Studies in Systems, Decision and Control

Volume 55

About this Series

The series "Studies in Systems, Decision and Control" (SSDC) covers both new developments and advances, as well as the state of the art, in the various areas of broadly perceived systems, decision making and control- quickly, up to date and with a high quality. The intent is to cover the theory, applications, and perspectives on the state of the art and future developments relevant to systems, decision making, control, complex processes and related areas, as embedded in the fields of engineering, computer science, physics, economics, social and life sciences, as well as the paradigms and methodologies behind them. The series contains monographs, textbooks, lecture notes and edited volumes in systems, decision making and control spanning the areas of Cyber-Physical Systems, Autonomous Systems, Sensor Networks, Control Systems, Energy Systems, Automotive Systems, Biological Systems, Vehicular Networking and Connected Vehicles, Aerospace Systems, Automation, Manufacturing, Smart Grids, Nonlinear Systems, Power Systems, Robotics, Social Systems, Economic Systems and other. Of particular value to both the contributors and the readership are the short publication timeframe and the world-wide distribution and exposure which enable both a wide and rapid dissemination of research output.

More information about this series at http://www.springer.com/series/13304

Georgi M. Dimirovski
Editor

Complex Systems

Relationships between Control, Communications and Computing

Springer

Editor
Georgi M. Dimirovski
School of Electrical Engineering
and Information Technology
Saints Cyril and Methodius University
of Skopje
Skopje
Macedonia

ISSN 2198-4182 ISSN 2198-4190 (electronic)
Studies in Systems, Decision and Control
ISBN 978-3-319-28858-1 ISBN 978-3-319-28860-4 (eBook)
DOI 10.1007/978-3-319-28860-4

Library of Congress Control Number: 2015960770

Printed on acid-free paper

This Springer imprint is published by SpringerNature
The registered company is Springer International Publishing AG Switzerland

Torrent of Changes Renews this World
Over and Over Again as the Uninterrupted
Flow of Time Brings Youth to the
Frontier-less Eternity.

Marcus Aurelius

Progress Has Been and Will Be Achieved
Always When Daring To Cross Borders,
Surpass Barriers and Exceed Limits.

Georgi M. Dimirovski

Foreword: Fascinating Ideas on Complexity and Complex Systems Control

Abstract The focus of this Foreword is to explore the fascinating ideas on complexity in conjunction with systems and control sciences. It is by no means an attempt to give final answers to what the control of complex systems may comprise. Nonetheless, it is believed to have gone far beyond "Complexity is in the eye of beholder" stated as the message of the similar 2001 Springer monograph. This essay is an exposition of personal visions of complexity, complex systems as well as control and supervision of complex systems, with certain complexity features as an essential system property. Though, it remains dedicated to explore the issues of integrated control and supervision within their possible interplay over complex plants and processes. It is thus hopped the justification for continuing the adventure of investigating complex systems and controls from a standpoint of physics and not solely mathematics shall emerge by itself. It has been found, the exploration road of complex networks and systems as well as their feasible controls still lies wide open before interested researchers from both theoretical and technological points of departure. For, complexity of dynamic networks and system is an evolving issue.

Keywords Dynamic networks and systems · Control and supervision · Controlled synchronization · Collective versus individual goals · Communications · Evolution adaptability · Evolving computation · Feedback · Information · Nonlinear dynamics

Prologue

At first, let us recall here what the famous Henry Poincare had emphasized in his own time: 'The thought must never submit itself nor to a dogma, nor to a party, nor to a passion, nor to a preconceived idea, not to anything that exists if it were not the facts themselves. For, the thought to submit itself, it shall be as to cease to exist'. In the sequel, with all due respect to a variety of thinking views complied from the literature, throughout Foreword *this one of Henry Poincare'* thoroughly is observed. In addition, since Whitehead and Russell [247] had derived the entire mathematical science (1927) on the grounds of Cantor's naïve definition of a set as the fundamental mathematical concept by observing the laws of logic, the discussion presented here also employs the set theory and the logic of natural sciences to its full. Long ago Leonardo da Vinci was the first to state: 'There is no much truth in sciences that do not use mathematics'. Furthermore, much later Immanuel Kant emphasized: 'In every department of physical science there is only so much science, properly so-called, as there is mathematics'. And even much more recently, Rudolph E. Kalman pointed out: 'First get the physics right. The rest is mathematics.' The underlying scientific background here comprises physics, set theory and its special branch on advanced graph theory [3, 25, 28, 89, 97, 128, 163, 206, 207, 210]. For, neither control nor systems sciences represent solely enterprise of mathematics, but rather they imply what one might call a kind of 'generalized' physical science handling objects and processes operating below the speed of light (for instance, see [3, 12, 13, 20, 52, 63, 64, 73, 91, 96, 112, 132, 154, 158, 171, 219, 233]) by virtue of *interfering and interacting energy, matter and information.*

Indeed, there exist in the literature a number of monographs with the ambition to give answers to what the control of complex networks and systems is all about; some of the most important ones are cited here. For instance, the author of one such recent book, Complexity Explained (Springer 2008), Peter Erdi, begins by arguing about the terms 'complexity', 'complex systems' and alike, and also by citing Stephen W. Hawking's statement: "I think the next century will be the century of complexity." Further, he points out that one aspect of the complexity is related to the structure of a system, and then proceeds to mention the underlying facts that all scientists recognize in omnipresent networks and webs. And yet he argues that one aspect of complexity is the structure of a system and further suggests that it is the fundamental organizational form of systems despite whether the adopted scientific approach is the holistic or the reductionist one. In contrast, we here argue that the fundamental form pertinent to some entity, object or process, to be considered a system is that of (some) organization in addition to a purpose functionality (e.g., see [29, 30, 31,35, 51, 57,66, 96,114,128, 135, 155, 157, 160, 164, 180, 204, 208, 223, 224, 231]). It is further argued here that a proper understanding of the concept of complex systems inevitably requires referring to the reality, including the living systems and also societal systems.

The present collective authored monograph has no ambition to give final answers but solely to shed deeper light into understanding complex systems in a wider setting, applications inclusive, hence via observing physics, in the first place. From a certain viewpoint on control and supervision, the present book may appear as if it were a reincarnation of the statement by Howard H. Rosenbrock: '…My own conclusion is that engineering is an art rather than science, and by saying this I imply a higher status…', see [189], in the present prospect of complex dynamic networks and systems. While it may well be true this Rosenbrock's paradigm of engineering as arts-rather-than-science to lie at the heart of the conceptual definition known as 'System of systems' given by Aiguier et al. in [1], dynamic networks and systems per se have a much deeper scientific foundation. Thus, I do not believe such pure mechanistic concept of 'System of systems', and the respective methodological approach too, could go beyond the ideas of interconnected and large-scale systems (e.g., see [5, 11, 16, 35, 39, 43, 48, 57, 65, 71, 76, 107, 109, 163, 167, 168, 192, 211, 227, 243, 260] and references therein). For, the entire history of natural sciences and of physics, in particular, gives a plethora of evidence that this kind of simplistic view cannot be scientifically viable.

A Look Back

It seems the idea and issues of complexity were first put forward in the year 1948 on the eve of emergent powerful computing machines as it appeared in the 1964 edited multi-authored book by Edwin F. Beckenbach [24], in which Warren Weaver was cited in the Preface to have authored them (American Scientist 36, 1948). But it was Herbert A. Simon [213] who in 1962 has managed to formulate complexity in scientific terms as well as to promote complexity as an exclusive direction of important research prospect (Proceedings of American Philosophical Society, 106).

A common belief at that time across all fields of science typically was: there exist two main categories of complexity—the ones of organized complexity associated with the applied combinatorial mathematics (also including graphs, matrix games and optimal control; e.g., see [2, 15, 26, 27, 47, 55, 67, 71, 72, 74, 94, 96, 101, 108, 121, 133, 153, 162, 177, 207, 214, 229, 233, 237, 246, 247, 249]), and the others of disorganized (or unorganized) complexity associated with the uncertainty, probabilities, and stochastic processes (e.g., see [6, 7, 14, 31, 37, 83, 95, 98,

Fig. 1 A macro-world complexity: Telescopic snapshot has recorded a cosmic moment in the tumultuous lives of large spiral galaxy NGC 3227 and smaller elliptical NGC 3226 in their close encounter. Spanning about 90,000 light-years and similar in size to the Milky Way, NGC 3227 is recognized as an active Seyfert galaxy with a central super-massive black hole

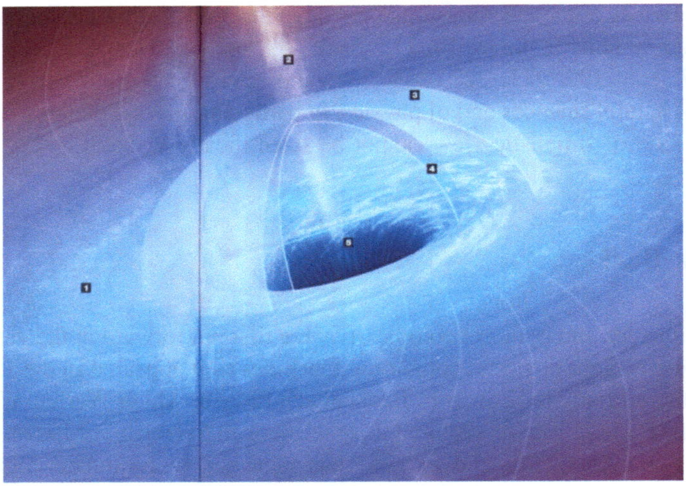

Fig. 2 In the center of Milky Way: Artistic perception by author M.A. Garlick of the phenomenology of black hole SgrA*, weighting 4 million times more than Sun and radiating powerful radio waves (National Geographic 102(4); 2014, p. 29): (1) superheated energy disk, (2) RO-radiation flow-stream, (3) static boundary virtual horizon, (4) horizon of events, (5) cosmic singularity. A. Einstein: "I would like to know God's thought. I would like to know how God created this world." (see brothers Bogdanov Igor and Grishka [32])

99, 132, 173, 184, 228]). It should be appreciated that both categories were indeed well conceptualized and clearly defined.

This dichotomy-like thinking, it may well be noticed, appears to follow the illustration in Fig. 1 that presents a contemporary telescope snapshot record of some part of the cosmos, which I name conditionally here as the macro-world complexity. It should be clearly noticed: certain elliptic- and spiral-shaped galaxies versus the vast background of seemingly disorganized ones. Furthermore, a recent insight into the sub-sub-world of the boson (Fig. 6) discovery by Peter Higgs and François Englert seems to emphasize over and over that similarity and symmetry [263] are quite natural properties in physical world [32, 53]; e.g., see both Figs. 2 and 3. Then why not (indeed, ?!?) accept the use of both similarity and symmetry when developing mathematical representation models and control system designs (Zhang 1994 [263]; Dimirovski 2011 [51]) for real-world objects and processes. Indeed, considerably early the logic of nature has yielded their use in systems and control studies, both the theoretical ones as well as the ones in engineering applications and technology [26, 35, 69, 109, 133, 195 , 227, 229, 238]. Thereby it appeared that exploiting the similarity and symmetry features have also given rise to results and some sort of trend to seek simplified representation of endowed control problems and then find relevant solutions, a process somewhat opposite to the large-scale systems. In particular, Prof. Si-Ying Zhang [264] and his many collaborators in due times have been rather instrumental for these developments and discoveries [264]; also, see works [44, 45, 47, 53, 95, 115].

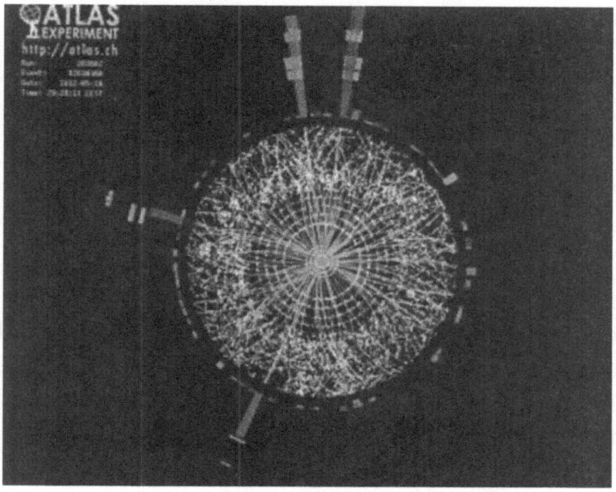

Fig. 3 Reported outlook of God's particle—boson of Peter Higgs and François Englert, which explained how other particle acquire mass and yet may not be the smallest one in the universe, following CERN collider experiment ATLAS: Fascinating properties of symmetry and similarity with respect to and transversal that may partition it into two halves

It was Prof. Zhang's encouragement and influence that have materialized in solutions to two different topics within this trend by Jing et al. [45, 115] on force resources deployment in battle [44, 115] and on warfare command systems [45]. In particular, a suitable modification of Lanchester square law was used in deriving force partitioning strategy in [44] to obtain simulations results that yielded precisely the historical outcome of the Trafalgar naval battle between English fleet (side x; Admiral Nelson) and French–Spanish fleet (side y; Admiral Villanueva), outlined further below. It is amazing to notice for this complex, command, control problem a rather simple solution was found albeit a game-based analysis was employed. The background is the following simple Lanchester-type attrition model [115]:

$$\begin{aligned} \dot{x}_{0i}(t) &= -\alpha y_{0i}, i = 1, 2, \cdots, \\ \dot{y}_{0i}(t) &= -\beta x_{0i}, i = 1, 2, \cdots, \end{aligned} \quad (1)$$

The derived resulting theorem stated: Assume employing the nonlinear logic of these Lanchester square equation of warfare and suppose that $x_{0i} > \frac{\alpha+\beta}{2\beta} y_{0i}$. If the conditions $\sqrt{\beta} x_{sh} > \sqrt{\alpha} y_{0,i+1}$ can be satisfied, then $y(T)$ will become zero before $x(T)$. This is to say, the defending side x will gain the final victory in the battle under those conditions by an appropriate force partitioning strategy among the feasible ones as shown in Table 1.

From Table 1, with respect to the influence of the maximum residual strength the minimum warfare time, the optimal partitioning strategy for the English fleet is (11, 16) and (7, 10, 16). The maximum residual strength, $x'_{sh} = 16$ battle units, and the minimum warfare, $T = 1.9678$ time units have been found. The relationship

Table 1 The partitioning of the force strength and the warfare time

(x_{01},x_{02})	(y_{01},y_{02},y_{03})	Warfare time	x'_{sh}
(9,18)	(6,11,16)	2.0380	16
(11,16)	(7,10,16)	1.9678	16
(12,15)	(7,10,16)	2.0380	16
(13,14)	(9,8,16)	2.1039	16

between the residual forces of the English fleet x and the number of force partitioning of the optimization model has been computed as in Fig. 4. The established relationship between the feasible force partitioning strategies and the actual number of force partitioning is depicted in Fig. 5. Similarly, the relationship between the minimum warfare time and the number of force partitioning is computed as in Fig. 6.

It can be readily seen from Fig. 4 that the residual force strength of the operational side reduces with the increase in the number of the force partitioning. There exists a limited value of the number of the residual force strengths. When the residual strength is less than the limited value, the strategy is not in line with the physical background of the battle. For instance, it should be supposed that $\alpha = \beta$ and the strategy of force partitioning should be chosen as (7, 20) and (2, 13, 18), then the residual strength on both sides becomes zero. This strategy was not used in that naval battle. As seen from Fig. 4, the residual force strength of the operational side x may also become zero if the number of the force partitioning is increased to 11 (Nelson had not used it).

It can be seen from Fig. 5, the warfare time shows the growth trend with the increase in the number of the force partitioning. Of course, it is well known that the warfare time must be influenced by the optimal force partitioning. There exists a limited value of the warfare time, which seems to have been intuitively well known

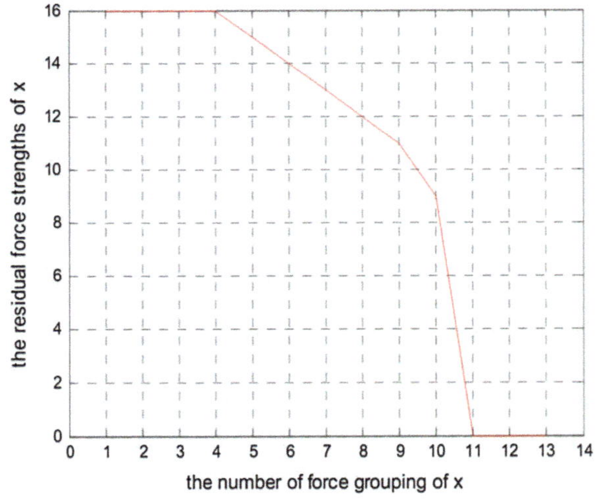

Fig. 4 The relationship of the residual force of side x and the number of force partitioning

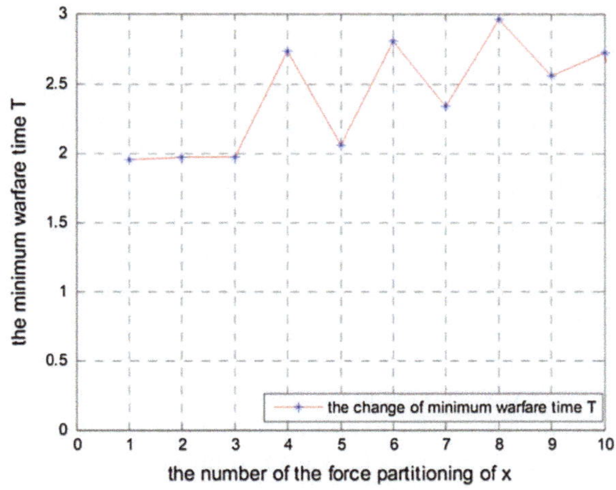

Fig. 5 The relationship of the minimum warfare time of side x and the number of force partitioning

to both admirals. It seems more than just interesting to note that a complex system dynamics as this naval battle had historically exhibited could have been so fairly and easily reproduced via model-based, computing simulation as Si-Ying Zhang [264] has argued in [265]. Furthermore, this study has emphasized how crucial in the real world are both role and place of information within the naturally existing *interplay of energy and matter with information* [2, 56, 63, 81, 91, 95, 154]. Indeed the information is the third fundamental quantity having the same rank and range as energy and matter do.

Author Peter Erdi in his 2008 Springer monograph [67] provided argument considerably in favor of importance about the similarities and symmetries being exploited in various topics on complexity. Indeed, given the facts from the world of the universals, such as in Fig. 2, and the world of the infinitesimals, such as in Fig. 6, the issue of similarity/symmetry features as properties of natural and

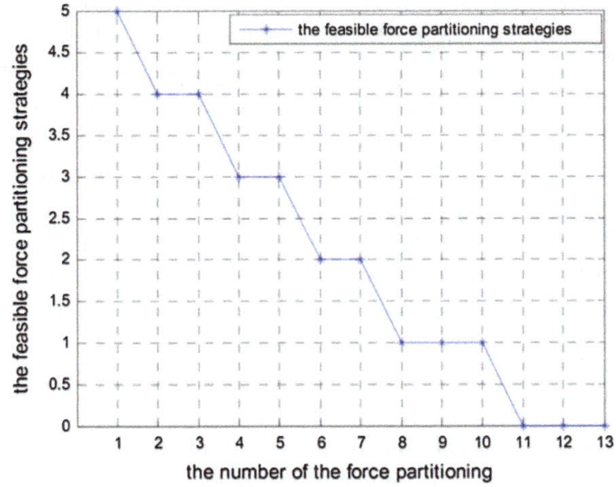

Fig. 6 The relationship of the feasible partitioning strategies of side x and the number of force partitioning

man-made systems cannot be ignored. It is therefore not surprising that, in his monograph, Peter Erdi has conceptually introduced several kinds of complexities, in general and in particular structural, functional, and dynamical complexities, in the first place. Only next he placed the algorithmic information complexity, largely attributed to Andrey N. Kolmogorov [134]; the cognitive complexity, largely attributed to Heinz von Foerester [234, 235]; the computing machine complexity, largely attributed to Allan M. Turing [230]; the organizational complexity, largely attributed to Ludwig von Bertalanffy [67] albeit it was Alexander A. Bogdanov [29] who had first invented, investigated, and made use of it in his theory of organizations called 'Tektology' (1925). Furthermore, even Erdi [67] himself has noted that certain field-of-application specific notions of complexities, namely ecological complexity; economic complexity; financial complexity; political complexity; and social complexity. Even in [4] Ahamdi and Jungers 2014 explored complexity of Lyapunov functions for switched linear systems,only for the linear ones.

The original Kolmogorov's definition, which reads "The algorithmic (descriptive) complexity of data sequence is the length of the shortest binary computer program that prints out the sequence and halts," is strikingly compatible to the fundamental theorems of computing languages, grammars, and machines [46, 128, 230]. More importantly in my opinion, however, it is compatible with the underlying algorithmic nature of all developments in sciences and engineering of computing, communication networks, and control systems alike. For, what always an operating feedback control system does after any disturbing action from the environment, in a sense, it is solving over and over the initial value problem through processing of the control error dynamics of the closed-loop system with certain newly triggered initial conditions. Therefore, Kolmogorov's algorithmic understanding of the complexity seems the most fundamental one from the viewpoint of the integrated control and supervision of complex networks and systems. Thus I do believe that Kolmogov's conceptualization precisely gave rise to the complexity theory. Later, it was embraced by Rissanen [184, 185] and developed to matured level by Li and Vitanyi [141]. These studies and the 1978 contribution by Rissanen [185], in turn, seem to have been instrumental for Gruenwald's 2007 study on the principle of *minimum description length* (MDL) in the model development for a given data set [84]. The latter is also strongly supported by Haykin in his 2009 monograph (pp. 109–112) within the context of learning machines and neural networks [89].

System properties of structural symmetry and/or similarity have been well exploited in the investigation of various kinds of interconnected and large-scale system representations, which have yielded an outstanding plethora of developments and discoveries. For instance, one should study the works by Dragoslav D. Siljak (1970–2010) and his many collaborators: the monographs [207, 210, 260] are particularly important. They have discovered most of structural features of decentralized control of such complex systems in the presence of information constraints and structural perturbations as well as uncertainty.

Systems and control theories, in particular those concerning the linear systems and linear-quadratic optimal control (as well as their applications), have had a blossom of developments following the fundamental discoveries by Rudolph E.

Kalman [120, 121, 123–125] on theory of systems and control. A similar impact could be seen for the fundamental discoveries of Vladimir A. Yakubovich [250–252] on matrix inequalities, S-procedure, and the celebrated Kalman–Popov–Yakubovich lemma [33, 81]. Practically, all these developments have been extended later to both large-scale systems and nonlinear systems in various ways by a number of authors (e.g., some are found in the references given). Yet the ideas and issues as well as concepts about control of complex systems continue to get enriched and to evolve until now such as most recently within the paradigm of complex networks and networked systems. Evolution happened somewhat gradually beginning with exploiting natural features of similarities and symmetries in both of the real worlds, the dead and the living ones.

In order to return to the issue of primary concern in this discussion, notice that there exists considerable knowledge about astrophysics perception of the cosmic complexity and birth, life, and death of celestial objects and galaxies (seen via Figs. 2 and 3). Yet the journey into the discovering the essence of life via studying its manifested features (for instance, see [170]) has started progressing not so long ago and was due largely to various visionary perceptions of what complexity and complex system dynamics may be all about (e.g., see [6, 15, 22, 34, 39, 41, 42, 49, 50, 57, 62, 66, 83, 96, 97, 101, 104, 112, 156, 166, 176, 181, 190, 193, 218, 234, 235]). On the other hand, our real-world experience via human perception of natural ones as well as of man-made dynamic processes demonstrates the existence of variation in "sheds" of complexity between these categories. Nonetheless, it is within this context precisely where Heinz von Foerester [233–235] within his thinking about cybernetics and systems in the wider setting of nature and society, seems to have highlighted considerable deeper roots involving not only the interplay of energy and matter with information [53, 55, 82, 135, 138, 154] but also the *cognition* process [235, 239]. It appears, his emphasis on cognition has emerged to be considerable crucial in contemporary developments of systems engineering design and automation technology. The cognition process, however, also implies some anticipation and maybe consciousness within the evolution dynamics in the nature. But can it indeed be so, then? There is no clear-cut answer as yet, and it does not seem one will appear in the near future.

Indeed, most of the consulted source literature suggested for quite long time there have existed considerable endeavors to explore the life itself. It is a strikingly old idea that actually has led to a deeper understanding of the issues of complexity and complex systems, as a matter of fact. This shift has taken place—it may well be argued so—due to the crucial impact of discoveries in genetic research within biological systems leading to the idea of systems biology. Indeed, considerably many aspects of complex systems on the way towards complex dynamic networks have been evolved into an innovated insight since the works by Darwin [49], Bogdanov [31], Pavlov [170], Dobzhansky [58] and Radcliffe-Brown [180]. Moreover, it appeared transcending these ideas into the evolution of life and the societal dynamics have had even greater stimulating impact towards the development of the science of complex dynamic networks and systems (e.g., see [1, 16, 51, 66, 77, 84, 101, 127, 157, 165, 167, 168, 169, 190, 193]) and had induced other insight into the issues of complexity and complex systems as well as their control strategies.

The evidence in the literature also suggests that the general trend during the past decades towards an ever increasing interconnection and cross-coupling in all societal and human activities caused a definite shift towards a unified comprehension of complex dynamic systems and complex dynamic networks in various realization forms. Furthermore, perhaps this is the most important paradigm of the current state-of-the-arts in complex networks as well as the whole of systems science. A brief scan of a small sample of research studies, such as [5, 6, 20, 102, 104, 144, 145, 150, 163, 168, 178, 179, 210, 212, 223, 224, 241, 243, 269, 270], clearly demonstrates the concluding remark. In addition, as Muscolesi and Mascolo [161] noted, the unprecedented recent advances in computer networks, communications, and information technologies [36, 59, 116, 201, 220, 255] as well as the world-wide expansion of mutually interconnected economies have only reinforced this major trend. Furthermore, this major trend is to expand even further in all domains of societal and human activities thus becoming the overwhelming one [20] nowadays. It seems however, the life itself plays a more essential role in the collective adaptability and the evolution dynamics than it has been admitted in so far [66, 114, 157, 188, 190], nonetheless.

Riding the Wave: A Forward–Backward Inquiry

Complex dynamic networks [219, 243], which comprise a set of nodes possessing locally controlled or uncontrolled dynamic systems, and a finite graph structure of finite capacity communications as well as interactions among those node systems, have already gained a growing attention from scientists and engineers world-wide. Nowadays, a dynamic network is commonly understood to be a large-scale systemic structure consisting of a graph that contains enumerable set of dynamic systems as its nodes and a finite set of signal-flow interconnections among them. Thus, in their very essence, I believe complex dynamical networks and systems, the roots of which can be traced back to Aizerman and his collaborators [3] and to Ford and Fulkerson [73], can be completely described by the dynamic graphs as proposed in [212] by Dragoslav D. Siljak. During the last couple of decades, research activities in exploring synchronous evolution or motion dynamics has been steadily growing and both spontaneous and controlled synchronization has been extensively addressed in a number of works (e.g., see [25, 39, 48, 51, 87, 146–148, 211, 223, 224, 241, 242, 270, 272, 273], and by many other authors as well). The importance of synchronized evolution or motion does not lie only in those situations in which synchronization can be found, but also where synchronous motion can be induced to ensure the proper functioning of particular collective devices such as multiple machine ensembles. It should be noted some of those works have supplied feasible control methods for reinforcing the synchronization in a few fairly wide classes of dynamic networks with regard to their topology.

The investigation of the aspects and issues related to the controlled dynamical networks has become a rather attractive research area for two main reasons. First, dynamical networks appear in various worlds of the nature and human societies on Earth as well as many events and processes in the real world can be modeled by dynamical networks. Second, a large number of important applications of dynamical networks can be identified in various disciplines, which include astrophysics biology, economics, engineering, life science, neuroscience, and sociology too [87, 168, 181].

In the sequel, building an argument is attempted in favor of observing Darwin's evolutional dynamics, as a point of centrality in the quest for a consistent theory of controlled complex networks and systems via the theory of Siljak's dynamic graphs as an adequate tool from applied mathematics and the physics behind systems science [55] is attempted. The crucial point of departure to notice [91, 92] is the *unique integrity* of the interplay of energy and matter with information in the universe; [92] hence also on Earth's nature [55, 56, 82, 176, 188].

It should be noted that reference to biology, life science, and sociology inevitably puts a special emphasis on the conceptually *fundamental category of organization* and its feasible evolution while not solely on the structure as such. Darwin's evolutional dynamics of species [49] involves interactive process of mutual adjustment and co-evolution that is matching the structural coupling up [58, 126, 127, 160, 176] while the organization of an "*autopoietic*," i.e., a self-reproducing circular nature, class identity "living organism" comes out from the feature of the organization possessing a certain structure determinism [57, 156, 166, 187]. In turn, reference to the living points out to essentially another but deeper understanding of complex dynamic network and systems that most often possess goal-seeking behavior.

It is in this regard that I do believe complex dynamic networks and systems, which employ combined control and supervision to enforce their organized integration, ought to be viewed as a kind of approximation to biological systems. Thus, they should be endowed with capacity of becoming autonomous, closed, circular dynamical processes as the life itself on Earth [5, 50, 55, 63, 86, 112, 130, 155, 160, 166, 181, 193, 215, 218, 232, 235]. Thus, the employed, combined control and supervision ought to possess higher qualities than known at present. In particular, the intuition suggests the supervisory level ought to have some properties of controlling cognition within the broader view on evolution process dynamics that is associated with the entire phenomenon of living biological systems. It is therefore that supervisory function should be sought to have implemented certain mentalist activities hence cognition within the wider set of decision and control activities. As such they must be compatible with and coherent to a set of decision and control activities within the complex networks and systems so as to guarantee that not only the system integrity and stability but also the goal-seeking behavior are simultaneously reinforced. Indeed these aspects may well open a new prospect but they also pose some tremendous complicated analytical problems both mathematically as well as from the viewpoint of systems and control sciences. The established insight that all complex networks and systems are indeed characterized with the feature of inseparable twin of structure determinism along with structural coupling is due to extensive internal communications and information sharing exchange mechanisms [26, 35, 52, 54, 155, 186, 192, 210, 232, 233, 272].

A comparative analysis of technologies and science of dynamic networks, systems, and their control at the times of the classics of engineering cybernetics ([15, 196, 229, 233, 237]; Bogdanov 1925 [29], [79]; Wiener 1948 [246]) with the recent ones readily yields not only too many but also the most important changes to notice. These changes, I believe, can be summarized in two entities: *incredibly powerful com-putational* technologies, on the one hand, and *unprecedented pervasive fast com-putational* and *transportation* networks, on the other hand. Of course, along with these technologies, the underlying scientific knowledge of the highest quality and in enormous quantity has been acquired as well. It is therefore that the real-world systems are abundant in all sorts of nonlinear and time-varying processes, which may not be amenable to mathematical representation entirely.

Still, as argued by Rosenbrock [189] in 1977, both mathematics and physics are indispensable to systems and control sciences, and therefore I believe a special attention should be paid to an important feature in modern physics. Namely, *modern quantum theory has abandoned to deal with strict causal determination for elementary atomic processes and has turned more towards statistical anticipation about those processes*. Should we now take a closer insight into the network control-theoretic developments and practical designs, then no special imagination is needed to see that such network systemic structures have parallels with the worlds of atoms and subatomic processes as well as with the molecular world. I believe the subsystem processes in networked control systems, and even more so, in complex dynamic networks are to a certain extent analogous to those elementary atomic processes, and indeed to various macro- and micro-worlds on Earth. Thus, it occurred to me, indeed we ought to try hard *to transcend some of modern quantum theory into the general framework of reasoning about the systems and control science* [51]; hence complex dynamic networks systems as well. For the time being, let leave off this issue and address what specific features can be recognized about control systems over communication networks as implementations, which represent pervasive existence nowadays. But notice nonetheless, these are by and large computer controlled hence computer science is indispensable.

In my opinion, the features of primary importance are the known kinds of *complexities* and of *uncertainties*, *information constraint* as well as *connecting/disconnecting sustainability and survivability*. The latter is being directly related to all variants of pertinent systems stability problems, now largely tractable by Linear Matrix Inequalities (LMI) based techniques due to original discoveries of Vladimir A. Yakubovich in fundamental theory of control [250–252] and in applications [72, 253]. Without a doubt adaptation and re-organization also play rather important roles. And even more so do adaptation and re-organizing adaptation and re-organization on a massive scale, not necessarily collective unless induced by some consensus strategy. By and large these all are control reinforced. However, I believe adaptation and re-organization come only second next to the previously pointed complexities.

Upon noting several kinds of complexities, let us first take into consideration the *organized* ones [244], because it is the control infrastructure precisely that can convert the unorganized complexities and uncertainties into the organized ones. It should be noted, solely the mathematical hence also the computer science, complexities are the ones precisely defined. These are referred to as NP-completeness and NP-tractability, usually. However, it has become apparent in the present day encountered complex dynamic networks that *architectural*, *structural*, and *topological* system complexities appear in addition to the *computational* one. Yet it is true that none of them is either completely or precisely defined as it is the NP-completeness in computer science and ICT. It should be noted further that the last couple of decades have yielded considerable advances in various issues about complex dynamic networks and systems, one of which is an improved understanding of topological complexity versus network sustainability seems to be instrumental for the further developments. However, it appeared much deeply involved for the present day existing knowledge and understanding. Nonetheless, some of the recently reported research [146–150, 223, 224, 268–271] has indicated that enormous control effort is needed in order to reinforce controlled synchronization, in general, and also to ensure connective sustainability, in particular.

The feature of information constraint is considerably well understood, thanks to the advances in large-scale systems theory and decentralized controls for large-scale systems. Following the 2010 Springer's monograph [260], Control of Complex Systems: Structural Constraints and Uncertainties, there still remain open issues in this regards as well as about the decentralized control of complex and large-scale systems—in Siljak's words—beyond the decentralized feedback (Fig. 7), in general. In particular, despite the existing knowledge on connective stability, this knowledge still remains to be transcended in the wider setting of connecting/disconnecting sustainability and survivability. One rather important pathway forward to be pointed out, I do believe, is precisely the one via dynamic graphs of Siljak [210]. In other words, complex dynamic networks are systemic structures that do possess features of dynamics requiring control mechanisms that transcend beyond the current knowledge on decentralized feedback control and supervision in functional synergy.

The issues of the uncertainty have been subject to study from the very beginnings of information theory as part of the theoretical studies in cybernetics and its applications. However, as with complexities, there appear various kinds of uncertainties albeit in systems and control most often the parametric and structural uncertainties are being accounted for. Should we recognize the unavoidable need for some integrative organizational strategy imposed even in largely decentralized complex dynamic networks, then the uncertainty too becomes as multifaceted one as the complexity is. At this point, it should point out the largely neglected need for developing sophisticated theory on supervisory control strategies that can guarantee survivability of the complex dynamic network under various ad hoc topological circumstances via some controlled reinforcement of system integration (see Fig. 7). In my opinion, for the issues of supervisory controls longer time is left for more empirical investigations than theoretical studies. It is in this regard that I believe

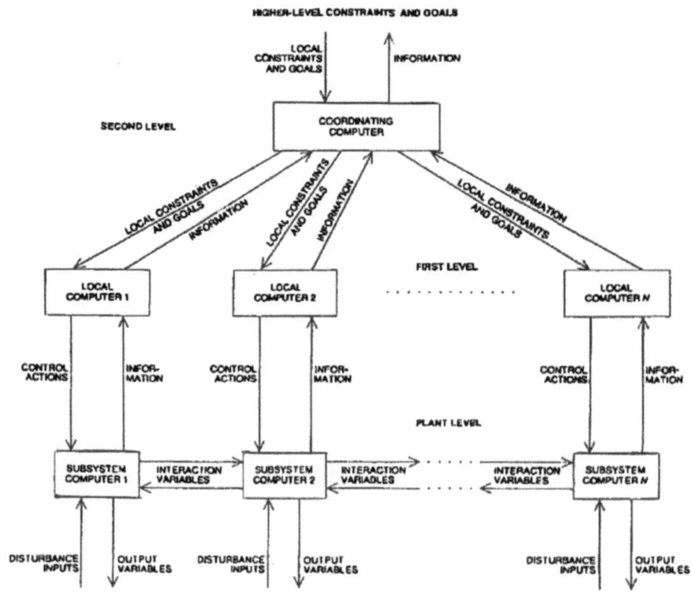

Fig. 7 The well-known, hierarchical, computer control based, implementation of integrated control and supervision for complex dynamic networks and systems in terms of dynamic sub-networks of nonlinear dynamic nodes, which in compliance with Siljak's concept of dynamic graphs [210, 212]

Siljak's concept of dynamic graphs does provide an innovative and sound path forward.

At this point , I would like to infer from the issue of exploiting some hints from modern quantum theory [91, 96], which seems to have been foreseen by Einstein his fellow-friends [63]. Namely, I think that sooner or later we have to place our main focus on how to mitigate consequences of strict use of the causality principle in system and control science. I incline to believe no major breakthrough is feasible without endowing the integrative supervisory control with some capacity for anticipation; only then combined control and supervision could deliver far beyond the present day knowledge on various predictive controls. Therefore I believe that the increasing precision with decreasing intelligence (IPDI) principle due to George N. Saridis, i.e., the analytical formulation of the principle of IPDI [194], albeit primarily it was driven for the needs in robotics, has been underestimated and left aside unjustified for that matter. In here and for the purpose of complex dynamic networks and systems, I would dare to re-interpret it in terms of *increasing intelligence with decreasing precision* hence and, on the grounds of entropy theory, come up with *supervisory control processing* albeit with some *incomplete anticipatory properties*. However, such supervisory controls cannot be achieved without employing *computational intelligence* techniques. Hence exploiting the *cognitive approach* in control imitating human intelligence capacities, which we may dare to envisage in terms of *evolving computational intelligence*, becomes as necessary as mathematics and physics are. It is within this context that the technological implementation of integrated control and supervision, which is illustrated in Fig. 7, seems to gain its full systems engineering value as emanated from control system theoretical endeavors of the large-scale systems theory [26, 35, 55, 77, 107, 158, 192, 207, 209, 210, 227].

The IPDI principle of Saridis [194], in my opinion, has made a rather stimulating impact towards exploiting computational intelligence methods [89, 113, 127, 259] for solving complicated tasks of control and supervision for nonlinear and complex systems as well as for complex dynamic networks. These methods, along with the respective techniques for representation modeling of processes to be controlled, include theories methodological techniques of fuzzy systems, neural networks, fuzzy-neural or neuro-fuzzy systems, and fuzzy-Petri nets as well as learning machines and systems in a broader sense [2, 7, 52–54, 60, 61, 90, 103, 104, 113, 127, 129, 133, 135, 152, 172, 174, 183, 191, 221, 255, 257–259, 262]. Furthermore, some of the most recent studies have clearly demonstrated how many mingling and mutually influencing are nonlinear theories of systems and control with interacting fuzzy system and neural network theories, from the viewpoint of stability in particular. In this regard, the interested reader is suggested, for instance, to consult recent articles [52, 89, 103, 104, 142, 221, 228, 254]. It is interesting to note that computational intelligence techniques appear to by and large another methodology to tackle mathematically complex nonlinear problems where analytical techniques fail to deliver technologically tangible results.

It should be noted further, in addition, during the last couple of decades significant scientific advances have been achieved in the theories of both *complex dynamic networks* as well as of *switched systems and switching-based controls*. Such dynamic networks and systems largely occur due to combined event-driven, state-driven, and time-driven control and supervision in man-made technologies [13, 16, 39, 40, 74, 80, 94, 107, 142, 145–151, 223, 224, 265–273]. It seems more than just coherent that these theoretical developments are also well coherent with technological implementation scheme that is presented in Fig. 7. Nonetheless, the complex dynamic networks involving switched topologies and switching base controls, I do believe, appeal for essentially modified conceptualizations on how we perceive processes that evolve due to the combined time-, event-, and state-driven dynamics. In fact, these two areas have opened a whole new prospects towards networked systems engineering creations of the future. It should be noted, although at infancy now, some encouraging results on the synergy of complex dynamic networks and switching control based modes have appeared.

For the sake of a certain scientific evidence of these issues mentioned above, let me put an overview of the following reasonably general and widely applicable model of complex dynamical networks, which involves realistic case of non-identical nodes with relative-degree-one nonlinear dynamic systems [150]. It is described by the following set of state and output equations:

$$\begin{aligned} \dot{x}_i &= f_i(x) + g_i(x)u_i, \\ y_i &= h_i(x),\ i = 1, \ldots, N, \end{aligned} \quad . \tag{2}$$

Here the symbols denote $x_i \in R^n$, $u_i \in R^m$, $y_i \in R^m$ are the states, the control inputs, and the controlled outputs of the i-th node, respectively; the vector functions $f_i\colon R^n \to R^n$, $g_i\colon R^n \to R^{n\times m}$, $h_i\colon R^n \to R^m$ are (assumed) to be twice differentiable or smooth mappings that possess the real-world properties $f_i(0) = 0$ and $h_i(0) = 0$.

For illustration purpose, at this instance, note that one such dynamic network having five nonlinear dynamic nodes is described as follows:

$$\begin{aligned} \dot{x}_{i1} &= x_{i1} + (l_i - 1)x_{i2} - \sin(x_{i1} + x_{i2}) - u_i, \\ \dot{x}_{i2} &= \sin(x_{i1} + x_{i2}) - u_i, \\ y_i &= x_{i2}. \end{aligned} \tag{3}$$

In here, $i = 1, \ldots, 5$, and in addition, values of l_i are defined as: $l_1 = 1.2$, $l_2 = 1.3$, $l_3 = 1.05$, $l_4 = 0.8$, $l_5 = 0.9$. Obviously, all nodes are not passive, but they all are nodes of relative-degree-one nonlinear dynamic systems, quite common in practical applications. Communication coupling topology for the network is defined by

$$v_i = \sum_{j=1}^{j=N} a_{ij}\Gamma y_j. \tag{4}$$

It has been shown [150] that this can be rewritten as $u = (A \otimes \Gamma)y$, where $A = \left(a_{ij}\right)_{NN}$ is a matrix, called the outer coupling matrix, Γ is an $m \times m$ diagonal matrix, called the inner coupling matrix. The quantity $A \otimes \Gamma$ is the Kronecker product between these two matrices, which is presented below.

$$A \otimes \Gamma = \begin{pmatrix} a_{11}\Gamma & a_{12}\Gamma & \ldots & a_{1N}\Gamma \\ a_{21}\Gamma & a_{22}\Gamma & \ldots & a_{2N}\Gamma \\ \vdots & \vdots & \ddots & \vdots \\ a_{N1}\Gamma & a_{N2}\Gamma & \ldots & a_{NN}\Gamma \end{pmatrix} \in R^{N_m \times N_m}, \tag{5}$$

where

$$A = \begin{bmatrix} -2 & 1 & 1 & 0 & 0 \\ 0 & -3 & 1 & 1 & 1 \\ 1 & 0 & -2 & 1 & 0 \\ 1 & 1 & 0 & -3 & 2 \\ 0 & 1 & 0 & 0 & -2 \end{bmatrix} \tag{6}$$

is the outer coupling matrix that does not possess the usually assumed property in the relevant literature; also, it is shown that Γ can be taken as the identity matrix [150].

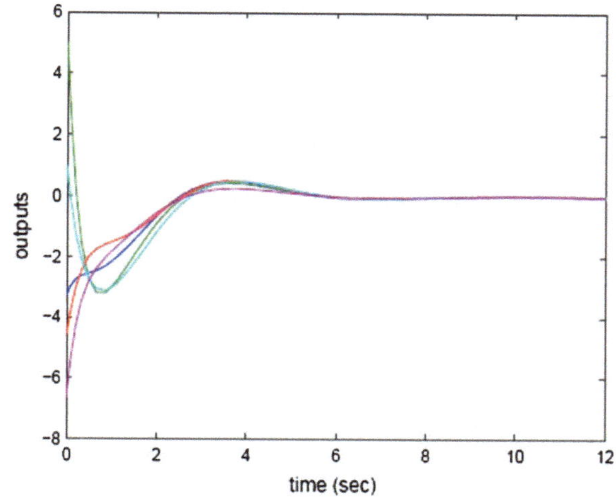

Fig. 8 The synchronized output responses of the network with communication coupling topology and nodes of nonlinear dynamic systems having relative-degree-one [150]

It may well be inferred from the above mathematical representation model that it implies complexity in several respects. Still, the controlled output synchronization can be well enforced as the simulation results in Fig. 8 demonstrate.

Indeed, synchronization is a collective behavior of dynamical networks, of course, but it is one of the crucial issues in studying potential behavior of complex dynamical networks. It has a direct relevance to and certain relations with the logic of life itself (Rosen 1991; [50,126,166, 186]). And the life itself indeed is the best teacher for both organization and self-organization in the universe [50, 127, 156]. The main issue of concern seems to appear in terms of how to accommodate within complex networks setting the need to employ some techniques of game theories and some anticipation features [186], because they are also crucial in comprehending the real-world living, ecology, and societal systems in their mutual interactive couplings [85,155,181,190, 218, 233].

The ideas about engineering implementation of integrating control and supervision of man-made complex industrial plants also deserve to be appropriate, but briefly addressed in this essay tractate. It should be noted here that, in the first place, the heritage from control developments aimed at interconnected and large-scale system complexes in terms of hierarchically organized computer control system architecture and the respective applications software technology. These were precisely the technologies that enabled the modern-day automated factories, power plants, as well as communication, power, and transportation systems, and even stimulate the exploration of the paradigm "factories of the future" further. For this purpose, from Romero et al. and the related works [119, 200, 231], respectively, the borrowed Fig. 9 is given as presented above [188].

It may well be considerably stimulating, and possibly instructive too, if in this context a few more points are made. Then this discussion inevitably involves the issues of operating symbiosis of human(s) and automation system [119]. It is therefore that also man–machine interactions [70, 111, 154, 156, 200] as well as

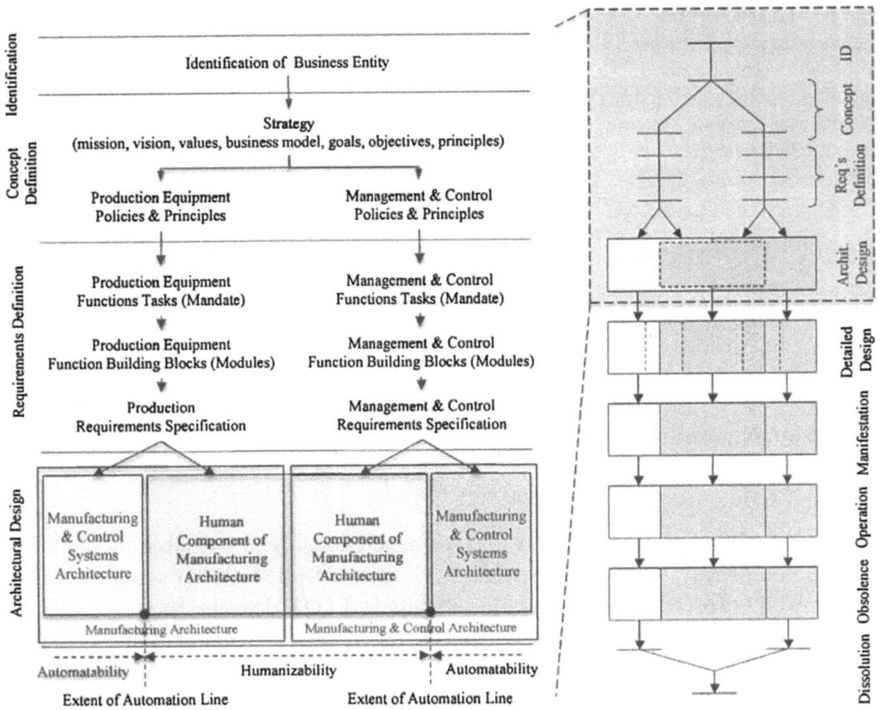

Fig. 9 An overview of the current state-of-the-art developments in human–automation symbiosis [188] within design variants of human-centered automation systems [70] in terms of the Purdue Enterprise Reference Architecture; the segment on the right points out the ongoing research tasks

features of humans in the role of control operator [182, 187, 232] ought to be placed within the scene and scenarios of control and supervision of complex dynamic networks and systems. It is in this regard that the supervisory level ought to have some properties of controlling cognition as a broader evaluation of controlled processes within the automated complex plant. Therefore that supervisory control level is to be designed to have implemented certain mentalist activities and hence cognition within the wider set of decision and control activities. As such they must be coherent to the entire feasible set of decision and control activities within the complex networks and systems in order to guarantee not only the system integrity and stability but also the overall adaptability and goal-seeking evolution. Furthermore, following a recent study [231] by Tzafestas this conceptualization seems to transcend beyond the technological complex system, since it does reflect upon the society and hence the nature[85, 180, 203, 218, 245].

Finally, within this discussion about complex dynamic networks and systems integrated via their control and supervision infrastructure, issues of stability ought to be addressed because these are sine-qua-non in systems and control sciences from the time of Poincare' [175] and classic analytical mechanics. It has been so since the days of developments in theory and engineering of linear dynamic systems and their control; for instance, Anderson [8, 9], Barnett [21], Bernusou, and Titli [26], Dimirovski et al. [55], Kalman [121, 122], Krasovskii and Pospelov [133], Mesarovic and Takahara [158], Rosenbrock [189], Siljak [202, 207], Tsien [229]. True, these aspects are considerably more difficult and more involved within this context of complex dynamic networks and systems. Moreover, there were much more involved than studies of nonlinear dynamic systems at the time of Liapunov [173] and Poincare' [175] within the realm of classic analytical mechanics. It is the extension towards the control of nonlinear dynamic systems and its optimization that made essential difference; for instance, see Anderson [8], Emelyanov [65], Filippov and Arscott [71], Isidori [110], Moylan [98], Kalman [120], Krasovskii [132], Lurye [153], Pontryagin et al. [177], Siljak [203, 207, 210]. These developments have given considerable understanding of the nature of Lyapunov functions and Lyapunov–Krasovskii functionals to the extent to become fundamental tools for explorations in the realm of complex dynamic networks and systems.

It should be noted first of all, the first considerable extensions of Lyapunov stability theory have been made rather early by Lurye [153] and by Krasovskii [132]. These extensions have paved way towards a deep and far-reaching study of the original conceptualization of Lyapunov function that yielded innovative modifications up to now to various Lyapunov-like functions in the search of constructing Lyapunov that are suitable for specific but nonlinear and/or time-delay phenomena; for instance, see Antsaklis [13], Basar and Bernhard [22], Fomin et al. [72], Fridman and Shaked [75], Ikeda and Siljak [106], Kalman [124], Kozyakin [138], Lee and Jiang [140], Liberzon et al. [142], Liu and Fridman [145], Long and Zhao [151], Tee et al. [225, 226], Zhang et al. [261], J. Zhao and Dimirovski [266], J. Zhao and Hill [267], Zhao et al. [148]. However, as recently pointed out by Ahamdi et al. (2013), a kind of complexity of Lyapunov functions for switched linear systems exists if these are sought in the tradition of quadratic stability, due to the underlying linear algebra [117, 118, 139] that seems even set-theoretic methods in control [28] could not be overcome.

Indeed, during the last couple of decades many new and/or innovated exiting stability results for complex dynamic networks and systems and for network control systems, all based on the ideas about Lyapunov functions and Laypunov–Krasovskii functionals, have emerged; for instance, see [13, 25, 27, 43, 51, 78, 86, 103, 104, 140, 145–151, 178, 179, 212, 224, 269–273]. Furthermore, even computational intelligence [46, 60, 90, 129, 191], based on either fuzzy systems or neural networks, recently have been handled as complex nonlinear systems by means of novel developments of either Lyapunov functions or Lyapunov–Krasovskii functionals [13, 52, 78, 103, 104, 152, 178, 183, 221, 254, 261]. In the view of these it may well be inferred that the derivation of sophisticated Lyapunov functions and/or Lyapunov–Krasovskii functionals in combination with the dynamic graphs and the principle of increasing precision with decreasing intelligence are the building blocks for future exploration of complex dynamic networks and systems under integrated supervision and control. However, as shown by Filippov and Arscott [71] and by Siljak [207, 210], the stability problem requires constructive use of techniques based on the theory of differential inclusions, e.g., see Aubin and Cellina [18], and based on the theory of functional differential equations, e.g., see Hale and Lumel [87]. The stability problem in such systems is by far more involved and subtle to handle in a clear-cut way.

Concluding Remarks

In this foreword, an attempt was made to bring many up-to-date facets of complexity and complex systems. It is only up to the interested readership to evaluate whether and to what extent it has been accomplished. To this author's belief, the concept of organization is the most fundamental one into the further quest for deeper comprehension of complex dynamic networks and systems if a theoretically sound, systematic, and yet reasonably pragmatic engineering design creativity has to be achieved. This seems to be an appropriate ahead perception for the future, which is largely argued for in the preceding sections as well as indicated in Figs. 7 and 9.

It is by intention, and by the permission of Prof. Ji-Feng Zhang, the Editor-in-Chief of AASC journal, that I selected and included here Fig. 10 as made of the two original figures from article [240]. Indeed, it may well be argued that Chinese colleagues promote the paradigm, control–computing–communication (C^3) as a dynamically evolving entity that, being integrated by means of control functions, decision and supervision essentially remain a simultaneous systemic information processing and communication mechanism [17, 39, 94, 179, 263, 266, 271]. Authors, L.Y. Wang and P. Zhao, have rightly entitled their article "Evolution of the feedback mechanism in information era"; in AASC 2(1), 70–76 (2014).

But it seems more than just interesting to notice: Was it just a coincidence with the decision by Springer Editorial to publish this book albeit documentation argued the same paradigm on the grounds of observing the evolution of life within the context of control of complex dynamic networks and systems control? Well, only time may provide some answer after all.

This monograph has materialized out of the efforts of a large international group of authors and matured over a period of a couple of years following a theme conference dedicated to the book's title, for which I am profoundly grateful to them all. It is therefore only in this foreword I dared to present my own perceptions, reasoning, and beliefs about complex networks and systems from a control perspective. Furthermore, I dared to place these on a prospect with regard to wider scientific knowledge and advanced engineering technologies available at present. Therefore, the responsibility is solely mine for the views expressed here. These views are expressed via symbolic paradigms as a look back and a look forward primarily on the background of physics although—to cite again words of late Prof. H.H. Rosenbrock—"…mathematics is indispensable…" [189], hence a sine-qua-non tool in systems and control sciences. In this control intellectual effort I did observe verbally the background mathematics of sets, set relations, and sets of

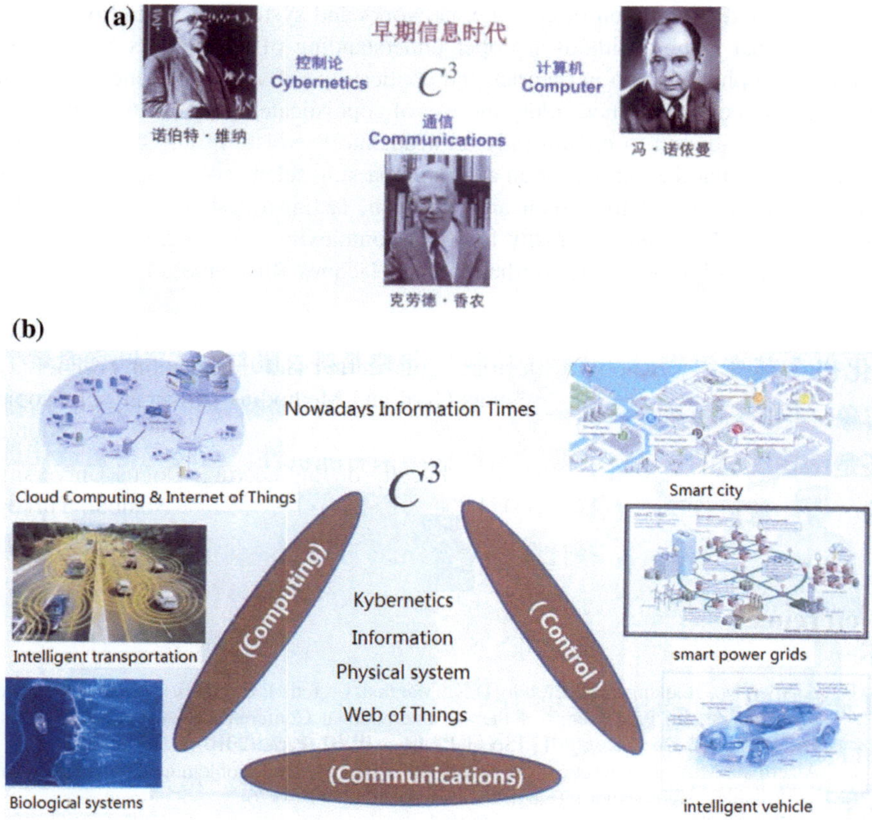

Fig. 10 The three celebrities of information-based sciences Norbert Weiner, John von Neuman, and Claude Shannon (**a**), and the C^3 paradigm in the journal, All About Systems & Control, AASC, founded in 2014 by the Academy of Mathematics & Systems Science of the Chinese Academy of Sciences (**b**)

sets as well as graphs and graphs following the pigeonhole principle of Church–Turing hypothesis [129] solely and in the first place. Till now only the computational complexity and the undecided-ability issue are formally and rigorously explained to the full via the NP-completeness or non-completeness [126]. Within the same context, also the potential emergence of paradoxes [62, 63, 191, 247] has been made a legitimate mathematical outcome event. In the view of these facts, one may argue that some paradoxes in the system-theoretic sense could also make legitimate outcome events if these are kept as questionable conjectures.

Finally, at this point I would like to emphasize again the famous remark put forward by Prof. Rudolph E. Kalman in one of his many plenary lectures: "First get the physics of considered problem right. The rest is mathematics." Indeed I do appreciate Kalman's advice deeply, despite my arguments favoring the necessity to involve cognitive and computational intelligence sciences as well into our perception and exploration of complex networks and systems. It should be widely noted for that matter, without a proper understanding of the physics behind any investigated phenomenon in theories and applications of systems science, decision science, and control science solely the use of sophisticated mathematical methods and models is bound to remain only an academic exercise. In any case, my own conclusion is that the buildup of an all-encompassing science of complex networks and systems is bound to remain an ever-open, lasting quest that appears to be correct, precisely because of many facets of complexity. "Who dares to say—this far man can go but not a step further."—Jean Jacques Russeaux [51, 56].

Georgi M. Dimirovski
School of Electrical Engineering and Information Technology
Saints Cyril and Methodius University of Skopje
Skopje, Macedonia

Faculty of Engineering, Dogus University
Istanbul, Turkey

References

1. Aiguier, M., Boulanger, F., Clotide, D.K., Marchal, C., (eds.): Complex systems design and management. In: Proceedings of the 4th International Conference on Complex Systems, Paris, FR, 4–6 December. 2013. Special Edition XXII, Paris (2014)
2. Aizerman, M.A.: Fuzzy sets, fuzzy proofs and some unsolved problem in automatic control (in Russian). Avtomatika i Telemekhanika **36**, 171–177 (1976)
3. Aizerman, M.A., Gusev, L.A., Petrov, S.V., Smirnov, I.M.: Dynamic approach to analysis of structures described by graphs (Foundations of graph dynamics; in Russian). Avtomatikai Telemekhanika **37**, 135–151 (1977)
4. Ahmadi, A.A., Jungers, R.M.: On complexity of Lyapunov functions for switched linear systems. In: Preprints of the 19th IFAC World Congress, Cape Town, ZA, 24–29 August. The IFAC and the SACAC, Pretoria, ZA, 5992–5997 (2014)
5. Albert, R., Barbasi, A.-L.: Topology of evolving networks: Local events and universality. Phys. Rev. Lett. **85**, 5234–5237 (2000)
6. Albert, R., Barbasi, A.-L.: Statistical mechanics of complex networks. Rev. Mod. Phys. **74**, 47–97 (2002)
7. Amari, S.: Field theory of self-organizing neural nets. IEEE Trans. Syst. Man Cybern. **13** (7), 741–748 (1983)
8. Anderson, B.D.O.: Stability of control systems with multiple nonlinearities. J. Franklin Inst. **282**, 155–160 (1966a)
9. Anderson, B.D.O.: Solution of quadratic matrix equations. Electron. Lett. **2**, 371–372 (1966b)
10. Anderson, B.D.O.: A system theory criterion for positive real matrices. SIAM J. Control **5**, 171–182 (1967b)
11. Andriyevsky, B.R., Matveev, A.S., Fradkov, A.L.: Control and estimation under information constraints: Toward a unified theory of control, computation and communications (in Russian). Avtomatika i Telemekhanika **71**(4), 34–99 (2010)

12. Anadriyevsky, B.R., Fradkov, A.L.: Control of chaos: Methods and applications; (i) Methods. Autom. Remote Control **64**(5), 673–713 (2003)
13. Antsaklis, P.J., (ed.): Special Issue on Hybrid Systems. Proc. IEEE, **88**(7), 879–1123 (2000)
14. Arrow, K.J., Hahn, F.H.: General Competitive Analysis. Holden-Day, San Francisco, CA (1971)
15. Ashby, W.R.: Design for a brain: The Origin of Adaptive Behaviour. Chapman & Hall, London, UK (1952)
16. Astroem, K., Albertos, P., Blanke, M., Isidori, A., Shaufelberger, W., Sanz, R., (eds.): Control of Complex Systems. Springer-Verlag Ltd., London (2001)
17. Akyol, E., Langbord, C., Basar, T.: Strategic compression and transmission of information: Crawford-Sobel meet Shannon. arXive paper pp. 1–10 (private communication), October (2015)
18. Aubin, J.-P., Cellina, A.: Differential Inclusions. Springer, Heidelberg (1984)
19. Azodomolky, S., Wieder, P., Yahyapour, R.: Cloud computing networking: Challenges and opportunities for innovations. IEEE Commun. Mag. **51**(7), 54–62 (2013)
20. Barbasi, A.: Linked: The New Science of Networks. Perseus Books, Reading, MA (2002)
21. Barnett, S.: Matrices in Control Theory. Van Nostrand Reinhold Co., London, UK (1971)
22. Basar, T., Bernhard, P.: H_{inf} Optimal Control and Related Minimax Design Problems (2nd edn.). Birkhaeser, Boston, MA (1995)
23. Basar, T., Olsder, G.J.: Dynamic Noncoopertive Game Theory. The SIAM, Philadelphia, PA (1999)
24. Beckenbah, E.F., (ed.): Applied Combinatorial Mathematics (Reprint). R.E. Krieger Pub. Co., Malabar, FL (1981)
25. Belykh, I., Belykh, V., Hasler, M.: Generalized connection graph method for synchronization in asymmetrical networks. Physica D **224**(1–2), 42–51 (2006)
26. Bersnusou, J., Titli, A.: Interconnected dynamical systems: Stability, Decomposition and Decentralization. North-Holland, Amsterdam, NL (1982)
27. Bernstien, D.: Feedback control: An invisible thread in the history of technology. IEEE Control Syst. Mag. **22**, 53–68 (2002)
28. Blanchini, F., Miani, S.: Set-Theoretic Methods in Control. Springer, Heidelberg (2007)
29. Bogdanov, A.A.: Tektology: The Universal Organizational Science (in Russian), Vol. 1 and 2 of Re-published 1925 Edition. Ekonomika, Moscow, RU (1989)
30. Bogdanov, A.: Essays in tektology: The General Science of Organization (2nd edn.). Intersystems Publishers, Seaside, CA (1984)
31. Bogdanov, A.A.: Philosophy of Living Experience (3rd edn.; in Russian). Kniga, Moscow USSR (1923)
32. Bogdanov, I., Bogdanov, G.: The Thought of God (in French). Bernard Grasset, Paris (2012)
33. Boyd, S., El Ghaoui, L., Feron, E., Balakrishnan, v.: Linear matrix inequalities in systems and control theory. The SIAM, Philadelphia, PA (1994)
34. Brown, J.C., Vincent, T.L.: Co-evolution as an evolutionary game. Evolution **41**, 66–79 (1987)
35. Buslenko, N.P.: Modeling of Complex Systems (2nd edn.; in Russian). Izdatelstvo "Nauka", Moskva, SSSR (1972)
36. Buttazzo, G.C.: Hard Real-Time Computing Systems: Predictable Scheduling Algorithms and Applications. Kluwer Academic Publishers, New York, NY (1997)
37. Callaway, D., Newman, M., Strogatz, S., Watts, D.: Network robustness and fragility: Percolation on random graphs. Phys. Rev. Lett. **85**, 5468–5471 (2000)
38. Casandras, C.G., Lafortune, S.: Introduction to Discrete Event Systems (2nd printing.) Kluwer Academic Pub., Norwell, MA (2001)
39. Casandras, C.G.: Event-driven control, communications, and optimization. In: Proceedings of the 32nd Chinese Control Conference (H.-F. Chen, General Chair, J.F. Zhang, General Co-Chair; and Pan, Q., Zhao, Q. Program Co-Chairs), Xi'an, July 26–28, Technical Committee on Control Theory, Chinese Association of Automation, Beijing, CN, 1–5 (2012)

40. Casandras, C.G., Lin, X., Ding, X.C.: An optimal control approach to the multi-agent persistent monitoring problem. IEEE Trans. Autom. Control **58**(4), 947–961 (2013)
41. Chen, D.: On finite potential game. Automatica **50**, 1793–1801 (2014)
42. Chen, D., Wang, Y., Liu, T.: A survey on potential evolutionary game and its applications. J. Control Decis. **2**(1), 26–45 (2015)
43. Chen, X., Dimirovski, G.M., Ma, R., Zhao, J.: Global stabilization of a class of switched cascade nonlinear systems. In: Chen, H.-F., General Chair, Zhang, J.-F., General Co-Chair; Xu, S., Zhao, Q., Program Co-Chairs, Proceedings of the 33rd Chinese Control Conference, Nanjing, CN, 28–30 July. Technical Committee on Control Theory, Chinese Association of Automation, Beijing, CN, 4023–4027 (2014)
44. Chen, X., Jiang, N., Jing, Y.: Differential game model and its solutions for force resource complementary via Lanchester squar law equation. In S. Bittanti, A. Cenedes, S. Zampieri, (eds.), Preprints of the 18th IFAC World Congress, Milano, IT, 28 August, 2 September. The IFAC Organization and CNR of Italy, Milano, IT, 14229–14233 (2011)
45. Chen, X., Jing, Y., Li, C., Dimirovski, G.M.: Effectiveness evaluation of warfare command systems with dissymmetrical warfare information. In: Proceedings of the 29th American Control Conference (Masada, G., General Chair; Braaz, R.D., Program Chair), Baltimore, M, 30 June–02 July. The IEEE, Piscataway, NJ, 5556–5560 (2010)
46. Churchland, P.S., Sejnowski, T.J.: The Computational Brain. MIT Press, Cambridge, MA (1992)
47. Colegrave, R.K., Hyde, J.M.: The Lanchester square-law model extended to a (2, 2) conflict. IMA J. Appl. Math. **51**(2), 95–109 (1993)
48. Cogill, R., Lall, S.: Control design for topology-independent stability of interconnected systems. In Proceedings of the American Control Conference, Boston, MA. The AACC and the IEEE, Piscataway, NJ, 3717–3722 (2004)
49. Darwin, C.: On the Origin of Species by Means of Natural Selection, or the Preservation of Favored Races in the Struggle for Life. Murray, London (1859)
50. Dennett, D.S.: Darwin's Dangerous Idea: Evolution and the Meanings of Life. Simon Schuster, New York, NY (1995)
51. Dimirovski, G.M.: System-of-systems, Large-scale Systems, Complex Systems or Complex Networks: Where do we stand? In Jorge Pereira, Open Consultation on Complex Systems Engineering towards System-of-Systems, Brussels, 4 July 2011. EC-INFSO, Brussels, BE, Position Paper II.9 (2011)
52. Dimirovski, G.M.: Complexity versus integrity solution in adaptive fuzzy-neural inference models. Int. J. Intell. Syst. **23**, 556–573 (2008)
53. Dimirovski, G.M., Jing, Y.: Kolmogorov networks and process characteristic input-output modes decomposition. In: Samad, T., Sgurev, V., Co-Chairs Proceedings of the 1st International IEEE Symposium Intelligent Systems, Varna, BG, 10–12 September. Federation of Technical Unions of Bulgaria and IEEE, Piscataway, NJ, I, 59–66 (2002)
54. Dimirovski, G.M.: Fuzzified Petri-Nets and Their Application to Organizing Supervisory Controller. In Kaynak, O., Zadeh, L.A., Turksen, B., Rudas, I.J., (eds.), Computational Intelligence: Soft Computing and Fuzzy-Neuro Integration with Applications. NATO ASI Series in Computer and Systems Sciences Vol. 162. Springer, Berlin, 260–282 (1998)
55. Dimirovski, G.M., Gough, N.E., Barnett, S.: Categories in systems and control theory. Int. J. Syst. Sci. **8**(9), 1081–1090 (1977)
56. Dimirovski, G.: Systems and Control: Energy-Matter-Information Interactions (Seminar 7; in Macedonian). Mathematical Institute with Numerical Centre, St. Cyril, St. Methodius University, Skopje, MK (1970)
57. Dooley, K.J., Ven, A.H.V. de: Explaining complex organizational dynamics. Organ. Sci. **10** (3), 358–372 (1999)
58. Dobzhansky, T.G.: Genetics and the Origin of Species. Columbia University Press, New York, NY (1937)

59. Duan, S., Sun, J.: A self-adaptive middleware for efficient routing in distributed sensor networks. UC Davis paper to appear in the Proceedings of the IEEE International Conference on Systems, Man, and Cybernetics SMC2015, Hong Kong, HK, 09–12 October. Retrieved through eScholarhsip of the University of California, pp. 1–6 (2015)
60. Dubios, D.M.: The Labyrinth of the Intelligence (2nd edn.; in French). Inter-Editions, Paris-Academia et Luoven-le-Neuve (1990)
61. Dubios, D.M., Resconi, G.: Mathematical foundation of non-linear threshold logic: A new paradigm for the technology of neural machines. Academie Royal de Belgique, Builletin de la Class des Sciences, 6-eme serie', Tome IV (1–6), 91–122 (1993)
62. Einstein, A., Podolsky, Rosen, N.: Can quantum-mechanical description of physical reality be considered complete? Phys. Rev. **47**, 777–780 (1935)
63. Einstein, A.: Remarks on Bertrand Russell's theory of knowledge. In: P.A. Schlipp, (ed.), The Philosophy of Bertrand Russell, Northwestern University, Evanston, IL, 211–232 (1944)
64. Elman, J.L.: Finding structure in time. Cognitive Sci. **14**, 179–211 (1990)
65. Emelyanov, S.V.: A method for obtaining complex regulation laws using only the error signal or the regulated coordinate and its first derivatives (in Russian). Avatomatikai Telemekhanika **18**(10), 873–875 (1957)
66. Epstein, J.: Nonlinear Dynamics, Mathematical Biology, and Social Science. Santa Fe Institute Studies in Sciences of Complexity. Addison Wesley, Reading, MA (1997)
67. Erdi, P.: Complexity Explained. Springer, Berlin (2008)
68. Evans, J.S.B.T.: In two minds: Dual-process accounts of reasoning. Trends Cognitive Sci. **7**, 454–459 (2003)
69. Fasol, K.H.: Hermann Schmidt: Pioneer in control and cybernetics. IEEE Control Syst. Mag. **22**(2), 24–38 (2002)
70. Fasth-Berglund, A., Stahre, J.: Cognitive automation strategy for reconfigurable and sustainable assembly systems. Assembly Autom. **33**, 294–303 (2013)
71. Filippov, A.F., Arscott, F.M.: Differential Equations with Discontinuous Right Hand Sides: Control Systems. Kluwer Springer, Dordrecht, NL (1988)
72. Fomin, V.N., Fradkov, A.L., Yakubovich, V.A.: Adaptive Control of Dynamical Objects (in Russian). Nauka, Moskva, SSSR (1981)
73. Ford, I.R., Fulkerson, D.R.: Flows in Networks. Princeton University Press, Princeton, NJ (1962)
74. Fradkov, A.L.: Cybernetic Physics: Principles and Examples (in Russian; SPb). Nuaka. Moskva, RU (2003)
75. Firdman, E., Shaked, U.: Delay-dependent stability and H_{inf} control: Constant and time-varying delays. Int. J. Control **76**(1), 48–60 (2003)
76. Fudenberg, D., Tirole, J.: Game Theory. MIT Press, Cambridge, MA (1991)
77. Gajic, Z., Shen, X.: Parallel Algorithm for Optimal Control of Large Scale Linear Systems. Springer, London (1993)
78. Gao, H., Chen, T., Lam, J.: A new delay system approach to network-based control. Automatica **44**(1), 39–52 (2008)
79. Gare, A.: Aleksandr Bogdanov and systems theory. Democracy Nature **6**(3), 341–359 (2000)
80. Garcia, E., Antsaklis, P.J.: Model-based event-triggered control for systems with quantization and time-varying network delays. IEEE Trans. Autom. Control **58**, 422–434 (2013)
81. Gelig, A.H., Leonov, G.A., Fradkov, A.L., Editors-in-Chief: In Honor of Vladimir A. Yakubovich—Frequency and Matrix Inequalities (in Russian). Gosudarstvennoe Izdatelstvo Fiziko-Matematicheskoy Literaturi, Moskva, RU (2008)
82. Gitt, W.: Information: The third fundamental quantity. Siemens Rev. **56**, 36–41 (1987)
83. Grandy, Jr., W.T.: Foundations of Statistical Mechanics: Equilibrium Theory. Reidel, Dodrecht, NL (1987)
84. Gruenwald, P.D.: The Minimum Description Length Principle. MIT Press, Cambridge, MA (2007)

85. Guastello, S.J.: Chaos, Catastrophe, and Human Affairs: Application of Nonlinear Dynamics to Work, Organization and Social Evolution. Mathway Pub., New Jersey, NJ (1995)
86. Haddad, W.M., Chellaboina, V.: Dissipativity theory and stability of feedback interconnections for hybrid dynamical systems. Math. Prob. Eng. **7**(4), 299–335 (2001)
87. Haddad, W.M., Chellaboina, V.: Stability and dissipativity theory for nonnegative dynamical systems: A unified analysis framework for biological and physiological systems. Nonlinear Analysis: Real World Applications **6**, 35–65 (2005)
88. Hale, J.K., Lumel, S.M.V.: Introduction to Functional Differential Equations. Springer-Verlag Ltd., New York, NY (1993)
89. Harary, F., Norman, R.Z., Cartwright, D.: Structural Models: An Introduction to the Theory of Directed Graphs. J. Wiley and Sons, New York, NY (1965)
90. Haykin, S.: Neural Networks and Learning Machines (3rd edn.). Pearson Prentice Hall, Upper Saddle River, NJ (2009)
91. Hawking, S.W.: Brief History of Time. Cambridge University Press, Cambridge, UK (1988)
92. Hawking, S.W.: Conservation of information and estimation of time for black holes. arXive paper pp. 1–3, January (2014)
93. Hebb, D.O.: The Organization of Behavior: A Neuropsychological Theory. J. Wiley, Sons, New York, NY (1949)
94. Heemels, W.P.M.H., Donkers, M.C.F.: Model-based periodic event-triggered control for linear systems. Automatica **49**(7), 698–711 (2013)
95. Helmbold, R.L.: A modification of Lanchester's equations. Oper. Res. **13**(2), 857–859 (1975)
96. Heisenberg, W.: Physics and Philosophy (Reprint of the 1962 book). Penguin Books Ltd., New York, NY (1990)
97. Hermann, R., Krener, A.: Nonlinear controllability and observability. IEEE Trans. Autom. Control **22**, 728–740 (1977)
98. Hill, D.J., Moylan, P.J.: Dissipative dynamical systems: Basic input-output and state properties. J. Franklin Inst. **309**(5), 327–357 (1980)
99. Ho, Y.C.: On stochastic approximations and optimal filtering. J. Math. Anal. Appl. **6**(1), 152–154 (1963)
100. Ho, Y.C., Lee, R.C.K.A.: Bayesian approach to problems of stochastic estimation and control. IEEE Trans. Autom. Control **9** (October), 333–339 (1964)
101. Holland, J.H.: Emergence—From Chaos to Order. Oxford University Press, Oxford (1998)
102. Holland, J.H.: Hidden Order—How Adaptation Builds Complexity. Perseus Books, Oxford (1995)
103. Hong, Y., Dimirovski, G.M., Zhao J.: Switched Fuzzy Systems: Representation Modeling, Stability Analysis, and Control. In J. Kacprzyk, Editor, Studies in Computational Intelligence **109**, 169–184 (2008)
104. Huang, C., Ho, D.W.C., Lu, J., Kurths, J.: Pinning synchronization of T-S fuzzy complex networks with partial and discrete-time couplings. IEEE Trans. Fuzzy Syst. **23**(4), 1274–1285 (2015)
105. Ikeda, M., Siljak, D.D.: Lotka-Volterra equations: Decomposition, stability and structure, Part I: Equilibrium analysis. J. Math. Biol. **9**, 56–83 (1980)
106. Ikeda, M., Siljak, D.D.: Hierarchical Lyapunov functions. J. Math. Anal. Appl. **111**, 110–128 (1985)
107. Ilic, M., Zaborsky, J.: Dynamics and Control of Large Electric Power Systems. J. Wiley, New York, NY (2000)
108. Isaacs, R.: Differential Games. J. Wiley, New York, NY (1965)
109. Isermann, R.: Theoretical Analysis of the Dynamics of Industrial Processes (in German). Bibliographische Institut, Manheim, DE—Wien, AT—Zuerich, CH (1967)
110. Isidori, A.: Nonlinear Control Systems (3rd edn.). Springer, Berlin, DE (1995)
111. ISO 14258: Industrial Automation Systems: Concepts and Rules for Enterprise Models (2005)

112. Jackson, E.: Exploring Nature's Dynamics. Wiley Series in Nonlinear Science. J. Wiley, New York, NY (2001)
113. Jang, J.-S.T., Sun, C.-T.: Neuro-fuzzy modeling and control. Proc. IEEE **83**, 378–406 (1995)
114. Jensen, H.: Self-Organized Criticality. Cambridge University Press, Cambridge, UK (1998)
115. Jing, Y., Chen, X., Li, C., Stojanovski, G.S., Dimirovski, G.M.: Optimal resource partitioning in warfare command systems using Lanchester equations: A simple solution to complex problem. In: Kolemisevska-Gugulovska, T., Stankovski, M.J., (eds.), Proceedings of International Conference on Complex Systems: Synergy of Control, Communications and Computing, Ohird, MK, 16–20 September. The ETAI Society of R. Macedonia, Skopje, MK, 229–235 (2011)
116. Jones, E.P.C., Li, L., Schmidtke, J.K., Ward, P.A.S.: Practical routing in delay-tolerant networks. IEEE Trans. Mob. Comput. **6**(8), 943–959 (2007)
117. Jungers, R.M., Blondel, D.V.: On fitness property for rational matrices. Linear Algebra Appl. **428**, 2283–2295 (2008)
118. Jungers, R.M., Protasov, Yu.V., Blondel, D.V.: Efficient algorithm for deciding the type of growth of products of integer matrices. Linear Algebra Appl. **428**, 2296–2311 (2008)
119. Kaber, D.B., Riley, J.M., Tan, K. Endsley, M.R.: On the design of adaptive automation for complex systems. Int. J. Cogn. Ergon. **5**(1), 37–57 (2001)
120. Kalman, R.E.: Physical and mathematical mechanisms of instability in nonlinear automatic control systems. Transactions ASME J. Basic Eng. **79**, D, 553–556 (1957)
121. Kalman, R.E.: Contributions to the theory of optimal control. Boletin de la Sociedad Matematica Mexicana **5** (segundaserie, 1), 102–119 (1960a)
122. Kalman, R.E.: New approach to linear filtering and prediction problems. Trans. ASME J. Basic Eng. **82**, D, 35–45 (1960b)
123. Kalman, R.E.: On the general theory of control systems. In Proceedings of the First International Congress of the IFAC, Moscow, 25 June–5 July 1960. Buthreworths & Company, London, UK, 481–492 (1961)
124. Kalman, R.E.: Lyapunov functions for the problem of Lurye in automatic control. Proc. Nat. Academy Sci. USA **49**(2), 201–205 (1963a)
125. Kalman, R.E.: Mathematical description of dynamical systems. SIAM J. Control **1**, 152–192 (1963b)
126. Kauffman, S.: The Origins of Order: Self-Organization and Selection in Evolution. Oxford University Press, Oxford, UK (1993)
127. Kauffman, S.: At Home in the Universe: The Search for Laws of Self-Organization and Complexity. Oxford University Press, Oxford, UK (1995)
128. Kinber, E., Smith, C.: Theory of Computing—A Gentle Introduction. Prentice Hall, Upper Saddle River, NJ (2001)
129. Kosko, B.: Neural Networks and Fuzzy Systems, a Dynamical Systems Approach to Machine Intelligence. Prentice Hall, Englewood Cliffs, NJ (1992)
130. Krapivsky, P.L., Redner, S., Leyvraz, F.: Connectivity of growing networks. Phys. Rev. Lett. **85**, 4629–4632 (2000)
131. Krapivsky, P.L., and Redner, S.: Organization of growing random networks. Phys. Rev. E **63**, 066123 (2001)
132. Krasovskii, N.N.: Some Problems of the Theory of Stability of Motion (in Russian). GosudarstvennoeI zdatelstvo Fiziko-Matematicheskoy Literaturi, Moskva, SSSR (1959)
133. Krasovskii, A.A., Pospelov, G.S.: Foundations of Automatic Control and Engineering Cybernetics (in Russian). Gosudarstvennoe Energeticheskoe Izdatelstvo, Moskva—Leningrad, SSSR (1962)
134. Kolmogorov, A.N.: Interpolation and extrapolation of stochastic sequences (in Russian). Izvestiya Akademii Nauk SSSR (Mathematics series) **5**, 3–14 (1941)
135. Kolmogorov, A.N.: On the representation of continuous functions of several variables by superposition of continuous functions of one variable and addition. Dokladi Akademii Nauk SSSR **114**, 953–956 (1957)

136. Kolmogorov, A.N.: Three approaches to the quantitative definition of information. Prob. Inf. Transm. **1**(1), 1–7 (1965)
137. Kurkova, V.: Kolmogorov's theorem and multi-layer neural networks. Neural Netw. **5**, 501–506 (1992)
138. Kozyakin, V.A.: Algebraic unsolvability of the problem of absolute stability of desynchronized systems. Autom. Remote control **51**, 754–759 (1990)
139. Lagariss, J.C., Wang, Y.: The fitness conjecture for the generalized spectral radius of a set of matrices. Linear Algebra Appl. **214**, 17–42 (1995)
140. Lee, T.C., Jiang, Z.P.: A generalization of Krasovskii-LaSalle theorem for nonlinear time-varying systems: Converse results and applications. IEEE Trans. Autom. Control **50**, 1147–1163 (2005)
141. Li, M., Vitanyi, P.: An Introduction to Kolmogorov Complexity and Its Applications. Springer, New York, NY (1993)
142. Liberzon, D., Nesic, D., Teel, A.R.: Lyapunov-based small-gain theorems for hybrid systems. IEEE Trans. Autom. Control **59**(6), 1395–1410 (2014)
143. Liapunov, A.M.: Problem general de la stabilite' du movement. Annals de la Faculte' des Sciences de l'Universite' de Toulouse, **9**, 203–474 (1907); translated into English and published as General Problem of the Stability of Motion. Annals of Mathematics Studies, Vol. 17, Princeton University Press, Princeton, NJ (1949)
144. Liu, K., Fridman, E., Johanson, K.H.: Dynamic quantization of uncertain linear networked control systems. Automatica **59**, 248–255 (2015)
145. Liu, K., Fridman, E.: Network-based stabilization via discontinuous Lyapunov functional. Int. J. Robust Nonlinear Control **22**(4), 420–436 (2012)
146. Liu, T., Dimirovski, G.M., Zhao, J.: Exponential synchronization of complex delayed dynamical networks with general topology. Physica A **387**(2–3), 643–652 (2008)
147. Liu, T., Zhao, J., Hill, D.J.: Exponential synchronization of complex delayed dynamical networkswith switching topology. IEEE Trans. Circuits Syst. I Regul. Pap. **57**(11), 2312–2322 (2010)
148. Liu, T., Hill, D.J., Zhao, J.: Output synchronization of dynamical networks with incrementally-dissipative nodes and switching topology. IEEE Trans. Circuits Syst. I Regul. Pap. **62**(9), 2312–2322 (2015)
149. Liu, Y.Y., Zhao, J., Dimirovski, G.M.: Pssivity, feedback equivalence and stability of switched nonlinear systems using multiple storage functions. In H.-F. Chen, General Chair, Zhang, J.-F., and Zhao, Q.-C., Program Co-Chairs, Proceedings of the 30th Chinese Control Conference, Yantai, CN, 22–24 July. Technical Committee on Control Theory, Chinese Association of Automation, Beijing, CN, 1805–1809 (2011)
150. Liu, Y., Dimirovski, G., Zhao, J.: Output synchronization of dynamical networks with nodes of relative degree one. In: Kolemisevska-Gugulovska, T., Stankovski, M.J., (eds.), Proceedings International Conference on Complex Systems: Synergy of Control, Communications and Computing, Ohird, MK, 16–20 September. The ETAI Society of R. Macedonia, Skopje, MK, 115–120 (2011)
151. Long, L., Zhao, J.: A small-gain theorem for switched interconnected nonlinear systems and its applications. IEEE Trans. Autom. Control **50**(4), 1082–1088 (2014)
152. Long, L., Zhao, J.: Adaptive output feedback neural control of switched uncertain nonlinear systems with average dwell time. IEEE Trans. Neural Netw. Learn. Syst. **26**(7), 1350–1374 (2015)
153. Lurye, A.I.: Some Nonlinear Problems in the Theory of Automatic Control (in Russian). Gosudarstvenoe Izdatelstvo Tekhnicheskoy Literaturi, Moscow–Leningrad, SSSR (1951)
154. MacFarlane, A.G.J.: Information, Knowledge, and Control. In: Trentelmann, H.L., Willems, J.C., (eds.), Essays on Control: Perspectives in the Theory and Applications. Birkhauser, Boston, MA, 1–28 (1993)
155. Maturana, H.R., Varela, F.J.: Autopoiesis and Cognition: The Realization of the Living. Dordrecht, NL, Reidel, (1980) and Kluwer Academic (1991)

156. Maturana, H.R., Varela, F.J.: The Tree of Knowledge: The Biological Roots of Human Understanding, Shambhala Pub., Boston, MA (1987)
157. Maynard-Smith, J.: Evolution and the Theory of Games. Cambridge University Press, Cambridge, UK (1982)
158. Mesarovic, M.D., Takahara, Y.: General Systems Theory: Mathematical Foundations. Academic Press, New York, NY (1975)
159. Monderer, D., Shapley, L.S.: Potential Games. Games and Economic Behavior **14**, 124–143 (1996)
160. Morel, B., Ramanujam, R.: Through the looking glass of complexity: The dynamics of organizations as adaptive evolving systems. Organ. Sci. **10**(3), 278–293 (1999)
161. Musolsi, M., Mascolo, C.: CAR: Context aware adaptive routing for delay-tolerant mobile networks. IEEE Trans. Mob. Comput. **8**(2), 246–260 (2009)
162. Nash, J.: Non-cooperative game. The Ann. Math. **54**, 286–295 (1951)
163. Newman, M.E.J.: Networks. An Introduction. Oxford University Press, Oxford, UK, and New York, NY (2010)
164. Newman, M.E.J.: Modularity and community structure in networks. Proc. Nat. Acad. Sci. U. S.A. **103**, 8577–8582 (2006)
165. Nillson, N.J.: Principles of Artificial Intelligence. Springer, New York, NY (1980)
166. Nowak, M.: Evolutionary Dynamics. Exploring the Equations of Life. The Belknap Press—Harvard University Press, New Haven, NJ (2006)
167. Olfati-Saber, R., Fax, J.A., Murray, R.M.: Consensus and cooperation in networked multi-agent systems. Proc. IEEE **95**(1), 215–233 (2007)
168. Palla, G., Derenyi, I., Farkas, I., Vicsek, T.: Uncovering the overlapping community structure of complex networks in nature and society. Nature **435**(7043), 814–818 (2005)
169. Pagels, H.: The Dreams of Reason: The Computer and the Rise of the Sciences of Complexity. Simon Schuster, New York, NY (1988)
170. Pavlov, I.P.: Conditional Reflexes: An Investigation of the Physiological Activity of Cerebral Cortex (Translation form Russian by G.V. Anreo). Oxford University Press, New York (1927)
171. Penrose, R.: Emperor's New Mind: Concerning Computers, Minds and Laws of Physics. Oxford University Press, Oxford, UK (1989)
172. Perreto, P.: Collective properties of neural networks: A statistical physics approach. Biol. Cybern. **50**, 51–62 (1984)
173. Perreto, P., Niez, J.-J.: Stochastic dynamics of neural networks. IEEE Trans. syst. Man Cybern. **16**(2), 73–83 (1986)
174. Poggio, T., Girosi, F.: Networks for approximation and learning. Proc. IEEE **78**, 1481–1497 (1990)
175. Poincare', H.: Sur le proble'me de trois corps et de les e'quations de la dinamique. Acta Mathematica **13**, 1–270 (1890)
176. Prigogine, I.: From Being to Becoming. Freeman, San Francisco, CA (1980)
177. Pontryagin, L.S., Boltyanskiy, V.G., Gamkrelidze, P.V., Mishchenko, E.F.: Mathematical Theory of Optimal Processes (in Russian). Gosudarstvennoe Izdatelstvo Fiziko-Matematicheskoy Literaturi, Moskva, SSSR (1961)
178. Qin, J., Gao, H., Zheng, W.X.: Exponential synchronization of complex networks of linear systems and nonlinear oscillators: A unified analysis. IEEE Trans. Neural Netw. Learn. Syst. **26**(3), 510–521 (2015)
179. Qin, J., Yu, C., Gao, H.: Coordination for linear multi-agent systems with dynamic interaction topology in the leader-following framework. IEEE Trans. Industr. Electron. **61** (5), 2412–2422 (2014)
180. Radcliffe-Brown, A.: On Social structure. J. Roy. Anthropol. Inst. Great Britain Ireland **70**, 1–12 (1940)

181. Rand, D.A., Wilson, H., and McGlade, J.M.: Dynamics and evolution: Evolutionary stable attractors, invasion exponents and phenotype dynamics. Philos. Trans. R. Soc. Lond. Ser. B Biol. Sci. **343**(1305), 261–283 (1994)
182. Rasmunsen, J.: Skills, rules, knowledge, signals, signs and symbols and other distinctions human performance models. IEEE Trans. Syst. Man Cybern. **13**, 257–266 (1983)
183. Ren, B., Ge, S.S., Tee, K.P., and Lee, T.H.: Adaptive neural control for output feedback nonlinear systems using barrier Lyapunov functions. IEEE Trans. Neural Netw. **21**(8), 1339–1345 (2010)
184. Rissanen, J.: Stochastic Complexity in Statistical Inquiry. World Scientific, Singapore (1989)
185. Rissanen, J.: Modeling by shortest data description. Automatica **14**, 465–471 (1978)
186. Rosen, R.: Anticipatory Systems: Philosophical, Mathematical and Methodological Foundations. Pergamon Press, Oxord, UK (1985)
187. Rosen, R.: Life Itself: A Comprehensive Inquiry into the Nature, Origin and Fabrication of Life. Columbia University Press, New York, NY (1991)
188. Romero, D., Noran, O., Stahre, J., Pernus, P., Fast-Berglund, A.: Towards human-centered reference architecture for next generation balanced automation systems: Human-automation symbiosis. In: Umeda, S., Nakano, M., Hibino, N., Kiritsis, D., von Cieminski, G., Eds., Proceedings APMS2015—Innovative Production Management Towards Sustainable Growth, Service, Manufacturing, and Resilient Value Chain. The IFIP, Part II, AICT 460, pp. 1–15 (2015)
189. Rosenbrock, H.H.: The future of control. Automatica **13**, 389–392 (1977)
190. Rosenbrock, H.H.: Machines with a Purpose. Oxford University Press, Oxford, UK (1990)
191. Russell, B.: Vagueness. Aust. J. Psychol. Philos. **1**, 84–92 (1923)
192. Salthe, S.N.: Evolving Hierarchical Systems. Columbia University Press, New York, NY (1985)
193. Samuelson, L.: Evolutionary Games and Equilibrium Selection. The MIT Press, Cambridge, MA (1997)
194. Saridis, G.N.: Analytical formulation of the principle of increasing precision with decreasing intelligence, Automatica **25**, 461–467 (1989)
195. Schmidt, H.: The development of technology as a phase of human adaptation (in German). Z. VDI **96**, 118–122 (1954).
196. Schmidt, H.: Origin and rise of cybernetics (in German). Z. VDI **106**, 749–753 (1964)
197. Sezer, S., Scott-Hayward, S., Chouhan, P. K., Fraser, B., Lake, D., Finnegan, J., Viljoen, N., Miller, M., Navneet, R.: Implementation challenges for software-defined networks. IEEE Commun. Mag. **52**, 36–43 (2013)
198. Shannon, C.E.: A mathematical theory of communication. Bell Syst. Tech. J. **27**, 379–423 and 623–656 (1948)
199. Shannon, C.E.: Communication in the presence of noise. Proc. Radio Eng. **37**(1), 10–21 (1949)
200. Sheridan, T., Parasuraman, R.: Human-automation interaction. Hum. Factors Ergon. **1**(1), 89–129 (2006)
201. Sher Decusatis, C.J., Caranzza, A., Decusatis, C.M.: Communication with clouds: Open standards and proprietary protocols for data center networking. IEEE Commun. Mag. **50**(9), 26–33 (2012)
202. Siljak, D.D.: On stability of large-scale system under structural perturbations. IEEE Trans. Syst. Man Cybern. **3**, 415–417 (1973)
203. Siljak, D.D.: Connective stability of competitive equilibrium. Automatica, **11**, 389–400 (1975)
204. Siljak, D.D., Vukcevic, M.B.: Large-scale systems: Stability, complexity, reliability. J. Franklin Inst. **301**, 49–69 (1976)
205. Siljak, D.D.: On reachability of dynamic systems. Int. J. Syst. Sci. **8**, 321–338 (1977a)
206. Siljak, D.D.: On pure structure of dynamic systems. Nonlinear Analysis: Theory, Methods, and Applications. **1**, 397–413 (1977b)

207. Siljak, D.D.: Large-Scale Dynamic Systems: Stability and Structure. North-Holland Books, New York, NY (1978)
208. Siljak, D.D.: Complex dynamic systems: Dimensionality, structure, and uncertainty. Large Scale Syst. **4**, 279–294 (1983)
209. Siljak, D.D.: Overlapping Decentralized Control. In: Singh, M.G., Editor-in Chief, Systems and Control Encyclopedia, pp. 3568–3572. Pergamon Press, London, UK, (1987)
210. Siljak, D.D.: Decentralized Control of Complex Systems. Academic Press, Cambridge, MA (1991)
211. Siljak, D.D., Zecevic, A.I.: Control of large-scale systems: Beyond decentralized feedback. Annu. Rev. Control **29**, 69–179 (2005)
212. Siljak, D.D.: Dynamic graphs. Nonlinear Analysis: Hybrid Systems **2**, 544–567 (2008)
213. Simon, H.D.: The architecture of complexity. Proc. Am. Philos. Soc. **106**, 467–482 (1962)
214. Stackelberg, H.V.: Theory of Games and Economic Behaviour. Oxford University Press, Oxford, UK (1952).
215. Steinhardt, P., Turok, N.: A cyclic model of the universe. Science **296**, 1436–1439 (2002)
216. Stipanovic, D.M., Siljak, D.D.: Stability of polytopic systems via convex M-matrices and parameter-dependent Lyapunov functions. Nonlinear Analysis **40**, 589–609 (2000)
217. Stipanovic, D.M., Siljak, D.D.: Connective stability of discontinuous dynamic systems. J. Optim. Theory Appl. **115**, 711–726 (2002)
218. Strogatz, S.: Nonlinear Dynamics and Chaos: With Applications to Physics, Biology, Chemistry and Engineering. Addison Wesley, Reading, MA (1994)
219. Strogatz, S.H.: Exploring complex networks. Nature **410**, 268–276 (2001)
220. Sun, J., Duan, S.: A self-adaptive middleware for efficient routing in distributed sensor networks. Appeared in Proceedings of the IEEE International Conference on Systems, Man, and Cybernetics SMC2015, Hong Kong, CN, 9–12 October. The IEEE, Piscataway, NJ, UCD-pp. 1–6 (2015)
221. Sun, Q., Li, R., Zhang, P.: Stable and optimal adaptive fuzzy control of complex systems using fuzzy dynamical model. Fuzzy Sets Syst. **133**, 1–17 (2003)
222. Taylor, J.G.: Lanchester Models of Warfare, Vols. 1 and 2. Operations Research Society of America, Military Applications Section, Arlington, VA (1983)
223. Tang, Y., Gao, H., Shi, Y., Kurths, J.: Sychronizability, controllability and observability of networked multi-agent systems (Editorial). J. Franklin Inst. **352**(9), 3405–3409 (2015)
224. Tang, Y., Gao, H., Zhou, W., Kurths, J.: Distributed synchronization of agent systems with nonlinearities and random switchings. IEEE Trans. Cybern. **43**(1), 358–370 (2013)
225. Tee, K.P., Ge, S.S., Tay, E.H.: Barrier Lyapunov functions for the control of output constrained nonlinear systems. Automatica **45**(4), 918–927 (2009)
226. Tee, K.P., Ge, S.S.: Control of state-constrained nonlinear systems using integral barrier Lyapunov functional. In: Proceedings of the 51st IEEE Conference on Decision and Control, Maui, Hawaii, 10–13 December. The IEEE, Piscataway, NJ, 3239–3244 (2012)
227. Titli, A., Editor-in-Chief: Analysis and Control of Complex Systems. Monograph AFCET (in French). Cepadeus, Toulouse, FR (1979)
228. Tseng, H.C., Siljak, D.D.: A learning scheme for dynamic neural networks: Equilibrium manifold and connective stability. Neural Netw. **8**, 853–864 (1995)
229. Tsien, H.-S.: Engineering Cybernetics. McGraw-Hill, New York, NY (1954)
230. Turing, A.M.: Computing machinery and intelligence. Mind **59**, 433–460 (1950)
231. Tzafestas, S.: Concerning human-automation symbiosis in the society and the nature. International Journal of Factory Automation, Robot. Soft Comput. **1**(3), 16–24 (2006)
232. Varela, F., Thompson, E., Rosch, E.: The Embodied Mind: Cognitive Science and Human Experience. The MIT Press, Cambridge, MA (1991)
233. Von Foerester, H.: Observing Systems. Intersystems Publishers, Salinas, CA (1981)
234. Von Foerester, H.: Cybernetics of Cybernetics (2nd edn.). Future Systems, Inc., Minneapolis, MN (1995)

235. Von Foerester, H.: Understanding Understanding: Essays on Cybernetics and Cognition. Springer, New York, NY (2002)
236. Von Neumann, J., Morgenstern, O.: Theory of Games and Economic Behavior. J. Wiley, New York, NY (1944)
237. Von Neumann, J.: The Computer and the Brain. Yale University Press, New Heaven, NJ (1958)
238. Wagner, R.: Biological mechanisms of control. Z. VDI **96**, 123–130 (1954)
239. Wan, B.: Second-order cybernetics and its founder Heinz von Foerester. All About Systems and Control **1**(2), 30–40 (2014)
240. Wang, L.Y., Zhao, P.: Evolution of the feedback mechanism in the information era. All About Systems and Control **2**(1), 70–76 (2015)
241. Wang, X.F., Chen, G.: Synchronization in small-world dynamical networks. Int. J. Bifurcat. Chaos **12**(1), 187–192 (2002a)
242. Wang, Y.W., Wang, H.O., Xiao, J.W., Guan, Z.H.: Synchronization of complex dynamical networks under recoverable attacks. Automatica **46**(1), 197–203 (2010a)
243. Watts, D.: Small Worlds: The Dynamics of Networks Between Order and Randomness. Princeton Studies in Complexity. Princeton University Press, Princeton, NJ (1999)
244. Weaver, W.: Science and complexity. Am. Sci. **36**, 538 (1948)
245. Weidlich, W.: Sociodynamics. A Systematic Approach to Mathematical Modeling in the Social Sciences. Harwood Academic Publishers, New Haven, NJ (2000)
246. Wiener, N.: Cybernetics or Control and Communication in the Animal and the Machine. The MIT Press, Cambridge, MA (1948)
247. Whitehead, A.N., Russell, B.: Principia Mathematicae (2nd edn.). Cambridge University Press, Cambridge, UK (1927)
248. Wolkenhauer, O.: Systems biology: The reincarnation of systems theory applied in biology? Briefings in Bioinformatics **2**, 258–270 (2001)
249. Wright, S.J.: Primal-Dual Interior-Point Methods. The SIAM, Philadelphia, PA (1997)
250. Yakubovich, V.A.: Solution to some matrix inequalities encountered in the theory of automatic control (in Russian). Dokladi Akadamii Nauk SSSR **143**(6), 1304–1307 (1962)
251. Yakubovich, V.A.: Solution to some matrix inequalities encountered in the nonlinear theory of automatic control (in Russian). Dokladi Akadamii Nauk SSSR **156**(2), 278–281 (1964)
252. Yakubovich, V.A.: S-procedure in the nonlinear theory of automatic control (in Russian). Vestnik Leningradskogo Gosudarsvenogo Universiteta, Seriya Matematika, Mehanika, Astronomiya **1**, 62–77 (1971)
253. Yakubovich, V.A.: Optimization and invariance in linear stationary control system (in Russian). Avtomatika i Telemekhanika **8**, 5–44 (1984)
254. Yang, B., Wang, R., Shi, P., Dimirovski, G.M.: New delay-dependent stability criteria for recurrent neural networks with time-varying delays. Neurocomputing **151**(P3), 1414–1422 (2015)
255. Yeganeh, S., Tootoonchian, A., Ganjali, Y.: On scalability of software defined networking. IEEE Commun. Mag. **51**, 136–141 (2013)
256. Zadeh, L.A.: Outline of a new approach to the analysis of complex systems and decision processes. IEEE Trans. Syst. Man Cybern. **3**(1), 28–44 (1973)
257. Zadeh, L.A.: Inference in fuzzy logic. IEEE Proc. **68**, 124–131 (1980)
258. Zadeh, L.A.: From computing with numbers to computing words—from manipulation of measurements to manipulation of perceptions. IEEE Trans. Circuits Syst. Part I, **4**(3), 105–111 (1999)
259. Zecevic, A.I., Siljak, D.D.: Control design with arbitrary information structure constraints. Automatica **44**, 2642–2647 (2008)
260. Zecevic, A.I., Siljak, D.D.: Control of Complex Systems—Structural Constraints and Uncertainty. Springer Science, New York Dordrecht Heidelberg London (2010)

261. Zhang, B., Lam, J., Xu, S.: Stability analysis of distributed delay neural networks based on relaxed Lyapunov-Krasovskii functionals. IEEE Trans. Neural Netw. Learn. Syst. **26**(7), 1480–1492 (2015)
262. Zhang, J., Feng. G.: Event-driven observer-based output feedback control for linear systems. Automatica **50**, 1852–1859 (2014)
263. Zhang, S.-Y.: The structures of symmetry and similarity of complex control systems. Control Theory Appl **11**(2), 231–237 (1994)
264. Zhang, S.-Y.: Seeking simplicity from complexity. All About Systems and Control **1**(1), 38–49 (2014)
265. Zhang, X., Han, Q.: Event-triggered dynamic output feedback control for networked control systems. IET Control Theory Appl. **8**, 226–234 (2014)
266. Zhao, J., Dimirovski, G.M.: Quadratic stability of switched nonlinear systems. IEEE Trans. Autom. Control **49**(4), 574–578 (2004)
267. Zhao, J., Hill, D.J.: On stability L_2-gain, and $H\text{-}_{inf}$ control for switched systems. Automatica **44**(5), 1220–1232 (2008)
268. Zhao, J., Hill, D.J.: Passivity and stability of switched systems: A multiple storage function method. Syst. Control Lett. **57**(2), 158–164 (2008)
269. Zhao, J., Hill, D.J.: Dissipativity theory for switched systems. IEEE Trans. Autom. Control **53**(4), 941–953 (2008)
270. Zhao, J., Hill, D.J., Liu, T.: Synchronization of complex dynamical networks with switching topology: A switched system point of view. Automatica **45**(11), 2502–2511(2009)
271. Zhao, X.-Q., Zhao, J., Dimirovski, G.M.: Output feedback control for continuous-time switched systems with actuator saturation. In: A. Xue, General Chair; Q. Zhao, International Program Chair, and S. Liu, International Program, Vice-Chair, Proceedings of the 34th Chinese Control Conference, Hangzhou, CN, 28–30 July. Technical Committee on Control Theory, Chinese Association of Automation, Beijing, CN, 5619–5623 (2015)
272. Zhong, W.S., Stefanovski, J.D., Dimirovski, G.M., Zhao, J.: Decentralized control and synchronization of time-varying complex dynamical network. Kybernetika **45**(1), 151–167 (2009a)
273. Zhong, W.S., Dimirovski, G.M., Zhao, J.: Decentralized synchronization of an uncertain complex dynamical network. J. Control Theory Appl. **7**(3), 225–230 (2009b)

Preface: An Overview of the Contributions

This book is a multi-authored monograph on advanced research in the field of complex dynamic network and dynamic systems involving both event-driven and time-driven evolution dynamics, and also some statistical mechanics, hence employing some kind of combined control and supervision. It does present a number of case studies and certain reflections in a broader prospective towards implementable system engineering creations along the lines of complex dynamic networks and systems. The book comprises five parts: 1. Control and supervision for complex networks and systems; 2. Machine intelligence and learning in complex control systems; 3. Control and supervision of complex mechanical and robotic structures; 4. Control and supervision in multi-agent and industrial systems; and 5. Novel ideas and variable-structure and switched systems control. This overview is presented consequently following the parts along with enumeration of chapters albeit individual chapters in the book are solely ordered accordingly, i.e., do not bear enumeration. However, the names of the authors of each chapter are included beneath the title of their respective contribution.

Initially, this collective monograph has emerged from the selected a set of selected contributions, but in expanded and revised versions, of a special international conference chaired by Prof. Mogens Blanke, one of the leaders of scientific research program COSY 1995–1999 of European Science Foundation (ESF), and also one of the editors of the 2001 book, Control of Complex Systems, by Springer. Several of the present contributing authors, including this editor, also participated in the ESF research program in complex systems control and wrote that monograph. Thus this book is an important step forward into both theoretical and technological issues of complex network and systems.

The main feature in common of the contributions authored in this book is a kind of highlighting certain existing synergies of control, computing, and communication in order to achieve a guaranteed stable and sustainable plant system operation with robust quality. In the individual chapters, there appear contributions that range from a generally applicable model-based diagnosis and systems engineering to medical, to communication, to power-grid and airport networks, to creating biological brains for control applications and safety-critical human–machine systems, to process industries and industrial complexes, to biped robots, to large space structures and UAVs, to precision servomechanisms, as well as to other advanced mechatronics technologies. Nonetheless, most of the contributions introduce certain novel theoretical techniques for hard-to-control networks and systems, which go beyond standard decentralized feedback and where compound system architectures of control and supervision are employed. It is this sense present book may be viewed as a kind of follow-up monograph to those collective research efforts summarized in the 2001 monograph by Springer.

Part I Control and Supervision for Complex Networks and Systems

Diagnosis for Control and Decision Support for Autonomous Vehicles
Mogens Blanke, Søren Hansen, and Morten Rufus Blas

In this chapter there are presented the theoretical foundations for design methodologies that now appear as enabling technology for a new area of system designs that are reliable in practice. Yet they are also affordable due to the use of fault-tolerant philosophies and tools that make engineering efforts minimal for their implementation. It also includes the examples for an autonomous aircraft and a baling system for agriculture to illustrate the generic design procedures and real-life results. It should be noted, diagnosis and, when possible, prognosis of faults are essential for safe and reliable operation. Developments of methodologies that cope with complex and nonlinear systems have considerably matured and methodology and associated tools have become available in the form of theory and software for design. Genuine industrial cases have also become available. Analysis of system topology, referred to as structural analysis, has proven to be unique and simple in use and a recent extension to *active structure* techniques have made fault isolation possible in a wide range of systems. Following residual generation using these topology-based methods, deterministic and statistical change detection has proven very useful for online prognosis and diagnosis. For complex systems, results from non-Gaussian detection theory have been employed with convincing results.

Integration of Supervisory Control Synthesis in Model-Based Systems Engineering
Jos C.M. Baeten, Joanna M. van de Mortel-Fronczak, and Jakobus E. Rooda

This chapter discusses the integration of recently developed supervisor synthesis techniques and tools in engineering plant processes based on suitable models. Formal models play an important role here because they enable the use of various model-based analyses for early integration techniques, and tools. Engineering processes based on formal models are shown to be able to cope with complexity. They also support time-to-market and development costs reduction. Moreover, application of supervisory control synthesis in the development of control systems can speed up the process considerably. The proposed approach is illustrated by examples of industrial cases, where supervisors synthesized have successfully been implemented and integrated in existing resource-control platforms.

Output Synchronization of Dynamical Networks Having Nodes with Relative-Degree-One Nonlinear Systems
Yanyan Liu, Georgi M. Dimirovski, and Jun Zhao

This chapter studies the output synchronization problem in complex dynamic networks that have nonlinear dynamic systems with relative-degree-one at their nodes. System property of passivity has been found to be a useful tool for solving the output synchronization problem in such dynamic networks. Although not all dynamic systems may be passive, however, it is nonetheless shown still passivity can be successfully used to output synchronize dynamic networks. If a node nonlinear system is weakly minimum phase and has relative-degree-one, then it is shown to be feedback equivalent to a passive system. The feedback passivation result is exploited in order to investigate the output synchronization of dynamic networks. The conditions are found under which the output synchronization of dynamic networks, having node nonlinear systems with relative-degree-one, is achieved without the need the negative definiteness property of the outer coupling matrix.

Mechanism Design for Incentive Compatible Control of Networks
Anil Kumar Chorppath and Tansu Alpcan

This chapter provides an overview of the recent results in the area of mechanisms and games for distributed control of networks and authors recent contributions. The methodology and algorithms developed are applied to diverse network control problems such as interference and spectrum management. The heterogeneous behavior of users, which ranges from altruistic to selfish and to malicious, is being modeled within the analytical framework of game theory. Network mechanism design aims to achieve system-level goals such as maximization of aggregate network performance using specific methods in networks, where users are strategic and selfish decision-makers with individual preferences. A mechanism design approach is adopted to quantify the effect of adversarial behavior, which ranges from extreme selfishness to destructive maliciousness. Differentiated pricing is proposed as a method to counter and mitigate adversarial behavior. An additional application to the location privacy in mobile commerce is also briefly discussed.

Building Smart Grid: Optimal Coordination of Consumption with Decentralized Energy Generation and Storage
Araz Ashouri, Sebastian Gaulocher, and Petr Korba

This work describes a timely implementation of an office and/or personal smart grid for environmentally friendly buildings. These can be equipped with a local energy source (e.g., photovoltaic panels or combined heat-power units), energy storage devices (batteries, electric hot water boilers, heating and ventilation systems including air conditioning), a building energy management system with sensors (e.g., providing the room temperatures), and household appliances acting as actuators (in general, split into groups of schedulable and non-schedulable ones). The idea behind this work is to develop an automatic control system which would optimally decide for the end-customer when to buy, sell, or store electric energy with the objective to minimize his total costs. At the same time, it fulfills all constraints in terms of the limits on power allowed to be taken from the grid. In this project, a model predictive control approach to the energy optimization problem in a building has been proposed based on utilization of a real-time pricing signal which reflects daily peaks in consumption to the energy management system. Different scenarios have been run and the results are discussed here.

Passivity-Based Switching Rule and Control Law Co-design of Networked-Switched Systems with Feedback Delays
Dan Ma, and Georgi M. Dimirovski

In this theoretical study, a class of switched linear systems under a hybrid state feedback controller with time-varying delays is studied. The main contribution is given on the issue of how to co-design switching rule and feedback control law so as to make the closed-loop system strictly input feed-forward output-feedback passive for all admissible time delays in the feedback channels. Sufficient conditions for strict input feed-forward, output-feedback 'passivitication' of switched systems with time-varying delays under some state-dependent switching rule by using the method of multiple storage functions is derived. The proposed switching rule can achieve strict input feed-forward, output-feedback passivity of the switched delay systems whose all subsystems can be input feed-forward output-feedback non-passive. The finite gain L_2 stability in closed-loop is guaranteed. Furthermore, under the proposed switching rule, the asymptotic stability can be guaranteed if the switched system is zero-state detectable when exogenous disturbance input is zero.

Part II Machine Intelligence and Learning in Complex Control Systems

Creating and Controlling Complex Biological Brains
Kevin Warwick

This chapter presents a closer look into how animal and/or human brain cells can be cultivated (grown) and given a robot physical body (as a controlling brain) in which they can move around and interact with the world. These observations are utilized to propose a design that is aimed at creating a specific kind of cyborg. This is realized as a new form of artificial intelligence in which the complexity of a highly nonlinear biological neural network is employed to uniquely control a real-world robot. An adequate presentation is given in which the communication/control feedback loop is described and considered in terms of learning, performance, long-term operation, and specialization within the neural structure. Experimental results are presented and also an outline of certain open philosophical arguments is given.

Iterative Learning Control as an Enabler for Robotic-Assisted Upper Limb Stroke Rehabilitation
Eric Rogers, Chris T. Freeman, Ann-Marie Hughes, Jane H. Burridge, Katie L. Meadmore, and Tim Excell

In this chapter, a recent research is surveyed where iterative learning control, developed initially for robots executing commonly encountered industrial tasks such as sequentially collecting objects from one location and transferring them to another, is used to control the assistive stimulation in robotic-assisted upper limb stroke rehabilitation. The presentation is accompanied with a number of both analytical and empirical results obtained in the course of this research endeavor. The results given include the outcomes of small-scale clinical trials with stroke patients, and areas for future research are also briefly discussed. This research is aimed at coping with an increased burden on health care and rehabilitation resources due to the number of people suffering a stroke and therefore novel approaches to rehabilitation are required, if the capacity of health services is to meet future demands.

Adaptive Fuzzy Modeling Based Quantitative Assessment of Operator Functional State in Complex and Safety-Critical Human–Machine Systems
Jian-Hua Zhang and Ru-Bin Wang

In this chapter the human operator functional state, OFS, is quantitatively estimated by using multiple sources of measured psycho-physiological data. In the data acquisition experiments, an automation-enhanced Cabin Air Management System (aCAMS) was employed to simulate with high fidelity a highly complex multitask platform of human–machine cooperative process control. Two types of adaptive fuzzy models, viz., adaptive-network-based fuzzy inference system and genetic algorithm based Mamdani fuzzy model, are constructed to estimate the temporal fluctuations of the OFS. The fuzzy models are used to reveal the complex unknown correlation between the psycho-physiological (i.e., electroencephalographic and cardiovascular) variables and the operator performance (i.e., primary-task-related performance). The adaptive fuzzy modeling paradigm was validated by using the data measured from a group of young healthy and well-trained male subjects (two trials for each), who were engaged in the manual control tasks under aCAMS experimental environment.

Space Independent Community Detection in Airport Networks
Emil Gegov, Maria Nadia Postorino, Alexander Gegov, and Boriana Vacthova

In this research contribution, the problem of community detection has been investigated within large networks with highly changeable but most often unpredictable flows. The given objects of exploration are the topology and passenger flows of the United States Airport Network (USAN) over two decades. The network model consists of a time-series of six network snapshots for the years 1990, 2000, and 2010, which capture bimonthly passenger flows among US airports. The volume of these flows is naturally affected by spatial proximity, and therefore, a model (recently proposed in the literature) accounting for this phenomenon is used to identify the communities of airports that have particularly high flows among them, given their spatial separation. This research results highlight the fact that some general techniques from network theory, such as network modeling and analysis, can be successfully applied for the study of a wide range of complex systems; while others, such as community detection, need to be tailored for a specific system. Thus a successful empirical study on the complex systems in air transportation also involving network modeling and aimed at community detection problem has been accomplished.

Decentralized Control of Complex Dynamic Systems Employing Function Emulation by Neural Networks
Yuanwei Jing, Yanxin Zhang, Vesna M. Ojleska, Tatjana D. Kolemisevska, and Gerogi M. Dimirovski

In this contribution, a novel robust adaptive control design synthesis is proposed for a class of mechatronic nonlinear systems possessing similarity property has been derived. The design employs both high-order neural networks and math-analytical results in a compatible way. This approach exploits adequate usage of the structural feature of composite similarity systems and of neural networks to solve the representation issue of uncertainty interconnections and subsystem gains by online updating the weights of the neural networks. Lyapunov stability theory and attraction domain analysis are used. The proposed design synthesis guarantees the practical real stability in closed-loop, but also requires skills to obtain larger attraction domains around the operating equilibrium. The benchmark example of elastically interconnected two inverted pendulums on carts, thus creating a complex nonlinear dynamic system possessing inherent uncertainties, is investigated and its decentralized control solved.

Neural Networks with Strong Anticipation and Some Problems of Complexity Theory
Oleksandr S. Makarenko

This chapter presents the study of one new type of models of neural networks, which takes into account certain anticipation property. As the base model in it, Hopfield-type of models with anticipation has been explored. The basic new qualities discovered in this research is the possibility of multi-valued solutions of given neural networks. Different types of behavior of such systems have been explored depending on parameters of networks. The problems of complex solutions and stored information have been considered, including the measures of complexity in deterministic and non-deterministic cases. Presumable applications of such models for living and social systems are discussed within the context of these new type models of neural networks, which take into account property of anticipation.

As the base background model, the Hopfield-type models with anticipation are found to be crucial. Different types of behavior of such systems have been explored depending on parameters of networks.

Part III Control and Supervision of Complex Mechanical Structures and Robots

How to Cope with Disturbances in Biped Locomotion?
Miomir Vukobratović, Branislav Borovac, Mirko Raković, and Milutin Nikolić

Nowadays walking humanoid-like robots have become an expanding reality. Furthermore, it is expected that the humanoid robots of the near future will "live" and work in a common environment with humans. This imposes the requirement that their operational efficiency ought to be close to that of humans. The main prerequisite to achieve this is to ensure the robot's efficient motion quality, that is, its ability to compensate for the ever-present disturbances. This work considers precisely the strategies of how to compensate for the disturbances of different intensities: small which are permanently present and large that jeopardize the robot's dynamic balance instantly. It is illustrated that those two classes of disturbances require quite different compensation approaches.

New Adaptive Algorithm for Flexible Spacecraft Control
Vladislav Y. Rutkovsky, Victor M. Glumov, and Victor M. Sukhanov

During the last couple of decades, the design, implementation, and deployment of orbit in space of large space structures have been expanded to unprecedented extent. All these large space structures are essentially flexible structures, which exhibit delicate dynamics even when motion on orbit is well settled. For the case of large space structure control, a new adaptation algorithm for system with reference model is proposed. Its operation does not depend on the intensity and spectral composition of the input actions and its realization does not require estimation of external disturbances. The proposed algorithm functioning is illustrated on the example of large space structure control. It is a new type of large-size spacecraft (space energy stations, large orbiting reflectors). Such an object cannot be inserted into orbit in assembled condition because of its big size. Therefore LSS is assembled in orbit and it is a discretely evolving structure. As the control object it is multi-frequency oscillating system with discretely time-varying parameters and number of freedom degrees. For the case of large space structure control, proposed algorithm is simplified; yet the designed control system performs high-precision operation.

State-Dependent Riccati Equation-Based Tracking Control of a Hydraulic Seismic Isolator Test Rig
Stefano Pagano, Ricardo Russo, Salvatore Strano, and Mario Terzzo

A novel design and implementation of hydraulic seismic isolator test rig which employs a nonlinear optimal tracking control based on the state-dependent Riccati equation (SDRE) technique has been developed. It is aimed at testing the devise and systems which are used mitigate effects due to either earthquake or sever storm winds. Earthquake and wind storm effects can be mitigated by means of base isolation strategies. The base isolation is typically effected using passive, semi-active, or active systems. These devices must be tested in order to obtain the horizontal force-displacement cycle that allows for deducing analytical description of their dynamic characteristics if they are to be practically used. The SDRE algorithm fully preserves system nonlinearities, bringing the nonlinear system to a linear structure with state-dependent coefficient (SDC) matrices. The linear quadratic (LQ) synthesis method has been be applied to this state-dependent state-space equation characterized by the SDC matrices and improved control design achieved. A dSPACE DS1103 hardware has been employed for the control implementation and also for the real-time resolution of the SDRE, which supported by the obtained results from real-time experiments.

Multi-Robot Navigation Using Market-Based Optimization
Rainer Palm, Abdelbaki Bouguerra, and Muhammad Abdullah

This contribution is dedicated to a thorough investigation of the artificial force potential fields for obstacle avoidance and their optimization by a market-based approach in scenarios where several robots are acting in a shared area. The potential field method has been enhanced by fuzzy logic, traffic rules, and the technique of market-based optimization (*MBa*). Fuzzy rules are used to deform repulsive potential fields in the vicinity of obstacles to produce smoother motions around them. Traffic rules are used to deal with situations where robots are crossing each other. The *MBa,* on the other hand, is used to strengthen or weaken repulsive potential fields generated due to the presence of other robots. For testing and verification, the navigation strategy is implemented and tested in simulation of more realistic vehicles. Extensive simulation experiments are performed to examine the improvement of the traditional potential field method by the *MBa* strategy and verify the performance achieved.

Fault-Tolerant Estimation of UAV Dynamics via Robust Kalman Filter
Chingiz Hajiyev and Halil Ersin Soken

This contributed chapter presents a novel robust adaptive algorithm for estimating sensor and actuator faults in unmanned aerial vehicles (UAV). A covariance scaling based robust adaptive Kalman filter (RAKF) algorithm has been developed for the case of sensor/actuator faults. The proposed RAKF uses variable scale factors for scaling the process and measurement noise covariances and eliminating the effect of the faults on the estimation procedure. First the existing covariance estimation based adaptation techniques are reviewed. After choosing the efficient adaptation method, an overall concept for the RAKF is proposed. In this concept, the filter initially isolates the fault, either in the sensors or actuators, and then it applies the required adaptation process such that the estimation characteristic is not deteriorated. The performance of the proposed filters is investigated via simulations for the UAV state estimation problem.

Guidance Laws and Navigation Systems for Quadrotor UAV: Theoretical and Practical Findings
Stojce Deskovski, Vasko Sazdovski, and Zoran Gacovski

This chapter presents a novel contribution towards the design of small-size, inexpensive, quadrotor-based unmanned aerial vehicle (UAV). Nowadays UAVs are becoming essential for many applications where human presence is considered unnecessary, dangerous, or impossible. These applications include variety of scientific, civilian, and military applications. This paper reflects the efforts that we are taking over the years toward a deeper understanding of these technologies. A presentation of a low-cost, small-size quadrotor UAV that we have modified for our experiments is given. Both practical and theoretical research activities in the guidance navigation and control algorithms for quadrotor UAVs are discussed here. These analytical, simulation, and experimental studies have yielded certain novel findings, which are reported in this contribution.

Part IV Control and Supervision in Multi-Agent and Industrial Systems

Distributed Supervisory Strategies for Multi-Agent Networked Systems
Allesandro Casavola, Emanuel Garone, and Francesco Tedesco

In this chapter, certain novel distributed supervisory strategies for multi-agent linear systems that are connected via data networks and subject to coordination constraints are presented Such a *coordination-by-constraint* paradigm is based on the online management of the prescribed set points and it is characterized by a set of spatially distributed dynamic systems, connected via communication channels, with possibly dynamical coupling among them which need to be supervised and coordinated in order to accomplish their overall objective. Two distributed strategies will be fully described and analyzed. First, a "sequential" distributed strategy is presented where only one agent per decision time is allowed to manipulate its own reference signal. Such a strategy is then instrumental to introduce a more effective "parallel" distributed strategy, in which all agents are allowed to modify their own reference signals simultaneously under certain conditions.

Petri Net-Based Synthesis of Agent Cooperation by Means of Modularity and Supervision Principles
František Čapkovič

This chapter presents an exploration study on the possibility and the means how the principles of modularity and of supervision can be utilized in the synthesis of the cooperation among a collective of agents. Subsystems modeling agents of different kinds are understood to be modules of discrete-event systems (DES). They are modeled by means of place/transition Petri nets (P/T PN). A desired strategy of the mutual behavior of agents during their cooperation is expressed by conditions for the DES-based supervisor synthesis. Then, the synthesized supervisor does obtrude the cooperation strategy on the agents at the realization of a common job. The supervisor synthesis is realized either by means of the P/T PN place invariants (P-invariants) or by the virtually extended method, where P-invariants are complemented by conditions imposed on P/T PN transitions and/or on the Parikh's vector, especially in order to express priorities.

Adaptive Internal Model-Based Distributed Output Agreement in a Class of Multi-Agent Dynamic Systems
Esma Gül and Veysel Gazi

This contribution presents a novel study of the agreement problem in a class of multi-agent dynamic systems that have uncertainties. In particular, the case of the distributed output agreement problem has been studied and novel solution is derived. The investigated problem is formulated as a nonlinear servomechanism problem, and then an adaptive internal model based controller has been employed to achieve agreement of the agent outputs using local information. Various agent neighborhood topologies have been considered and the overall performance has been verified using fairly simple numerical simulations. Thus a novel solution to the output agreement problem in multi-agent dynamic systems has been found that tolerates presence of uncertainties.

An Example of Fault Detection and Reconfiguration-Based Tolerance Within Distributed Embedded Control Systems
Matjaž Colnarič, Domen Verber and Matej Šprogar

This contributed chapter introduces certain novel, recently devised solutions for the fault detection within embedded control systems. These represent a kind of follow-on elaboration on the successful IST FW5 project IFATIS, which has been carried out at the authors' Laboratory for Real-Time Systems of the Faculty of Electrical Engineering and Computer Science in Maribor. The topic is first re-elaborated and the overall results of the original project presented to some extent. Then, in continuation, certain later enhancements and improvements are shown all together with original implementations of specific parts. In particular, the discrete FPGA- and PSoC-based fault monitoring cells are given proper attention. All the novel improvements are discussed via properly emphasized presentation.

Predictive Control of Thermal Processes in Complex Industrial Furnaces
Goran S. Stojanovski, and Mile J. Stankovski

This chapter presents a thorough investigation of advanced predictive control methods for industrial thermal processes that have been developed and practically implemented in the ASE Institute Laboratory of the Faculty of Electrical-Electronics Engineering and Information Technologies in Skopje. This research is largely carried out on the grounds of identified representation models of two high-power, industrial furnaces that are operated in our country. Such industrial thermal processes typically require high fuel consumption, and therefore the optimization of the fuel costs is always needed. It is widely known that reducing those costs yields dramatically reduced costs of the final product delivered by the industrial plant. For this purpose, the advanced predictive control methods appear especially tailored for furnace thermal processes, since employing predictive control techniques enforces, at the same time, the plant to achieve both faster response and increased robustness. This is clearly supported by both experimental and simulation results are obtained.

Closed-Loop Control with Evolving Gaussian Process Models
Juš Kocijan and Dejan Petelin

This contribution presents a novel development in the design of control systems that is based on employing evolving Gaussian process (GP) models. The GP models are known to provide a probabilistic, nonparametric modeling approach for black-box identification of nonlinear dynamic systems. They can highlight areas of the input space where prediction quality is poor, by indicating the higher variance around the predicted mean, which may occur due to either the lack of data or the underlying complexity. While the GP models are Bayesian models, the output has normal distribution, expressed in terms of mean and variance. The evolving GP model is the conceptual approach within which various ways of model adaptations can be used. If the prior knowledge about the system to be controlled is scarce or the plant system varies with either the time or the operating region, then this control problem can be solved with an iterative method that adapts the model by means of information obtained with streaming data and thus concurrently optimizes hyper-parameter values.

Part V: Novel Control Ideas and Variable-Structure Systems Control

Attenuation of Uncertain Disturbances Through Fast Control Inputs
Alexander B. Kurzhanski and Alexander N. Daryin

In this chapter, there is presented a new class of controls that ensure an effect similar to the one produced by conventional matching conditions between control and disturbance inputs in a linear system. However, in this study a broader class of such inputs has been obtained. This is due to an application of piecewise-constant control functions with varying amplitudes, generated by approximations of "ideal controls" which are linear combinations of delta-functions and their higher order derivatives. Such a class allows for calculation of feedback control solutions by solving problems of open-loop control, thus reducing the overall computation burden. It is believed that this control approach does open a new prospect for future developments of control techniques.

Sliding Manifold Design for Linear Systems with Scalar Unmatched Disturbances
Boban Veselić, Branislava Draženović and Čedomir Milosavljević

This chapter presents an efficient sliding manifold design method that minimizes the impact of unmatched disturbances on sliding mode (SM) dynamics in variable-structure control systems. Although variable-structure control systems are known to be insensitive to so-called matched disturbances in ideal sliding mode, nonetheless they are vulnerable to the unmatched ones. The system sensitivity upon an unmatched constant external disturbance is evaluated through the steady state vector norm. An infinite set of the sliding hyper-planes that minimize the chosen optimization criterion is determined. A way of selecting a manifold out of that set that provides adopted SM dynamics is also suggested. The proposed approach has been demonstrated on several numerical examples and investigated by means of computer simulations.

Sliding Mode Based Anti-Lock Braking System Control
Dragan S. Antić, Darko B. Mitić, Zoran D. Jovanović, Staniša Lj. Perić, Marko T. Milojković and Saša S. Nikolić

This chapter presents the results of a thorough investigation of the anti-lock braking system control by means of sliding mode control. There are considered different continuous- and discrete-time sliding mode control (SMC) techniques in the control of anti-lock braking system (ABS). The SMC is found a right choice for its control because of its robust characteristics in the view that inherently the ABS is characterized by nonlinear and uncertain dynamics. The survey of continuous-time SMC algorithms based on nonlinear models of ABS is given first. Then, the discrete-time nonlinear model of ABS is derived, and the overview of existing discrete-time SMC techniques is presented. The experimental results are given to verify the effectiveness of the investigated SMC methods.

Switching Frequency Optimization of DC/AC Inverters Using Sliding Mode
Khalifa Al-Hoseni and Vadim I. Utkin

This chapter investigates the application of siding mode in order to achieve the switching frequency optimization of DC/Ac invertors. It is common that a DC/AC converter for three-phase load is designed for controlling two variables such as speed and flux of an AC motor. An additional degree of freedom can be utilized to minimize the switching frequency, which depends on the voltage of the load neutral point. A methodology of switching frequency minimization is proposed in the framework of the modified hysteresis control. The load neutral point voltage is selected as the third variable to be controlled. First, the tracking system algorithm is developed and then optimization with the switching frequency as a criterion is performed by a proper choice of the reference input for the neutral point voltage. The system accuracy is determined by the width of hysteresis loop and is the same for any switching frequency.

Discrete-Time Sliding-Mode Servo Systems Design with Disturbance Compensation Approach
Čedomir Milosavljević, Branislava Draženović and Boban Veselić

In this chapter, there is presented a novel discrete-time sliding mode control design employing a new disturbance compensator. This novel contribution is an essentially chattering-free, discrete-time, sliding mode, control algorithm with a new combined disturbance compensator. It is based on switching function measurement only. The overall system behaves as a high accuracy tracking system with an excellent compensation of matched disturbances. Thus the proposed servo system design represents a new design synthesis solution to this fundamental control engineering problem. Properties of the proposed design method are demonstrated on a velocity and a positional servo system. Analytically derived results as well as the experimental ones have demonstrated a superior performance in comparison with the existing designs.

Pragmatic Design Methods Using Adaptive Controller Structures for Mechatronic Applications with Variable Parameters and Working Conditions
Stefan Preitl, Radu-Emil Precup, Zsuzsa Preitl, Alexandra-Iulia Stînean, Claudia-Adina Dragoş and Mircea-Bogdan Rădac

This chapter presents an exploration study of two pragmatic design methods for controllers dedicated to mechatronic applications working under variable conditions. Adaptive structure of the control algorithms are known to be rather important for such applications. Basically, the design is founded on two extensions of the modulus optimum method and of the symmetrical optimum method (SO-m): the extended SO-m and the double parameterization of the SO-m (2p-SO-m). Both methods, which are attributed to the authors, make use of specific PI(D) controllers that are capable of ensuring high control performance in terms of: increased value of the phase margins, improved tracking performance, and efficient disturbance rejection. A short and systematic presentation of the methods and digital implementation aspects using an adaptive structure of the algorithms for industrial applications are given. Application deals with a cascade speed control structure for driving systems with continuously variable reference input, moment of inertia, and disturbance.

Acknowledgments

First of all, I would like to express my gratitude to all the contributing authors. It was their dedication that made this multi-authored monograph to materialize and come into being. My thanks are due for the patience with which they responded to various request of mine during the last stage of this common endeavor. In particular, I do remain indebted to Profs. Mogens Blanke, Alexander B. Kurzhanskiy, Jakobus E. Rooda, Vladislav Y. Rutkovskiy, Branislava Drazenovic, and Kevin Warwick for their encouraging support throughout all the activities involved in producing this book. However, my special gratitude goes to late Prof. Miomir K. Vukobratovic, my personal teacher of robotics, who passed away at the very early stage of this working endeavor, and thus I missed his inspiring counsel so much.

At this moment, I am glad to extend my sincere thanks to both Ms. Karin de Bie, Springer's production manager, and Mr. Oliver Jackson, Springer's engineering editor, for their understanding about my difficulties and cooperating with me during these years. Indeed their cooperation and help were instrumental in bringing this work to its fruitful end.

In addition, by all means I owe my profound gratitude to my entire family for being very patient as to tolerate often missing me from home and hence to carry on without a husband, a father, and a grandfather. Thank you so much my dearest ones.

Contents

Part II Machine Intelligence and Learning Control in Complex Systems

Editor and Contributors

About the Editor

Dr. Georgi M. Dimirovski was born on December 20, 1941 in Greece, in village Nestorion—Nestram in Aegean Macedonia. Currently, he is a Research Professor (retired life-time professor) of Automation and Systems Engineering at St. Cyril and St. Methodius University of Skopje, Republic of Macedonia, and of Computer Science and Information Technologies at Dogus University of Istanbul, Republic of Turkey. In 2015, he was elected an Active Member of the European Academy of Sciences in Salzburg, Austria, while in 2004 he was elected a Foreign Member of the Academy of Engineering Sciences in Belgrade, Serbia. Dr. Dimirovski is an invited guest professor on the graduate studies at Istanbul Technical University (Aeronautical Engineering) and before he was at Dokuz Eylul University of Izmir (Mechatronics Engineering) too; part-time teaching. In 2011, he was awarded the honor 'Pro Universitas' Professor of Obuda University in Budapest, Hungary, and included in the academic staff of the Doctoral School there. Since 2011, he has been teaching in summer graduate schools at either of universities in Dalian, Nanjing, Shanghai, and Shenyang. He received his Ph.D. in ACC in 1977 from the University of Bradford, UK. He has obtained his degrees: Dipl.-Ing. in EE from St. Cyril and St. Methodius University of Skopje, Macedonia, and M.Sc. in EEE from University of Belgrade, Serbia—then both within the former S.F.R. of Yugoslavia—in years 1966 and 1974, respectively. As a young engineer, Dimirovski spent 3 years in the industry before joining the academia in 1969. At University of Bradford, he has held a postdoctoral position in 1979, and was a Visiting Professor in 1984, 1986, and 1988, respectively. He has been a Visiting Scientist to Wolverhampton University in 1990 and 1992. Dr. Dimirovski was a Senior Research Fellow and Visiting

Professor of the Free University of Brussels, Belgium, in 1994, and also of Johannes Kepler University of Linz, Austria, in 2000. During the last three decades, he has paid long- or short-term academic visits with seminars at the universities in Aalborg, Ankara, Belgrade, Bradford, Bochum, Brussels, Coimbra, Covilha, Duisburg, Grenoble, Hannover, Istanbul, Izmir, Linz, Lisbon, Ljubljana, London, Maribor, Nis, Portsmouth, Sarajevo, Sevastopol, Sofia, Split, Valencia, Wien, Wolverhampton, Zagreb, and Kaohsiung (TW); Beijing, Dalian, Nanjing, Shanghai, and Shenyang (CN). He participated in and served on the Steering Committee of European Science Foundation's Scientific Programme on Control of Complex Systems (COSY) 1995–1999 under the leadership of Profs. Karl Astroem and Manfred Thoma, along with his postdoctoral research fellow and four graduate students. He has successfully supervised three postdoctoral, 16 Ph.D., and 27 M.Sc. as well as more than 300 graduation students projects, and advised more than 30 graduate students at the NEU of Shenyang within the framework of his academic cooperation with Profs. Yuanwei Jing and Jun Zhao. He has served on Ph.D. evaluation juries and/or panels for several universities including Bradford and Wolverhampton in the UK, Izmir in Turkey, Ljubljana in Slovenia, Chennai in India, Zagreb in Croatia, Sofia, Varna in Bulgaria, and Craiova in Romania. He was editor of six proceedings volumes of the IFAC series and of one volume of the IEEE Series. Internationally, he has contributed one edited research monograph by Springer, nine invited theme papers (chapters) in monographs, and more than 100 journal articles as well as more than 350 conference papers in the IFAC and the IEEE proceedings series alone. Professor Dimirovski, as an Associate Editor (AE), now serves on Editorial Boards of the Journal of the Franklin Institute (The FI, USA), Asian J. of Control (Wiley, TW), AI & Society (Springer, UK), Intl. J. of Automation & Computing (CAS, CN) as well as regularly reviews for a number of archival journals refereed in the SCI and SCI-Expanded. Before he served as an AE of the IEEE Transactions on Systems, Man & Cybernetics (The IEEE, USA), Proceedings of the Institution of Mechanical Engineers Pt. I Systems & Control Eng. J. (Institution Mechanical Engineers., UK), Facta Universitatis J. Electronics & Energetics (University of Nis, RS), J. Electrical & Electronics Engineering (University of Istanbul, TR), Information Technologies & Control J. (Bulgarian Union of Automation and Informatics, Sofia), J. of Control Engineering & Applied Informatics (SRAIT, RO), Automatika (former S.F.R. Yugoslavia), and was Editor-in-Chief of Engineering J. (MK). In due times, he was awarded grants for five national projects in automation and control of industrial processes, and two in complex decision and control systems by the Ministry of Education and Science of Macedonia. He founded the Institute of Automation and Systems Engineering at the Faculty of Electrical Engineering of St. Cyril and St. Methodius University in 1985. He has initiated and subsequently in 1981 founded the ETAI Society—Macedonian IFAC NMO, was elected as its first president, and served several terms as president. In 1996, he joined Dr. Goce Arsov in founding the IEEE Republic of Macedonia Section, and in 2002 joined Dr. Okyay Kaynak in founding the IEEE Computational Intelligence Chapter of Turkey. He has developed a number of undergraduate and/or graduate courses in control and automation as well as others in robotics, in applied numerical, fuzzy system and neural-network computing, and in operations research at his home university and at universities in Bradford, Istanbul, Izmir, Linz, and Zagreb, respectively. During 1985–1991, he served three terms as President of the Yugoslav Association for ETAN, then Yugoslav IFAC NMO. During 1988–1993, he served the European Science Foundation on its Executive Council and in other capacities as well. He served the International Federation of Automatic Control (IFAC) in the capacities of: Vice-Chair of TC 9.3 on Developing Countries, during 1996–2002, and of Chair during 2002–2005; Chair of Coordination Committee CC9 and Technical Board Member of the IFAC (TB) during 2005–2011. He has served on the IPC's for many IFAC, IEEE, ECC, and WAC conferences, and for others. He was Co-Chairman of the IFAC SWIIS 2000 and of the IFAC DECOM-TT 2001, 2003, 2004, and 2007, and of the IFAC CEFIS 2007 as well as the Program Chair of the IEEE SMC-2010 and IEEE CCA-2003, and of the 1993 and 2003 International AAS ETAI symposia. Dr. Dimirovski received two Yugoslav ETAN Best Paper Awards: the 1974 in Automatic Control and the 1991 in Robotics & Automation. In 2009, the ETAI Society awarded him the ETAI Merits Award. In 2011, G.M. Dimirovski received the IFAC Outstanding Service Award from the IFAC Organization. In 2009, along with Dr. King-Qui Li and Prof. Jun Zhao, Prof. G.M. Dimirovski received the 2009 IET Premium Award from the UK Institution of Engineering and Technology for the best article of the IET-CTA journal in 2008. His publications have received more than 1280 citations in Scopus (Elsevier, NL), more than 790 in Web-of-Science (Thomson Reuters, USA), and more than 1790 citations in Google Scholar, until now. *BiSvP!*

Current address/contact means: Dogus University of Istanbul, Acibadem, Zeamet Sk. 21, Kadikoy, Istanbul 34722, Turkey

e-mail: gdimirovski@dogus.edu.tr; Phone: +90-216-544-5555, Extension: 1202; Fax: +90-216-544 5535

Contributors

Muhammad Abdullah Örebro University, Örebro, Sweden

Khalifa Al Hosani Electrical Engineering Department, The Petroleum Institute, Abu Dhabi, United Arab Emirates

Tansu Alpcan Department of Electrical and Electronic Engineering, The University of Melbourne, Melbourne, Australia

Dragan S. Antić Department of Control Systems, Faculty of Electronic Engineering, University of Niš, Niš, Republic of Serbia

Araz Ashouri Department of Mechanical and Process Engineering, ETH Zurich, Zurich, Switzerland

Jos C.M. Baeten Centrum Wiskunde and Informatica, Amsterdam, The Netherlands; Eindhoven University of Technology, Eindhoven, The Netherlands

Mogens Blanke Automation and Control Group, Department of Electrical Engineering, Technical University of Denmark, Kongens Lyngby, Denmark; AMOS CoE, Institute for Technical Cybernetics, Norwegian University of Science and Technology, Trondheim, Norway

Morten Rufus Blas CLAAS Agrosystems, Nivå, Denmark

Branislav Borovac Faculty of Technical Sciences, University of Novi Sad, Novi Sad, Serbia

Abdelbaki Bouguerra Örebro University, Örebro, Sweden

Jane H. Burridge Faculty of Health Sciences, University of Southampton, Southampton, UK

František Čapkovič Institute of Informatics, Slovak Academy of Sciences, Bratislava, Slovakia

Alessandro Casavola Università della Calabria, Rende, Italy

Anil Kumar Chorppath Technical University of Munich Lehrstuhl Für Theoretische Informationstechnik, Munich, Germany

Matjaž Colnarič Faculty of Electrical Engineering and Computer Science, University of Maribor, Maribor, Slovenia

Alexander N. Daryin Department of Computational Mathematics and Cybernetics, Moscow State (Lomonosov) University, Moscow, Russia

Stojche Deskovski Faculty of Technical Sciences, University St. Kliment Ohridski, Bitola, Republic of Macedonia

Georgi M. Dimirovski School FEEIT, St. Cyril and St. Methodius University, Skopje, Republic of Macedonia; Computer and Control Department, Dogus University, Acibadem, Istanbul, Republic of Turkey

Claudia-Adina Dragoş Department of Automation and Applied Informatics, "Politehnica" University of Timisoara, Timisoara, Romania

Branislava Draženović Faculty of Electrical Engineering, University of Sarajevo, Sarajevo, Bosnia and Herzegovina

Tim Exell Department of Electronics and Computer Science, University of Southampton, Southampton, UK

Chris T. Freeman Department of Electronics and Computer Science, University of Southampton, Southampton, UK

Zoran Gacovski Faculty for Information and Communication Technologies, FON University, Skopje, Republic of Macedonia

Emanuele Garone Université Libre de Bruxelles, Brussel, Belgium

Sebastian Gaulocher Institute of Automation, University of Applied Sciences and Arts Northwestern Switzerland, Basel, Switzerland

Veysel Gazi Department of Electrical and Electronics Engineering, Istanbul Kemerburgaz University, Bağcılar, Istanbul, Turkey

Alexander Gegov University of Portsmouth, Portsmouth, UK

Emil Gegov Brunel University, Uxbridge, UK

Victor M. Glumov Institute of Control Sciences, Russian Academy of Science, Moscow, Russia

Esma Gül Turkish Aerospace Industries (TAI), Fethiye Mahallesi, Kazan, Ankara, Turkey

Chingiz Hajiyev Aeronautics and Astronautics Faculty, Istanbul Technical University, Maslak, Istanbul, Turkey

Søren Hansen Automation and Control Group, Department of Electrical Engineering, Technical University of Denmark, Kongens Lyngby, Denmark

Ann-Marie Hughes Faculty of Health Sciences, University of Southampton, Southampton, UK

Yuanwei Jing College of Information Science and Engineering, Northeastern University, Shenyang, Liaoning, People's Republic of China

Zoran D. Jovanović Department of Control Systems, Faculty of Electronic Engineering, University of Niš, Niš, Republic of Serbia

Juš Kocijan Jožef Stefan Institute, Ljubljana, Slovenia; University of Nova Gorica, Nova Gorica, Slovenia

Tatjana D. Kolemisevska-Gugulovska School FEEIT, St. Cyril and St. Methodius University, Skopje, Republic of Macedonia

Petr Korba Institute of Energy Systems and Fluid Engineering, Zurich University of Applied Sciences, Zürich, Switzerland

Alexander B. Kurzhanski Department of Computational Mathematics and Cybernetics, Moscow State (Lomonosov) University, Moscow, Russia

Yanyan Liu State Key Laboratory of Synthetic Automation for Process Industries, Northeastern University, Shenyang, Liaoning, People's Republic of China

Dan Ma State Key Laboratory of Synthetical Automation for Process Industries, College of Information Science and Engineering, Northeastern University-Shenyang, People's Republic of China

Oleksandr S. Makarenko Institute for Applied System Analysis, National Technical University of Ukraine (KPI), Kyiv, Ukraine

Katie L. Meadmore Department of Electronics and Computer Science, University of Southampton, Southampton, UK

Marko T. Milojković Department of Control Systems, Faculty of Electronic Engineering, University of Niš, Niš, Republic of Serbia

Čedomir Milosavljević Faculty of Electrical Engineering, University of Istočno Sarajevo, Istočno Sarajevo, Bosnia and Herzegovina

Darko B. Mitić Department of Control Systems, Faculty of Electronic Engineering, University of Niš, Niš, Republic of Serbia

Joanna M. van de Mortel-Fronczak Eindhoven University of Technology, Eindhoven, The Netherlands

Milutin Nikolić Faculty of Technical Sciences, University of Novi Sad, Novi Sad, Serbia

Saša S. Nikolić Department of Control Systems, Faculty of Electronic Engineering, University of Niš, Niš, Republic of Serbia

Vesna M. Ojleska School FEEIT, St. Cyril and St. Methodius University, Skopje, Republic of Macedonia

Stefano Pagano Department of Industrial Engineering, University of Naples Federico II, Naples, Italy

Rainer Palm Örebro University, Örebro, Sweden

Staniša Lj. Perić Department of Control Systems, Faculty of Electronic Engineering, University of Niš, Niš, Republic of Serbia

Dejan Petelin Jožef Stefan Institute, Ljubljana, Slovenia

Maria Nadia Postorino Mediterranea University of Reggio Calabria, Reggio Calabria, Italy

Radu-Emil Precup Department of Automation and Applied Informatics, "Politehnica" University of Timisoara, Timisoara, Romania

Stefan Preitl Department of Automation and Applied Informatics, "Politehnica" University of Timisoara, Timisoara, Romania

Zsuzsa Preitl Siemens A.G., Erlangen, Germany

Mircea-Bogdan Rădac Department of Automation and Applied Informatics, "Politehnica" University of Timisoara, Timisoara, Romania

Mirko Raković Faculty of Technical Sciences, University of Novi Sad, Novi Sad, Serbia

Eric Rogers Department of Electronics and Computer Science, University of Southampton, Southampton, UK

Jacobus E. Rooda Eindhoven University of Technology, Eindhoven, The Netherlands

Riccardo Russo Department of Industrial Engineering, University of Naples Federico II, Naples, Italy

Vladislav Yu. Rutkovsky Institute of Control Sciences, Russian Academy of Science, Moscow, Russia

Vasko Sazdovski Faculty of Electrical Engineering, University Goce Delchev, Shtip, Republic of Macedonia

Halil Ersin Soken Institute of Space and Astronautical Science (ISAS), Japan Aerospace Exploration Agency (JAXA), Sagamihara, Kanagawa, Japan

Matej Šprogar Faculty of Electrical Engineering and Computer Science, University of Maribor, Maribor, Slovenia

Mile Stankovski Department of Automatics and System Engineering, Faculty of Electrical Engineering and Information Technology, SS Cyril and Methodius University Skopje, Skopje, Macedonia

Goran Stojanovski Department of Automatics and System Engineering, Faculty of Electrical Engineering and Information Technology, SS Cyril and Methodius University Skopje, Skopje, Macedonia

Salvatore Strano Department of Industrial Engineering, University of Naples Federico II, Naples, Italy

Alexandra-Iulia Stînean Department of Automation and Applied Informatics, "Politehnica" University of Timisoara, Timisoara, Romania

Victor M. Sukhanov Institute of Control Sciences, Russian Academy of Science, Moscow, Russia

Francesco Tedesco Università della Calabria, Rende, Italy

Mario Terzo Department of Industrial Engineering, University of Naples Federico II, Naples, Italy

Vadim I. Utkin Department of Electrical and Computer Engineering, The Ohio State University, Columbus, OH, USA

Boriana Vatchova Bulgarian Academy of Sciences, Sofia, Bulgaria

Domen Verber Faculty of Electrical Engineering and Computer Science, University of Maribor, Maribor, Slovenia

Boban Veselić Faculty of Electronic Engineering, University of Niš, Niš, Serbia

Miomir Vukobratović Mihajlo Pupin Institute, Belgrade, Serbia

Rubin Wang Institute of Cognitive Neurodynamics, East China University of Science and Technology, Shanghai, People's Republic of China

Kevin Warwick School of Systems Engineering, University of Reading, Whiteknights, Reading, UK

Jianhua Zhang Department of Automation, East China University of Science and Technology, Shanghai, People's Republic of China

Yanxin Zhang School of Electronic and information Engineering Haidian, Beijing Jiaotong University, Beijing, People's Republic of China

Jun Zhao State Key Laboratory of Synthetic Automation for Process Industries, Northeastern University, Shenyang, Liaoning, People's Republic of China

Part I
Control and Supervision for Complex Networks and Systems

No doubt, information is the third fundamental quantity of our world next to energy and matter. But only information has the capacity to reshape, subordinate, and govern energy and matter.

Georgi Dimirovski

Chapter 1
Diagnosis for Control and Decision Support for Autonomous Vehicles

Mogens Blanke, Søren Hansen and Morten Rufus Blas

Abstract Diagnosis and, when possible, prognosis of faults are essential for safe and reliable operation. The area of fault diagnosis has emerged over three decades. The majority of studies are related to linear systems but real-life systems are complex and nonlinear. The development of methodologies coping with complex and nonlinear systems have matured and even though there are many unsolved problems, methodology and associated tools have become available in the form of theory and software for design. Genuine industrial cases have also become available. Analysis of system topology, referred to as structural analysis, has proven to be unique and simple to use and a recent extension to *active structural* techniques has made fault isolation possible in a wide range of systems. Following residual generation using these topology-based methods, deterministic and statistical change detection has proven very useful for online prognosis and diagnosis. For complex systems, results from non-Gaussian detection theory have been employed with convincing results. The chapter presents the theoretical foundation for design methodologies that now appear as enabling technology for a new area of design of systems that are reliable in practice. Yet, they are also affordable due to the use of fault-tolerant philosophies and tools that make engineering efforts minimal for their implementation. The chapter includes examples for an autonomous aircraft and a baling system for agriculture to illustrate the generic design procedures and real-life results.

M. Blanke (✉) · S. Hansen
Automation & Control Group, Department of Electrical Engineering,
Technical University of Denmark, Kongens Lyngby, Denmark
e-mail: mb@elektro.dtu.dk

S. Hansen
e-mail: sh@elektro.dtu.dk

M. Blanke
AMOS CoE, Institute for Technical Cybernetics, Norwegian University
of Science and Technology, Trondheim, Norway

M.R. Blas
CLAAS Agrosystems, Møllevej 11, Nivå, Denmark
e-mail: rufus.blas@claas.com

G.M. Dimirovski (ed.), *Complex Systems*, Studies in Systems,
Decision and Control 55, DOI 10.1007/978-3-319-28860-4_1

1.1 Introduction

Diagnosis of faults and active accommodation of faults are tools to prevent faults from developing into failure. Diagnosis is needed for fault-tolerant control, where the diagnostic information is used without operator intervention to handle a fault or is used by a human supervisor in support of fault-tolerant operation.

The theory of fault diagnosis has a long history. Early papers that helped the directions in this field included seminal results on generation of residuals using parity space approaches [20], combined overview and research articles [40, 41, 52, 55]. Early applications appeared in [79] for diagnosis of flight control systems, in [27] for chemical processes, in [54]. The theory for fault-diagnostic observers was pioneered in [32, 33], and generation of robust residuals in [18, 35]. Extension to fault-tolerant control emerged in [7, 9, 100] and theories for integration of diagnosis and control for linear systems appeared [73]. Diagnosis for nonlinear systems using geometric theory was pursued in [81]. Robust methods matured [59, 85, 92], also for Fuzzy and neural network approaches [61, 78]. The crucial issue of threshold selection for diagnosis was treated in papers by [29] for the uncertain deterministic case and for the stochastic case by [3]. Extension to discrete event and quantised systems appeared in [70] and [69]. Early edited books helped develop the area [80], and several textbooks now show the maturity of the area [2, 5, 19, 21, 22, 42, 47, 56, 62, 71, 72, 76, 101].

The above methods rely mostly on detailed mathematical models of the plant. When complex systems are in focus, detailed models are very difficult or at least very expensive to obtain, and a different approach has become widely appreciated. This is based on graph-analysis of system properties and this is the subject of this chapter. The graph-based techniques do not replace the above analytical methods, but for complex systems they have proven to be very attractive. Several aspects of control and diagnosis of complex systems appeared in [1]. Applications are plenty and to mention a few from recent years, omitting many important contributions: [8, 13, 17, 44, 46, 65, 84, 93].

Coping with complex systems is a challenge. Dimensionality of the problems met in real life is one challenge, complexity and nonlinearity are others. The ideas of using graph-analysis concepts to help solving complex set of equations were studied early in the applied mathematics community with the result of [25] being instrumental for the area. Further theoretical developments in [26, 51] made the analysis of the structure of a set of equations or of a system described by such equations a feasible task, even for large systems of nonlinear equations. Structural analysis as this area is called, has been used intensively in Chemical Engineering for solving large sets of equations and issues on solvability have been pursued in a number of publications, see [68, 96] and the references herein. The structural approach and the features it offers for analysing monitoring and diagnosis problems was first introduced in [89] and further developed in [88]. Extensions to analysis of reconfigurability and fault-tolerance emerged in [87] and [90]. The structural analysis approach was brought into a digested form in [5, 95]. Structural analysis has hence evolved during several decades. However, the salient features of the theory and the possibilities it offers have only become apparent to a larger community in the field of automation and automatic control over the last few years. Reasons for the slow penetration into applications origin mainly in the lack of widely available tools to support the structural analysis method for automated industrial systems. Software tools appeared in [23] and in [6]. An approach to highly efficient algorithms was developed in [63].

When considering diagnosis for control, the safety, from a structural point of view, depends on services offered by a system not only in normal operation but more important, after reconfiguration of the system to accommodate one or more faults. Blanke and Staroswiecki [11] considered the safety of fault-tolerant control schemes when multiple faults could be present. It was shown how structural analysis could be applied to analyse cases of multiple faults and to synthesise residual generators. Fault isolation, which is instrumental for correct fault handling, was addressed and active isolation was introduced from a structural point of view.

This chapter revisits the theory for graph-based analysis of systems and introduces the notion of structural active isolability. It first reviews the concept of behaviours and shows how the behaviour of a system is equally well applied on the services offered by hardware and software components. It then interprets the impact on safety of a system that is supposed to work under conditions of multiple faults. While it is well established that structural analysis is very useful for residual generation of technical processes [24], it is less obvious that the generic technique is also very applicable on complex system, even generation of residuals for diagnosis in a natural environment. The combination with change detection techniques is highlighted, and the techniques needed for nonlinear systems, when non-Gaussian residuals occur, are demonstrated. Two real cases are included for illustration of the complete design procedure, an aeroplane speed sensor fault diagnosis problem and a case of baler control for agriculture. The latter includes vision sensor techniques and results show how the diagnostic concepts can also be applied to enhance the robustness of a vision-based control system.

1.2 Graph-Based Analysis

Fault-tolerant control (FTC) uses control or sensor reconfiguration to accommodate failures in instruments, plant components or actuators. Aiming at utilising existing redundancy in instrumentation and control devices as far as possible, fault-tolerant control can be applied to minimise the hazards associated with malfunction, even when several sensors or actuators fail, but several modifications need be made to the usual single fault FTC schemes in order to achieve the necessary level of safety.

1.2.1 Reconfigurability and Safety

The topology (structure) approach that is pursued in this context considers a system as consisting of a set of components which each offer a service and performs this service through defined normal behaviours. A component can offer different versions of services and command to the component can define which version of a service is made available. Within a component, fault-tolerant techniques can use fault-diagnosis and fault-handling to switch between services or offer a service in a version with degraded performance if local malfunction should make this necessary.

1.2.2 Subsystem Services

A system breakdown in Fig. 1.1 shows three different topologies, by which we mean the arrangement of the system components and their interaction. Component k has input u_k, output y_k, parameters θ_k and a behaviour $c_k(y_k, u_k, \theta_k) = 0$. The behaviour may be constructed from a set of constraints $\left\{c_{k1}, c_{k2}, \ldots, c_{kn}\right\}$ associated with the subsystem and the exterior behaviour of the component is the union of internal behaviours $c_k = c_{k1} \cup c_{k2} \cup \cdots \cup c_{kn}$ or for brevity, $c_k = \left\{c_{k1}, c_{k1}, \ldots, c_{kn}\right\}$. Following the generic component definition in [5], the service $S^{(k)}$ offered by component k is to deliver *produced variables* (output), based on *consumed variables* (input) and available *resources*, according to the specified behaviour $S^{(k)}_{(v)}$ where $v \in \{1, 2, 3, \ldots\}$ is the version of the service. Clearly, the exterior behaviour is associated with the service offered by the component, we denote this behaviour by $c_k^{(v)}$.

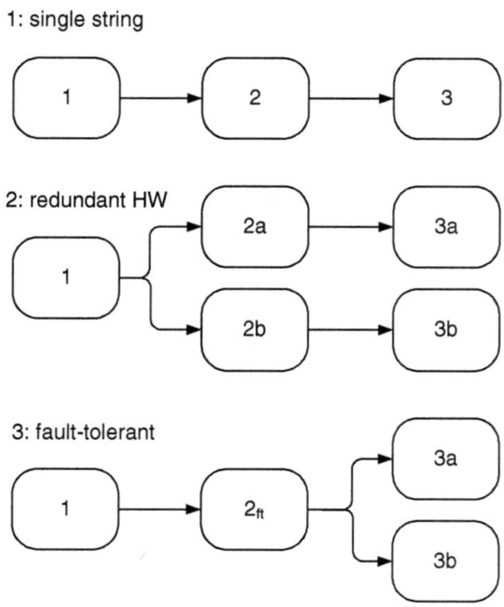

Fig. 1.1 Three architectures, *single line* with no redundancy (*1*), hardware redundancy (*2*) and combined fault-tolerance and redundancy (*3*)

In this context we particularly wish to consider versions of the same service that follow from the condition of the component, from normal over degraded to none. If a component has an internal failure, fault-tolerant techniques may still provide a version of the service with degraded performance $\left(S^{(k)}_{(d)}\right)$ or the service may not be available at all $\left(S^{(k)}_{(o)}\right)$. Hence, we consider the set of versions $v \in \{n, d_1, d_2, \ldots, o\}$ where n : *normal*; d_1 : *degraded*$_1$; d_2 : *degraded*$_2$; o : *none*.

1.2.3 Service at System Level

The service obtained by the system as an entirety is a function of the component architecture $\mathcal{A}$ and the versions for the present condition k_i of components. With m components in a system, each component in one out of p conditions, $k_i \in \mathbb{N}^{p_i}$, we have a versions vector $\mathbf{v} = [v_1(k_1), v_2(k_2), \ldots, v_m(k_m)]$, and the set of available behaviours $C_v = \left\{c_1^{v(k_1)}, c_2^{v(k_2)}, \ldots, c_m^{v(k_m)}\right\}$.

Definition 1.1 (*Overall Service*) The overall service available from a system is $S^{(s)}(c_i^{v(k_i)}) = \mathcal{A}(C_v|\mathbf{v}(\mathbf{k})), i = 1, \ldots, m$.

With a single string architecture from Fig. 1.1, we obtain

$$S^{(s)} = S^{(1)} \cap S^{(2)} \cap S^{(3)}. \tag{1.1}$$

With redundancy in the system, the hardware configuration with two parallel, totally redundant lines with only one component in common,

$$S^{(s)} = S^{(1)} \cap \left(\left(S^{(2a)} \cap S^{(3a)}\right) \cup \left(S^{(2b)} \cap S^{(3b)}\right)\right). \tag{1.2}$$

This solution is expensive as it requires two completely redundant subsystems. A cost-effective solution would be to have some components intrinsically safe $S^{(1)}$, have others equipped with fault-tolerant properties so their service $S^{(2)}_{v(2)}$ will be available but in degraded version when local faults occur, and just have hardware redundancy for few essential components $(3a, 3b)$. The fault tolerant architecture shown in part C of Fig. 1.1 is based on this idea. The service at system level is

$$S^{(s)} = S^{(1)} \cap S^{(2)}_{v(2)} \cap \left(S^{(3a)} \cup S^{(3b)}\right). \tag{1.3}$$

The paradigm in this architecture is that component failures should be detectable and control be switched to obtain a fault-tolerant service or reconfigure the system bypassing faulty components. This should be achieved by controlling the signal flow in the software of the system.

1.2.4 Availability and Safety

The plant at the system level is *available* as long as the predefined normal service is offered in some version, normal or degraded. A fault-tolerant version of the service is obtained when one or more of the component services are offered in a fault-tolerant version. A fail-operational version of the service is obtained when hardware reconfiguration has been made to bypass a failure in the redundant component.

When multiple local failures are present, the service at system level is

$$S^{(s)} = \mathcal{A}\left(C_v|\mathbf{v}\right).$$

Definition 1.2 (*Availability*) A system is available when $S^{(s)} \subseteq \mathcal{S}$ where $\mathcal{S} = \left\{S_1^{(s)}, S_2^{(s)}, \ldots, S_n^{(s)}\right\}$ is the set of admissible services that meet specified overall objectives for behaviour $\mathcal{O}$ of the system:

$$\forall S_i^{(s)}(c_i^{v(k_i)}) : C_v \subseteq \mathcal{O}.$$

Definition 1.3 (*Structural fault*) Fault. A fault in the structural description of behaviours is a deviation from normal behaviour,

$$\exists i : c_i \neq 0.$$

Definition 1.4 (*Critical fault*) A fault in c_i is critical $c_i \in C_{crit}$ if it will cause the system's behaviour to be outside the set of admissible behaviours

$$c_i \in C_{crit} \text{ iff } c_i \neq 0 \Rightarrow C_v \nsubseteq \mathcal{O}.$$

Definition 1.5 (*Useability*) Useability for reconfiguration. A faulty system is usable for reconfiguration, from the structure point of view, if all critical faults are structurally detectable,

$$\forall c_i \in C_{crit} : c_i \in C_{detectable}.$$

Assumption 1.1 (*Intrinsical safety*) It is a natural assumption that shutdown of the system is intrinsically safe and that the system can be shutdown to a safe mode from any condition where $S^{(s)} \subseteq \mathcal{S}$.

Definition 1.6 (*Structural reconfigurability*) A system is structurally reconfigurable if

$$c_i \neq 0 \Rightarrow \exists j \neq i, \mathbf{v}(j) \neq \mathbf{v}(i) : C_{v(j)} \subseteq \mathcal{O}.$$

The task of fault-tolerant control is to find an appropriate $\mathbf{v}(j)$ when the fault c_i is detected and isolated and bring the system from version $v(i)$ to $v(j)$.

Having defined the system properties in terms of behaviours, it is natural to employ structural analysis where behaviours are defined in terms of constraints between variables and graph theory methods offer rapid and rigorous analysis.

1.2.5 Structure Graph

A structural model of a system can be represented as a bipartite graph that connects constraints and variables. The structure graph [89] of a system (C, Z) is a bipartite graph $G = (C, Z, E)$ with two set of vertices whose set of edges $E \subseteq C \times Z$ is defined by $(c_i, z_i) \in E$ iff the variable z_i appears in constraint c_i.

The variables in Z are divided into known K and unknown variables X. Similarly, the constraints C are divided into constraints C_K that only apply to the known variables and C_X that involve at least one unknown variable. An incidence matrix S describes the structure graph where each row in the matrix represents a constraint and each column a variable. $S(i,j) = 1$ means that variable x_j appears in constraint c_i, $S(i,j) = x$ denotes a directed connection.

1.2.6 Constraints

Constraints represent the functional relations in the system, i.e., originating in a physical model using first principles. The constraints needed for structural analysis are far simpler. Instead of using the explicit system equations, structural analysis needs to know whether a certain constraint makes use of a particular variable. Parameters that are known from the physics of the plant or from properties of the automation system, e.g. a control gain, are treated as part of the constraint in which the particular parameter is used. A constraint can be directed. This implies that a variable on the left-hand side of the constraint cannot be calculated from the right-hand side of the constraint.

1.2.7 Variables

There are three different kinds of variables: *Input variables* are known, externally defined; *Measured variables* are entities measured in the system; *Unknown variables* are internal physical variables. Input and measured variables both belong to the set K but are separated for calculation of controllability.

1.2.8 Matching and Results

The central idea in the structure graph approach is to match all unknown variables using available constraints and known variables, if possible. If successful, the matching will identify overdetermined subgraphs that can be used as analytical redundancy relations in the system.

Use of a complete matching on unknown variables is not a necessary prerequisite to find analytic redundancy relations. Finding *minimal structural overdetermined* subgraphs within a structure graph, referred to as MSO sets, was introduced in [64], and their very efficient minimal structurally overdetermined (MSO) set algorithm makes structural analysis feasible for real-life complex systems.

Results of the structural analysis are:

- List of parity relations that exist
- Auto-generated suggestion of residual generators
- List of detectable behavioural faults
- List of isolable behavioural faults

The term *behavioural faults* is used to emphasise that the faults determined by the structural analysis are in violation of a normal behaviour.

When a matching has been found, backtracking to known variables will give a suggestion for parity relations that could be used as residual generators. A system with m constraints and n parity relations will give a relation showing which residuals depend on which constraints.

One view on these relations is the Boolean mapping

$$\mathcal{F} : r \leftarrow M \otimes \left(c_i \neq 0\right) \tag{1.4}$$

from which structural detectability and isolability can be found.

Definition 1.7 (*Structural detectability*) A fault is structurally detectable iff it has a non-zero Boolean signature in the residual

$$c_i \in C_{detectable} \ \textit{iff} \ \exists j : c_i \neq 0 \Rightarrow r_j \neq 0.$$

Definition 1.8 (*Structural isolability*) A fault is structurally isolable iff it has a unique signature in the residual vector, i.e., column m_i of M is independent of all other columns in M,

$$c_i \in C_{isolable} \ \textit{iff} \ \forall j \neq i : m_i \neq m_j.$$

1.2.9 Active Isolation

In some cases faults are group-wise isolable, i.e., within the group individual faults are detectable but not isolable. This implies that with the given architecture of the system, these faults are only group-wise isolable. This does not necessarily imply that individual isolation cannot be achieved in other ways. Indeed, although the same set of residuals will be "fired" when either one or the other of non-structurally isolable constraints is faulty, the time response of the residuals may be different under the different fault cases. Exciting the system with an input signal perturbation may therefore make it possible to discriminate different responses of the same residual set when different constraints within the group are faulty. The analytical idea of applying test signals to isolate faults is not new. Zhang [101] designed test signals for diagnosis. A sophisticated set-up for active diagnosis was made in [74] and controller switching made for active diagnosis was shown in [83]. The structural analysis approach was first suggested in [11] and extended to different use modes in [67].

Proposition 1.1 *Active structural isolation is possible if and only if both a structural condition and a quantitative condition are true.*

Structural condition: the known variables in the set of residuals associated with a group of non-structurally isolable constraints include at least one control input.

Quantitative condition: the transfer from control inputs to residuals and or to output is affected differently by faults on different constraints.

Definition 1.9 (*Input to output reachability*) Let $p^{(i,j)} = \{c_f, c_g, \ldots, c_h\}$ be a path through the structure graph from input u_i to z_j, where z is a residual or an output and $\prod^{(i,j)}$ the union of valid paths from u_i to z_j. Let $C^{(i,j)}_{reach} = \left\{ c_g \mid c_g \in \prod^{(i,j)} \right\}$. A constraint c_h is input reachable from input u_i if a path exists from u_j to any output (or residual) z_k and the path includes the constraint, $c_h \in C^{(i,k)}_{reach}$.

Proposition 1.2 (Active structural isolability) *Two constraints c_g and c_h are actively structurally isolable from output signatures if $\exists i, j, k, l : c_g \in C^{(i,j)}_{reach}, c_h \in C^{(k,l)}_{reach}$ and $\left\{c_g, c_h\right\} \notin C^{(i,j)}_{reach} \cap C^{(k,l)}_{reach}$.*

These structural properties easily reveal which possibilities there are for active isolation of faults in a system described by its behaviours and associated topology (structure). The detailed design of which test signals are feasible and how test signals are detected efficiently are subjects of the signal-based design that follows the analysis based on structure.

1.2.10 Analysis of Scenarios with Multiple Faults

Scenarios of multiple faults are dealt with, in the structural analysis context, by removing one or more constraints that represent the faulty parts of the system. Should

c_6 be subject to a local failure, the remaining system $\mathbf{S}_f = \mathbf{S} \setminus \{c_6\}$ needs to be reanalysed. The results can show which residual generators exist for the faulty system, and which further faults could be isolated or detected. An application to a marine control system was treated in [8] where analysis of multiple faults was demonstrated as part of a fault-tolerant design.

1.3 Design Procedure Based on Analysis of Behavioural Relations

Having introduced the formal background of the structure-based approach, the generic procedure for set-up and design of the diagnostic system is now presented, including a brief overview of the signal-based design steps that are needed to convert the results from analysis based on structure to detection filters and change detection algorithms. This section will outline the generic design procedure [5] based on a system described by dynamic and algebraic constraints

$$\begin{aligned} \dot{\mathbf{x}} &= \mathbf{g}(\mathbf{x}, \mathbf{z}, \mathbf{u}, \mathbf{d}, \theta) \\ \mathbf{0} &= \mathbf{m}(\mathbf{x}, \mathbf{z}, \mathbf{u}, \mathbf{d}, \theta) \\ \mathbf{y} &= \mathbf{h}(\mathbf{x}, \mathbf{z}, \mathbf{u}, \mathbf{d}, \theta) \\ \dot{\mathbf{x}} &= \frac{d\mathbf{x}}{dt}, \end{aligned} \tag{1.5}$$

where $\mathbf{x}$ are states, $\mathbf{z}$ are variables determined by algebraic constraints, $\mathbf{u}$ is input, $\mathbf{y}$ is measured output, $\mathbf{d}$ is unknown disturbances and θ are system parameters.

1.3.1 Tools for Analysis of System Structure Properties

Let a system be given by the set of constraints of the form (1.5). The basic tool for analysis of system structure and generation of analytical redundancy relations is the matching of the unknown variables $(\mathbf{x}, \mathbf{z}, \mathbf{d})$ in a bipartite graph to the known ones, $(\mathbf{y}, \mathbf{u})$ through the constraints, as outlined above.

A first approach to matching in structural analysis was to generate a single complete causal matching on the over-specified part of the system [89] from which structural detectability and isolability properties could be shown. For nonlinear systems, isolability can be enhanced by the use of combinations of residuals for a diagnosis. Therefore, all matchings were generated in [23, 28] to investigate isolability properties. Finding all possible matchings is computationally heavy or even impossible for large industrial scale systems and Krysander [63, 64] instead finds MSO subgraphs within a graph. An MSO set is a subgraph where only one constraint can be removed to get a just-determined part of the graph. An MSO set contains a complete matching plus one unmatched constraint, and one redundancy relation is available from each MSO set. The algorithm by [63] is very efficient and was found to be about 10^5 times faster for a nine-bus electrical power distribution system [60] than the earlier algorithms [28]. In cases where the system topology might change, further techniques exist to also explore the different modes of operation and deduct isolability from the operation in different *use modes* [67].

1.3.2 Transformation to Signal Space and Analysis

Having found the matchings or MSO sets from the structural representation, where the unmatched constraints are used for diagnosis, mapping to analytical form is obtained by *backtracking* of unknown variables to known ones. Solving for unknown variables is done in the order specified by the particular matching. Tools are available that can compute the residuals from specification of the constraints [6], from a Simulink® simulation of a system [38] or from a bond graph of the system, see [30] and references herein.

FDI properties from the structural analysis should be interpreted with a bit of care, since structural and analytical properties in detectability and isolability are not identical. The isolability relations were analysed in [66] who introduced *checking models* to better approximate analytical isolability from structural isolability.

In summary, analytical redundancy relations (ARR) are generated from unmatched constraints, and the residuals are the ARR functions of time. The general form of the analytical redundancy relation obtained is the vector $r(t)$, which is *zero* when there is no deviation from normal, the $\mathcal{H}_0$ condition, or *non-zero* when triggered by a fault, the $\mathcal{H}_1$ condition.

The set of constraints $\mathbf{c}_{arr}$ that form the ARRs obtained from structural analysis will be linear or nonlinear, according to how the system is modelled,

$$\mathbf{c}_{arr}(t) = \mathbf{g}_{arr}(y, u, x, \dot{x}, \theta), \tag{1.6}$$

where θ are system parameters. Backtracking through a matching will express the unknown variables by the known ones, hence the set of residuals, represented by the vector $\mathbf{r}(t)$ will be functions only of known variables and their derivatives,

$$\mathbf{r}(t) = \mathbf{g}_r(\mathbf{y}(t), \mathbf{u}(t), \dot{\mathbf{y}}(t), \dot{\mathbf{u}}(t), \ddot{\mathbf{y}}(t), \ddot{\mathbf{u}}(t), \ldots, \theta). \tag{1.7}$$

1.3.2.1 Residuals for Diagnosis

In terms of structural analysis, $\mathbf{r}(t)$ will be different from zero for some t if a constraint is violated. In order to analyse the detailed properties of residual generators, we introduce faults $\mathbf{f}$ in an explicit way in the system being analysed and can then deduct properties such as sensitivity of residuals to particular faults and estimation of magnitude of faults. The system is then described by

$$\begin{aligned}\dot{\mathbf{x}} &= \mathbf{g}(\mathbf{x}, \mathbf{z}, \mathbf{u}, \mathbf{d}, \theta, \mathbf{f})\\ \mathbf{0} &= \mathbf{m}(\mathbf{x}, \mathbf{z}, \mathbf{u}, \mathbf{d}, \theta, \mathbf{f})\\ \mathbf{y} &= \mathbf{h}(\mathbf{x}, \mathbf{z}, \mathbf{u}, \mathbf{d}, \theta, \mathbf{f})\\ \dot{\mathbf{x}} &= \frac{d\mathbf{x}}{dt},\end{aligned} \tag{1.8}$$

and the associated residual vector by

$$\mathbf{r}(t) = \mathbf{g}_r\left(\mathbf{y}(t), \dot{\mathbf{y}}(t), \ldots, \mathbf{u}(t), \dot{\mathbf{u}}(t), \ldots, \mathbf{f}(t), \dot{\mathbf{f}}(t), \ldots, \theta\right). \tag{1.9}$$

In the linear time invariant setting, using s as the Laplace transform variable, the residual vector can be written in the form

$$\mathbf{r}(s) = \mathbf{V}_{ru}\mathbf{u}(s) + \mathbf{V}_{ry}(s)\left(\mathbf{H}_{yu}(s, \theta)\mathbf{u}(s) + \mathbf{H}_{yd}(s, \theta)\mathbf{d}(s) + \mathbf{H}_{yf}s, \theta\mathbf{f}(s)\right). \tag{1.10}$$

Decoupling from input $\mathbf{u}$ and disturbances $\mathbf{d}$ are obtained as part of the structural analysis since unknown disturbances are matched. Hence, $\mathbf{V}_{ru}$ and $\mathbf{V}_{ry}$ obtain the property

$$\mathbf{0} = \left(\mathbf{V}_{ry}(s)\ \mathbf{V}_{ru}(s)\right)\begin{pmatrix}\mathbf{H}_{yu}(s) & \mathbf{H}_{yd}(s)\\ \mathbf{I} & \mathbf{0}\end{pmatrix}. \tag{1.11}$$

This could also be obtained through a left nullspace design in the frequency domain. For linear systems, the decoupling property could also be obtained by unknown input observer design [34]. The structural analysis, the left nullspace approach and the unknown input observer design are equivalent for stable systems. Structural analysis followed by an observer-based implementation is preferred for unstable systems.

Real-life residuals should not comprise pure derivatives and no pure integrations, so filtering is required. This is a natural part of the design that follows the basic calculation of residuals. Linear methods have been thoroughly analysed in [97] and references herein.

Ideal residual generators leave the residual perfectly decoupled from input and disturbances. The residuals are only sensitive to faults, but the decoupling requirement can sometimes mask particular faults. When a non-masked fault is present, the residual will change, and detection of the change makes room for diagnosis of the fault. The logical steps include first *detection* that a fault is present, then *isolation* of which fault is present and then *estimation* of the magnitude of the fault. Faults can be *strongly detectable* in residuals if the change in residual persists as long as the fault is present, or it can be *weakly detectable*. The definitions of these terms are provided in the standard literature on fault diagnosis, including textbooks as [5, 21, 42, 80].

1.3.3 Robust Residuals

The ideal performance of residual generators (1.9) and (1.10) is obtained assuming perfect knowledge of system dynamics. With uncertainty being present, robustness needs to be assured. The structural analysis approach can force residual generators to be robust to selected parameters by declaring these parameters as unknown variables in the analysis. In this way, the parameters are treated as completely uncertain. If robustness is desired against more generic uncertainty bounds, not related to one or more particular parameters, principles of robust design using $\mathcal{H}_\infty$ optimal design methods were thoroughly treated in [19, 72]. The principle is to define residual generation as an optimisation problem, weighing sensitivity for faults against sensitivity to input and to unknown disturbances. Robustness to variations and uncertainty in parameters θ can be obtained using $\mathcal{H}_\infty$ methods. The robust residual generator design has been extended to incorporate active diagnosis techniques [74].

Another approach is to use interval methods where an observer calculates the interval, a polygon in the output space, where output could be expected given the bounds on the θ vector. A prediction can also be made assuming the occurrence of particular faults and associated intervals [53]. Hypothesis about faults can be made by rejection of models that are not in conformance with observed behaviour [16, 86], and robust active diagnosis methods have been suggested [94].

1.3.4 Evaluation of Residuals

In most real applications, random noise in residuals requires that stochastic change detection is employed for their evaluation and testing of hypotheses about presence of faults. Using the notation $\mathcal{H}_0$ for nominal system (no faults) and $\mathcal{H}_1$ for occurrence of a particular fault. Two models are distinguished based on discrete time observation of the residual

$$\begin{aligned} \mathcal{H}_0 &: \mathbf{r}(k) = \mathbf{w}(k) \\ \mathcal{H}_1 &: \mathbf{r}(k) = \mathbf{a}(k) + \mathbf{w}(k), \end{aligned} \tag{1.12}$$

where $\mathbf{a}(k)$ is a change in the residual of known or unknown magnitude or time-wise profile, and $\mathbf{w}(k)$ is a random component that can have a Gaussian or other distribution. The increments of each component $w_i(k)$ of $\mathbf{w}(k)$ would ideally be independent and from identical distributions (IID). Change detection is often based on testing a log-likelihood ratio of the distribution of residuals $p(r|\mathcal{H}_0)$, assuming $\mathcal{H}_0$ and $p(r|\mathcal{H}_1)$, assuming $\mathcal{H}_1$. Taken over a window of size M, the log-likelihood

$$S(k) = \frac{1}{M} \sum_{j=k-M}^{k} ln \left(\frac{p(r(j)|\mathcal{H}_0)}{p(r(j)|\mathcal{H}_1)} \right). \tag{1.13}$$

Theoretical methods to detect various change profiles have been developed, see [2, 75] for Gaussian change detection, [58] for non-Gaussian cases and [47] for combined signal processing and change detection, and improved formulations are constantly evolving, see [31].

A test statistic $g(k)$ is computed based on $S(k)$ using cumulative sum (CUSUM) algorithms for known changes, generalised likelihood (GLR) methods or others to detect an unknown change. A popular detection scheme is to compare $g(k)$ with a threshold h and decide $\mathcal{H}_1$ when $g(k) > h$. Key features offered by change detection theory include the possibility to predict the probability of false detection P_F under $\mathcal{H}_0$ and the probability of detection P_D under $\mathcal{H}_1$. Theoretical results to determine h commonly assume IID to calculate a detection threshold

Real-life residuals often are neither Gaussian nor IID so theoretic thresholds fall short in relating practical choice of h with P_F and P_D. Instead, the distribution of $g(k)$ can be estimated from data, and h is determined from the estimated distribution. Estimating over a sliding window was pursued in [39, 48, 99] for different domains of applications.

1.4 Aeroplane Diagnosis and Fault Handling

A case of diagnosis and fault handling of air speed sensor (pitot tube) faults on a small Unmanned Aerial Vehicle (UAV) was first studied in [49] and a comprehensive solution was presented in [48]. Focusing on faults in the pitot tube, that easily causes a crash if not diagnosed and handled in time, residuals are generated that allow both isolation and handling of a single sensor failure. Recorded telemetry data of an actual event with a pitot tube defect illustrate the efficacy of the diagnostics.

1.4.1 Airspeed Sensor Problem

Defects on sensors can have catastrophic consequences for aircraft, especially smaller UAVs, which do not have the same sensor redundancy that is available on a larger aircraft. It is therefore important to be able to detect whether a sensor defect has occurred. One of the vital sensors for an UAV is the pitot tube which measures the airspeed of the vehicle. This sensor is very exposed because of its position in the airstream and can easily be clogged by dust or water particles that freeze at higher altitudes.

The solution to these clogging problems usually employed on larger aircrafts is to install several pitot tubes with built-in heating devices to have a redundant system that can accommodate icing. Because of weight and space limitations, adding more sensors is usually not an option on smaller UAVs. Therefore, a different approach must be taken to diagnose and accommodate faults. One way is to have artefact readings detected and replaced with estimated values. Detection of faults and fault-tolerance for UAVs has a lot of focus and, as described in [22], many parts of the aircraft control and operation can benefit from using fault tolerant methods. A systematic approach to fault detection is described in [36] and some of the applications of these methods are. detection of mechanical defects, like stuck control surfaces. These were studied in [4, 77] where active methods were used to isolate faults. Observer-based fault diagnosis was investigated in [50]. Nonlinear models that describe the aircraft can also be used in fault diagnosis, this was demonstrated on small helicopters in [37]. A comprehensive overview for the state of art in diagnosis in aerospace was given in [103], and [43, 46] covered uses in the aircraft industry.

1.4.2 Model for Diagnosis

Aircraft are usually modelled by the dynamic and kinematic equations that describe their motion through the air. For some aircraft very detailed models are available, which includes nonlinear terms caused by aerodynamic effects. As the equations of motion are inherently nonlinear, these models give a very accurate aircraft behavioural description. Several textbooks, including [91] describe the detailed models.

A formulation of the generic behaviours for an aircraft [36] showed that the structural analysis approach is also well suited for generating a model for fault diagnosis. The behavioural formulation reduces the complexity to precisely what is required for diagnosis and the details of particular nonlinearities need be scrutinised only when needed for particular residuals. Focusing on the speed sensor fault diagnosis, a simplified behaviour description, related only to vehicle speed, is shown in Table 1.1 where $\mathbf{v}_n$ is speed over ground (navigation frame), v_t is air speed through propeller disc if the propeller was not present, $\mathbf{v}_a$ is air speed vector, $\mathbf{v}_w$ is the wind velocity vector, v_p the pitot tube measured air speed and $\mathbf{v}_g$ the velocity estimated by the Global Positioning System (GPS).

In Table 1.1, the function g_1 is an estimator of air speed $\mathbf{v}_a$ using the propeller thrust relations, where n is known rotational speed, g is gravity acceleration constant, m i mass of the aircraft, T_{nn} and T_{nv} are parameters determined from propeller characteristics, and F_A is aerodynamic drag,

Table 1.1 Velocity-related behaviours

Constraint	Behaviour
c_1 :	$\mathbf{v}_n = \mathbf{v}_a + \mathbf{v}_w$
c_2 :	$\hat{v}_t = g_1(n, \lvert\mathbf{v}_a\rvert, \theta)$
c_3 :	$\hat{v}_t = \lvert\mathbf{v}_a\rvert$
c_4 :	$\hat{\mathbf{v}}_w = g_2(\mathbf{v}_n, \mathbf{v}_a)$
c_5 :	$\hat{\mathbf{v}}_w = \mathbf{v}_w$
m_1 :	$\mathbf{v}_g = \mathbf{v}_n$
m_2 :	$v_p = \lvert\mathbf{v}_a\rvert$

$$g_1 = \left(T_{nv}n\right)^{-1}\left(T_{nn}n^2 - F_A(v_a, \theta) - mg\sin(\theta)\right). \tag{1.14}$$

The function g_2 is an estimator that provides a fairly uncertain estimate of the wind velocity vector.

The sets $\mathcal{K}$ and $\mathcal{X}$ of known and unknown variables in this problem are

$$\begin{aligned} \mathcal{K} &= \{n, \mathbf{v}_g, v_p, \theta\} \\ \mathcal{X} &= \{\hat{v}_t, \hat{\mathbf{v}}_w, \mathbf{v}_n, \mathbf{v}_a, \mathbf{v}_w\}. \end{aligned} \tag{1.15}$$

With 5 unknowns and 7 constraints, there are up to 2 unmatched constraints that can be used as parity relations. When solving the set of matched constraints in Table 1.1, calculability needs to be accounted for since a vector length can be determined from the components of the vector, but not reversely. A representation of the structure graph associated with the constraints is shown in the form of the incidence matrix in Table 1.2. A 1 in the matrix shows that the associated variable can be calculated from the constraint, a -1 shows that it cannot.

One complete matching on the unknown variables is achieved as follows:

$$c_1 \to \mathbf{v}_a; c_4 \to \hat{\mathbf{v}}_w; c_5 \to \mathbf{v}_w; m_1 \to \mathbf{v}_n; c_3 \to \hat{\mathbf{v}}_t; \tag{1.16}$$

The set of unmatched constraints are $\{c_2, m_2\}$. Backtracking to known variables along the matching leads to a symbolic form of the residuals and insertion of the analytical constraints from Table 1.1 gives the residuals

$$R_1 = |\mathbf{v}_g - \hat{\mathbf{v}}_w| - v_p = |\mathbf{v}_g - g_2(\mathbf{v}_n, \mathbf{v}_a)| - v_p \tag{1.17}$$

$$R_2 = \quad \hat{v}_t - v_p = g_1(n, |v_p|, \theta) - v_p. \tag{1.18}$$

This result is quite intuitive but the formal procedure assures that all possible redundancies have been explored.

Finding all minimal structurally overdetermined subsystems in a set of constraints is the basis of an extremely efficient algorithm developed by Krysander et al. [64].

Table 1.2 Graph for aeroplane velocity case

	Known				Unknown				
	n	$\mathbf{v}_g$	v_p	θ	$\hat{v}_t$	$\hat{v}_w$	$\mathbf{v}_n$	$\mathbf{v}_a$	$\mathbf{v}_w$
c_1							1	1	1
c_2	-1			-1	1			-1	
c_3					1			-1	
c_4						1	-1	-1	
c_5						1			1
m_1		1					1		
m_2			1					-1	

The result is a table similar to the backtracking shown above that is a recipe to find an analytic redundancy relation. This algorithm is integrated in the latest version of SaTool, an open source software [6] that analyses detectability and isolability, and provides symbolic and analytic expressions of residuals from a set of constraints that express the dynamic behaviours of a system.

While the simple estimate g_1 was found to suffice for the purpose of fault diagnosis by [49], more sophisticated and precise nonlinear estimators have been suggested in the literature. Zhou and Blanke [102] described a way to estimate state and parameters in nonlinear systems with a structure similar to the thrust equation here, [10] applied an adaptive observer scheme and [82] showed nonlinear observer designs for thrust estimation. Several estimation techniques were integrated in determining airspeed sensor faults in [48] that also assessed detection and false alarm probabilities from real data.

1.4.3 *Signal Analysis*

Figure 1.2 shows a time history and a histogram for R_1 and R_2 from UAV flight data in the fault free case.

The power spectral densities of the residual signals are not white. Since white noise is one of the requirements for most statistical change detectors to perform opti-

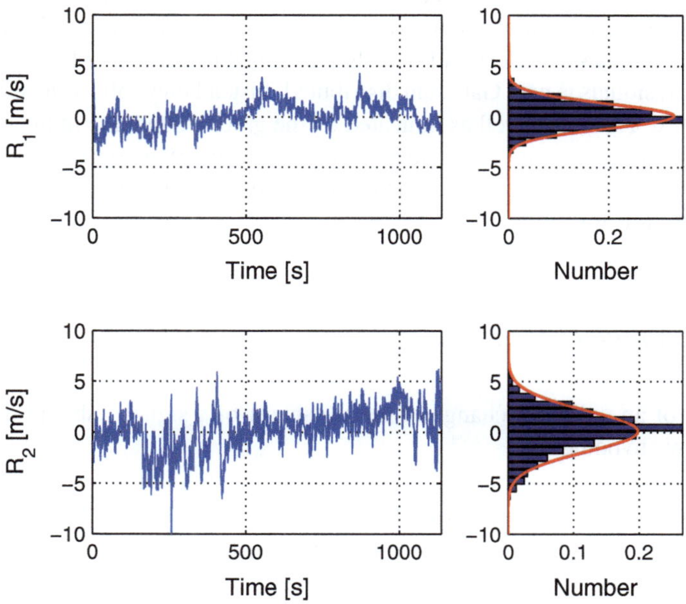

Fig. 1.2 Time development and histogram for residual R_1 and R_2 in the fault free case

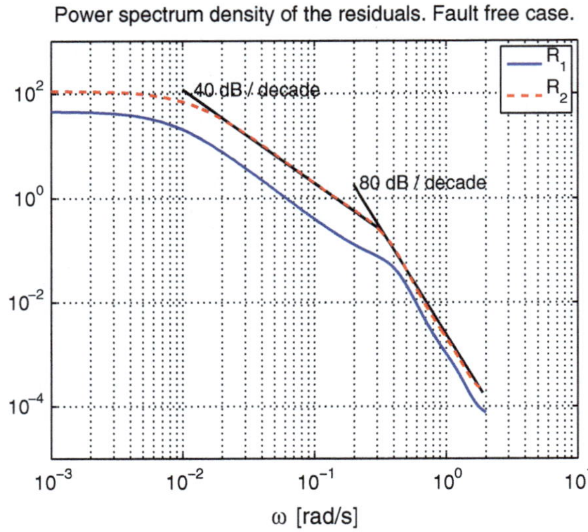

Fig. 1.3 Power spectrum densities for residual R_1 and R_2

mally, the low-pass filtered nature of the noise should be removed. One solution of dealing with coloured noise is to filter white noise through a suitable filter function to take into account the correlations present in the coloured noise. This can be created from a large record of data where all the signals properties are present. As indicated on Fig. 1.3 the power spectrum density of the two residuals consists of a part which decreases with 40 dB/decade and a part which decreases with 80 dB/decade. A whitening filter can be implemented as any stable filter, including as a Kalman filter.

Figure 1.4 shows time series of the whitened residuals and their histograms. Neither of the residuals is now Gaussian distributed. Instead they follow the Cauchy distribution (Eq. 1.20) very well as indicated by the green dotted line in the histogram. The Gaussian nature that the residuals had before whitening apparently came from the effects of low-pass filtering. When removed during the whitening process, the Gaussianity was lost.

1.4.4 Change Detection

Detection of an unknown change of mean value A of a signal x should distinguish between two hypotheses

$$\begin{aligned} \mathcal{H}_0 &: x[n] = w[n] \\ \mathcal{H}_1 &: x[n] = A + w[n]. \end{aligned} \tag{1.19}$$

The variance of the noise w is σ_w^2. The Generalised Likelihood Ratio Test (GLRT) is a standard way to solve such a problem. The standard solution assumed Gaussian

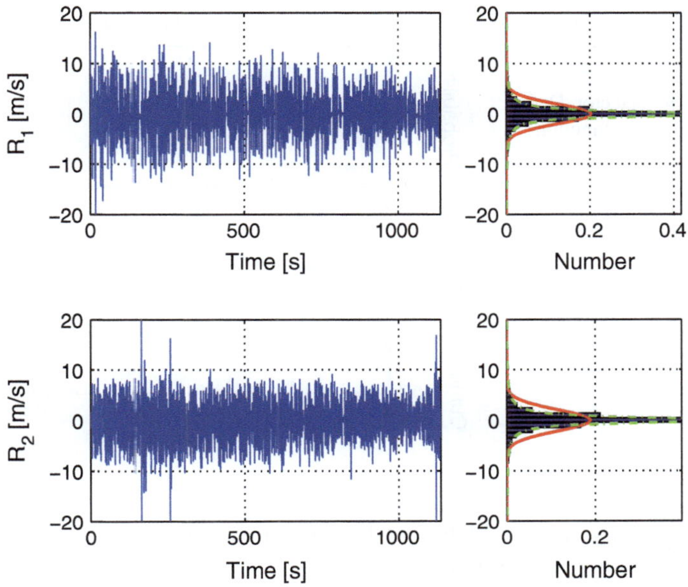

Fig. 1.4 Time development and histogram for the whitened residual R_1 and R_2 in the fault free case

noise. When the residuals are distributed according to a Cauchy distribution (as seen from Fig. 1.4) the following probability distribution function must be used:

$$p(x; x_o, \beta) = \frac{\beta}{\pi \left(x - x_o\right)^2 + \beta^2}, \tag{1.20}$$

where the two parameters are the half-width half-maximum scaling, β, and the offset x_o. The GLRT test statistic becomes

$$L_G(\mathbf{x}) = \frac{\prod_{i=1}^{N} p\left(x_i; \hat{x_o}, \hat{\beta}\right)}{\prod_{i=1}^{N} p\left(x_i; 0, \hat{\beta}\right)} > \gamma_c. \tag{1.21}$$

The maximum likelihood estimate (MLE) of $\hat{\beta}$ and $\hat{x_o}$ is found by fitting the data to Eq. 1.20.

Detection of change according to Eq. 1.21 requires that the threshold γ is determined. If the distribution of $L_G(\mathbf{x}|\mathcal{H}_o)$ is known, the probability of false alarm is

$$P_F(\mathbf{x}|\mathcal{H}_0, \gamma) = \int_{\gamma}^{\infty} p_{L_G}(x|\mathcal{H}_0)dx. \tag{1.22}$$

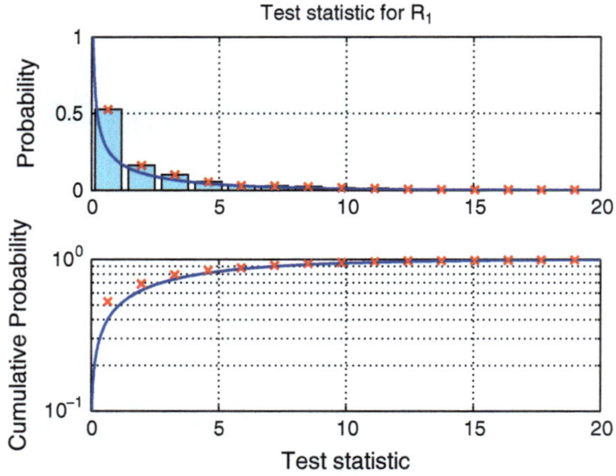

Fig. 1.5 Histogram and cumulative density for the $L_G(\mathcal{H}_0)$ test statistics for the GLRT Cauchy detector. Data and approximating function

The probability density (PDF) and the cumulative density (CDF) for $L_G(\mathcal{H}_0)$ are shown using flight data in Fig. 1.5.

The approximating function to the CDF in the figure is a Gamma distribution.

$$G(x; a, b) = \frac{1}{b^a \Gamma(a)} x^{a-1} \exp\left(-\frac{x}{b}\right) \qquad x > 0, \tag{1.23}$$

where Γ denotes the Gamma function. The distribution is fitted with an MLE and the following parameters are found:

	a	b
R_1	0.46	5.58
R_2	0.43	4.91

Selecting a threshold of $\gamma = 50$ gives a probability of false alarm of $1 - G(50; 0.46, 5.58) \cong 2.0\ 10^{-5}$ for R_1.

The theoretical performance could also be calculated using formulas from [49] but since theory and practice sometimes differ, it is advisable to check ones threshold value using an observed CDF from the test statistics of real data. This is particularly the case when the assumption of independent and identically distributed (*IID*) samples is not valid. The threshold required to achieve a certain false alarm probability can be very much different from its theoretical value [58] when the *IID* assumption is not valid.

1.4.5 Results

To test the detection performance, data from a real event are used where a UAV crashed, caused by a pitot tube defect and propagated effects on the flight control. The detection results are shown together with the residuals in Fig. 1.6.

As seen an alarm is raised, indicated by a value of 1, about 14 s after assumed instant when the incident started. Timely detection is hence obtained and with yet another 120 s elapsing until the crash happened, there would have been ample time to execute remedial actions had this detection system been available.

The simple detection methods in this example are sufficient to avoid simple accidents with simple equipped UAVs. Pitot tube defects have happened also on larger aircraft despite redundancy to a fail-operational level and the fault-diagnosis techniques could be part of fault-tolerant solutions and have a wider area of application.

1.5 Fault-Tolerant Guidance Using Vision

This second case from agriculture brings diagnosis and fault-tolerant control techniques in operation within advanced computer vision-based sensors. In agriculture, the typical tasks are to follow structures in the field to plow, seed, spray or harvest. The specific harvesting task of baling involves to follow rows of cut straw or grass (swath, see Fig. 1.7) in order to pick it up and process it into bales. This is a labour

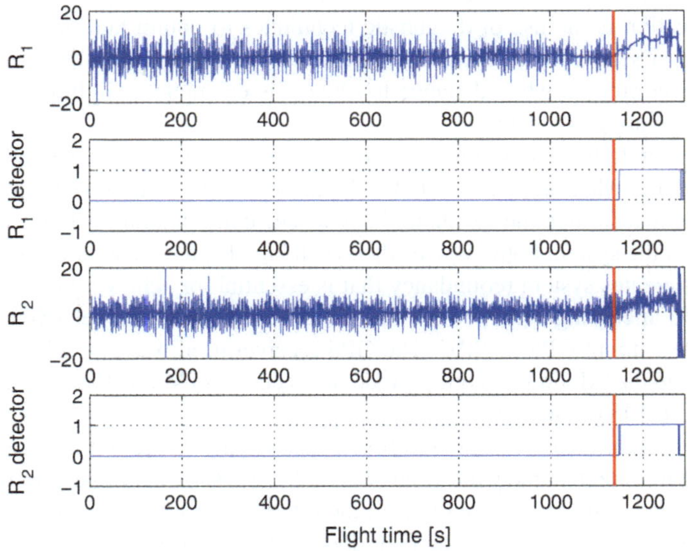

Fig. 1.6 Residuals and detector output. At approximately $t = 1140$ s the clogging of the pitot tube occurs (marked with the *red vertical line*)

Fig. 1.7 Swath in the field to be collected

intensive and repetitive task, which is of interest to automate. The difficulties pertaining to automating this task are similar to the difficulties in automating a large range of agricultural tasks. The ability to track this structure using 3D shape information from a stereo camera and/or GPS information was demonstrated in [15] and a detailed presentation of the baling problem was presented in [13] where a classifier was employed based on online learning of texture information about the swath and the surroundings. This was then coupled with shape information to extract the swath position and a mapping kept track of measured swath positions. The map was used to guide the vehicle along the swath by steering the tractor's front wheels while a driver controlled the throttle and brakes. The novelty with respect to diagnosis in this application is the use of diagnostic techniques to obtain fault tolerance in the stereo vision sensor itself, avoiding typical reasons for artefacts that occur in the stereo vision process when distance to objects is calculated. Highlights from this real-life case are given below.

The system to be analysed is equipped with stereo-vision, a single antenna GPS and an Inertial Measurement Unit (IMU), in one configuration. GPS positions of the vehicle that formed the swath are known. The combination of stereo-vision and GPS allows the system both to "see" the swath but also navigate based on the given map. This creates system redundancy that is essential for achieving fault tolerance. A visual odometry algorithm on the stereo-camera allows for the relative position of the vehicle without GPS or IMU. The GPS receiver used was ground-station compensated and the IMU was a tactical grade (low accuracy) Micro-Electro-Mechanical Systems (MEMS)-based unit.

The two main ideas presented is first a behavioural model for representing the natural environment, namely the swath. Secondly, it is shown how parts of this model (the swath location) can be used in conjunction with sensor inputs to create a fault-tolerant sensor fusion system. The fault diagnosis is illustrated using real data.

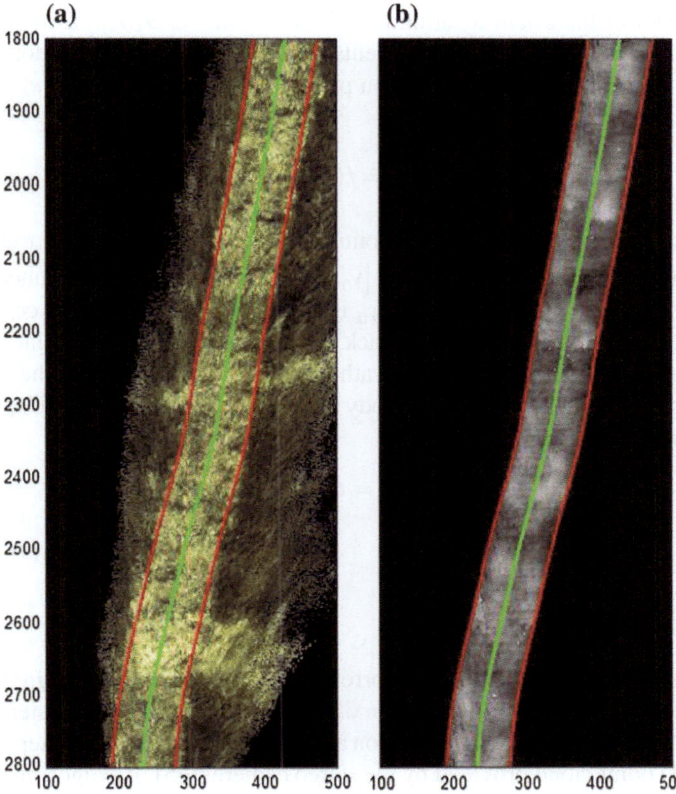

Fig. 1.8 **a** RGB *topdown-view* of the swath with swath location illustrated with a *green spline* and estimated swath width superimposed with *red borders*. **b** The same swath model but with grayscale values illustrating the height of the swath in a given cell

1.5.1 Modelling the Natural Environment

A model of the swath requires extracting the salient features of the environment required for field operation and storing them in the model representation. The salient features are the location of the swath and the distribution of the swath material across the swath. Swath location, width and height are modelled as a map using splines to follow the centre of the swath in a metric map. The model is illustrated in Fig. 1.8.

The swath location is defined as being in a 2D coordinate system on the ground plane. A function f represents the lines down the middle of the swaths. Given coordinate pairs (x, y) then f is

$$y = f(x). \tag{1.24}$$

The model of the swath location is then $s(x)$ with $s \in S_3(\mathbf{k}_{0:n})$, where $\mathbf{k}_{0:n}$ are the spline knots and the spline coefficients and S_3 is the cubic spline domain. Then the model is equal to the swath location plus the approximation error e_a of fitting a spline to f:

$$s(x) = f(x) + e_a. \tag{1.25}$$

Based on the concept of having a controller that allows the vehicle to follow the swath location, the position error $\varepsilon_x = \left[y_0, \phi, C_0\right]^T$ can be found as a function of the tractor position and the spline. Here, y_0 is the lateral offset from the control point to the spline (also known as the X-track error), angle ϕ is the angular difference between the vehicle attitude and the swath, and C_0 is the curvature of the swath near the vehicle. Defining the spline s_b in body coordinates a function $\mathcal{E}$ can be set to find the position error:

$$\varepsilon_x = \mathcal{E}(s_b). \tag{1.26}$$

1.5.2 Stereo Camera

A stereo algorithm is used to find the correspondence between features in the left and right image sensors (i_l, i_r). The position of the features relative to the stereo camera can then be inferred in 3D. Modern vision algorithms then exist to register 3D models with the 3D point cloud provided by the stereo camera: [45]. An algorithm has been constructed that allows such registration between the swath model and the 3D points. The stereo algorithm and registration will be denoted by the function $\mathcal{V}_{reg}$. Thus, given the two images a measurement of the swath location s_c can be computed for the part of the swath in the image.

$$s_c = \mathcal{V}_{reg}(i_l, i_r). \tag{1.27}$$

These measurements are stored in a map representation for an individual swath and s_m is the spline formed by combining N previous measurements

$$\mathbf{s}_m(k) = \min_{spline} \left(s_c(j), j = k - N, k\right). \tag{1.28}$$

1.5.3 Robust Stereo Enhancement by Texture

Stereo vision uses correlation of textures to determine distance to objects in view of both cameras. When misclassification of texture happens, gross errors may occur in the stereo calculated distance to objects if artefact objects occur. To robustify the stereo against such artefacts, [13] introduced statistical validation of texture using so-called *textons*. Texture properties can include colour distribution, intensity, shape and patterns. Textons are basis vectors extracted from the local descriptors of texture. Following [98], who showed that small local texture neighbourhoods may be better than using large filter banks, [13, 14] employed statistical classification to the problem and used the texture-based classification of swath as a supplement to the geometrical mapping described above.

1.5.4 Texton Labelling

Given a colour image as input, pixel neighbourhoods in the image are classified into a texton type, which belongs to a set of basis textons υ_i obtained by prior learning from a training image. This is done by first extracting a descriptor in the form of a vector from each pixel location in the image. For each location the p_i vector is

$$p_i = \begin{bmatrix} w_1 * l_c \\ w_2 * a_c \\ w_2 * b_c \\ w_3 * (l_1 - l_c) \\ \vdots \\ w_3 * (l_8 - l_c), \end{bmatrix}, \tag{1.29}$$

where $[l_c, a_c, b_c]$ is the colour of the pixel at this location in CIE*LAB colour-space. $(l_1 - l_c), \ldots, (l_8 - l_c)$ are the intensity differences between the pixel at this location and the 8 surrounding pixels in a 3×3 neighbourhood. The vector elements are then weighted using $\{w_1 = 0.5, w_2 = 1, w_3 = 0.5\}$. A K-means algorithm is then run on all these descriptors to extract cluster centres which we refer to as textons. The fast K-means algorithm [57] was used in [13] to find the set of textons υ_j that partitions the descriptors into κ sets $\mathcal{T} = \mathcal{T}_0, \mathcal{T}_1, \ldots, \mathcal{T}_\kappa$ by trying to minimise:

$$\mathcal{T}_{reg} = \underset{\mathcal{T}}{\operatorname{argmin}} \sum_{j=1}^{\kappa} \sum_{p_i \in \mathcal{T}_j} \left\| p_i - \upsilon_j \right\|^2 . \tag{1.30}$$

As a final step, processing of all labelled pixels in an image at time k gives an estimate of swath density using the textons that are classified as belonging to the swath Y_{swath},

$$s_t(k) = \max_{density} \mathcal{T}_{reg}(p_i \in Y_{swath}, i = k - N \ldots k). \tag{1.31}$$

Each pixel location in the image is then labelled by finding the nearest texton in Euclidean space. An example of the result of such a classification is shown in Fig. 1.9.

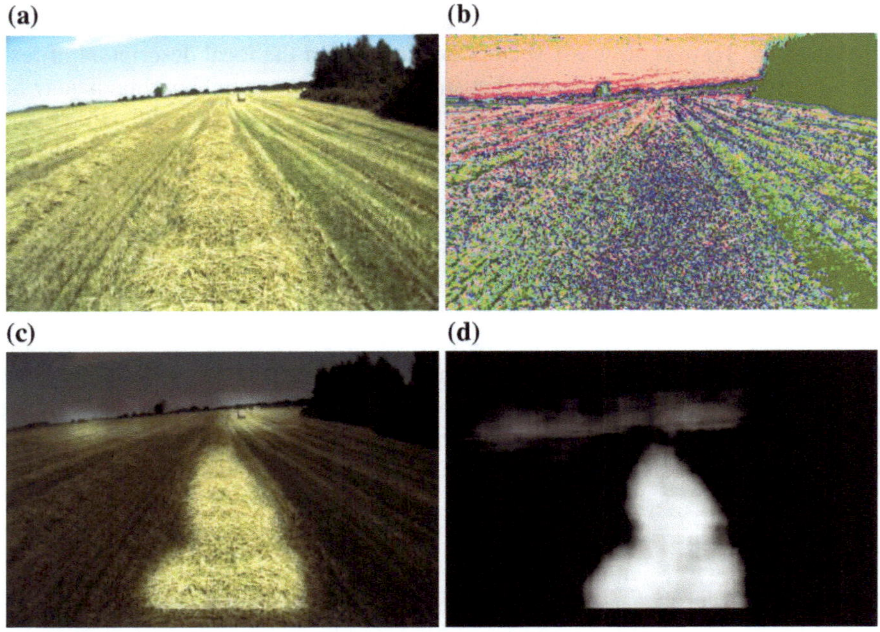

Fig. 1.9 **a** *Left image* from the stereo camera. **b** Texton classification. Pixel colour shows class of texton. **c** The stereo camera image with a transparency mask based on the swath classification. **d** Swath classification based on texture with intensity representing the strength of classification

One way of abstracting the texture processing could be in the form of a spline function describing the texture-classified swath

$$\mathbf{s}_t(k) = \min_{spline} \left(s_t(j), j = k - N, k \right). \tag{1.32}$$

Another use of the texton classification could be to robustify the stereo estimate (1.27) by,

$$\left[s_{ct}, g_{ct}, h_{ct} \right]^T = \mathcal{V}_{reg}(i_l, i_r, \mathcal{T}_{reg}). \tag{1.33}$$

Using the *intelligent sensor* capability of (1.33) represents a fault-tolerant processing within the stereo algorithms that could be used after learning of the set Y_{swath} has been obtained. This is an alternative to using (1.32) and (1.28) separately, but the latter could have benefits during unsupervised learning.

Table 1.3 Behaviours for Baler control example

$c_1 : s^b = \mathbf{R}_b^n(\Theta, \lambda)s + \mathbf{p}^n$
$c_2 : \varepsilon_x = \mathcal{E}(s^b)$
$d_1 : \dot{\mathbf{p}}^n = \frac{d}{dt}\mathbf{p}^n$
$m_1 : \mathbf{v}^b = \mathbf{R}_b^n(\Theta, \lambda)\dot{\mathbf{p}}^n$
$m_2 : \mathbf{a}^b = \frac{d}{dt}\mathbf{R}_b^n(\Theta, \lambda)\dot{\mathbf{p}}^n$
$m_3 : \mathbf{p}_1^n = \mathbf{p}^n$
$m_4 : s_g = s$
$m_5 : s_c = s^b$
$m_6 : s_m = s^b$
$m_7 : s_t = s^b$

1.5.5 Structural Model

The constraints in this case describe the structured natural environment of the field with swath, the baler and the available sensors. Let the constraints be composed of those from measurements (m), differential (d), and the "system" constraints (c). Using $\mathbf{v}^b$ for visual odometry measured velocity vector over ground seen in body coordinates; $\mathbf{a}^b$ for IMU measured acceleration vector; $\mathbf{p}^n$ the position in (North, East) coordinates with $\mathbf{p}_1^n$ being the position measurement from the GPS; $\mathbf{R}_b^n$ is the rotation matrix from body to navigation frame, which is a function of Θ, the attitude vector (Euler angles roll, pitch and yaw) and of λ, the latitude;. s^b is the swath position spline seen in body coordinates. In this analysis, the $\mathbf{R}_b^n$ matrix is assumed to be known.

With variables defined above, the sets of known and unknown variables are

$$\begin{aligned} \mathcal{K} &= \{\mathbf{v}^b, \mathbf{a}^b, \mathbf{p}_1^n, \mathbf{s}_g, \mathbf{s}_c, , \mathbf{s}_m, \mathbf{s}_t, \mathbf{R}_b^n(\Theta, \lambda)\} \\ \mathcal{X} &= \{\mathbf{p}^n, \dot{\mathbf{p}}^n, s^b, s, \varepsilon_x\}. \end{aligned} \tag{1.34}$$

The constraints are summarised in Table 1.3

1.5.6 Residuals for Fault Diagnosis

The number of constraints available depends on: the *use mode*, e.g. m_7 is only available when the tractor is in the field with a swath and texture learning has been completed; the *configuration*, e.g. whether an IMU is mounted; on *faults*, e.g. failure of an instrument. Structural analysis under different use modes was presented in [67] and aspects of analysis under simultaneous faults was discussed in [11]. A structural analysis of the baler example, without the texture element, was presented in [12]. Analysing the constraints listed in Eq. 1.3, the following residuals are obtained:

$$\begin{aligned} r_1 &= \mathbf{v}^b - \mathbf{R}_b^n(\Theta, \lambda)\frac{d}{dt}\mathbf{p}_1^n \\ r_2 &= \frac{d}{dt}\mathbf{v}^b - \mathbf{a}^b \\ r_3 &= \mathcal{E}(s_c) - \mathcal{E}(\mathbf{R}_b^n(\Theta, \lambda)s_g + \mathbf{p}_1^n) \\ r_4 &= \mathcal{E}(s_c) - \mathcal{E}(s_m) \\ r_5 &= \mathcal{E}(s_m) - \mathcal{E}(s_t). \end{aligned} \tag{1.35}$$

1.5.7 Field Tests

The properties of residuals were investigated based on recorded data. The position of the swath was first logged by following the middle of the swath manually—emulating the vehicle forming the swath. This was repeated for a second pass emulating the vehicle that should pick up the swath. This provides some form of limited ground truth. The position error of the driver is bounded between the runs as he constantly steers relative to the swath.

Experience with driving with balers puts the error associated with not driving exactly over the centre of the swath to under ± 0.2 m as this is required to pick up the swath successfully. In the data examined the GPS has a false offset in the second pass relative to the first pass of approximately 0.6 m for the first approx. 70 s before it corrects its position estimate to bring it to about 0.15 m of the swath location (See Fig. 1.10). This offset is acceptable for normal operation. Field tests enabled calculation of residuals r_1, r_3 and r_4 as an instrumentation issue prevented data reception

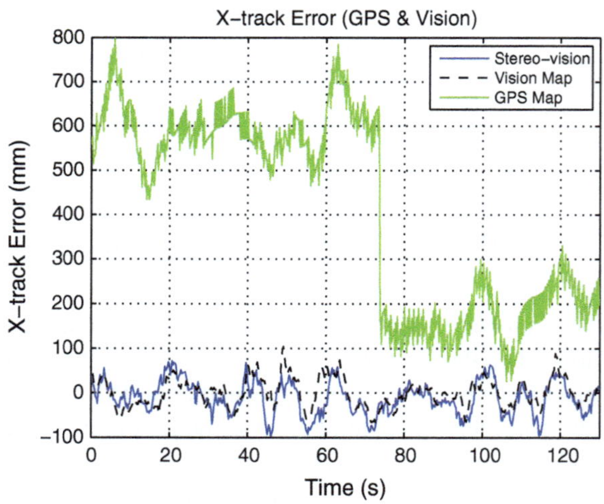

Fig. 1.10 The vehicle was driven manually over a swath. The driver centred the vehicle over the *middle* of the swath and drove for 2 min while maintaining this centered position. The x-track errors from the subsystems were recorded

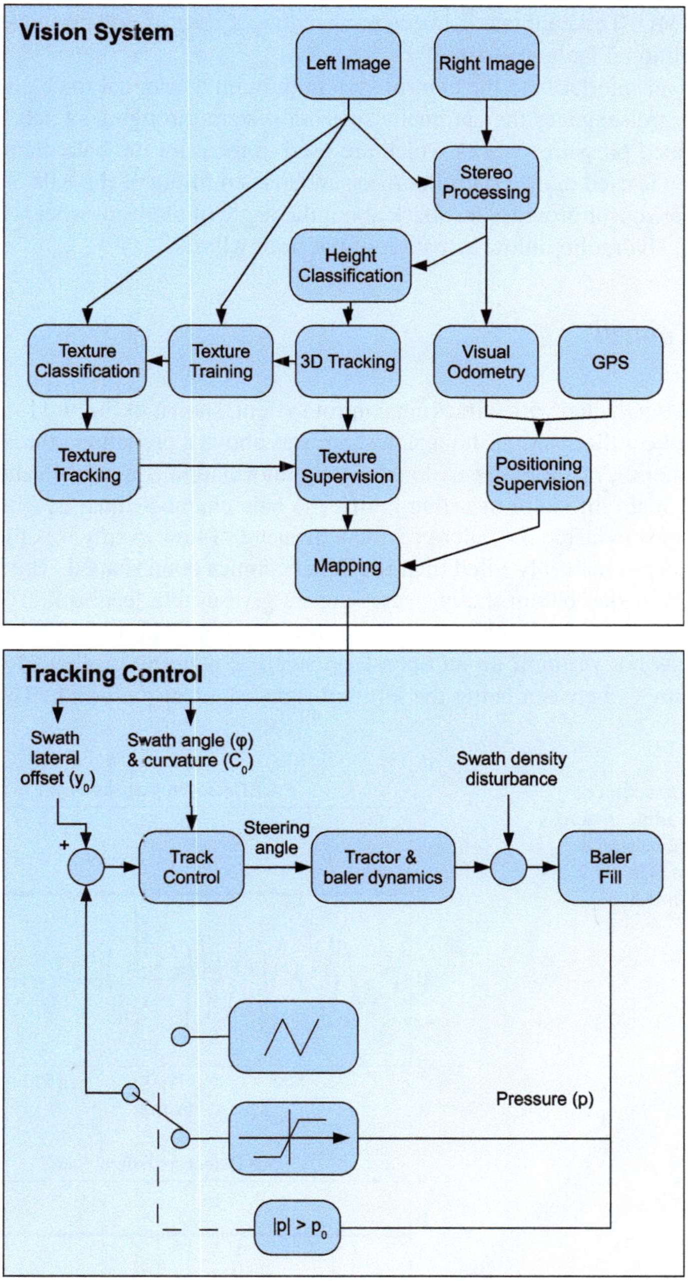

Fig. 1.11 Block diagram of vision system and tracking control for baling. Supervised classification and positioning provides mapping of field structures, which are fed to the steering controller

from the IMU. The field test is hence representing a case of one permanent failure and an additional fault occurring.

The driver interfaces to the control system through a terminal to change settings and engages/disengages the automatic steering system through a switch. The baler has integrated pressure sensors which are used to measure the bale diameter. This information is used in the controller to assure an even filling of the bale chamber. A wheel angle sensor provides feedback about the angle of the front wheels relative to the tractor. Hydraulics allow actuation of the front wheels.

1.5.8 Control

Tests were conducted with a tracking control system, shown in Fig. 1.11, which was made to collect the swath if the match score was above a predefined threshold. The control system remains active as long as map information is available ahead of the vehicle. A main difficulty in baling is that the bale chamber must be filled evenly. Pressure sensors inside the baler provide a measure of how evenly it is filled. If the bale chamber is unevenly filled then the bale becomes cone shaped. The bale must have a certain size before the pressure sensors give usable feedback. To compensate for this lack of feedback the controller has two states. In the initial state where pressure has not yet built up an open-loop steering pattern is followed where the vehicle changes between being the left and right edges of the swath. The steering

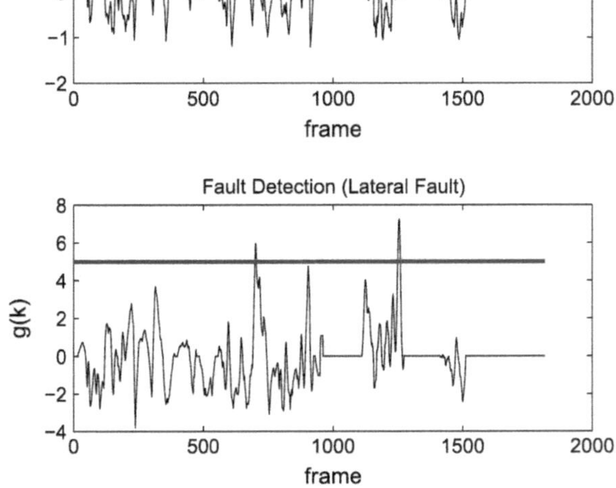

Fig. 1.12 Detection of deviation in angle or across the swath track using a CUSUM test. The *thick line* defines the threshold

system changes mode when the bale size reaches a threshold and sensor feedback is used. The sensor feedback spans an interval from −1 to 1 indicating how cone shaped the bale is. This information and the relative location of the swath (computed from a fused spline map) is used for control. The swath parameters y_0, ϕ, and C_0 of the spline relative to the vehicle are then fed to a generic curve tracking controller. Fault-tolerant control described in this example focused on the complex parts, i.e., the natural environment. Extension with the baling control part would be simple.

Figure 1.12 shows detection of deviations in steering angle and cross-track based on a CUSUM detection. The statistical change detection robust and provides a robust diagnosis.

1.6 Conclusions

This chapter has introduced generic principles of diagnosis and fault-tolerant control based on the formal description of *behaviours* and *services*. The steps of system analysis, residual generation and change detection were discussed and exemplified through two large case studies. One dealt with UAV sensor fault diagnosis and fault-handling, another with vision-based baling for agriculture. It was shown that the generic methods are indeed applicable to the complexity met in a natural environment, and the chapter showed how such generalisation was obtained in both theory and practice.

Acknowledgments Research support from CLAAS Agrosystems, from the Danish Forces Joint UAV Team and the Danish Ministry for Science, Development and Innovation are gratefully acknowledged.

References

1. Åström, K.J., Albertos, P., Blanke, M., Isidori, A., Schaufelberger, W., Sanz, R. (eds.): Control of Complex Systems. Springer (2000)
2. Basseville, M., Nikiforov, I.V.: Detection of Abrupt Changes: Theory and Application. Information and System Science. Prentice Hall, New York (1993)
3. Basseville, M.: Detecting changes in signals and systems—a survey. Automatica **24**(3), 309–326 (1988)
4. Bateman, F., Noura, H., Ouladsine, M.: Fault tolerant control strategy based on the DoA: Application to UAV. In: 7th IFAC Symposium on Fault Detection Supervision and Safety of Technical Processes (2009)
5. Blanke, M., Kinnaert, M., Lunze, J., Staroswiecki, M.: Diagnosis and Fault-tolerant Control, 2nd edn. Springer (2003, 2nd edn. 2006)
6. Blanke, M., Lorentzen, T.: SaTool—a software tool for structural analysis of complex automation systems. In: 6th IFAC Symposium on Fault Detection, Supervision and Safety of Technical Processes SAFEPROCESS. Beijing (2006)
7. Blanke, M.: Consistent design of dependable control systems. Control Eng. Pract. **4**(9), 1305–1312 (1996)

8. Blanke, M.: Fault-tolerant sensor fusion with an application to ship navigation. In: Proceedings of IEEE Mediterranean Control Conference, pp. 1385–1390, June 2005
9. Blanke, M., Izadi-Zamanabadi, R., Bøgh, S.A., Lunau, C.P.: Fault-tolerant control systems—a holistic view. Control Eng. Pract. **5**(5), 693–702 (1997)
10. Blanke, M., Izadi-Zamanabadi, R., Lootsma, T.: Fault monitoring and re-configurable control for a ship propulsion plant. J. Adapt. Control Signal Process. **12**, 671–688 (1998)
11. Blanke, M., Staroswiecki, M.: Structural design of systems with safe behavior under single and multiple faults. In: 6th IFAC Symposium on Fault Detection, Supervision and Safety of Technical Processes SAFEPROCESS, pp. 511–516. Beijing, P. R. China, Sept 2006
12. Blas, M.R., Blanke, M.: Natural environment modeling and fault-diagnosis for automated agricultural vehicle. In: Proceedings 17th IFAC World Congress, Seoul, Korea, pp. 1590–1595 (2008)
13. Blas, M.R., Blanke, M.: Automatic baling using stereo vision and texture learning. Comput. Electron. Agric. **75**, 159–168 (2011)
14. Blas, M.R., Blanke, M., Madsen, T.E.: A method of detecting a structure in a field, a method of steering an agricultural vehicle and an agricultural vehicle. Patent, Mar 2010, eP10157313.7
15. Blas, M.R., Blanke, M., Rusu, R.B., Beetz, M.: Fault-tolerant 3d mapping with application to an orchard robot. In: Proceedings of 7th IFAC International Symposium on Fault Detection, Supervision and Safety of Technical Processes (SAFEPROCESS'09), pp. 893–898 (2009)
16. Blesa, J., Puig, V., Saludes, J.: Robust fault detection using polytope-based set-membership consistency test. IET Control Theory Appl. **6**(12), 1767–1777 (2012)
17. Caponetti, F., Blas, M.R., Blanke, M.: Stochastic automata for outdoor semantic mapping using optimised signal quantisation. Control Eng. Pract. **19**, 223–233 (2011)
18. Chen, J., Patton, R.J.: Optimal filtering and robust fault-diagnosis of stochastic-systems with unknown disturbance. IEE Proc. Control Theory Appl. **143**(1), 31–36 (1996)
19. Chen, J., Patton, R.J.: Robust Model-based Fault Diagnosis for Dynamic Systems. Kluwer Academic Publishers (1999)
20. Chow, E.Y., Willsky, A.S.: Analytical redundancy and the design of robust failure detection systems. IEEE Trans. Autom. Control **AC-29**(7), 603–614 (1984)
21. Ding, S.M.: Model-based Fault Diagnosis Techniques: Design Schemes, Algorithms, and Tools. Springer (2008)
22. Ducard, G.J.J.: Fault-tolerant Flight Control and Guidance Systems. Springer (2009)
23. Düştegör, D., Cocquempot, V., Staroswiecki, M.: Structural analysis for fault detection and identification: an algorithmic study. In: Proceedings of the 2nd Symposium on System Structure and Control 2004 (SSSC'04). Oaxaca, Mexico, 8–10 Dec 2004
24. Düştegör, D., Frisk, E., Cocquempot, V., Krysander, M., Staroswiecki, M.: Structural analysis of fault isolability in the damadics benchmark. Control Eng. Pract. **14**(6), 597–608 (2006)
25. Dulmage, A.L., Mendelsohn, N.S.: A structure theory of bi-partite graphs. Trans. R. Soc. Can. Sec. **3**(53), 1–13 (1959)
26. Dulmage, A.L., Mendelsohn, N.S.: Two algorithms for bipartite graphs. J. Soc. Ind. Appl. Math. **11**(1), 183–194 (1963)
27. Dunia, R., Qin, S.J., Edgar, T.F., McAvoy, T.J.: Identification of faulty sensors using principal component analysis. AIChE J. **42**, 2797–2812 (1996)
28. Düştegör, D.: Structural analysis for fault detection and identification: algorithmic Issues. Ph.D. thesis, Universite des Sciences et Technologies de Lille—France, Automatic, Computer and System Engineering, Dec 2005
29. Emami-Naeini, A., Akhter, M., Rock, S.M.: Effect on model uncertainty on failure detection; the threshold selector. IEEE Trans. Autom. Control **AC-33**(12), 1106–1115 (1988)
30. Feki, M.E., Jardin, A., Marquis-Favre, W., Krähenbühl, L., Thomasset, D.: Structural analysis by bond graph approach: duality between causal and bicausal procedures. J. Syst. Control Eng. **226**, 82–100 (2012)
31. Fouladirad, M., Freitag, L., Nikiforov, I.: Optimal fault detection with nuisance parameters and a general covariance matrix. Int. J. Adapt. Control Signal Process. **22**(5), 431–439 (2008)

32. Frank, P.M.: Fault diagnosis in dynamic systems using analytical and knowledge-based redundancy. Automatica **26**(3), 459–474 (1990)
33. Frank, P.M.: On-line fault detection in uncertain nonlinear systems using diagnosticobservers; a survey. Int. J. Syst. Sci. **25**(12), 2129–2154 (1994)
34. Frank, P.M., Ding, S.X., Marcu, T.: Model-based fault diagnosis in technical processes. Trans. Inst. Meas. Control **22**(1), 57–101 (2000)
35. Frank, P.M., Ding, X.: Frequency domain approach to optimally robust residual generation and evaluation for model-based fault diagnosis. Automatica **30**(5), 789–804 (1994)
36. Fravolini, M.L., Brunori, V., Campa, G., Napolitano, M.R., La Cava, M.: Structured analysis approach for the generation of structured residuals for aircraft FDI. IEEE Trans. Aerosp. Electron. Syst. **45**(4), 1466–1482 (2009)
37. Freddi, A., Longhi, S., Monteriu, A.: A model-based fault diagnosis system for unmanned aerial vehicles. In: 7th IFAC Symposium on Fault Detection, Supervision and Safety of Technical Processes (2009)
38. Frisk, E., Krysander, M., Nyberg, M., Åslund, J.: A toolbox for design of diagnosis systems. In: Proceedings of IFAC Safeprocess'2006 (2006)
39. Galeazzi, R., Blanke, M., Poulsen, N.K.: Early detection of parametric roll resonance on container ships. IEEE Trans. Control Syst. Technol. **21**(2), 489–503 (2013)
40. Gertler, J.J.: Survey of model-based failure detection and isolation in complex plants. IEEE Control Syst. Mag. **8**(6), 3–11 (1988)
41. Gertler, J.J.: Fault detection and isolation using parity relations. Control Eng. Pract. **5**(5), 653–661 (1997)
42. Gertler, J.: Fault Detection and Diagnosis in Engineering Systems. Marcel Dekker (1998)
43. Gheorghe, A., Zolghadri, A., Cieslak, J., Goupil, P., Dayre, R., Berre, H.L.: Model-based approaches for fast and robust fault detection in an aircraft control surface servo loop: from theory to flight tests. IEEE Control Syst. Mag. **33**(3), 20–30 (2013)
44. Gálvez-Carrillo, M., Kinnaert, M.: Sensor fault detection and isolation in doubly-fed induction generators accounting for parameter variations. Renew. Energy **36**, 1447–1457 (2011)
45. Goshtasby, A.A.: 2-D and 3-D Image Registration: For Medical, Remote Sensing, and Industrial Applications. Wiley (2005)
46. Goupil, P.: Airbus state of the art and practices on FDI and FTC in flight control systems. Control Eng. Pract. **19**(6), 524–539 (2011)
47. Gustafsson, F.: Adaptive Filtering and Change Detection. Wiley (2000, 2nd edn. 2001)
48. Hansen, S., Blanke, M.: Diagnosis of airspeed measurement faults for unmanned aerial vehicles. IEEE Trans. Aerosp. Electron. Syst. **50**(1) (2014)
49. Hansen, S., Blanke, M., Adrian, J.: Diagnosis of UAV pitot tube defects using statistical change detection. In: 7th IFAC Symposium on Intelligent Autonomous Vehicles, vol. 7, no. 1. IFAC PapersOnLine (2010)
50. Heredia, G., Ollero, A., Bejar, M., Mahtani, R.: Sensor and actuator fault detection in small autonomous helicopters. Mechatronics **18**, 90–99 (2008)
51. Hopcroft, J.E., Karp, R.M.: An n/sup 5/2 algorithm for maximal matchings in bipartite graphs. SIAM J. Comput. 225–231 (1973)
52. Horak, D.T.: Failure detection in dynamic systems with modeling errors. J. Guidance Control Dyn. **11**(6), 508–516 (1988)
53. Ingimundarson, A., Bravo, J.M., Puig, V., Alamo, T., Guerra, P.: Robust fault detection using zonotope-based set-membership consistency test. Int. J. Adapt. Control Signal Process. **23**(4), 311–330 (2009)
54. Isermann, R.: On fuzzy logic applications for automatic control, supervision, and faultdiagnosis. IEEE Trans. Syst. Man Cybern. Part A-Syst. Hum. **28**(2), 221–235 (1998)
55. Isermann, R.: Supervision, fault-detection and fault-diagnosis methods—an introduction. Control Eng. Pract. **5**(5), 639–652 (1997)
56. Isermann, R.: Fault-Diagnosis Systems: An Introduction from Fault Detection to Fault Tolerance. Springer (2005)
57. Jain, A., Dubes, R.: Algorithms for Clustering Data. Prentice-Hall, Englewood Cliffs (1988)

58. Kay, S.M.: Fundamentals of Statistical Signal Processing: Detection Theory. Prentice-Hall PTR (1998)
59. Kinnaert, M.: Robust fault detection based on observers for bilinear systems. Automatica **35**, 1829–1842 (1999)
60. Knüppel, T., Blanke, M., Østergaard, J.: Fault diagnosis for electrical distribution systems using structural analysis. Int. J. Robust Nonlinear Control (2014)
61. Korbicz, J.: Fault diagnosis of non-linear dynamical systems using analytical and soft computing methods. J. Autom. Mobile Robot. Intell. Syst. **1**, 7–23 (2007)
62. Korbicz, J., Koscielny, J.M., Kowalczuk, Z.: Fault Diagnosis: Models, Artificial Intelligence, Applications. Springer (2004)
63. Krysander, M.: Design and analysis of diagnosis systems using structural methods. Ph.D. thesis, Linköping University, June 2006
64. Krysander, M., Åslund, J., Nyberg, M.: An efficient algorithm for finding minimal overconstrained subsystems for model-based diagnosis. IEEE Trans. Syst. Man Cybern. Part A: Syst. Hum. **38**(1), 197–206 (2008)
65. Krysander, M., Frisk, E.: Leakage detection in a fuel evaporative system. Control Eng. Pract. **17**(11), 1273–1279 (2009)
66. Krysander, M., Nyberg, M.: Predicting fault isolability properties using structural and analytical information. In: Proceedings of IFAC World Congress (2005)
67. Laursen, M., Blanke, M., Düştegör, D.: Fault diagnosis in a water for injection system using enhanced structural isolation. Int. J. Appl. Math. Comput. Sci. **18**(4), 593–603 (2008)
68. Leitold, A., Hangos, K.M.: Structural solvability analysis of dynamic process models. Comput. Chem. Eng. **25**, 1633–1646 (2001)
69. Lunze, J.: Diagnosis of quantised systems. In: Fault Detection, Supervision and Safety for Technical Processes, vol. 1, no. 1 (2000). Elsevier Science, June 2001
70. Lunze, J., Schröder, J.: Process diagnosis based on a discrete-event description. Automatisierungstechnik **47**(8) (1999)
71. Mahmoud, M.M., Jiang, J., Zhang, Y.: Active Fault Tolerant Control Systems: Stochastic Analysis and Synthesis. Springer (2003)
72. Mangoubi, R.S.: Robust Estimation and Failure Detection. Springer (1998)
73. Niemann, H.H., Stoustrup, J.: An architecture for fault-tolerant controllers. Int. J. Control **78**(14), 1091–1110 (2005)
74. Niemann, H.: A setup for active fault diagnosis. IEEE Trans. Autom. Control **51**(9), 1572–1578 (2006)
75. Nikiforov, I.V.: Two strategies in the problem of change detection and isolation. IEEE Trans. Inf. Theory **43**(2), 770–776 (1997)
76. Noura, H., Theilliol, D., Ponsart, J.C., Cham, A.: Fault-Tolerant Control Systems—Design and Practical Applications. Springer (2009)
77. Park, W., Lee, S.H., Song, J.: Fault detection and isolation of DURUMI-II using similarity measure. J. Mech. Sci. Technol. **23**, 302–310 (2009)
78. Patan, K., Witczak, M., Korbicz, J.: Towards robustness in neural network based fault diagnosis. Int. J. Appl. Math. Comput. Sci **18**(4), 443–454 (2008)
79. Patton, R.J.: Fault detection and diagnosis in aerospace systems using analytical redundancy. IEE Comput. Control Eng. J. **2**(3), 127–136 (1991)
80. Patton, R.J., Frank, P.M., Clark, R.N. (eds.): Fault Diagnosis in Dynamic Systems, Theory and Application. Prentice Hall (1989)
81. de Persis, C., Isidori, A.: A geometric approach to non-linear fault detection and isolation. IEEE Trans. Autom. Control **45**(6), 853–865 (2001)
82. Pivano, L., Johansen, T.A., Smogeli, O.N.: A four-quadrant thrust estimation scheme for marine propellers: theory and experiments. IEEE Trans. Control Syst. Technol. **17**(1), 215–226 (2009)
83. Poulsen, N.K., Niemann, H.: Active fault diagnosis based on stochastic tests. Int. J. Appl. Math. Comput. Sci. **18**(4), 487–496 (2008)

84. Quevedo, J., Puig, V., Cembrano, G., Blanch, J., Aguilar, J., Saporta, D., Benito, G., Hedo, M., Molina, A.: Validation and reconstruction of flow meter data in the Barcelona water distribution network. Control Eng. Pract. **18**(6), 640–651 (2010)
85. Romano, D., Kinnaert, M.: Robust design of fault detection and isolation systems. Qual. Reliab. Eng. Int. **22**(5), 527–538 (2006)
86. Seydou, R., Raissi, T., Zolghadri, A., Efimov, D.: Actuator fault diagnosis for flat systems: A constraint satisfaction approachr. Int. J. Appl. Math. Comput. Sci. **23**(1), 171–181 (2013)
87. Staroswiecki, M., Attouche, S., Assas, M.L.: A graphic approach for reconfigurability analysis. In: Proceedings of DX'99, June 1999
88. Staroswiecki, M., Cassar, J.P., Cocquempot, V.: Generation of optimal structured residuals in the parity space. In: 12th IFAC World Congress, vol. 8, pp. 299–305, July 1993
89. Staroswiecki, M., Declerck, P.: Analytical redundancy in nonlinear interconnected systems by means of structural analysis. In: Proceedings of IFAC AIPAC'89 Symposium, vol. 2, pp. 23–27. Elsevier—IFAC (1989)
90. Staroswiecki, M., Gehin, A.L.: Control, fault tolerant control and supervision problems. In: IFAC Safeprocess'2000. Budapest, Hungary (2000)
91. Stevens, B.L., Lewis, F.L.: Aircraft Control and Simulation, 2nd edn. Wiley (2003)
92. Stoorvogel, A.A., Niemann, H.H., Saberi, A., Sannuti, P.: Optimal fault signal estimation. Int. J. Robust Nonlinear Control **12**(8), 697–727 (2002)
93. Svärd, C., Nyberg, M.: Residual generators for fault diagnosis using computation sequences with mixed causality applied to automotive systems. IEEE Trans. Syst. Man Cybern. Part A: Syst. Hum. **40**(6), 1310–1328 (2010)
94. Tabatabaeipour, S.M.: Active fault detection and isolation of discrete-time linear time-varying systems: a set-membership approach. Int. J. Syst. Sci. (2014)
95. Travé-Massuyès, L., Escobet, T., Olive, X.: Diagnosability analysis based on component supported analytical redundancy relations. IEEE Trans. Syst. Man Cybern. Part A: Syst. Hum. **36**(6), 1146–1160 (2006)
96. Unger, J., Kröner, A., Marquardt, W.: Structural analysis of differential-algebraic equation systems—theory and applications. Comput. Chem. Eng. **19**(8), 867–882 (1995)
97. Varga, A.: A numerically reliable approach for the synthesis of periodic FDI filters. In: Proceedings of IFAC SAFEPROCESS 2012 (2012)
98. Varma, M., Zisserman, A.: Texture classification: Are filter banks necessary? In: Proceedings of the IEEE Conference on Computer Vision and Pattern Recognition, vol. 2, pp. 691–696 (2003)
99. Willersrud, A., Blanke, M., Imsland, L., Pavlov, A.: Fault diagnosis of downhole drilling incidents using adaptive observers and statistical change detection. J. Process Control (2015)
100. Wu, N.E., Klir, G.J.: Optimal redundancy management in reconfugurable control systems based on normalised nonspecificity. Int. J. Syst. Sci. **31**, 797–808 (2000)
101. Zhang, X.: Auxiliary Signal Design in Fault Detection and Diagnosis. Springer (1989)
102. Zhou, W.-W., Blanke, M.: Identification of a class of non-linear state space models using RPE techniques. IEEE Trans. Autom. Control **34**(3), 312–316 (1989)
103. Zolghadri, A.: Advanced model-based FDIR techniques for aerospace systems: today challenges and opportunities. Prog. Aerosp. Sci. **53**, 18–29 (2012)

Chapter 2
Integration of Supervisory Control Synthesis in Model-Based Systems Engineering

Jos C.M. Baeten, Joanna M. van de Mortel-Fronczak and Jacobus E. Rooda

Abstract Increasing system complexity, time to market and development costs reduction place higher demands on engineering processes. Formal models play an important role here because they enable the use of various model-based analyses and early integration techniques and tools. Engineering processes based on formal models are able to cope with complexity. They also support time to market and development costs reduction. Moreover, application of supervisory control synthesis in the development of control systems can speed up the process considerably. This paper discusses the integration of recently developed supervisor synthesis techniques and tools in engineering processes. To illustrate this approach, examples of industrial cases are presented, where supervisors synthesized have successfully been implemented and integrated in existing resource control platforms.

2.1 Introduction

In current industrial practice, it is very difficult to deal with high-tech multidisciplinary system development due to system complexity, market pressure, and resource limitations. To overcome the difficulties, various kinds of models are used increasingly often in the development process. Specifically, formal and executable models built and employed in the design phase can be used to assess functional correctness and performance of component designs and overall system design. Formal verification, in particular model checking [3], is employed when a high degree of confidence in functional correctness of a design is required. To assess design performance, one

J.C.M. Baeten
Centrum Wiskunde & Informatica, Amsterdam, The Netherlands
e-mail: jos.baeten@cwi.nl

J.C.M. Baeten · J.M. van de Mortel-Fronczak (✉) · J.E. Rooda
Eindhoven University of Technology, Eindhoven, The Netherlands
e-mail: j.m.v.d.mortel@tue.nl

J.E. Rooda
e-mail: j.e.rooda@tue.nl

G.M. Dimirovski (ed.), *Complex Systems*, Studies in Systems, Decision and Control 55, DOI 10.1007/978-3-319-28860-4_2

can use simulation directly on the model, or analyze an associated Markov chain or queueing model.

Successful use of models is, however, not limited to the design phase. Models are nowadays also used in the test and integration phases. The effort required to integrate components and to test the resulting high-tech multidisciplinary system against the initial requirements increases significantly with the increase in complexity. The integration and test phases have a large and growing influence on time to market and product quality, being the main business drivers for system developers. High-tech multidisciplinary systems are systems in which cutting edge technologies from multiple engineering disciplines are integrated to meet strict quality requirements set by the customer. A characteristic of such systems is that they are integration intensive, which means that integrating components and testing the resulting system against its requirements consumes a large portion of total system development effort. In some cases, the integration and test effort may even exceed the already considerable amount of development effort required to deal with the complexity of components. Examples of high-tech multidisciplinary systems are commercial systems such as wafer scanners, electron microscopes, and high-speed printers; medical systems such as magnetic resonance imaging (MRI) scanners; as well as aerospace systems such as satellites, airplanes, and spacecraft.

To wit, it is shown in [6, 7] that models can successfully be used for early integration of wafer scanners, where the idea is to start integrating and testing as soon as possible, using models as replacements of components as long as actual component realizations are not yet available. Using an appropriate integration infrastructure that implements the designed and modeled interaction behavior, realized components can be integrated with models of not yet realized components. The resulting model-based integrated system can then be tested on system level before complete system realization is available. Another application is in model-based testing [8], where a model is used to automatically derive and execute tests on an implementation, as shown for wafer scanners in [24].

The use of models enables application of model-based techniques and tools for thorough system analysis and systematic testing, helping to improve system overview for engineers. Additionally, models of components and requirements enable synthesis of supervisory controllers essential for correct functioning of systems. Supervisor synthesis procedures are based on supervisory control theory originating in [26].

Figure 2.1 schematically illustrates a high-tech system with the focus on control. At the bottom, the main structure is depicted, containing the physical hardware components. Sensors and actuators are mounted on these hardware components to monitor their position or state and to actuate them. The resource control layer assures that actuators reach a desired position in a desired way (feedback control). The supervisory control layer coordinates individual components and gives the desired functionality to the system.

The high-tech industry is also confronted with problems related to systematic upgrades of complex control software, necessary either for new systems or machines or for extending functionality of the existing ones. Also, in this context, system component and requirement models can be used to speed up the supervisory control

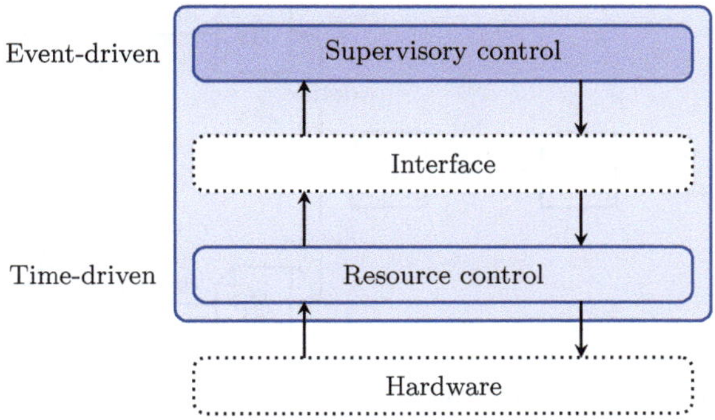

Fig. 2.1 Positioning supervisory control

design process. To embed supervisor synthesis in model-based engineering, distributed synthesis techniques are needed to handle the complexity frequently encountered in high-tech systems and to support the evolvability in the development process. High-tech companies are often challenged to increase the functionality and quality of a product, while simultaneously time to market and product costs should be reduced. Current practice shows that this is not straightforward. As a result, there is a need for new engineering processes. The purpose of this paper is to show how supervisory control theory introduced by Ramadge and Wonham [26] can contribute to system development and how it can be integrated in an engineering process. To this end, the applicability of recently developed supervisor synthesis techniques to industrial cases is discussed.

The paper is structured as follows. Section 2.2 gives a short overview of engineering processes that are based on models. In Sect. 2.3, we informally explain the basic supervisor synthesis procedure. Section 2.4 focusses on the role of model transformations in support of model-based engineering processes, especially for industrial applications. An overview of a few industrial cases in which a synthesis-based engineering method is applied is presented in Sect. 2.5. Section 2.6 concludes the paper by showing what impact supervisory control synthesis has on the system development process.

2.2 Overview

To support structured system development, several engineering processes are introduced originating from *systems architecting* [27] and *systems engineering* [19]. More recently, [20] proposed a conceptual model specifically supporting embedded systems architecting. In general, a system engineering process starts with a global definition of the requirements that the system should fulfill. Based on these requirements, the global system design is defined, usually partitioned into subsystems or modules. For each module, the requirements are derived from the global design, and a design is defined and then built. Every module is tested separately to assure that its requirements are satisfied. Subsequently, modules are integrated and the complete system

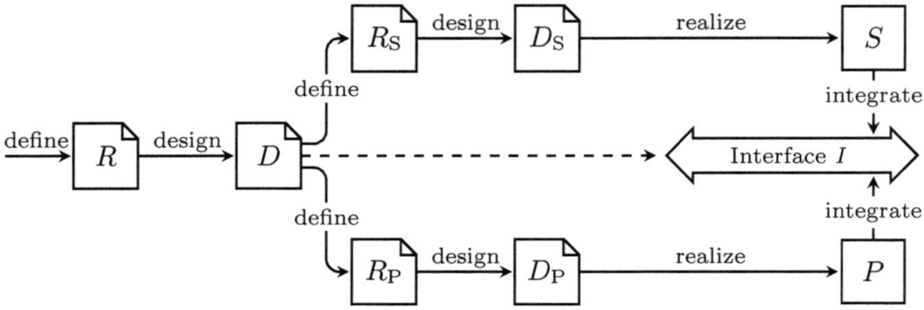

Fig. 2.2 Traditional engineering process

is also tested to assure that it complies to its requirements. For a system consisting of two modules, component *P* and its controller *S*, this is schematically depicted in Fig. 2.2.

Thanks to the introduction of microprocessors, it became possible to develop systems with substantially extended functionality. The extended functionality often implies a multidisciplinary character because mechanical and control engineers needed to cooperate with electrical, electronic, and computer science engineers. Especially for this kind of systems, traditional engineering processes are often not optimal in terms of system quality, time to market, and costs.

In the last decades, engineers started to use executable software models to test the designs before they are actually built. An advantage of such models is that they can be used not only to analyze designed system behavior but also to investigate how components or modules that are already built interact with the rest of the system that is not yet built. One can think of, for instance, checking if the system is nonblocking or estimating system performance. In [6], evidence is provided that executable models help in improving system quality and in decreasing time to market. Models also support evolvability, as they can easily be reused for similar systems. Integration of executable models in engineering processes is rendered in the scheme of Fig. 2.3.

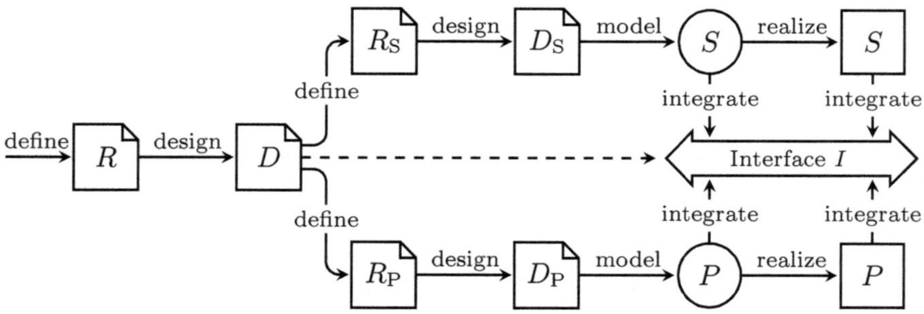

Fig. 2.3 Model-based engineering process

Certainly for complex systems, model-based systems engineering (MBSE) methods are state-of-the-art. However, all current MBSE approaches still carry a significant contribution to the overall system cost. The industry faces increasing complexity and is challenged to reduce development time and cost. To meet these challenges, a next-generation MBSE approach is required that reduces human errors and shortens the development cycle. An overview of the state-of-the-art commercially available MBSE methodologies, such as IBM Telelogic Harmony-SE, INCOSE Object-Oriented Systems Engineering Method (OOSEM), IBM Rational Unified Process for Systems Engineering (RUP SE) for Model-Driven Systems Development (MDSD), and Vitech MBSE Methodology, can be found in [10]. They include processes, methods, and tools supporting manual development of systems, therefore helping in achieving a paradigm shift from traditional document-based approach to model-based approach. However, in the context of control software design, they focus on design model *formulation*. To cope with industrial challenges, more powerful techniques and tools are required. The MBSE process proposed in [7] enables the use of various analysis techniques and tools based on formal models to support system development. Significant development cost and cycle time reduction can be gained by the incorporation of methods and tools for supervisor synthesis. To this end, as opposed to traditional engineering, an explicit separation of concerns related to the plant and the supervisor is necessary. This also means that requirements for supervisory control must be separated from requirements for regulative control. Supervisor synthesis techniques replace manual design of control logic by automatical *derivation* from component and requirement specifications. The synthesized control logic is nonblocking and by construction correct with respect to safety properties. In general, liveness properties still need to be verified. The position of supervisor synthesis in the development process is shown in Fig. 2.4 [28].

This setting partly corresponds to Fig. 2.3. As previously, requirements R of the system under supervision are defined first. Based on these requirements, design D of the system and a decomposition into the uncontrolled plant and the supervisor are defined. After decomposition, requirements R_S for the supervisor and R_P for the uncontrolled plant are specified. From the plant requirements, design D_P and one or more plant models P can be defined. However, instead of making a design and a model of the supervisor, the requirements for the supervisor are formally modeled.

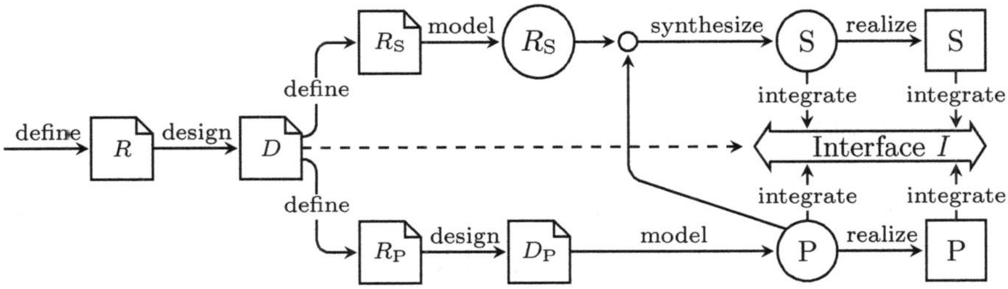

Fig. 2.4 Engineering process with supervisory control synthesis

A discrete-event model of the plant and the formal model of R_S can be used to *synthesize* a supervisor in the framework of supervisory control theory. Plant models can also be used to analyze the behavior of the uncontrolled plant under supervision of the supervisor. Formal and executable models built and employed in the design phase can be used to assess functional correctness and performance of the design of components and the overall system. One example of commercially available tool used for correctness checking is the Communicating Sequential Processes (CSP)-based tool of Verum (http://www.verum.com). To assess performance of the design, one can use simulation directly on the model, or use analytical or numerical methods to analyze an associated stochastic model.

In synthesis-based engineering, required system properties are used as input for generation of a design of a supervisor that is correct by construction. As a consequence, in the analysis phase verification can be eliminated to a large extent. This changes the development process from implementing and debugging the design and the implementation, to defining the requirements.

2.3 Synthesis-Based Supervisory Control Engineering

Supervisory control theory (SCT) of [26] allows to synthesize a model of the supervisor from formal models of the uncontrolled system and of the requirements. First, the uncontrolled system (plant P) is formally specified in terms of automata. A plant automaton describes the physically possible behavior of the system to be controlled. Then, the requirement specifications (R_S) for the system under control are formally defined in terms of automata. A model of the supervisor (S) is generated from these formal models.

The resulting supervisor can be used to supervise an uncontrolled plant, which is schematically illustrated in Fig. 2.5. The supervisor can only react to observable events that are generated by the uncontrolled plant. The supervisor influences the behavior of the plant by disabling certain controllable events. The method guarantees that the system consisting of the derived supervisor and the uncontrolled plant fulfills the requirements. If the supervised model does not contain the desired functionality, the models of the uncontrolled plant or the requirements are inadequate.

Supervisory control problem is formulated as follows. For plant P and requirement R, a supervisor S is required such that:

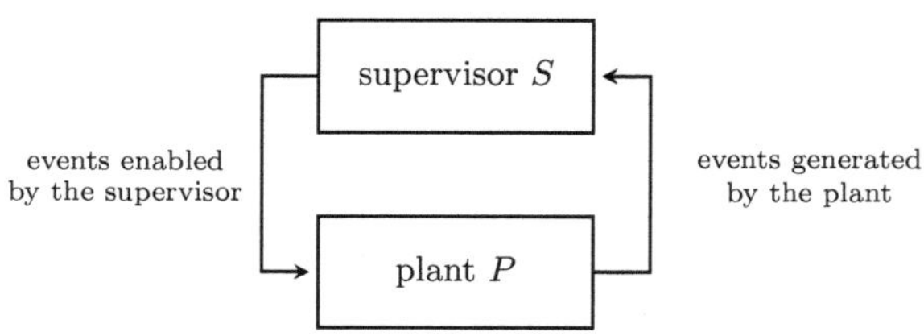

Fig. 2.5 The feedback loop of supervisory control

1. P under control of S, denoted by S/P, satisfies requirement R.
2. S does not disable uncontrollable events.
3. Output of S only depends on observable outputs of P.
4. S/P is nonblocking.
5. S is optimal, that is, maximally permissive.

Supervisory control theory provides means to synthesize S. It is a conceptually simple framework based on discrete-event system models. Different methods have been developed that allow for an automatic synthesis of a supervisor. As computational complexity is high for systems of industrial size, several advanced techniques have been introduced to reduce synthesis complexity, such as modular [25], hierarchical [35], interface-based hierarchical [14], aggregative [11, 17], distributed [9, 13, 32], coordinated distributed [29], and aggregated distributed [30]. They allow for splitting a system in a number of modules for which local supervisors can be synthesized. If needed, also a coordinator can be synthesized on top of them. The local supervisors together with the coordinator compose a nonblocking supervisor for the total system. A different approach is the state-based supervisory control framework of [16]. In this framework, discrete-event systems are represented by state tree structures (STS), a representation that allows for computationally efficient monolithic supervisor synthesis. Unlike the event-based framework, the state-based framework allows to formulate requirements as conditions over states. In [2], a first step is reported in the direction of defining supervisor synthesis techniques for a process-theoretic setting. Additionally, [31] describes how to compute a nonblocking supervisor that not only complies with the prescribed requirements but also achieves a time optimal performance such as throughput.

In the context of supervisory control, system component models are defined by automata. Often, each component and its resource controller are modeled together by one automaton. Automata consist of relevant states of each resource and transitions labeled by controllable and uncontrollable events. Controllable events represent relevant discrete commands to the resource control that can be enabled or disabled by the supervisor. Uncontrollable events represent messages from the resource control to the supervisor that cannot be disabled by the supervisor. The conditions the supervisor must satisfy are formally defined by expressions on languages generated by the automata.

To illustrate the basic supervisor synthesis procedure, a simple example from [36] is worked out below. A workcell consists of two machines M_1 and M_2, and an automated guided vehicle (*AGV*). The *AGV* can load a workpiece at M_1 or M_2 and unload it at M_2 or at the infinite buffer B, respectively. This is schematically shown in Fig. 2.6.

In this paper, we use italic event labels (e.g., c) and bold state labels (e.g., **Busy**). In the graphical automaton representation, initial states are denoted by an unconnected incoming arrow and marker states are denoted by an unconnected outgoing arrow. Controllable and uncontrollable events are denoted by solid and dashed edges, respectively.

Fig. 2.6 Workcell

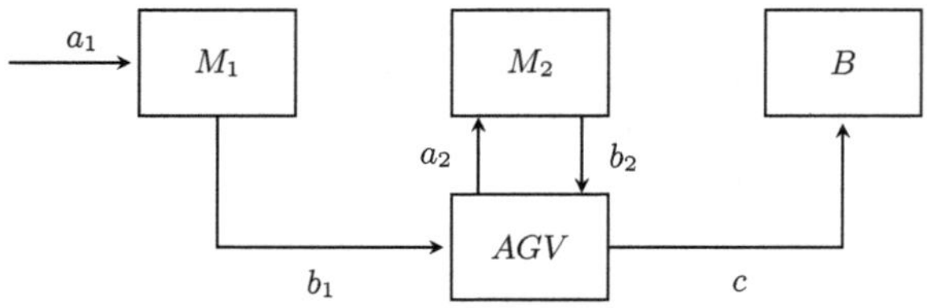

Fig. 2.7 Workcell automata

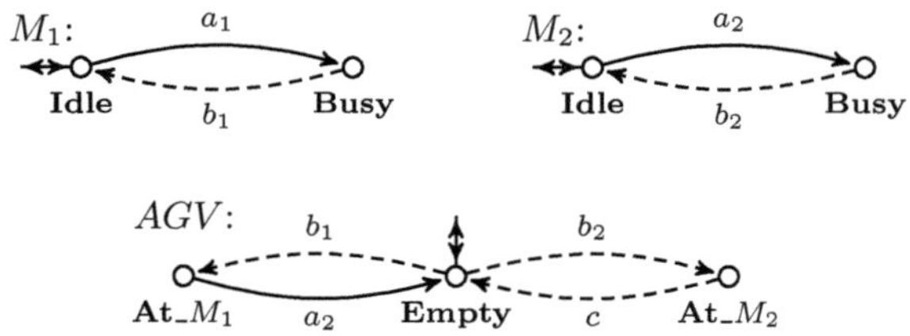

Fig. 2.8 Uncontrolled system

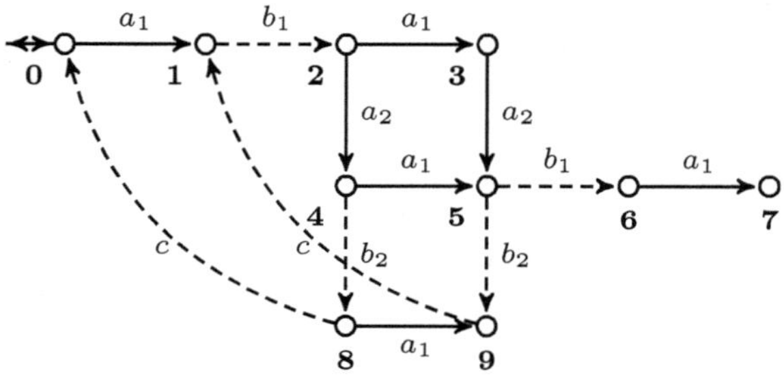

Components M_1, M_2, and AGV are modeled by automata shown in Fig. 2.7. The behavior of the uncontrolled system P is represented by the synchronous product of M_1, M_2, and AGV drawn in Fig. 2.8.

We informally explain the conditions that must be satisfied by the supervisor for this uncontrolled system. For simplicity, we assume the requirement does not impose any restrictions on admissible system behavior and all events are observable. This implies that conditions 1 and 3 are satisfied. Hence, we can focus on the remaining conditions. In the basic supervisor synthesis approach, the supervisor is derived from the intersection of the two languages: the uncontrolled plant behavior and the admissible behavior defined by the requirement. Because of the assumption, in our case it suffices to take the first one into account. Hence, we must check whether the plant is nonblocking. As can be seen above, the absence of control results in a blocking situation (deadlock in state 7). The supervisor we are constructing should influence the plant under control to avoid the blocking situation. Only controllable events may be disabled. To avoid blocking, controllable event a_1 must be disabled in state 4 and controllable event a_2 must be disabled in state 3. This results in Fig. 2.9a, where these supervisory control actions introduce a new blocking situation, state 3. This situation can be avoided by disabling controllable event a_1 in state 2 resulting

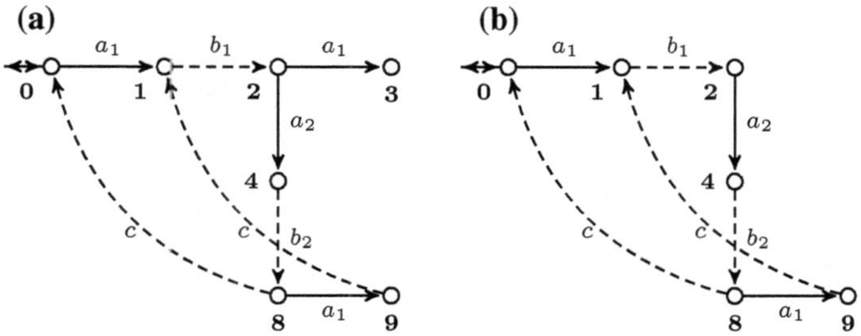

Fig. 2.9 Derivation of the supervisor. **a** Step 1. **b** Step 2

in Fig. 2.9b. Finally, by construction this supervisor gives a proper optimal control to the plant: conditions 1 through 5 are satisfied.

Supervisors can be synthesized according to the procedure sketched above. In a similar way, a self-made supervisor can be verified by checking the enumerated conditions.

In Sect. 2.5, a few synthesis techniques from the event-based and state-based frameworks are applied in industrial case studies.

2.4 Model Transformations

The main difference between the traditional engineering and the model-based engineering processes is the inclusion of models in the latter. System components are modeled before being realized. This extra step in the development process has several advantages, as mentioned in [7]:

- Abstract models give a systematic approach to specify component and system behavior with more consistency and less ambiguity than documents.
- Models make it easier to analyze dynamic behavior as well as the performance of a component of a system.
- Using various model-based analysis techniques indicated in Fig. 2.10, errors can be encountered in an early stage of the system development process when no component is realized, which decreases risk.

By simulating and validating models, the actual behavior can be analyzed in an early stage of the development process. However, simulation can only show that models might have correct behavior. It cannot guarantee model correctness. If models are validated and simulated and the conclusion is drawn that the system has the correct functionality, components can be realized. Formal verification techniques can be used to prove model properties.

Available components can be tested with hardware-in-the-loop simulation and testing. Composition of models and realizations can be tested as a whole to check for

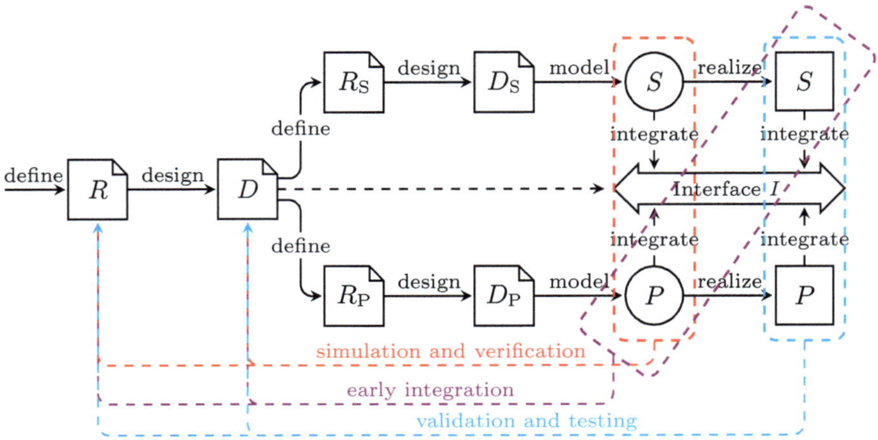

Fig. 2.10 Model-based analysis techniques

the correct functionality, which is called early integration. Also in this step, design errors can be encountered in an early stage, which decreases risk. If all components are realized, all models can be replaced by their realizations. The final implementation can be tested to verify if the system as a whole fulfills its requirement *R*. In this phase, model-based testing can be applied, where models are used to automatically derive and execute tests on the implementation.

In spite of the numerous advantages, model-based analysis techniques can only help the engineer with discovering errors in an early stage. There is no formal link between models and implementations. In other words, model correctness does not guarantee implementation correctness. Supervisory control theory explained in Sect. 2.3 provides this formal link and eliminates the manual design of the supervisor. This theory allows the generation of a supervisor that is mathematically correct with respect to formal plant models and requirements. As a consequence, design and implementation do not need to be tested against the requirements, the verification can be eliminated to a large extent, and the engineer can focus on system validation.

To support model-based systems engineering in a consistent way, model transformations are needed. The Chi language [4] has been developed to provide a process algebraic generic modeling formalism for hybrid systems. The language supports hierarchy and modularity to deal with large-scale systems, by providing operators for model reuse, parallelism, and nesting. It is based on processes that interact by shared variables, by communication via shared channels, and by synchronization by means of shared actions. Furthermore, arbitrary differential and algebraic equations are supported, for modeling of continuous time behavior. The Chi toolset provides translations to various verification tools, discrete events, and hybrid simulators, and supports hardware-in-the-loop simulation and control prototyping. Additionally, it facilitates code generation for various platforms.

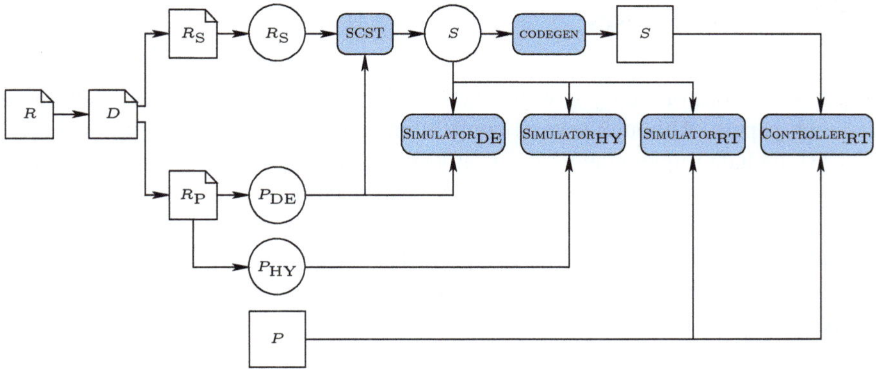

Fig. 2.11 Toolchain for model-based systems engineering

In Fig. 2.11, the Chi toolchain is depicted that supports model-based systems engineering with supervisor synthesis as described in [28]. In this figure, tools are depicted in blue, models are circles, and realizations are rectangles. The uncontrolled discrete-event behavior of the plant is formally modeled by automata represented by P_{DE}. Control requirements R_S are formally modeled by automata or logical expressions. From these models, a supervisor can be synthesized using a supervisor control synthesis tool (SCST) and translated to the Chi language, resulting in a model of the supervisor S. Three tools are used for supervisor synthesis and translations exist to and from the associated file formats:

- Translation Communication Tool (TCT) of [36] (http://www.control.utoronto.ca/cgi-bin/dldes.cgi) and Supervisor Synthesis Package (SSP) of Systems Engineering Group from Eindhoven University of Technology (http://se.wtb.tue.nl/sewiki/supcon/start) for the event-based supervisory control framework.
- NBC tool for the state-based supervisory control framework of [16].

The Chi simulator can be used to simulate the model of the supervisor S together with the discrete-event plant model P in order to analyze its behavior with respect to the control requirements. Furthermore, the discrete-event plant model can be replaced by the hybrid plant model P_{HY} or by real hardware in order to test whether it complies with the models. The supervisor model can also be used for code generation (CODEGEN). It can be implemented on a Personal Computer (PC) or Programmable-Logic-Controller (PLC) real-time control platform connected to the actual plant hardware.

In the next section, the model-based systems engineering approach supported by the toolchain described above is applied to three industrial cases in order to obtain a model of the supervisor, perform model-based system analysis, derive a supervisor implementation, and test it on the existing system.

2.5 Industrial Cases

Supervisory control theory is well established and received substantial attention in research papers. Although several applications reported in the literature, e.g., [15, 22, 23], include implementation of synthesized supervisors on a PC or a PLC, they are usually put in practice in an academic setup. To our knowledge, the industrial applications described in [12, 18, 34] are ones of the few that actually refer to an existing industrial product. Recently, a project is started in the automotive application area, where supervisor synthesis techniques are used in the design of the discrete part of a cruise control system.

2.5.1 MRI Scanner

In [34], supervisor synthesis is applied to a support system used to position patients in an MRI scanner of Philips Healthcare (http://www.medical.philips.com). MRI stands for magnetic resonance imaging, which is a medical imaging technique used in radiology to visualize detailed internal structures. An MRI scanner is used for noninvasive medical diagnosis. It uses a powerful magnetic field to align the magnetization of some atoms in the body, and nonionizing radio frequency fields to systematically alter the alignment of this magnetization. This causes the nuclei to produce a rotating magnetic field detectable by the scanner. This information is recorded to construct an image of the scanned area of the body. In clinical practice, MRI is used to distinguish pathologic tissue (such as a brain tumor) from normal tissue. One advantage of an MRI scan is that it is harmless to the patient, unlike Computed Tomography (CT) scans and traditional X-rays, which both use ionizing radiation. The traditional MRI scanner is a large cylinder-shaped tube surrounded by a circular magnet, called the bore. The patient lies on a moveable examination table, called the patient support system, which slides into the center of the magnet, see Fig. 2.12.

To properly control the patient support system, several actuators and sensors, as well as the extended user interface, must be taken into account. The modeled part

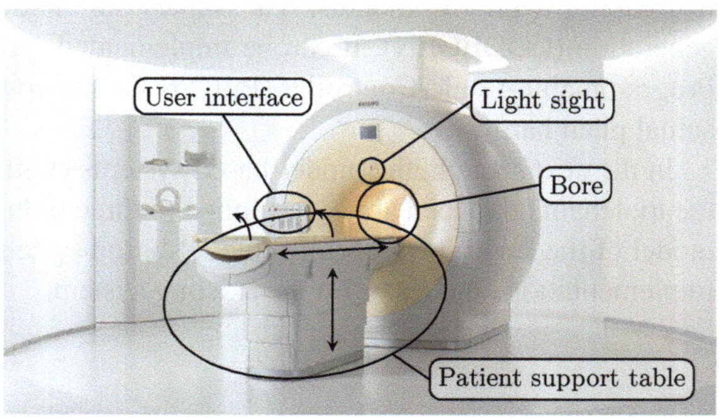

Fig. 2.12 The MRI scanner

of the system is divided into three modules: the vertical axis, the horizontal axis, and the user interface. The vertical axis consists of the lift driven by the vertical motor drive and associated end-position sensors. The horizontal axis contains the tabletop that can be moved into and out of the bore, either manually or by applying the horizontal motor drive depending on the state of the clutch. It also contains sensors detecting the presence and the position of the tabletop. The tabletop can only be placed or removed in the maximally out position. The clutch connects the motor to the tabletop. When the clutch is applied, the motor controls the movement of the tabletop. Otherwise, the tabletop can be moved freely by the operator. The tabletop release (TTR) is a hardware safety system that releases the clutch independently of the controller. If TTR is active, the table can be moved freely, even if the controller enables the clutch. The user interface allows the operator to use the manual button and the tumble switch. The manual button is used to switch between the manual and automatic mode of the tabletop movements. The tumble switch determines whether and how the patient support table should move. The patient support system model does not include the hardware safety systems, the light visor for marking, and automated positioning of the scan plane in the bore.

If no tabletop is present, horizontal movement is not allowed and the associated motor should be stopped. The tabletop may only be moved by the horizontal motor if the clutch is applied and the TTR is not active. If the clutch is not applied, the motor may not move the table. While the motor is moving the table, the clutch may not be released. The horizontal movement is only allowed to start when TTR is off. It cannot be prevented that TTR is turned on while moving. Whenever TTR is turned on, the table should be stopped. Only if TTR is turned off, the clutch may be turned on or off.

Control requirements are related to tabletop handling. Four safety-related aspects can be distinguished:

- Ensure that the tabletop does not move beyond its vertical and horizontal end positions.
- Prevent collisions of the tabletop with the magnet.
- Define the conditions for manual and automatic movements of the tabletop.
- Enable the operator to control the system by means of the manual button and the tumble switch.

The components and requirements described above are modeled by automata. The plant model consists of 672 states and 14.384 transitions. The total requirement model consists of 4.128 states and 34.884 transitions. Based on these models, the centralized supervisor was synthesized using the TCT tool of [36]. The automaton representing the supervisor consists of 2.976 states and 22.560 transitions. The plant under control of the supervisor was validated using simulation and the supervisor was tested on the real system. Subsequently, the experiment was extended to investigate system evolvability in case of changing functionality. In the original setup, if the table is not maximally up and not maximally out it does not move when the user switches the tumble switch up. As this behavior appeared counter-intuitive for the user, a new specification was introduced stating that when the tumble switch is up, the table should first go maximally out and then move up if the tumble switch is still up. This functional change introduced only a few small changes to formal requirements and a new supervisor was successfully synthesized. Approximately, four-hour work was needed to produce this new result. Implementing the same functional change in the existing design process was estimated for a week, approximately.

In a related project, supervisor synthesis has also successfully been applied to the patient communication system of the MRI scanner.

2.5.2 *Océ printer*

In the Océ printer case study, reported in [18], supervisory control synthesis is applied to the scheduling of maintenance operations in a high-end printer of Océ-Technologies (http://www.oce.com) schematically shown in Fig. 2.13. Such a printer consists of three modules: module 1 takes care of paper input, module 2 performs the printing, and module 3 provides stacking and stapling capabilities. The printing process consists of several components as depicted in Fig. 2.13, part *2b*. In this process, the toner image is applied onto the toner transfuse belt and it is fused onto a paper sheet. To maintain high printing quality, several maintenance operations need to be carried out, such as

1. Toner transfuse belt jittering, which displaces the transfuse belt to prolong its lifespan.
2. Black image operation, which prints completely black pages to remove paper dust from the toner transfuse belt.
3. Coarse toner particles removal operation, which removes leftover particles from the toner.

In general, a maintenance operation is scheduled after a given number of prints, unless there is an active print job. There is also a hard threshold of the number of prints after which the maintenance operation must be carried out even if an active print job needs to be interrupted. To perform a maintenance operation, the printing process has to change the power mode: from Run mode used for printing to Standby mode required for maintenance. The change of the power mode can trigger pending

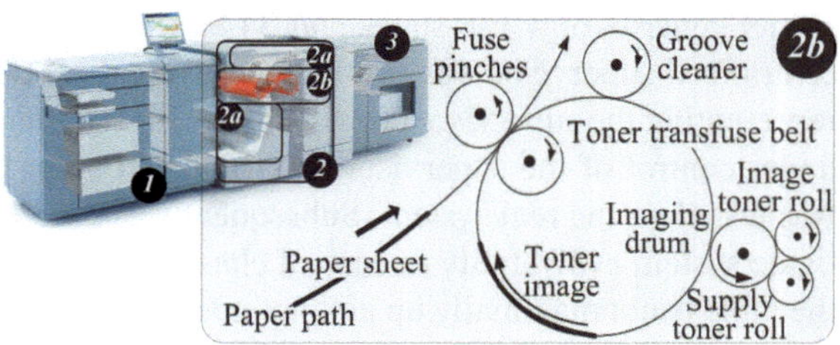

Fig. 2.13 High-end printer

maintenance operations, which might prolong the waiting time of the interrupted print job.

The purpose of control requirements is to coordinate the maintenance procedures with the rest of the printing process. Specifically,

- Maintenance operations may only be performed if the power mode of the printing process is Standby.
- Maintenance operations should be scheduled if their soft deadline is reached and no print jobs are in progress or if their hard deadline is reached.
- Only scheduled maintenance operations can be started.
- The power mode of the printing process should conform to the mode determined by the print job managers unless it is overridden by a pending maintenance operation.

The system components are modeled by automata placed in a state tree structure as defined in [16]. To simplify the specification process, the requirements related to coordination and scheduling of maintenance operations are modeled by generalized state-based expressions. The plant model consists of 25 automata with 2–24 states. The requirements are represented by 23 generalized state-based expressions, which translate to more than 500 standard state-based expressions. Using the synthesis tool of [16], the centralized supervisor is derived. To analyze the controlled behavior, both the plant model and the supervisor are converted to an executable simulation model. After simulation-based validation, the supervisor is converted to C++ for execution on the existing control platform.

In another project, described in [5], supervisor synthesis has also successfully been applied to exception handling in printers.

2.5.3 Theme Park Vehicle

In the theme park vehicle case study, reported in [12], supervisory control synthesis is applied to the multimovers of ETF (http://www.etf.nl), as shown in Fig. 2.14a. Multimovers are automated guided vehicles that can follow an electrical wire integrated in the floor. This track wire produces a magnetic field that can be measured by track sensors. Next to the track wire, floor codes are positioned that can be read by a metal detector. These floor codes give additional information about the track, such as the start of a certain scene program, a switch, a junction, or a dead end. The scene program, which is read by the scene program handler, defines when the vehicle should ride at what speed, when it should stop, rotate, play music and in which direction the vehicle should move if it has arrived at a junction.

An operator is responsible for powering up the vehicle and deploying it into the ride manually. The operator also controls the dispatching of the vehicles in the passenger boarding and outboarding area. The vehicle can receive messages from Ride Control. Ride Control coordinates all vehicles and sends start/stop commands to these vehicles. This communication is either wireless or by means of the track wire.

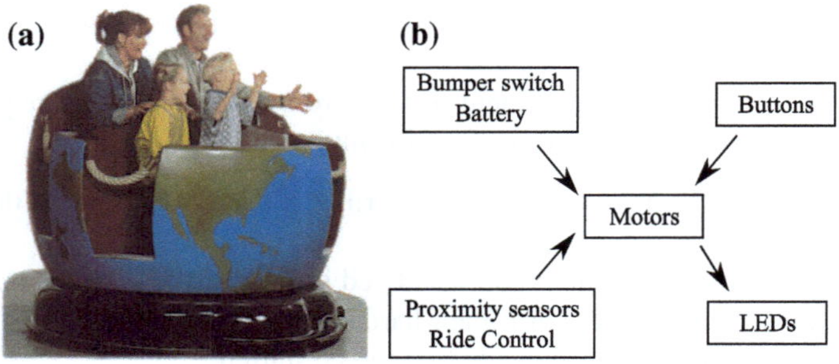

Fig. 2.14 Theme park vehicle. **a** The multimover. **b** Interaction between components

Multimovers are not able to communicate with other vehicles. As safety is an important aspect of this vehicle, several sensors are integrated in it to avoid collisions. First, proximity sensors are integrated in the vehicle to avoid physical contact with other objects. Two types of proximity sensors are used: long-range proximity sensors to detect obstacles in the vicinity of six meters and short-range proximity sensors to detect obstacles in the vicinity of one meter. Second, a bumper switch is mounted on the vehicle that can detect physical contact with other objects. The interactions between vehicle components are shown schematically in Fig. 2.14b.

The main requirement for supervisory control synthesis is safety. Three safety-related aspects can be distinguished:

- **Proximity handling** The supervisor has to assure that the multimover does not collide with other vehicles or obstacles. To this end, proximity sensors are integrated at the front and back which can detect an obstacle in the vicinity of the multimover. To avoid collisions, the multimover should drive at a safe speed and stop if the obstacle is too close to it.
- **Emergency handling** The system should stop immediately and should be powered off when a collision occurs. To detect collisions, a bumper switch is mounted on the multimover. The same applies when the battery level is too low. The Light Emitting Diode (LED) interface should give a signal when an emergency stop has been performed. The multimover should be deployed back into the ride by an operator manually.
- **Error handling** When a system failure occurs (e.g., a malfunction of a motor), the system should stop immediately and should be powered off to prevent any further problems. The LED interface should give a signal that an emergency stop has been performed. The multimover should be deployed back into the ride by an operator manually.

The multimover control problem cannot be solved in the centralized event-based setting because of the large state-space size. To enable the application of distributed synthesis methods, it is divided into five smaller subproblems that are solved separately.

- **LED actuation** An operator must be able to check in which state the multimover is by looking at the interface LEDs. This means that the states of the LEDs represent the current state of the multimover. It is a task of the supervisor to actuate the LEDs according to the state of the multimover.
- **Motor actuation** The drive motor, steer motor, and scene program handler have to be switched on and off according to the state of the multimover. If the multimover is in the state **Active**, all motors can be switched on. If the multimover is in the state **Reset** or **Emergency**, all motors have to be switched off.
- **Button handling** The user interface of the multimover contains three buttons. The reset button is used to reset the vehicle if the multimover is active and deployed into the ride or it is in the state **Emergency**. The forward and the backward buttons are used to deploy the vehicle into the corresponding direction. The supervisor has to assure that the corresponding state is reached after a button is pushed.
- **Proximity and Ride Control handling** On each side of the multimover, two proximity sensors are mounted: one long range and one short range. If a long-range proximity sensor detects an object in the traveling direction, the multimover should react by slowing down to a safe driving speed. If an obstacle is detected by a short-range proximity sensor, the multimover should stop in order to prevent a collision. When the short-range proximity sensor does not detect an object any more, the vehicle should start moving automatically. If the multimover receives a stop command from Ride Control, it should stop as in the case of short-range proximity handling. If Ride Control sends a start command, the multimover should automatically start with the speed depending on the state of the proximity sensors related to the current driving direction.
- **Emergency and error handling** In order to guarantee the safety of the passengers, the multimover should be deactivated immediately when an emergency situation occurs. It should not be possible to reset the multimover if the bumper switch is still activated or the battery power is still too low. A control task of the supervisor is to enter the **Emergency** state of the multimover when an emergency situation occurs.

The plant model is defined by automata representing an event-based abstraction of the actual behavior of the physical components and their resource control. Requirements described above are also defined by automata. Based on the plant and requirement models, an optimal supervisor is synthesized, validated, and implemented. The plant model consists of 17 automata with 2–4 states. The requirements are represented by 30 automata with 2–7 states.

Both a centralized supervisor and a distributed supervisor are synthesized for the supervisory control problem of the multimover. A centralized supervisor has been synthesized with the state-based framework based on state tree structures of [16]. The state-based synthesis produces binary decision diagrams (BDD) for each controllable event. The maximum BDD size is 15 and the minimum BDD size is 1. Furthermore, a distributed supervisor has been synthesized with an aggregated approach of [30]. The results of the aggregated synthesis are shown in Table 2.1.

Table 2.1 Distributed supervisors

Module	Order	# States	# Trans.	Order	# States	# Trans.
LED actuation	1	25	77	5	41	125
Motor actuation	2	41	222	2	257	1428
Button handling	3	465	3477	4	177	765
Emergency handling	4	89	626	3	118	609
Proximity handling	5	225	1953	1	481	4513

Synthesized supervisors were evaluated to check whether the models of the controlled system are consistent with the intended behavior. For this purpose, discrete-event simulation was used persistently. Specifically, the state-space stepper was used to check whether the supervisor disables the right transitions in the right states when evaluating the closed-loop system behavior. The toolset described in [1] was used for discrete-event simulation. Currently, a new version of the toolset supporting MBSE and synthesis is being developed based on CIF [21, 33].

To deliver a proof of concept for synthesis-based engineering, a prototype of a supervisory controller with the synthesized supervisors is implemented in the existing control software of the multimover. This implementation is first tested by means of simulation and then on the existing implementation platform. Subsequently, the number of proximity sensors is extended. The engineering process used presently requires approximately 2 days for making necessary changes to the control system. The synthesis-based engineering process described in this paper requires approximately four hours to cope with the same change.

2.6 Conclusions

Formal models and associated model transformations provide consistent support for synthesis-based engineering. If consequently used from the beginning of the product development process, they constitute a vehicle that:

- Helps in specifying component and system behavior in a structured, systematic, and unambiguous way.
- Forces to clarify all relevant aspects of the system being developed.
- Shortens the time needed to deliver a good design and correctly functioning control software.

The synthesis phase requires modeling of the uncontrolled system by finite-state machines and modeling of the requirements by finite-state machines and logical expressions. This kind of modeling is intuitive and usually results in (relatively) simple and easy to maintain models as input to the synthesis procedure. Because the synthesis is automatic, the focus in the control software development process changes from designing and debugging controller code to designing and analyzing requirements, if the input models are correct. The downside is, however, that there is not sufficient tool support for validation of these models. This is especially problematic for larger applications. In the industrial case studies described in this paper, simulation was used extensively for this purpose, which consumes too much development time. In a subsequent project, another analysis techniques, like model checking, are investigated in this context.

The results of the case studies show another advantage of synthesis-based engineering that manifests itself when the structure or the requirements of the system under development change due to customer demands. In traditional engineering, all changes have to be made in the software design manually, which is difficult to do without introducing errors or inconsistencies. In the synthesis-based approach, only component models or requirement models have to be adapted and a new supervisor can again automatically be synthesized. Since the desired behavior is specified in models instead of in the control software, it is easier to validate it with respect to the original informal specifications. This also contributes to system evolvability.

References

1. Baeten, J., van Beek, D., Cuijpers, P., Reniers, M., Rooda, J., Schiffelers, R., Theunissen, R.: Model-based engineering of embedded systems using the hybrid process algebra Chi. ENTCS **209**, 21–53 (2008)
2. Baeten, J., van Beek, D., Luttik, B., Markovski, J., Rooda, J.: A process-theoretic approach to supervisory control theory. In: Proceedings of ACC, pp. 4496–4501. IEEE (2011)
3. Baier, C., Katoen, J.: Principles of Model Checking. MIT Press (2008)
4. van Beek, D., Man, K., Reniers, M., Rooda, J., Schiffelers, R.: Syntax and consistent equation semantics of hybrid Chi. JLAP **68**(1–2), 129–210 (2006)
5. Bertens, E., Fabel, R., Petreczky, M., van Beek, D., Rooda, J.: Supervisory control synthesis for exception handling in printers. In: Proceedings of the Philips Conference on Applications of Control Technology (2009)
6. Braspenning, N., Boumen, R., van de Mortel-Fronczak, J., Rooda, J.: Estimating and quantifying the impact of using models for integration and testing. Comput. Ind. **62**(1), 65–77 (2011)
7. Braspenning, N., van de Mortel-Fronczak, J., Rooda, J.: A model-based integration and testing method to reduce system development effort. ENTCS **164**(4), 13–28 (2006)
8. Brinksma, E., Tretmans, J.: Testing transition systems: an annotated bibliography. In: Modeling and Verification of Parallel Processes, Lecture Notes in Computer Science, vol. 2067, pp. 187–195. Springer (2001)
9. Cai, K., Wonham, W.: Supervisor localization: a top-down approach to distributed control of discrete-event systems. IEEE Trans. Autom. Control **55**(3), 605–618 (2010)
10. Estefan, J.: Survey of Model-Based Systems Engineering (MBSE) methodologies. Technical report, INCOSE (2008). http://www.incose.org
11. Flordal, H., Malik, R., Fabian, M., Akesson, K.: Compositional synthesis of maximally permissive supervisors using supervisor equivalence. Discrete Event Dyn. Syst. **17**(4), 475–504 (2007)
12. Forschelen, S., van de Mortel-Fronczak, J., Su, R., Rooda, J.: Application of supervisory control theory to theme park vehicles. Discrete Event Dyn. Syst. **22**(4), 511–540 (2012)
13. Hill, R., Tilbury, D., Lafortune, S.: Modular supervisory control with equivalence-based conflict resolution. In: Proceedings of ACC, pp. 491–498 (2008)

14. Leduc, R., Lawford, M., Wonham, W.: Hierarchical interface-based supervisory control-part II: parallel case. IEEE Trans. Autom. Control **50**(9), 1336–1348 (2005)
15. Liu, J., Darabi, H.: Ramadge-Wonham supervisory control of mobile robots: lessons from practice. Proc. ICRA **1**, 670–675 (2002)
16. Ma, C., Wonham, W.: Nonblocking supervisory control of state tree structures. IEEE Trans. Autom. Control **51**(5), 782–793 (2006)
17. Malik, R., Flordal, H.: Yet another approach to compositional synthesis of discrete event systems. In: Proceedings of WODES, pp. 16–21 (2008)
18. Markovski, J., Jacobs, K., van Beek, D., Somers, L., Rooda, J.: Coordination of resources using generalized state-based requirements. In: Proceedings of WODES, pp. 297–302 (2010)
19. Martin, J.: Systems Engineering Guidebook. CRC Press (1996)
20. Muller, G.: Coupling enterprise and technology by a compact and specific architecture overview. In: Proceedings of INCOSE (2007)
21. Nadales Agut, D.E., van Beek, D.A., Rooda, J.E.: Syntax and semantics of the compositional interchange format for hybrid systems. JLAP **82**(1), 1–52 (2012)
22. Noorbakhsh, M., Afzalian, A.: Design and PLC based implementation of supervisory control for under-load tap-changing transformers. In: Proceedings of ICCAS, pp. 901–906 (2007)
23. Nourelfath, M., Niel, E.: Modular supervisory control of an experimental automated manufacturing system. Control Eng. Pract. **12**(2), 205–216 (2004)
24. van Osch, M.: Automated model-based testing of hybrid systems. Ph.D. thesis, Eindhoven University of Technology, The Netherlands (2009)
25. de Queiroz, M., Cury, J.: Modular supervisory control of composed systems. In: Proceedings of ACC, pp. 4051–4055 (2000)
26. Ramadge, P., Wonham, W.: Supervisory control of a class of discrete event processes. SIAM J. Control Optim. **25**(1), 206–230 (1987)
27. Rechtin, E., Maier, M.: The Art of Systems Architecting. CRC Press (1997)
28. Schiffelers, R., Theunissen, R., van Beek, D., Rooda, J.: Model-based engineering of supervisory controller using CIF. In: Proceedings of the 3rd International Workshop on Multi-Paradigm Modeling. Electronic Communication of the EASST, vol. 21, pp. 1–10 (2009)
29. Su, R., van Schuppen, J., Rooda, J.: Synthesize nonblocking distributed supervisors with coordinators. In: Proceedings of the 17th Mediterranean Conference on Control and Automation, pp. 1108–1113 (2009)
30. Su, R., van Schuppen, J., Rooda, J.: Aggregative synthesis of distributed supervisors based on automaton abstraction. IEEE Trans. Autom. Control **55**(7), 1627–1640 (2010)
31. Su, R., van Schuppen, J., Rooda, J.: The synthesis of time optimal supervisors by using heaps-of-pieces. IEEE Trans. Autom. Control **57**(1), 105–118 (2012)
32. Su, R., Thistle, J.: A distributed supervisor synthesis approach based on weak bisimulation. In: Proceedings of WODES, pp. 64–69 (2006)
33. Systems Engineering Group TU/e: CIF toolset. http://cif.se.wtb.tue.nl (2013)
34. Theunissen, R., Schiffelers, R., van Beek, D., Rooda, J.: Supervisory control synthesis for a patient support system. In: Proceedings of ECC, pp. 4647–4652 (2009)
35. Wong, K., Wonham, W.: Hierarchical control of discrete-event systems. Discrete Event Dyn. Syst.: Theory Appl. **6**(3), 241–273 (1996)
36. Wonham, W.: Supervisory control of discrete-event systems. Technical report, University of Toronto (2010)

Chapter 3
Output Synchronization of Dynamical Networks Having Nodes with Relative-Degree-One Nonlinear Systems

Yanyan Liu, Georgi Dimirovski and Jun Zhao

Abstract In this study, the output synchronization problem in dynamical networks consisting of nodes that have nonlinear dynamic systems with relative degree one has been explored and solved. The system property of passivity has been found to be a useful tool in solving the output synchronization problem in complex dynamical networks. It is well known, however, that not all dynamic systems possess the passivity property. Nonetheless, it has been shown here how to use passivity in order to output synchronize complex dynamical networks. It is also known that if a nonlinear system is weakly minimum phase and has relative degree one, then it is feedback equivalent to a passive system. Thus, the feedback passivation result has been exploited to investigate and solve the output synchronization problem in dynamical networks. The conditions are found, which do not require the assumption of the negative definiteness property of the outer coupling matrix under which output synchronization of dynamical networks having nodes with relative degree one systems is achieved. Furthermore, it has been found that, when all nodes are feedback equivalent to a strictly passive system, the output synchronization criterion is accomplished with less conservative conditions. An illustrative example along with numerical and simulation results is given to demonstrate the effectiveness of the theoretical findings in this study.

Y. Liu · J. Zhao
State Key Laboratory of Synthetic Automation for Process Industries, Northeastern University, Shenyang 110819, Liaoning, People's Republic of China
e-mail: yanyanliu11@yahoo.cn

J. Zhao
e-mail: zhaojun@mail.neu.edu.cn

G. Dimirovski
Faculty of Engineering, Dogus University, Acibadem, 34722 Kadikoy, Istanbul, Turkey

G. Dimirovski (✉)
School FEEIT, SS Cyril and Methodius University, 1000, Skopje, Republic of Macedonia
e-mail: gdimirovski@dogus.edu.tr

G.M. Dimirovski (ed.), *Complex Systems*, Studies in Systems, Decision and Control 55, DOI 10.1007/978-3-319-28860-4_3

3.1 Introduction

In the past couple of decades, dynamical networks that [1] comprise a finite set of nodes, with locally controlled or uncontrolled dynamic systems [2–4] at the nodes, and a certain communication coupling topology among them have gained growing attention by scientists and engineers worldwide. Nowadays, a dynamical network is commonly understood to be a large-scale systemic structure [5] consisting of a graph that contains enumerable set of very many dynamic systems as its nodes and a large number of signal-flow interconnections among them; for instance, see works by Boccaletti et al. [6], Jiang et al. [7], Newman [8], Palla et al. [9], Strogatz [10], and Wang and Chen [15, 20], Watts [1]. Thus, in their very essence, dynamical networks are completely described and represented by dynamic graphs [11].

The investigation of the aspects and issues concerning or related to the controlled dynamical networks has become one of the attractive, hot research areas for two main reasons. First reason, dynamical networks appear in the world of nature, in human societies on Earth, and very many events in the real world can be modeled by dynamical networks. More precisely, there exist real-world networks and systems such as food web, computer and communication networks, metabolic network systems, power grids, and the World Wide Web. For instance, see Bertsekas and Gallager [12], Dimirovski [13], Ilic and Zaborski [14], Palla et al. [9], Siljak [11], Strogatz [10], Watts [1]. Second one, there can be identified large numbers of important applications of dynamical networks in various disciplines including biology, economics, engineering, neuroscience, and sociology as pointed out by Wang and Chen [15].

Synchronization is a collective behavior of dynamical networks, and is one of the major issues in studying dynamical networks. Synchronous evolution or motion dynamics is of particular interest and synchronization has been extensively addressed in Arenas et al. [16], Das and Lewis [17], Lu and Chen [18], Olfati-Saber et al. [19], Wang and Chen [20], Wang et al. [21], Wang et al. [22]; and others. The importance of synchronized motion does not lie only in those situations in which synchronization can be found, but also where synchronous motion can be induced to ensure the proper functioning of a particular device such as multi-robot ensemble in Rodriguez-Angeles and Nijmeijer [23]. An interconnected multi-robot system with synchronous motion can carry out tasks that are difficult to an individual robot because the collective behavior of all robots offers more flexibility and maneuverability. Therefore, a number of research studies on synchronization have been reported in recent years as seen in DeLellis et al. [24], Dimirovski [13], Estrada et al. [25], Wang et al. [26], Zhao et al. [27], Zhu and Hu [28]. Several works have supplied good control methods on the synchronization problem as shown in Liu et al. [29]; Liu et al. [30]; Zhong et al. [31, 32].

On the other hand, the passivity first introduced by Willems in 1972 has been shown to be an important system property, because it means the energy dissipated inside a dynamic system is no more than the energy supplied outside. That is, it determines that the energy supply rate is dependent on the input and the output; see

Willems [33]. Thus, the passivity concept offers an effective tool to stabilize nonlinear systems because the storage function induced by passivity is often a good candidate Lyapunov function as pointed out in Kokotovic and Arcak [34]; Haddad and Chellaboina [35]; Zhao and Hill [36].

Since the passivity property is considerably meaningful from the viewpoint of physics, naturally, it should also be employed for controlled synchronization of dynamical networks. The answer to this task has been found positive and it is further elaborated in this contribution. In fact, by and large the nodes of networks are typical nonlinear dynamic systems. It is therefore that the passivity property appeared rather useful in the study of the output synchronization problem for dynamical networks, which is clearly demonstrated in Chopra and Spong [37]; Igarashi et al. [38]; Listmann and Woolsey [39]; Zhao et al. [40].

Recently, several works on passivity-based synchronization for networks have been reported. Most of them deal with the output synchronization problem. Reference work by Chopra and Spong [41] investigated the output synchronization of multi agents via passivity for the first time, and treated the exchange information as interconnected graph. Furthermore, these results have been extended to: the synchronization of nonlinear dynamic systems with time delay in communications in Chopra and Spong [37], of networked Euler–Lagrange systems in Yu and Antsaklis [42], and in general dynamical networks with nonidentical nodes in Zhao et al. [40]. In addition, these results have been extended to the output synchronization of nonlinear dynamic systems with relative degree one in Chopra and Spong [43] and to the output synchronization problem in the Special Euclidean Group of dimension three in Igarashi et al. [38]. Moreover, passivity-based control and synchronization of complex dynamical networks were addressed in Yao et al. [44], while passivity-based output synchronization of networks with switching topologies was studied in Liu et al. [30], Zhao et al. [40], and Zhu et al. [45].

The aforementioned papers all deal with the output synchronization of network using the passivity property. Passivity proved to be an efficient tool for solving the synchronization problem of dynamical networks with passive nodes. Nonetheless, when not all the nodes are passive the issue of whether passivity can be used to output synchronize dynamical networks has arisen. The idea of feedback passivation offers a good method to solve this problem too. It is well known that a weakly minimum phase system with relative degree one is the feedback equivalent to a passive system, which was proved in Byrnes et al. [46]. Based on this finding, the network node satisfying such conditions can be feedback equivalent to a passive system. Thus, output synchronization of dynamical networks can be achieved by feedback passivation. On the other hand, Zhao et al. [40] in their study thoroughly investigated the passivity-based output synchronization of the dynamical network. Their criteria are given by means of relaxing the commonly used assumptions that the outer coupling topology is symmetric and have the zero row sum property, as argued in Jiang et al. [7].

In this chapter, a revised version of the work by Liu et al. [30] is presented that extends further the results of Zhao et al. [40] to investigate the output synchronization problem of dynamical networks with relative degree one via the passivity property following the direction of local feedback passivation of nodes by Chopra and Spong [43]. First, we derive the conditions under which the nodes can be feedback equivalent to a passive system and then the output synchronization criterion of the dynamical network using passivity. This result is extended to the output synchronization problem via the output strict passivity. By using the output strict passivity (OSP), the negative semi-definiteness of the outer coupling matrix is removed, and the positive definiteness of the inner coupling matrix is not assumed either. Thus, the results in this manuscript represent certain new contributions to the investigated problem topic.

The rest of this paper is organized in the conventional way. Section 3.2 gives the necessary preliminaries. Section 3.3 presents the main results. An illustrative example of a relevant dynamical network along with its numerical and computer simulation results is included. The simulation results that demonstrate the effectiveness of the main theoretic results of this contribution are presented in Sect. 3.4. In Sect. 3.5, the concluding remarks finalize this study.

3.2 Problem Statement and Preliminaries

Consider a dynamical network with N nonidentical nodes described by means of state and output, Eq. (3.1):

$$\begin{aligned} \dot{x}_i &= f_i(x) + g_i(x)u_i, \\ y_i &= h_i(x),\ i=1,\ \dots, N. \end{aligned} \tag{3.1}$$

Here in (3.1) the symbols denoting $x_i \in R^n$, $u_i \in R^m$, $y_i \in R^m$ are the state, control input, and output vectors of the ith node, respectively; vector functions f_i: $R^n \to R^n$, g_i: $R^n \to R^{n\times m}$, h_i: $R^n \to R^m$ are smooth mappings satisfying the standard real-world properties $f_i(0)=0$, and $h_i(0)=0$.

The objective of this paper is to investigate the output synchronization problem of dynamical networks via the passivity property of node dynamic systems. If all nodes of the dynamical network are passive, output synchronization can be achieved via passivity. If the nodes of the dynamical network do not satisfy the passivity property, but each node can be feedback equivalent to a passive system, one can still use passivity to achieve the output synchronization. The obtained results here, as will be seen in the sequel, shed novel insight into the underlying relationship between the network's outer coupling matrix and the nodes passivity or strict passivity.

First, we give the definitions of passivity and feedback passivation of each node in the dynamical network.

Definition 3.1 [47] System (3.1) is said to be passive if there exists a positive definite continuous differentiable function $V_i(x_i)$, such that the inequality

$$V_i \le h_i^T u_i, \tag{3.2}$$

is satisfied, where $V_i(x_i)$ is called the storage function.

Definition 3.2 [47] System (3.1) is said to be output strictly passive if there exists a positive definite continuous differentiable function $V_i(x_i)$, such that the inequality

$$\dot{V}_i \le h_i^T u_i - \rho_i h_i^T h_i, \tag{3.3}$$

is satisfied for some constant $\rho_i > 0$, where ρ_i is called the output strict passivity (OPI, in short) index.

Definition 3.3 [47] System (3.1) is said to be feedback equivalent to a passive system if there exists a feedback controller $u_i = \alpha_i(x_i) + \beta_i(x_i)\upsilon_i$ such that

$$\begin{aligned} \dot{x}_i &= f_i(x_i) + g_i(x_i)\alpha_i(x_i) + g_i(x_i)\beta_i(x_i)\upsilon_i, \\ y_i &= h_i(x_i), \end{aligned} \tag{3.4}$$

satisfies the passivity inequality with the new input υ_i and the output y_i.

Also, the known definition of output synchronization of dynamical networks is recalled next.

Definition 3.4 [8] A dynamical network is said to be output synchronization if

$$\lim_{t \to \infty} \left(y_i(t) - y_j(t) \right) = 0, \; \forall i, j = 1, \ldots, N. \tag{3.5}$$

In the sequel the investigation is focused on the output synchronization of the investigated class of complex dynamic networks (3.1).

3.3 The Output Synchronization

First, we review the conditions under which each nonlinear node is feedback equivalent to a passive system.

As in Byrnes et al. [46], suppose the following specifications hold as follows:

Hypothesis H1 The matrix $L_{g_i} h_i(x)$ is nonsingular for each $x_i \in R^n$;

Hypothesis H2 The vector fields $\tilde{g}_{i1}(x), \ldots, \tilde{g}_{im}(x)$ are complete where

$$[\tilde{g}_{i1}(x) \ldots \tilde{g}_{im}(x)] = g_i(x) \left[L_{gi} h_i(x) \right]^{-1};$$

Hypothesis H3 The vector fields $\tilde{g}_{i1}(x), \ldots, \tilde{g}_{im}(x)$ commute; Then there exist state variables $z_i = \eta_i(x_i)$ and $y_i = h_i(x_i)$, such that each node (3.1) is transformed into the normal form

$$\begin{aligned} \dot{z}_i &= q_i(z_i, y_i), \\ \dot{y}_i &= b_i(z_i, y_i) + a_i(z_i, y_i)u_i, \end{aligned} \tag{3.6}$$

where $a_i(z_i, y_i)$ is nonsingular, hence invertible, and $\dot{z}_i = q_i(z_i, 0)$ is the zero dynamics of the ith node.

If each node of the network is weakly minimum phase, there exists a positive definite smooth function $W_i(z_i) \leq 0$ with $w_i(0) = 0$, such that $L_{q_i(z_i,0)}W_i(z_i) \leq 0$. The first equation of the normal form (3.6) can be rewritten as

$$\dot{z}_i = q_i(z_i, 0) + p_i(z_i, y_i)y_i. \tag{3.7}$$

The feedback controller is chosen as

$$u_i = a_i^{-1}(z_i, y_i)\left[-\left(L_{p_i(z_i,y_i)}W_i(z_i)\right)^T - b_i(z_i, y_i) + v_i\right]. \tag{3.8}$$

The Lyapunov function of the ith node is chosen as

$$V_i(z_i, y_i) = W_i(z_i) + \frac{1}{2}y_i^T y_i. \tag{3.9}$$

Then, the derivative of V_i along the state trajectory is

$$\begin{aligned} \dot{V}_i(z_i, y_i) = L_{q_i(z_i,0)}W_i(z_i) + L_{p_i(z_i,y_i)}W_i(z_i)y_i \\ + y_i^T(b_i(z_i, y_i) + a_i(z_i, y_i)u_i) \leq y_i^T v_i. \end{aligned} \tag{3.10}$$

Thus, the ith node is a nonlinear dynamic system that is feedback equivalent to a passive system.

Therefore, we can use the passivity property of each node to accomplish output synchronization of the network. The coupling topology is given as

$$v_i = \sum_{j=1}^{N} a_{ij}\Gamma y_j. \tag{3.11}$$

It also can be rewritten as $u = (A \otimes \Gamma)y$, where $A = \left(a_{ij}\right)_{NN}$ is a matrix, called the outer coupling matrix and Γ is an $m \times m$ diagonal matrix, called the inner coupling matrix. The quantity $A \otimes \Gamma$ is called a Kronecker product between two matrices, and is defined as

$$A \otimes \Gamma = \begin{pmatrix} a_{11}\Gamma & a_{12}\Gamma & \dots & a_{1N}\Gamma \\ a_{21}\Gamma & a_{22}\Gamma & \dots & a_{2N}\Gamma \\ \vdots & \vdots & \ddots & \vdots \\ a_{N1}\Gamma & a_{N2}\Gamma & \dots & a_{NN}\Gamma \end{pmatrix} \in R^{N_m \times N_m}. \tag{3.12}$$

If $a_{ii} = -\sum_{j=1, j\neq i}^{N} a_{ij}$, the coupling matrix A is called to be zero row sum matrix. This is a well-known property, which is usually assumed for the synchronization problem investigations in the literature. In this study, we do not assume the output coupling matrix A to be a zero row sum matrix, which contributes to its generality.

If the coupling topology is chosen as (3.11), the network representation can be rewritten as

$$\begin{aligned} \dot{z}_i &= q_i(z_i, 0) + p_i(z_i, y_i)y_i, \\ \dot{y}_i &= -\left(L_{p_i(z_i, y_i)} W_i(z_i)\right)^T + \sum_{j=1}^{N} a_{ij}\Gamma y_j. \end{aligned} \tag{3.13}$$

Theorem 3.1 *Consider the dynamical network consisting of the nodes described by* (3.1). *Suppose all nodes satisfy H1–H3 and are globally weakly minimum phase. If the output coupling matrix A satisfies that* $A + A^T$ *is negative semidefinite with a zero eigenvalue and is zero row sum, the inner coupling matrix* Γ *is positive definite, then the dynamical network is achieved to be output synchronization.*

Proof If each node satisfies hypothesizes H1–H3 and is globally weakly minimum phase, then each node is feedback equivalent to a passive system with the new input v_i. In addition, the control inputs $v_i (i = 1, \dots, N)$ are chosen as (3.11), thus the network can be rewritten as (3.13). Choose a candidate Lyapunov function as

$$V = \sum_{i=1}^{N} \left(2W_i(z_i) + y_i^T y_i\right). \tag{3.14}$$

Then the first time derivative of (3.14) along trajectories generated by system (3.13) is given as follows:

$$\begin{aligned} \dot{V} &= 2\sum_{i=1}^{N} \left(\dot{W}_i(z_i) + y_i^T \dot{y}_i\right) \\ &= 2\sum_{i=1}^{N} \left(L_{q_i(z_i,0)} W_i(z_i) + L_{p_i(z_i,y_i)} W_i(z_i) y_i\right) \\ &\quad + 2\sum_{i=1}^{N} y_i^T \left(-\left(L_{p_i(z_i,y_i)} W_i(z_i)\right)^T + \sum_{j=1}^{N} a_{ij}\Gamma y_j \right) \\ &\leq 2\sum_{i=1}^{N} y_i^T \sum_{j=1}^{N} a_{ij}\Gamma y_j \\ &= y^T \left(A + A^T\right) \otimes \Gamma y. \end{aligned} \tag{3.15}$$

In expression (3.15), for the time derivative $\dot{V}$ there are $y=(y_1, \ldots, y_N)^T$ and $A=(a_{ij})_{NN}$. Since $A+A^T$ is a symmetric matrix, a unitary matrix $\Phi=(\Phi_1, \ldots, \Phi_N)$ with $\Phi_i=(\phi_{1i}, \ldots, \phi_{Ni})^T$ can be chosen such that

$$\Phi^T(A+A^T)\Phi = diag\{\lambda_1, \lambda_2, \ldots, \lambda_N\}. \tag{3.16}$$

□

On the other hand, $A+A^T$ is negative semidefinite with one zero eigenvalue. Without loss of generality, we suppose that $0=\lambda_1>\lambda_2\geq\cdots\geq\lambda_N$.

Let $\omega=(\Phi^T\otimes I_m)y$. Then using (3.13) it yields

$$\dot{V}\leq\omega^T(\Phi^T\otimes I_m)((A+A^T)\otimes\Gamma)(\Phi\otimes I_m)\omega=\sum_{i=2}^{N}\omega_i^T\lambda_i\Gamma\omega_i. \tag{3.17}$$

Thus, $\dot{V}\leq 0$, and $\dot{V}=0$ if and only if $(\omega_2^T, \ldots, \omega_N^T)^T=0$. Therefore,

$$\begin{pmatrix} y_1 \\ y_2 \\ \vdots \\ y_N \end{pmatrix} = (\Phi\otimes I_m)\omega = \begin{pmatrix} \phi_{11}I_m & * & \cdots & * \\ \phi_{21}I_m & * & \cdots & * \\ \vdots & \vdots & \ddots & \vdots \\ \phi_{N1}I_m & * & \cdots & * \end{pmatrix}\begin{pmatrix} \omega_1 \\ 0 \\ \vdots \\ 0 \end{pmatrix} = \begin{pmatrix} \phi_{11}\omega_1 \\ \phi_{21}\omega_1 \\ \vdots \\ \phi_{N1}\omega_1 \end{pmatrix}. \tag{3.18}$$

Then, $\dot{V}=0$ if and only if $\phi_{j1}\omega_i=\phi_{i1}\omega_j, \forall i,j=1,\ldots N$. Furthermore, the zero row sum property of $A+A^T$ ensures that $\Phi_1=\frac{1}{\sqrt{N}}(1, \ldots, 1)^T$ can be chosen. Thus, output synchronization of the dynamical network is achieved.

Remark 3.1 If the output coupling matrix A satisfies that $A+A^T$ is negative definite, and the inner coupling matrix Γ is positive definite, then the dynamical network is achieved to be output synchronization and x(t) converges to the set $\ker(h)=\{x|h_i(x_i)=0, i=1,2,\ldots,N\}$. In addition, if each node is zero state detectable, state synchronization is achieved and $x(t)\rightarrow 0$.

Next, we design another feedback control and make each node to be feedback equivalent to an output strictly passive system. We investigate output synchronization of dynamical networks via output strict passivity.

If hypotheses H1–H3 for each of the nodes in the network hold true, then each node can be represented to a normal form (3.6). In addition, should all the nodes be globally weakly minimum phase with a positive function W_i, the feedback control of the ith node is given as

$$u_i=a_i^{-1}(z_i,y_i)[-(L_{p_i(z_i,y_i)}W_i(z_i))^T-b_i(z_i,y_i)-\rho_i y_i+v_i], \tag{3.19}$$

where ρ_i is a positive constant. Choose storage function as (3.14), and find the derivative of Vi to be

$$\dot{V}_i(z_i, y_i) \leq y_i^T v_i - \rho_i y_i^T y_i. \tag{3.20}$$

Then the ith node can be feedback equivalent to an output strictly passive system.

The communication among nodes is described by (3.11). The output synchronization criterion is given as follows.

Theorem 3.2 *Consider the dynamical network consisting of nodes described by (3.1). Suppose all nodes satisfy H1–H3 and are globally weakly minimum phase. Let λ_i be eigenvalues of $A + A^T$ and μ_i be eigenvalues of Γ. If $\lambda_i\mu_j < 2\rho = 2\min_{k=1..N}\{\rho_k\}$, $i = 1, \ldots, N$, $j = 1, \ldots, m$, then the dynamical network is achieved to be output synchronization and $y_i(t) \to 0$ as $t \to \infty$.*

Proof As shown before, the node of the dynamical network is represented by

$$\begin{aligned} \dot{z}_i &= q_i(z_i, 0) + p_i(z_i, y_i)y_i, \\ \dot{y}_i &= -(L_{p_i(z_i, y_i)}W_i(z_i))^T - \rho_i y_i + \sum_{j=1}^{N} a_{ij}\Gamma y_j. \end{aligned} \tag{3.21}$$

Again, we choose the Lyapunov function as

$$V = \sum_{i=1}^{N} (2W_i(\lambda_i) + y_i^T y_i). \tag{3.22}$$

□

Then the time derivative of V is

$$\begin{aligned} \dot{V} &= 2\sum_{i=1}^{N} (\dot{W}_i(z_i) + y_i^T \dot{y}_i), \\ &\leq 2\sum_{i=1}^{N} y_i^T (\sum_{j=1}^{N} a_{ij}\Gamma y_j - \rho y_i), \\ &= y^T((A + A^T) \otimes \Gamma - 2P \otimes I_m)y, \end{aligned} \tag{3.23}$$

where $P = diag\{\rho, \ldots, \rho\}$.

In fact, $\lambda_i\mu_j$ ($i = 1, \ldots, N$, $j = 1, \ldots, m$) are the eigenvalues of $(A + A^T) <$ g) Γ.

Since $\lambda_i\mu_j < 2\rho$, $(A + A^T)(E)\Gamma - 2P(E)I_m$ is negative definite. Thus, $V < 0$, and $V = 0$ if and only if $y = 0$. Then output synchronization is achieved and $y_i(t) \to 0$ as $t \to 0$.

3.4 An Illustrative Example

In this section, we give a numerical example to show the effectiveness of the main results. Consider a dynamical network consisting of five nonlinear systems, which are described as

$$\begin{aligned} \dot{x}_{i1} &= x_{i1} + (l_i - 1)x_{i2} - \sin(x_{i1} + x_{i2}) - u_i, \\ \dot{x}_{i2} &= \sin(x_{i1} + x_{i2}) - u_i, \\ y_i &= x_{i2}, \end{aligned} \tag{3.24}$$

where $i = 1, \ldots, 5$. In addition, let us define values of l_i as follows: $l_1 = 1.2$, $l_2 = 1.3$, $l_3 = 1.05$, $l_4 = 0.8$, $l_5 = 0.9$. Obviously, all nodes are not passive, but they all are nodes of relative-degree-one nonlinear dynamic systems.

Nonetheless, computer simulation has revealed the outputs of the isolated nodes with $u_i = 0.0$ that are shown in Fig. 3.1. It is clearly seen that the output synchronization has not been achieved.

Next, let us choose the following coordinate transformation:

$$\begin{aligned} z_i &= x_{i1} + x_{i2}, \\ y_i &= x_{i2}. \end{aligned} \tag{3.25}$$

Then the nodes described with (3.24) are represented as

$$\begin{aligned} \dot{z}_i &= -z_i + l_i y_i, \\ \dot{y}_i &= \sin z_i + u_i. \end{aligned} \tag{3.26}$$

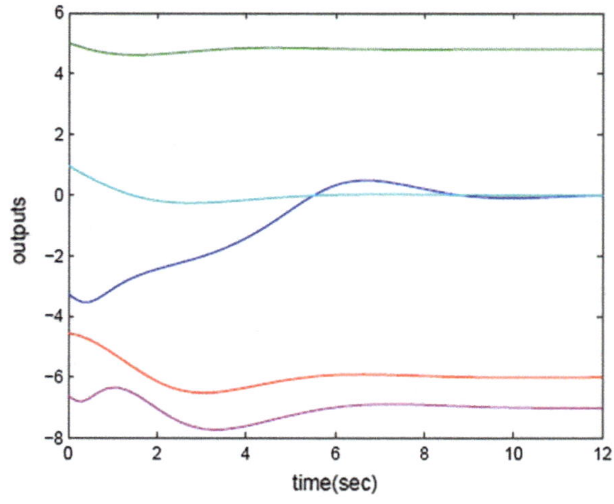

Fig. 3.1 The output responses of the isolated nodes having relative-degree-one nonlinear dynamic systems

The storage function of each node is chosen as $V_i = (1/2)z_i^2 + (1/2)y_i^2$, and the control input is chosen as $u_i = \sin z_i - l_i z_i + v_i$, where v_i is the new input. Then the derivative of along (3.26) is

$$\dot{V}_i = -z_i^2 + l_i y_i z_i - y_i \sin + y_i u_i \leq y_i v_i. \tag{3.27}$$

The node (3.26) is represented as

$$\begin{aligned} \dot{z}_i &= -z_i + l_i y_i, \\ \dot{y}_i &= -l_i z_i + v_i. \end{aligned} \tag{3.28}$$

Therefore, each node is made feedback equivalent to a passive system, hence their passivity property can be used for the purpose of the output synchronization problem.

It should be noted further, the communication coupling topology for the network is chosen as defined by (3.11), where

$$A = \begin{bmatrix} -2 & 1 & 1 & 0 & 0 \\ 0 & -3 & 1 & 1 & 1 \\ 1 & 0 & -2 & 1 & 0 \\ 1 & 1 & 0 & -3 & 2 \\ 0 & 1 & 0 & 0 & -2 \end{bmatrix}. \tag{3.29}$$

In addition, it should be also noted in addition that the simplest matrix $\Gamma = I$ was adopted.

The coupling matrix satisfies the condition of Theorem 3.1, and therefore the output synchronization of network (3.28) with (3.11) is achieved. The respective computer simulation results of the synchronized outputs in this case are depicted in Fig. 3.2. As it may well be noticed, these demonstrate that a rather quick

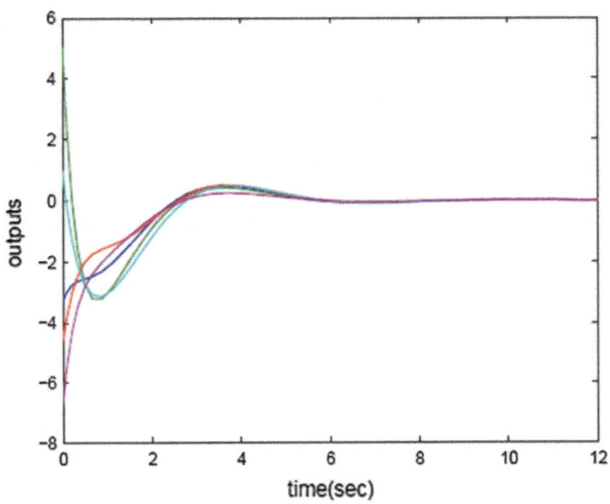

Fig. 3.2 The synchronized output responses of the network with communication coupling topology and nodes having relative-degree-one nonlinear dynamic systems

synchronization of all the nodes has been achieved. Thus, indeed a new solution to the output synchronization for complex dynamic networks with nonlinear nodes having relative degree one has been found and its favorable performance is confirmed.

3.5 Conclusion

This paper has investigated and solved the output synchronization problem for dynamical networks having in all nodes the relative-degree-one nonlinear dynamic systems. First, the conditions are given under which each node of the dynamical network is made feedback equivalent to a passive system. Second, by representing all the nodes as nonlinear dynamic systems in normal form, the output synchronization criterion is obtained without assuming the negative definiteness property and the zero row sum property of the outer coupling matrix.

Furthermore, the appropriate control inputs have been designed so that each of the nodes was made feedback equivalent to an output strictly passive system. On the grounds of this result, the output synchronization has been achieved via the output strict passivity property of node systems.

It should be noted, however, that there remains an open issue for future research to establish when in general the passivity-based synchronizability may be achieved. It is thus believed that also a deeper understanding of when such synchronizability conditions exist shall also be gained. The output synchronization in complex networks having nodes with nonlinear systems of relative degree higher than, naturally, is also another topic for future research.

References

1. Watts, D.: Small worlds: the dynamics of networks between order and randomness In: Princeton Studies in Complexity, Princeton University Press. Princeton
2. Anadriyevsky, B.R., Matveerv, A.S., Fradkov, A.L.: Control and estimation under information constraints: toward a unified theory of control, computation and communications (in Russian). Avtomatika i Telemechanika **71**(4), 34–99 (2010)
3. Dimirovski, G.M., Gough, N.E., Barnett, S.: Categories in systems and control theory. Int. J. Syst. Sci. **8**(9), 1081–1090 (1977)
4. Mesarovic, M.D., Takahara, Y.: General Systems Theory: Mathematical Foundations. Academic Press, New York (1975)
5. Siljak, D.D.: Large-Scale Dynamic Systems: Stability and Structure. North-Holland Books, New York (1978)
6. Boccaletti, S., Latora, V., Moreno, Y., Chavez, M., Hwang, D.U.: Complex networks: structure and dynamics. Phys. Rep. **424**(4–5), 175–308 (2006)
7. Jiang, Y., Lozada-Cassou, M., Vinet, A.: Synchronization and symmetry-breaking bifurcations in constructive networks of coupled chaotic oscillators. Phys. Rev. E: Stat., Nonlin, Soft Matter Phys. **68**(6), 1–4 (2003)

8. Newman, M.E.J.: The structure and function of complex networks. SIAM Rev. **45**(2), 167–256 (2003)
9. Palla, G., Derenyi, I., Farkas, I., Vicsek, T.: Uncovering the overlapping community structure of complex networks in nature and society. Nature **435**(7043), 814–818 (2005)
10. Strogatz, S.H.: Exploring complex networks. Nature **410**(6825), 268–276 (2001)
11. Siljak, D.D.: Dynamic graphs. Nonlinear Analysis: Hybrid Systems **2**, 544–567 (2008)
12. Bertsekas, D., Gallager, R.: Data Networks. Prentice Hall, Englewood Cliffs (1987)
13. Dimirovski, G.M.: System-of-systems, Large-scale Systems, Complex Systems or Complex Networks: Where do we stand? In Jorge Pereira, Open Consultation on Complex Systems Engineering towards System-of-Systems, Brussels, 4 July 2011. The EC-INFSO, Brussels, BE, Position Paper II.9 (2011).
14. Ilic, M., Zaborsky, J.: Dynamics and Control of Large Electric Power Systems. Wiley, New York (2000)
15. Wang, X.F., Chen, G.R.: Synchronization in scale-free dynamical networks: robustness and fragility. IEEE Trans. Circuits Syst. I: Fundam. Theory Appl. **49**(1), 54–62 (2002)
16. Arenas, A., Daz-Guilera, A., Kurths, J., Moreno, Y., Zhou, C.: Synchronization in complex networks. Phys. Rep. **469**(3), 93–153 (2008)
17. Das, A., Lewis, F.L.: Distributed adaptive control for synchronization of unknown nonlinear networked systems. Automatica **46**(12), 2014–2021 (2010)
18. Lu, J., Chen, G.: A time-varying complex dynamical network model and its controlled synchronization criteria. IEEE Trans. Autom. Control **50**(6), 841–846 (2005)
19. Olfati-Saber, R., Fax, J.A., Murray, R.M.: Consensus and cooperation in networked multi-agent systems. Proc. IEEE **95**(1), 215–233 (2007)
20. Wang, X.F., Chen, G.: Synchronization in small-world dynamical networks. International J. Bifurcat. Chaos **12**(1), 187−192 (2002)
21. Wang, L., Jing, Y., Kong, Z., Dimirovski, G.M.: Adaptive exponential synchronization of uncertain complex dynamical networks with delay coupling. Neuro-Quantology: Int. J. Neurol. Quantum Phys. **6**(4), 397–404 (2008)
22. Wang, Z., Wang, Y., Liu, Y.: Global synchronization for discrete-time stochastic complex networks with randomly occurred nonlinearities and mixed time delays. IEEE Trans. Neural Netw. **21** (1), 11–25 (2010)
23. Rodriguez-Angeles, A., Nijmeijer, H.: Mutual synchronization of robots via estimated state feedback: a cooperative approach. IEEE Trans. Control Syst. Technol. **12**(4), 542–554 (2004)
24. DeLellis, P., diBernardo, M., Garofalo, F.: Novel decentralized adaptive strategies for the synchronization of complex networks. Automatica **45**(5), 1312–1318 (2009)
25. Estrada, E., Gago, S., Caporossi, G.: Design of highly synchronizable and robust networks. Automatica **46**(11), 1835–1842 (2010)
26. Wang, Y.W., Wang, H.O., Xiao, J.W., Guan, Z.H.: Synchronization of complex dynamical net-works under recoverable attacks. Automatica **46**(1), 197–203 (2010a)
27. Zhao, J., Hill, D.J., Liu, T.: Synchronization of complex dynamical networks with switching topology: a switched system point of view. Automatica **45**(11), 2502–2511 (2009)
28. Zhu, Z.Q., Hu, H.P.: Robust synchronization by time-varying impulsive control. IEEE Trans. Circuits Syst. II Express Briefs **57**(9), 735–739 (2010)
29. Liu, T., Dimirovski, G.M., Zhao, J.: Exponential synchronization of complex delayed dynamical networks with general topology. Physica A **387**(2–3), 643–652 (2008)
30. Liu, Y., Dimirovski, G.M., Zhao, J.: Output synchronization of dynamical networks having nodes with relative degree one nonlinear systems. In: Kolemisevska-Gugulovska, T.D., Stankovski, M.J. (eds.) Proceedings of COSY-2011 Papers—Special International Conference on Complex Systems: Synergy of Control, Communications and Computing, Ohrid, The Society of ETAI, Skopje, pp. 115−120, 16–20 September 2011 (2011)
31. Zhong, W.S., Stefanovski, J.D., Dimirovski, G.M., and Zhao, J.: Decentralized control and synchronization of time-varying complex dynamical network. Kybernetika **45**(1), 151–167 (2009)

32. Zhong, W.S., Dimirovski, G.M., Zhao, J.: Decentralized synchronization of an uncertain complex dynamical network. J. Control Theory Appl. **7**(3), 225–230 (2009)
33. Willems, J.C.: Dissipative dynamical systems part I: general theory. Arch. Ration. Mech. Anal. **45**(5), 321–351 (1972)
34. Kokotovic, P., Arcak, M.: Constructive nonlinear control: a historical perspective. Automatica **37**(5), 637–662 (2001)
35. Haddad, W.M., Chellaboina, V.: Dissipativity theory and stability of feedback interconnections for hybrid dynamical systems. Math. Prob. Eng. **7**(4), 299–335 (2001)
36. Zhao, J., Hill, D.J.: Dissipativity theory for switched systems. IEEE Trans. Autom. Control **53** (4), 941–953 (2008)
37. Chopra, N., Spong, M.W.: Output synchronization of nonlinear systems with time delay in communication. In: Proceedings of the 45th IEEE Conference on Decision and Control, pp. 4986–4992 (2006)
38. Igarashi, Y., Hatanaka, T., Fujita, M., Spong, M.W.: Passivity-based output synchronization in SE(3). In: Proceedings of the 27th American Control Conference, pp. 723–728 (2008)
39. Listmann, K.D., Woolsey, C.A.: Output synchronization of systems in chained form. In: Proceedings of the 48th IEEE Conference on Decision and Control CDC2009 held jointly with the 28th Chinese Control Conference, pp. 3341–3346 (2009)
40. Zhao, J., Hill, D.J., Liu, T.: Passivity-based output synchronization of dynamical networks with non-identical nodes. In: Proceedings of the 49th IEEE Conference on Decision and Control, pp. 7351–7356 (2010)
41. Chopra, N., Spong, M.W.: Passivity-based control of multi-agent systems In: Advances in Robot Control. Springer, New York (2006)
42. Yu, H., Antsaklis, P.J.: Passivity-based output synchronization of networked Euler-Lagrange systems subject to nonholonomic constraints. In: Proceedings of the 29th American Control Conference, pp. 208–213 (2010)
43. Chopra, N., Spong, M.: Output Synchronization of Nonlinear Systems with Relative Degree One Recent Advances in Learning and Control. Springer, Berlin (2008)
44. Yao, J., Guan, Z.H., Hill, D.J.: Passivity-based control and synchronization of general complex dynamical networks. Automatica **45**(9), 2107–2113 (2009)
45. Zhu, Y., Cheng, D., Hu, X.: Synchronization of a class of networked passive systems with switching topology. In: Proceedings of the 46th IEEE Conference on Decision and Control, pp. 2271–2276 (2007)
46. Byrnes, C.I., Isidori, A., Willems, J.C.: Passivity, feedback equivalence, and the global stabilization of minimum phase nonlinear systems. IEEE Trans. Autom. Control **36**(11), 1228–1240 (1991)
47. Van der Schaft, A.J.: L2-Gain and Passivity Techniques in Nonlinear Control. Springer, Berlin (2000)

Chapter 4
Mechanism Design for Incentive Compatible Control of Networks

Anil Kumar Chorppath and Tansu Alpcan

Abstract Network mechanism design aims to achieve system-level goals such as maximization of aggregate network performance using specific methods in networks, where users are strategic and selfish decision-makers with individual preferences. By imposing certain rules and pricing schemes on users, the mechanism designer aligns the system-wide objectives with those of the users, and achieves the targeted goals while ensuring user incentive compatibility. This chapter provides an overview of recent results in the area of mechanisms and games for distributed control of networks. The methodology and algorithms developed are applied to diverse network control problems such as interference and spectrum management. We model the heterogeneous behavior of users, which ranges from altruistic to selfish and to malicious, within the analytical framework of game theory. A mechanism design approach is adopted to quantify the effect of adversarial behavior, which ranges from extreme selfishness to destructive maliciousness. Differentiated pricing is proposed as a method to counter and mitigate adversarial behavior. An additional application to location privacy in mobile commerce is also briefly discussed.

4.1 Introduction

The theory of mechanism design is concerned with obtaining outcomes that meet certain system-level goals, despite the fact that individuals in the system pursue only their self-objectives. In this setting a social planner or designer faces the problem

This work is supported by Deutsche Telekom Laboratories, Berlin, Germany.

A.K. Chorppath (✉)
Technical University of Munich Lehrstuhl Für Theoretische Informationstechnik, 80333 Munich, Germany
e-mail: anil.chorppath@tum.de

T. Alpcan
Department of Electrical and Electronic Engineering,
The University of Melbourne, VIC 3010, Melbourne, Australia
e-mail: tansu.alpcan@unimelb.edu.au

G.M. Dimirovski (ed.), *Complex Systems*, Studies in Systems, Decision and Control 55, DOI 10.1007/978-3-319-28860-4_4

of using the reported values of strategic multiple users to obtain a system-level outcome when the actual values are private to users. Mechanism design can be considered as a subbranch of game theory, where the rules of the induced game in the mechanism are designed to achieve a socially desirable outcome. Thus, mechanism design can be viewed alternatively as reverse engineering of games. Recently, there has been widespread interest in using mechanism design for modeling, analyzing, and solving problems in network resource allocation and network economics that are decentralized in nature.

We present in this chapter an overview of applications of mechanism design and game theory to revenue maximization as well as privacy and security, based on our earlier work [1–4].

4.1.1 Revenue Maximization in Wireless Networks

In addition to pricing user transmit powers for obtaining social goals, the designer may like to maximize the revenue obtained from these prices. We next introduce pricing mechanisms for designer **revenue maximization** which may lead to nonoptimal social welfare. Myerson [5] introduced optimal auctions in which the designer knowing the distribution of private values of players maximizes the expected revenue. In [6], for a wideband wireless network that employs Code Division Multiple Access (CDMA) as the spectrum access mechanism, the revenue maximization problem is formulated as a Stackelberg game. The optimal prices are obtained for the Nash equilibrium points. For revenue maximization in a similar setting, suboptimal constant distributed pricing scheme is proposed in [7]. In [8], for a general delay network, a two-stage dynamic pricing-congestion game in which the service provider sets a price anticipating demand of users and users chose their flow vectors given the prices is analyzed. An optimal revenue maximizing pricing is proposed for networks with several competing oligopolists and the extent of inefficiency loss is lower bounded. In [9], a lower bound for the ratio between the revenue from flat entre fee pricing rule and revenue maximizing rule is provided, which they refer to as the Price of Simplicity (PoS). A price discrimination scheme is also studied and Price of Simplicity is obtained for it.

Here, we present an overview of revenue maximizing pricing mechanisms for **interference coupled** multi-carrier wireless systems which manipulate the Nash equilibrium with optimal prices [1]. The revenue maximizing mechanism is analyzed in Sect. 4.3.

4.1.2 Network Mechanism Design with Malicious Users

The behavior of different users on networks may range from altruistic on the one end to malicious (adversarial) on the other end of a wide spectrum. While a selfish player strategize to maximize her throughput by getting the proportional share of resources, a malicious user, however, tries to get a disproportionate share of network resource, and in addition may disrupt the whole network. Well-known examples of this adversarial behavior in networks include jamming in wireless networks and denial-of-service (DoS) attacks [10, 11].

In networked systems with selfish users, a loss in overall social welfare was identified and referred to as *Price of Anarchy* in [12, 13]. In [14], with the presence of malicious users this concept was extended and *Price of Byzantine Anarchy* and *Price of Malice* were first introduced and obtained bounds on these metrics, which are parameterized by the number of malicious users for a virus inoculation game in social networks. A modified definition was proposed in [15] for congestion games based on the delay experienced at Nash equilibrium point with and without the presence of a malicious player. Both of these works observed a *Windfall of Malice*, where malicious behavior actually improves the social welfare of non-oblivious selfish users due to the better cooperation resulting because of the 'fear factor' or effects similar to Braess's paradox [15]. In [16], a more general definition of Price of Malice was given with weaker assumptions than above-mentioned works in the presence of Byzantine players and using a no-regret analysis. A game theoretic model for the strategic interaction of legitimate and malicious players was introduced in [17], where the authors derived a bound on the damage caused by the malicious players. In [18], partial altruism of some of the users was analyzed and a bound on Price of Anarchy was obtained as a function of the altruism parameter. In [19], the degree of cooperation of a user as a vector of values was used to obtain a convex combination of other user utilities and to model altruistic behavior in the context of network routing games. The value of unilateral altruism (VoU) was defined to be the ratio of the equilibrium utility of the altruistic user to the equilibrium utility she would have received in Nash equilibrium if she was selfish and was calculated for routing games in [20].

In order to circumvent *Price of Anarchy*, a pricing scheme for price-taking users [21–23] and auctions for price anticipating users [24, 25] were developed. In [26], the effect of spiteful behavior of some of the users was analyzed in the context of first and second price auctions and the revenues obtained from each were compared. A Bayesian Nash equilibrium is obtained. A similar analysis was carried out in [27]. In this chapter, we analyze the effect of malicious behavior in the case of Vickrey-Clarke-Groves (VCG) mechanism for divisible resource allocation in a non-Bayesian framework. In addition, we define and quantify the Price of Malice of the mechanisms proposed for network resource allocation including VCG.

To counter the adversarial behavior, Micali and Valiant in [28], developed a modified VCG mechanism, taking into account collusive, irrational, and adversarial user behavior for combinatorial auctions. In the proposed mechanism, the price charged to an agent is increased from VCG price by a scaled factor of the maximum social welfare of other agents. The first price auction was modified to make it incentive compatible to adversarial behavior and other externality effects in [29]. Following a similar direction, we modify the pricing in the proportional fair allocation mechanisms to punish the malicious users and incentivize them to adopt a selfish behavior instead.

We model the coexistence of selfish and malicious players using a noncooperative game theoretic formulation and adopt a mechanism design approach [30]. Here, we assume that malicious users mainly stay within the rules of the system but exhibit adversarial behavior. We model them by assigning different utility functions than selfish players, such as own selfish utility minus the sum of utility of other users in the system or a convex one in contrast to the usually concave utility functions of selfish users. Thus, we map their destructive behavior such as jamming other players and launching denial-of-service (DoS) attacks to rational incentives. The model and the mechanisms are elaborated in Sects. 4.4 and 4.5, respectively.

4.1.3 Privacy in Mobile Commerce

In mobile commerce, a company provides location-based services to a set of mobile users. The users report to the company their location with a level of granularity to maintain a degree of anonymity, depending on their perceived risk, and receive in return monetary benefits or better services from the company. This chapter formulates a quantitative model in which information theoretic metrics such as entropy, quantify the anonymity level of the users. The individual perceived risks of users and the benefits they obtain are considered to be linear functions of their chosen location information granularity. The interactions between the mobile commerce company and its users are investigated using mechanism design techniques as a privacy game. The user's best responses and optimal strategies for the company are derived under budgetary constraints on incentives, which are provided to users in order to convince them to share their private information at the desired level of granularity.

We consider a mobile commerce environment in which the users or customers get benefits from a company (service provider) by disclosing their location with certain degree of accuracy. At the same time, disclosing their location information brings users certain risks and compromises their privacy. Therefore, users have a motivation to maintain anonymity by giving less granular information about their location or no information at all. In this chapter, we propose a *mechanism design* [31] approach in which the company acts as a designer and properly motivates its users through the benefits in terms of payment [32] provided to them, in order to obtain desired *granularity of location information* from all the users. We refer to the mechanisms in this setting as *privacy mechanisms* [33].

The benefits offered by the company to the users can be the location-based service resources, discount coupons, or monetary awards. It is assumed that the more accurate the information, the more valuable it is for the company. For example, street-level information leads to contextual advertisements while city-level granularity is less valuable. Concurrently, by being less anonymous, the users take a privacy risk. We take a commodity view of the privacy here, where the users can trade their privacy to obtain benefits from the company in an individual risk aware way. The model and the privacy mechanisms are given in Sect. 4.6.

We propose mechanisms for multi-carrier systems, where the designer without knowing users' utility functions achieve different designer objectives through appropriate pricing. The pricing mechanisms obtained here can be implemented through a distributed iterative algorithm rather than existing heuristic suboptimal or centralized algorithms. The general model is given next.

4.2 General System Model

Let us consider an uplink multiple access system with spectrum divided into N orthogonal carriers shared among K users. We assume the base station acts as a designer $\mathcal{D}$ who manages the resource sharing among the users. In this scenario, let us define a K-player strategic noncooperative game, $\mathcal{G}$, where each player $i \in \mathcal{A}$ has a respective **decision variable** x_i such that

$$x = [x_1, \ldots, x_K] \in \mathcal{X} \subset \mathbb{R}^{K \times N},$$

where $\mathcal{X}$ is the decision space of all users. Let

$$x_{-i} = [x_1, \ldots, x_{i-1}, x_{i+1}, \ldots x_K] \in \mathcal{X}_{-i} \subset \mathbb{R}^{K-1 \times N},$$

be the profile of decision variable of users other than ith player and $\mathcal{X}_{-i}$ is the respective decision space. This chapter assumes vector decision variables and a compact and convex decision space. Due to the inherent coupling between the users, the decisions of users directly affect each other's performance as well as the aggregate allocation of limited resources.

Each user decides on the power level over the N carriers. Therefore,

$$x_i = [x_i^{(1)}, \ldots, x_i^{(N)}],$$

where $x_i^{(n)} = h_i^{(n)} p_i^{(n)}$ denote the received power level over carrier n as a product of uplink transmission power $p_i^{(n)}$ and channel loss $0 < h_i^{(n)} < 1$ of player i. If linear interference is assumed, then the signal-to-interference ratio (SIR) of the received signal on channel n is

$$\gamma_i^{(n)} = \frac{x_i^{(n)}}{\sum_{j \neq i} x_j^{(n)} + \sigma}, \tag{4.1}$$

where σ represents the background noise. Let us denote the interference at receiver for user i over channel n as $I_i^{(n)} = \sum_{j \neq i} x_j^{(n)} + \sigma$.

The **preferences** of the users are captured by utility functions and for multi-carrier wireless systems it is given by

$$\sum_n U_i(\gamma_i^{(n)}(x)) : \mathcal{X} \to \mathbb{R}, \quad \forall i \in \mathcal{A},$$

which are usually chosen to be continuous and differentiable for analytical tractability. We assume that $U_i(.)$ is any concave nondecreasing function.

The designer $\mathcal{D}$ devises a **pricing mechanism** $\mathcal{M}$, which can be represented by the mapping $\mathcal{M} : \mathcal{X} \to \mathbb{R}^{K\times N}$, implemented by introducing incentives in the form of *prices* to the users. The latter can be formulated by adding it as a cost term such that the player i has the cost function

$$J_i(x) = c_i(x) - \sum_n U_i(\gamma_i^{(n)}(x)). \tag{4.2}$$

The designer imposes prices $P_i^{(n)}(x)$ per unit power over channel n for user i to align the strategic users with the global objective. The total payment by user i is then $\sum_n x_i^{(n)} P_i^n(x)$ and the individual cost of users will be

$$J_i(x) = \sum_n (x_i^{(n)} P_i^n(x) - U_i(\gamma_i^{(n)}(x))). \tag{4.3}$$

The **user objective** is to solve the following individual optimization problem in the strategic game

$$\min_{x_i} J_i(x), \tag{4.4}$$

under the given constraints of the strategic game, and prices imposed by the designer. We differentiate between two kinds of mechanisms, auctions and pricing, which differ in the assumption on nature of the users and the interaction rules.

In auction mechanisms, the designer $\mathcal{D}$ imposes on a player $i \in \mathcal{A}$ a user-specific

- resource allocation rule, $Q_i(x)$,
- resource pricing, $c_i(x)$,

based on their bids x. The price anticipating users decide on their bid, minimizing their individual cost. In this case, the cost function is

$$J_i(x) = P_i(x)Q_i(x) - U_i(Q(x)).$$

Similar to player preferences, the **designer objective**, e.g., maximization of aggregate user utilities or social welfare, can be formulated using a smooth objective function V for the designer:

$$V(x, U_i(\gamma_i(x)), c_i(x)) : \mathcal{X} \to \mathbb{R},$$

where $c_i(x)$ are user-specific pricing terms and player utilities, respectively. Hence, the global optimization problem of the designer is simply $\max_x V(x, U_i(x), c_i(x))$, which it solves *indirectly* by setting prices. In some cases, the objective function V

characterizes the desirability of an outcome x from the designers perspective. In other cases when the designer objective is to satisfy certain minimum performance constraints such as users achieving certain quality-of-service levels, the objective can be characterized by a region (a subset of the game domain $\mathcal{X}$). The main problem the designer has is that he has no information about the utility function of the users except that it belongs to a class of concave nondecreasing (in this case logarithmic) functions.

The players share and compete for limited resources in the given environment under its information and communication constraints. We focus on two basic types of resource sharing and coupling, which are often encountered in a variety of problems in networking:

1. *Additive resource sharing*: the players share a finite resource C such that

$$\sum_{i=1}^{K}\sum_{n=1}^{N} x_i^{(n)} = C.$$

   This type of coupling is encountered in bandwidth sharing and rate control in networks.
2. *Interference coupling* (linear interference): the resource allocated to player i, γ_i, is inversely proportional to interference generated by others such that

$$\gamma_i(x) = \frac{h_i x_i}{\sum_{j \neq i} h_j x_j + \sigma},$$

   where h_i $\forall i$ and σ denote some system parameters. Interference coupling occurs in wireless networks where γ represents signal-to-interference ratio.

The formal definitions of the game theoretical and mechanism design concepts used in this chapter are given in Appendix.

4.3 Pricing Mechanism for Designer Revenue Maximization

We consider a multi-carrier system, where the transmit power is allocated across multiple orthogonal channels as in Orthogonal Frequency Division multiplexing (OFDM). Each user receives a different price for power consumption over different channels and the prices influence the user best responses (IWF solution) [34]. In this section we extend the results obtained earlier for single carrier system to multiple carrier systems.

In many practical scenarios the designer will be interested more in maximizing her revenue than sum of utilities of users. In this section, the global objective of the designer aims to maximize her total revenue as a monopolistic entity, while trying to limit the user power levels to $P_{\max}$. The total revenue of the designer will be,

$$V(x) = \sum_j \sum_n P_j^{(n)}(x) x_j^{(n)},$$

The user best response obtained from first-order derivative is

$$\frac{dU_i(\gamma_i^{(n)}(x))}{dx_i^{(n)}} - P_i^{(n)} = 0, \ \forall \, i \in \mathcal{A}, n. \tag{4.5}$$

and

$$\frac{dU_i}{d\gamma_i^{(n)}} = P_i^{(n)} I_i^{(n)}, \ \forall \, i \in \mathcal{A}, n. \tag{4.6}$$

Also,

$$\frac{dU_j}{dx_i^{(n)}} = \frac{x_j^n P_j^n}{(\sum_{k \neq j} x_k^n + \sigma)} = \gamma_j^n P_j^n, \ \forall \, j, i \in \mathcal{A}, n. \tag{4.7}$$

The designer $\mathcal{D}$ solves the constrained optimization problem

$$\max_x V(x) \text{ such that } \sum_n \frac{x_i^{(n)}}{h_i^{(n)}} \leq P_{\max} \ \forall i, n,$$

and

$$\sum_j \sum_n \frac{x_j^{(n)}}{h_j^{(n)}} \leq P_{total}.$$

By substituting Eq. (4.5), we obtain

$$V(x) = \sum_j \sum_n \frac{dU_j}{dx_j^{(n)}} x_j^{(n)},$$

and this objective function is not guaranteed to be convex. To find a local maximum we form the Lagrange function given by

$$L = V(x) - \sum_j \mu_j (\sum_n \frac{x_j^{(n)}}{h_j^{(n)}} - P_{\max}), \tag{4.8}$$

$$+ \rho \sum_j \sum_n \frac{x_j^{(n)}}{h_j^{(n)}} - P_{total}, \tag{4.9}$$

where μ_i's and ρ are Lagrange multipliers. The resulting Karush-Kuhn-Tucker (KKT) conditions for optimality are

$$P_i^{(n)} + \sum_j x_j^{(n)} \frac{d^2 U_j^{(n)}}{dx_i^{(n)} dx_j^{(n)}} - \mu_i - \rho = 0, \ \forall\, i, n,$$

$$\mu_i(\sum_n \frac{x_i^{(n)}}{h_i^{(n)}} - P_{\max}) = 0, \ \forall i,$$

$$\rho(\sum_j \sum_n \frac{x_j^{(n)}}{h_j^{(n)}} - P_{total}) = 0.$$

Let us analyze this problem for the wireless channel, for example,

$$U_i(x) = \alpha_i \sum_n \log(1 + \gamma_i^{(n)}).$$

Substituting from Eq. (4.7), the above equation becomes

$$P_i^{(n)} - \sum_j \frac{x_j^{(n)} \alpha_j}{(\sum_k x_k^{(n)} + \sigma)^2} - \frac{\mu_i}{h_i^{(n)}} - \frac{\rho}{h_i^{(n)}} = 0, \ \forall i, n.$$

By aligning the problems we can rewrite it as

$$P_i^{(n)} - \sum_j \frac{x_j^{(n)} P_j^{(n)}}{(\sum_k x_k^{(n)} + \sigma)} - \frac{\mu_i}{h_i^{(n)}} - \frac{\rho}{h_i^{(n)}} = 0, \ \forall i, n. \tag{4.10}$$

Let us denote $F_j^{(n)} = \frac{x_j^{(n)}}{(\sum_k x_k^{(n)} + \sigma)}$, and note that $F_j^{(n)} = \frac{\gamma_i^{(n)}}{1+\gamma_i^{(n)}}$ is already available as side information. Then, the above equation can be again rewritten as

$$P_i^{(n)} - \sum_j F_j^{(n)} P_j^{(n)} = \frac{\mu_i}{h_i^{(n)}} + \frac{\rho}{h_i^{(n)}}, \ \forall i, n. \tag{4.11}$$

As in previous cases we obtain a matrix form as

$$C^{(n)} \cdot P^{(n)} = D^{(n)} \cdot K,$$

where $C^{(n)}$ and $D^{(n)}$ are defined accordingly and $K = [\mu_1, \ldots, \mu_N, \rho]^T$.

4.4 Mechanism Design and Game Model with Malicious Users

We consider here a mechanism design having heterogeneous users in the induced game, in which one subset of users has ‘abnormal’ utility function compared to the class of regular selfish users. The utility function of the class of malicious users can be very different depending on their nature and goals. The disrupting nature of malicious users where they want to create loss to other users even at the cost of their benefit can be captured with a modified utility function. One such modified utility function can be obtained by a convex combination of user utilities

$$U_i^m = U_i + \theta_i \sum_j U_j, \tag{4.12}$$

where θ_i is the parameter between -1 and 1 which captures the range of behavior of a user. This utility function can be modified by taking the average of the utilities of all the users in the second term [18]. The table below lists the values of θ and corresponding user behavior.

θ	Behavior
$\theta = 0$	Selfish
$-1 < \theta < 0$	Malicious

Let us define the set of selfish users be $\mathcal{S} \subset \mathcal{A}$. Also, the set of malicious users, i.e., users with $\theta_i \neq 0$, is defined as $\mathcal{B}$ and $\mathcal{B} = \mathcal{A} \backslash \mathcal{S}$. The utility function of malicious users can be modified as

$$U_i^m = U_i + \theta_i \sum_{j \in \mathcal{S}} U_j, \ \forall i \in \mathcal{B}. \tag{4.13}$$

An alternate utility function,

$$U_i^m = (1 - |\theta_i|)U_i + \theta_i \sum_{j \neq i} U_j, \tag{4.14}$$

models the user behavior with a gradual decrease in the self-utility when $|\theta|$ increases. In the case of network resource allocation, the malicious users take disproportionate higher share of resources and thereby reduce the utility of other users. This model does not capture such a malicious behavior because it will not lead to a disproportionate higher share of resource to malicious user. This observation is demonstrated for a specific example later. In the case of network resource allocation, Eq. (4.14) is not appropriate for modeling malicious behavior and therefore the malicious modeling in Eq. (4.13) is adopted in this chapter. The extreme selfishness or greedy nature of malicious users can be also captured with a convex utility function.

In this case they will take the maximum possible share of the resource constrained above by either physical limits or a level that leads to immediate detection.

In this chapter, for tractability purposes, we model user i's utility function as logarithmic, parameterized by her private value α_i. In this case, the aim of the designer in the auction setting will be to make the users report their true private value, i.e., $x_i = \alpha_i$ and carry out an efficient allocation based on that. We assume that the malicious users have information about the utility function of other selfish users but the regular selfish users do not have the information about the existence of malicious users and their identities.

4.5 Price of Malice in Mechanisms

In this section, we quantify the robustness of mechanisms described in the above setting, against malicious players. For this purpose, we first redefine the metric Price of Malice ($PoM(M)$) of mechanism M suitable for mechanisms in resource sharing setting.

Definition 4.1 The metric Price of Malice of a mechanism M is defined as

$$PoM(M) := \frac{\sum_{j \in S} U_j(Q_j(x)) - \sum_{j \in S} U_j(Q'_j(x))}{\sum_{j \in S} U_j(Q_j(x))},$$

where Q is the allocation at the Nash equilibrium when none of the users are malicious and Q' is the allocation at the Nash equilibrium in the presence of malicious users.

Now, we estimate the value of Price of Malice parameter for two networks which differ in user coupling and resource sharing as described in the previous section.

4.5.1 Price of Malice in VCG Mechanism:

Consider the case where only user k is malicious and reports U_k^m to the designer to reduce the share of resource allocation to the other selfish users. Then, the optimal allocation becomes

$$Q^* = \arg\max_Q \, (U_k(Q_k) + (1 + \theta_k) \sum_{j \neq k} U_j^m(Q_j(x))).$$

Again, in the case of logarithmic user utility functions, we have

$$Q^{m*} = \arg\max_Q \; \alpha_k \log\left(Q_k\right) + (1 + \theta_k) \sum_{j \neq k} \alpha_j \log\left(Q_j(x)\right).$$

The individual user allocations are given in the malicious case as

$$Q_k^{m*} = \frac{\alpha_k C}{\alpha_k + (1+\theta_k)\sum_{j\neq k}\alpha_j}, \tag{4.15}$$

$$Q_j^{m*} = \frac{\alpha_j(1+\theta_k)C}{\alpha_k + (1+\theta_k)\sum_{j\neq k}\alpha_j}, \forall j \neq k. \tag{4.16}$$

We observe that the allocation to malicious user increases and that of other users reduce when θ_k decreases from 0 toward -1 compared to the case where none of the users are malicious. Therefore, the malicious user is able to destroy the efficiency property of the VCG mechanism.

Furthermore, in the malicious case when user i is not in the system, we have

$$Q_k^i = \frac{\alpha_k C}{\alpha_k + (1+\theta_k)\sum_{j\neq k,i}\alpha_j}, i \neq k, \tag{4.17}$$

$$Q_j^i = \frac{\alpha_j(1+\theta_k)C}{\alpha_k + (1+\theta_k)\sum_{j\neq k,i}\alpha_j}, \forall j \neq k, i, \tag{4.18}$$

$$Q_j^k = \frac{\alpha_j C}{\sum_{m\neq k}\alpha_m}. \tag{4.19}$$

In the case where user k is malicious and users have logarithmic utility function, we could obtain analytical expression for PoM in VCG mechanism which can be generalized to other cases. $PoM(VCG)$ is derived directly by substituting Eqs. (4.18) and (4.17) into the Definition 4.1 given above. For the additive resource sharing case, the Price of Malicious of VCG mechanism $PoM(VCG)$ is

$$PoM(VCG) = \frac{\sum_{j\neq k}\alpha_j \log\left(\frac{\alpha_k + (1+\theta_k)\sum_{j\neq k}\alpha_j}{(1+\theta_k)\sum_m \alpha_m}\right)}{\sum_{j\neq k}\alpha_j \log\left(\frac{\alpha_j C}{\sum_m \alpha_m}\right)}.$$

For the case where users are symmetric $\alpha_i = \alpha,\ \forall i$, and only one user is malicious or all the malicious user coordinate to form one entity, this simplifies to

$$PoM(VCG) = \frac{\log\left(\frac{N-1+\frac{1}{1+\theta_k}}{N}\right)}{\log\left(\frac{C}{N}\right)}.$$

We could observe that when the maliciousness of users increase, i.e., as θ decreases from 0 to -1, we can see that the Price of Malice increases. We could observe that the $PoM(VCG)$ can be bounded for different possible values of θ_k and is unbounded when θ_k reaches -1.

4.5.2 Price of Malice in Indirect Auction Mechanisms

Here we present indirect auction mechanisms [31] for two network coupling schemes, rate control in wired networks and power allocation in interference coupled wireless networks, and quantify the Price of Malice for both cases. In the indirect mechanisms, instead of reporting their utility function to the designer, the players take a best response to the actions of other players and to the allocation and pricing rules set by the designer. Therefore, the allocation and pricing rules are not a function of utility functions unlike direct VCG mechanism, but rather fixed functions of the player strategies. We consider indirect auction mechanisms with scalar bid here since they have only one-dimensional communication requirement which is suitable for network resource allocation. The malicious behavior considered in this section is that the malicious players take maximum possible share of the resources according to their θ value. By this way the malicious players aim to disrupt other players by denying their fair share of resources.

Additive Sharing (Rate Control in Networks)
We consider the rate sharing problem with users having separable utility function of their allocation and quantify the effect of the adversarial behavior on it. Let users with utilities $U_i(Q_i)$ share a fixed bandwidth C such that $\sum_{i=1}^{N} Q_i(x) \leq C$, where $x_i \in (0, x_{\max})$. The vector x in this case denotes player flow rates and Q the capacity allocated to them [32, 35]. Consider the utility function given in (4.13) and the cost of ith user is then given by

$$J_i^m = c_i - U_i - \theta_i \sum_{j \in S} U_j, \forall i. \tag{4.20}$$

We consider the efficient proportional allocation auction mechanism M_a introduced in [24] which is an indirect mechanism where the users submit a scalar bid. The proportional allocation which is defined based on the bid vector of players x is

$$Q_i := \frac{x_i}{\sum_j x_j + \omega} C, \tag{4.21}$$

where ω can be seen as the reserve bid [36] and it removes the singularity of the function. For $\omega = 0$, we could see that the resource is completely utilized, i.e., $\sum_i Q_i = C$.

We next briefly show how the pricing rule/function is designed with the use of a generator function, as in [24]. Let us define $t = \sum_j x_j + \omega$ as a measure of demand for the resource and which allows us to characterize agent optimal responses with respect to a parameter which is identical for all agents at equilibrium. The generator function $g(.)$ is a function of t to $\mathbb{R}^+$ and plays the role of Lagrange multiplier to generate the optimal pricing function. The total payment of ith user has several choices, depending on the choice of generator function.

For $g(t) = t^2$,the payment function is derived in [24] as

$$c_i = x_i \sum_{j \neq i} x_j + \omega, \tag{4.22}$$

which is convex payment function in x_i and is sufficient to guarantee a unique Nash equilibrium.

Being oblivious to the presence of malicious users, the designer employs the same allocation rule and payment to ith user as the one obtained above for mechanism $\mathcal{M}_a$ assuming all the users are just selfish. First, we characterize the modified Nash equilibrium if some of these users are malicious or altruistic. Let us check for the special case of logarithmic utilities and the mechanism $\mathcal{M}_a$.

For the allocation in (4.21) the strategy space constrained by the set $0 \leq x_i \leq x_{\max}\ \forall i$ satisfies the assumption the strategy space $\mathcal{X}$ has scalar decision variables, is compact and convex, and has a nonempty interior. Then by a standard theorem of game theory (Theorem 4.4, p.176, in [37]), the network game admits a *Nash equilibrium* (NE).

For the payment given in (4.22), allocation in (4.21) and logarithmic utility function, the cost functions are twice continuously differentiable in all its arguments and strictly convex in x_i for $\theta_i < 0$. For altruistic case, i.e., $\theta_i > 0$, the cost functions are strictly convex in x_i only for

$$\theta_i \leq \alpha_i\left(\frac{1}{x_{\max}} + \frac{t}{x_{\max}^2}\right).$$

This is obtained by checking for $\dfrac{d^2 J_i^m}{dx_i^2} \geq 0$. We consider in the game only altruistic users satisfying this condition, in order to obtain an equilibrium. Since the cost function is twice continuously differentiable in all its arguments and strictly convex in x_i for all the users with $-1 < \theta_i \leq \alpha_i(\dfrac{1}{x_{\max}} + \dfrac{t}{x_{\max}^2})$, the best response points obtained from first-order conditions gives a Nash equilibrium.

The best response of user becomes

$$\frac{\partial J_i^m}{\partial x_i} = 0 \Longrightarrow x_i^* = \frac{\alpha_i}{t(1+\theta_i)C},$$

by using the fact that selfish users will have the Nash equilibrium point $x_i^* = \dfrac{\alpha_i}{tC}$, from the incentive compatibility property of the mechanism $\mathcal{M}_a$.

The mechanism $\mathcal{M}_a$ defined by (4.21) and (4.22) with users having logarithmic utilities admits several Nash equilibria and one NE point is given as

$$x_i^* = \frac{\alpha_i}{t(1+\theta_i)C}, \tag{4.23}$$

where $-1 < \theta_i \le \alpha_i(\frac{1}{x_{\max}} + \frac{t}{x_{\max}^2})$.

Remark 4.1 In [38], the conditions for existence of a unique NE for an $N-$person game are given. In addition to assumption that cost functions are twice continuously differentiable in all its arguments and strictly convex in x_i, the cost functions should satisfy diagonal strict concavity of the weighted nonnegative sum of the cost functions as given in Theorem 2 of [38]. The cost function does not necessarily satisfy this condition in our case. Therefore, the NE is not unique.

We can observe that malicious users having $-1 < \theta < 0$ will have the Nash equilibrium point as $x_i^* > \frac{\alpha_i}{tC}$. Therefore, the malicious users bid higher than the selfish users and obtain a disproportionate higher share of resource.

If we use the modeling in Eq. (4.14), the Nash equilibrium point of the mechanism $\mathcal{M}_a$ with logarithmic utilities is obtained in similar way as above is

$$x_i^* = \frac{(1-|\theta_i|)\alpha_i}{t(1+\theta_i)C}. \tag{4.24}$$

We can observe that for malicious users having $-1 < \theta < 0$, the NE point is same as the all-selfish case ($x_i^* = \frac{\alpha_i}{tC}$), i.e., no malicious effect. But it can be observed that there is a higher effect of altruistic users on selfish users in the case of (4.24) compared to (4.23). For example, with $\theta = 1$, NE point of altruistic user is $x_i^* = \frac{\alpha_i}{2tC}$ according to (4.23) but $x_i^* = 0$ according to (4.24), i.e., the altruistic user leaves the entire resource for the selfish users' usage. Therefore, modeling in Eq. (4.14) is useful to capture extreme altruistic behavior.

The allocation for the regular selfish users, i.e., users with $\theta_i = 0$ in the presence of malicious users, can be written as

$$Q_i' = \frac{\frac{\alpha_i}{tC}C}{\sum_{j\in S}\frac{\alpha_j}{tC} + \sum_{k\in B}\frac{\alpha_k}{t(1+\theta_k)C} + \omega}. \tag{4.25}$$

Let

$$r_i = \frac{Q_i}{Q_i'} = \frac{\sum_{j\in S} \alpha_j + \sum_{k\in B} \dfrac{\alpha_k}{(1+\theta_k)} + \omega}{\sum_j \alpha_j + \omega}, \tag{4.26}$$

be the ratio of allocation of selfish users before and after the presence of malicious users. Now we obtain the value of PoM of the mechanism $\mathcal{M}_a$ at the NE point given above in Eq. (4.23). For the additive resource sharing case, the Price of Malicious $PoM(\mathcal{M}_a)$ is

$$PoM(\mathcal{M}_a) = \frac{\sum_{j\in S} \alpha_j \log(r_j)}{\sum_{j\in S} \alpha_j \log\left(\frac{\alpha_j C}{\sum_i \alpha_i + \omega}\right)}.$$

For the case where users are symmetric $\alpha_i = \alpha,\ \forall i$ and only one user is malicious or all the malicious user coordinate to form one entity, this simplifies to

$$PoM(\mathcal{M}_a) = \frac{\log\left(\frac{\alpha(N-1+\frac{1}{1+\theta_k})+\omega}{N\alpha+\omega}\right)}{\log\left(\frac{\alpha C}{N\alpha+\omega}\right)}.$$

Remark 4.2 We could see that $PoM(\mathcal{M}_a)$ is equal to $PoM(VCG)$ when $\omega = 0$ for the special case of the utility function considered. It is because the proportional allocation coincides with the VCG allocation for this case. But we get very different $PoM(\mathcal{M}_a)$ and $PoM(VCG)$ in the case of the other utility functions, for example $U_i(Q_i) = \alpha_i \log(1 + Q_i)$.

Interference Coupled Systems (CDMA Power Control)

Consider an auction mechanism in the context of a wireless network and uplink power control setting [36, 39] where due to the interference coupling the resource sharing is inherently competitive. Let the user utilities be defined as $U_i(x) = \alpha_i \log \gamma_i(Q(x))$ and the individual power levels, Q, satisfy $\sum_{i=1}^N Q_i \le C$, where the signal-to-interference ratio (SINR) received by the base station is

$$\gamma_i = \frac{Q_i(x)}{\sum_{j\neq i} Q_j(x) + \sigma},$$

and $x_i \in (0, x_{\max})$.

An auction-based mechanism, $\mathcal{M}_b$, can be defined based on the bid of player i, with the resource allocation rule

$$Q_i := \frac{x_i}{\sum_j x_j} C, \tag{4.27}$$

which is proportional allocation as first analyzed in [24]. We can see that using this proportional allocation, full utilization of resource is attained, i.e., $\sum_i Q_i = C$. Now we decouple the user utilities by rewriting γ_i as

$$\gamma_i(Q_i) = \frac{Q_i(x)}{C - Q_i(x) + \sigma}, \tag{4.28}$$

using the full utilization property. For the allocation given in (4.27), the SINR is

$$\gamma_i(x) = \frac{x_i}{\sum_j x_j(C + \sigma) - x_i}. \tag{4.29}$$

Then, we obtain $PoM(M_b)$ as [30]

$$PoM(M_b) = \frac{\sum_{j\in S} \alpha_j \log(\frac{\gamma_j}{\gamma_j'})}{\sum_{j\in S} \alpha_j \log(\frac{\alpha_j C}{\sum_k \alpha_k (C+\sigma)-\alpha_j})}.$$

In the symmetric case and only one user is malicious, the PoM becomes

$$PoM(M_b) = \frac{\log(\frac{(N-1+\frac{\alpha}{\alpha+\theta_k})(C+\sigma)-1)}{N(C+\sigma)-1})}{\log(\frac{C}{N(C+\sigma)-1})}.$$

A similar behavior of $PoM(M_b)$ as in the case of additive sharing can be observed for different values of θ. The variation of $PoM(M_b)$ for different values of θ is given in the simulation section for a specific set of parameters.

4.5.3 *Price of Malice in Pricing Mechanisms*

A counterpart of the Price of Malice metric in Definition 4.1 for pricing mechanisms [21], which differ from auctions by their lack of an explicit resource allocation scheme, can be obtained by replacing $Q(x)$ and $Q'(x)$ with the action vector without malicious users x and with malicious users x', respectively.

In the case of additive resource sharing, the users with utilities $U_i(x_i) = \alpha_i \log x_i$ share the fixed resource $\sum_{i=1}^{N} x_i = C$, and $x_i \in (0, x_{\max})$. Consider an efficient mechanism M_c, which can be implemented in an iterative way. The efficient allocation is

$$x_i = \frac{\alpha_i}{\lambda},$$

where λ is the Lagrange multiplier. In the case of all- selfish users $\lambda = \sum_i \alpha_i / C$ and it will be set as price to the users.

Let each malicious user takes a share x_m which can be $x_{\max}$, the maximum share they can use without detection, according to their utility function, in order to disrupt others. Let λ' be the Lagrange multiplier in this case which will be a different point rather than $\lambda = \sum_i \alpha_i / C$. The remaining resource $(C - \sum_B x_m)$ will be shared among good users, under the efficient mechanism M_c. In the additive sharing case $PoM(M_c)$ is

$$PoM(M_c) = \frac{\sum_{j \in S} \alpha_j \log(\frac{C\lambda'}{\sum_i \alpha_i})}{\sum_{j \in S} \alpha_j \log(\frac{\alpha_j C}{\sum_i \alpha_i})}.$$

For symmetric case, where $\alpha_i = \alpha \ \forall i$, it becomes

$$PoM(M_c) = \frac{\log(\frac{C\lambda'}{N\alpha})}{\log(\frac{C}{N})}.$$

The counterpart of auction in the interference coupled case for pricing can be obtained in a similar way and the mechanism.

4.5.4 Differentiated Pricing

We propose a softer response scheme than blocking toward malicious users after explicit detection based on any well-known (threshold) detection scheme. There are numerous methods of detection already available as given in PART IV of [40]. The response mechanism is implemented by the designer by deploying a differentiated pricing. A necessary assumption we make in this subsection is that malicious users stay within the system and do not have any means to evade the pricing mechanisms imposed by the designer. This assumption is relaxed in the next subsection.

Auctions for Additive Sharing

We consider the network mechanism $\mathcal{M}_a$ proposed for network rate sharing above and modify it with a new payment function. We propose now a differentiated payment function to counter the malicious behavior of users and propose a new mechanism using this payment function. We first assume here that the designer knows the value of θ of malicious user. In practical problems, this is not realistic and the designer needs to make the decision on payment function entirely based on user bids. Therefore, we assume that after detecting the malicious user using a threshold detection scheme based on the bids, the designer punishes the malicious users with a price function assuming $\theta = -1$, i.e., extreme maliciousness. Alternatively, one can couple this parameter with the confidence of the detection scheme used, i.e., low θ values for high probability of malicious behavior and vice versa. We propose mechanism $\mathcal{M}_m$ in the following proposition which is efficient in the presence of malicious users, i.e., $PoM(\mathcal{M}_m) = 1$. The cost function of users from Eq. (4.20) for the proportional allocation given in (4.21) with $\omega = 0$ and logarithmic utility function is

$$J_i^m = c_i - \alpha_i \log\left(\frac{x_i}{\sum_k x_k}\right) - \theta_i \sum_{j\neq i} \alpha_j \log\left(\frac{x_j}{\sum_k x_k}\right), \forall i. \tag{4.30}$$

The first derivative of the cost in terms of the signal or bid to be sent, to calculate the best response of the ith, user gives

$$\frac{\partial U_i(Q_i)}{\partial Q_i} = \frac{(\sum_k x_k)^2}{\sum_{j\neq i} x_j}\left(\frac{\partial c_i}{\partial x_i} + \theta_i \sum_{j\neq i} \frac{\alpha_j}{x_j \sum_k x_k}\right).$$

Let us denote $t = \sum_j x_j$, then $x_i = \dfrac{tQ_i}{C}$ and

$$\sum_{j\neq i} x_j = t - x_i = t\left(1 - \frac{Q_i}{C}\right).$$

Doing the substitutions,

$$\begin{aligned}\frac{\partial U_i(Q_i)}{\partial Q_i} &= \frac{t}{1 - \dfrac{Q_i}{C}}\left(\frac{\partial c_i(Q_i, t)}{\partial x_i} + \theta_i \sum_{j\neq i} \frac{1}{t}\right),\\ &:= f(Q_i, t).\end{aligned} \tag{4.31}$$

The designer should solve the constrained optimization problem

$$\max_Q V(Q) \Leftrightarrow \max_Q \sum_i U_i(Q_i) \text{ such that } \sum_i Q_i = C, \tag{4.32}$$

in order to find a globally optimal allocation Q that satisfies this **efficiency criterion**. The associated Lagrangian function is then

$$L(Q) = \sum_i U_i(Q_i) + \lambda\left(C - \sum_i Q_i\right),$$

where $\lambda > 0$ is a scalar Lagrange multiplier. Under the convexity assumptions made, this leads to

$$\frac{\partial L}{\partial Q_i} \Rightarrow U_i'(Q_i) = \lambda, \ \forall i \in \mathcal{A}, \tag{4.33}$$

and the efficiency constraint

$$\frac{\partial L}{\partial \lambda} \Rightarrow \sum_i Q_i = C \tag{4.34}$$

and $Q_i = 0$ for users with $U_i'(Q_i) < \lambda$.

When we compare (4.31) and (4.33), we can see that $f(Q_i, t)$ is equal to the Lagrange multiplier λ. Since $f(Q_i, t)$ is a function of Q_i, there will be unequal marginal valuations at equilibrium. For efficient allocation we need to obtain a price function which will induce a $f(Q_i, t)$ which will give identical marginal valuations at equilibrium [24]. For this we make $f(Q_i, t)$ independent of Q_i and derive the corresponding price function. Let $f(Q_i, t) = g(t)$ where $g(t)$ is the generator function and by integrating over x_i, we obtain

$$\begin{aligned} c_i(x) = & \int_0^{x_i} \frac{g(s + \sum_{j \neq i} x_j)}{(s + \sum_{j \neq i} x_j)^2} ds \sum_{j \neq i} x_j \\ & - \theta_i \int_0^{x_i} \frac{ds}{s + \sum_{k \neq j} x_k} \sum_{j \neq i} \frac{\alpha_j}{x_j}. \end{aligned} \tag{4.35}$$

For $g(t) = t^2$, we obtain

$$c_i(x) = x_i \sum_{j \neq i} x_j - \theta_i \log\left(1 + \frac{x_i}{\sum_{j \neq i} x_j}\right) \sum_{j \neq i} \frac{\alpha_j}{x_j}. \tag{4.36}$$

Let us assume that the users except ith user are merely selfish due to the payment function of the mechanism they report $x_i = \frac{\alpha_i}{tC}$. Then, we obtain (4.37) as the payment function which corresponds to the efficient allocation. If the malicious user takes best response using the payment (4.37) in (4.20), the best response is obtained as $x_i = \frac{\alpha_i}{tC}$. The mechanism $\mathcal{M}_m$ defined by the allocation in (4.21) with $\omega = 0$ and the payment

$$c_i(x) = x_i \sum_{j \neq i} x_j - \theta_i (N-1) tC \log\left(1 + \frac{x_i}{\sum_{j \neq i} x_j}\right), \tag{4.37}$$

is efficient in the presence of malicious users and makes the malicious user take the strategy $x_i^* = \frac{\alpha_i}{tC}$ for network rate sharing with users having logarithmic utility functions.

Remark 4.3 If the designer punishes the users who are detected as malicious with a payment in which $\theta_i = -1$, without knowing the exact θ value in a more realistic situation, the pricing function becomes

$$c_i(x) = x_i \sum_{j \neq i} x_j + \log\left(1 + \frac{x_i}{\sum_{j \neq i} x_j}\right)(N-1). \tag{4.38}$$

For this cost function to be convex, in order to take the best response, from the second-order conditions we get

$$N \leq \frac{\sum_{j \neq i} \alpha_j}{C^2} + 1.$$

Note that in this differentiated pricing scheme, the malicious users who will try to bid something higher than its private value will have to pay an additional amount proportional to their bid as in (4.38). Even if the cost function is not convex, it does not affect the equilibrium; since anticipating the additional payment, the malicious user will bid taking the best response according to the cost with payment given by Eq. (4.22) which is convex.

Such a differentiated pricing scheme is widely used today in various settings, such as network access. For example, if some users of an Internet service provider (ISP) are creating burden to the network using much higher amount of resources above a pre-determined cap, they are priced differentially higher compared to other users. This reality is captured in our model since the higher usage above a threshold is punished even if it is not coming from the disproportionate use due to malicious nature.

In a similar way, a differentiated pricing mechanism can be also derived for interference coupled CDMA systems.

4.6 Privacy Mechanism Model

Consider a mobile network composed of a set of mobile users with cardinality N. Around user i at any time t, let a group of $n_i(t)$ mobile users, $\mathcal{A}$, are in close proximity in an area. The service provider gives location-based applications to the mobile users. Therefore, it asks for the location information from the mobiles.

We use an information theoretic approach to quantify the anonymity level of the individual mobile users while giving the location information. The uncertainty of service provider about the location information of user i is defined using the entropy term

$$A_i = \sum_{i=1}^{n_i(t)} p_i \log_2 \frac{1}{p_i}.$$

where probability p_i corresponds to the probability that a user is in a location. The parameter A_i concurrently quantifies the anonymity level of a users i. We can see that $p_i = \frac{1}{\log_2 n_i(t)}$. Then A_i simply boils down to

$$A_i = \log_2 n_i(t).$$

Table 4.1 Values of $n_i(t)$, N and g

$n_i(t)$	N	g
10^1	10^3	$\frac{2}{3}$
10^3	10^6	$\frac{1}{2}$
10^6	10^9	$\frac{1}{3}$

We next define a metric called *granularity of location information*, g_i, for the ith user as

$$g_i = 1 - \frac{A_i}{\log_2 N}.$$

The value of g_i is between zero and one for each user. The anonymity level obtained by user i by reporting with a granularity level g_i is

$$A_i = (1 - g_i)\ \log_2 N.$$

Here, $g_i = 0$ means the user i keeps its location completely private and $g_i = 1$ means the user gives exact location to the mobile company. We can see that the more the value of g, the less anonymous are the users. With a given value of g_i the users specify the size of the crowd it belongs to, i.e., $n_i(t)$. Table 4.1 gives values of g for different combinations of $n_i(t)$ and N. We can see that as the size of the population N increases the more anonymous become the users.

The users decide on the value of g which they report to the company. In the scenario considered in this model, the users have a continuous decision space resulting from a risk-benefit trade-off optimization, i.e., the allowed decisions are not just full or null information. This allows the designer to provide benefit based on the level of information given by the users.

There is a cost of perceived risk c_i associated with the user's privacy when they give location information, which linearly increases with the granularity of information, i.e.,

$$c_i = r_i g_i,$$

where r_i is the risk factor. The risk factor may result from disclosing your daily routine or behavior to unknown parties. For example, the users may not like others to know when they are in their office or home or they may simply care about their privacy on principle. The users estimate or learn about their risk level from past experiences or from reliable sources or by exchanging information with users like how much level of g with which they report to the designer.

While gaining on location privacy, each user loses on the benefits of location-based applications/services due to the anonymity. For example, while depending on whether users are in office, home, or a particular street or city, they might be targeted with different kinds of offers and services. When they give wrong information they are given wrong services and offers. The total benefit obtained by user i can be quantified as

$$s_i = b_i(g)\log(1 + g_i),$$

where $b_i(g) \in R^+$ is the benefit or subsidy factor provided by the company. Note that the benefit factor b_i provided for user i is designed based on the granularity level chosen by all the users. In other words, the company provides benefits based on the total available information in the actual "information market." We model that the total benefit increases logarithmically with the granularity level, since for low granularity level marginal increase in the value of location information is higher. The logarithmic assumption in this chapter can be generalized to any nondecreasing, concave function.

We now summarize the definitions of some of the terms discussed so far.

1. **(Location) Privacy**: (Location) privacy of an individual user refers to how she discloses and controls the dissemination of her personal (location) data.
2. **Anonymity (location)**: Anonymity of a user i, A_i, is the uncertainty of the service provider about the users location.

$$A_i = \sum_{i=1}^{n_i(t)} p_i \log_2 \frac{1}{p_i}.$$

3. **Granularity of Information**: Granularity of information is the level of granularity with which a user i reports its location.

$$g_i = 1 - \frac{A_i}{\log_2 N}.$$

4. **Perceived risk (cost)**: It is the total cost perceived by user i as a result of reporting her location with a certain level of granularity of information, which is modeled as linear in g_i,

$$c_i = r_i g_i.$$

5. **Benefit**: The total subsidy or reward user i obtains from the mobile commerce company by disclosing her location with a certain level of granularity of information,

$$s_i = b_i(g)\log(1 + g_i).$$

In a mechanism design setting, there is a *designer* $\mathcal{D}$ at the center who influences N *players* participating in a **strategic (noncooperative) game**. Let us define the interaction of the users in the close proximity in the above setting as an N-player strategic game, $\mathcal{G}$, where each player $i \in \mathcal{A}$ has a respective **decision variable** g_i such that

$$g = [g_1, \dots, g_N] \in \mathcal{X} \subset \mathbb{R}^N,$$

where $\mathcal{X}$ is the decision space of all players. The cost of each mobile user i will be the risk it perceives minus the benefits it obtains from the company, i.e.,

$$J_i(g) = r_i g_i - b_i \log(1 + g_i)\ \forall i.$$

Each mobile user then solves her own optimization problem

$$\min_{g_i} J_i(g). \tag{4.39}$$

Note that from the user perspective the benefit b_i is a constant designed by the company, since each user has an information constraint to know the granularity level of other users and calculate its benefit. The users just take best response given the benefit provided by the company. The NE is a widely accepted and useful solution concept in strategic games, where no player has an incentive to deviate from it while others play according to their NE strategies. The NE is at the same time the intersection point of players' best responses obtained by solving user problems individually.

The company acts here as the mechanism designer and has the goal of obtaining a desired level of location information granularity from the users. In this work, the designer has an unconventional objective compared to other works in mechanism design where the designer usually looks for social welfare or designer revenue maximization. The designer or company here wants to improve the precision of location information from each user, which is captured by a designer objective function that takes granularity of information of all the users as its argument. The designer objective we consider here is

$$\max_b V = \max_b \sum_{i=1}^{N} w_i \log(1 + g_i(b_i)), \tag{4.40}$$

subject to a budget or resource constraint

$$\sum_{i=1}^{N} b_i \leq B,$$

where w_i is the weight given to individual users as desired by the designer and B is the total budget. The weights depend on how much the company values the location information from different types of users.

It is important to note here that the designer (the mobile commerce company) tries to achieves its objective indirectly by providing benefits to users b as it naturally does not have control on their behavior, i.e., g. Essentially, the company tries to move the NE point vector of g of the resulting game to a desirable point using the benefits provided to the users.

4.6.1 Privacy Mechanism

In a privacy mechanism, each user decides on the location privacy level to be reported, i.e., g_i, depending on its risk-level perception as a best response to the benefit set by the company by minimizing individual cost. The underlying game may converge to a Nash equilibrium, which may not be desirable to the service provider because the required level of location information is not obtained. Therefore, the designer employs a pricing or subsidy mechanism to motivate the users by properly selecting the benefits delivered to each user by solving a global objective. We obtain the optimum benefit for each user by aligning user problems and designer problem as

$$b_i^* = \frac{w_i}{\nu^* + \lambda_i^* - \mu_i^*}, \quad \forall i \in \mathcal{A}, \tag{4.41}$$

where $\nu^*, \lambda_i^*, \mu_i^*$ are Lagrange multipliers. Then, the optimal granularity level of each user will be

$$g_i = \begin{cases} 0, & \text{if } b_i \le r_i \\ \dfrac{w_i}{(\nu^* + \lambda_i^* - \mu_i^*) r_i} - 1, & \text{if } r_i \le b_i \le 2r_i \\ 1, & \text{if } b_i \ge 2r_i. \end{cases}$$

The designer can obtain desired granularity of information from each user by properly selecting the functions in the global objective and the weights in the function. Note that to formulate the objective and for imposing the constraints on the global problem, the designer needs to know the user r's. This she can obtain using a learning method which will be considered next.

4.7 Discussion and Open Problems

In this chapter, solutions for network control problems with strategic users are obtained using a mechanism design approach. First, an uplink power control game of multi-carrier wireless systems with strategic users is analyzed. The users have scalar-parameterized logarithmic utility functions that are unknown to the designer (base station). Distributed and iterative pricing mechanisms for uplink power and spectrum allocation are proposed for revenue maximization of the designer.

Next, we have studied adversarial behavior in network resource allocation schemes including pricing and auctions by adopting a mechanism design approach. We have considered two types of coupling of resource sharing, additive sharing, and interference coupling. We have analyzed the robustness of the existing network mechanisms to adversarial behavior using a quantitative metric Price of Malice. Next, we have presented one method to counter such adversarial behavior which is a differentiated pricing to punish the aggressive user.

Finally, the interaction of a mobile commerce company with its users who obtain location-based services is a strategic game. A privacy mechanism is designed where the company motivates users to report their location information at a granularity level desired by the company. In return, the benefits obtained by a user depend on the weight the designer gives for her in the global objective. The users report their location with nonzero granularity of information when the subsidy by the company exceeds their perceived risk factor. As expected, the granularity of location information selected by the users decreases with the risk factor.

Future research directions include obtaining bounds on the parameters dealt in this paper and detailed study of collusion by malicious users. Another direction is to study Bayesian mechanisms in this context. It is also an interesting direction to analyze the effect of altruism or partial altruism of some of the users in this context, as in the work [18].

Appendix

Definitions:
The properties of mechanisms considered in this chapter can be formally defined as follows.

Definition 4.2 *Efficiency*: Efficient mechanisms maximize designer objective, i.e., they solve the problem $\max_x V(x, U_i(x), c_i(x))$.

Definition 4.3 *Nash Equilibrium*: The strategy profile $x^* = [x_1^*, \ldots, x_N^*]$ is in Nash Equilibrium if the cost of each player is minimized at the equilibrium given the best strategies of other players.

$$J_i(x_i^*, x_{-i}^*) \leq J_i(x_i, x_{-i}^*), \forall i \in \mathcal{A}, x_i \in \mathcal{X}_i.$$

Definition 4.4 *Strategy proofness or Incentive Compatibility*: If the players do not gain anything by reporting a value other than their true value, i.e.,

$$J_i(x_i, x_{-i}) \leq \tilde{J}_i(\tilde{x}_i, x_{-i}), \forall i \in \mathcal{A}, \tilde{x}_i \in \mathcal{X}_i, x_{-i} \in \mathcal{X}_{-i},$$

where x is the original value vector, and $\tilde{x}_i$ is the "misrepresented" value or action, then the mechanism is strategy proof.

References

1. Chorppath, A.K., Alpcan, T., Boche, H.: Pricing mechanisms for multi-carrier wireless systems. In: Proceedings of IEEE International Dynamic Spectrum Access Networks (DySPAN) Symposium, Aachen, Germany (May 2011)
2. Chorppath, A.K., Alpcan, T., Boche, H.: Adversarial behavior in network games. Dyn. Games Appl. (June 2013)
3. Chorppath, A.K., Alpcan, T.: Learning user preferences in mechanism design. In: Proceedings of 50th IEEE Conference on Decision and Control and European Control Conference, Orlando, Florida (December 2011)
4. Chorppath, A.K., Alpcan, T.: Trading privacy with incentives in mobile commerce: a game theoretic approach. Pervasive Mob. Comput. **9**(4), 598–612 (2013)
5. Myerson, R.B.: Optimal auction design. Math. Oper. Res. **6**(1), 58–73 (1981)
6. Daoud, A.A., Alpcan, T., Agarwal, S., Alanyali, M.: A stackelberg game for pricing uplink power in wide-band cognitive radio networks. In: 47th IEEE Conference on Decision and Control, 2008. CDC 2008, pp.1422–1427, 9–11 December 2008
7. Ren, S., van der Schaar, M.: Revenue maximization and distributed power allocation in cognitive radio networks. In: Proceedings of the 2009 ACM workshop on Cognitive radio networks. CoRoNet'09, pp. 43–48. New York, NY, USA, ACM (2009)
8. Acemoglu, D., Ozdaglar, A.: Competition and efficiency in congested markets. Math. Oper. Res. **32**(1), 1–31 (2007)
9. Shakkottai, S., Srikant, R., Ozdaglar, A., Acemoglu, D.: The price of simplicity. In: Forty-First Asilomar Conference on Signals, Systems and Computers, ACSSC 2007, pp. 1450–1454. Pacific Grove, CA (2007)
10. Xu, W., Trappe, W., Zhang, Y., Wood, T.: The feasibility of launching and detecting jamming attacks in wireless networks. In: MobiHoc'05 Proceedings of the 6th ACM international symposium on Mobile ad hoc networking and computing, pp. 47–56 (2005)
11. Avrachenkov, K., Altman, E., Garnaev, A.: A Jamming Game in Wireless Networks with Transmission Cost. Lecture Notes in Computer Science, vol. 4465, pp. 1–12. Springer, Berlin (2007)
12. Koutsoupias, E., Papadimitriou, C.: Worst-case Equilibria. Lecture Notes in Computer Science, pp. 404–413. Springer, Berlin (1999)
13. Roughgarden, T.: The price of anarchy is independent of the network topology. In: Proceedings of the 34th Annual ACM Symposium on the Theory of Computing (May 2002)
14. Moscibroda, T., Schmid, S., Wattenhofer, R.: When selfish meets evil: byzantine players in a virus inoculation game. In: Proceedings of the twenty-fifth annual ACM symposium on Principles of distributed computing, Denver, Colorado (2006)
15. Babaioff, M., Kleinberg, R., Papadimitriou, C.H.: Congestion games with malicious players. In: Proceedings of the 8th ACM Conference on Electronic Commerce, pp. 103–112. San Diego, California (2007)
16. Roth, A.: The price of malice in linear congestion games. In: WINE'08: Proceedings of the 4th International Workshop on Internet and Network Economics, pp. 118–125 (2008)
17. Theodorakopoulos, S., Baras, J.S.: Game theoretic modeling of malicious users in collaborative networks. IEEE J. Sel. Areas Commun. **26**(7), 1317–1327 (2008). August
18. Chen, P.A., Kempe, D.: Altruism, selfishness, and spite in traffic routing. In: Electronic Commerce EC08, pp. 8–125. Illinois, Chicago (July 2008)
19. Azad, A.P., Altman, E., Azouzi, R.E.: From altruism to non-cooperation in routing games. CoRR (2008). arXiv:abs/0808.4079
20. Azad, A.P., Musacchio, J.: Unilateral altruism in network routing games with atomic players. CoRR (2011). arXiv:abs/1108.1233
21. Kelly, F.P., Maulloo, A.K., Tan, D.: Rate control in communication networks: shadow prices, proportional fairness and stability. J. Oper. Res. Soc. **49**, 237–252 (1998)
22. Alpcan, T., Pavel, L.: Nash Equilibrium Design and Optimization. In: Proceedings of International Conference on Game Theory for Networks (GameNets 2009), Istanbul, Turkey (May 2009)

23. Boche, H., Naik, S., Alpcan, T.: A unified mechanism design framework for networked systems. Technical report (September 2010). arXiv:1009.0377 [cs.GT]
24. Maheswaran, R.T., Basar, T.: Social welfare of selfish agents: motivating efficiency for divisible resources. In: 43rd IEEE Conference on Decision and Control (CDC), pp. 1550–1555. Paradise Island, Bahamas (December 2004)
25. Johari, R., Mannor, S., Tsitsiklis, J.: Efficiency loss in a network resource allocation game: the case of elastic supply. IEEE Trans. Autom. Control **50**(11), 1712–1724 (2005). November
26. Brandt, F., Sandholm, T., Shoham, Y.: Spiteful bidding in sealed-bid auctions. In: IJCAI'07 Proceedings of the 20th international joint conference on Artifical Intelligence, pp. 1207–1214. Hyderabad, India (2007)
27. Steiglitz, K., Morgan, J., Reis, G.: The spite motive and equilibrium behavior in auctions. Contrib. Econ. Anal. Policy, **2**(5) (2003)
28. Micali, S., Valiant, P.: Revenue in truly combinatorial auctions and adversarial mechanism design. Technical report, MIT-Computer Science and Artificial Intelligence Laboratory (June 2008)
29. Netzer, N.: An externality-robust auction. Working Paper (2012)
30. Chorppath, A.K., Alpcan, T.: Adversarial behavior in network mechanism design. In: Proceedings of 4th International Workshop on Game Theory in Communication Networks (Gamecomm), ENS, Cachan, France (May 2011)
31. Krishna, V.: Auction theory (2nd ed.). Academic Press (2010)
32. Srikant, R.: The Mathematics of Internet Congestion Control. Systems & Control: Foundations & Applications. Birkhauser, Boston (2004)
33. Chorppath, A.K., Alpcan, T.: A privacy mechanism for mobile commerce. In: Proceedings of 2nd IEEE International Workshop on Data Security and Privacy in wireless Networks, Lucca, Italy (June 2011)
34. Yu, D.D., Cioffi, J.M.: Iterative water-filling for optimal resource allocation in OFDM multiple-access and broadcast channels. In: IEEE Global Telecommunications Conference (GLOBECOM'06), San Francisco, CA (2006)
35. Alpcan, T., Başar, T.: A utility-based congestion control scheme for internet-style networks with delay. IEEE Trans. Netw. **13**(6), 1261–1274 (2005). December
36. Huang, J., Berry, R., Honig, M.: Auction-based spectrum sharing. ACM Mob. Netw. Appl. J. **24**(5), 405–418 (2006). June
37. Başar, T., Olsder, G.J.: Dynamic Noncooperative Game Theory, 2nd edn. SIAM, Philadelphia (1999)
38. Rosen, J.B.: Existence and uniqueness of equilibrium points for concave n -person games. Econometrica **33**(3), 520–534 (1965)
39. Chorppath, A.K., Bhashyam, S., Sundaresan, R.: A convex optimization framework for almost budget balanced allocation of a divisible good. IEEE Trans. Autom. Sci. Eng. **8**(3), 520–531 (2011). July
40. Alpcan, T., Basar, T.: Network Security: A Decision and Game Theoretic Approach. Cambridge University Press, Cambridge (2010)

Chapter 5
Building Smart Grid: Optimal Coordination of Consumption with Decentralized Energy Generation and Storage

Araz Ashouri, Sebastian Gaulocher and Petr Korba

Abstract This work describes an implementation of an office and/or personal smart grid for environmentally friendly buildings. These can be equipped with a local energy source (e.g., photovoltaic panels or combined heat-power units), energy storage devices (batteries, electric hot water boilers, heating, and ventilation systems including air conditioning), a building energy management system with sensors (e.g., providing the room temperatures), and household appliances acting as actuators (in general, split into groups of schedulable and non-schedulable ones). The idea behind this work has been to develop an automatic control system which would optimally decide for the end customer when to buy, sell, or store electric energy with the objective to minimize his total costs. At the same time, it fulfills all constraints in terms of the limits on power allowed to be taken from the grid. For the user, besides the optimal scheduling of household appliances and selling energy to the grid whenever it becomes profitable, the developed prototype gives also the opportunity to monitor the local generation, storage, and consumption in real time. On the other side, the electric utility can produce a real-time pricing signal which reflects daily peaks in consumption. Making such a price signal available for this building energy management system will lead to a decrease of power consumption from the grid. In this project, a model predictive control approach to the energy optimization problem in a building has been proposed. Different scenarios have been run and the results are discussed here.

A. Ashouri (✉)
Department of Mechanical & Process Engineering, ETH Zurich,
Sonneggstrasse 3, 8092 Zurich, Switzerland
e-mail: ashouri@alumni.ethz.ch

S. Gaulocher
Institute of Automation, University of Applied Sciences
and Arts Northwestern Switzerland, Basel, Switzerland
e-mail: sebastian.gaulocher@fhnw.ch

P. Korba
Institute of Energy Systems and Fluid Engineering,
Zurich University of Applied Sciences, Zürich, Switzerland
e-mail: petr.korba@zhaw.ch

G.M. Dimirovski (ed.), *Complex Systems*, Studies in Systems,
Decision and Control 55, DOI 10.1007/978-3-319-28860-4_5

5.1 Abbreviations

The following abbreviations are used in the text:

BEMS	Building Energy Management System
CHF	Swiss Francs
CHP	Combined Heat and Power
DG	Distributed Generation
MILP	Mixed-Integer Linear Programming
MLD	Mixed-Logical Dynamical
MPC	Model Predictive Control
NMPC	Nonlinear MPC
RTP	Real-Time Pricing

5.2 Introduction

Regulated electric systems were built and operated around the philosophy "supply follows demand" where the customer—who is in fact only a passive consumer—has the right to demand any amount of energy and pays a (often very infrequently updated, e.g., 1–2 times per year) constant, pre-specified price per consumed unit of energy [kWh]. This approach may be criticized for many reasons such as follows: *(technical reasons)* (1) Need for fast load tracking and large spinning reserves leads to inefficient use of fuel or (2) large ratio between peak and average load implies extra system capacity required for peak demands. There are also other *(economical reasons)* such as (3) fixed nature of prices is not supporting any forms of energy conservation or load leveling/peak shaving or (4) isolation of customers from the problems of the supply system and utilities from the effects of competition.

5.2.1 Smart Grid Philosophy

All energy costs have risen in the past and they are expected to continue to rise, and the markets for electric systems have been deregulated. With upgrades and improvements of information and communication technologies applied in electric power systems resulting in so-called smart grids, one can expect that demand-sensitive pricing of electricity will soon become a standard pricing mechanism. It is well known from other industries/services (like airlines, cinemas, car rentals, etc.) that demand-sensitive pricing is a proved way for the philosophy "demand follows supply" [1]. The following can then be expected with regard to the existing power system infrastructure: (1) It will be used more efficiently, in general and (2) large capital costs to meet short peak demands will be reduced.

What is meant here is that power utilities have to invest a lot into new generation, transmission, and distribution of electricity to meet all requirements only for the short periods of peaks of the electricity demand, whereas the idea of smart grids is to go the way of pushing stronger emphasis on energy efficiency and using demand response to reduce and to "flatten" the peaks in electricity consumption. In general, this consumption has predictable, cyclical patterns, with profiles and trends observable in periods of days, weeks, and years. With no feedback (or incentives like real-time pricing, fees for exceeding limits of power consumption, etc.) from the electric grid, utilities, or politics, there can hardly be expected any changes in these patterns. However, all consumers of electricity have certain degree of flexibility in the timing of their electricity usage. With a flat rate, there is no incentive to motivate any changes in the behavior of electricity consumption. With the new smart grid infrastructure enabling two-way communication, reading meters in short intervals, etc., utilities will soon have the means to enforce changes in the consumption patterns, for example, using the real-time price (RTP) which can vary from simple 2 or 4 tariffs to a real function of time reflecting the actual conditions on the electric grid as it is already happening on the high voltage level today. The time-varying electricity price can be seen as a good monetary incentive for all customers to revise their behavior with regard to their consumption of electricity.

The objective of introducing RTP for the utilities is to manage better the peak demand through expected load leveling and achieving a flatter demand curve this way. Using the proposed approach, RTP can also be combined with maximal power consumption limits: This P_{max} value characterizing the maximal power in-feed from the grid can be a given constraint subject to which the proposed optimization is periodically carried out so that this value is not exceeded at any time. As a result of shaving the demand peaks, the existing portfolio of generation sources and transmission &distribution systems could meet the expected demand growth for much more years resulting then in a delay of required additional costs for building new generation, transmission, and distribution.

Yet, another point of view on the same problem is the integration of distributed generation: The electric power from intermittent and weather-dependent sources (non-dispatchable energy) such as wind turbines, photovoltaic panels, etc. cannot be neglected in future anymore when the installed capacity increases above a certain level. The time-varying nature of these energy sources requires their balancing by fast dispatchable generation (such as gas turbine plants) which are expensive, not ecological, and additionally operated inefficiently (they are also optimized for a constant output). The alternative would be again RTP providing the incentive for consumers to modify their electricity demand pattern so that the variability of these non-dispatchable sources is (at least partially) balanced by price variations. The idea behind is that balancing this variability locally would lead to smoother, flatter profile of the overall electricity demand.

5.2.2 Building Energy Management Systems in Smart Grids

Practical feasibility of a smart grid approach as discussed above will depend mainly on the consumers who have to be able to respond to the real-time conditions in the electric power system—without substantial losses of comfort in the preferred lifestyle. Having distributed generation (DG) units, storage, and controllable appliances in the own household, the average consumer will have a wide range of options related to the optimal management such as to decide when to use, store, or sell electricity in response to current and forecasted grid and ambient conditions. This can only be done using a building energy management system (BEMS) that will efficiently (and optimally subject to given constraints) replace the consumers' own decision making process and operate automatically in the background. Considering the same problem from the electric grid point of view, the balancing problem between demand and supply over time, it is questionable whether RTP can become the only successful solution to this concept. Assuming a large number of the considered BEMS have been installed, there is a danger that too many automated responses to a varying price signal carried out at the same time may lead to instability of the entire electric grid or to certain oscillations. Therefore, the considered BEMS will probably need to respond also to utility (demand response) signals within a predefined agreement framework between the utility and the customers. Examples of such an agreement could include incentives like further compensations of the customers for their willingness to shed a certain amount of their actual load (disconnect some running appliances) in order to prevent the demand level to exceed the actually available supply; for example, during periods with lack of sun for photovoltaic panels, and vice versa, the BEMS should also be able to coordinate an increase of demand (e.g., early charging of hot water boilers or electric cars) during periods with unexpected increase of wind or sun radiation. For these purposes, the possibility to store the generated electricity in local storages such as stationary batteries or connected plug-in vehicles would be the best solution to maintain a reliably working electric grid and still to make an efficient usage of the generated electric power and simplify the grid management.

The basic idea goes back to [2]. As a consumer, one can think of four different approaches how to handle the discussed problem [3]:

(1) do nothing (makes only a sense if the electricity price never varies with time),
(2) use a timer (possible only if there are not too many different price tariffs and if the switching happens during a priori known time),
(3) rule-based system, or
(4) optimization-based approach such as model-based predictive control developed and discussed in this work.

This paper is organized as follows: Sect. 5.1 has given a generic introduction to the problems being solved here. Section 5.2 describes the methodology for the mathematical modeling and predictive control employed in this work. The developed models for optimization and the controller are discussed in more detail in Sect. 5.3. Some obtained results are shown in Sect. 5.4. Section 5.5 concludes this work.

5.3 Methodology

In this section, the methodology underlying the presented BEMS will be presented. In a first step, the framework used for modeling the physical components of a BEMS will be described. Afterward, it will be shown how the task of automatically making optimal decisions can be formulated as a mathematical optimization problem, more specifically model predictive control.

5.3.1 Modeling Framework and Paradigms

The future behavior of a dynamical system can only be predicted based on a reliable dynamic model of the relevant aspects of the system. The approach used here relies on the mixed-logical dynamical (MLD) framework which has the advantage of incorporating both real and Boolean variables as well as constraints. Furthermore, MLD systems offer great flexibility (e.g., piecewise linear dynamics) compared to traditional approaches (e.g., state-space systems). The definition of an MLD reads as follows:

$$\begin{aligned} x(t+1) &= A_t x(t) + B_1 u(t) + B_2 \delta(t) + B_3 z(t) + B_5, \\ y(t) &= C_t x(t) + D_1 u(t) + D_2 \delta(t) + D_3 z(t) + D_5, \\ E_2 \delta(t) + E_3 z(t) &\leq E_1 u(t) + E_4 x(t) + E_5 \;. \end{aligned} \tag{5.1}$$

where $x(t+1)$, $u(t)$, and $y(t)$ are the real state, input, and output variables at time step t, respectively. MLD systems are formulated in discrete time. Therefore, t has to be considered a time index rather than absolute time, and the time steps t and $t+1$ are separated by the sampling interval Δt. $\delta(t)$ and $z(t)$ are boolean and real auxiliary variables, respectively. Equation set (5.1) is an extended version of the MLD formulation described in [4]. It is worthwhile to note that an equality constraint can be implemented using two inequality constraints of opposite signs, or also by extending the above-mentioned framework. The discrete-time state-space representation is contained in the MLD framework as a special case. MLDs can be defined using the hybrid system description language (HYSDEL, see [5] for more details), which significantly simplifies the modeling process. The two following modeling paradigms have been paramount to formulating the model of a BEMS:

5.3.1.1 Modularity

The first paradigm is modularity. This means that, in order to provide the maximum ability to examine different possible scenarios for the BEMS, component-based modeling is utilized instead of resorting to a monolithic model. This approach enables the user to include and exclude any component or service of the building at any time with little effort. Furthermore, the complexity of the individual building blocks can be hidden from the end user, and only the high-level structure of the system is of concern and can be understood easily. The assembly of the overall model from basic MLD modules can be achieved using the constraints, cf. [6]. The resulting model again has the form of an MLD.

5.3.1.2 First principles

The second paradigm is to resort to first-principles models, which is a natural choice for the physical components of a BEMS. First principles stands for making use of basic physical principles in order to formulate the system equations. For instance, the energy conservation principle states that the time derivative of the temperature in a room is proportional to the heat balance of this room, i.e., the difference between heat supply (e.g., heating) and heat removal (e.g., air conditioning or loss through walls and windows). Once the physical principles have been identified, the process of casting this statement into equations is straightforward. An alternative approach that has not been considered because it would require experimental data is the use of empirical models, e.g., black-box models.

5.3.2 Model Predictive Control

The core of the presented solution is model predictive control (MPC), a multivariable control technique whose fundamental principle is to predict the process behavior during a finite horizon in the future using a dynamic model [4, 7, 8].

Traditionally, MPC was used in the process industries [9], but rarely in power systems. However, its ability to cope with constraints and to accommodate scheduling problems—in particular in combination with the MLD modeling framework—comes in handy for energy management applications [10, 11]. Figure 5.1 shows a graphical representation of the concept underlying MPC.

Using a prediction (dashed blue line with circles) of the system states and outputs, a sequence of future control moves (solid red line) is computed which complies with specified constraints and optimizes an objective function that may express goals such as reaching and maintaining a desired set-point. The optimal sequence is the solution of an optimization problem, but only its first control move is carried out afterward (gray circle at time t). Repeating this procedure at each sampling time yields closed-loop control. One possible mathematical formulation of the optimization problem at time t reads as follows:

Here, $y_{cost}(\tau)$ is the vector of cost outputs which can be freely defined as long as they can be cast into the MLD framework. For instance, a frequently chosen formulation for a cost output is the difference between a variable and a (possibly time-varying) reference trajectory for this variable. In general, only a subset of the variables defined in the system will appear in the above-mentioned cost function, namely those variables that are in some way related to the optimization objective.

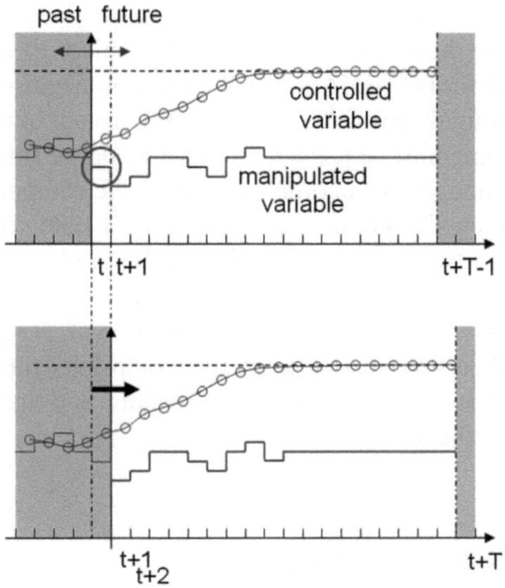

Fig. 5.1 Principle of model predictive control illustrated using two subsequent points in time t (*top*) and $t + 1$ (*bottom*)

$\|.\|_1$ is the 1-norm of a vector. An alternative formulation that has not been pursued here is to use the two-norm $\|.\|_2$ for the cost function. The optimization step in MPC thus consists of minimizing a specific cost function translating the objectives to achieve. There exist various methods to formulate MPC problems, e.g., linear and nonlinear MPC (NMPC, see e.g., [12]). If—as above—an MLD model is used in combination with a linear cost function and linear constraints, a mixed-integer linear programming (MILP) problem is the outcome.

$$\min_{\{u(t),\ldots,u(t+T-1)\}} J \tag{5.2}$$

where

$$J = \sum_{\tau=t}^{t+T-1} \|y_{cost}(\tau)\|_1$$

subject to

$$\begin{aligned}
&E_2\delta(\tau) + E_3 z(\tau) \leq E_1 u(\tau) + E_4 x(\tau) + E_5 \\
&x(\tau+1) = Ax(\tau) + B_1 u(\tau) + B_2\delta(\tau) \\
&\qquad\qquad +B_3 z(\tau) + B_5 \\
&y(\tau) = Cx(\tau) + D_1 u(\tau) + D_2\delta(\tau) \\
&\qquad\qquad +D_3 z(\tau) + D_5 \\
&y_{cost}(\tau) = C_{cost} x(\tau) + D_{cost,1} u(\tau) + D_{cost,2}\delta(\tau) \\
&\qquad\qquad +D_{cost,3} z(\tau) + D_{cost,5} \\
&\forall \tau = t, \ldots, t+T-1 \\
&x(t) = x_{initial}\,.
\end{aligned} \tag{5.3}$$

In order to determine the optimal sequence of inputs $\{u^*(t), \ldots, u^*(t+T-1)\}$, the MILP described in (5.2) subject to constraints in (5.3) must be solved (as described in [4]). Ideally, the optimization horizon T is long enough in order to roughly cover the longest time constant of the system to be controlled. Furthermore, care must be taken to obtain a model of sufficiently low complexity such as to allow enough time in order to perform the computations for solving the optimization problem. The knowledge of the current state vector $x_{initial}$ is of great importance. In this contribution, it will be assumed that all states can be measured directly. In case, however, this is not possible for all states, state estimation techniques such as the Kalman filter or moving horizon estimation cf. [13] can be employed. The three main ingredients for model predictive control are thus a dynamic model, the combination of objective function and constraints, and the knowledge of the current dynamic state of the process.

5.4 Modeling

A list of components which have been modeled is provided in Table 5.1. A controllable load means a consumer device that its action is uncritical and could be interrupted by the controller. As an example, the heating system is considered to be controllable, since the room temperature is allowed to fluctuate between two predefined comfort levels. This implies that the heating system could be possibly delayed to avoid a peak in electricity tariff. When it is not desirable that the controller manipulates the operation of a device, it is categorized as any uncontrollable load. Devices such as PC and notebook, TV, and lamps, and kitchen appliances like oven and coffee maker are in this group. Using the same terminology, generator unit can be controllable (e.g., micro CHP and gas boiler) and uncontrollable (e.g., photovoltaic panel

Table 5.1 Building components with the most important monitored property

Building components		
Device	Type	Parameter
Heating system	Controllable load	Input power
AC	Controllable load	Input power
Washing machine	Controllable load	Input power
Dish washer	Controllable load	Input power
Fridge	Controllable load	Input power
Lamps	Uncontrollable load	Input power
Battery	Storage	Capacity
Micro CHP	Generator	Output power
Photovoltaic (PV) panel	Generator	Output power
Any	Uncontrollable load	Input power

and wind turbine). In the second case, the stochastic energy output of the device is monitored and either used instantly or stored for future use. Furthermore, controllable loads can fall into two general groups of set-point load (e.g., Heating system and fridge/freezer) and schedulable loads (e.g., washing machine and dish washer). The term "schedulable" means that the controller is free to shift the running time of the device as long as all user-defined deadlines are satisfied.

In order to keep the simulation structure close to a real power grid, it is stated that a *power equilibrium* is required to be hold at every sampling time. This condition follows directly from the Kirchhoff's current law and is guaranteed by introducing positive and negative signs for powers consumed and produced, respectively. As shown in Fig. 5.2, storage units can behave as a load when being charged and as a generator when becoming discharged to the grid. Figure 5.3 is an alternative representation of the power equilibrium node given by (5.4).

$$
\begin{aligned}
&(P_G - P_C) + (P_{BD} - P_{BC}) + (P_B - P_S) = 0 \\
&P_{BD} + P_G \leqslant P_S \,.
\end{aligned}
\tag{5.4}
$$

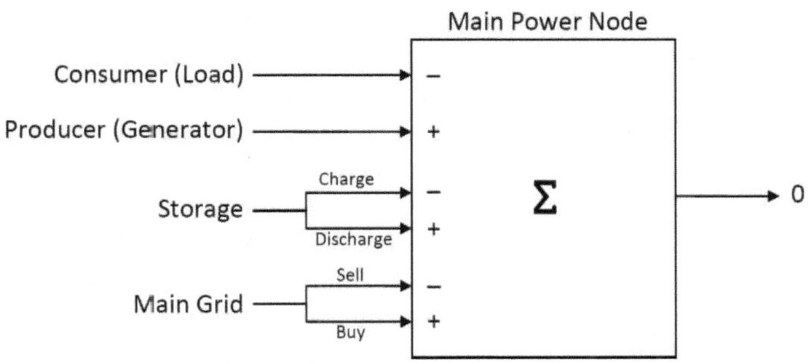

Fig. 5.2 A symbolic demonstration of power equilibrium as proposed in the optimization implemented in MATLAB

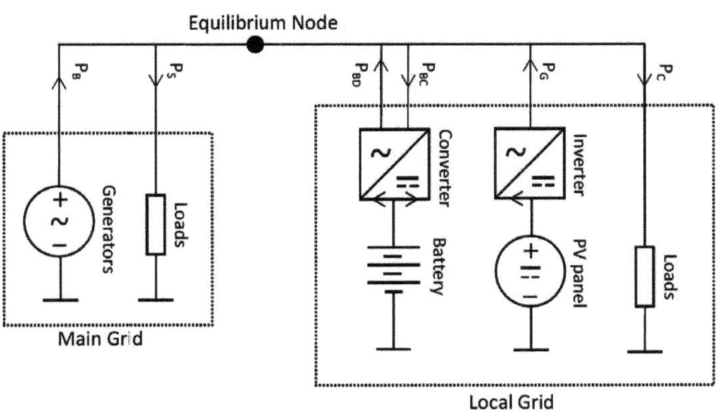

Fig. 5.3 Power equilibrium shown as an electrical circuit

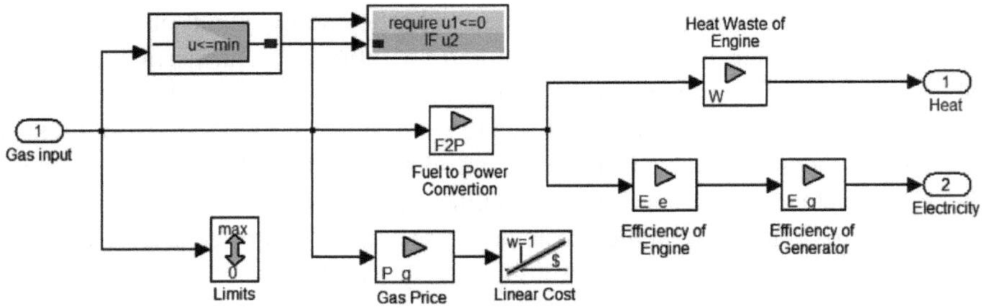

Fig. 5.4 Model of a combined heat and power (CHP) unit

Here, P stands for power and the variables are defined according to Fig. 5.3. The battery interaction is shown with battery charging (P_{BC}) and discharging (P_{BD}) powers, while the power generation and consumption are denoted using P_G and P_C powers. The non-equality corresponds to the fact that one cannot sell (P_S) at any node more energy than one is generating (P_G) and discharging from the battery. An advantage of this approach is the ease in handling the power flow to/from the main grid. On one hand, if the required power demand at each time step is not met by the amount provided from electricity generators and the instant electricity price is less than the one used previously to charge the battery, the difference is compensated by drawing power from the grid. On the other hand, if there is an excess production of electricity and selling tariff is high (compared to that of buying), energy is sold back to the grid instead of being stored in the battery. The individual component models, as well as the overall model, are set up using a graphical environment. More specifically, an MLD toolbox [14] developed by ABB Corporate Research and based on MATLAB® and Simulink® significantly speeds up the modeling process.

Figure 5.4 illustrates one exemplary building block used in the overall model: a combined heat and power (CHP) unit with a graphical representation.

The MLD matrices for a CHP unit are given in (5.5). The omitted matrices from (5.1) are zero matrices of appropriate dimensions.

$$D_1 = \begin{bmatrix} (1-\eta_{mech}) \cdot H_c \\ \eta_{el} \cdot \eta_{mech} \cdot H_c \end{bmatrix} = \begin{bmatrix} 1200 \\ 760 \end{bmatrix},$$

$$D_{cost,1} = P_{fuel} = 1.5,$$

$$E_1 = \begin{bmatrix} 1 \\ -1 \\ -1 \\ 1 \\ -1 \end{bmatrix},$$

$$E_2 = \begin{bmatrix} 0 \\ 0 \\ u_{thr} - u_{max} - \varepsilon \\ u_{thr} - u_{min} \\ u_{max} - u_{thr} \end{bmatrix} = \begin{bmatrix} 0 \\ 0 \\ -0.8 - 10^{-6} \\ 0.2 \\ 0.8 \end{bmatrix}, \tag{5.5}$$

$$E_5 = \begin{bmatrix} -u_{min} \\ u_{max} \\ u_{min} - \varepsilon \\ -u_{min} \\ u_{max} \end{bmatrix} = \begin{bmatrix} 0 \\ 1 \\ -10^{-6} \\ 0 \\ 1 \end{bmatrix}.$$

The only input to the component model is $u(t)$, the relative engine load of the CHP unit. The two outputs $y_1(t)$ and $y_2(t)$ are the generated heating and electrical power, respectively. The block entitled "Limits" is a constraint that keeps this input between a minimum value u_{min} and a maximum value u_{max}. In addition, the two gray blocks near the top left corner prohibit that the engine be operated below a threshold value for the relative engine load u_{thr}. These two blocks require an auxiliary boolean variable $\delta(t)$ indicating whether the relative engine load is below u_{min} or not. The model also contains a cost output entitled "Linear Cost" that indicates the price which has to be paid for the fuel consumed. The outputs of the model are heating and electrical power generated by the CHP unit, respectively. The following numerical values (without units) have been used:

(1) $\eta_{mech} = 0.4$ (efficiency of combustion engine), $\eta_{el} = 0.95$ (efficiency of generator), $H_c = 2000$ (specific heat of combustion of the fuel), $P_g = 1.5$ (specific price of fuel), $u_{max} = 1$ (maximum relative engine load), $u_{min} = 0$ (minimum relative engine load), $u_{thr} = 0.2$ (threshold for relative engine load), and $\varepsilon = 10^{-6}$ (very small constant for numerical reasons).

(2) The different component models are interconnected in line with the actual structure of the building. One important feature is the summation of all power signals using appropriate signs (positive for energy production and negative for energy consumption). As a result, the conservation of energy holds at all times. The sampling time Δt has been chosen to be 15 min. This period is sufficiently short so that relevant processes, e.g., operation of a washing machine or temperature

dynamics of a building, can be subdivided into multiple time steps, and sufficiently long so that the optimization problem remains tractable. The optimization horizon has been chosen to be initially 5 h, corresponding to $T = \frac{5\,\mathrm{h}}{15\,\mathrm{min}} = 20$ steps, and modified later as a function of the available CPU power up to 12 h.

5.5 Results

The above-mentioned control strategy is used for automatic energy management in the forms of load shifting, set-point tracking, and peak power reduction. The overall objective described in the cost function is the minimization of the electrical energy bill. The control strategy is intended to be implemented on off-the-shelf hardware and to run in a real-time environment. It forms part of the *Smart Grid Demo Lab*, a demonstrator developed within ABB Corporate Research in Switzerland. It incorporates various sources of electrical energy, various consumer devices, and various ways of influencing the allocation of electrical energy.

In addition to the knowledge of the price of electrical energy in function of daytime, the controller is provided with information on the weather forecast (used for predicting the generation of solar and wind power) and on the predicted consumption of electrical energy over the day.

5.5.1 Load Shifting

MPC uses the prediction on future electricity prices to find the optimal time for controllable devices to be running [15]. The illustration in Fig. 5.5 should be considered together with the corresponding daily electricity tariffs, which in this case is the one shown in Fig. 5.6. Knowing that there would be a jump in the buying price in the morning, the controller commands the battery to get fully charged before the jump happens. Another decision taken here is to shift the schedulable load to the late evening, when the tariff is low again. An important problem boundary is the maximum

permitted power drawn from the grid. In this case, there is no limit applied. The effect of having such limit is investigated in the next section.

Regardless of the load shifting actions, set-point devices (e.g., the fridge or heating system) will follow the desired user-defined (set-point) values in an optimal way decided by the MPC algorithm, as illustrated in Fig. 5.7

It is also interesting to investigate the possible difference results while a more dynamic electricity tariff is applied. Figure 5.8 depicts a 4-tariff buying price. In order to make the comparison meaningful, the new tariff has the same mean value of 0.15 CHF/kWh as the previous one. The resulting optimal behavior is then shown in Fig. 5.9. The most noticeable difference compared to the previous scenario is the change in the running time of schedulable loads which are now delayed until the final low tariff (valid after 22:00).

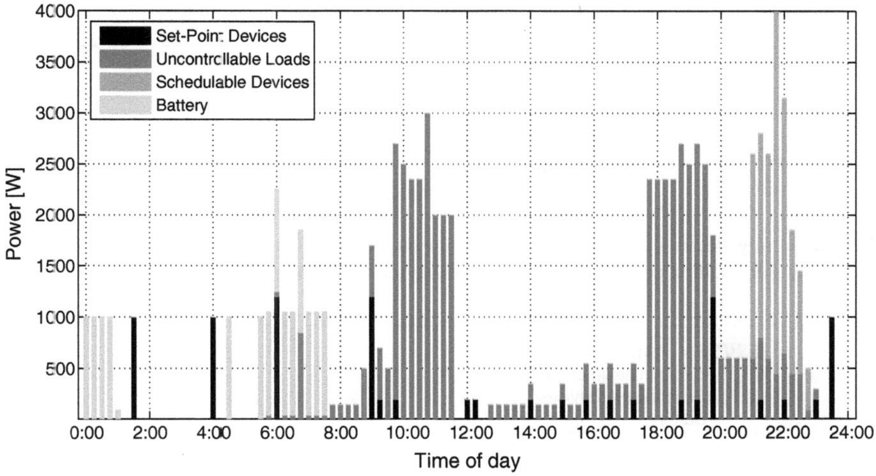

Fig. 5.5 Load shifting with the electricity tariffs of Fig. 5.6. The operating devices include battery (ivory), schedulable loads (*light gray*), uncontrollable loads (*dark gray*), and set-point loads (*black*). No power limit is applied

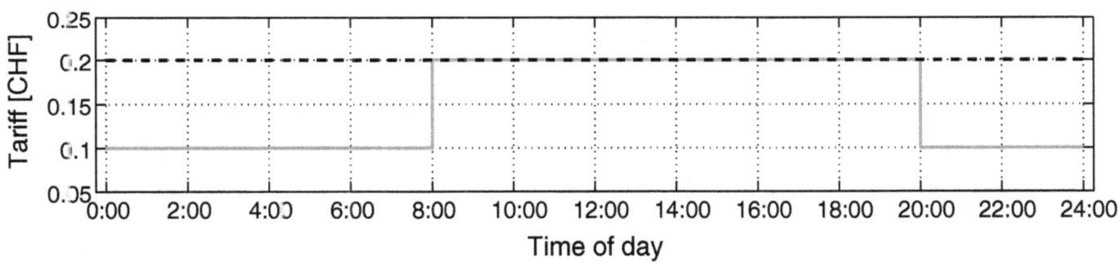

Fig. 5.6 Daily electricity tariffs: 2-tariff price for buying from (*solid gray*) and selling to (*dashed black*) the grid. Selling price is higher than or equal to the buying price

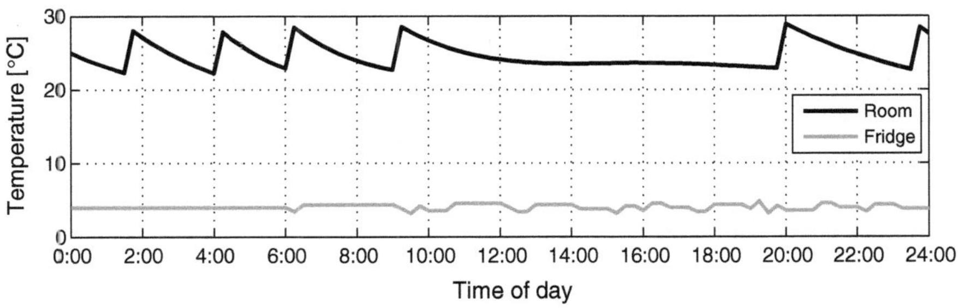

Fig. 5.7 Continuous states and set-point tracking: Room temperature (*black*/set-point 25 °C ± 2) and internal fridge temperature (*gray*/set-point 4 °C ± 1) during a day

It should be mentioned that the reason the battery is not charged during the first low tariff (valid from 0:00 to 4:00) is that the controller is still not aware of the price

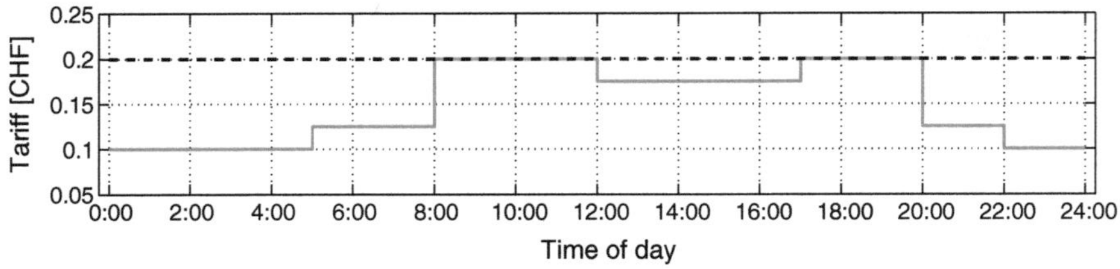

Fig. 5.8 Daily electricity tariffs: 4-tariff price for buying from (*solid gray*) and selling to (*dashed black*) the grid. Selling price is higher than or equal to the buying price

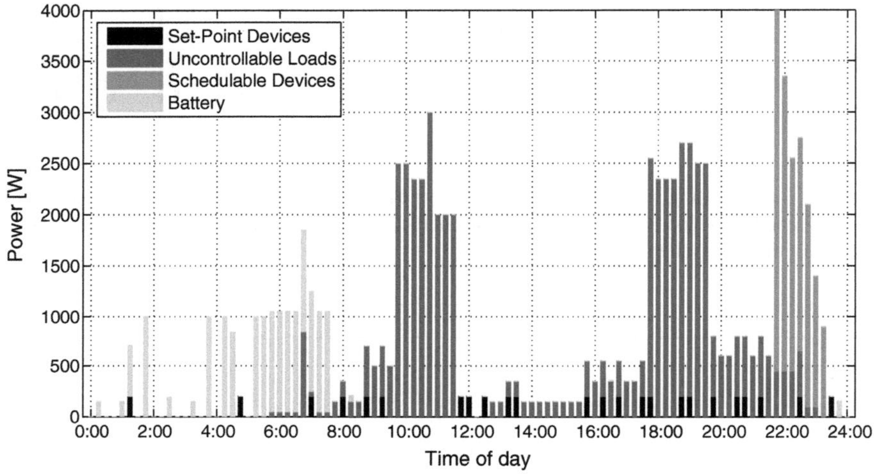

Fig. 5.9 Load shifting with the electricity tariffs of Fig. 5.8. The operating devices include battery (ivory), schedulable loads (*light gray*), uncontrollable loads (*dark gray*), and set-point loads (*black*). No power limit is applied

jump of 8 o'clock. This is a result of limited control horizon of the controller (seeing only 5 h ahead). Therefore, increasing the horizon would result in a new optimal solution when this situation is avoided.

5.5.2 *Power Limiting*

Another interesting capability of MPC is observed when limits are applied to the maximum permitted power draw from the grid. This is a realistic scenario and is already implemented in some European countries. Figure 5.10 demonstrates how MPC handles this situation.

When compared to Fig. 5.5, the most noticeable difference is the behavior of schedulable loads. This time, the controller expands the run time for these devices, avoiding the high power peak which was created in the previous case.

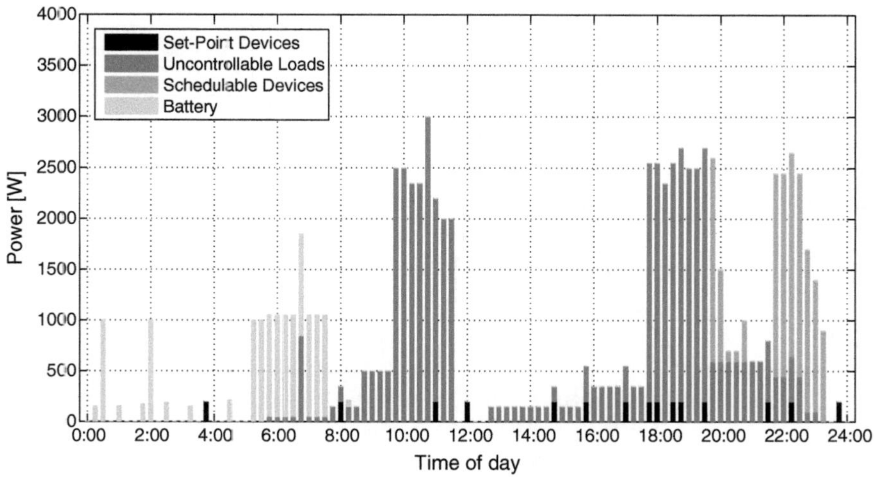

Fig. 5.10 Load shifting with the electricity tariffs of Fig. 5.6. The operating devices include battery (ivory), schedulable loads (*light gray*), uncontrollable loads (*dark gray*), and set-point loads (*black*). A power limit of 3 kilowatts is applied

5.5.3 *Effect of Tariffs*

The discussions above suggested how an advance knowledge of electricity buying tariffs may influence the optimal behavior of building services. However, this is also true for the selling tariffs. The main difference is that here, the main affected devices are not loads, but the battery. In order to investigate these effects, Fig. 5.11 depicts the daily charging status of the battery, together with the instant rate of electricity production and consumption.

In this first scenario, the electricity tariffs of Fig. 5.6 are used. Since the selling tariff is higher than or equal to the buying price at all times, the production from the PV panels is directly sold out to the grid. Therefore, the battery is charged from the grid during the low tariff hours to provide energy to the loads later. Figure 5.6 shows this situation.

In the second scenario, a different electricity tariff of Fig. 5.12 is used instead. The modified behavior of the battery is shown in Fig. 5.13. The decreased selling prices made the produced electricity less valuable. Therefore, the loads are fed with the output of PV panels when possible. Also, the battery is only charged with the locally generated energy. After the sunset (around 20:00 in this situation), the grid power is used to supply the loads' demand if there is no energy left at the battery. It should be mentioned that in all scenarios, battery charge level must stay above a minimum level of 1 kWh to reduce the aging factor.

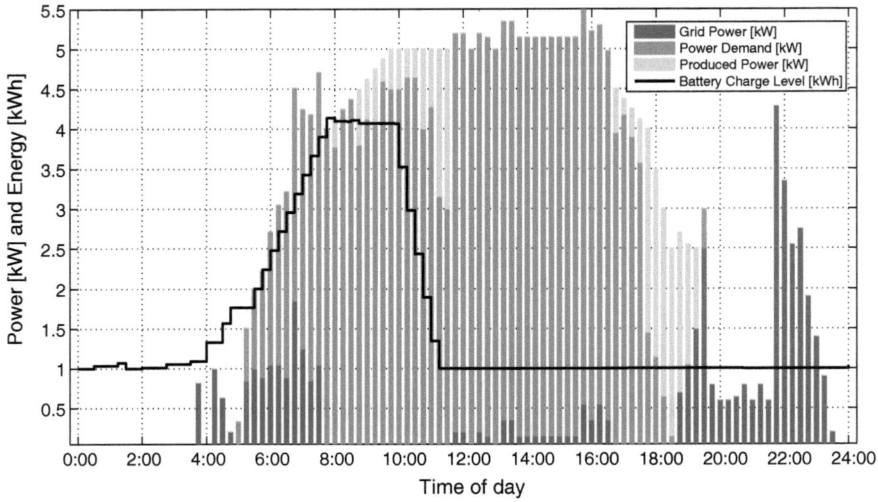

Fig. 5.11 Battery status with the electricity tariffs of Fig. 5.6. The unit for the PV panels produced power (ivory), loads' demand power (*light gray*), and the power draw from the grid (*dark gray*) is kW. The unit for the battery charge level (*black*) is kWh

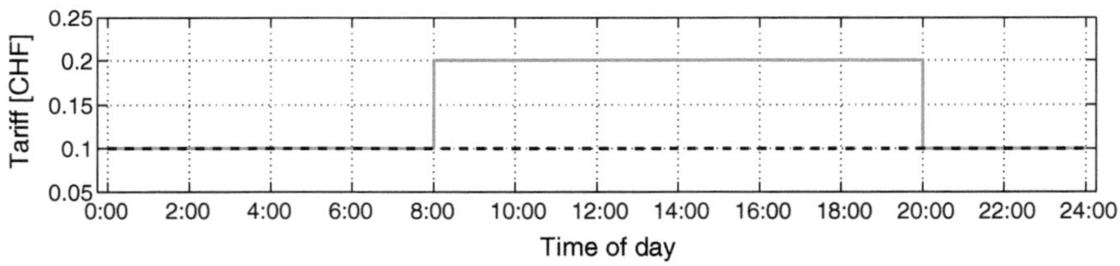

Fig. 5.12 Daily electricity tariffs: Price for buying from (*solid gray*) and selling to (*dashed black*) the grid. Selling price is lower than or equal to the buying price

5.5.4 Some Practical Considerations

The mentioned capabilities discussed in the last three sections (a) load shifting, (b) power limiting, and (c) adapting to different tariffs are the most important advantages of using a model-based optimal control.

In a large number of obtained results the following could be observed in terms of the achieved cost savings for the case of 2- or 4-tariff prices (thus reflecting today most common situation on the electricity markets): About 15 % lower energy costs in the optimized cases can be achieved. In the same setup, considering the real-time price signal (when the energy price varies, e.g., every 15 min), the cost savings can be even higher.

The optimization approach based on the model-based predictive controller implemented in MATLAB considering a half-a-day ahead prediction horizon and all mod-

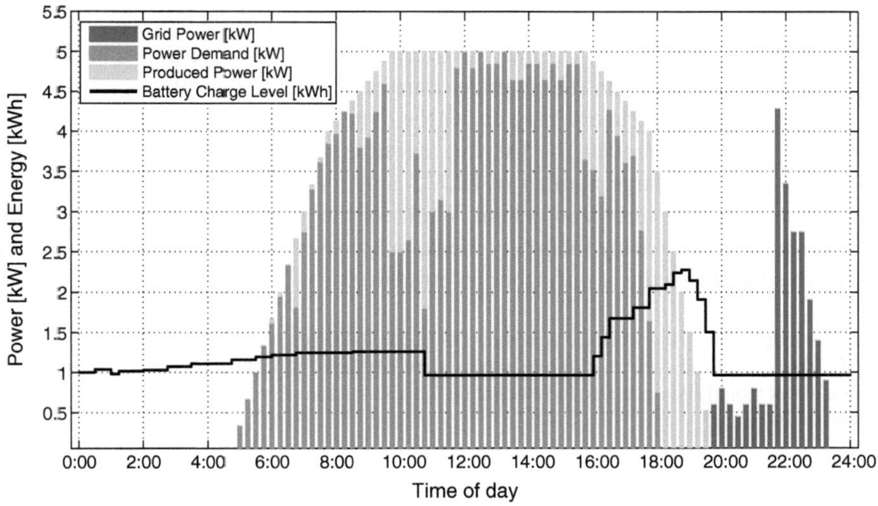

Fig. 5.13 Battery status with the electricity tariffs of Fig. 5.12. The unit for the PV panels produced power (Ivory), loads' demand power (*Light Gray*), and the power draw from the grid (*Dark Gray*) is kW. The unit for the battery charge level (*Black*) is kWh

eled appliances given in Table 5.1 takes up to 5 min on a standard 2 GHz Laptop PC with a dual core processor. Thus, for the considered optimization step of 15 min, it is feasible to implement this controller in real time.

5.6 Conclusions

It has been shown that the described problem (an optimization-based demand response algorithm for energy management of buildings) can technically be solved using a model-based predictive control approach and existing hardware. In case of 2 or 4 tariffs (most common on today electricity markets), savings of about 15 % (compared to a case without any optimization) can typically be achieved through the optimized demand response proposed in this work, i.e., without any loss of comfort for the end user. In case of real-time pricing (when the price for electric energy can vary several times per hour), the cost savings will further increase. Independently on the form of time-varying electricity prices, the proposed optimization also takes into consideration important constraints such as the limits on the maximum power in-feed from the main grid.

From the consumer point of view, the price benefit ratio plays an important role. Thus, the value of money this system saves should cover its cost within a reasonable amount of time. Today, electricity prices may be too low to justify economically an investment into the required information and communication infrastructure without any other incentives. However, this can change in future.

From the transmission and distribution grid management point of view, there can be few challenges to think of in case of a high penetration of the proposed BEMS installed in a power system: (a) Many BEMS (all default wise delivered and programmed the same way) will connect and disconnect at the same time a large number of appliances when the energy price changes in order to optimize our electricity bill. This sudden load change may excite the dynamics of the entire power system, including its critical modes. (b) Varying real-time price signal followed closely by the reactions of the large number of BEMS may introduce oscillations in demand across the whole synchronized system. While for the former concern a simple solution might be just to delay randomly the response of every single BEMS, the latter problem should be subject of further investigations.

Acknowledgments This work describes results of the R&D project "Smart Grid" carried out at ABB Switzerland Ltd. where all three authors were employed at that time in different positions. The authors would like to thank Jan Poland of ABB Corporate Research for all the constructive discussions.

References

1. Livengood, D., Larson, R.: The energy box—locally automated optimal control of residential electricity usage. Serv. Sci. **1**, 1–16 (2009)
2. Schweppe, F.C., Tabors, R.D., Kirtley, J.L., Outhred, H.R., Pickel, F.H., Cox, A.J.: Homeostatic utility control. IEEE Trans. Power Apparatus Syst. PAS-99,1151-1163 (1980)
3. Korba, P., Oudalov, A.: Personal Smart Grid, Technical report C6-066-07/00, ABB Switzerland Ltd., Corporate Research Ltd. (2009)
4. Bemporad, A., Morari, M.: Control of systems integrating logic, dynamics, and constraints. Automatica **35**, 407–427 (1999)
5. Torrisi, F.D., Bemporad, A.: HYSDEL—a tool for generating computational hybrid models for analysis and synthesis problems. IEEE Trans. Control Syst. Technol. **12**, 235–249 (2004)
6. Gallestey, E.A., Castagnoli, D., Stothert, A.: Method of generating optimal control problems for industrial processes. European Patent, EP1607809B1 (2008)
7. Garcia, C.E., Prett, D.M., Morari, M.: Model predictive control: theory and practice—a survey. Automatica **25**, 335–348 (1989)
8. Maciejowski, J.: Predictive Control with Constraints. Pearson Education Limited, Prentice-Hall, Essex (2002)
9. Stadler, K.S., Wolf, B., Gallestey, E.: Precalciner Control in the Cement Production Process Using MPC. IFAC MMM, Québec (2007)
10. Fazlollahi, S., Becker, G., Ashouri, A., Marechal, F.: Multi-objective, multi-period optimization of district energy systems: IV-A case study. Energy **84**, 365–381 (2015)
11. Ashouri, A., Petrini, F., Bornatico, R., Benz, M.J.: Sensitivity analysis for robust design of building energy systems. Energy **76**, 264–275 (2014)
12. Findeisen, R., Allgöwer, F.: An introduction to nonlinear model predictive control. Summerschool on The Impact of Optimization in Control, pp. 3.1-3.45 (2001)
13. Rawlings, J.B.: Model predictive control and moving horizon estimation: assessment of the current status. In: Danish Automation Society Model Predictive Control Conference, Skaerbaek, Fredericia, Denmark (2007)
14. Poland, J., Bauer, M.: Building, Estimating and Optimizing Models using MAMPF, NOMOT, and Expert Optimizer. Technical report 9ADB001834 – 012, ABB Switzerland Ltd (2009)
15. Manjunatha, A.P., Korba, P., Stauch, V.: Integration of large battery storage system into distribution grid with renewable generation, powertech 2013. Grenoble France, pp.16-20, June 2013

Chapter 6
Passivity-Based Switching Rule and Control Law Co-design of Networked Switched Systems with Feedback Delays

Dan Ma and Georgi M. Dimirovski

Abstract In this chapter, a class of switched linear systems under a hybrid state feedback controller with time-varying delays is studied. The main contribution is given on the issue of how to co-design switching rule and feedback control law so as to make the closed-loop system strictly input feed-forward output-feedback passive for all admissible time delays in the feedback channels. We derive the sufficient conditions for strict input feed-forward, output-feedback passification of switched systems with time-varying delays under some state-dependent switching rule by using the method of multiple storage functions. The proposed switching rule can achieve strict input feed-forward, output-feedback passivity of the switched delay systems, whose subsystems are all allowed to be input feed-forward output-feedback nonpassive. The finite–gain L_2 stability in closed-loop is achieved in this way. Furthermore, under the proposed switching rule the asymptotic stability can be guaranteed if the switched system is zero-state detectable when exogenous disturbance input is zero. Finally, we present an algorithm on how to co-design switching rule and feedback control law to maintain strict input feed-forward output-feedback passivity of switched systems with time-varying delays.

D. Ma (✉)
State Key Laboratory of Synthetical Automation for Process Industries, College of Information Science and Engineering, Northeastern University, Shenyang 110819, People's Republic of China
e-mail: madan@mail.neu.edu.cn

G.M. Dimirovski
School FEIT, St Cyril and St Methodius University, Karpos 2, 1000 Skopje, Republic of Macedonia
e-mail: gdimirovski@dogus.edu.tr

G.M. Dimirovski
Computer and Control Department, Dogus University, Acibadem, 34722 Istanbul, Republic of Turkey

G.M. Dimirovski (ed.), *Complex Systems*, Studies in Systems, Decision and Control 55, DOI 10.1007/978-3-319-28860-4_6

6.1 Introduction

It is well known nowadays that a switched system consists of a family of continuous- or discrete-time subsystems and a rule that orchestrates the switching between them. It belongs to a special subcategory of the category of hybrid systems [3, 7, 12]. Most research on switched systems has been focused on finding stability/stablizability conditions under an arbitrary switching rule or on finding the constraints that should be imposed to the switching signal to guarantee the stability of the system; see [6, 9, 11–13, 15, 17, 25, 27, 28, 32–37]. The common Lyapunov function (CLF) technique for all subsystems guarantees stability under arbitrary switching [12, 27]. The average dwell time (ADT) technique [11, 33] is an effective tool for achieving stability under a class of slow switching signals. The ADT technique is able to deal with the problem on stability property of switched systems consisting of both stable and unstable subsystems [33], and it can also be extended to the stability analysis of systems with controller failure [28]. Most of the above methods can only be used to achieve the stability/stabilization of switched systems that contain stable subsystems (at least one when using the ADT method). The technique of multiple Lyapunov functions (MLF), first proposed by Peletis and DeCarlo in [25] then elaborated further by Branicky [3], and extended and promoted by Zhao and Hill in [36], is a powerful tool for finding a switching rule to guarantee the stability of switched systems, even when none of the subsystems is stable. However, in practice, due to the switched and hybrid nature of many physical processes and the growing use of communication networks in control of physical plants, inevitably switched systems are subjected to time delays in feedbacks channels. Therefore, the existing effective methods for switched systems without time delays must be reevaluated before they become applicable to switched systems with hybrid feedback delays. Furthermore, if subsystems cannot be stabilized by its candidate delayed hybrid feedback control law, how to co-design switching rule and feedback control law to maintain the performance of switched systems for all time-varying delays in feedback control has posed a theoretical control problem.

Although there exist in the literature some results on the stability of switched systems with time delays, the control synthesis of hybrid feedback delays has not been deeply investigated. In [9], a class of hybrid multi-rate control models with a constant time delay and switching controllers is formulated and robust passivity analysis for the discrete system under an arbitrary switching signal is investigated. Since most switched systems in practice do not possess a common Lyapunov function and, consequently, an arbitrary switching law, the stability conditions for switched system with delays under an arbitrary switching signal is obviously limited and conservative. In [34] is investigated the asymptotic stability and stabilization of switched systems with state delays. A piecewise quadratic Lyapunov function is constructed and stability and stabilization conditions are derived. However, the hybrid time-varying delay feedback in switched systems is different from the state delay in switched systems [34]. On one hand, it occurs within the feedback channels, which might be caused by transmission-induced delay of measurements [19, 20, 21]. On the other hand, it is much more complicated with respect to the state delays [8, 23], especially in design of hybrid control law. In [6], a class of convex combination of switched time-delay systems is presented and a delay-dependent stability criterion under a state-driven switching law is derived.

The dissipativity and passivity theory [30, 31] has provided a framework for both design and analysis of control systems using an input–output description based on energy-related considerations [4, 10, 18]. Passivity provides considerable properties that can be used to design stable systems. For instance, passive systems are Lyapunov stable. Additionally, if other stability results are desired, such as L_2 stability and asymptotic stability, it is possible to use more restrictive conditions to get such results. The passivity theory is intimately related to the circuit analysis method and plays an important role in both electrical–electronic networks and control systems. This is mainly due to the fact that passivity and stability are closely related, thus supplying a new methodological approach to solve the stabilization problem [10, 26]. Moreover, passivity-based control, as an energy-based analysis approach, also has important robustness properties [18].

The study of passivity of switched systems has drawn considerable attention in recent years [1, 5,13,14,16, 29, 35, 37]. A common storage function for all subsystems is often adopted to characterize passivity under an arbitrary switching rule [1, 5]. However, this kind of passivity property is much too restrictive for a switched system. In [35], a notion of passivity for general nonlinear continuous-time switched systems with multiple positive definite storage functions has been successfully used for the stability analysis and design of a state-dependent switching law that renders the closed-loop system passive. In [37], a framework of dissipativity theory for switched systems is given, and the L_2 gain and passivity properties are also addressed. In [13] is focused the passivity and feedback passification of switched discrete-time linear systems. In [16] is investigated the making of strict dissipative system for continuous switched systems with state delays under the average dwell time switching rule. The proposed conditions to obtain strict dissipativity require that each subsystem be made strictly dissipative. Most recent passivity study for a class of discrete-time switched nonlinear systems has appeared in [29].

To the authors' best knowledge, however, the issue of input feed-forward, output-feedback (IF-OF) passification of switched systems with time-varying delays in the area of hybrid feedback control has not been investigated. If each subsystem is not dissipative, how to co-design the switching rule and the feedback control law which takes time-varying delays into account to obtain the strict IF–OF passivity and passification of switched systems is an open problem. Consequently, so are the asymptotic stability and the finite–gain L_2 stability for such systems, which are of great importance, of course. This motives our present work.

This chapter is organized as follows: a switched system with hybrid feedback delays is formulated in Sect. 6.2. Multiple storage functionals are given to analyze the strict input feed-forward, output-feedback passivity of switched time-varying delay systems. Delay bound-dependent criteria on strict input feed-forward, output-feedback passification of the system are given in Sect. 6.3. Conditions on finite–gain L_2 stability and asymptotic stability are also presented there. Section 6.4 gives the switching rule and the feedback controller co-design algorithm. An example is given to illustrate our main results in Sect. 6.5. Finally, Sect. 6.6 contains the conclusions.

6.2 Preliminaries

A networked switched control system can be depicted in Fig. 6.1. A switched plant, via actuators and sensors, interacts with hybrid controllers through network transmission channels. Thus, the closed-loop system can be modeled as a continuous-time linear switched system

$$\begin{aligned} \dot{x}(t) &= A_{\sigma(t)}x(t) + B_{\sigma(t)}u_{\sigma(t)}(t) + \Gamma_{\sigma(t)}, \omega(t) \\ z(t) &= C_{\sigma(t)}x(t) + D_{\sigma(t)}\omega(t), \end{aligned} \tag{6.1}$$

with hybrid state feedback controllers

$$u_{\sigma(t)}(t^+) = K_{\sigma(t)}\hat{x}(t_k),\ t \in [t_k, t_{k+1}), \tag{6.2}$$

where $x(t) \in R^n$ denotes the state, $\sigma(t): R_+ = [0, \infty) \to \Upsilon = \{1, 2, \ldots . m\}$ is the switching signal which is a piecewise constant function depending on time t and/or state $x(t)$, while $z(t)$ and $\omega(t) \in L_2[0, \infty)$ are the output and exogenous disturbance input vectors.

When $\sigma(t) = i$, $u_i(t) \in R^q$, $i \in \Upsilon$ is the control input of the ith subsystem, $A_i, B_i, C_i, D_i, \Gamma_i, \forall i \in \Upsilon$, are constant matrices; $K_i, \forall i \in \Upsilon$ are the feedback controller gains to be designed. The switching signal $\sigma(t)$ can be characterized by the following switching sequence:

$$\Sigma = \{x_{t_0}; (i_0, t_0), \ldots, (i_k, t_k), \ldots, | i_k \in \Upsilon, k \in N\},$$

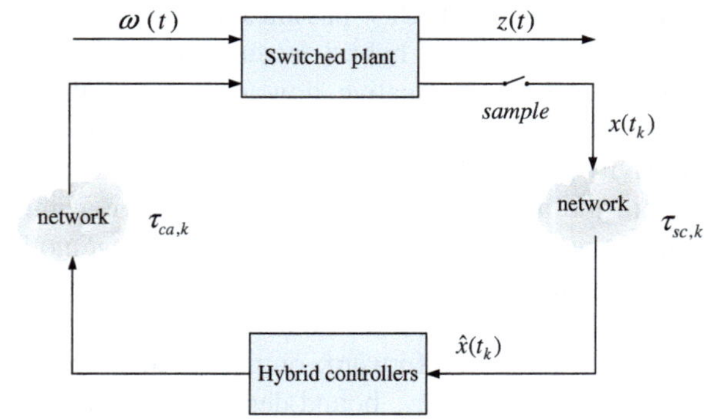

Fig. 6.1 Architecture of networked switched control systems

in which x_{t_0} is the initial state, t_0 is the initial time, and N is the set of nonnegative integers. The i_k th subsystem is activated when $t \in [t_k, t_{k+1})$, $\sigma(t) = i_k$. Time t_k is also the sampling instant. $u_i(t^+)$, $\forall i \in \Upsilon$, are piecewise continuous functions, and only change the value at t_k.h is the sampling period, which is a positive scalar; and $\hat{x}(t_k)$ denotes the state signal received via network transmission at the sampling instant t_k.

As depicted in Fig. 6.1, if network congestion or node failure occurs, depending on the network protocol employed, the network-induced delays and packet dropout happen inevitably. For the purpose of analysis, the whole delay at each sampling period is denoted by $\tau_k = \tau_{sc,k} + \tau_{ca,k}$, where $\tau_{sc,k}$ is the sensor-to-controller delay, and $\tau_{ca,k}$ is the controller-to-actuator delay. The received signals by the hybrid controllers can be conveniently described as follows:

$$\begin{cases} \hat{x}(t_k) = x(t_k), & \text{if no packet dropout at time } t_k; \\ \hat{x}(t_k) = x(t_k - h), & \text{if one packet dropout at time } t_k; \\ & \vdots \\ \hat{x}(t_k) = x(t_k - n(k)h), & \text{if } n(k) \text{ packets dropout at time} t_k. \end{cases}$$

Therefore, network-induced delays and packet dropout can be lumped into a generalized description

$$\hat{x}(t_k) = x(t_k - n(k)h - \tau_k). \tag{6.3}$$

Replacing $t - (t_k - n(k)h - \tau_k)$ by a new variable $\tau(t)$, we have

$$\hat{x}(t) = x(t - \tau(t)), \tag{6.4}$$

where $\tau(t)$ is a time-varying delay satisfying

$$0 < \tau(t) = t - t_k + n(k)h + \tau_k \le \tau_{\max},\ t \in [t_k, t_{k+1}) \tag{6.5}$$

$\tau_{\max} > 0$, is the upper bound of delay.

The closed-loop NSCS can be described as

$$\begin{aligned} \dot{x}(t) &= A_{\sigma(t)}x(t) + B_{\sigma(t)}K_{\sigma(t)}x(t - \tau(t)) + \Gamma_{\sigma(t)}\omega(t),\ t \in [t_k, t_{k+1}), \\ z(t) &= C_{\sigma(t)}x(t) + D_{\sigma(t)}\omega(t), \end{aligned} \tag{6.6}$$

where $x(\theta) = \phi(\theta)$, $\theta \in [-\tau_{\max}, 0)$, $\phi(\theta) \in C_n$ is a differentiable vector-valued initial function on $[-\tau_{\max}, 0]$.

From the above modeling procedure, we know that networked switched systems can be described by switched systems with feedback delays. Therefore, in what follows, we will focus on the analysis and synthesis of such a system based on the passivity. Here we take the general case into account that each feedback delay of switched subsystem is different, which means the hybrid feedback controllers with time-varying delays can be given as

$$u_{\sigma(t)}(t) = K_{\sigma(t)}\, x(t - \tau_{\sigma(t)}(t)), \tag{6.7}$$

$\tau_i(t)$ is the time-varying feedback delay satisfying $0 < \tau_i(t) \le \tau_{i,\max}$, where $\tau_{i,\max} > 0$ represents the upper bound of time-varying delay of each subsystem. Further, $x(\theta) = \phi(\theta)$, $\theta \in [-\tau_{i,\max}, 0]$, where $\phi(\theta) \in C_n$ is a differentiable vector-valued, initial-state function on interval $[-\tau_{i,\max}, 0]$.

Thus, the closed-loop system to be studied is described as

$$\begin{aligned} \dot{x}(t) &= A_{\sigma(t)}x(t) + B_{\sigma(t)}K_{\sigma(t)}x(t - \tau_{\sigma(t)}(t)) + \Gamma_{\sigma(t)}\omega(t), \\ z(t) &= C_{\sigma(t)}x(t) + D_{\sigma(t)}\omega(t). \end{aligned} \tag{6.8}$$

In what follows, we give the definitions on dissipativity, strict dissipativity, and $(\tilde{Q}, \tilde{S}, \tilde{R})$-strict dissipativity for systems (6.8) first. Second, we study the related properties, including passivity, finite–gain L_2 stability and asymptotic stability. Finally, we propose an algorithm on how to co-design switching rule and feedback control law in order to guarantee the passivity of system (6.8).

We assume the state of system (6.8) is continuous at the switching time instant. Also, switching function $\sigma(t)$ is assumed to have a finite number of switchings on any finite interval of time. Notice that these assumptions are standard in the literature on switched systems and control [11, 12, 27].

Definition 6.1 System (6.8) with a specific switching rule $\sigma(t)$ is said to be *dissipative* if there exists a positive-definite continuous differentiable function $V_{\sigma(t)}(x_t)$, called *storage function*, a locally integrable function $s_{\sigma(t)}(\omega, z)$, called *supply rate*, such that for all $\omega(t) \in L_2[0, \infty)$, $x(\theta) = \phi(\theta)$, $\theta \in [-\tau_{\sigma(t),\max}, 0]$

$$\dot{V}_{\sigma(t)}(x_t) \le s_{\sigma(t)}(\omega, z), \tag{6.9}$$

holds, where $x_t \in R^n$ is the solution to the system state equation. Furthermore, if there exist some positive definite function $\Psi_{\sigma(t)}(x_t)$ satisfying

$$\dot{V}_{\sigma(t)}(x_t) + \Psi_{\sigma(t)}(x_t) \le s_{\sigma(t)}(\omega, z), \tag{6.10}$$

then system (6.1) is called *strictly dissipative*.

Remark 6.1 Similar to Definition 3.3 in [37], in order to induce stability for switched systems here, we also need positive definite storage functions. Yet, unlike Definition 3.3 of [37], here we need to focus only on the active subsystem of system (6.8), but not on the inactive one. This means a subsystem is dissipative when it is active, but not all the subsystems are dissipative. This definition includes strict dissipativity.

When $\sigma(t) = i$, the supply rates $s_i(\omega, z)$ can be a general function of the state x, e.g. $s_i(x, \omega, z)$, which implies internal stability and input-to-state stability. For convenience of analysis, we introduce a quadratic form of the supply rates and then the $(\tilde{Q}, \tilde{S}, \tilde{R})$-dissipativity is defined as follows:

Definition 6.2 System (6.8) with a specific switching rule $\sigma(t)$ is said to be $(\tilde{Q}, \tilde{S}, \tilde{R})$-*dissipative* if there exists a positive-definite continuously differentiable function $V_{\sigma(t)}(x_t)$ satisfying (6.3) with respect to the following supply rate:

$$s_{\sigma(t)}(\omega, z) = \begin{pmatrix} z \\ \omega \end{pmatrix}^T \begin{pmatrix} \tilde{Q}_{\sigma(t)} & \tilde{S}_{\sigma(t)} \\ * & \tilde{R}_{\sigma(t)} \end{pmatrix} \begin{pmatrix} z \\ \omega \end{pmatrix} = z^T \tilde{Q}_{\sigma(t)} z + 2z^T \tilde{S}_{\sigma(t)} \omega + \omega^T \tilde{R}_{\sigma(t)} \omega, \tag{6.11}$$

where matrices $\tilde{Q}_{\sigma(t)}$, $\tilde{S}_{\sigma(t)}$, $\tilde{R}_{\sigma(t)}$ have appropriate dimensions, and $\tilde{Q}_{\sigma(t)}$ and $\tilde{R}_{\sigma(t)}$ are symmetric ones. Consequently, if condition (6.10) holds with respect to (6.11), then system (6.8) is said to be $(\tilde{Q}, \tilde{S}, \tilde{R})$-*strictly dissipative*.

It should be noted, Definition 6.2 is a natural extension of (Q, S, R)-dissipativity notion for non-switched systems proposed in [2]. If there exists a common storage function for switched systems such that (6.9) holds, then the system is (Q, S, R)-dissipative. Reference [24] introduced (Q, S, R)-dissipativity for a discrete-time switched system, in which each inactive subsystem should be considered.

It is well known that passivity is one of the most useful forms of dissipativity, and it is directly related to the stability of systems. Assuming ω and $\mathbf{z}$ have the same dimensions, we introduce the following passivity definition.

Definition 6.3 System (6.8) with a specific switching rule $\sigma(t)$ is said to be *strictly input feed-forward output-feedback passive* (IF–OFP) if it is strictly dissipative with respect to

$$s_{\sigma(t)}(\omega, z) = \omega^T z - \gamma \omega^T \omega - \beta z^T z, \tag{6.12}$$

for some $\gamma \geq 0$, $\beta \geq 0$. System (6.1) is called *strictly input (output) passive* (SIP, SOP) if $\gamma > 0$ $(\beta > 0)$.

Remark 6.2 The constants $\gamma \geq 0, \beta \geq 0$ are called passivity induces of system (6.8), since they characterize how passive the system is.

In order to obtain the dissipativity, passivity, and stability results of system (6.8), we need to find relevant multiple storage functionals and to construct the switching rule that generates the switching sequence Σ. In what follows, we will give the conditions to guarantee the strict input feed-forward output-feedback passivity of system (6.8) under certain state-dependent switching rule, which is equivalent to the $(\tilde{Q}, \tilde{S}, \tilde{R})$-strict dissipativity of system (6.8) under the switching rule. On the grounds of the proposed conditions, we know that system (6.8) is asymptotically stable if it is asymptotically zero-state detectable whenever the exogenous disturbance input is zero. Before giving the main results, we recall the definition of asymptotic zero-state detectability of system (6.8).

Definition 6.4 [37]: System (6.8) is said to be *asymptotically zero-state detectable* with degree $\alpha > 0$ if there exists $\delta > 0$ such that when $\|z(t+s)\| \leq \delta$ holds for some $t \geq t_0, \tau > 0$ and $0 < s \leq \tau$, we have $\|x_t(t+s)\| \leq ce^{-\alpha\tau}\|x_t\|$.

In the sequel, we derive the conditions and the switching rule under which system (6.8) is strictly input feed-forward output-feedback passive (IF–OFP).

6.3 Passivity-Based Switching Rule Design

In this section, we propose the conditions on strictly input feed-forward output-feedback passification of system (6.8) by using the constructive multiple storage functionals, and the asymptotical stability and finite-gain L_2 stability conditions are also given under some state-dependent switching rule.

Theorem 6.1 *For given constants* $\alpha > 0$ $\tau_{i,\max} > 0$, $\beta \geq 0$, $\gamma \geq 0$, $\alpha_{ij} \geq 0$, $\forall i, j \in \Upsilon$ *and controller gain matrices* K_i, *if there exist symmetric positive matrices*

$$P_i, Q, Z, X_i = \begin{pmatrix} X_i^{11} & X_i^{12} & X_i^{13} \\ * & X_i^{22} & X_i^{23} \\ * & * & X_i^{33} \end{pmatrix}, \text{ and matrices } N_i = \begin{bmatrix} N_i^1 \\ N_i^2 \\ N_i^3 \end{bmatrix}, \; M_i = \begin{pmatrix} M_i^1 \\ M_i^2 \\ M_i^3 \end{pmatrix}, \; i \in \Upsilon,$$

such that the following inequalities:

$$\begin{pmatrix} \Phi_i^{11} + \beta C_i^T C_i & \Phi_i^{12} & \Phi_i^{13} & -M_i^1 & \tau_{i,\max} A_i^T Z \\ * & \Phi_i^{22} & \Phi_i^{23} & -M_i^2 & \tau_{i,\max} K_i^T B_i^T Z \\ * & * & \Phi_i^{33} & -M_i^3 & \tau_{i,\max} \Gamma_i^T Z \\ * & * & * & -e^{-\alpha\tau_{i,\max}} Q & 0 \\ * & * & * & * & -\tau_{i,\max} Z \end{pmatrix} < 0, \tag{6.13a}$$

$$\begin{pmatrix} X_i^{11} & X_i^{12} & X_i^{13} & N_i^1 \\ * & X_i^{22} & X_i^{23} & N_i^2 \\ * & * & X_i^{33} & N_i^3 \\ * & * & * & e^{-\alpha\tau_{i,\max}} Z \end{pmatrix} \geq 0, \tag{6.13b}$$

$$\begin{pmatrix} X_i^{11} & X_i^{12} & X_i^{13} & M_i^1 \\ * & X_i^{22} & X_i^{23} & M_i^2 \\ * & * & X_i^{33} & M_i^3 \\ * & * & * & e^{-\alpha\tau_{i,\max}} Z \end{pmatrix} \geq 0, \tag{6.13c}$$

hold for $\forall i, j \in \Upsilon$, *then system (6.8) is strictly input feed-forward output-feedback passive (IF-OFP) under the switching rule*

$$\sigma(t) = \arg\min_{i \in \Upsilon} \{x^{\mathrm{T}} P_i x\}, \tag{6.14}$$

where

$$\Phi_i^{11} = P_iA_i + A_i^TP_i + Q + \alpha P_i + (N_i^1)^T + N_i^1 + \tau_{i,\max}X_i^{11} + \sum_{j=1}^{m} \alpha_{ij}(P_j - P_i),$$
$$\Phi_i^{12} = P_iB_iK_i + (N_i^2)^T - N_i^1 + M_i^1 + \tau_{i,\max}X_i^{12},$$
$$\Phi_i^{13} = P_i\Gamma_i + (N_i^3)^T + \tau_{i,\max}X_i^{13} + \beta C_i^TD_i - C_i^T/2,$$
$$\Phi_i^{22} = -(N_i^2)^T - N_i^2 + (M_i^2)^T + M_i^2 + \tau_{i,\max}X_i^{22},$$
$$\Phi_i^{23} = -(N_i^3)^T + (M_i^3)^T + \tau_{i,\max}X_i^{23},$$
$$\Phi_i^{33} = \tau_{i,\max}X_i^{33} + \beta D_i^TD_i - D_i + \gamma I.$$

Proof Consider the multiple storage functionals candidate, which is constructed as follows:

$$V_{\sigma(t)}(x_t) = V_i(x_t) = V_i^1(t) + V_i^2(t) + V_i^3(t), \tag{6.15}$$

where

$$V_i^1(t) = x^T(t)P_ix(t),$$
$$V_i^2(t) = \int_{-\tau_{i,\max}}^{0}\int_{t+\theta}^{t} \dot{x}^T(s)e^{\alpha(s-t)}Z\dot{x}(s)\mathrm{d}s\mathrm{d}\theta,$$
$$V_i^3(t) = \int_{t-\tau_{i,\max}}^{t} x^T(s)e^{\alpha(s-t)}Qx(s)\mathrm{d}s,$$

and where P_i, Q, and Z are positive matrices for any $i \in \Upsilon$. The time derivatives of each $V_i(x_t)$ along the state trajectory of system (6.8) become:

$$\begin{aligned}\dot{V}_i^1(t) &= 2x^T(t)P_i\dot{x}(t)\\ &= x^T(t)(P_iA_i + A_i^TP_i)x(t) + 2x^T(t)P_iB_iK_ix(t-\tau_i(t)) + 2x^T(t)P_i\Gamma_i\omega(t),\end{aligned} \tag{6.16}$$

$$\begin{aligned}\dot{V}_i^2(t) = &-\alpha\int_{-\tau_{i,\max}}^{0}\int_{t+\theta}^{t} \dot{x}^T(s)e^{\alpha(s-t)}Z\dot{x}(s)\mathrm{d}s\mathrm{d}\theta\\ &+\tau_{i,\max}\xi^T(t)\begin{pmatrix} A_i^T \\ (B_iK_i)^T \\ \Gamma_i^T \end{pmatrix} Z(A_i \quad B_iK_i \quad \Gamma_i)\xi(t) - \int_{t-\tau_{i,\max}}^{t} \dot{x}^T(s)e^{\alpha(s-t)}Z\dot{x}(s)\mathrm{d}s,\end{aligned} \tag{6.17}$$

$$\dot{V}_i^3(t) = -\alpha\int_{t-\tau_{i,\max}}^{t} x^T(s)e^{\alpha(s-t)}Qx(s)\mathrm{d}s + x^TQx - x^T(t-\tau_{i,\max})e^{-\alpha\tau_{i,\max}}Qx(t-\tau_{i,\max}). \tag{6.18}$$

Next, let us define $\xi(t) = [\,x^T(t) \quad x^T(t-\tau_i(t)) \quad \omega^T(t)\,]^T$. From the well-known Newton–Leibniz formula, we have

$$2[x^T N_i^1 + x^T(t-\tau_i(t))N_i^2 + \omega^T(t)N_i^3] \cdot \left(x(t) - \int_{t-\tau_i(t)}^{t} \dot{x}(s)\mathrm{d}s - x(t-\tau_i(t)) \right) = 0, \tag{6.19}$$

$$2[x^T M_i^1 + x^T(t-\tau_i(t))M_i^2 + \omega^T(t)M_i^3] \cdot \left(x(t-\tau_i(t)) - x(t-\tau_{i,\max}) - \int_{t-\tau_{i,\max}}^{t-\tau_i(t)} \dot{x}(s)\mathrm{d}s \right) = 0. \tag{6.20}$$

For any matrices

$$X_i = \begin{pmatrix} X_i^{11} & X_i^{12} & X_i^{13} \\ * & X_i^{22} & X_i^{23} \\ * & * & X_i^{33} \end{pmatrix} \geq 0,$$

it follows that this equation holds true:

$$\tau_{i,\max}\xi^T(t)X_i\xi(t) - \int_{t-\tau_i(t)}^{t} \xi^T(t)X_i\xi(t)\mathrm{d}s - \int_{t-\tau_{i,\max}}^{t-\tau_i(t)} \xi^T(t)X_i\xi(t)\mathrm{d}s = 0. \tag{6.21}$$

Therefore, we can obtain

$$\begin{aligned} &\dot{V}_i(x_t) + \alpha V_i(x_t) + \sum_{j=1}^{m} \alpha_{ij}(V_j - V_i) - \omega^{\mathrm{T}}(t)z(t) + \beta z^{\mathrm{T}}(t)z(t) + \gamma\omega^{\mathrm{T}}(t)\omega(t), \\ &= \dot{V}_i^1(x_t) + \dot{V}_i^2(x_t) + \dot{V}_i^3(x_t) + \alpha V_i(x_t) + \sum_{j=1}^{m} \alpha_{ij}(V_j - V_i) - \omega^{\mathrm{T}}(t)z(t) + \beta z^{\mathrm{T}}(t)z(t) + \gamma\omega^{\mathrm{T}}(t)\omega(t), \\ &= v^T(t)\Phi_i v(t) - \int_{t-\tau_i(t)}^{t} \eta^T(t)\Theta_i^1\eta(t)\mathrm{d}s - \int_{t-\tau_{i,\max}}^{t-\tau_i(t)} \eta^T(t)\Theta_i^2\eta(t)\mathrm{d}s, \end{aligned} \tag{6.22}$$

where

$$v(t) = [\,x^T(t) \quad x^T(t-\tau_i(t)) \quad \omega^T(t) \quad x^T(t-\tau_{i,\max})\,]^T,$$

$$\eta(t) = \left[\,x^T(t) \quad x^T(t-\tau_i(t)) \quad \omega^T(t) \quad \dot{x}^T(s)\,\right]^T,$$

$$\Phi_i = \begin{pmatrix} \Phi_i^{11} + \tau_{i,\max} A_i^T Z A_i + \beta C_i^T C_i & \Phi_i^{12} + \tau_{i,\max} A_i^T Z B_i K_i & \Phi_i^{13} + \tau_{i,\max} A_i^T Z \Gamma_i & -M_i^1 \\ * & \Phi_i^{22} + \tau_{i,\max} K_i^T B_i^T Z B_i K_i & \Phi_i^{23} + \tau_{i,\max} K_i^T B_i^T Z \Gamma_i & -M_i^2 \\ * & * & \Phi_i^{33} + \tau_{i,\max} \Gamma_i^T Z \Gamma_i & -M_i^3 \\ * & * & * & -e^{-\alpha\tau_{i,\max}} Q \end{pmatrix},$$

$$\Theta_i^1 = \begin{pmatrix} X_i^{11} & X_i^{12} & X_i^{13} & N_i^1 \\ * & X_i^{22} & X_i^{23} & N_i^2 \\ * & * & X_i^{33} & N_i^3 \\ * & * & * & e^{-\alpha\tau_{i,\max}} Z \end{pmatrix},$$

$$\Theta_i^2 = \begin{pmatrix} X_i^{11} & X_i^{12} & X_i^{13} & M_i^1 \\ * & X_i^{22} & X_i^{23} & M_i^2 \\ * & * & X_i^{33} & M_i^3 \\ * & * & * & e^{-\alpha\tau_{i,\max}} Z \end{pmatrix}.$$

From (6.13a)–(6.13c) and (6.22), we know that

$$\dot{V}_i(x_t) + \alpha V_i(x_t) + \sum_{j=1}^{m} \alpha_{ij}(V_j - V_i) - \omega^{\mathrm{T}}(t)z(t) + \beta z^{\mathrm{T}}(t)z(t) + \gamma\omega^{\mathrm{T}}(t)\omega(t) \leq 0.$$

Since $V(x_t) \geq 0, \alpha > 0$, it holds true that

$$\dot{V}_i(x_t) + \alpha V_i(x_t) + \sum_{j=1}^{m} \alpha_{ij}(V_j - V_i) - \omega^{\mathrm{T}}(t)z(t) + \beta z^{\mathrm{T}}(t)z(t) + \gamma\omega^{\mathrm{T}}(t)\omega(t) \leq 0. \tag{6.23}$$

If we choose a switching rule satisfying (6.14), then we find that

$$\dot{V}_{\sigma(t)}(x_t) + \Psi_{\sigma(t)}(x_t) \leq s_{\sigma(t)}(\omega, z), \tag{6.24}$$

where

$$s_{\sigma(t)}(\omega, z) = \omega^{\mathrm{T}}(t)z(t) - \beta z^{\mathrm{T}}(t)z(t) - \gamma\omega^{\mathrm{T}}(t)\omega(t), \quad \Psi_{\sigma(t)}(x_t) = \alpha V_{\sigma(t)}(x_t).$$

Thus on the grounds of Definitions 6.1 and 6.3, we can say that system (6.8) is strictly dissipative with respect to (6.11) under the switching rule (6.14), and also it is strictly input feed-forward output-feedback passive (IF-OFP) under (6.14). This completes the proof. □

Remark 6.3 Given feedback control gains $K_i, i \in \Upsilon$, Theorem 6.1 gives sufficient conditions for strictly input feed-forward output-feedback passification of system (6.8) under switching rule (6.14). It should be noted that each subsystem of system (6.8) is not required to be strictly input IF–OFP. The inequalities (6.13a)–(6.13c) hold true while depending on $\tau_{i,\max}$, the important parameter which describes the upper bound of time-varying delays of each subsystem. Therefore, the proposed conditions (6.13a)–(6.13c) are delay bound-dependent ones.

Remark 6.4 In Theorem 6.1, if $\gamma > 0,\ \beta = 0$, and (6.13a)–(6.13c) are satisfied, then system (6.8) is strictly input passive (SIP); and if $\gamma = 0,\ \beta > 0$, and (6.13a)–(6.13c) are satisfied, then system (6.8) is strictly output passive (SOP).

Corollary 6.1 *For given constants* $\alpha > 0,\ \beta \geq 0,\ \gamma \geq 0,\ \alpha_{ij} \geq 0$ *and* $u_i = 0,\ \forall i, j \in \Upsilon$, *if there exist symmetric positive matrices* P_i, *such that the following inequalities*:

$$\begin{pmatrix} \Phi_i^{11} + \beta C_i^T C_i & \beta C_i^T D_i - C_i^T/2 \\ * & \beta D_i^T D_i - D_i + \gamma I \end{pmatrix} < 0, \tag{6.25}$$

where $\Phi_i^{11} = P_i A_i + A_i^T P_i + \alpha P_i + \sum_{j=1}^{m} \alpha_{ij}(P_j - P_i)$, *hold for* $\forall i, j \in \Upsilon$, *then the unforced system (6.8)* $(u_i = 0,\ \forall i, j \in \Upsilon)$ *is strictly input feed-forward output-feedback passive (IF-OFP) under the switching rule (6.14).*

Corollary 6.2 *System (6.8) is finite-gain* L_2*-stable if it is strictly output passive under the switching rule (6.14).*

Proof From Theorem 6.1, if system (6.8) is strictly output passive under the switching rule (6.14), we have

$$\dot{V}_{\sigma(t)}(x_t) - \omega^{\mathrm{T}}(t)z(t) + \beta z^{\mathrm{T}}(t)z(t) < 0. \tag{6.26}$$

Combining with (6.14), we know that $V_{\sigma(t)}(x_t)$ is continuous in $[0, +\infty)$, and thus we can use $V(x_t)$ to denote $V_{\sigma(t)}(x_t)$, which leads to:

$$\begin{aligned} \dot{V}(x_t) = V_{\sigma(t)}(x_t) &< \omega^{\mathrm{T}}(t)z(t) - \beta z^{\mathrm{T}}(t)z(t), \\ &= -\frac{1}{2\beta}(\omega(t) - \beta z(t))^{\mathrm{T}}(\omega(t) - \beta z(t)) + \frac{1}{2\beta}\omega^{\mathrm{T}}(t)\omega(t) - \frac{\beta}{2}z^{\mathrm{T}}(t)z(t), \\ &< +\frac{1}{2\beta}\omega^{\mathrm{T}}(t)\omega(t) - \frac{\beta}{2}z^{\mathrm{T}}(t)z(t). \end{aligned} \tag{6.27}$$

Integrating the above inequality from 0 to T, we obtain

$$\int_0^T z^{\mathrm{T}}(s)z(s)ds < \frac{1}{\beta^2}\int_0^T \omega^{\mathrm{T}}(s)\omega(s)ds - \frac{2}{\beta}[V(x(T)) - V(x(0))], \tag{6.28}$$

which is equivalent to

$$\|z\|_{L_2} < \frac{1}{\beta}\|\omega\|_{L_2} + \sqrt{\frac{2}{\beta}V(x(0))}. \tag{6.29}$$

□

Corollary 6.3 *When $\omega(t)=0$, system (6.8) is asymptotically stable under (6.8) if system (6.8) is asymptotically zero-state detectable.*

Remark 6.5 It should be noted that Corollary 6.1 gives sufficient conditions on strict input feed-forward output-feedback passivity of switched system (6.8) with $u_i=0$: in this case, the feedback delay is zero. Corollary 6.2 says system (6.8) is finite-gain L_2-stable if it is strictly output-feedback passive under the switching rule (6.8). Furthermore, by virtue of Corollary 6.3, we can deduce that system (6.8) with $\omega(t)=0$, is asymptotically stable under the switching rule (6.14).

Remark 6.6 From the proof procedure of Theorem 6.1, if system (6.8) is strictly input feed-forward output-feedback passive (IF-OFP) under the switching rule (6.14), we can deduce system (6.8) is $(\tilde{Q},\tilde{S},\tilde{R})$- strictly dissipative by choosing

$$z(t)=\left[x^T(t)\quad x^T(t-\tau_i(t))\quad x^T(t-\tau_{i,\max})\right]^T,$$

and thus it follows that

$$\tilde{Q}=\begin{pmatrix} \Phi_i^{11}+\tau_{i,\max}A_i^TZA_i+\beta C_i^TC_i & \Phi_i^{12}+\tau_{i,\max}A_i^TZB_iK_i & -M_i^1 \\ * & \Phi_i^{22}+\tau_{i,\max}K_i^TB_i^TZB_iK_i & -M_i^2 \\ * & * & -e^{-\alpha\tau_{i,\max}}Q \end{pmatrix}, \tag{6.30a}$$

$$\tilde{S}=\begin{pmatrix} \Phi_i^{13}-\tau_{i,\max}A_i^TZ\Gamma_i \\ \Phi_i^{23}+\tau_{i,\max}K_i^TB_i^TZ\Gamma_i \\ -M_i^3 \end{pmatrix},\quad \tilde{R}=\Phi_i^{33}+\tau_{i,\max}\Gamma_i^TZ\Gamma_i. \tag{6.30b}$$

6.4 Switching Rule and Feedback Control Law Co-design

In this section, we are concerned with co-design of the switching rule and the state feedback controllers, which together render the closed-loop system strictly IF–OFP for all admissible time-varying delays.

Theorem 6.2 *For given scalars $\alpha>0,\ \tau_{i,\max}>0,\beta\geq 0,\ \gamma\geq 0,\ \alpha_{ij}\geq 0\ \forall i,j\in\Upsilon$ suppose there exist positive definite matrices $\bar{P}_i,\bar{Q},\bar{Z},\bar{X}_i=\begin{pmatrix} \bar{X}_i^{11} & \bar{X}_i^{12} & \bar{X}_i^{13} \\ * & \bar{X}_i^{22} & \bar{X}_i^{23} \\ * & * & \bar{X}_i^{33} \end{pmatrix}$, and matrices $\bar{N}_i=\begin{pmatrix} \bar{N}_i^1 \\ \bar{N}_i^2 \\ \bar{N}_i^3 \end{pmatrix},\quad \bar{M}_i=\begin{pmatrix} \bar{M}_i^1 \\ \bar{M}_i^2 \\ \bar{M}_i^3 \end{pmatrix},\quad Y_i$ such that the following linear matrix inequalities*

$$\begin{pmatrix} \psi_{i11} & \psi_{i12} & \psi_{i13} & -\bar{M}_i^1 & \tau_{i,\max}\bar{P}_iA_i^T & \bar{P}_iC_i^T & \sqrt{\alpha_{i1}}\bar{P}_i & \sqrt{\alpha_{i2}}\bar{P}_i & \cdots & \sqrt{\alpha_{im}}\bar{P}_i \\ * & \psi_{i22} & \psi_{i23} & -\bar{M}_i^2 & \tau_{i,\max}Y_i^TB_i^T & 0 & 0 & 0 & \cdots & 0 \\ * & * & \psi_{i33} & -\bar{M}_i^3 & \tau_{i,\max}\Gamma_i^T & 0 & 0 & 0 & \cdots & 0 \\ * & * & * & -e^{-\alpha\tau_{i,\max}}\bar{Q} & 0 & 0 & 0 & 0 & \cdots & 0 \\ * & * & * & * & -\tau_{i,\max}\bar{Z} & 0 & 0 & 0 & \cdots & 0 \\ * & * & * & * & * & -\beta^{-1} & 0 & 0 & \cdots & 0 \\ * & * & * & * & * & * & -\bar{P}_1 & 0 & \cdots & 0 \\ * & * & * & * & * & * & * & -\bar{P}_2 & \cdots & 0 \\ * & * & * & * & * & * & * & * & \ddots & 0 \\ * & * & * & * & * & * & * & * & * & -\bar{P}_m \end{pmatrix} < 0, \tag{6.31a}$$

$$\begin{pmatrix} \bar{X}_i^{11} & \bar{X}_i^{12} & \bar{X}_i^{13} & \bar{N}_i^1 \\ * & \bar{X}_i^{22} & \bar{X}_i^{23} & \bar{N}_i^2 \\ * & * & \bar{X}_i^{33} & \bar{N}_i^3 \\ * & * & * & e^{-\alpha\tau_{i,\max}}(2\bar{P}_i-\bar{Z}) \end{pmatrix} \geq 0, \tag{6.31b}$$

$$\begin{pmatrix} \bar{X}_i^{11} & \bar{X}_i^{12} & \bar{X}_i^{13} & \bar{M}_i^1 \\ * & \bar{X}_i^{22} & \bar{X}_i^{23} & \bar{M}_i^2 \\ * & * & \bar{X}_i^{33} & \bar{M}_i^3 \\ * & * & * & e^{-\alpha\tau_{i,\max}}(2\bar{P}_i-\bar{Z}) \end{pmatrix} \geq 0, \tag{6.31c}$$

hold true for $\forall i,j \in \Upsilon$. *Then the hybrid state feedback controllers (6.7) and the switching rule*

$$\sigma(t) = \arg\min_{i\in\Upsilon}\{x^{\mathrm{T}}\bar{P}_i^{-1}x\}, \tag{6.32}$$

guarantee the strictly input feed-forward, output-feedback passification of system (6.8) for all admissible time-varying delays. Moreover, the hybrid state feedback controller gains are given by means of

$$K_i = Y_i\bar{P}_i, \quad \forall i \in \Upsilon, \tag{6.33}$$

where

$$\psi_{i11} = A_i\bar{P}_i + \bar{P}_iA_i^T + \bar{Q} + \alpha\bar{P}_i + (\bar{N}_i^1)^T + \bar{N}_i^1 + \tau_{i,\max}\bar{X}_i^{11} - \sum_{j=1}^{m}\alpha_{ij}\bar{P}_i,$$

$$\psi_{i12} = B_iY_i + (\bar{N}_i^2)^T - \bar{N}_i^1 + \bar{M}_i^1 + \tau_{i,\max}\bar{X}_i^{12},$$

$$\psi_{i13} = \Gamma_i + (\bar{N}_i^3)^T + \tau_{i,\max}\bar{X}_i^{13} + \bar{P}_i(\beta C_i^TD_i - C_i^T/2),$$

$$\psi_{i22} = -(\bar{N}_i^2)^T - \bar{N}_i^2 + (\bar{M}_i^2)^T + \bar{M}_i^2 + \tau_{i,\max}\bar{X}_i^{22},$$

$$\psi_{i23} = -(\bar{N}_i^3)^T + (\bar{M}_i^3)^T + \tau_{i,\max}\bar{X}_i^{23},$$

$$\psi_{i33} = \tau_{i,\max}\bar{X}_i^{33} + \beta D_i^TD_i - D_i + \gamma I.$$

Proof Let us pre-multiply and post-multiply (6.13a) in Theorem 6.1 by matrices $\mathrm{diag}\{P_i^{-1}, P_i^{-1}, I, P_i^{-1}, Z^{-1}\}$ and their transpose, respectively. Simultaneously, let us pre-multiply and post-multiply Θ_i^1 and Θ_i^2 by $\mathrm{diag}\{P_i^{-1}, P_i^{-1}, P_i^{-1}, P_i^{-1}\}$ and its transpose, respectively. Furthermore, let

$$\begin{aligned}
&\bar{P}_i = P_i^{-1}, \bar{Z} = Z^{-1}, \bar{Q} = P_i^{-1}QP_i^{-1}, \bar{N}_i^1 = P_i^{-1}N_i^1P_i^{-1},\ \bar{N}_i^2 = P_i^{-1}N_i^2P_i^{-1}, \bar{N}_i^3 = P_i^{-1}N_i^3\\
&\bar{M}_i^1 = P_i^{-1}M_i^1P_i^{-1}, \bar{M}_i^2 = P_i^{-1}M_i^2P_i^{-1}, \bar{M}_i^3 = P_i^{-1}M_i^3 \quad Y_i = K_iP_i^{-1},\\
&\bar{X}_i^{11} = P_i^{-1}X_i^{11}P_i^{-1}, \bar{X}_i^{12} = P_i^{-1}X_i^{12}P_i^{-1}, \bar{X}_i^{13} = P_i^{-1}X_i^{13},\\
&\bar{X}_i^{22} = P_i^{-1}X_i^{22}P_i^{-1}, \bar{X}_i^{23} = P_i^{-1}X_i^{23}, \bar{X}_i^{33} = X_i^{33}.
\end{aligned}$$

By noting that $\bar{Z} > 0$, we have $(\bar{Z} - \bar{P}_i)\bar{Z}^{-1}(\bar{Z} - \bar{P}_i) \geq 0$, which is equivalent to the inequality $\bar{P}_i\bar{Z}^{-1}\bar{P}_i \geq 2\bar{P}_i - \bar{Z}$. Therefore, we can obtain (6.31a)–(6.31c) from (6.13a)–(6.13c), which completes this proof. □

Remark 6.7 From Theorem 6.2, given scalars $\alpha > 0, \tau_{i,\max} > 0, \beta \geq 0, \gamma \geq 0, \alpha_{ij} \geq 0$, $\forall i, j \in \Upsilon$, first solving the inequalities (6.31a)–(6.31c), if they are feasible, we can obtain $\bar{P}_i, \bar{Q}, \bar{Z}, \bar{X}_i$, and matrices $\bar{N}_i, \bar{M}_i, Y_i$. Then we get the switching rule (6.32) and the feedback controller gain (6.33). Assume that we know each $\tau_{i,\max} > 0, i \in \Upsilon$, for given parameters $\alpha > 0$, $\beta \geq 0$, $\gamma \geq 0$, $\alpha_{ij} \geq 0, \forall i, j \in \Upsilon$. Then, if (6.31a)–(6.31c) have no feasible solution, we can adjust α to be smaller or constants α_{ij} to be larger.

The following algorithm is proposed to compute the switching rule and the controller law that guarantee strict input feed-forward output-feedback passification of system (6.8).

Algorithm 6.1

Step 1. For given matrices $A_i, B_i, C_i, D_i, \Gamma_i$ and the exogenous disturbance input $\omega(t) \in L_2$, take $\alpha > 0$, $\tau_{i,\max} > 0$, $\beta \geq 0$, $\gamma \geq 0$, $\alpha_{ij} \geq 0$, and try to find feasible solutions to satisfy (6.31a)–(6.31c)

Step 2. If (6.31a)–(6.31c) are feasible, write down the feasible solutions

$$\bar{P}_i, \bar{Q}, \bar{Z}, \bar{X}_i = \begin{pmatrix} \bar{X}_i^{11} & \bar{X}_i^{12} & \bar{X}_i^{13} \\ * & \bar{X}_i^{22} & \bar{X}_i^{23} \\ * & * & \bar{X}_i^{33} \end{pmatrix}, \ \bar{N}_p = \begin{pmatrix} \bar{N}_i^1 \\ \bar{N}_i^2 \\ \bar{N}_i^3 \end{pmatrix}, \ \bar{M}_i = \begin{pmatrix} \bar{M}_i^1 \\ \bar{M}_i^2 \\ \bar{M}_i^3 \end{pmatrix}, Y_i,$$

and then compute the switching rule (6.32) and the feedback controller gains (6.33). Otherwise, adjust α, α_{ij}, and return to Step 1).

6.5 An Illustrative Example

In this section, we investigate the application to an example of switched linear systems with feedback delays in order to illustrate our main results.

Consider the following two switched subsystems:
Subsystem 1

$$\dot{x}=\begin{pmatrix}0.3 & 0\\ 0 & 0.1\end{pmatrix}x+\begin{pmatrix}0\\ 1\end{pmatrix}u_1+\begin{pmatrix}-0.1\\ 0\end{pmatrix}\omega(t),\quad z=(\,0.1\quad 0\,)x+\omega(t);$$

Subsystem 2

$$\dot{x}=\begin{pmatrix}1.5 & 0\\ 0 & 0.2\end{pmatrix}x+\begin{pmatrix}0\\ 0.1\end{pmatrix}u_2+\begin{pmatrix}-0.5\\ 0\end{pmatrix}\omega(t),\quad z=(\,0\quad 0.1\,)x+\omega(t);$$

where $\omega(t)\in L_2$, and $u_i=K_ix(t-\tau_i(t))$, $i=1,2$.

By virtue of Theorem 6.1, for the given positive constants $\tau_{1,\max}=0.1$, $\tau_{2,\max}=0.15$, $\alpha_{12}=\alpha_{21}=10$, $\gamma=0.1$, $\alpha=0.3$, $\beta=0.3$ solving the LMIs (6.31a)–(6.31c) yields:

$$\bar{P}_1=\begin{pmatrix}6.2026 & -0.1042\\ -0.1042 & 75.4354\end{pmatrix},\ \bar{P}_2=\begin{pmatrix}7.1453 & -0.1290\\ -0.1290 & 72.6838\end{pmatrix},$$

$$Y_1=(\,-0.1139\quad -157.3196\,),\ Y_2=(\,1.900\quad -139.68\,),$$

$$\bar{N}_1=\begin{pmatrix}-21.9413 & 0.0197\\ 0.0197 & -129.8055\\ 20.6361 & 0.0417\\ 0.0417 & 56.9148\\ 0.0997 & 0.0010\end{pmatrix},\ \bar{N}_2=\begin{pmatrix}-91.2923 & 0.4059\\ 0.4059 & -121.3558\\ 126.8874 & -0.1012\\ -0.1012 & 57.1093\\ 2.1877 & 1.0280\end{pmatrix},$$

$$\bar{M}_1=\begin{pmatrix}-0.3325 & 0.0178\\ 0.0178 & -5.8782\\ -2.0106 & 0.0486\\ 0.0486 & -36.9515\\ 0.0003 & -0.0014\end{pmatrix},\ \bar{M}_2=\begin{pmatrix}-0.1060 & 0.0962\\ 0.0962 & -6.5343\\ -2.0961 & 0.0900\\ 0.0900 & -37.9476\\ -0.0047 & 0.2549\end{pmatrix}.$$

Thus, the respective hybrid state feedback controllers have gains

$$K_1=Y_1\bar{P}_1=(\,-0.0534\quad -2.0856\,),\quad K_2=Y_2\bar{P}_2=(\,-0.0836\quad -19.2183\,).$$

Notice that the eigenvalues of each closed-loop subsystem is $\lambda_{1,1}=-1.9856$, $\lambda_{1,2}=0.3000$,, and $\lambda_{2,1}=-1.7218$, $\lambda_{2,2}=1.500$, respectively. Obviously, they indicate that neither of each subsystem is passificated by their candidate controllers since both of them have eigenvalues located in the right half-plane. However, by using Theorem 6.2, we find that the switching signal for guaranteeing the strictly input feed-forward, output-feedback passification of the considered switched system is obtained by the switching law

$$\sigma(t) = \arg\min_{i \in \Upsilon}\{x^{\mathrm{T}} \bar{P}_i^{-1} x\}.$$

6.6 Conclusions

We studied strictly input feed-forward, output-feedback passification for a class of switched systems by synthesizing hybrid time-varying feedback delays. We derived delay bound-dependent conditions on the strict input feed-forward, output-feedback passivity of switched delay systems under a class of state-dependent switching signals and the respective generating control law. Consequently, also the finite-gain L_2 stability and asymptotic stability of such system are also studied. Finally, we give an algorithm that implements the know-how to co-design the switching rule and the feedback control law. The foreseen future work is dedicated to bring general results on co-design of switching law and feedback passification for nonlinear switched systems with feedback delays.

Acknowledgment This work was supported in part by the National Natural Science Foundation of China (No. 61233002), and the Fundamental Research Funds for the Central Universities (No. N120404019).

References

1. Ahn, C.K.: Switched exponential state estimation of neural networks based on passivity theory. Nonlinear Dyn. **67**, 573–586 (2012)
2. Boyd, S., El Ghaoui, L., Feron, E., Balakrishnan, V.: Linear Matrix Inequalities in System and Control Theory. The SIAM, Philadelphia (1994)
3. Branicky, M.S.: Multiple Lyapunov functions and other analysis tools for switched and hybrid systems. IEEE Trans. Autom. Control **43**(4), 475–482 (1998)
4. Brogliato, B., Lozano, R., Maschke, B., Egeland, O.: Dissipative Systems Analysis and Control: Theory and Applications, 2nd edn. Springer, London (2007)
5. Chen, W.T., Saif, M.: Passivity and passivity based controller design of a class of switched control systems. In: Proceedings of the 16th IFAC World Congress, pp. 143–147 (2005)
6. Chioua, J.S., Wang, C.J., Cheng, C.M.: On delay-dependent stabilization analysis for the switched time-delay systems with the state-driven switching strategy. J. Franklin Inst. **348**(2), 261–276 (2011)
7. Dimirovski, G.M., Gough, N.E., Barnett, S.: Categories in systems and control theory. Int. J. Syst. Sci. **7**(9), 1081–1090 (1977)
8. Fridman, E., Seuret, A., Richard, J.P.: Robust sampled-data stabilization of linear systems: an input delay approach. Automatica **40**, 1441–1446 (2004)

9. Guan, Z.H., Zhang, H., Yang, S.H.: Robust passive control for internet-based switching systems with time-delay. Chaos, Solitons Fractals **36**, 479–486 (2008)
10. Hill, D.J., Moylan, P.J.: The stability of nonlinear dissipative systems. IEEE Trans. Autom. Control **21**(5), 708–711 (1976)
11. Hespanha, J.P., Morse, A.S.: Stability of switched systems with average dwell-time. In: Proceedings of the 38th IEEE Conference on Decision and Control, pp. 2655–2660. Phoenix, AZ, Dec 1999
12. Liberzon, D.: Switching in Systems and Control. Birkhauser, Boston (2003)
13. Li, J., Zhao, J.: Passivity and feedback passification of switched discrete-time linear systems. Syst. Control Lett. **62**, 1073–1081 (2013)
14. Lian, J., Shi, P., Feng, Z.: Passivity and passification for a class of uncertain switched stochastic time-delay systems. IEEE Trans. Cyber. (IEEE Soc. Syst. Man Cybern.) **43**(1), 3–13 (2013)
15. Lin, H., Antsaklis, P.J.: Stability and stabilizability of switched linear systems: a survey of recent results. IEEE Trans. Autom. Control **54**(2), 308–322 (2009)
16. Liu, J., Liu, X.Z., Xie, W.C.: Input-to-state stability of impulsive and switching hybrid systems with time-delay. Automatica **47**(5), 899–908 (2011)
17. Liu, Y.Y., Zhao, J.: Stabilization of switched nonlinear systems with passive and non-passive subsystems. Nonlinear Dyn. **67**(3), 1709–1716 (2012)
18. Lozano, R., Brogliato, B., Landon, I.D.: Passivity and global stabilization of cascaded nonlinear systems. IEEE Trans. Autom. Control **37**(9), 1386–1388 (1992)
19. Ma, D., Liu, J.C.: Robust exponential stabilization for network-based switched control systems. Int. J. Control Autom. Syst. **8**(1), 67–72 (2010)
20. Ma, D., Dimirovski, G.M., Zhao, J.: Hybrid state feedback controller design of networked switched systems with packet dropout. In: Proceedings of the 29th American Control Conference, pp. 1368–1373. Baltimore, MD, June–July 2010
21. Ma, D., Guo, Z.F., Dimirovski, G.M., Zhao, J.: Passive control for networked switched systems with network-induced time-delays and dropout. In: Proceedings of the Joint 48th IEEE Conference on Decision and Control & 28th Chinese Control Conference, pp. 4258–4263. Shanghai, P.R. China, Dec 2009
22. Mahmoud, M.S.: Delay-dependent dissipativity analysis and synthesis of switched delay systems. Int. J. Robust Nonlinear Control **21**(1), 1–20 (2011)
23. Moon, Y.S., Park, P., Kwon, W.H., Lee, Y.S.: Delay-dependent robust stabilization of uncertain state-delay systems. Int. J. Control **74**, 1447–1455 (2001)
24. McCourt, M.J., Antsaklis, P.J.: Stability of interconnected switched systems using QSR dissipativity with multiple supply rates. In: Proceedings of the 31st American Control Conference, pp. 4564–4569. Montréal, Canada, June 2012
25. Peleties, P., DeCarlo, R.: Asymptotic stability of m-switched systems using Lyapunov-like functions. In: Proceedings of the 10th American Control Conference. pp. 1679–1684, Boston, MA, June 1991
26. Sepulchre, R., Jankovic, M., Kokotovic, P.V.: Constructive Nonlinear Control. Springer, London (1997)
27. Sun, Z.D., Ge, S.S.: Switched Linear Systems-Control and Design. Springer, New York (2004)
28. Sun, X.M., Liu, G.P., Rees, D., Wang, W.: Stability of systems with controller failure and time-varying delay. IEEE Trans. Autom. Control **53**(10), 2391–2396 (2008)
29. Wang, Y., Gupta, V., Antsaklis, P.J.: On passivity of a class of discrete-time switched nonlinear systems. IEEE Trans. Autom. Control **59**(3), 692–702 (2014)
30. Willems, J.C.: Dissipative dynamical systems part I: general theory. Arch. Ration. Mech. Anal. **45**(5), 321–351 (1972)
31. Willems, J.C.: Dissipative dynamical systems Part II: Linear systems with quadratic supply rates. Arch. Ration. Mech. Anal. **45**(5), 352–393 (1972)
32. Ye, H., Michel, A.N., Hou, L.: Stability theory for hybrid dynamical systems. IEEE Trans. Autom. Control **43**(4), 461–474 (1998)

33. Zhai, G.S., Hu, B., Yasuda, K., Michel, A.: Disturbance attenuation properties of time-controlled switched systems. J. Franklin Inst. **338**, 765–779 (2001)
34. Zhang, L.X., Wang, C.H., Gao, H.J.: Delay-dependent stability and stabilization of a class of linear switched time-varying delay systems. J. Syst. Eng. Electron. **18**(2), 320–326 (2007)
35. Zhao, J., Hill, D.J.: Passivity and stability of switched systems: a multiple storage function method. Syst. Control Lett. **57**(2), 158–164 (2008)
36. Zhao, J., Hill, D.J.: On stability L_2-gain and H_∞ control for switched systems. Automatica **44** (5), 1220–1232 (2008)
37. Zhao, J., Hill, D.J.: Dissipativity theory for switched systems. IEEE Trans. Autom. Control **53** (4), 941–953 (2008)

Part II
Machine Intelligence and Learning Control in Complex Systems

Learning without thinking is labour lost; thinking without learning is perilous.

Confucius

Chapter 7
Creating and Controlling Complex Biological Brains

Kevin Warwick

Abstract In this contribution, a look is taken at how animal and/or human brain cells can be cultivated (grown) and given a robot physical body (as a controlling brain) in which they can move around and interact with the world. This is realised as a new form of Artificial Intelligence in which the complexity of a highly non-linear biological neural network is employed to uniquely control a real-world robot. The communication/control feedback loop is described and considered in terms of learning, performance, long-term operation and specialisation within the neural structure. Experimental results are presented and philosophical arguments opened up, e.g. can the robot be considered to be a living, conscious entity?

7.1 Introduction

Considerable progress has been made towards hybrid systems in which biological neurons are integrated with electronic components. As an example, Reger et al. [1] demonstrated the use of a lamprey brain to control a small-wheeled robot's movements; meanwhile others were successfully able to send control commands to the nervous system of cockroaches [2] or rats [3] as if they were robots. These studies inform us about information processing and encoding in the brains of living animals [4], however, they do pose ethical questions and can be technically problematic since access to the brain is limited by barriers such as the skin and skull, and data interpretation is complicated due to the number of neurons present in the brain of even the simplest animal. Also, approaches which involve recording the activity of individual neurons or small populations of neurons are limited by their invasive, and hence destructive, nature. As a result, neurons cultured under labo-

K. Warwick (✉)
School of Systems Engineering, University of Reading, Whiteknights, PO Box 225, Reading RG6 6AY, UK
e-mail: k.warwick@reading.ac.uk

G.M. Dimirovski (ed.), *Complex Systems*, Studies in Systems, Decision and Control 55, DOI 10.1007/978-3-319-28860-4_7

ratory conditions on a planar array of non-invasive electrodes provide an attractive alternative with which to probe the operation of biological neuronal networks.

In the past few years, research has focussed on culturing networks of some tens of thousands of brain cells grown in vitro [5]. These cultures are created by enzymatically dissociating neurons obtained from foetal rodent cortical tissue and then culturing them in a specialised chamber, in doing so providing suitable environmental conditions and nutrients. An array of electrodes is embedded in the base of the chamber (a multielectrode array; MEA) providing an electrical interface to the neuronal culture [6–9]. The neurons in such cultures spontaneously begin to branch out and within an hour of placement, even without external stimulation, they begin to reconnect with other nearby neurons and commence electrochemical communication. This propensity to spontaneously connect and communicate demonstrates an innate tendency to network. Studies of neural cultures demonstrate distinct periods of development defined by changes in activity which appear to stabilise after 30 days and, in terms of useful responses, last for at least 2–3 months [10–12]. The cultures of neurons form a monolayer on the MEA, making them both amenable to optical microscopy and accessible to physical and chemical manipulation [9].

An aim of the ongoing project described here is to investigate the use of cultured neurons for the control of mobile robots. In order to produce useful processing, the disembodied biological network must develop in the presence of meaningful input/output relationships as part of closed-loop sensory interaction with the environment. This is evidenced by animal and human studies, which show that development in a sensory-deprived environment results in poor or dysfunctional neural circuitry [13, 14]. To this end the overall closed-loop hybrid system involving a primary cortical culture on an MEA and a mobile robot body must exist within a sufficiently rich and reasonably consistent environment. This then constitutes an interesting and novel approach to examining the computational capabilities of biological networks [15].

Typically, in vitro neuronal cultures consist of thousands of neurons generating highly variable, multidimensional signals. In order to extract components and features representative of the network's overall state from such data, appropriate pre-processing and dimensionality reduction techniques must be applied. Several schemes have till now been constructed. Shkolnik et al. [16] created a control scheme for a simulated robot body in which two channels of an MEA were selected and an electrical stimulus consisting of a ±600 mV, 400 μs biphasic pulse was delivered at varying inter-stimulus intervals. Information coding was formed by testing the effect of electrically induced neuronal excitation with a given time delay termed the inter-probe interval (IPI) between two stimulus probes. This technique gave rise to a characteristic response curve which formed the basis for deciding the robot's direction of movement using basic commands (forward, backward, left and right).

DeMarse and Dockendorf investigated the computational capacity of cultured networks by implementing the control of a 'real-life' problem, namely controlling a simulated aircraft's flight path (e.g. altitude and roll adjustments) [17]. Meanwhile

Shahaf and Marom [18] reported one of the first experiments to achieve desired discrete output computations by applying a simple form of supervised learning to disembodied neuronal cultures. Recently, in [19] a Learning classifier system was used to manipulate culture activity towards a goal level using simple input signals. In both of these latter experiments, the desired result was only achieved in about one-third of the cases, indicating some of the difficulties in achieving repeatability.

This should not really come as a surprise as this is a field of study very much in its infancy. There are bound to be difficulties, however, there is much to be learnt. It is apparent that even at this early stage, such re-embodiments (real or virtual) have an important role to play in the study of biological learning mechanisms and neurological behaviour in general. Our physical embodied robots provide the starting point for creating a proof-of-concept control loop around the neuronal culture and a basic platform for future—more specific—reinforcement learning experiments. The fundamental problem is the coupling of the robot's goals to the culture's input–output mapping, the design of the robot's architecture discussed in this paper therefore emphasises the need for flexibility and the use of machine learning techniques in the search of such coupling.

In the section which follows, the general procedure for laying out the neural culture (the biological component) is described; this is followed by a description of the main elements of the closed-loop control system, including the culture as an important element in the feedback loop. Details of the current system's architecture are given in Sect. 7.3. Section 7.4 includes a description of our initial tests and preliminary results. Section 7.5 meanwhile provides a discussion of possible biological underpinnings providing the motivation and context for the use of the machine learning (ML) methods, and Sect. 7.6 concludes with an overview of current progress. Finally, Sect. 7.7 discusses new ongoing research and planned future extensions.

7.2 Culture Preparation

To realise the cultured neural network, cortical tissue is dissected from the brains of embryonic rats and neuronal cells enzymatically dissociated before seeding onto planar multielectrode arrays (MEAs). The cells are restricted to lie within the recording horizon of the electrode array by means of a template placed on the MEA prior to seeding and removed immediately after cells have settled (~1 h). The MEA is also filled with a conventional cell culture medium containing nutrients, growth hormones and antibiotics of which 50 % is replaced twice weekly. Within the first hour after seeding, neurons appear to extend connections to nearby cells (even within the first few minutes this has been observed) and within 24 h, a thick matt of neuronal extensions is visible across the seeded area.

The connectivity between seeded cells increases rapidly over subsequent days. After 7 days, electrical signals are observed in the form of action potentials which, in the 'disembodied culture' (not connected within the closed loop), over the following week transform into dense bursts of simultaneous electrical activity across the entire network. This bursting feature subsequently continues through to maturity (30 days in vitro and onwards). It is not well understood what the bursting actually means and how much it is part of normal neural development. However, such continued behaviour, after this initial development phase, may subsequently be representative of an underlying pathological state resulting from impoverished sensory input and may differ from the activity of a culture developing within a closed loop [20]. This is something which remains to be studied further.

Cultures usually remain active until approximately 3 months of age. During this time, they are sealed with Potter rings [21] to maintain sterility and osmolarity, and are maintained in a humidified, 37 °C, 5 % CO_2 incubator. Recordings are undertaken in a non-humidified 37 °C, 5 % CO_2 incubator for between 30 min and 8 h dependent on environmental humidity and the resulting stability of activity.

7.3 Experimental Platform

The multielectrode array enables voltage fluctuations in the culture (relative to a reference ground electrode outside the network) to be recorded in real-time at 59 sites out of 64 in an '8 × 8' array (Fig. 7.1). This allows for the detection of neuronal action potentials within a 100 μm radius (or more) around an individual electrode. Using spike sorting algorithms [12], it is then possible to separate the firings of multiple individual neurons, or small groups of neurons, as monitored on a single electrode. As a result, multielectrode recordings across the culture permit a picture of the global activity of the entire neuronal network to be formed. It is possible to electrically stimulate via any of the electrodes to induce focussed neural activity. The multielectrode array therefore forms a functional and non-destructive bidirectional interface to the cultured neurons.

Electrically evoked responses and spontaneous activity in the culture (the neuronal network) are coupled to the robot architecture, and thence on to the physical robot, via a machine learning interface, which maps the features of interest to specific actuator commands. Sensory data fed back from the robot is associated with a set of appropriate stimulation protocols and is subsequently delivered to the culture, thereby closing the robot-culture loop. Thus, signal processing can be broken down into two discrete sections (a) 'culture to robot', in which an output machine learning procedure processes live neuronal activity, and (b) 'robot to culture', which involves an input mapping process, from robot sensor to stimulus.

It is important to realise that the overall system employed in this experiment has been designed based on a closed-loop, modular architecture. As neuronal networks exhibit spatiotemporal patterns with millisecond precision [22], processing of these signals necessitates a very rapid response from neurophysiological recording and robot control systems. The software developed for this project runs on Linux-based workstations communicating over the Ethernet via fast server–client modules, thus providing the necessary speed and flexibility required.

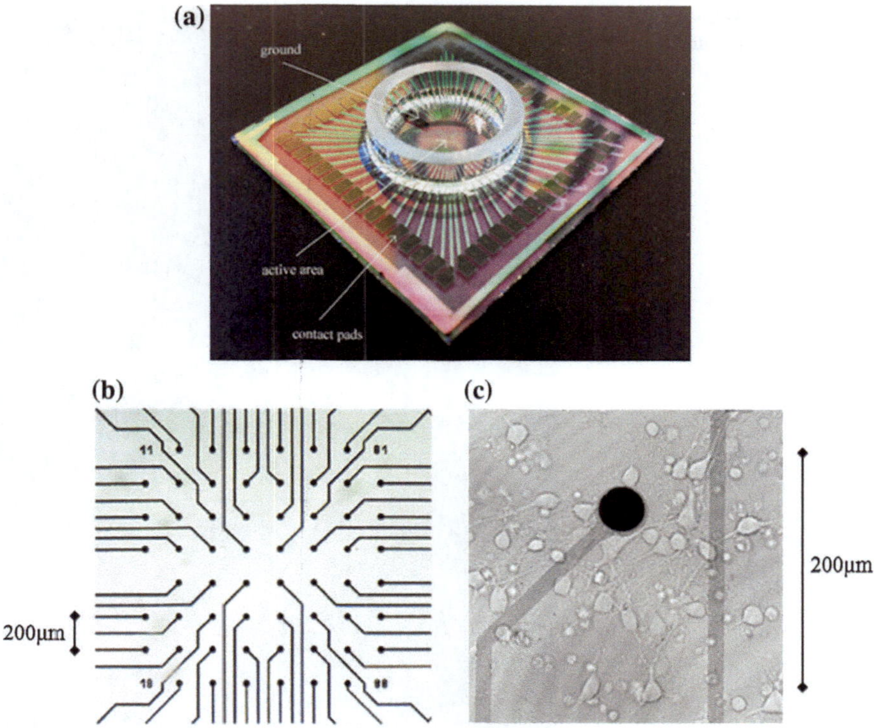

Fig. 7.1 **a** An MEA, showing the 30 μm electrodes which lead to the electrode column–row arrangement, **b** Electrode arrays in the centre of the MEA seen under an optical microscope, **c** An MEA at ×40 magnification, showing neuronal cells in close proximity to an electrode, with visible extensions and interconnections

In recent years, the study of neuronal cultures has been greatly facilitated by commercially available planar MEA systems. These consist of a glass specimen chamber lined with an 8 × 8 array of electrodes as shown in Fig. 7.1. It is just such one of these MEAs that we have employed in our overall robot system.

A standard MEA (Fig. 7.1a) measures 49 × 49 × 1 mm and its electrodes provide a bidirectional link between the culture and the rest of the system. The associated data acquisition hardware includes a head-stage (MEA connecting interface), 60 channel amplifier (1200 × gain; 10–3200 Hz band-pass filter), stimulus generator and PC data acquisition card.

To this point, we have successfully created a modular closed-loop system between a (physical) mobile robotic platform and a cultured neuronal network using a multielectrode array, allowing for bidirectional communication between the culture and the robot. It is estimated that the cultures employed in our studies consist of approximately (on average) 100,000 neurons. The actual number in any one specific culture depends on natural density variations in proliferation post-seeding and experimental aim. The spontaneous electrochemical activity of the culture

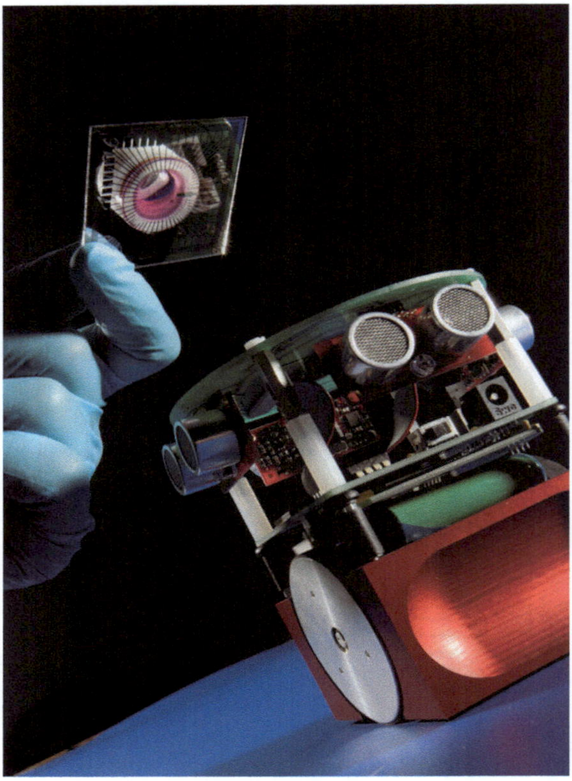

Fig. 7.2 The Miabot robot with a cultured neural network

realising signals at certain of the electrodes is used as input to the robot's actuators and the robot's (ultrasonic) sensor readings are (proportionally) converted into stimulation signals received by the culture, effectively closing the loop.

We are using a versatile, commercially available, Miabot robot (Fig. 7.2) as our physical platform. This exhibits accurate motor encoder precision (~0.5 mm) and has a maximum speed of approximately 3.5 m/s. Hence it can move around quite quickly in real time. Recording and stimulation hardware is controlled via open-source MEABench software [23].

The overall closed-loop system therefore consists of several modules including the Miabot robot, an MEA and stimulating hardware, a directly linked workstation for conducting computationally expensive neuronal data analyses and a separate machine running the robot control interface; a network manager routing signals directly between the culture and the robot body.

The Miabot is wirelessly controlled via Bluetooth. Communication and control are performed through custom C++ server code and TCP/IP sockets and clients running on the acquisition PC which has direct control of the MEA recording and stimulating software. The server sends motor commands and receives sensory data via a virtual serial port over the Bluetooth connection, while the client programs contain the closed-loop code which communicates with and stimulates the MEA culture. The client code also performs text logging of all important data during an experiment run.

This modular approach to the architecture has resulted in a system with easily reconfigurable components. The obtained closed-loop system can efficiently handle the information-rich data that is streamed via the recording software. A typical sampling frequency of 25 kHz of the culture activity demands large network, processing and storage resources. Consequently, on-the-fly streaming of spike-detected data is the preferred method when investigating real-time closed-loop learning techniques.

7.4 Experimental Results

First an existing appropriate neuronal pathway was identified by searching for strong input/output relationships between pairs of electrodes. Suitable input/output pairs were defined as those electrode combinations in which neurons proximal to one electrode responded to stimulation of the other electrode at which the stimulus was applied (at least one action potential within 100 ms of stimulation) more than 60 % of the time and responded no more than 20 % of the time to stimulation on any other electrode. An input–output response map was then created by cycling through all preselected electrodes individually with a positive-first biphasic stimulating waveform (600 mV; 100 μs each phase, repeated 16 times). By averaging over 16 attempts, it was ensured that the majority of stimulation events fell outside any inherent culture bursting that might have occurred. In this way, a suitable input/output pair could be chosen, dependent on how the cultures had developed, in order to provide an initial decision making pathway for the robot.

Thus, in the initially developed culture, we found by experimentation a reasonably repeatable pathway in the culture from stimulation to response. We then employed this pathway to achieve desired robot control—for example, for the obstacle avoidance task—if the ultrasonic sensor was active we mapped the culture response onto the robot actuators in order to cause the robot to turn away from the object being located ultrasonically. This ensures constant movement of the robot, without bumping into any obstacles.

The robot followed a forward path within its corral confines until it reached a wall, at which point the front sonar value decreased below a threshold (set at approximately 30 cm), triggering a stimulating pulse as shown in Fig. 7.3. If the responding/output electrode registered activity following the input pulse then the robot turned to avoid the wall. Essentially, activity on the responding electrode was interpreted as a command for the robot to turn in order to avoid the wall. It was apparent that, in fact, the robot turned spontaneously whenever activity was registered on the response/output electrode. The most relevant result for the experiment was the occurrence of the chain of events: wall detection stimulation response. From a philosophical and neurological perspective, it is of course also of interest to

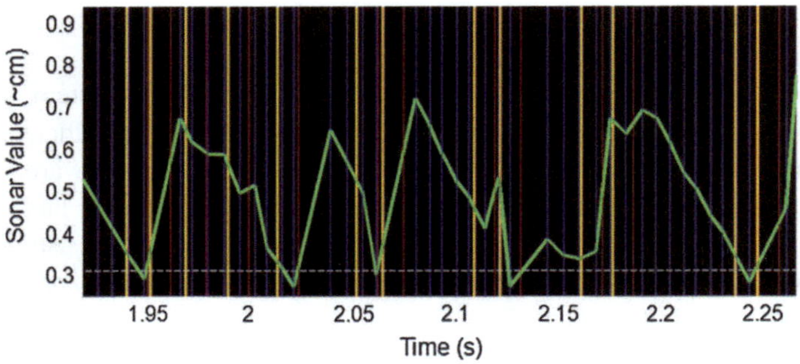

Fig. 7.3 Analysis of the robot's activity during a simple wall detection/right turn experiment

speculate why there was activity on the response electrode when no stimulating pulse had been applied.

The typical behaviour in the cultures studied was generally a period of inactivity (or low-frequency activity) prior to stimulus, followed by heightened network activity induced almost immediately (within few millseconds) after stimulus, which decayed (typically after ~100 ms) to baseline pre-stimulus activity. The study opens up the possibility of investigating response times of different cultures under different conditions and how they might be affected by external influences such as electrical fields and pharmacological stimulants [24]. At any one time we have typically 25 different cultures available and hence such comparative developmental studies are now being conducted.

With the sonar threshold set at approximately 30 cm from a wall, a stimulation pulse was applied to the culture, via its sensory input, each time this threshold was breached–effectively when the robot's position was sufficiently close to a wall. An indication of the robot's typical activity during a simple wall detection/right turn experiment is shown in Fig. 7.3. The green trace indicates the front sonar value. Yellow bars indicate stimulus pulse times and blue/red bars indicate sonar timing/actuator command timing.

As can be witnessed, these response events may occur purely spontaneously or due to electric stimulation as a result of the sensor threshold being breached. Such events are deemed 'meaningful' only in the cases when the delay between stimulation and response is less than 100 ms. In other words, such an event is a strong indicator that the electric stimulation on one electrode caused a neural response on the recording electrode. The red vertical lines indicate the time that a rotation command is sent to the robot. These events are always coupled (the first one starts the right turn rotation and the second simply ends the rotation). Only the second signals of each pair can be clearly seen here as the rotation initiation commands are overlaid by the yellow electrode firing bars (as a result of electrode firing which instantly initiates a rotation command). A 'meaningful' event chain would be for example at 1.95 s, where the sonar value drops below the threshold value (30 cm) and a stimulation-response subsequently occurs.

Table 7.1 Basic statistics from a wall avoidance experiment

Results	Simulation	Live culture
Wall - > Stimulation event	100 %	100 %
Stimulation - > Response event	100 %	67 %
Total closed loop time	0.075 s	0.2–0.5 s
Run time	240 s	140 s
Meaningful turns	41	22
Spontaneous turns	41	16

Table 7.1 shows typical results from a live culture test in comparison with a 'perfect' simulation. If the live culture acted 'perfectly', making no mistakes, then the two columns would be identical. This of course raises the question as to what a 'perfect' response actually is. In this case it could be regarded as a programmed exercise—which some might refer to as 'machine-like'. In a sense therefore the culture is asserting its own individuality by not being 'perfect'.

To explain Table 7.1 further—'Total closed loop time' refers to the time between wall detection and a response signal witnessed from the culture. 'Meaningful turns' refer to the robot turning due to a 'wall detection-stimulation-response' chain of events. A 'wall to stimulation' event corresponds to the 30 cm threshold being breached on the sensor such that a stimulating pulse is transmitted to the culture. Meanwhile a 'stimulation to response' event corresponds to a motor command signal, originating in the culture, being transmitted to the wheels of the robot to cause it to change direction. It follows that for the culture some of the 'stimulation to response' events will be in 'considered' response to a recent stimulus–termed meaningful, whereas other such events–termed spontaneous—will be either spurious or in 'considered' response to some thought in the culture, about which we are unaware.

By totalling the results of a series of such trials carried out (over 100), considerable differences (as typically indicated in Table 7.1) are observed between the ratio of expected and spontaneous turns between the simulation and the live culture. Under the control of the simulation 95 $\pm$ 4 % (Mean $\pm$ SD) meaningful turns were observed whilst the remaining spontaneous turns (5 $\pm$ 4 %) were easily attributable to aspects of thresholding spike activity. In contrast the live culture displayed a relatively low number of meaningful turns (46 $\pm$ 15 %) and a large number of spontaneous turns 54 $\pm$ 19 % as a result of intrinsic neuronal activity. Such a large number of spontaneous turns was perhaps only to be expected in an uncharacterised system and current work aims to both quiet the level of on-going spontaneous, reminiscent of epileptiform, activity present in such cultures and to discover more appropriate input sites and stimulation patterns.

As a follow up closed-loop experiment, the robot's individual (right and left separately) wheel speeds were controlled by using the spike firing frequency recorded from the two chosen motor/output electrodes. The frequency is actually calculated by means of the following simple principle: A running mean of spike rate from both the output electrodes was computed from the spike detector. The detected spikes for each electrode were separated and divided by the signal acquisition time to give a frequency value. These frequencies were linearly mapped (from their typical range of 0–100 Hz) to a range of 0–0.2 m/s for the individual wheel linear velocities. Meanwhile, collected sonar information was used to directly control (proportionally) the stimulating frequency of the two sensory/input electrodes. The typical sonar range of 0–100 cm was linearly re-scaled into the range 0.2–0.4 Hz for electrode stimulation frequencies (600 mV voltage pulses).

The overall setup can be likened to a simple Braitenberg model [25]. However, in our case, sensor-to-speed control is mediated by the cultured network acting as the sole decision-making entity within the overall feedback loop. One important aspect being focussed on is the evocation of long-term potentiation (LTP), i.e., directed neural pathway changes in the culture, thereby effecting plasticity between the stimulating-recording electrodes. Although this was not a major initial target in carrying out this part of the experiment, it has been noted elsewhere that a high-frequency burst time can induce plasticity very quickly [27, 28]. As a result, we are now investigating spike timing dependent plasticity based on the coincidence of spike and stimulus.

7.5 Learning

Inherent operating characteristics of the cultured neural network have been taken as a starting point to enable the physical robot body to respond in an appropriate fashion—to get it started. The culture then operates over a period of time within the robot body in its corral area. Experimental duration, e.g. how long the culture is operational within its robot body, is merely down to experimental design. Several experiments can therefore be completed within a day, whether on the same or differing cultures. The physical robot body of course can operate 24/7.

In our studies thus far, learning and memory [26] investigations are at an early stage. However, we were able to observe that the robot appeared to improve its performance over time in terms of its wall avoidance ability. We are currently investigating this and examining whether it can be repeated robustly and subsequently quantified. What we have witnessed could mean that neuronal structures/pathways employ some form of adaptive processes which allow the system to adjust to the closed-loop stimulation environment. In fact, such adaptation may underlie some forms of plasticity and hence learning. Typically, learning is considered to involve creating representations of the relationships between different objects or inputs. This is the so called associative learning paradigm and in its broadest form it includes auto associative and hetero associative paradigms such as classical or Pavlovian conditioning, operant conditioning and many others. However, psychologists also recognise other broad class of learning paradigms which do not involve association (at least not explicitly). Forms of such non-associative learning include sensitisation and habituation. The latter involves gradual adjustment of responses to repetitive stimuli and has been well documented in the psychological literature. Putative biological mechanisms of habituation have been well characterised in invertebrates.

We believe, that the experiment involving the robot wall avoidance task may constitute a natural setup for a form of habituation to take place. The observed improvement of the robot performance due to repetition of a satisfactory action may then be a result of the strengthening of neural pathways through a process of adaptation–learning due to habit.

This hypothesis opens a plethora of further interesting questions which we are addressing. What processes, structures or pathways in the cortical neural cultures support such habituation?

What is the 'cognitive' capacity of such form of learning? The issue involves characterisation of behavioural repertoire that can be supported by habituation. This includes the complexity of behaviours that can be exhibited as well as the number of different behaviours that can be simultaneously acquired by this form of learning. What are the dynamics of habituation? The length of time over which the habituation takes place, the level of robustness of the acquired behaviours, and what are the properties of degradation of them over time, if the behaviour is not maintained.

Another, more practical but far from trivial, problem that needs to be addressed is how to characterise the habituation and elucidate all the problems and issues highlighted above from data. The signals collected from a culture are highly dimensional, noisy and non-stationary. Moreover, operation in the closed loop requires a quick turnaround time which means that, in spite of the potentially long times over which the experiments are run; the data are of finite and short duration. Hence, the issue of reliable recognition and characterisation of neural responses in such circumstances comes to the fore.

We are encouraged by the fact that such habituation processes in animals have been reported on elsewhere, e.g. [29, 31], and experimentation has been carried out to investigate the effects of sensory deprivation on subsequent culture development. In our case we are monitoring changes and attempting to provide a quantitative characterisation relating plasticity to experience and time. The potential number of confounding variables is however, considerable, as the subsequent plasticity process, which occurs over quite a period of time, is (most likely) dependent on such factors as initial seeding and growth near electrodes as well as environmental transients such as feed rate, temperature and humidity.

On completion of these first phases of the infrastructure setup, a significant research contribution, it is felt, lies in the application of machine learning (ML) techniques to the hybrid system's closed-loop experiments. These techniques may be applied in the spike sorting process (dimensionality reduction of spike data profiles, clustering of neuronal units), the mapping process between sensory data and culture stimulation as well as the mapping between the culture activity and motor commands, and the application of learning techniques on the controlled electrical stimulation of the culture, in an attempt to exploit the cultured networks' computational capacity.

7.6 Conclusions

We have successfully realised a closed-loop adaptive feedback system involving a (physical) mobile robotic platform and a cultured neuronal network using a multielectrode array (MEA), which necessitates real-time bidirectional communication between the culture and the robot. A typical culture being employed presently consists of approximately 100,000–150,000 neurons, although at any one time only a small proportion of these neurons are actively firing.

Trial runs have been carried out with the overall robot and comparisons have been made with an 'ideal' simulation which responds to stimuli perfectly as required. It has been observed that the culture on many occasions responds as expected, on other occasions, however, it does not, and in some cases it provides a motor signal when it is not expected to do so.

The concept of an 'ideal' response is difficult to address because a biological network is involved, and it should not be seen in negative terms when the culture does not achieve (what is in our eyes) an ideal. We know very little about the fundamental neuronal processes that give rise to meaningful behaviours, particularly where learning is involved, we therefore need to retain an open mind as to a culture's performance.

The culture preparation techniques employed are constantly being refined and have led to stable cultures that exhibit both spontaneous and induced spiking/bursting activity which develops in-line with the findings of other groups, e.g. [15, 32].

A stable robotic infrastructure has been set up, tested and is in place for future culture behaviour and learning experiments. This infrastructure could be easily modified in order to investigate culture-mediated control of a wide array of alternative robotic devices, such as a robot head, an 'autonomous' vehicle, robotic arms/grippers, mobile robot swarms and multi-legged walkers.

In terms of robotics, this study and others like it, show that a robot can have a biological brain to make its 'decisions'. The 100,000–150,000 neuron size is due to present day limitations—clearly this will increase. Indeed it is already the case that 3-dimensional structures are being investigated [20]. Simply increasing the complexity from 2-dimensions to 3-dimensions (on the same basis) realises a figure of 30 million neurons (approximately) for the 3-dimensional case. The whole area of research is therefore a rapidly expanding one as the range of sensory inputs is expanded and the number of cultured neurons encapsulated rises. The potential capabilities of such robots, including the range of tasks they can perform, therefore needs to be investigated.

Understanding neural activity becomes a much more difficult problem as the culture size is increased. Even the present neuron cultures are far too complex at present for us to gain an overall insight. When they are grown to sizes such as 30 million neurons and beyond, clearly the problem is significantly magnified, particularly with regard to neural activity in the centre of a culture volume, which will be (effectively) hidden from view. On top of this, the nature of the neurons may be diversified. To this point rat neurons are employed in our studies–however, any animal neurons could be used. Human neurons, which are readily available are particularly suitable due to the fact that any results obtained could be more immediately relevant to human studies. The author wishes to record his feelings here that it is important to stress the need for ethical concerns to be listened to in such circumstances.

7.7 Future Research

There are a number of ways in which the current research programme is being taken forward. First the robot is being extended to include additional sensory devices such as extra sonar arrays, audio input, mobile cameras and other range-finding hardware such as an on-board infrared sensor. This will provide an opportunity to investigate sensory fusion in the culture and to perform more complex behavioural experiments, possibly even attempting to demonstrate links between behaviour and culture plasticity, along the lines of [29], as different sensory inputs are integrated.

Provision of a powered-floor for the robot's corral will provide the robot with relative autonomy for a longer period of time while different learning techniques are applied and behavioural responses monitored. For this the Miabot must be adapted to operate on an in-house powered-floor, providing the robot with an unlimited power supply. This feature, which is based on an original design for displays in museums [30] is necessary since learning and culture behaviour tests will be carried out for hours at a time.

Current hardcoded mapping between the robot goals and the culture input/output relationships can be extended using learning techniques to eliminate the need for an a priori choice of the mapping. In particular, reinforcement learning techniques can be applied to various mobile robot tasks such as wall following and maze navigation, in an attempt to provide a formal framework within which the learning capabilities of the neuronal culture will be studied [35, 36].

To increase the effectiveness of culture training beyond the ~30 % success rate seen in previous work, biological experiments are currently being performed to identify physiological features which may play a role in cellular correlates of learning processes. These experiments also investigate possible methods of inducing an appropriate receptive state in the culture that may allow greater control over its processing abilities and the formation of memories [26] involving specific network activity changes, which may allow identification of the function of given network ensembles. In particular, in terms of cholinergic influences, the possible effect of acetylcholine (ACh) [33] in coordinating the contributions of different memory systems is being investigated.

The learning techniques employed and the results obtained from the culture need to be benchmarked. In order to achieve this we are developing a model of the cultured neural network, based on experimental data about culture density and activity. In doing so we hope to gain a better understanding of the contribution of culture plasticity and learning capacity to the observed control proficiency. At present we are investigating hidden markov models (HMMs) as a technique for uncovering dynamic spatiotemporal patterns emerging from spontaneously active or stimulated neuronal cultures [34, 37]. The use of hidden markov models enables characterisation of multichannel spike trains as a progression of patterns of underlying discrete states of neuronal activity.

Key for the future, however, will be the possible employment of 30 million human neurons as the brain of a robot body. Through technical development, it is envisioned that the number of neurons will subsequently increase further still [38]. As the 100 billion typical number of neurons is approached, other issues then become apparent—as the sheer size of the robot brain tends to approach that of a human—e.g. what, if any, rights should the robot have?

Acknowledgements This work is funded by the UK Engineering and Physical Sciences Research Council (EPSRC) under Grant No. EP/D080134/1.

References

1. Reger, B., Fleming, K., Sanguineti, V., Alford, S., Mussa-Ivaldi, F.: Connecting brains to robots: An artificial body for studying the computational properties of neural tissues. Artif. Life **6**, 307–324 (2000)
2. Holzer, R., Shimoyama, I., Miura, H.: Locomotion control of a bio-robotic system via electric stimulation. In: Proceedings of International Conference on Intelligent Robots and Systems, Grenoble, France (1997)
3. Talwar, S., Xu, S., Hawley, E., Weiss, S., Moxon, K., Chapin, J.: Rat navigation guided by remote control. Nature **417**, 37–38 (2002)
4. Chapin, J., Moxon, K., Markowitz, R., Nicolelis, M.: Real-time control of a robot arm using simultaneously recorded neurons in the motor cortex. Neuroscience **2**, 664–670 (1999)
5. Bakkum, D.J., Shkolnik, A., Ben-Ary, G., DeMarse, T. Potter, S.: Removing Some `A' from AI: Embodied Cultured Networks, Lecture Notes in Computer Science, pp. 130-145 (2004)
6. Thomas, C., Springer, P., Loeb, G., Berwald-Netter, Y., Okun, L.: A miniature microelectrode array to monitor the bioelectric activity of cultured cells. Exp. Cell Res. **74**, 61–66 (1972)
7. Gross, G.: Simultaneous single unit recording in vitro with a photoetched laser deinsulated gold multimicroelectrode surface. IEEE Trans. Biomed. Eng. **26**, 273–279 (1979)
8. Pine, J.: Recording action potentials from cultured neurons with extracellular microcircuit electrodes. J. Neurosci. Methods **2**, 19–31 (1980)
9. Potter, S., Lukina, N., Longmuir, K., Wu, Y.: Multi-site two-photon imaging of neurons on multi-electrode arrays. SPIE Proc. **4262**, 104–110 (2001)
10. Gross, G., Rhoades, B., Kowalski, J.: Dynamics of burst patterns generated by monolayer networks in culture. In: Neurobionics: An Interdisciplinary Approach to Substitute Impaired Functions of the Human Nervous System, pp. 89–121 (1993)
11. Kamioka, H., Maeda, E., Jimbo, Y., Robinson, H., Kawana, A.: Spontaneous periodic synchronized bursting during the formation of mature patterns of connections in cortical neurons. Neurosci. Lett. **206**, 109–112 (1996)
12. Lewicki, M.: A review of methods for spike sorting: the detection and classification of neural action potentials. Network (Bristol) **9**(4), R53 (1998)

13. Saito, S., Kobayashik, S., Ohashio, Y., Igarashi, M., Komiya, Y., Ando, S.: Decreased synaptic density in aged brains and its prevention by rearing under enriched environment as revealed by synaptophysin contents. J. Neurosci. Res. **39**, 57–62 (1994)
14. Ramakers, G.J., Corner, M.A., Habets, A.M.: Development in the absence of spontaneous bioelectric activity results in increased stereotyped burst firing in cultures of dissociated cerebral cortex. Exp. Brain Res. **79**, 157–166 (1990)
15. Chiappalone, M., Vato, A., Berdondini, L., Koudelka-Hep, M., Martinoia, S.: Network dynamics and synchronous activity in cultured cortical neurons. Int. J. Neural Syst., **17**(2), 87–103 (2007)
16. Shkolnik, A.C.: Neurally controlled simulated robot: applying cultured neurons to handle an approach/avoidance task in real time, and a framework for studying learning in vitro, in mathematics and computer science. Masters thesis, Department of Computer Science, Emory University, Georgia (2003)
17. DeMarse, T.B., Dockendorf, K.P.: Adaptive flight control with living neuronal networks on microelectrode arrays. In: Proceedings of the 2005 IEEE International Joint Conference on Neural Networks, pp.1549–1551 (2005)
18. Shahaf, G., Marom, S.: Learning in networks of cortical neurons. J. Neurosci. **21**(22), 8782–8788 (2001)
19. Bull, L., Uruokov, I.: Initial results from the use of learning classifier systems to control in vitro neuronal networks. In: Proceedings of the 9th Annual Conference on Genetic and Evolutionary Computation. (GECCO).pp. 369–376. The ACM, London, England (2007)
20. Hammond, M., Marshall, S., Downes, J., Xydas, D., Nasuto, S., Becerra, V., Warwick, K. Whalley, B.J.: Robust methodology for the study of cultured neuronal networks on MEAs. In: Proceedings 6th International Meeting on Substrate-Integrated Micro Electrode Arrays, pp. 293–294 (2008)
21. Potter, S.M., DeMarse, T.B.: A new approach to neural cell culture for long-term studies. J. Neurosci. Methods **110**, 17–24 (2001)
22. Rolston, J.D., Wagenaar, D.A., Potter, S.M.: Precisely timed spatiotemporal patterns of neural activity in dissociated cortical cultures. Neuroscience **148**, 294–303 (2007)
23. Wagenaar, D.A.D., T.B.; Potter, S.M.: MEABench: a toolset for multi-electrode data acquisition and on-line analysis. In: Proceedings of the 2nd International IEEE EMBS Conference on Neural Engineering, pp. 518–521 (2005)
24. Xydas, D., Warwick, K., Whalley, B., Nasuto, S., Becerra, V., Hammond, M., Downes, J.: Architecture for Living Neuronal Cell Control of a Mobile Robot. In: Proceedings of the European Robotics Symposium EUROS08, pp. 23–31. Prague, CZ, (2008)
25. Hutt, B., Warwick, K., Goodhew, I.: Emergent Behaviour in Autonomous Robots. In: Bryant, J., Atherton, M. and Collins, M. (eds.) Chapter 14 in Information Transfer in Biological Systems. Design in Nature Series, vol. 2, The WIT Press (2005)
26. Hasselmo, M.E.: Acetycholine and learning in a cortical associative memory source. Neural Comput. Arch. **5**, 32–44 (1993)
27. Cozzi, L., Chiappalone, M., Ide, A., Novellino, A., Martinoia, S., Sanguineti, V.: Coding and Decoding of Information in a Bi-directional Neural Interface. Neurocomputing **65**(66), 783–792 (2005)
28. Novellino, A., Cozzi, L., Chiappalone, M., Sanguinetti, V., Martinoia, S.: Connecting neurons to a mobile robot: an in vitro bi-directional neural interface, Comput. Intell. Neurosci. (2007)
29. Karniel, A., Kositsky, M., Fleming, K., Chiappalone, M., Sanguinetti, V., Alford, T., Mussa-Ivaldi, A.: Computational Analysis In Vitro: Dynamics and Plasticity of a Neuro-Robotic System. J. Neural Eng. **2**, S250–S265 (2005)
30. Hutt, B., Warwick, K.: Museum robots: multi-robot systems for public exhibition. In: Proceedings of the 35th International Symposium on Robotics, p.52. Paris, France (2004)
31. Marks, P.: Rat-brained robots take their first steps. New Sci. **199**(2669), 22–23 (2008)
32. DeMarse, T., Wagenaar, D., Blau, A., Potter, S.: The neurally controlled animat: Biological brains acting with simulated bodies. Auton. Robots **11**, 305–310 (2001)

33. Chang, Q., Gold, P.: Switching memory systems during learning: Changes in patterns of brain acetylcholine release in the hippocampus and striatum in rats. J. Neurosci. **23**, 3001–3005 (2003)
34. Xydas, D., Downes, J., Spencer, M., Hammond, M., Nasuto, S., Whalley, B., Becerra, V., Warwick, K.: Revealing ensemble state transition patterns in multi-electrode neural recordings using hidden Markov models. IEEE Trans. Neural Syst. Rehabil. Eng. **19**(4), 345–355 (2011)
35. Warwick, K., Xydas, D., Nasuto, S., Becerra, V., Hammond, M., Downes, J., Marshall, S., Whalley, B.: Controlling a mobile robot with a biological brain. Def. Sci. J. **60**(1), 5–14 (2010)
36. Warwick, K., Nasuto, S., Becerra, V., Whalley, B.: Experiments with an in-vitro robot brain. In: Cai Y., (ed.) Computing with Instinct, Lecture Notes in Computer Science, vol. 5897, pp. 1–15, Springer, Berlin (2011)
37. Spencer, M., Downes, J., Xydas, D., Hammond, M., Becerra, V., Warwick, K., Whalley, B., Nasuto, S.: Multiscale evolving complex network model of functional connectivity in neuronal cultures, IEEE Trans. Biomed. Eng., **59**(1), 30–34 (2012)
38. Warwick, K.: Implications and consequences of robots with biological brains. Ethics Inf. Technol. **12**(3), 223–234 (2010)

Chapter 8
Iterative Learning Control as an Enabler for Robotic-Assisted Upper Limb Stroke Rehabilitation

Eric Rogers, Chris T. Freeman, Ann-Marie Hughes, Jane H. Burridge, Katie L. Meadmore and Tim Exell

Abstract An increased burden on health care and rehabilitation resources is due to the number of people suffering a stroke and if the capacity of health services is to meet future demand novel approaches to rehabilitation are required. In this chapter recent research is surveyed where iterative learning control, developed initially for robots executing commonly encountered industrial tasks such as sequentially collecting objects from one location and transferring them to another, is used to control the assistive stimulation in robotic-assisted upper limb stroke rehabilitation. The results given include the outcomes of small-scale clinical trials with stroke patients and areas for future research are also briefly discussed.

8.1 Introduction

Stroke is a leading cause of death and disability in humans. Every year 15 million people worldwide suffer a stroke and 5 million are left permanently disabled. In the UK, as one example, approximately 50 % of people who survive a stroke require some form of rehabilitation to reduce impairment and assist with activities of daily living [1]. The upper limb is particularly important in regaining independence following stroke as impairments impact on daily living and well-being [2]. A very large volume of literature exists on the various aspects of stroke discussed in this section and the references cited are indicative only.

A stroke is usually caused when a blood clot blocks a vessel in the brain and acts like a dam stopping the blood reaching the regions downstream. Alternatively, it may be caused by a hemorrhage where a vessel ruptures and leaks blood into the

E. Rogers (✉) · C.T. Freeman · K.L. Meadmore · T. Exell
Department of Electronics and Computer Science, University of Southampton,
Southampton SO17 1BJ, UK
e-mail: etar@ecs.soton.ac.uk

A.-M. Hughes · J.H. Burridge
Faculty of Health Sciences,University of Southampton,
Southampton SO17 1BJ, UK
e-mail: a.hughes@soton.ac.uk

G.M. Dimirovski (ed.), *Complex Systems*, Studies in Systems,
Decision and Control 55, DOI 10.1007/978-3-319-28860-4_8

surrounding areas. As a result, some of the connecting nerve cells die and the person commonly suffers partial paralysis on one side of the body, termed hemiplegia. Cells killed in this way cannot regrow, but the brain has some spare capacity and hence new connections can be made. Moreover, brain is continually and rapidly changing as new skills are learned, new connections are formed, and redundant ones disappear. A person who relearns skills after a stroke goes through the same process as someone learning to play tennis or a baby learning to walk, requiring sensory feedback during repeated practice of a task. Unfortunately, the problem is that they can hardly move and therefore do not receive feedback on their performance.

Research [3] has consistently identified treatment intensity and goal-oriented strategies as critical elements for successful therapeutic outcomes. To further maximize rehabilitation effects, novel therapeutic and cost-effective rehabilitation interventions need to be developed and may combine different methodological techniques. For example, the combined use of electrical stimulation, as referred to as Functional Electrical Stimulation (FES) in this chapter, robot-aided therapy and virtual reality (VR) environments have been suggested to be particularly promising with respect to upper limb rehabilitation in chronic stroke [4].

Following stroke, robotic and FES therapies have been demonstrated to reduce upper limb motor impairments [5]. Furthermore, these techniques have been identified as a way to facilitate the intensity of the training received [3] and allow training despite muscle weakness and without the aid of a therapist. In addition, when used with a real-time system which displays the participants' arm and hand movements in a VR environment, the practiced movements can be very task-specific [4]. These types of technologies may be more easily transferred into patients' homes, increasing the intensity and task specificity of the training and reducing the time and expense constraints on therapists. The therapeutic effect of FES in rehabilitation is known to increase when associated with a person's voluntary effort [6]. However, a disadvantage of many FES approaches is that they fail to encourage voluntary contribution.

The research described in this chapter is from a program that is investigating the feasibility and effectiveness of novel rehabilitation systems that combine robotic support, FES, and VR. These systems allow patients to receive the benefits of muscle-specific targeted FES within a tightly controlled, safe, and motivating environments. The FES is mediated by iterative learning control (ILC), a technology transferred from industrial robotics which is applicable to systems which repeatedly perform a finite duration tracking operation [7, 8]. After each repetition, termed a trial, ILC uses data gathered on previous executions of the task, in combination with a model of the underlying dynamics, to update the FES signal that is applied on the next trial. This control law design method allows strict regulation of the amount of FES used, its trial-to-trial variation, and the resulting movement error.

The next section gives a brief overview of how the 'technology transfer' to stroke rehabilitation has been achieved by considering the first work on a planar reaching task and also describes the measures health professionals use to assess the progress of a patient in a rehabilitation program for the upper limb. Currently, research is focused on 3D rehabilitation but the initial work was on planar tasks that are also essential to daily living, such as reaching out over a table top to an object. This research, including the outcomes of a small-scale clinical trial are covered in Sect. 8.3. Section 8.4 considers 3D rehabilitation, again with clinical trial results, and final section gives the overview of the progress achieved together with discussion of ongoing and planned research. A preliminary version of this chapter is given in [9].

8.2 ILC Applied to Stroke Rehabilitation

Iterative Learning Control (ILC) emerged from industrial applications where the system involved, such as a robot, executes the same operation many times over a fixed time interval. When each trial is complete, resetting to the starting location takes place and the next one can commence immediately or after a stoppage time. A particular example is a gantry robot undertaking a pick and place operation in synchronization with a moving conveyor or assembly line. The sequence of the operations is: (a) the robot collects a payload from a fixed location, (b) transfers it over a finite duration, (c) places it on the moving conveyor, (d) returns to the original location for the next object, and then (e) repeats the previous four steps for as many payloads as is required or can be transferred before it is required to stop.

To operate it is necessary to supply the robot with a trajectory to follow. The task for a control law is to ensure that the robot follows the prescribed trajectory exactly or, more realistically, to within a specified tolerance. In addition to controlling the robots movement and that of the payload, the control law must prevent other effects, such as disturbances and signal noise, from degrading tracking and forcing it outside of the tolerance bound. If the robot begins to operate outside permissible limits, the control task is to bring it back as quickly as required or is physically possible. This must be achieved without causing damage to, for example, the sensing and actuating technologies used.

In the ILC literature, each completion or execution of the task is described as a pass, iteration or trial, but in this work the latter term is exclusively used. Similarly, the finite time each trial takes will be referred to as the trial length. Once a trial is complete, all data used and generated during its completion is available for use in computing the control action to be applied on the next trial. The use of such data is a form of learning and is the essence of ILC, embedding the mechanism through which performance may be improved by past experience.

Evidence from the application of conventional therapy and motor learning theory provide evidence that intensity of practice of a task and feedback are important, see, for example, [10] in stroke rehabilitation. This knowledge is motivating the development of novel treatments such as robotic therapy that provide the opportunity for repetitive movement practice. Although the use of robotic therapy in upper limb rehabilitation is relatively recent, reviews of the literature suggest that robot-aided therapy improves motor control of the proximal upper limb and may improve functional outcomes [11]. Moreover, electromechanical and robotic devices may have an advantage over conventional therapies in the frequency of movement repetitions because of an increased motivation to train and also the opportunity for independent exercise [11]. Based on existing evidence, use of rehabilitation robots is recommended in the UK stroke guidelines [12].

Clinical evidence exists to support the therapeutic use of functional electrical stimulation, denoted as FES, to improve motor control [13]. This form of stimulation makes muscles work by causing electrical impulses to travel along the nerves in much the same way as electrical impulses from the brain, and if stimulation is carefully controlled, a useful movement can be made. Theoretical results from neurophysiology [14] and motor learning research support clinical research with the conclusion that the therapeutic benefit of stimulation is maximized when applied coincidently with a patient's own voluntary intention to move [15]. A hypothesis has been proposed in [16] to explain the added benefit of increased recovery when FES is used to mimic a weak or paralyzed movement.

A variety of FES model-based control methods, surveyed in [17] have been employed to control movement, but the majority are intended for spinal cord injury subjects, which is reflected in the number of approaches focused on the lower limb. Advanced techniques, such as those referenced above, have rarely transferred to clinical practice, as discussed in [17] with supporting references. See also [18] for a control-related treatment of FES applied to muscles.

A major reason for the lack of model-based methods finding application in a program of patient trials is the difficulty in obtaining reliable biomechanical models. In the clinical setting there is minimal set-up time, reduced control over environmental constraints, and little possibility of repeating any one test in the program of treatment undertaken; controllers are required to perform to a minimum standard on a wide number of subjects and conditions. These problems are often exacerbated in the case of stroke because hemiplegic subjects exhibit both voluntary and involuntary responses to applied stimulation. The limited number of model-based approaches that have been used in stroke rehabilitation therefore provide limited scope to adapt the applied stimulation to changes in the underlying system due to fatigue or spasticity, leading to reduction in performance and an inability to fully exploit the therapeutic potential.

This chapter describes the application of ILC in stroke rehabilitation, including clinical trials that constitute the first major stage toward eventual transfer into practice. In contrast to the other approaches employed to control FES, ILC exploits the repeating nature of the patients' tasks in order to improve performance by learning from past experience. By updating the control input using data collected over previous attempts at the task, ILC is able to respond to physiological changes in the system, such as spasticity and the presence of patient's voluntary effort, which would otherwise erode performance. Moreover, ILC can closely regulate the amount of stimulation supplied, ensuring that minimum assistance is provided, thereby promoting the patient's maximum voluntary contribution to the task completion. As the treatment progresses this control action encourages patients to exert increasing voluntary effort with each trial and therefore a corresponding decrease in the level of FES applied.

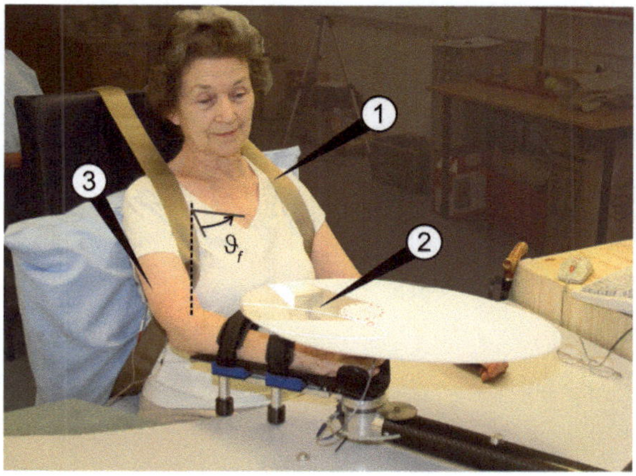

Fig. 8.1 A frontal view of a patient using the planar robotic workstation: showing (*1*) shoulder strapping, (*2*) tracking task, and (*3*) surface electrodes

The first research in this area concentrated on a planar task that replicates the everyday task of reaching out to, for example, a cup, where the aim was to establish the basic feasibility and hence the exclusion of more complicated tasks, such as reaching out and then extending the arm. This research is described in the next section where Fig. 8.1 shows a stroke patient using the system designed for this purpose.

The patient in this picture is seated with her impaired arm supported by the robot, and elliptical trajectories are projected onto a target above the hand and FES is applied to her triceps, using the surface electrodes, in order to assist tracking of a point that moves along the reference trajectory. At the end of the task, the arm is returned to the starting position in preparation for the next trial. During the reset time an ILC control law is used to calculate the stimulation to be applied on the subsequent trial. The stimulation applied to the triceps muscle produces a torque about the elbow and the control problem is equivalent to controlling the angle ϑ_f. The shoulder strapping is to prevent forward movement by the patient's trunk during the trials, which would conflict with the desired objective of reaching out with the arm.

Figure 8.2 shows a plain view of the motion of the patients' arm in Fig. 8.1 and the link with the operation of systems for which ILC was developed is clear. The patient aims to track the supplied reference signal, in the form of a light path beamed down from above, with the aid of FES applied to the triceps muscle and the error between the path followed by the patient and the reference is measured. Once the end is reached, the arm is returned to the starting location and in this reset time, plus a rest time to help counter the effects of muscle fatigue, the measured previous trial error is used to compute the level of stimulation to be applied on the next trial.

If the patient is improving with each successive trial then the level of stimulation should decrease from one trial to the next and the patients' voluntary effort increase. This is an essential property if recovery to completing the task is to be achieved and retained and this was detected in the clinical trial discussed in the next section.

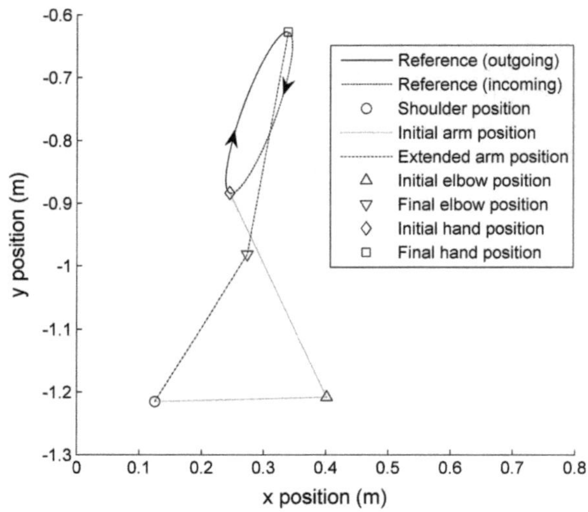

Fig. 8.2 Block diagram representation of the ILC control scheme in the planar case

The reference signal for the patients in this trial was determined by health professionals based on an assessment of each patient. If this signal is one the patient can already complete then no benefit is gained and if it is too far beyond current unaided capability this risks demotivating the patient. An open research question is the provision of tools based on the measurements of current capability to assist in the formulation of reference signals in this application area.

8.2.1 Measurement in Neurorehabilitation

The purpose of measurement in neurorehabilitation is twofold. First to design therapy (to make initial decisions and decide changes to therapy programs) and second to measure progress. The World Health Organisation's International Classification of Functioning, Disability and Health (ICF) is a framework for measuring both health and disability [19]. It consists of domains which are 'health' and 'health related' described as two lists: body functions and structures, and activity and participant. Impairments are defined as problems in body function or structure such as a significant deviation or loss, whereas activity is the execution of a task or action by an individual and participation is involvement in a life situation (society).

Active assisted or partially facilitated exercises are recommended for stroke patients who are unable to move by themselves [20]. To measure the effectiveness of such techniques, healthcare professionals are more likely to used activity or participant-based outcome measures which explain how effective an intervention has been (which may be more relevant to the patient) than impairment-based measures, which normally require more equipment to explain why these changes are being seen.

Conventionally, healthcare professionals use standardized validated clinical outcome measures to address changes in body functions and structures or activities. There is little consensus within the literature regarding the best motor performance outcome measure for stroke patients, although it has been suggested that trials should use valid instruments that measure upper limb skills specifically, such as the Action Research Arm Test (ARAT) or Structural Myofascial Therapy (SMFT) to assess improvement in activities of daily living [21].

In the research considered in this chapter, the primary measure for upper limb function used was the ARAT, for which reliability and validity have been established. Movement, coordination and sensation of the upper limb have also been measured using the Fugl-Meyer Assessment (FMA) [22], a valid and reliable measure of post-stroke impairments. Both of these ordinal measures have been extensively used in the clinical robotic and electrical stimulation literature, but still have limitations including floor and ceiling effects, experimenter bias, and inter-rater reliability across clinical trial sites.

The ARAT [23] was developed to monitor function related to everyday tasks and uses a hierarchical measure of grasp, grip, pinch, and gross movement. The reaching and grasping movements are rated on quality and speed in three dimensions. The ARAT assesses primarily activity limitations, i.e., a patient's functional loss when interacting with the environment by means of the upper limb. Reliability and validity have been established, however, the measure shows both floor and ceiling effects, i.e., limitations for people who have either very low or very high function. The test is widely used in the electrical stimulation literature.

The primary outcome used to detect changes in the upper limb impairment is the FMA, which primarily assesses impairment in terms of loss or abnormality of movement, i.e., the ability to perform movements in accordance with specified joint motion pattern. It provides an adequate, reproducible, and fairly standardized picture of a patient's sensorimotor and joint characteristics. It is an ordinal scale testing gross movement, coordination, and sensation of the upper limb. The motor part of the scale scores a maximum of 66 points: section A (shoulder, elbow, and forearm 36 points), B (wrist 10 points), C (hand 14 points), and D (coordination/speed 6 points). This test is appropriate for severe to mildly affected patients and has high reliability and validity.

The widespread use of the FMA in research involving rehabilitation robots, together with its utility and standardized procedure, meant that it was the obvious choice of impairment outcome measure. In terms of resolution, the FMA could in contrast to the ARAT, detect differences throughout the spectrum of motor dysfunction of the study population and is less affected by floor or ceiling effects. Patient Reported Outcome Measures (PROMs) which provide validated evidence of health from the user or patient perspective are increasingly being used. This is due in part to their use as evidence of outcomes of services for comparative performance evaluation, quality assurance, and audit.

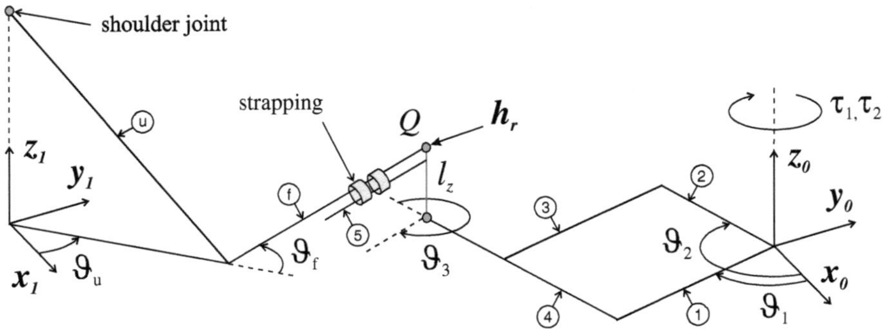

Fig. 8.3 Geometry of combined human arm and robotic arm system

8.3 ILC for Rehabilitation of Planar Tasks

The robotic system used was especially designed for this research, with full details in [24]. Figure 8.3 shows the geometry of the dual human and robotic system, where Q denotes where the subject's hand grasps the robot. The robot joint angle vector is $\boldsymbol{q}_r = \left[\vartheta_1 \ \vartheta_2\right]^T$, where ϑ_1 and ϑ_2 are the joint angles corresponding to links 1 and 2 respectively. The torque supplied by the motors is given by $\boldsymbol{\tau}_r = \left[\tau_1 \ \tau_2\right]^T$, where τ_1 and τ_2 are applied to actuate joints 1 and 2 respectively.

It is assumed that the subject interacts with the robot by applying a vector of forces and torques at the point Q, which has a z_0 component of l_z. This corresponds to the vertical distance of the subject's hand above the force/torque sensor. The vector of the components of these forces applied in the $\boldsymbol{x}_0$ and $\boldsymbol{y}_0$ directions is given by $\boldsymbol{h}_r$. The subject's arm is strapped to the fifth link of the robot and the human arm model therefore includes the properties of this link, which will be neglected from the model of the robot. To ensure the robot's safe interaction with an unknown environment, a form of impedance control, detailed in [25], is used to govern the torque demand supplied to the motors. This is a strategy that has previously been used to control robotic therapy devices [26], although alternative approaches could instead be used, such as force or compliant control [27].

The controller compensates for the inertial and damping properties of the robotic arm and at point Q, the full derivation is in [27]

$$\boldsymbol{h}_r = \boldsymbol{K}_{K_x}\tilde{\boldsymbol{x}}_r + \boldsymbol{K}_{B_x}\dot{\tilde{\boldsymbol{x}}}_r + \boldsymbol{K}_{M_x}\ddot{\tilde{\boldsymbol{x}}}_r, \tag{8.1}$$

where $\hat{\boldsymbol{x}}_r$ is the reference position, $\tilde{\boldsymbol{x}}_r = \hat{\boldsymbol{x}}_r - \boldsymbol{x}_r$, $\boldsymbol{x}_r = \boldsymbol{k}_r(\boldsymbol{q}_r)$, $\dot{\boldsymbol{x}}_r = \boldsymbol{J}_r(\boldsymbol{q}_r)\dot{\boldsymbol{q}}_r$ and $\ddot{\boldsymbol{x}}_r = \boldsymbol{J}_r(\boldsymbol{q}_r)\ddot{\boldsymbol{q}}_r + \dot{\boldsymbol{J}}_r(\boldsymbol{q}_r, \dot{\boldsymbol{q}}_r)\dot{\boldsymbol{q}}_r$. The direct kinematics equation for the robotic system is $\boldsymbol{x}_r = \boldsymbol{k}_r(\boldsymbol{q}_r)$, and $\boldsymbol{J}_r^T(\boldsymbol{q}_r)$ is the Jacobian of this system. When the robot is moved freely by the subject in the absence of assistance, the gain matrices are set as $\boldsymbol{K}_{K_x} = \boldsymbol{0}$, $\boldsymbol{K}_{B_x} = K_{B_x}\boldsymbol{I}$ and $\boldsymbol{K}_{M_x} = K_{M_x}\boldsymbol{I}$, where $\boldsymbol{I}$ is the identity matrix and the values of $K_{B_x} > 0$ and $K_{M_x} > 0$ are tuned to create a 'natural' feel. An additional requirement is that $\hat{\boldsymbol{x}}_r$

equals a constant value. When the robot is required to move the subject's arm along the predefined trajectories, it is necessary to set $\boldsymbol{K}_{K_x} = K_{K_x}\boldsymbol{I}$ with $K_{K_x} > 0$ and the three gain matrices are tuned to produce the required tracking performance.

8.3.1 Human Arm Model: Passive System

In the system considered, the forearm must, due to the constraint, lie in the horizontal plane, and rotation is possible about the axis along the upper arm. Only forces and torques along the unconstrained directions are shown (only rotation about the axis parallel to $\boldsymbol{z}_1$ is unconstrained at this point as rotation is not possible about the two orthogonal axes). The triceps has been selected for stimulation since stroke patients typically experience problems with shoulder and elbow extension during reaching tasks [28].

Actuation of the triceps is modeled as a torque, $T_\beta \geq 0$, acting about an axis orthogonal to both the upper arm and forearm. Components of the forces in the $\boldsymbol{x}_1$ and $\boldsymbol{y}_1$ directions applied by the subject's hand at the point of interaction with the robot are denoted by F_{x_1} and F_{y_1} respectively. Following the derivation in [29], the dynamic model of the constrained arm can be written as

$$\boldsymbol{B}_a(\boldsymbol{q}_a)\ddot{\boldsymbol{q}}_a + \boldsymbol{C}_a(\boldsymbol{q}_a, \dot{\boldsymbol{q}}_a)\dot{\boldsymbol{q}}_a + \boldsymbol{F}_a(\boldsymbol{q}_a, \dot{\boldsymbol{q}}_a) = \boldsymbol{\tau}_a - \boldsymbol{J}_a^T(\boldsymbol{q}_a)\boldsymbol{h}, \tag{8.2}$$

where B_p, $C_p \in \mathscr{R}^{2\times 2}$ are the inertial and Coriolis matrices, respectively, $q_p = \left[\vartheta_u, \ \vartheta_f\right]^T$ is the joint angle vector, $J_p(q_p)$ is the system Jacobian and the external force vector satisfies $h = h_r$ as a consequence of the connection with the robotic arm. The moment vector about the forearm axis produced by the applied FES is

$$\tau_p = \begin{bmatrix} 0 \\ \tau_e \sigma\left(\vartheta_f\right) \end{bmatrix}, \qquad \sigma\left(\vartheta_f\right) = \frac{-s_f c_{\bar{\gamma}},}{\sqrt{1 - c_f^2 c_{\bar{\gamma}}^2}}. \tag{8.3}$$

The decoupled form of the nonconservative force

$$F_p(q_p, \dot{q}_p) = \left[F_u(\vartheta_u, \dot{\vartheta}_u), \ \ F_f(\vartheta_f, \dot{\vartheta}_f)\right]^T,$$

provides a satisfactory compromise between repeatability and the accuracy of the overall model and includes effects such as spasticity.

Various forms of the friction term $\boldsymbol{F}_a(\boldsymbol{q}_a, \dot{\boldsymbol{q}}_a)$ have been considered in order to arrive at a compromise between repeatability and the accuracy of the overall model. The most general form considered is

$$\boldsymbol{F}_a(\boldsymbol{q}_a, \dot{\boldsymbol{q}}_a) = \left[\ F_{a1}(\vartheta_u, \dot{\vartheta}_u) \ \ F_{a2}(\vartheta_f, \dot{\vartheta}_f)\ \right]^T, \tag{8.4}$$

where $F_{a1}(\cdot)$ and $F_{a2}(\cdot)$ are piecewise linear functions.

To account for the action of the triceps, an established model of the torque, T_β, generated by electrically stimulated muscle acting about a single joint is given by

$$T_\beta(\beta,\dot{\beta},u,t) = g(u,t) \times F_{ma}(\beta,\dot{\beta}) + F_{mp}(\beta,\dot{\beta}), \tag{8.5}$$

where u denotes the stimulation pulse width applied, and β is the joint angle [30]. A Hammerstein structure incorporating a static nonlinearity, $h_{\mathrm{IRC}}(u)$, representing the isometric recruitment curve, cascaded with linear activation dynamics, $h_{\mathrm{LAD}}(t)$, produces the first term, $g(u,t)$. The activation dynamics can be adequately captured using a critically damped second-order system [31]. The term $F_{ma}(\beta,\dot{\beta})$ models the multiplicative effect of the joint angle and joint angular velocity on the active torque developed by the muscle and the term $F_{mp}(\beta,\dot{\beta})$ accounts for the passive properties of the joint. Since γ is invariant, $F_{mp}(\beta,\dot{\beta})$ is accounted for when using the most general form of $\boldsymbol{F}_a(\boldsymbol{q}_a,\dot{\boldsymbol{q}}_a)$. It has therefore been decided not to add additional forms of friction to the existing model. Details of the procedures used to establish the parameters in the model are given in [32]. Further results on the use of system identification to determine a model for the muscle response are given in [33, 34].

The robotic arm must make the task a feasible yet productive one and the following points concern the choice of trajectory and role of the robot when the task is performed:

1. The trajectories will be elliptical reaching tasks for each subject's dominant arm, and should be achievable given their identified arm model.
2. The triceps muscle will provide the only actuating torque about the elbow, and the robotic arm will use the control scheme given by (8.1) to make the dynamics about the elbow feel 'natural' to the subject.
3. The robotic arm will provide a torque acting about the subject's shoulder in order to track the reference in a manner which is entirely governed by the angle of the forearm. This makes the task feasible without lessening the role played by the triceps.

8.3.2 Robotic Control Scheme

Combining the human arm model described by (8.2) and the end effector dynamics (8.1), gives

$$\boldsymbol{B}_a(\boldsymbol{q}_a)\ddot{\boldsymbol{q}}_a + \boldsymbol{C}_a(\boldsymbol{q}_a,\dot{\boldsymbol{q}}_a)\dot{\boldsymbol{q}}_a + \boldsymbol{F}_a(\boldsymbol{q}_a,\dot{\boldsymbol{q}}_a) = \boldsymbol{\tau}_a + \boldsymbol{J}_a^T(\boldsymbol{q}_a)(\boldsymbol{K}_{K_x}\tilde{\boldsymbol{x}}_a + \boldsymbol{K}_{B_x}\dot{\tilde{\boldsymbol{x}}}_a + \boldsymbol{K}_{M_x}\ddot{\tilde{\boldsymbol{x}}}_a), \tag{8.6}$$

where setting $\hat{\boldsymbol{x}}_a = \hat{\boldsymbol{x}}_r - [l_x\ l_y\ l_z]^T$ gives $\tilde{\boldsymbol{x}}_a = \tilde{\boldsymbol{x}}_r$. To separate the dynamics of the end effector in the directions corresponding to the human arm joint angles, it is necessary to set

$$\begin{aligned}&\boldsymbol{K}_{K_x}\tilde{\boldsymbol{x}}_a + \boldsymbol{K}_{B_x}\dot{\tilde{\boldsymbol{x}}}_a + \boldsymbol{K}_{M_x}\ddot{\tilde{\boldsymbol{x}}}_a = \\ &\boldsymbol{J}_a^{-T}(\boldsymbol{q}_a)\Big(\boldsymbol{K}_{K_q}\tilde{\boldsymbol{q}}_a + \boldsymbol{K}_{B_q}\dot{\tilde{\boldsymbol{q}}}_a + \boldsymbol{K}_{M_q}\ddot{\tilde{\boldsymbol{q}}}_a\Big),\end{aligned} \tag{8.7}$$

where $\tilde{\boldsymbol{q}}_a = \hat{\boldsymbol{q}}_a - \boldsymbol{q}_a$ and $\hat{\boldsymbol{q}}_a = \boldsymbol{k}_a^{-1}(\hat{\boldsymbol{x}}_a)$,. Hence

$$\boldsymbol{B}_a(\boldsymbol{q}_a)\ddot{\boldsymbol{q}}_a + \boldsymbol{C}_a(\boldsymbol{q}_a, \dot{\boldsymbol{q}}_a)\dot{\boldsymbol{q}}_a + \boldsymbol{F}_a(\boldsymbol{q}_a, \dot{\boldsymbol{q}}_a) = \boldsymbol{\tau}_a + \boldsymbol{K}_{K_q}\tilde{\boldsymbol{q}}_a + \boldsymbol{K}_{B_q}\dot{\tilde{\boldsymbol{q}}}_a + \boldsymbol{K}_{M_q}\ddot{\tilde{\boldsymbol{q}}}_a \tag{8.8}$$

and to satisfy the requirements of points 2 and 3 of the previous subsection, requires that

$$\boldsymbol{K}_{K_q} = \mathrm{diag}\{K_{K_1}, 0\},\ \boldsymbol{K}_{B_q} = \mathrm{diag}\{K_{B_1}, K_{B_2}\},\ \boldsymbol{K}_{M_q} = \mathrm{diag}\{K_{M_1}, K_{M_2}\} \tag{8.9}$$

and $\hat{\boldsymbol{q}}_a = \begin{bmatrix}\hat{\vartheta}_u & c\end{bmatrix}^T$ where c is a constant and $K_{K_1}, K_{B_1}, K_{B_2}, K_{M_1}, K_{M_2} \geq 0$. This allows a choice of arbitrary second-order dynamics to be imposed about the shoulder and the damping and inertia about the elbow to be prescribed.

The right-hand side of (8.8) now is

$$\boldsymbol{\tau}_a + \begin{bmatrix} K_{K_1}\tilde{\vartheta}_u + K_{B_1}\dot{\tilde{\vartheta}}_u + K_{M_1}\ddot{\tilde{\vartheta}}_u \\ -K_{B_2}\dot{\vartheta}_f - K_{M_2}\ddot{\vartheta}_f \end{bmatrix}, \tag{8.10}$$

and provides the required dynamic relationship for both components of the torque on the assumption that $\hat{\vartheta}_f = c$ and therefore $\dot{\hat{\vartheta}}_f = \ddot{\hat{\vartheta}}_f = 0$. This last assumption is unsuitable for the desired references and hence this control scheme is not appropriate for the intended tracking task. If instead it is assumed that $\dot{\hat{\vartheta}}_u$ and $\ddot{\hat{\vartheta}}_u$ are sufficiently small, the controller can still ensure asymptotic stability to externally applied forces by using

$$\boldsymbol{\tau}_a + \begin{bmatrix} K_{K_1}\tilde{\vartheta}_u - K_{B_1}\dot{\vartheta}_u - K_{M_1}\ddot{\vartheta}_u \\ -K_{B_2}\dot{\vartheta}_f - K_{M_2}\ddot{\vartheta}_f \end{bmatrix}, \tag{8.11}$$

instead of (8.10).

This last requirement can be produced by choosing $\hat{\boldsymbol{q}}_a = \begin{bmatrix}\hat{\vartheta}_u & \hat{\vartheta}_f\end{bmatrix}^T$ and replacing (8.1) by

$$\boldsymbol{h}_r = \boldsymbol{K}_{K_x}\tilde{\boldsymbol{x}}_r - \boldsymbol{K}_{B_x}\dot{\boldsymbol{x}}_r - \boldsymbol{K}_{M_x}\ddot{\boldsymbol{x}}_r \tag{8.12}$$

and then (8.7) can be replaced by

$$\begin{aligned}&\boldsymbol{K}_{K_x}\tilde{\boldsymbol{x}}_a - \boldsymbol{K}_{B_x}\dot{\boldsymbol{x}}_a - \boldsymbol{K}_{M_x}\ddot{\boldsymbol{x}}_a = \\ &\boldsymbol{J}_a^{-T}(\boldsymbol{q}_a)\Big(\boldsymbol{K}_{K_q}\tilde{\boldsymbol{q}}_a - \boldsymbol{K}_{B_q}\dot{\boldsymbol{q}}_a - \boldsymbol{K}_{M_q}\ddot{\boldsymbol{q}}_a\Big).\end{aligned} \tag{8.13}$$

To obtain the required values of $\hat{\vartheta}_u$ and $\hat{\vartheta}_f$, components of (8.13) are compared and hence

$$\boldsymbol{K}_{K_x} = \frac{K_{K_1}\left(\hat{\vartheta}_u - \vartheta_u\right)}{\left|\hat{\boldsymbol{x}}_a - \boldsymbol{x}_a\right|(l_{u1} + l_{u2})c_\gamma s_f}\boldsymbol{I} \tag{8.14}$$

and

$$\hat{\boldsymbol{x}}_a = \boldsymbol{x}_a + \left|\hat{\boldsymbol{x}}_a - \boldsymbol{x}_a\right| \begin{bmatrix} c_{uf} \\ s_{uf} \end{bmatrix}. \tag{8.15}$$

Consequently $\hat{\boldsymbol{x}}_a$ is not uniquely defined, but can be any point lying on a line extending along the forearm and passing through $\boldsymbol{x}_a$. To achieve the tracking task it must therefore be set equal to the point of intersection with the trajectory. Hence the reference point is then defined formally as

$$\begin{aligned} \hat{\boldsymbol{x}}_a = \boldsymbol{\Omega}\left(\boldsymbol{x}_a, \Psi(\cdot)\right) &:= \boldsymbol{k}_a\left(\begin{bmatrix} \Psi(\hat{\vartheta}_f) \\ \hat{\vartheta}_f \end{bmatrix}\right), \\ \left|\boldsymbol{k}_a\left(\begin{bmatrix} \Psi(\hat{\vartheta}_f) \\ \hat{\vartheta}_f \end{bmatrix}\right) \right. &= \boldsymbol{x}_a + \lambda \begin{bmatrix} c_{uf} \\ s_{uf} \end{bmatrix}, \end{aligned} \tag{8.16}$$

where λ is a scalar.

The resulting control system is shown in Fig. 8.4, and the necessary controller matrices are

$$\boldsymbol{K}_{B_x} = \boldsymbol{J}_a^{-T}(\boldsymbol{q}_a)\boldsymbol{K}_{B_q}\boldsymbol{J}_a^{-1}(\boldsymbol{q}_a) \tag{8.17}$$

and

$$\boldsymbol{K}_{M_x}\left(\boldsymbol{J}_a(\boldsymbol{q}_a)\ddot{\boldsymbol{q}}_a + \dot{\boldsymbol{J}}_a(\boldsymbol{q}_a, \dot{\boldsymbol{q}}_a)\dot{\boldsymbol{q}}_a\right) = \boldsymbol{J}_a^{-T}(\boldsymbol{q}_a)\boldsymbol{K}_{M_q}. \tag{8.18}$$

Using (8.16) the remaining controller matrix (8.14) can be written explicitly as

$$\boldsymbol{K}_{K_x}\left(\boldsymbol{x}_a, \Psi(\cdot)\right) = \frac{K_{K_1}\left(\Psi(\hat{\vartheta}_f) - \vartheta_u\right)}{\lambda(l_{u1} + l_{u2})c_\gamma s_f}\boldsymbol{I}. \tag{8.19}$$

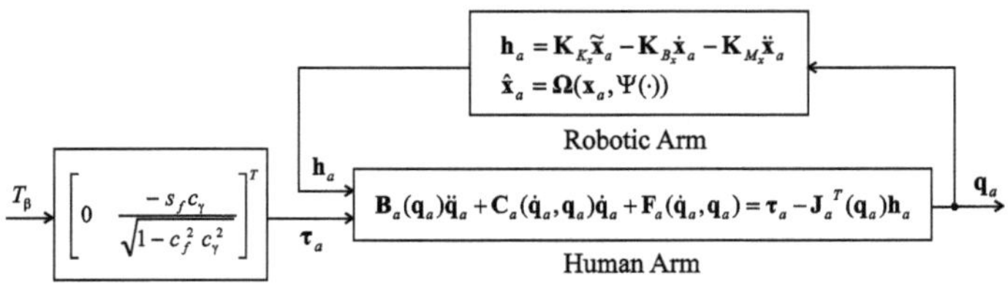

Fig. 8.4 Human arm system with robotic assistance

8.3.3 Trajectory Selection

Given the robotic control system, the next step is to select the trajectories used in the ILC design. How the trajectories feel to the patient is critical and impacts on the selection of K_{B_2} and K_{M_2} as follows:

1. For $\Psi(\vartheta_f)$ to be a one-one continuous function, both $\vartheta_u^*(t)$ and $\vartheta_f^*(t)$ must be monotone.
2. With the use of robotic assistance, it is shown in [35] that the behavior of the electrically stimulated forearm can be approximated by the system represented schematically in Fig. 8.5. The approximations used are based on experimentally confirmed properties of the human arm model, and the ability of the robotic control system to provide accurate tracking of $\hat{\vartheta}_u$ by ϑ_u.

   Given $T_\beta \geq 0$, point 1 above requires that $\vartheta_f^*(t)$ is monotone decreasing. Hence from the geometry of the task, the line joining $\boldsymbol{x}_a^*(T)$ and $\boldsymbol{x}_a^*(0)$ must pass through the origin of the arm system, since at both these points the direction of the elliptical trajectory is orthogonal to the major axis.
3. Under the assumption that $\vartheta_f^*(t)$ is tracked perfectly, the torque that must be applied to the arm system shown in Fig. 8.5 is given by

$$T_\beta^*(t) \approx \left(\frac{-\sqrt{1-c_{f^*(t)}^2 c_\gamma^2}}{s_{f^*(t)} c_\gamma} \right) \Big\{ K_{B_2} \dot{\vartheta}_f^*(t) + \left(K_{M_2} + b_{a3}\right) \ddot{\vartheta}_f^*(t) + F_{a2}\left(\vartheta_f^*(t), \dot{\vartheta}_f^*(t)\right) \Big\}, \tag{8.20}$$

   and the stimulation required, $u^*(t)$, must satisfy

$$\int_0^t h_{\mathrm{IRC}}\left(u^*(\tau)\right) h_{\mathrm{LAD}}\left(t-\tau\right) d\tau = \frac{T_\beta^*(t)}{F_{ma}(\beta^*(t), \dot{\beta}^*(t))}, \tag{8.21}$$

   where $\beta^*(t)$ corresponds to the reference trajectory and hence

$$\beta^*(t) = \beta(\vartheta_f^*(t), t). \tag{8.22}$$

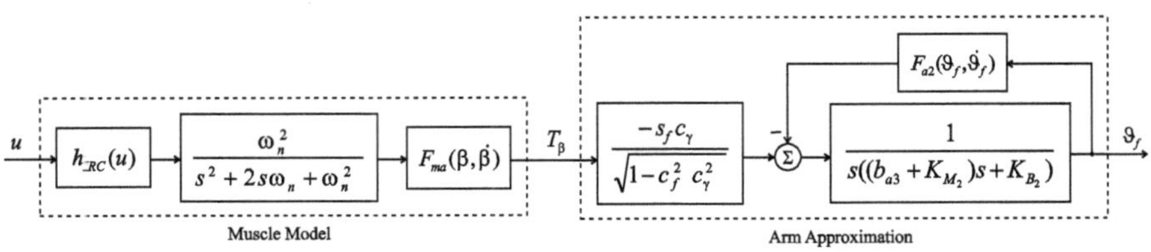

Fig. 8.5 Approximate model of stimulated human arm

This last equation clearly limits the magnitude and rate of change of any achievable torque trajectory and the existence of a solution to (8.21) has been ensured by selecting slow trajectories that comprise half ellipse segments that are comfortably within both the robot's and the subject's workspace. The start and end points are chosen such that they can be reached by a smooth extension of the elbow, and are individually calculated for each subject depending on their maximum reach capability. The K_{B_2} and K_{M_2}, are selected in order to mimic a realistic activity, provide a high level of stability in response to sudden stimulation inputs, and to require a moderate level of work from the muscles in order to track the trajectory.

8.3.4 ILC Structure and Design

The control strategy consists of a linearizing controller in a feedback control arrangement to which ILC is added as a feedforward signal, where the first component of the linearizing controller is the inverse of the isometric recruitment curve, $h_{\mathrm{IRC}}^{-1}(\cdot)$, that has been identified for each subject. The controller's remaining action is then motivated by the form of the remaining nonlinear terms $-s_f c_\gamma / \sqrt{1 - c_f^2 c_\gamma^2}$, $F_{ma}(\beta, \dot{\beta})$ and $F_{a2}(\vartheta_f, \dot{\vartheta}_f)$ appearing in the arm model of Fig. 8.5. The value of all the three functions can be shown to vary only slowly when the trajectories considered in this chapter are followed perfectly [32].

The control action attempts to remove the effect of these functions to produce a system that approximates the linear activation dynamics in series with the linear arm dynamics (the transfer functions appearing in the left and right subsystems, respectively, of Fig. 8.5). To achieve this, the controller next applies the gain term

$$\frac{-\sqrt{1 - c_f^2 c_\gamma^2}}{s_f c_\gamma F_{ma}(\beta, \dot{\beta})} = \left(-s_f c_\gamma / \sqrt{1 - c_f^2 c_\gamma^2}\right)^{-1} \left(F_{ma}(\beta, \dot{\beta})\right)^{-1}, \tag{8.23}$$

to address the multiplicative effect of the first two nonlinear terms and then $F_{a2}(\vartheta_f, \dot{\vartheta}_f)$ is added to the input in order to cancel the additive effect of the third nonlinear term. The resulting linearizing controller is

$$u = h_{\mathrm{IRC}}^{-1}\left(\left(F_{a2}(\vartheta_f, \dot{\vartheta}_f) + w\right) \frac{-\sqrt{1 - c_f^2 c_\gamma^2}}{s_f c_\gamma F_{ma}(\beta, \dot{\beta})}\right). \tag{8.24}$$

and the validity of this approach will now be investigated. For conciseness, attention is restricted to the case where $F_{a2}(\cdot)$ and $F_{ma}(\cdot)$ are functions of their first argument only. The resulting system is shown in Fig. 8.6 in which

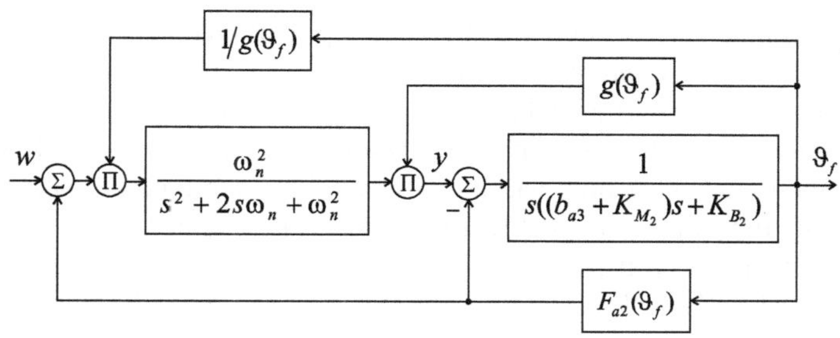

Fig. 8.6 Stimulated arm system and linearizing controller

$$g(\vartheta_f) = \frac{-s_f c_\gamma F_{ma}(\beta(\vartheta_f))}{\sqrt{1 - c_f^2 c_\gamma^2}} \tag{8.25}$$

and Π denotes multiplication of inputs.

In [35] an equivalent state-space expression for this system is derived and then linearized at time $\bar{t}$. When the first derivatives of $F_{a2}(\vartheta_f)$ and $g(\vartheta_f)$ are zero for all t, this linearized system corresponds with the desired relationship

$$\frac{\vartheta_f}{w}(s) = \frac{\omega_n^2}{s^2 + 2s\omega_n + \omega_n^2} \cdot \frac{1}{s\left((b_{a3} + K_{M_2})s + K_{B_2}\right)} = G(s). \tag{8.26}$$

To ensure that (8.26) adequately approximates the system behavior when the derivatives are nonzero, trajectories must be chosen over which $g_1(t)$, $g_2(t)$ and $F'_{a2}(\cdot)$ are small. Also, as detailed in [35], their effect can also be reduced by ensuring that the dynamics of the arm system are slower than the activation dynamics. This also reduces the effect of the terms involving $F'_{a2}(\vartheta_f(\bar{t}))$, which can be more clearly seen by examining Bode plots with $g_1(\bar{t})$, $g_2(\bar{t}) = 0$ using various values of $F'_{a2}(\vartheta_f(\bar{t}))$. The same goal could, of course, be achieved if it were possible to achieve more rapid activation dynamics in the patients tested.

To examine whether the choice of trajectories and arm dynamics correspond to a model that is well approximated by (8.26), simulations were conducted in which its output is compared with that of the linearizing controller applied to the full model shown in Fig. 8.5. The applied input, w, has been chosen to result in approximate tracking of the more rapid of the demands used. The model outputs used were in close agreement which supports the use of $G(s)$ to approximate the combined linearizing controller and arm system in the remainder of this chapter. Care must be taken, however, to ensure that the control schemes subsequently considered are robust to the modeling error that is present in the system.

Linearization along a trajectory is a well-established approach which allows the linearized dynamics to be used to infer properties of the nonlinear system when the state variables and input are close to those of the linearized system. In particular, if the resulting time-varying system is stable then the nonlinear system also has that property in the neighborhood of the trajectory. This technique can be applied to assess the local stability and robustness of the proposed control system by using trajectories comprising experimental test data or those resulting from simulations using the system model. Stability of the linearized system can then be assessed using well-known methods for linear time-varying systems. This can then be repeated whilst varying the model parameters in order to gain a broader picture of the system robustness and performance properties. The linearized system variation with respect to F_{a2} and F_{ma} depends only on F'_{a2}, and the values of F_{ma}, F'_{ma}, and F''_{ma} respectively. The stability of the system can therefore be examined through variation of these quantities, together with variation of the remaining model parameters, in order to provide a measure of the controller's robustness. This approach can then be extended to provide local convergence properties of the ILC algorithms used.

The next stage is to choose a feedback controller to supply the torque demand, w, necessary for the system to track the intended references. A Proportional Derivative (PD) controller has been selected and its transfer function approximated using

$$C(s) = \frac{K_d s + 1}{\varepsilon K_d s + 1}. \tag{8.27}$$

with $\frac{1}{6} \geq \varepsilon \geq \frac{1}{20}$. The level of stimulation that first produces a response from the triceps, u_m, is used to supply an offset such that the feedback system operates within the torque generating capabilities of the muscle. The feedback controller is then tuned for each subject with an emphasis on robustness, since stability is of greater concern than accurate tracking, and therefore a conservative bandwidth and high gain and phase margins are desired.

In common with other forms of control, the design of ILC laws can be based on tuning a small number of parameters in a simple structure without the use of the model of the dynamics or else be model based. One critical difference in ILC is that information that would be non-causal in the standard case can be used. A phase-lead ILC algorithm in the discrete-time case with sample index p has the structure

$$v_{k+1}(p) = v_k(p) + Le_k(p + \lambda), \tag{8.28}$$

where v is the input signal, e is the error between the supplied reference and output, L is the control law gain and $\lambda > 0$ represents the ILC phase lead in samples. The term $Le_k(p + \lambda)$ is allowed ILC as the complete previous trial has been generated before the current one begins. Also the Ziegler Nichols tuning rules do not apply but still phase-lead ILC has been used to effect in many application areas, some of which are referenced in the survey papers [7, 8]. In the stroke rehabilitation area both phase-lead and model-based ILC have been used where in the planar case the former has been used as detailed next.

In z transfer function terms, the phase-lead ILC law (8.28) can be written as, see [7] for the details of how the z transform can be applied despite the finite trial length,

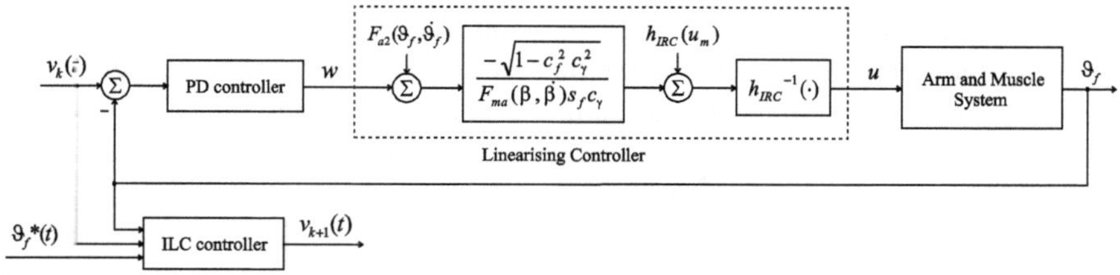

Fig. 8.7 Block diagram of the ILC system

$$v_{k+1}(z) = v_k(z) + Lz^\lambda e_k(z). \tag{8.29}$$

The overall control scheme is shown in Fig. 8.7 where the feedback controller is commonly used in many applications [7, 8] of ILC to increase stability and noise rejection. Using (8.29) gives

$$e_{k+1}(z) = e_k(z) - Lz^\lambda P(z)e_k(z) \tag{8.30}$$

with

$$P(z) = \frac{C(z)G(z)}{1-C(z)G(z)}, \tag{8.31}$$

where $G(z)$ and $C(z)$ are the discretized representations of (8.26) and (8.27) respectively.

The relationship

$$e_{k+1}(z) = e_k(z)\left(1 - LP(z)z^\lambda\right) \tag{8.32}$$

yields the monotonic convergence criterion

$$\left|1 - LP(e^{j\omega T_s})e^{j\omega\lambda T_s}\right| < 1, \tag{8.33}$$

for ω up to the Nyquist frequency. Satisfying this at a given frequency is a sufficient condition for monotonic convergence at that frequency. Furthermore the convergence speed is dictated by the magnitude of the left-hand side; if it is close to zero, convergence will occur in a single trial, while if it is greater than one, divergence is likely to occur at that frequency. Therefore L and λ are chosen such that the left-hand side is minimized to provide the greatest convergence over those frequencies present in the reference. Frequencies above this value, or those at which plant uncertainty may cause the criterion to be violated, are removed through the use of a non-causal zero-phase filter applied to the error. The sample time, $T_s = \frac{1}{40}$ s, will be used since this corresponds to the frequency at which stimulation pulses are applied to the patient. This frequency is synchronized with the robotic control system and each pulse is produced with a delay of less than 10 μs.

A value of $L = 0.2$ has been chosen in both cases in order to produce a robust system at the expense of convergence speed. Reducing L can be shown to increase robustness to model uncertainty over all frequencies. It can also be verified that choosing the phase lead to maximize the convergence at a given frequency also achieves a maximum robustness to gain and phase uncertainty at that frequency, which means that there is no compromise between robustness and convergence speed with respect to choosing this parameter. Therefore the phase lead will be selected to maximize the convergence over a suitable frequency range, and a zero-phase filter implemented to ensure stability at all higher frequencies (this being advisable even if the convergence criterion is satisfied). For general application, it is natural for this frequency range to correspond to the bandwidth of the system. It is routine to verify that the system with $\omega_b = 0.45$ Hz has maximum convergence over its bandwidth using a phase lead of 30 samples (although a phase lead of 20 samples produces greater stability over higher frequencies, which is an important factor if a filter is not used). Similarly, the system with $\omega_b = 0.53$ Hz has a maximum convergence over its bandwidth of 20 samples. When using these phase leads the system with the higher bandwidth has the property of monotonic convergence at greater frequencies than the system with lower bandwidth.

8.3.5 *Clinical Assessment*

The ILC control scheme of Fig. 8.7 was applied, after testing on unimpaired volunteers and obtaining the necessary ethical approval, in a clinical trial with 5 stroke patients. The evaluation of any rehabilitation intervention is critical to establish effect, i.e., the benefit or otherwise to the patient. The 5 participants, 3 male and 2 female, recruited gave written consent and their demographic characteristics ranged in age from 38 to 77 years with a mean age of 52 years. Participants had suffered strokes ranging 8 months to 8.4 years, mean 4 years, prior to recruitment to the study; 3 were affected on the right side and 2 on the left side. Patients one year beyond the stoke occurring are not likely to recover all lost functionality but improvement to a lower level could also have benefits and, for example, enable them to live independently.

The intervention forming the clinical trials consisted of either 18 or 25 treatment sessions where participants practiced planar reaching tasks augmented by responsive FES of the triceps brachii muscle. Results from applying the stimulation to a stroke patient to assist tracking of the required trajectory are given in Fig. 8.8a and b, which show typical changes in the angle of the shoulder and elbow, respectively, over the duration of the supplied trajectory. Figure 8.8c shows the FES pulse width that is applied using ILC to produce the assisted movements.

For each participant, the change in tracking error data over the four different unassisted tracking trajectories performed at the beginning of each treatment session is shown in Fig. 8.9, these vary in orientation, duration, and length [36]. During these tasks the patient received neither FES nor assistance from the robot. Although the trend for participants is a decrease in tracking error over time, the change is not monotonic, that is, adjacent values may increase. The tracking error decreased most for participants three and five across all trajectories over the intervention. The

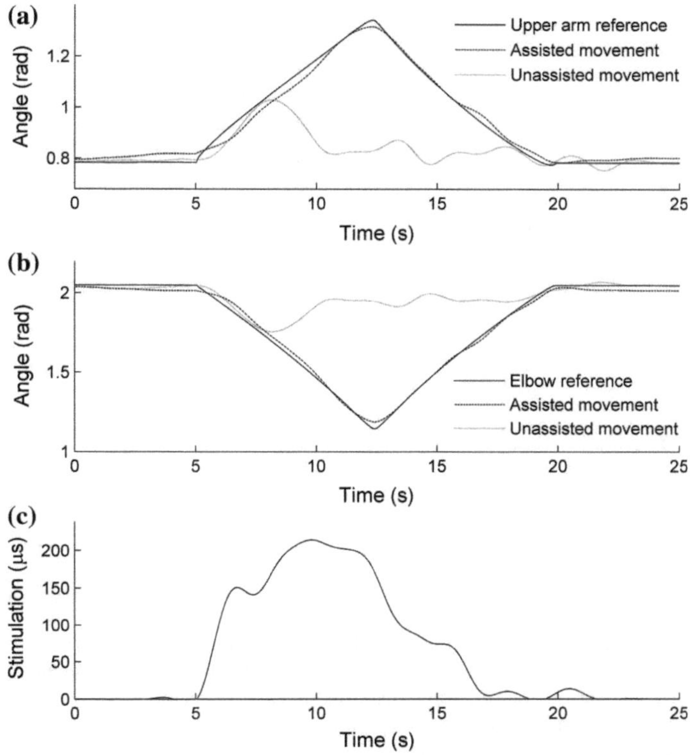

Fig. 8.8 Results from applying the stimulation to a stroke patient to assist tracking of a required trajectory, where (**a**), and (**b**) show the changes in the angle of the shoulder, and elbow, respectively, over the duration of the supplied trajectory. In the plots, the *solid line* shows the ideal movements that would be required to complete the trajectory, the *dotted line* represents unassisted movement, and the *dashed line* shows movement assisted by FES. Figure 8.8c shows the FES pulse width that is applied using ILC to produce the assisted movements. During the 5-s countdown period, before the target movement starts, there is minimal stimulation. On the reach component of the trajectory, from 5 to 12.5-s, stimulation increases rapidly, there is a delay period of approximately 2-s between the stimulation peak and the peak shoulder and elbow angle, associated with the biomechanical response to stimulation

improvement in unassisted tracking is statistically significant in three of the four tasks [36].

The clinical evaluation of this trial [36] exhibits significant changes in the outcome measures, error tracking, isometric muscle force, percentage maximum level of stimulation required to correct error, and muscle activation patterns, respectively, for the 5 chronic stroke participants taking part in the intervention. The ILC system takes into account that the effect of FES is enhanced when associated with the participant's intention to move and that to maximize plasticity stroke participants need to work at their maximum effort in planning and executing tasks during rehabilitation interventions. Although systems have been developed where electrical stimula-

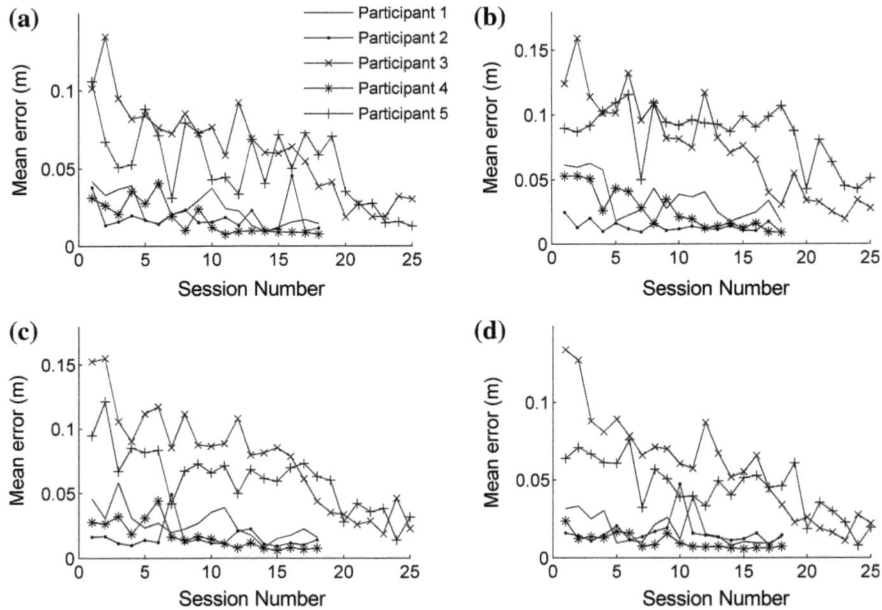

Fig. 8.9 Clinical intervention results for planar ILC system: Tracking error data with no FES assistance as a measure of patient improvement following treatment. For each participant, the change in tracking error data over the four different unassisted tracking trajectories performed at the beginning of each treatment session is shown. Although the trend for participants is a decrease in tracking error over time, the change is not monotonic, that is, adjacent values may increase. The tracking error decreased most for participants three and five across all trajectories over the intervention

tion is triggered by muscle activity, they do not allow feedback to adjust stimulation parameters during tasks, which is a drawback compared with the ability of the training modalities available during robotic assistance to promote voluntary activity [37].

By modifying the gain applied to the ILC update in accordance with the observed error, the ILC system adjusted the level of assistance in response to the users' performance and hence provided only the minimum level of stimulation needed to assist the participant in performing the task to a specified level of accuracy. During the intervention, the error tracking remained within a limited range, specifically, less than 15 mm, and the FES required to achieve this error tracking reduced over the course of the intervention. Thus, the balance of FES and voluntary effort required to perform the reaching task changed, with the participants proportionally contributing greater voluntary movement, indicating that motor learning had occurred [36].

The effectiveness of any rehabilitation critically depends on participant perspectives of the procedures and equipment used [13]. Currently, there is no generic evaluation tool available to be used across different rehabilitation robot systems. Consequently, participant comments were recorded during the intervention sessions and subsequently a question set developed to explore effectiveness, acceptability, and usability of the ILC system. The question set [38] was administered by a health psychologist to the five stroke participants, and found to be easy to interpret. The findings from this study using a robotic workstation and FES are congruent with other studies, referenced in [36]. Patients' comments on the best aspects of the study could be separated into physical and psychological benefits, research interaction, being involved, feedback, and enjoyment.

These results add to the growing body of evidence that suggests that robotic or FES interventions can be used both to provide objective assessments, before, during, and after an intervention, in addition to being an accepted and well tolerated treatment that results in changes in impairment levels. In engineering terms, the next stage is to determine if this approach can be extended to more complex 3D movements, as discussed in the next section.

8.4 Iterative Learning Control of the Unconstrained Upper Limb

In the work described in the previous section, the patient's forearm is constrained to lie in a horizontal plane. The resulting simplification to the underlying biomechanical model enables the development of a simple linearizing controller and the subsequent implementation of phase-lead ILC. To maximize the treatment's potential for rehabilitation, it is necessary to use a wider range of more functional movements, which more closely resemble the tasks necessary for daily living and are aligned with the activity-based ARAT measures. This section develops the previous model of the arm to remove the planar forearm constraint, permitting unconstrained movement, and applies ILC to the resulting system, where model-based ILC design is also used.

8.4.1 Stimulated Arm Model

When providing assistive FES during unconstrained 3D upper limb reaching movements, FES must, as in the previous section, be applied using a controlled environment to reduce fatigue, and ensure safety and comfort across a broad spectrum of patient abilities. There exist many exoskeletal robotic systems capable of providing such support, although very few have been combined with FES, and fewer still with model-based FES control schemes. In the research covered in this section, a commercially available device is selected, but the ILC algorithms may be applied to a range of such supports.

The mechanical exoskeleton employed is shown in Fig. 8.10 and has two springs incorporated in the mechanism to provide support to overcome gravity. This passive unweighing device allows patients to focus practice on the impaired muscles rather than those acting against gravity. Whilst it is supplied with its own broad range of

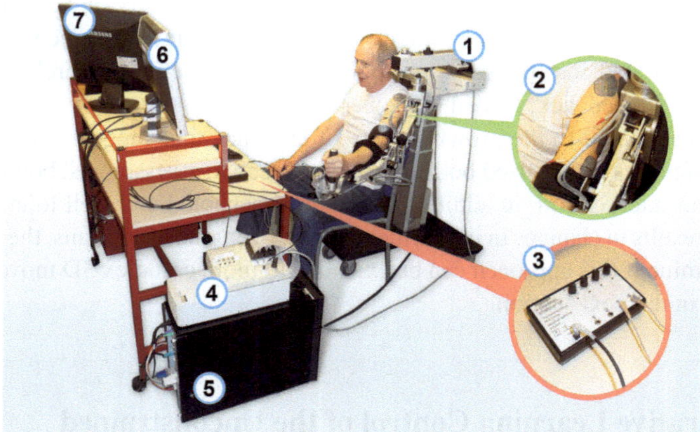

Fig. 8.10 3D ILC system components: (*1*) unweighing robotic device, (*2*) surface electrodes on triceps and anterior deltoid, (*3*) FES module, (*4*) real-time processor and interface module, (*5*) PC, (*6*) monitor displaying task, and (*7*) operator monitor

virtual reality tasks, these are not suitable for controller evaluation, and a custom task display system has instead been developed. The patient is seated with their impaired arm strapped into the mechanical unweighing device, whose segmental lengths and degree of antigravity support are adjusted for each user. Joint positional data provided by resolvers mounted on the support mechanism are transferred to the interface module, which connects external devices to the real-time hardware, see [39] for the details of the development of these components of the resulting system.

To increase the potential for rehabilitation, FES is applied to both the anterior deltoid and triceps muscles in accordance with the clinical objective of providing more assistance, a greater degree of feedback, enabling more accurate movements, and ensuring more muscles are usefully activated.

The biomechanical system comprises the human arm and exoskeleton mechanical support system shown in Fig. 8.11a. Figure 8.11b shows the kinematic structure of the exoskeleton support, where the joint variables $q_a = \left[\vartheta_1,\ \vartheta_2,\ \vartheta_3,\ \vartheta_4,\ \vartheta_5\right]^T$ correspond to the measured joint angles. Also the parallelogram structure of the upper arm section results in $\vartheta_3 = -\vartheta_3'$. The human arm is shown in Fig. 8.11c, and since it is strapped to the support, within the necessary joint ranges there exists a unique bijective transformation between their coordinate sets, given by $q_u = f_a\left(q_a\right)$, where $q_u = \left[\vartheta_a,\ \vartheta_b,\ \vartheta_c,\ \vartheta_d,\ \vartheta_e\right]^T$ contains the anthropomorphic variables shown in Fig. 8.11c. Using this relationship, application of Lagrangian analysis produces a dynamic model of the combined robotic and human arm systems given in terms of anthropomorphic coordinates as

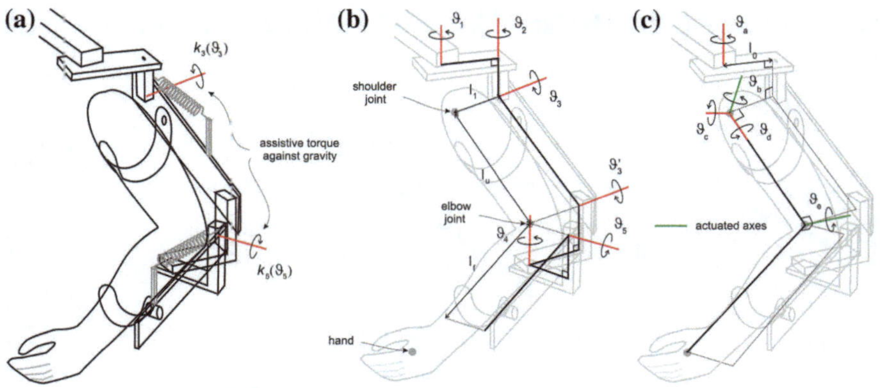

Fig. 8.11 Kinematic system relationships, **a** combined system, **b** unweighing robotic device, and **c** human arm

$$B_u\left(q_u\right)\ddot{q}_u + C_u\left(q_u, \dot{q}_u\right)\dot{q}_u + F_u\left(q_u, \dot{q}_u\right) + G_u\left(q_u\right) + K_u\left(q_u\right) = \tau_u\left(u, q_u, \dot{q}_u\right) - J_u^T\left(q_u\right)h \tag{8.34}$$

where $B_p, C_p \in \mathscr{R}^{5\times5}$ are the inertial and Coriolis matrices, respectively, for the 3D system, $G_u \in \mathscr{R}^5$ is the vector of moments produced by gravity, and $K_u \in \mathscr{R}^5$ is the vector of moments produced by the unweighing action where [39] gives a full description of the individual components.

The nonconservative forces assume the same form as in the planar case and hence

$$F_u\left(q_u, \dot{q}_u\right) = \left[F_a\left(\vartheta_a, \dot{\vartheta}_a\right), \quad \dots \quad F_e\left(\vartheta_e, \dot{\vartheta}_e\right)\right]^T$$

where the entries in this vector incorporate friction and spasticity. The vector due to stimulated muscle $\tau_u\left(u, q_u, \dot{q}_u\right) = \left[0, \ \tau_b\left(u_b, \vartheta_b, \dot{\vartheta}_b\right), \ 0, \ 0, \ \tau_e\left(u_e, \vartheta_e, \ \dot{\vartheta}_e\right)\right]^T$, contains elements of the form (8.5), where $u_b(t)$ and $u_e(t)$ are the electrical stimulation sequences applied to the triceps and anterior deltoid muscles, respectively, and $u = \left[u_a, \ u_b, \ u_c, \ u_d, \ u_e\right]^T$. The term h represents a vector of external forces and torques applied by the therapist using a handle mounted on a sensor attached to the robotic support. This is only used during identification tests and the model (8.34) is used by the FES control system to calculate a control signal that results in accurate tracking of a reference trajectory. Since assistive torque is applied about the ϑ_b and ϑ_e axes only, the system is underactuated. When applied during the treatment of patients, the controller assists tracking about ϑ_b and ϑ_e alone and, in response to clinical guidelines, it is assumed that the patient has sufficient control over the remaining axes to adequately perform the task.

Unlike the planar case of the previous section, it is not feasible to display a real-world tracking task in 3D and therefore a virtual task is displayed to the patient to ensure clarity, with provision for additional visual instructions and performance feedback. The patient's screen runs a 3D virtual reality environment which displays

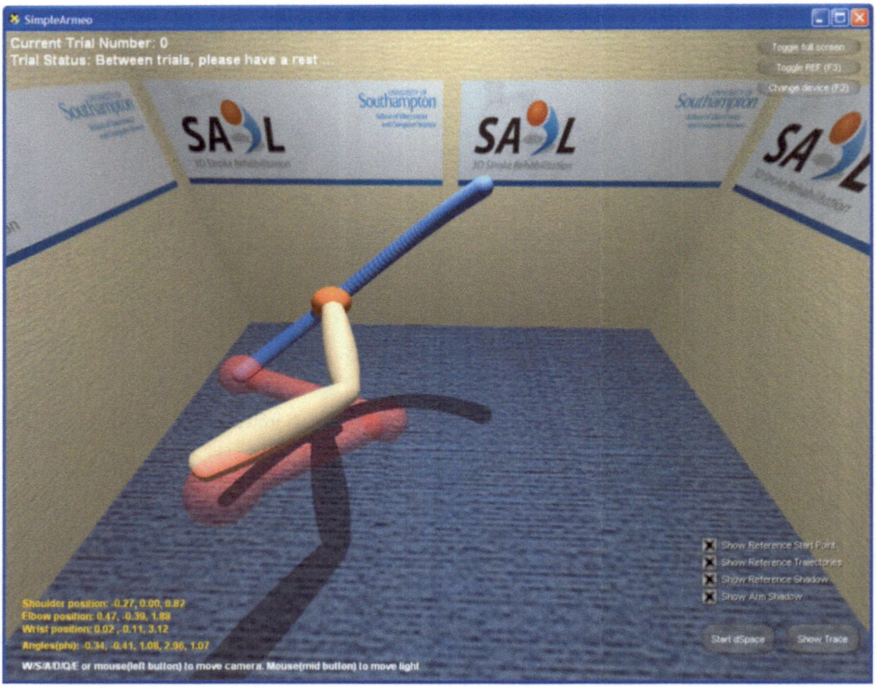

Fig. 8.12 3D virtual reality environment for patients, showing reaching trajectory comprising a sphere moving along a path and a graphic of the arm. Visual feedback provided by changing sphere color

a graphic of their arm in real time, together with the trajectory tracking task and is shown in Fig. 8.12. The aim of the tracking task is for the patient to follow a sphere which travels along the trajectory at various speeds, whilst FES controllers designed using the biomechanical system (8.34) assist their completion. The graphic of the patient's hand changes color to indicate their current error level. Feedback of performance is also given by an error percentage score displayed after each set of trials. A graphic of the initial arm position is displayed to ensure accurate resetting of the system at the start of each trial.

8.4.2 3D Control Schemes

Figure 8.13 shows the final control system in this case and it is possible to extend the phase-lead ILC design of the previous section [17, 39]. The more complicated dynamics in this case also warrants consideration of model-based ILC.

Newton's method-based ILC [40] uses the full model in the calculation of the new input. First, the closed-loop connection of the system (8.34) and feedback controller is written as the following discrete-time state-space model

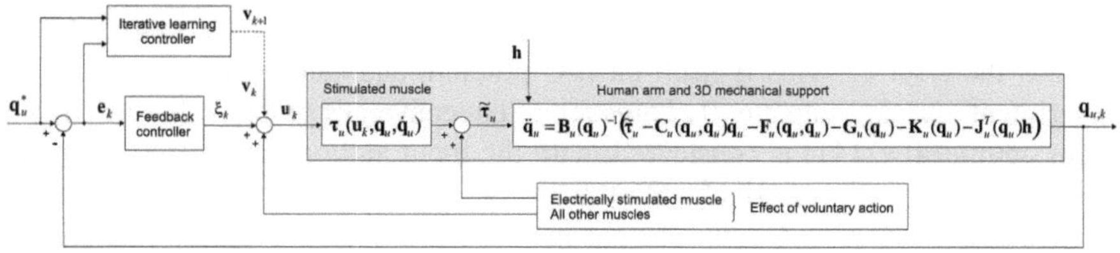

Fig. 8.13 Block diagram representation of the ILC control scheme for the 3D rehabilitation system

$$
\begin{aligned}
x_k(t+1) &= f[x_k(t), v_k(t)],\\
q_{u,k}(t) &= h[x_k(t)],
\end{aligned}
\tag{8.35}
$$

where $t \in [0, 1, 2, \ldots, N-1]$ is the sample number, $x_k(t)$ is the state vector, and $N = T/T_s + 1$ with T_s the sampling frequency. Introducing the vectors

$$
\begin{aligned}
v_k &= [v_k(0)^T, v_k(1)^T, \ldots, v_k(N-1)^T]^T,\\
q_{u,k} &= [q_{u,k}(0)^T, q_{u,k}(1)^T, \ldots, q_{u,k}(N-1)^T]^T,
\end{aligned}
\tag{8.36}
$$

and the reference vector

$$
q_u^* = [q_u^*(0)^T, q_u^*(1)^T, \ldots, q_u^*(N-1)^T]^T,
\tag{8.37}
$$

the Newton's method-based ILC update takes the form

$$
v_{k+1} = v_k + g'(v_k)^{-1} e_k,
\tag{8.38}
$$

where again $e_k = q_u^* - q_{u,k}$ is the tracking error on trial k. The term $g'(v_k)$ is equivalent to system linearization around v_k, with $\tilde{q}_u = g'(v_k)\tilde{v}$ corresponding to the linear time-varying, denoted LTV, system

$$
\begin{aligned}
\tilde{x}(t+1) &= A(t)\tilde{x}(t) + B(t)\tilde{v}(t)\\
\tilde{q}_u(t) &= C(t)\tilde{x}(t) + D(t)\tilde{v}(t), \qquad t = 0, 1, \ldots, N-1
\end{aligned}
\tag{8.39}
$$

with

$$
\begin{aligned}
A(t) &= \left(\frac{\partial f}{\partial x}\right)_{v_k(t), x_k(t)}, \quad & B(t) &= \left(\frac{\partial f}{\partial v_k}\right)_{v_k(t), x_k(t)},\\
C(t) &= \left(\frac{\partial h}{\partial x}\right)_{v_k(t), x_k(t)}, \quad & D(t) &= \left(\frac{\partial h}{\partial v_k}\right)_{v_k(t), x_k(t)}.
\end{aligned}
\tag{8.40}
$$

The term $g'(v_k)^{-1}$ in (8.38) is computationally expensive and may be singular or contain excessive amplitudes and high frequencies. To overcome this difficulty, introduce

$$
e_k = g'(v_k)\Delta v_{k+1}
\tag{8.41}
$$

and then $\Delta v_{k+1} = v_{k+1} - v_k$ equals the input that forces the LTV system (8.39) to track the error e_k. This is itself an ILC problem and can be solved in between experimental trials using any ILC algorithm that converges globally. In this chapter norm optimal ILC [41] is considered, with the input and output on trial j denoted by $e_{k,j}$ and $\Delta v_{k+1,j}$, respectively. On trial $j+1$, the trade-off between minimizing the tracking error, $e_k - e_{k,j}$, and the change in control input, $\Delta v_{k+1,j+1} - \Delta v_{k+1,j}$, is represented by the cost function

$$
\begin{aligned}
J_{j+1} &= \sum_{t=0}^{N-1} (e_k - e_{k,j})(t)^T Q (e_k - e_{k,j})(t), \\
&+ \sum_{t=0}^{N-1} (\Delta v_{k+1,j+1} - \Delta v_{k+1,j})(t)^T R (\Delta v_{k+1,j+1} - \Delta v_{k+1,j})(t), \qquad (8.42)
\end{aligned}
$$

where Q and R are symmetric positive definite weighting matrices of compatible dimension. The ILC computation is stopped after 100 trials or after the error reaches a preset threshold. The input obtained, $\Delta v_{k+1,j}$, is then used to approximate Δv_{k+1} in (8.38) to generate the control input for the next trial.

Nine custom reference trajectories are generated for each participant, producing tracking tasks which extend the arm out in front of the patient in response to clinical need. These are calculated using their identified workspace to establish the maximal arm extension directly in front of them and out to their affected side. By interpolating these 2 points a third intermediate point is then generated. Each reference starts from an initial point close to the patient's body, and extends 60, 80 and 100 % of the distance to one of these points. The task comprises reaching out to one of the end point locations, with the fixed trajectories for each of the five joints generated by scaling a third-order ramp signal of 10 s duration and adding an offset, so that it smoothly connects the required start and end joint angles. This results in the vector of reference trajectories $q_u^*(t) = \left[\vartheta_a^*(t),\ \vartheta_b^*(t),\ \vartheta_c^*(t),\ \vartheta_d^*(t),\ \vartheta_e^*(t)\right]^T$ where the presence of non-fixed $\vartheta_a^*(t)$, $\vartheta_c^*(t)$, and $\vartheta_d^*(t)$ components makes the task more natural to the patients, who can use their remaining voluntary effort to move these joints (Fig. 8.14).

In common with the planar case of the previous section, the phase lead and Newton designs were tested on unimpaired volunteers, see [39] for the phase-lead results. Figure 8.15 shows error norm results over 10 trials of the Newton's method-based ILC algorithm for two of the volunteers, using a long off-center and a medium off-center trajectory. The results are representative of all the tasks tested and confirm that accurate tracking is achieved within very few trials. These results are for $Q = 30I$ and $R = I$ respectively. Figure 8.16 shows representative input, output, and error signals recorded on trial 8 for one participant and confirm that a high level of tracking performance is achieved with an input signal that is not excessive. The theoretical quality of rapid monotonic convergence to zero tracking error is degraded due to inaccuracies in the human arm model which deteriorate performance, motivating development of more accurate identification procedures, and future use of online and recursive

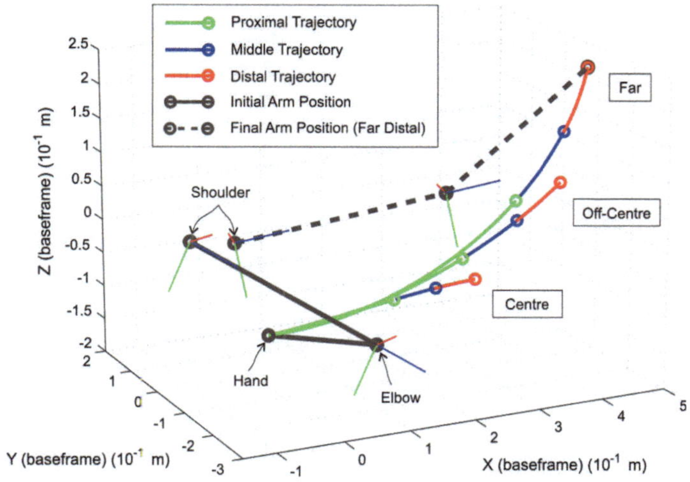

Fig. 8.14 Reference trajectories used for the 3D rehabilitation system

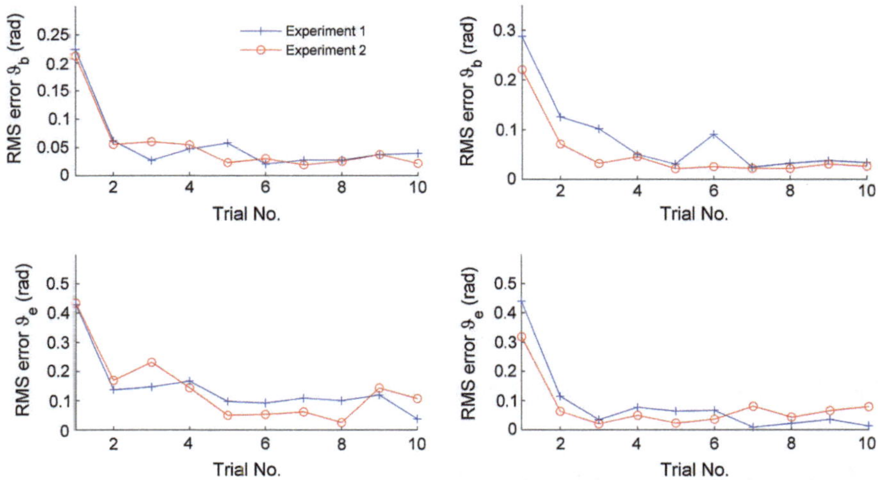

Fig. 8.15 Root mean square error plots for Newton's method-based ILC, subject 1, *left* hand plots, subject 2, *right* hand plots

techniques. Such identifications routines, however, must be suitable for application within the restrictive conditions of clinical trials, where there is limited set-up time, little opportunity to repeat measurements, and where techniques must yield satisfactory results for a wide range of subjects and changing physiological conditions.

Results from a clinical trial, again with 5 stroke patients, are also available [17, 42]. In summary, these provide strong evidence as to the beneficial role of ILC in 3D stroke rehabilitation and provide a basis for further research as discussed in the next section.

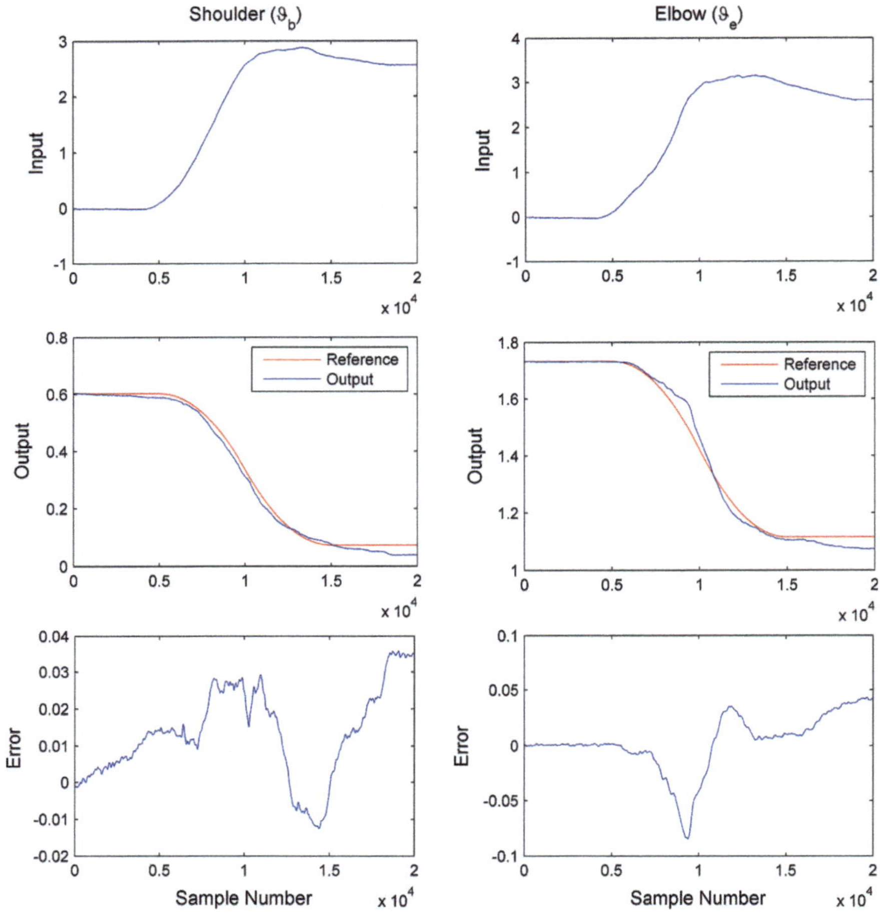

Fig. 8.16 Tracking performance of the Newton's method-based ILC algorithm

8.5 Conclusions and Further Research

Driven next generation health care needs there is currently significant interest in both rehabilitation robotics and therapeutic application of FES, with a clinically supported need to combine technologies whilst developing controllers for the latter that enable precise control of movement during functional tasks. This chapter has given an overview of research that established how robotic and FES controllers may be integrated to supply complementary assistance driven by clinical need.

The progress reported in this chapter provides a critical first step in the development of FES controllers for full reach and grasp movements involving shoulder, elbow, wrist, and hand stimulation during fully functional movements and there are many areas for further research in both engineering and clinical assessment before uptake by practicing medical professionals is possible. In the latter aspect, the use of any new medical intervention or treatment will require randomized control trials. The low number of patients used in the clinical trials described in this chapter show feasibility and a sample size calculation can be conducted to establish exactly how many patients would be needed to conduct a randomized controlled trial.

In engineering terms, further development of the algorithms used is required. One area for short- to medium-term research is the question of how to select the reference signal for each patient. At present this is done by a physiotherapist from an assessment of the patient without any decision support tools and if the task given to patients is already within their capability then no rehabilitation takes place as a result of the treatment. Conversely, a task which is far beyond current capabilities can lead to serious patient demotivation. An open question then is whether the region of effective operation can be estimated from pre-assessment and measurement data.

The ability to use the hand is an integral component of the ARAT and other functional outcome measures and strongly motivates the development of ILC for functional control of the hand and wrist. As the movement complexity increases, there is more emphasis on model-based approaches to provide optimal performance that maximizes effectiveness of therapy. However there are virtually no model-based control approaches for the hand and wrist which have been applied clinically, making this an important area for further research. Some recent progress is reported in [43].

Effective FES-based rehabilitation demands that patients are assisted during functional tasks in a manner that mimics their performance in the absence of impairment. There exists extensive literature on motor control for both unimpaired and stroke subjects which provide kinematic analysis of movement, for example, framing tasks as constrained optimization problems [44]. Such exploitation requires development of ILC laws which do not comprise the tracking of fixed reference trajectories defined over the task duration, but rather incorporate point-to-point movements with embedded input and output constraints [45]. These constraints may also reflect explicit coordination acting between joint variables, and the presence of muscle synergies. It is also necessary to adapt the task objective in response to identified system knowledge and ongoing performance, incorporating expert knowledge, clinical data, prior performance relationships, and established task progression models. Lastly, the robust stability of the overall system, incorporating adaptive model identification, ILC learning, and task adaption loops, must be established in a general framework.

References

1. Carroll, K., Murad, S., Eliahoo, J., Majeed, A.: Stroke incidence and risk factors in a population-based propsective cohort study. Health Stat. Q. **12**, 18–26 (2001)
2. Barreca, S., Wolf, S.L., Fasoli, S., Bohannon, R.: Treatment interventions for paretic upper limb of stroke survivors: a critical review. Neurorehabilitation and Neural Repair **17**, 220–226 (2003)
3. Oujamaa, L., Rlave, I., Froger, J., Mottet, D., Pelissier, J.-Y.: Rehabilitation of arm function after stroke: literature review. Ann. Phys. Rehabil. Medicene **52**, 269–293 (2009)
4. Dobkin, B.H.: Strategies for stroke rehabilitation. Lancet Neurol. **3**, 528–536 (2004)

5. Kwakkel, G., Kollen, B.J., Krebs, H.I.: Effects of robot-assisted theraphy on upper limb recovery after stroke: a systematic review. Neurorehabilitation and Neural Repair **22**, 111–121 (2008)
6. De Groon, J.R., Ijzerman, M.J., Chae, J., Lankhorst, G.J., Zilvold, G.: Relation between stimulation characteristics and clinical outcome of the upper extremity in stroke. Rehabil. Medicene **37**, 65–74 (2005)
7. Bristow, D.A., Tharayil, M., Alleyne, A.G.: A survey of iterative learning control: a learning-based method for high-performance tracking control. IEEE Control Syst. Mag. **26**(3), 96–114 (2006)
8. Ahn, H.S., Chen, Y., Moore, K.L.: Iterative learning control: brief survey and categorization. IEEE Trans. Syst. Man Cybern. Part C **37**(6), 1109–1121 (2007)
9. Rogers, E., Freeman, C.T., Cai, Z., Meadmore, K.L., Hughes, A.-M., Burridge, J.H.: Iterative learning control for robotic-assisted 3D stroke rehabilitation: from algorithms to clinical trials. In: Kolemeisevska-Gugulovska, T., Stankovski, M. (eds.) Complex Systems: Synergy of Control Communication and Computing—Selected COSY 2011 Papers, pp. 17–22. Society of ETAI, Skopje, R. Macedonia (2011)
10. Pomeroy, V.M., King, L., Pollack, A., Baily-Hallon, A., Longhorne, P.: Electrostimulation for promoting recovery of movemnent or functional ability after stroke. Cochrane Database of Syst. Rev. Issue 2. Art. No.: CD003241 (2006). doi:10.1002/14651858.CD003241.pub2
11. Prange, G.B., Jannick, M.A., Groothuis-Oudshoorn, C.G.M., Hermens, M., Ijzerman, M.J.: Systematic review of the effect of robot-aided therapy on recovery of the hemiparetic arm after stroke. J. Rehabil. Res. Devel. **43**, 171–184 (2006)
12. Intercollegiate Working Party for Stroke: National Clincial Guidelines for Stroke. Royal College of Physicians, London (2004)
13. de Kroon, J.R., van der Lee, J.H., Izerman, M.J., Lankhorst, G.J.: Therapeutic electrical stimulation to improve motor control and functional abilities of the upper extremity after stroke: a systematic review. Clin. Rehabil. **16**, 350–360 (2002)
14. Burridge, J.H., Ladouceur, M.: Clinical and therapeutic applications of neuromuscular stimulation: a review of current use and speculation into future developments. Neuromodulation **4**(4), 147–154 (2001)
15. de Kroon, J.R., Ijzerman, M.J., Lankhorst, G.J., Zilvold, G.: 'Relation between stimulation characteristics and clinical outcome in studies using electrical stimulation to improve motor control of upper extremity in stroke. J. Rerhabil. Med. **35**, 65–74 (2005)
16. Rushton, D.N.: Functional electrical stimulation and rehabilitation—an hypothesis. Medicial Eng. Phys. **1**, 75–78 (2003)
17. Freeman, C.T., Rogers, E., Hughes, A.-M., Burridge, J.H., Meadmore, K.L.: Iterative learning control in health care: electrical stimulation and robotic-assisted upper-limb stroke rehabilitation. IEEE Control Syst. Mag. **32**(1), 18–43 (2012)
18. Lynch, C.L., Popovic, M.R.: Functional electrical stimulation: closed-loop of induced muscle contractions. IEEE Control Syst. Mag. **28**(2), 41–50 (2008)
19. Internaitonal Classification of Functioning, Disability and Health (ICF): World Health Organisation (2001)
20. Jackson, J.: Specific treatment techniques. In: Stokes, M. (ed.) Physical Management in Neurological Rehabilitation. Elsevier (2004)
21. Kwakkel, G., Kollen, B., Krebs, H.I.: Effects of robot-assisted therapy on upper limb recovery after stroke: a systematic review. Stroke **34**, 2181–2186 (2008)
22. Fugl-Meyer, A.R., Jaasko, L., Leyman, I., Olsson, S., Steglind, S.: The post-stroke hemiplegic patient. 1. a method for evaluation of physical performance. Scandinavian Journal of Rehabilitation Medicine **7**, 13–31 (1975)
23. Carroll, D.: A quantitative test for upper extremity function. J. Cornic Disabil. **18**, 479–491 (1965)
24. Freeman, C.T., Hughes, A.-M., Burridge, J.H., Chappell, P.H., Lewin, P.L., Rogers, E.: A robotic workstation for stroke rehabilitation of the upper extremity using FES. Med. Eng. Phys. **31**(3), 364–373 (2009)

25. Colgate, J.E., Hogan, N.: Robust control of dynamically interacting systems. Int. J. Control **48**(1), 65–88 (1988)
26. Krebs, H.I., Hogan, N., Aisen, M.L., Volpe, B.T.: Robot-aided neurorehabilitation. IEEE Trans. Rehabil. Eng. **6**, 75–87 (1998)
27. Siciliano, B., Villani, L.: Robot Force Control. Kluwer Academic Publishers, Boston (1999)
28. Lum, P.S., Burgar, C.G., Shor, P.C.: Evidence for improved muscle activation patterns after retraining of reaching movements with the MIME robotic system in subjects with post-stroke hemiparesis. IEEE Trans. Neural Syst. Rehabil. Eng. **12**(2), 186–194 (2004)
29. Freeman, C.T., Hughes, A.-M., Burridge, J.H., Chappell, P.H., Lewin, P.L., Rogers, E.: A model of the upper extremity using FES for stroke rehabilitation. J. Biomech. Eng. **131**(3), 031011–031023 (2009)
30. Shue, G., Crago, P.E., Chizeck, H.J.: Muscle-joint models incorporating activation dynamics, moment-angle, and moment-velocity properties. IEEE Trans. Biomed. Eng. **42**(2), 213–223 (1995)
31. Baratta, R., Solomonow, M.: The dynamic response model of nine different skeletal muscles. IEEE Trans. Biomed. Eng. **37**, 243–251 (1990)
32. Freeman, C.T., Hughes, A.-M., Burridge, J.H., Chappell, P.H., Lewin, P.L., Rogers, E.: An experimental facility for the application of iterative learning control as an intervention aid to stroke rehabilitation. Meas. + Control: J. Inst. Meas. Control **40**(1), 20–23 (2007)
33. Le, F., Markovsky, I., Freeman, C.T., Rogers, E.: Identification of electrically stimulated muscle models of stroke patients. Control Eng. Pract. **18**(4), 396–407 (2010)
34. Le, F., Markovsky, I., Freeman, C.T., Rogers, E.: Recursive identification of Hammerstein systems with application to electrically stimulated muscle. Control Eng. Pract. **20**(4), 386–396 (2012)
35. Freeman, C.T., Hughes, A.-M., Burridge, J.H., Chappell, P.H., Lewin, P.L., Rogers, E.: Iterative learning control of FES applied to the upper extremity for rehabilitation. Control Eng. Pract. **17**, 368–381 (2009)
36. Hughes, A.-M., Freeman, C.T., Burridge, J.H., Chappell, P.H., Lewin, P.L., Rogers, E.: Feasibility of iterative learning control mediated by functional electrical stimulation for reaching after stroke. Neurorehabilitation and Neural Repair **23**(6), 559–568 (2009)
37. Huang, V.S., Krakauer, J.W.: Robotic neurorehabilitation: a computational motor learning perspective. J. NeuroEng. Rehabil. **6**(5), 1–13 (2009)
38. Hughes, A.-M., Freeman, C.T., Chappell, P.H., Lewin, P.L., Rogers, E., Burridge, J.H., Donovan-Hall, M., Dibb, B.: Stroke participants perceptions of robotic and electrical stimulation therapy: a new approach. Disab. Rehabil.: Technol. **6**(2), 130–138 (2011)
39. Freeman, C.T., Tong, D., Meadmore, K.L., Cai, Z., Rogers, E., Hughes, A.-M, Burridge, J.H.: Phase-lead iterative learning control algorithms for functional electrical stimulation based stroke rehabilitation. Proc. Inst. Mech. Eng. Part I **225**, 850–859 (2011)
40. Lin, T., Owens, D.H., Hatonene, J.: Newton method based iterative learning control for discrete non-linear systems. Int. J. Control **79**(10), 1263–1276 (2006)
41. Amann, N., Owens, D.H., Rogers, E.: Iterative learning control for discrete-time with exponential rate of convergence. Proc. Instution Electr. Eng., Part D Control Theory Appl. **143**(2), 217–224 (1996)
42. Meadmore, K.L., Hughes, A.-M., Freeman, C.T., Cai, Z., Tong, D., Burridge, J.H., Rogers, E.: Function electrical stimulation mediated by iterative learning control and 3D robotics reduces motor impairment in chronic stroke. J. Neuroengineering Rehabil. (2012). doi:10.1186/1743-0003-9-32
43. Soska, A., Freeman, C.T., Exell, T., Rogers, E.: Surface electrode array based control of the wrist and hand. In: Proceedings of the IFAC International Workshop on Adaptation and Learning in Control and Signal Processing, pp. 164–169 (2013)
44. Wolpert, D.M., Ghahramani, Z.: Flanagan, J.R.: Perspectives and problems in motor learning. Trends Cog. Sci. **11**(1), 487–494 (2001)
45. Freeman, C.T., Cai, Z., Rogers, E., Lewin, P. L.: Iterative learning control for multiple point-to-point tracking application. IEEE Trans. Control Syst. Technol. **19**(3), 590–600 (2011)

Chapter 9
Adaptive Fuzzy Modeling Based Assessment of Operator Functional State in Complex Human–Machine Systems

Jianhua Zhang and Rubin Wang

Abstract The quantitative analysis of human operator functional state (OFS) plays a crucial role in modeling and adaptive control of a large class of complex and safety-critical human–machine systems arising from such diverse fields as manned aerospace, air traffic control, and nuclear power plant. In this chapter, the OFS is quantitatively estimated using multiple sources of measured psychophysiological data. In the data acquisition experiments, an automation-enhanced cabin air management system (aCAMS) was employed to simulate with high fidelity a highly complex multitask platform of human–machine cooperative process control. Two types of adaptive fuzzy models, i.e., adaptive-network-based fuzzy inference system (ANFIS) and genetic algorithm (GA)-based Mamdani fuzzy model, are constructed to estimate the temporal fluctuations of the OFS. The fuzzy models are used to reveal the complex unknown correlation between the psychophysiological (i.e., electroencephalographical and cardiovascular) variables and operator performance (i.e., primary-task-related performance). The adaptive fuzzy modeling paradigm was validated using the data measured from 11 young healthy and well-trained male subjects (2 trials for each), who were engaged in the manual control tasks under aCAMS experimental environment. The fuzzy modeling methods proposed may provide an objective and quantitative way to accurately estimate the OFS related to mental or cognitive workload (stress) of the human operator.

J. Zhang (✉)
Department of Automation, East China University of Science and Technology, Shanghai 200237, People's Republic of China
e-mail: zhangjh@ecust.edu.cn

R. Wang
Institute of Cognitive Neurodynamics, East China University of Science and Technology, Shanghai 200237, People's Republic of China
e-mail: rbwang@163.com

G.M. Dimirovski (ed.), *Complex Systems*, Studies in Systems, Decision and Control 55, DOI 10.1007/978-3-319-28860-4_9

9.1 Introduction

The human operator's role in a human–machine system has become more crucial with increasing operational demands, stress, and fatigue, with a resultant threat to safety and reliability, in particular, in many safety-critical applications [1]. The operator functional state (OFS) approach was proposed to address this problem [2]. From the OFS perspective, during a period of time before the system performance breakdown occurs, the operator is likely to be in a vulnerable state able to manage predictable demands but not unexpected or difficult ones, because he/she has reduced spare capacity to respond to emergencies [3]. If this state can be detected by any means, we would be able to predict when (exactly at what periods of time) operational risk is heightened with an aim to prevent serious failure of human—machine systems.

The OFSs can be characterized by a battery of psychophysiological markers [4, 5]. To provide for all these possibilities, we adopted a hybrid approach in which we derived electroencephalogram (EEG) and cardiovascular indicators in the present investigation. Although OFS modeling is likely to be more reliable when multiple psychophysiological measures are used, there is a fundamental problem of data fusion—how to combine those appropriate OFS markers as the model inputs? Wilson and others [6–8] have demonstrated the feasibility of artificial neural networks (ANNs) in precisely classifying the operator states of high and low workload conditions. For our purposes, the ANN has serious limitations on the opaqueness of its model output result in that the physiologically meaningful (or plausible) interpretation of the modeled (or predicted) states of the human–machine system operator is not available in the approach. In contrast, fuzzy models are well-known to be transparent (i.e., interpretable), due to its structured knowledge representation and utilization technique by means of a set of fuzzy linguistic rules and certain fuzzy inference mechanisms, without loss of modeling accuracy. Over the past three decades, fuzzy-logic-based techniques have been successfully applied to a wide range of engineering and medical problems [9–11], including workload management, target monitoring, [12] and prediction of the heart rate variability (HRV) changes under physical stress [13]. Nevertheless, the use of them in assessing OFS has been rarely reported, except for the preliminary studies described in [14–18].

The primary objective of this work is to build a model that describes the input–output (I/O) relationship between the psychophysiological measures (e.g., cardiovascular and EEG activities) and functional (i.e., cognitive, mental, or psychological) state of the human operator under a simulated process control environment. In the present investigation, the identification of the OFSs is achieved using adaptive fuzzy modeling approach which requires the measured psychophysiological and primary-task performance data only. Two types of fuzzy models, using HRV and EEG indices as the inputs and the primary-task performance marker (i.e., *time-in-range:* percentage of the in-range parameter values) as the output, are constructed and optimized automatically. The proposed modeling approach is then shown by pertinent simulation results to be capable of effectively exploiting the information reflected by the measured physiological and performance data.

Using this model, the OFS may be identified or predicted in real-time by monitoring the momentary changes of the psychophysiological and performance data. The vulnerable states identified can then be used to trigger an adaptive automation (AA) interface between human operator and machine (computer), both playing an equally important role in maintaining the operational performance of human–machine system under study. Although a large body of empirical work has substantiated the general benefits of AA [19], there has been very little systematic work on modeling OFSs in a human–machine system [5]. The fuzzy models developed in this chapter can be used as reference models for control design, permitting simulations for design validation prior to real-time implementation.

9.2 Data Acquisition Experiments

9.2.1 Subjects

As process control operators in this experiment, a total of 11 volunteering participants (all are healthy male postgraduate students, aged at 23–29 years old) were selected from a subject pool on the basis of level of training (>10 h) and domain expertize (e.g., process engineering, chemistry) on a simulated process control environment–automation-enhanced cabin air management system (aCAMS) originally developed by Hockey and his colleagues and modified in later studies by Lorenz and others [20–22]. All subjects were fit and healthy, had normal or corrected-to-normal visual acuity, did not have any diseases, and did not take any medicine that could have affected their performance at the time of experiment. Subjects in the CAMS experiments had been informed that the experiment was concerned with the test of OFS in process control operations and informed about the series of changes in level of automation (LOA) they underwent under the experimental operations. In addition to six general training sessions, they carried out a specific training session on the cyclic loading task, followed by two formal experimental sessions.

9.2.2 Experimental Equipments and Environment

For each subject, two experimental sessions were scheduled in an air-conditioned laboratory (3.75 $\times$ 4.25 m) without windows, illuminated with two pairs of high-frequency fluorescent lamps and furnished with desks for the subject and the experimenter and movable walls for separation. The process control software was running on a PC and presented on a 19'' Thin Film Transistor (TFT) monitor on the subject's desk at about 50 cm visual distance. A PC keyboard and a mouse were used as hardware controls. The Active Two Base System (BioSemi, The Netherlands) was used for continuously acquiring psychophysiological data from 45 channels for ECG (electrocardiogram; Nehb's triangle), respiration (nasal/oral thermistor for 3-point measurement), EMG (electromyographic activity from the dominant forearm), EOG (vertical, horizontal electrooculogram), reference (left and right mastoids), and EEG (32 sites on a headcap arranged in the international 10–20 system [23]). Data acquisition for all signals sampled at 2048 Hz was controlled by an interface (ActiView 5.33, BioSemi, The Netherlands). The ActiView interface displayed on a 19'' TFT monitor provided the experimenter with signal preprocessing, marking specific events (environmental disturbances or other artifacts), and storing all physiological data and markers in BioSemi data files.

9.2.3 Experimental Tasks

The aCAMS served as a simulated process control environment, which required manually monitoring and control of the atmospheric environment (e.g., air quality, temperature, pressure, etc.) of a closed system (such as a spaceship capsule or submarine). The primary tasks contained in the aCAMS simulation environment required monitoring of an autonomous air management system and performing fault management in case of system malfunction (or failures). The potential faults vary in complexity and demand high-level knowledge-based decisions about (1) appropriate immediate safety actions (changing settings of automated control subsystems or assuming manual control of subsystems) and (2) diagnosing (or localizing) the possible causes of the fault.

The mutitask environment enabled us to analyze strategies in information acquisition as well as in secondary tasks (e.g., alarm reaction time and prospective memory) complementing primary-task performance data (percentage of the time duration when the five key system variables fall within their targeted ranges). In a modified version of aCAMS, the process control task was performed either by the computer when in automatic mode or by the subject when in manual mode. Operators interacted with a dynamic display that provides data on controlled variables and system functions via a range of control and automation tools. In the experiment, five conditions, each with different levels of task difficulty (LTD) and hence, different intensities of mental workload, were introduced: (1) C5–absence of automation: without any kind of system support, the subject had to manually control all the five technical subsystems simultaneously; (2) notification: the system provided information about what happened; (3) suggestion: the system notified and suggested the appropriate action to take; and (4) action: the system resolved the problem and consequently notified this to the subject. All subjects encountered all five LTD in each session of experiment.

In order to vary the level of task load that an operator underwent, different automation failures were defined for the process control operation. When automatic control components failed, operators had to maintain several key variables within their target ranges by assuming manual control. This was the primary task of the human–machine system. Workload in individual conditions (A, B, C, D) was increased by the number of variables (1, 3, 4, 5) requiring to be under manual control, i.e., failures of the controllers for oxygen flow, nitrogen flow, CO_2 scrubber, dehumidifier, and temperature cooler. Thus a new 'cyclic loading' approach was designed to assess OFS in both loading (conditions A→D) and unloading (conditions D→A) phase of an experimental session of process control operation. The goal of such experimental paradigm was to induce artificially a performance breakdown in the compensatory control mechanism, allowing the detection of OFSs corresponding to the breakdown and pre-breakdown phases of process control operation.

Prior to formal experimental sessions, each subject (with sound engineering background) had been trained on the manual process control tasks for more than 10 h to get familiar with aCAMS environment and the experimental tasks pre-specified. In each experimental session of process control operations, a subject interacted with aCAMS while his/her EEG and ECG responses were recorded simultaneously.

9.2.4 Experimental Procedure

The subjects worked for two sessions, each lasting about 2 h. In order to avoid the unwanted circadian effects, the two sessions were performed at the same time of two days spanning an interval of one month. Immediately after filling out the health questionnaire and subjective ratings, the subject started process control operations. Each process control condition lasted for 15 min and was interrupted by filling in subjective ratings for about 20 s. The time taken to complete the ratings was used to reset aCMAS and change the file for continuous recording of psychophysiological data. The data analysis includes two sessions, each containing 2 × 4 conditions different according to two factors: phase (loading/unloading) and level (1, 3, 4, 5 variables demanding manual control) of task load.

The psychophysiological measures, especially Heart Rate (HR), HRV, and EEG, have usually been used as typical markers of the mental workload (MWL)-related OFSs [4, 5]. HR is one of the most practical techniques for the assessment of MWL imposed on operator. However, it is only one of several possible parameters of the complex homeostatic cardiovascular system reflecting energetic, thermoregulatory, respiratory, and/or emotional processes. Consequently, the HRV measures in both time and frequency domain have been adopted to decompose different processes that influence the cardiovascular regulation. In recent years, the HRV measure has been widely used to analyze MWL since it allows differentiation of levels and types of MWL in information processing, for instance automated versus controlled information processing (latter requiring more effort and thus resulting in a suppression of HRV). Among the available HRV measures, the suppression of spectral power of the 0.1 Hz component of HRV has been considered as a sensitive and diagnostic indicator of MWL under real or simulated human–machine interaction [24–28].

The ECG signal was composed of signals recorded from Nehb's triangle and segmented in 10 s epochs using Brain Vision Analyzer (Brain Products, Germany). After baseline correction (epoch length) the R-peak in the signal was triggered by a level indicator (1 mV) and time-stamped by a marker. Instantaneous HR and the 0.1 Hz component of HRV were calculated using the procedure described in [28] with the LabVIEW^{TM} virtual instruments (National Instruments, USA). With Brain Vision Analyzer (Brain Products, Germany), the EEG signals were recorded at 32 electrode sites defined by international 10-20 system. The EEG electrodes were re-referenced to two linked mastoids. The EEG, sampled at a rate of 2048 Hz, was preprocessed with a band-pass filter between 1.6 and 55 Hz. The epoched EEG signals (epochs of 2 s) were analyzed via fast fourier transform (FFT; 10 % Hanning window) and normalized to obtain power spectra (0.5–50 Hz) in 0.5 Hz resolution comparable across EEG segments. The primary-task-related control system performance data, *time-in-range*, were simultaneously recorded during process control operations.

9.3 Adaptive Fuzzy Modeling Methods

Fuzzy set and fuzzy logic theory were originally proposed by Zadeh in [29, 30] to formalize the representation and management of imprecise (or approximate) knowledge in complex systems. Many optimization techniques, such as artificial neural networks (ANN) and genetic algorithms (GA) have been employed to tune fuzzy system in an automatic manner. The fundamental advantage of fuzzy models is that they can combine human knowledge and expertize with sensory numeric measurements within a unified mathematical framework [31, 32].

The criteria for OFS identification or prediction must allow for overlapping classification of the operator states. The most appropriate approach for this is fuzzy modeling, since it recognizes that classification of many real-world variables cannot be captured using strict logic-based techniques. Fuzzy models are particularly suited to problems such as operator performance, where the process is so complex and ill-defined that precise mathematical models cannot be derived and if at all derived they would prove to be either inaccurate or less interpretable. This important feature may lead to easy fusion of information extracted from multiple physiological measures.

The OFS modeling problem may be roughly stated as: Given *n* independent or dependent physiological measures (variables), build a model to accurately assess or predict the OFS. With this description of the problem at hand, the psychophysiological data and observer's perceptions (or visual examination) can be formalized in terms of linguistic variables characterized by their membership functions (MFs) defined over certain domain of discourse. As will be shown by this work, fuzzy modeling approach can provide concise (parsimonious) and robust description of OFSs. By utilizing fuzzy logic, two types of OFS models were developed based on the HR, HRV, and EEG data in this work. Using the developed fuzzy models, a state metric (either crisp value or class) can be computed based on multiple physiological measures only.

9.3.1 OFS-Related Psycho-physiological Markers

Candidate markers will be selected by briefly analyzing their reliability and sensitivity for assessment of risky OFS under process control operation. Cardiovascular indices, notably HR and HRV, have been found to respond reliably to changes in workload and mental effort [33], particularly in operating settings where executive problem-solving is involved [34]. Current analysis of HRV uses spectral analysis of the cardiac interval signal to separate effects of effort on different components. As with HRV, power spectrum analysis of the EEG is usually necessary to reveal effects of mental stress. The most sensitive markers of general alertness are usually based on various ratios between the power in higher and lower frequency bands.

The first HRV index, HRV_1, is defined as the average of the 0.1 Hz components of HRV for time windows of 128 s. The calculation was performed for inter-beat intervals (normalized to their mean value) using windows of 128 s, which were shifted forward by steps of 1 s. Thus spectral densities were obtained for 128 s intervals across the whole session, containing about 99 % overlapping with adjacent intervals. The second HRV index, HRV_2, is defined as the HR variation coefficient and given by:

$$HRV_2 = \frac{\sigma_{\text{HRV}}}{\mu_{\text{HRV}}}, \tag{9.1}$$

where σ and μ denote the standard deviation and average of HRV segment (7.5 min).

We also include an EEG measure referred to as the 'Task Load Index' (TLI) calculated using different EEG band powers [35]. This is based on the presence of high levels of theta activity at frontal midline sites, with concomitant attenuation of alpha power [theta/alpha]. The reduced frontal-midline theta power may reflect strategic disengagement from the frequent executive requirements of aCAMS [36]. This subtle shift of task focus may be a more sensitive marker of operational risk. The TLI indices, TLI_1 and TLI_2, used in this chapter are given by:

$$\begin{cases} \text{TLI}_1 = \dfrac{P_{\theta, Fz}}{P_{\alpha, Pz}} \\ \text{TLI}_2 = \dfrac{P_{\theta, AF_z}}{P_{\alpha, CP_z PO_z}}, \end{cases} \quad (9.2)$$

where P_θ and P_α denote the theta- and alpha-band power, respectively; the frequency bands of EEG signals recorded on different sites are defined in order as: *θ, Fz*: 6–7 Hz; *α, Pz*: 10–12 Hz; *θ, AFz*: 5–7 Hz; *α, CPz*: 8–10.5 Hz; *α, POz*: 10–13.5 Hz; and *Fz, Pz, AFz, CPz*, and *POz* are the five EEG sites introduced in standard 10-20 system. The extracted data for each frequency band was exported to separate files and the indices were calculated using Office Excel (Microsoft, USA).

9.3.2 Basics of Fuzzy Modeling Paradigm

The construction of fuzzy models of OFS consists of the following steps:
Step 1-Determination of model structure.
The candidate inputs of the fuzzy model may include HR, HRV, and EEG markers, namely HR, HRV1, HRV2, TLI1, TLI2, which were found to be most sensitive to the changes in mental workload [4, 5, 37]. The optimal number of inputs were selected from the above five candidate inputs based on visual inspection and linear correlation analysis of the input and output data. As the number of candidate inputs used for the static model is small, one can alternatively select the parsimonious model structure by comparing the accuracy of models using different combinations of the candidate inputs. The single output of the model is *time- in-range* related to the primary-task performance.
Step2-Fuzzy reasoning mechanism.
Apply the membership grades to individual rules to produce a set of rule-firing strengths. The most popular fuzzy rules processing techniques are the Mamdani- and TSK-type [38]. The general forms of their *i*th rule are given by:

$$R^i\text{: If } x_1 \text{ is } A_{i1} \text{ and } x_2 \text{ is } A_{i2} \text{ and } \ldots \text{ and } x_m \text{ is } A_{im}, \text{ Then } y_i \text{ is } B_i, \quad (9.3)$$

$$R^i\text{: If } x_1 \text{ is } A_{i1} \text{ and } \ldots \text{ and } x_m \text{ is } A_{im}, \text{ Then } y_i = f(x_1, \ldots, x_m), \quad (9.4)$$

where $\mathbf{x} = [x_1, \ldots, x_m]^T \in \mathbf{U} \subset \mathfrak{R}^m$ is the input vector, y the output, $A_{i1}, A_{i2}, \ldots, A_{im}$, B_i are linguistic labels;$f(x_1, \ldots, x_m) = c_0 + c_1 x_1 + \cdots + c_m x_m$ is a linear function; and $R^i, i = 1, 2, \ldots, M$ denotes the *i*th rule in the rule-base containing a total of M rules.

In the ANFIS model using TSK-type fuzzy rules [39], the output $f(\mathbf{x}) \in \mathbf{V} \subset \mathfrak{R}$ is given by

$$\begin{cases} f(\mathbf{x}) = \sum_{i=1}^{M} w_i \cdot y_i / \sum_{i=1}^{M} w_i \\ w_i = \prod_{j=1}^{m} \mu C_j^i(x_j), \end{cases} \tag{9.5}$$

where M denotes the number of rules and m the number of inputs.

The ANFIS and Mamdani models were tuned by hybrid learning [39] and genetic algorithms (GA) [40, 41], respectively. The optimal MF parameters were determined when the learning or optimization process reaches the minimum of mean squared error (MSE) defined by

$$\mathrm{MSE} = \frac{1}{N} \sum_{k=1}^{N} [y(k) - y_M(k)]^2, \tag{9.6}$$

where $y(k)$ is the actual output and $y_M(k)$ is the calculated model output at the sampling instant k, respectively.

9.3.3 Mamdani-Type Fuzzy Model

When designing Mamdani-type fuzzy models, we use MIN and MAX for connectives 'and' and 'or' used in the fuzzy rules, MIN for implication operator, MAX for aggregation operator, singleton fuzzification, and centroid defuzzification. As in the ANFIS models, all MFs are of a Gaussian type. The composite rule of inference [29] is adopted to perform fuzzy reasoning.

A. I/O data normalization

To facilitate the GA optimization procedure, the universes of input and output variables need to be normalized to the range [0, 1]. The linear transformation used for this normalization is as follows:

$$\begin{cases} z' = \dfrac{z - z_{\min}}{L_z} \\ L_z = z_{\max} - z_{\min}, \end{cases} \tag{9.7}$$

where $z' \in [0, 1]$ denotes the normalized data, z the original data, and $z_{\min}$ and $z_{\max}$ the minimal and maximal values among the original data.

B. GA optimization

The optimization of the fuzzy rules is achieved using GA. In the first instance, all the input and output data are normalized within the interval [0, 1] according to Eq. (9.7). The fitness function chosen for the GA is the mean squared error (MAE) between the actual and model output. GA is used to search for the minimum of this fitness function with respect to the parameters of the MFs in the Mamdani model.

The input and output MFs are initialized with Gaussian functions. Three MFs (c. f. Table 9.1), namely '*small,*' '*medium,*' and '*large,*' are assigned to each input, while four MFs: '*low,*' (indicating an unacceptable or a very vulnerable state) '*normal,*' (indicating a normal state) '*high,*' (indicating a good physiological state) and '*very high,*' (indicating a very good physiological state) are assigned to the output.

There are 20 MF parameters and 9 rule strengths (weights) to be optimized by the GA. The population size is fixed at 100 in all the simulation examples given below. The rank technique is used for fitness rescaling. After extensive trials, the GA operators found to produce the best results for our problem were preset as follows: stochastic uniform selection, scattered crossover with a probability of $P_c = 0.8$, and uniform mutation with a probability of $P_m = 0.02$. At the end of the GA run, the optimized parameters of the MFs involved in the fuzzy model are provided and stored.

9.4 OFS Estimation Results and Discussion

In all the simulations presented here, the signal data sampling interval is taken to be 7.5 min. Gaussian MFs were used for both fuzzy models. The training error profiles have been examined by varying the number of fuzzy partitions for each input from 2 to 5. It was found that assigning three fuzzy partitions to each input produces the best modeling accuracy. The three initial fuzzy partitions are identical in shape, i.e., the width parameters of the initial Gaussian MFs are taken to be the same. The variables HRV2 and TLI2 among the five candidates were selected as the inputs to both fuzzy models. The training and checking data sets were obtained from the first and second session of experiment, respectively.

9.4.1 OFS Estimation Results

The input/output data measured from two experimental sessions for P11 (healthy and fit male, 25 years old, 70 kg, 170 cm) is shown in Fig. 9.1. It is indicated by

Table 9.1 Fuzzy rule lookup table for P11

		HRV_2		
		S	M	La
TLI_2	S	**N ($\mathbf{w_{R1}}$** = 0.5346)	Lo (w_{R4} = 0.0034)	**H ($\mathbf{w_{R7}}$** = 0.7905)
	M	H (w_{R2} = 0.2978)	H (**$\mathbf{w_{R5}}$** = 0.9262)	**H ($\mathbf{w_{R8}}$** = 0.9687)
	La	**Lo($\mathbf{w_{R3}}$** = 0.4804)	H (w_{R6} = 0.1201)	**H ($\mathbf{w_{R9}}$** = 0.8986)

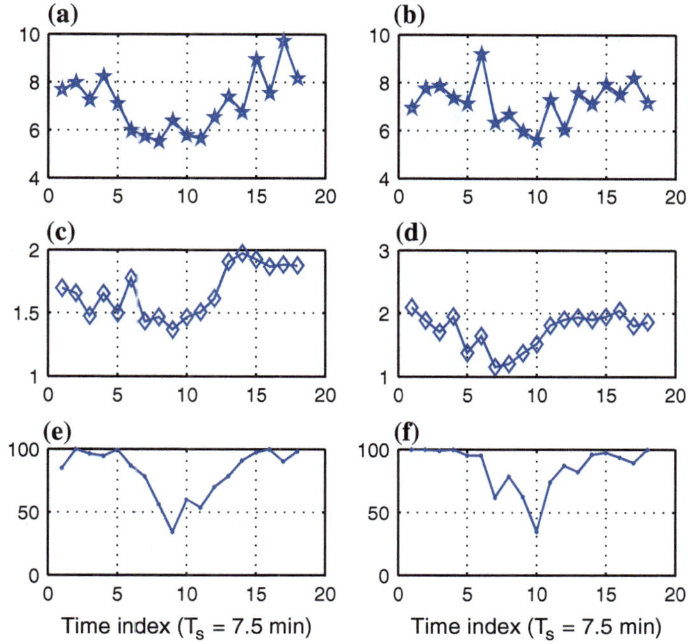

Fig. 9.1 The I/O data measured from experimental session 1 (*left column*) and session 2 (*right column*) for P11. **a** HRV2 of session 1. **b** HRV2 of session 2. **c** TLI2 of session 1. **d** TLI2 of session 2. **e** Time in range (%) of session 1. **f** Time in range (%) of session 2

Fig. 9.1e, f that control performance was very high in low task-load conditions (i.e., A and B) and was degraded in high task-load conditions (i.e., C and D). Under low level of monitoring and control tasks, the operator maintained the total system in range for more than 90 % of the time across the complete set of conditions in both sessions of data acquisition experiment. However, as expected there was some evidence of control performance breakdown for more demanding conditions: the parameter *time-in-range* falling abruptly below 50 % when all five variables had to be controlled manually. According to the primary-task performance data, we can thus assume certain level of strain and high operational risk at high levels of loading conditions.

The ANFIS modeling result for P11 in the case of C1 is shown in Fig. 9.2. For the Mamdani model, the initial fuzzy rule-base for each subject is constructed by visually examining the qualitative if-then relationship underlying the measured data. As a result, the fuzzy rule-bases are different from subject to subject (in other words, each subject has a unique rule-base, reflecting the cross-individual difference). The optimal MFs and degrees of belief in each rule are searched using a GA approach. The evolution of best and mean fitness value in each generation of the GA approach is demonstrated in Fig. 9.3. In the lower panel of Fig. 9.3, the first, second ten variables, and the last nine ones denote the optimized spread widths, center positions

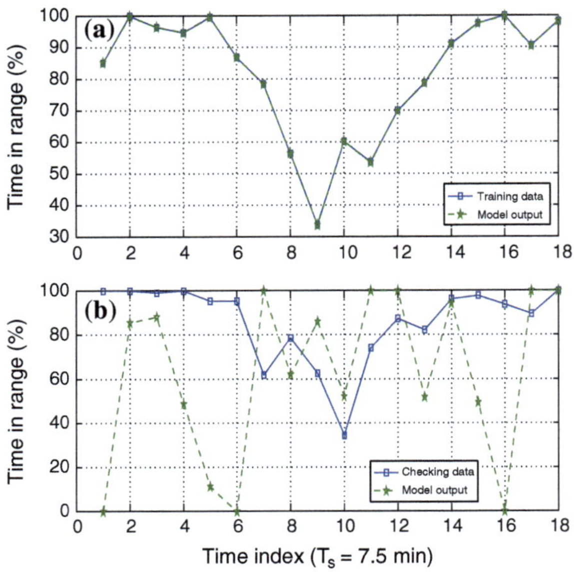

Fig. 9.2 ANFIS modeling result for P11. **a** ANFIS model training result. **b** ANFIS model checking result

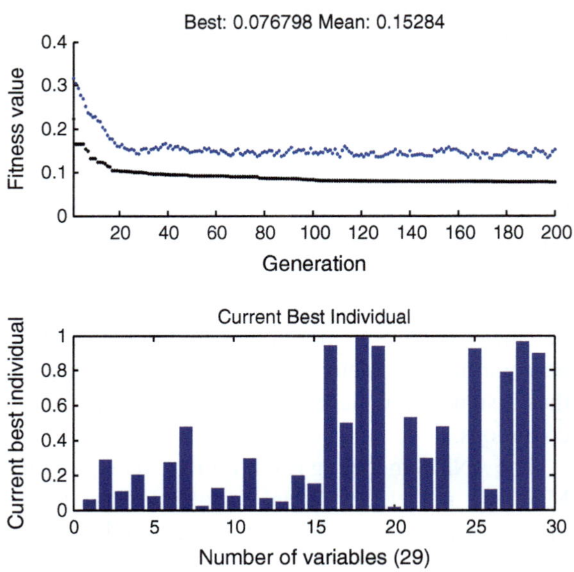

Fig. 9.3 GA optimization solution corresponding to optimal model parameters

of the input and output MFs, and rule weights (used to indicate the strength or importance of individual rules), respectively. The structure of the optimized Mamdani model is schematically shown in Fig. 9.4. The rules of the optimized Mamdani-type fuzzy model are illustrated in Table 9.1, where the labels S, M, La,

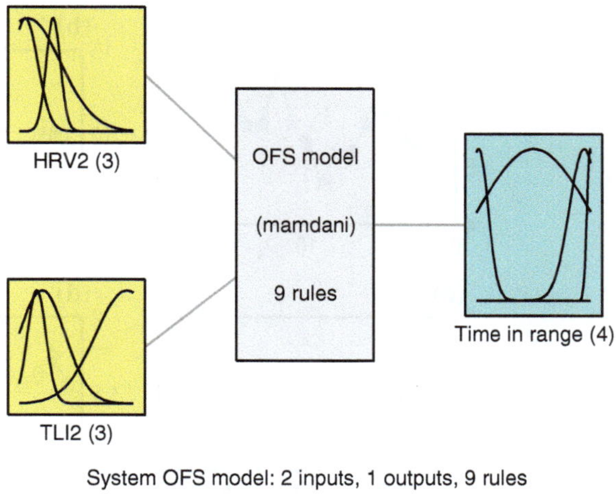

Fig. 9.4 The structure of GA optimized Mamdani-type OFS model

Lo, N, H, VH denote the linguistic terms 'small,', 'medium,' 'large,' 'low,' 'normal,' 'high,' and 'very high,' respectively. It is noted that the first, third, fifth, seventh, eighth, and ninth rule (Table 9.1 in bold) are more important (in terms of the larger weights to be applied to each rule) in comparison with the other rules in Table 9.1 after applying the GA-based fuzzy rule optimization procedure. The comparison of the model output and desired output is shown in Fig. 9.5.

The results for subject P9 (very healthy and fit male, 23 years old, 75 kg, 173 cm) were shown in Fig. 9.6, 9.7, 9.8, 9.9 and 9.10 and Table 9.2 (a different set of fuzzy rules with different weights in this case when compared with those for P11). By comparing Figs. 9.1 and 9.6, it was clearly seen that the individual

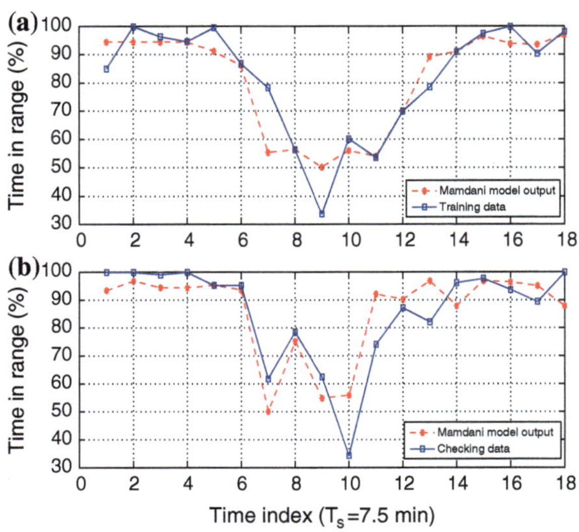

Fig. 9.5 Modeling results via GA-based Mamdani model for P11. **a** Comparison between model output and training data. **b** Comparison between model output and checking data

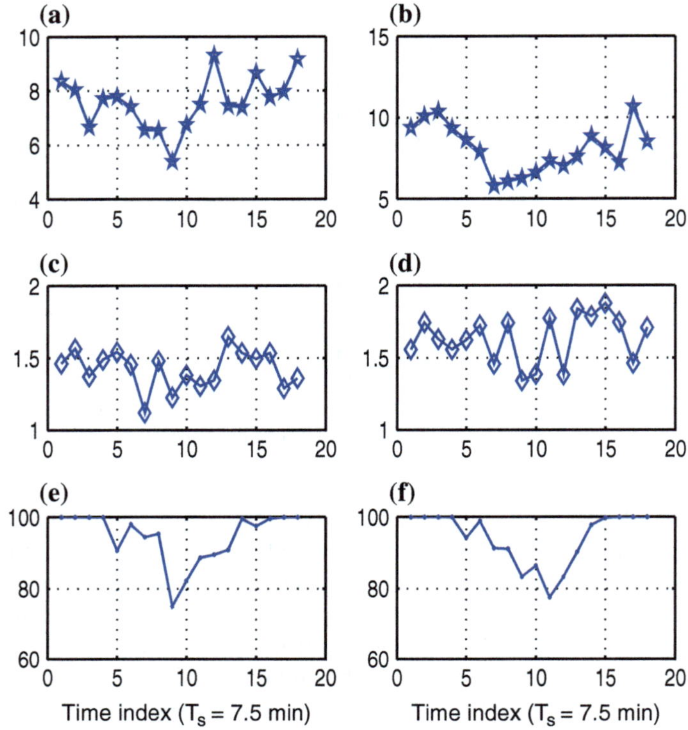

Fig. 9.6 The I/O data measured from experimental session 1 (*left column*) and session 2 (*right column*) for P9. **a** HRV2 of session 1. **b** HRV2 of session 2. **c** TLI2 of session 1. **d** TLI2 of session 2. **e** Time in range (%) of session 1. **f** Time in range (%) of session 2

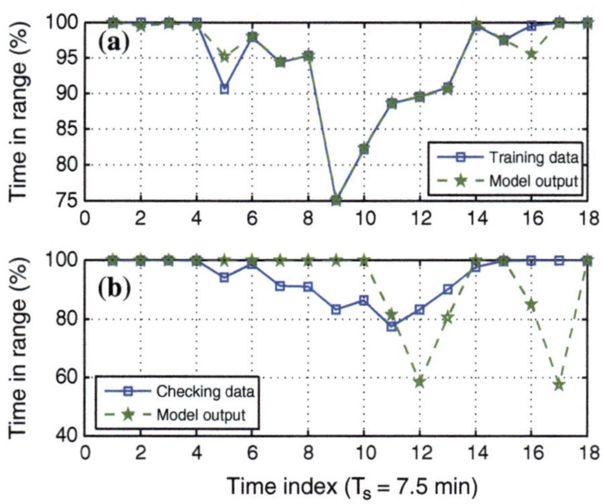

Fig. 9.7 ANFIS modeling result for P9. **a** Comparison of ANFIS output and training data. **b** Comparison of ANFIS output and checking data

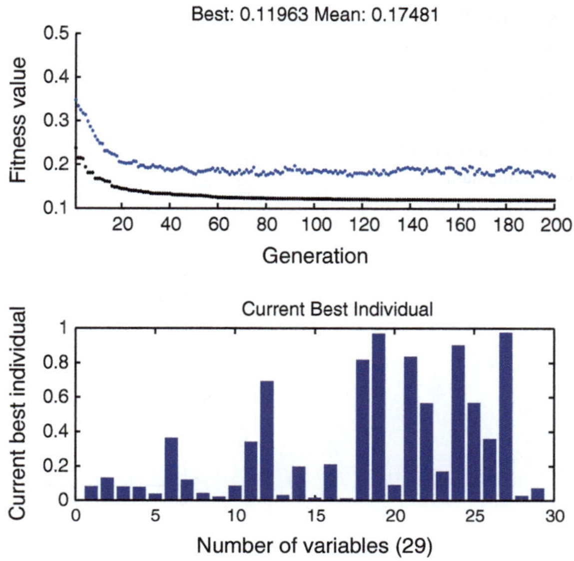

Fig. 9.8 The GA optimization solution corresponding to optimal model parameters

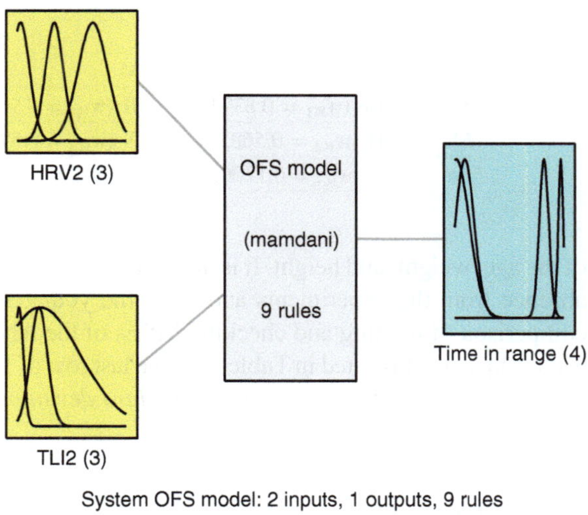

Fig. 9.9 The structure of GA optimized Mamdani-type OFS model

difference exists, P9 showing less significant decrement of performance even with a load of all five: average performance began to show impairments when four variables had to be controlled manually and fell below 80 % (50 % for the subject P11) with a load of all five.

Likewise, we constructed the ANFIS and Mamdani models for the remaining 9 subjects in the same experimental group. The characteristic information of all 11 well-trained subjects is given in Table 9.3, where the last row summarizes the statistics

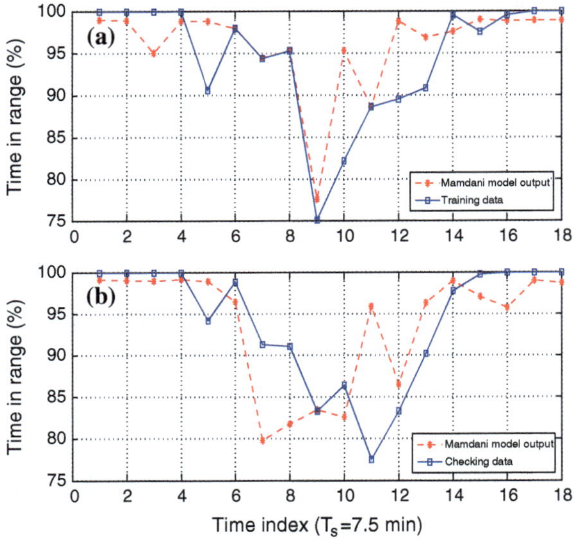

Fig. 9.10 Modeling results via GA-based Mamdani model for P9. **a** Comparison between model output and training data. **b** Comparison between model output and checking data

Table 9.2 Fuzzy rule lookup table for P9

		HRV_2		
		S	M	La
TLI_2	S	**Lo ($\mathbf{w_{R1}}$ = 0.8374)**	**H ($\mathbf{w_{R4}}$ = 0.9040)**	**Lo ($\mathbf{w_{R7}}$ = 0.9775)**
	M	**H ($\mathbf{w_{R2}}$ = 0.5683)**	**N ($\mathbf{w_{R5}}$ = 0.5702)**	H (w_{R8} = 0.0286)
	La	N (w_{R3} = 0.1706)	H (w_{R6} = 0.3596)	VH (w_{R9} = 0.0793)

of the age, weight, and height. It is noted that P7 is missing because of his unexpected absence from the experiments and P9 is the youngest and very fit physically. The comparison of training and checking MSEs of the ANFIS and Mamdani models for each subject is illustrated in Table 9.4. The last row of Table 9.4 shows the statistics of the MSEs in the format of '*mean* ± *standard deviation*'.

9.4.2 Discussions

The robustness and generalizability of the fuzzy models suggested are improved by providing optimization capacity, leading to hybrid intelligent system architectures like ANFIS and GA-based Mamdani model adopted in this work. In terms of generalization performance, particularly with respect to the operator state prediction (i.e., use some initial data during a task to estimate or predict the rest of the task as suggested by one of the reviewers), the ANFIS models for P2 and P8 ∼ 10 work better than for other subjects, while Mamdani model works best for P6 ∼ 11. This

Table 9.3 Subject information

Subject no.	Age (years)	Weight (kg)	Height (cm)
P1	27	56	168
P2	27	62	172
P3	27	65	175
P4	27	60	170
P5	26	90	180
P6	26	65	173
P8	**28**	**86**	**175**
P9	**23**	**75**	**173**
P10	**29**	**67**	**178**
P11	25	70	170
P12	29	85	180
	26.72 ± 1.74	71 ± 11.046	174 ± 4.05

implies that we may need to use *different models* for different subjects in order to predict the individual OFSs well. In comparison, among all models the Mamdani model tailored to P9 achieved the best balance between training and validation. The ANFIS model for P9 also works best. By analysis, this is because that the psychophysiological patterns of P9 already settled down in the second session of experiment (i.e., after training) and thus he performed in a similar way to manage the mental workload in both experimental sessions (refer to Fig. 9.6). It is interesting to see from Table 9.2 that P9 is the youngest in the group. Also we are aware that he is very fit because of regular exercise (for example, cycling everyday). It may be intuitively expected that he has a better ability to adapt to the task-load situations demanding a lot of mental resources and efforts. The modeling results are well consistent with this intuition.

Surprisingly it seems that the ANFIS training is still too good even if we reduce the number of training iterations to a few epochs. This observation inspired us to consider the use of Mamdani model to improve the generalizability of the OFS model. It was clearly shown by Table 9.4 that the GA-based Mamdani model has better generalization performance using the checking (or validation) data set than ANFIS model. More specifically, from Figs. 9.5b and 9.10b we see that the Mamdani-type models capture the output trends more effectively than the ANFISs, an interesting observation which strengthens the viewpoint that OFSs are indeed subjective and difficult to be quantitatively identified using conventional mathematical models even ANFIS. In the present case, the ANFIS model is essentially an optimized mathematical model in terms of function approximation, but it does not actually exploit the available qualitative knowledge about the model I/O relationship. On the other hand, the I/O information contained in the training data set (i.e., the first experimental session) and the checking data set (i.e., the second experimental session) may not be consistent or stable, but complementary to each other. This serious paucity of necessary numerical information would mainly account for the marked discrepancy between the training and checking error for the ANFIS

Table 9.4 Training and checking RMSEs (root mean squared errors) of fuzzy models for each of 11 subjects

	ANFIS model		GA-based Mamdani model	
	Training	Checking	Training	Checking
P1	4.2263	59.4475	7.7597	40.9133
P2	**0.3748**	**15.9376**	6.4588	14.9934
P3	0.1345	76.8967	11.3326	11.9885
P4	7.68e-002	42.5335	6.755	20.5058
P5	0.04	44.308	12.6905	25.7907
P6	5.55e-003	40.2976	**6.077**	**7.8643**
P8	**3.9326**	**17.7885**	**5.9657**	**7.5489**
P9	***1.4392***	***13.7408***	***4.7382***	***6.1433***
P10	**1.5067e-003**	**14.3889**	**3.489**	**9.3083**
P11	0.0938	49.8096	**8.054**	**9.3657**
P12	0.6745	71.4031	13.9648	17.0875
	1 ± 1.5821	40.5956 ± 22.9367	7.935 ± 3.3384	15.5918 ± 10.3771

model. For the Mamdani model, it would not cause severe degradation of the generalization performance since the model was effectively built upon the optimized fuzzy rule-base which may be expected to be a set of generic (or universal) law governing the model I/O dependency in both sessions of experiment. This is why the two models display very different generalizability for the same set of checking data.

From a more pragmatic point of view, we also need to consider the practicality of the approach especially for real-time assessment of OFS. The small number of fuzzy rules (9 in total in the present case) elicited by the Mamdani model suggests that it is possible to reconcile accuracy, interpretability (transparency), and computational complexity of the model, which will greatly help when the model structure is augmented with more input dimensions, especially when more EEG features (i.e., other frequency band power or power ratio indices different from that introduced in Eq. (9.2)) are taken into account. In regard to the real-time constraints of the Mamdani model, since a system for online analysis of the psychophysiological recordings in real-time is under construction, the continuously recorded data stored in the BioSemi data files were analyzed only off-line in this work. Nonetheless it should be noted that the proposed fuzzy models include a generic structure (outer shell) common or at least very similar to all subjects, and an individual-tailored inner shell that can be readily tuned using individual baseline data. In general, the choice of the model parameter optimization techniques depends on the nature of the problem domain, the available information and the realistic requirement for computational complexity. In other words, we need to consider the above three aspects and try to achieve a trade-off among them when determining the appropriate optimization paradigm. For the current problem of model-based quantitative identification and prediction of the OFSs under investigation, this can be accomplished using multi-objective GAs. Such model parameter optimization can be performed either off-line, using our operator models, or on-line as the operator is performing the task. In the latter case, incremental GAs can be used for the sake of required real-time performance [41]. Considering the moderate computing overhead of the GA-based Mamdani model, we firmly believe that it would be feasible for online data modeling and analysis to be carried out as the next step of this research.

In summary, the GA-based Mamdani models may constitute a more powerful solution to the OFS prediction problem in terms of both generalizability and practical usefulness for real-time evaluation of OFS.

9.5 Conclusion and Future Work

In this chapter, we developed adaptive fuzzy models of the OFS under operational risk (associated with increased cognitive strain through stepwise changes in task load) based on the experimentally measured psychophysiological and performance data. Several psychophysiological input variables of the fuzzy models were selected (extracted). Based on the training data set, two types of adaptive fuzzy modeling techniques, viz., ANFIS and Mamdani model were developed to estimate the OFS. The OFS-related data analysis results on 11 subjects have demonstrated that the proposed technique is robust with respect to both initialization of model parameters and noises contained in the measured training data and thus can be used effectively for identification and prediction of high-risk OFS and performance breakdown. Using the modified aCAMS, we investigated the effect of dynamic LTD changes on operator performance and found quantitatively the complex nonlinear relationship between physiological indices and operator performance in human–machine systems. The work has potential to be used as the basis for future implementation of adaptive automation of human–machine systems.

In summary, the salient features of the proposed fuzzy models include:

(a) They may provide a parsimonious, transparent (interpretable by means of fuzzy if-then rules) and robust characterization of OFS.
(b) They can be easily extended to accommodate additional inputs, MFs and fuzzy rules.
(c) They may lead to 'traffic-light'-like decision-making approach to trigger adaptive control of aCAMS system. In this case, the fuzzy subsets of '*red,*' ('*low*' used in the chapter instead) '*yellow,*' ('*normal*') and '*green,*' ('*high*' and '*very high*') may correspond to risky, slightly risky, and acceptable OFSs. This decision can be used to trigger adaptive task allocation between human operator and computer-based automatic controller.

Considering the practical significance of the current work, the estimated OFS can be used to design a closed-loop control system to trigger control task allocation between human operator and computer, thus allowing for design and implementation of adaptive automation of human–machine systems. To this end, the following problems should be further studied in depth in the future work:

(1) The model structural identification problem (i.e., how to select an optimal number of model inputs) needs to be resolved. In addition to the physiological and performance variables used in this work, other OFS-related indices, such as eye movements (e.g., point-of-gaze), EEG, and HRV indices [42–48], should also be examined.
(2) We still need to investigate whether or not the temporary reintroduction of manual control, found to be beneficial in [49] and the present investigation, can also benefit the fault management missions in aCAMS.
(3) More experimental subjects and more trials for each subject will have to be used since high inter-individual differences in sensitivity to the level of task difficulty and task type were already reported [50, 51]. Therefore, the possible generalizability of the developed fuzzy models across subjects with similar physiological characteristics should be further studied. On the other hand, it may be necessary to establish individual-specific OFS indices for a reliable prediction of risky OFS. The future development of intelligent control architecture for adaptive man–machine functional allocation will take into account the idiosyncrasies of individual subjects.

Acknowledgments This work was supported by the National Natural Science Foundation of China under Grant No. 61075070 and Key Grant No. 11232005. The authors would also like to thank Professor D. Manzey, Technical University of Berlin, Germany for providing the AUTO-CAMS software which was used in the data acquisition experiments. The invitation from Professor G. M. Dimirovski, for the chapter contribution is also gratefully acknowledged.

References

1. Hancock, P.A., Desmond, P.A. (eds.): Stress, Workload and Fatigue. Erlbaum, Mahwah (2001)
2. Hockey, G.R.J., Gaillard, A.W.K., Burov, O. (eds.): Operator Functional State: The Assessment and Prediction of Human Performance Degradation in Complex Tasks. IOS, Amsterdam (2003)
3. Hockey, G.R.J.: Compensatory control in the regulation of human performance under stress and high workload: a cognitive energetical framework. Biol. Psychol. **45**, 72–93 (1997)
4. Nickel P., Roberts A.C., Hockey G.R.J.: Assessment of high risk operator functional state markers in dynamical systems – preliminary results and implications. In Proceeding Human Factors and Ergonomics Society Europe Chapter Annual Meeting 2005, Turin, Italy, 26–28 October 2005
5. Nickel P., Hockey G.R.J., Roberts A.C., Roberts M.H.: Markers of high risk operator functional state in adaptive control of process automation. In: Proceeding of IEA (2006)
6. Wilson, G.F., Russell, C.A.: Performance enhancement in an uninhabited air vehicle task using psychophysiologically determined adaptive aiding. Hum. Factors **49**, 1005–1018 (2007)

7. Wilson G.F., Russell C.A.: Operator functional state classification using multiple psychophysiological features in an air traffic control task, Hum. Factors, **45**, 381–389 (2003)
8. Wilson G.F., Russell C.A.: Real-Time assessment of mental workload using psychophysiological measures and artificial neural networks, Hum. Factors, **45,** 635–644 (2003)
9. Cox, E.: Fuzzy fundamentals, IEEE Spectrum, 58–61 (1992)
10. de Silva, C.W.: Intelligent Control and Fuzzy Logic Applications. CRC Press, Boca Raton (1995)
11. Mendel, J.M.: Uncertainty, fuzzy logic and signal processing. Signal Process. **80**, 913–933 (2000)
12. Parasuraman, R., Masalonis, A.J., Hancock, P.A.: Fuzzy signal detection theory: basic postulates and formulas for analyzing human and machine performance. Hum. Factors **42**, 635–659 (2000)
13. Mahfouf M., El-Samahy E., Linkens D.A.: Development of a grey-box closed-loop model relating to volunteers subjected to physical stress. In: Proceeding 15th IFAC World Congress, Barcelona (2002)
14. Fraser, W.D., Human Factors of CC-130 Operations: Future Aircraft Performance Measures, Technical Report., R-98–16, Toronto: DCIEM (1998)
15. Moon, B.S., Lee, H.C., Lee, Y.H., Park, J.C., Oh, I.S., Lee, J.W.: Fuzzy systems to process ECG and EEG signals for quantification of the mental workload. Inf. Sci. **142**, 23–35 (2002)
16. Hockey G.R.J., Nickel P., Roberts A., Mahfouf M., Linkens D.A.: Implementing adaptive automation using on-line detection of high risk operator functional state. In: Proceeding 9th International Symposium of the ISSA,Nice, France, 1–3, March 2006
17. Zhang J., Mahfouf M., Linkens D.A. et al.: Adaptive fuzzy model of operator functional state in human-machine system: a preliminary study. In: Proceeding IASTED International. Conference. on Biomedical Engineering, Innsbruck, Austria, 14–16, February 2007
18. Ntuen C.A., Estimating (human) cognitive states in a human-machine system: a fuzzy modeling approach. In: Proceeding 4th Symposium on Human Interaction with Complex Systems (HICS), p. 209 (1998)
19. Kaber, D.B., Riley, J.M., Kheng-Wooi, T., Endsley, M.R.: On the design of adaptive automation for complex systems. Int. J. Cogn. Ergon. **5**, 37–57 (2001)
20. Hockey, G.R.J., Wastell, D., Sauer, J.: Effects of sleep deprivation and user interface on complex performance: a multi-level analysis of compensatory control. Hum. Factors **40**, 233–253 (1998)
21. Sauer, J., Wastell, D.G., Hockey, G.R.J.: A conceptual framework for designing micro-worlds for complex work domains: a case study on the Cabin Air Management System. Comput. Hum. Behav. **16**, 45–58 (2000)
22. Lorenz, B., Di Nocera, F., Röttger, S., Parasuraman, R.: Automated fault-management in a simulated spaceflight micro-world. Aviat. Space Environ. Med. **73**, 886–897 (2002)
23. Jasper, H.H.: Report of the committee on methods of clinical examination in electroencephalography. Electroencephalogr. Clin. Neurophysiol. **10**, 370–375 (1958)
24. Luczak, H., Laurig, W.: An analysis of heart rate variability. Ergonomics **16**, 85–98 (1973)
25. Kitney, R.I., Rompelman, O. (eds.): Study of Heart-Rate Variability. Clarendon Press, Oxford (1980)
26. Boucsein, W., Backs, R.W.: Engineering psychophysiology as a discipline: Historical and theoretical aspects. In: Backs, R.W., Boucsein, W. (eds.) Engineering Psychophysiology: Issues and Applications, pp. 3–30. Erlbaum, Mahwah (2000)
27. Izsó, L.: Developing Evaluation Methodologies for Human-Computer Interaction. Delft University Press, Delft (2001)
28. Nickel, P., Nachreiner, F.: Sensitivity and diagnosticity of the 0.1-Hz component of heart rate variability as an indicator of mental workload. Hum. Factors **45**, 575–590 (2003)
29. Zadeh, L.A.: Fuzzy sets. Inf. Control **8**, 338–353 (1965)
30. Zadeh, L.A.: An outline of a new approach to the analysis of complex systems and decision processes. IEEE Trans. Syst. Man Cybern. **3**, 28–44 (1973)

31. Nie, J., Linkens, D.A.: Fuzzy-Neural Control: Principles, Algorithms and Applications. Prentice-Hall, New York (1995)
32. Wang, L.X.: A Course on Fuzzy Systems and Control. Prentice-Hall, Englewood Cliffs (1997)
33. Fahrenberg, J., Wientjes, C.W.J.: Recording methods in applied environments. In: Backs, R. W., Boucsein, W. (eds.) Engineering Psychophysiology: Issues and Applications, pp. 111–136. Erlbaum, Mahwah (2000)
34. Tattersall, A.J., Hockey, G.R.J.: Level of operator control and changes in heart rate variability during simulated flight maintenance. Hum. Factors **37**, 682–698 (1995)
35. Pope, A.T., Bogart, E.H., Bartolome, D.S.: Biocybernetic system evaluates indices of operator engagement in automated task. Biol. Cybern. **40**, 187–195 (1995)
36. Lorenz, B.: Detection and prediction of an automation-induced state of impaired operator competence. In: Proceeding NATO ARW on Operator Functional State (2002)
37. Zhang J., Nassef A., Mahfouf M., Linkens D.A. et al.: Modeling and analysis of HRV under physical and mental workloads. In: Proceeding. 6th IFAC Symposium on Modeling and Control in Biomedical Systems, (2006) pp.189–194. Reims, France, 20–22 September 2006
38. Tagaki, T., Sugeno, M.: Fuzzy identification of systems and its application to modeling and control. IEEE Trans. Syst. Man Cybern. **15**, 116–132 (1985)
39. Jang, J.: ANFIS: Adaptive-network-based fuzzy inference system. IEEE Trans. Syst. Man Cybern. **23**, 665–685 (1993)
40. Goldberg, D.E.: Genetic Algorithms in Search, Optimization and Machine Learning, Addison-Wesley (1989)
41. Linkens, D.A., Nyongesa, H.O.: Genetic algorithms for fuzzy control (Part 1: Offline system development and application and Part 2: Online system development and application). IEE Proc. Control Theory Appl. **142**, 161–185 (1995)
42. Harding, R.M.: Human Respiratory Responses During High Performance Flight. AGARD/NATO, AG 312, Neuilly-sur-Seine, France (1987)
43. Schvaneveldt, R.W., Reid, G.B., Gomez, R.L., Rice, S.: Modeling mental workload. Cogn. Technol. **3**, 19–31 (1998)
44. Parasuraman, R., Sheridan, T.B., Wickens, C.D.: A model for types and levels of human interaction with automation. IEEE Trans. Syst. Man Cybern. **30**, 286–297 (2000)
45. Scerbo, M., Freeman, F.G., Mikulka, P.J., Parasuraman, R., Di Docero, F., Prinzel, L.J.: The Efficacy of Psychophysiological Measures for Implementing Adaptive Control, NASA – Technical Paper (2001-211018). Langley Research Center, Hampton (2001)
46. Laine T.I., Bauer K.W., Lanning J.J. W., Russell C.A., Wilson G.F.: Selection of input features across subjects for classifying crewmember workload using artificial neural networks, IEEE Trans. Syst. Man Cybern. – Part A: Syst. Hum. **32** 691–704 (2002)
47. Berka, C., Levendowski, D.J., Lumicao, M.N., Yau, A., Davis, G., Zivkovic, V.T., Olmstead, R.E., Tremoulet, P.D., Craven, P.L.: EEG correlates of task engagement and mental workload in vigilance, learning, and memory tasks. Aviat. Space Environ. Med. **78**, B231–B244 (2007)
48. Di Nocera F., Camilli M., Terenzi M.: A random glance at the flight deck: Pilots' scanning strategies and the real-time assessment of mental workload, J. Cogn. Eng. Decis. Mak. **1** 271–285 (2007)
49. Parasuraman, R., Mouloua, M., Molloy, R.: Effects of adaptive task allocation on monitoring of automated systems. Hum. Factors **38**, 665–679 (1996)
50. Gevins, A., Smith, M.E., McEvoy, L., Yu, D.: High-resolution EEG mapping of cortical activation related to working memory: effects of task difficulty, type of processing, and practice. Cereb. Cortex **7**, 374–385 (1997)
51. Smith, M.E., Gevins, A., Brown, H., Karnik, A., Du, R.: Monitoring task loading with multivariate EEG measures during complex forms of human-computer interaction. Hum. Factors **43**, 366–380 (2001)

Chapter 10
Space-Independent Community Detection in Airport Networks

Emil Gegov, Maria Nadia Postorino, Alexander Gegov and Boriana Vatchova

Abstract This research explores the topology and passenger flows of the United States Airport Network (USAN) over two decades. The network model consists of a time-series of six network snapshots for the years 1990, 2000 and 2010, which capture bi-monthly passenger flows among US airports. Since the network is embedded in space, the volume of these flows is naturally affected by spatial proximity, and therefore, a model (recently proposed in the literature) accounting for this phenomenon is used to identify the communities of airports that have particularly high flows among them, given their spatial separation. This research results highlight the fact that some general techniques from network theory, such as network modelling and analysis, can be successfully applied for the study of a wide range of complex systems, while others, such as community detection, need to be tailored for a specific system.

10.1 Introduction

Complex networks can be usually described by their community structure, which is also a prominent feature in many biological [24], social [5] and technological [5] complex systems. Community structure is defined as the presence of highly

E. Gegov (✉)
Brunel University, Uxbridge, UK
e-mail: e.gegov@gmail.com

M.N. Postorino
Mediterranea University of Reggio Calabria, Reggio Calabria, Italy
e-mail: npostorino@unirc.it

A. Gegov
University of Portsmouth, Portsmouth, UK
e-mail: alexander.gegov@port.ac.uk

B. Vatchova
Bulgarian Academy of Sciences, Sofia, Bulgaria
e-mail: boriana.vatchova@gmail.com

G.M. Dimirovski (ed.), *Complex Systems*, Studies in Systems, Decision and Control 55, DOI 10.1007/978-3-319-28860-4_10

intra-connected modules of nodes that are loosely inter-connected to the rest of the network. In other words, nodes are organised in clusters and most links are inside those clusters. The reason for this phenomenon is that nodes that share functional similarity and/or dependency tend to interact more and therefore they should be more connected. There are two main advantages of this community architecture: the first is efficiency, as most interactions are within modules which are internally well-connected, thereby reducing the path length (the number of links that separate a pair of nodes); and the second is robustness, as entire modules may fail autonomously, without severely affecting the operation of other modules, and hence, the function of the entire network. Therefore, the emergence of community structures in complex networks has implications for their efficiency and robustness, as well as their particular characteristics.

In recent years, research on complex networks has proposed many community detection methods [21] that aim to discover the most sensible partition of a network into communities. Most of them work on the principle of modularity [25] optimisation, aiming to maximise the modularity benefit function describing the quality of a network partition. The more links that fall within a community compared to an ensemble of benchmark random networks with the same community structure, then the more bias there is for links to connect to nodes belonging to the same community, and therefore the higher the modularity Q (Eq. 10.1). In essence, modularity measures how sharply the modules are defined.

$$Q = (\text{fraction of links within communities}) - (\text{expected fraction of such links}). \tag{10.1}$$

The expected fraction of links within communities is calculated from an ensemble of random networks that resemble the network under scrutiny in terms of its strength (total weight on all adjacent links) distribution. In addition, it is necessary to quantify the average level of interaction between a pair of nodes, and this is achieved by defining a null model matrix P_{ij} that describes the expected weight of a link between nodes i and j, over the ensemble. The standard choice for P_{ij}, defined by [25], preserves the strength of nodes in the random networks:

$$P_{ij}^{NG} = \frac{k_i k_j}{2m}, \tag{10.2}$$

where k_i is the strength of node i and m is the total weight in the network. A limitation of this null model, and of community detection methods that use it, is that only network topology and traffic are considered, but this is insufficient for networks embedded in space, such as the USAN. The reason for this is that most spatial networks (excluding the Internet for example) are very biased towards short-range connections due to the cost involved in long-range interactions in physical space. In terms of topology, an airport network is not a typical spatial network, as long-range connections are common. However, in terms of traffic, the higher financial and temporal costs involved in long-range travel play an important role for passengers, thereby affecting the flow on the network. Hence, standard community detection methods (typically based on the NG null model) discover communities of nodes that are spatially close, as opposed to communities that have particularly strong internal interactions [1, 3, 11, 16, 22]. To address this, Expert (2011) proposed an alternative null model for P_{ij} that takes into account the effect of space by favouring communities of nodes i and j that are more connected than expected, given the physical distance d_{ij} between them:

$$P_{ij}^{Spa} = N_i N_j f(d_{ij}), \quad (10.3)$$

where N_i is the importance (typically the strength) of node i and $f(d_{ij})$ is the function that incorporates the effect of space. This so-called deterrence function describes the expected level of interaction between nodes i and j that are separated by some distance d_{ij}. In other words, the function defines how interaction decays, analogous to gravity, as distance between objects increases.

Transportation networks are a good example of spatial networks. Their network topology is not only characterised by spatial aspects such as the location of nodes and the length of links but also by the association of a transport cost to the link length; implying that longer links are typically balanced by some benefit, such as connecting to a high-degree node, or a node in an attractive location. Transportation networks typify the specific nature of spatial networks particularly with regard to issues such as congestion, fast-growing urban sprawl and disease propagation. Network structure and dynamics play a key role in most, if not all, of these challenges. Transportation networks can be planar, as in road and rail networks, or non-planar, as in airport networks. In addition, transportation networks are usually weighted, where the link weight describes the intensity of some form of interaction, e.g. the link travel cost, the amount of traffic. Air transportation networks are an important example of spatial networks. Here, nodes identify airports and links represent the existence of a direct air service among them. Weights on links may represent the number of passengers flying on that connection, and the distribution of weights is an initial indication of the existence of possible strong heterogeneities.

The existence of links among airports depends on factors related to both airline strategies and passenger demand. Airlines decide to operate at a given airport on the basis of a significant demand, allowing them to reach satisfactory load factors. Location and socio-economic characteristics of the airport catchment area are the key factors generating air traffic demand. The airport choice made by both airlines and travellers depends on factors that can be ultimately reduced to time and monetary costs. For example, reduced airport charges may help airlines to offer lower air fares to potential travellers, and hence to induce more flights. The airport network is an example of a heterogeneous network where the hubs have high connectivity, high weight (in terms of traffic) and long-distance links [4].

In recent years, the analysis of complex transport networks has received considerable attention, mainly in terms of commuting networks [15, 27, 28]. Airport networks have also been studied to characterise their level of degree correlations and clustering, their evolution in time, and their potential scale-free properties and D'Souza, [2, 18, 31]. In terms of network robustness, network failure due to external factors such as bad weather conditions, volcanic eruptions, and political or security issues, may have significant impact on the air traffic depending on the criticality of the involved nodes and the extent of their influence. In terms of socio-economic characteristics, the emergence of community structure depends on the location and distribution of relevant activities. Concentration of activities in a given area generally means concentration of short trips in that area, and this is a typical commuting pattern. For medium long-distance trips, the main contributing factor is mass migration rather than commuting, and air transportation plays an important role in facilitating easier migration of workers. Within larger countries, such as the United States, a new kind of commuting by air can be identified, as people working in different parts of the country during the week return home at weekends. The changes in the availability, frequency and cost of air travel facilitate trips for migrants located far from traditional gateways (large airports with hub functions (hub-and-spoke) and inter-continental links) [10].

10.2 Research Methodology

This section presents a case study of a continuously developing air transportation network that is vital for the mobility of millions of passengers per day. The USAN was chosen for several reasons. First, it is large and growing, so it is clearly a good candidate for studying network evolution. Second, there are few detailed models that trace the network for more than a few years. Third, there is a large quantity of available data, dating back to 1990, when the network looked very different to what it is today.

Over the past few decades air travel in the US has changed considerably. Apart from the obvious increase in the number of airports, connections and passengers, the structure (topology) of the USAN has been transformed, thereby affecting all aspects of air travel. Up to the 1970s the USAN had mainly a hub-and-spoke architecture: flights coming from many origins (spokes) converge to the airport (hub) from which new flights start toward other destinations (spokes). The hub-and-spoke architecture is characterised by a high spatial network concentration, and a time co-ordination of flights at the hub according to a flight wave concept [8]. The ideal wave is the set of arriving and departing flights such that for each arriving flight there is a departing one allowing travellers to get an easy transfer to the final destination, and the integration of air services at the hub (e.g. baggage transfer). The main disadvantage of the hub-and-spoke architecture for passengers is that they would have to change flights at the hub, taking more time to reach their final destination. Furthermore, passengers travelling between other destinations may experience poor service, including infrequent flights and many changes [19]. As a result, a number of low-cost airlines emerged in the 1980s, providing point-to-point direct services between poorly connected destinations. One example is JetBlue, which is still considered very successful even when compared against larger airlines, such as American Airlines and United Airlines [6]. Consequently, the resulting USAN topology is a combination of both hub-and-spoke and point-to-point architectures.

In large countries, such as the US, the domestic airport network is a key factor in facilitating domestic migration, i.e., the movement of people within the United States. Particularly, migrants are defined as people moving among states (inter-state migration). Incoming migration (in-migration) is defined as movements into an area during a given period, while outgoing migration (out-migration) is defined as movements out of an area during the same period.

Migration can be thought of as population redistribution within a country or between countries. It is often linked to an asymmetric distribution of employment and affluence: people are attracted to areas with better job markets, services and quality of life. These aspects relate to the concept of city competitiveness, in other words, attractive cities (or regions) are efficient, accessible and offer economic opportunities to both investors and workers [7, 12–14]. In terms of accessibility, attractive areas have efficient transport systems mainly in terms of external connections linking those areas to other parts of a large territory [26]. In this context, air services can play an important role because they provide fast links among distant locations, even though there may be alternative forms of transportation. The fusion of air transportation with migration is an important novelty in the field of transportation since migration is a driving factor behind many passenger flows in an airport network.

First, it is necessary to decide which specific interactions in the airport network are of particular interest. For example, these can be the number of passengers flying between airports, the number of aircraft flying between airports, or quite possibly any other metric describing the link between a pair of airports. Then, a long enough time interval is chosen, such that there are available data to be modelled, and the scale of the observed evolution is maximised. The chosen interval is partitioned into equal time slices, depending on the required level of granularity. In the case where a long interval and high granularity result in an unfeasible number of time slices, a sample of those can be selected for the actual modelling.

The number of passengers flying from an origin to a destination airport was chosen as the variable for this study, because it is the common choice in the literature, and it is perhaps the most influential factor in the expansion and organisation of the network. The longest possible time period—from 1990 to 2010—was selected, based on the availability of data for this period. To investigate seasonal variation within a given year and to build more precise models of the network, time slices of length 2 months offer a good balance, so a year is divided into six equal parts. To reduce the huge amount of modelling (120 networks) without losing too much information, only 3 years are modelled in this study: 1990, 2000 and 2010. These years capture the oldest, the intermediate, and the newest, open source states of the network. Data were obtained from the Bureau of Transportation Statistics (BTS) of the US Department of Transport [9]. More specifically, data contained monthly records of origin-destination pairs of domestic airports and the number of passengers carried.

The methodology consists of two parts: Network Modelling and Network Analysis. The former describes how the USAN time-series model is developed and the latter presents the network analysis techniques that are used to quantify various properties of the networks.

10.2.1 Network Modelling

To investigate the evolution of the USAN from 1990 to 2010 the network is modelled in a discrete time-series consisting of three stages: 1990, 2000 and 2010. Each of those is further split into six bi-monthly intervals, in order to capture finer temporal detail and to explore seasonal variations in the network. Hence, the network model consists of 18 network snapshots depicting topology and traffic for a 2-month time-slice. Each network is defined by a set of nodes (the airports) and a set of directed, weighted links (the flight connections) representing topology. Link directionality reflects the difference in passengers flying from A to B and vice versa. Links weight represents the total number of passengers that flew on that connection within the specified time-slice. Each network includes a number of isolated nodes and self-loops. Isolated nodes denote airports that handled aeroplane departures and/or arrivals, but no actual passengers. Self-loops occur when an aeroplane takes off and lands at the same airport for some reason, such as an emergency.

Using network modelling, both dynamics on the network in terms of traffic fluctuations and dynamics of the network in terms of topology fluctuations are studied. The more recent structure of the network (reference year: 2010) is compared with migration patterns among the four US macro-regions (West, Midwest, North–east and South), in order to identify possible relationships. Figure 10.1 shows a map of the US regions and states, including the locations of the main airports in terms of 2010 passenger flows.

10.2.2 Network Analysis

The analysis of the USAN involves simple statistical parameter analysis and more complex community structure analysis. The idea of the former is to identify general network properties of the USAN as a whole, such as the average number of airport connections $<k>$. The latter exposes specific traffic patterns at the airport level, thereby revealing deeper individual characteristics. For example, if New York and Los Angeles happen to be members of the same community, then this implies that there is significantly more air traffic between them than expected, given their distance apart.

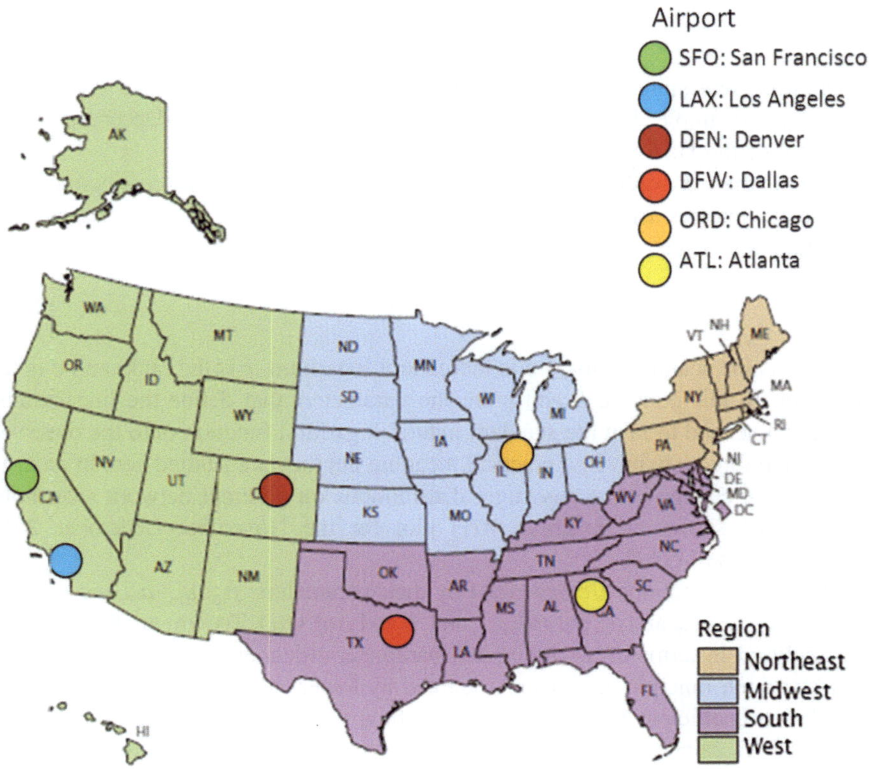

Fig. 10.1 US macro-regions and major airports in 2010

10.2.2.1 Network Parameters

In the USAN model, N is the total number of US airports; E is the total number of one-way domestic connections; Gulf Cooperation Council (GCC) is the number of airports in the largest connected subnetwork; $<k>$ is the average number of domestic connections per airport; L is the average number of flights that need to be taken to get from A to B; and C is the expected proportion of airport neighbours (all connected to the airport) that are connected themselves. The latter two of those are calculated for a simple (unweighted and undirected) version of the network due to computational complexity but most connections are bidirectional anyway so the results should be fairly accurate. $P(k_{\text{in}})$ and $P(k_{\text{out}})$ are the probability distributions of a randomly chosen airport having k_{in} incoming and k_{out} outgoing connections, respectively. By extracting the first two data points (0 and 1 connection) and taking them as separate parameters p and q, the degree distributions are well-approximated by a power–law fitting function of the form:

$$W(r) = b\mathrm{Ln}(r) + c, \tag{10.4}$$

where a is the scaling factor, k is in/out-degree, and n is the exponent. Figure 10.2 shows an example in-degree distribution for the November–December 2010 snapshot of the USAN.

$W(r)$ is the rank-ordered passenger distribution on all network connections. For systematic analysis across all networks, $W(r)$ is normalised to be in the range (0, 1]. This function is well-approximated by a logarithmic fit of the form:

$$W(r) = b\mathrm{Ln}(r) + c, \tag{10.5}$$

where b is the scaling factor, Ln is the natural logarithm, r is the rank, and c is the intercept. In this context, b and c are the parameters that define the linear transformation needed to map the standard natural logarithm function onto the observed data. Therefore, they have no practical meaning but they are studied here in order to measure the change in the passenger distributions on different network snapshots. Figure 10.3 shows an example $W(r)$ plot for the November–December 2010 snapshot of the USAN.

Hence, the functions are described by their parameters: p_{in}, q_{in}, a_{in} and n_{in} of $P(k_{\mathrm{in}})$; p_{out}, q_{out}, a_{out} and n_{out} of $P(k_{\mathrm{out}})$; and, b and c of $W(r)$. To sum up, the networks are analysed in terms of six individual parameters (denoted by capital letters and $\langle k \rangle$), and ten function parameters (denoted by lower case letters). In addition, the correlation coefficient R—which measures how well the best-fit approximates the real data—is calculated. The individual parameters are calculated using Network Workbench; the degree distributions are fitted using the EzyFit toolbox for MATLAB; and the ranked weight distributions are fitted in SPSS. For each parameter and for each of the 3 years (1990, 2000, and 2010), the mean parameter value and the Standard Error of the Mean (SEM) of all six network snapshots were calculated. The SEM indicates the amount of bi-monthly variation.

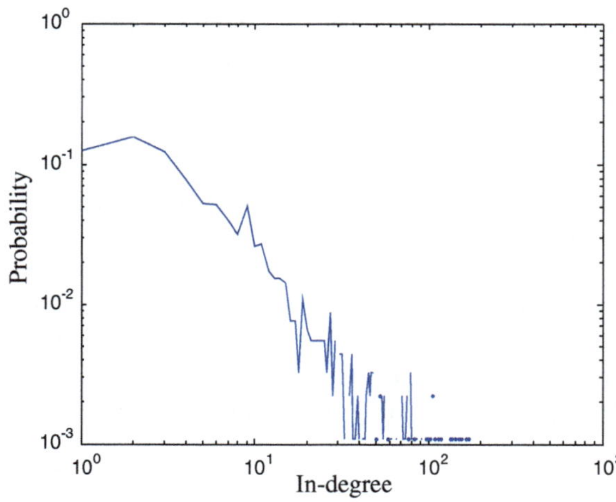

Fig. 10.2 In-degree probability distribution for November–December 2010 snapshot of the USAN

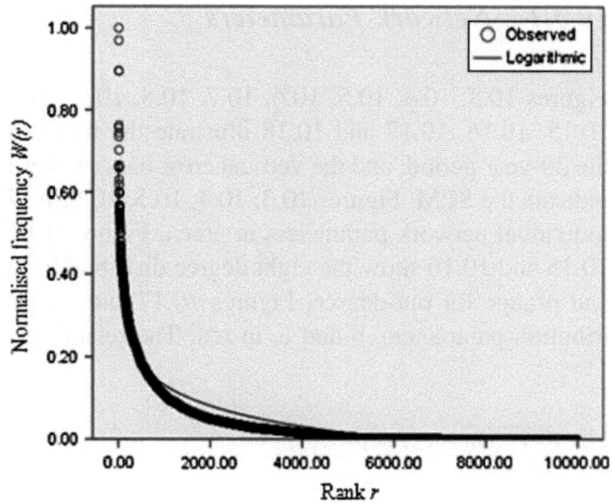

Fig. 10.3 Ranked weight (normalised frequency of passengers) $W(r)$ for November–December 2010 snapshot of the USAN

10.2.2.2 Community Structure

Since distance is expressed in terms of degrees of arc length where one degree is approximately 60 miles, the largest distance in the distance matrix is 149. The bin populations and the deterrence function were checked for bin sizes 0.1, 1, 2 and 3, and 1 was chosen as it provided balanced bin populations and a smooth deterrence function.

There is potentially a large number of nearly optimal partitions, and therefore, a non-deterministic implementation of the algorithm is applied twice to each USAN network snapshot, in order to discover better partitions and to check their stability (similar partitions for the same snapshot). This is achieved using Normalised Variation of Information (NVI) (Meila, 2003), which measures the distance between two partitions in the range 0–1 (0 if they are identical, approaching 1 if they are very different). The average NVI values across the six snapshots for the years 1990, 2000 and 2010 are 0.40, 0.34 and 0.26, respectively. These values indicate that the community detection is considerably stable.

10.3 Simulation Results

Over the past 20 years, the USAN experiences dramatic growth: airports triple from about 350 to over 1,100, and direct connections double from 5,000 to 10,000.

10.3.1 Network Parameters

Figures 10.3, 10.4, 10.5, 10.6, 10.7, 10.8, 10.9, 10.10, 10.11, 10.12, 10.13, 10.14, 10.15, 10.16, 10.17 and 10.18 illustrate the trend of each parameter average over the 20-year period, and the vertical error bars (where visible, due to higher variance) indicate the SEM. Figures 10.3, 10.4, 10.5, 10.6, 10.7, 10.8 and 10.9 present the six individual network parameters in green. Figures 10.10, 10.11, 10.12, 10.13, 10.14, 10.15 and 10.16 show the eight degree distribution parameters in blue for in-degree and orange for out-degree. Figures 10.17 and 10.18 report the ranked weight distribution parameters, b and c, in red. The results are discussed in Sect. 10.4.1.

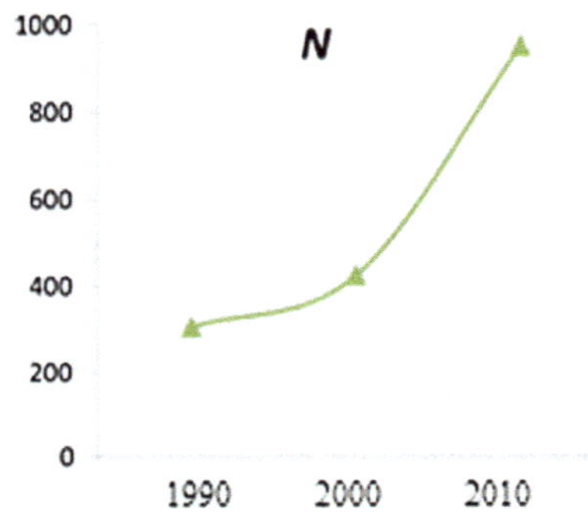

Fig. 10.4 Number of airports as a function of time

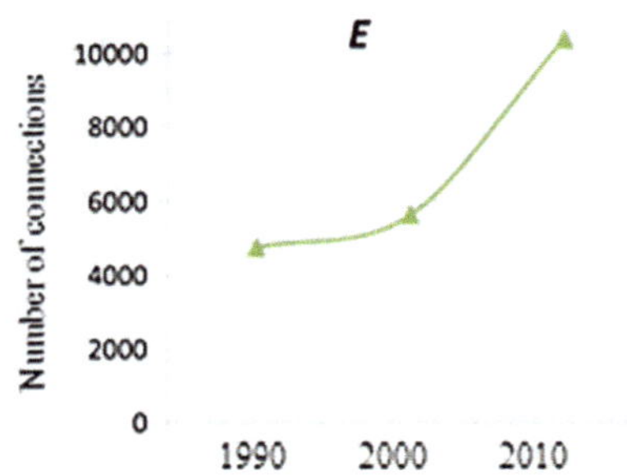

Fig. 10.5 Number of connections as a function of time

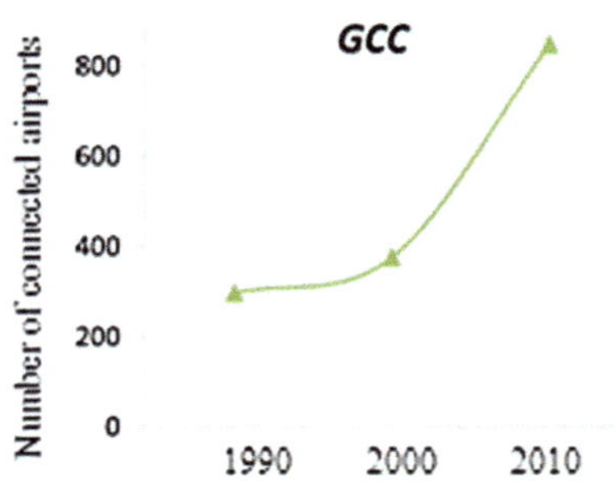

Fig. 10.6 Number of connected airports as a function of time

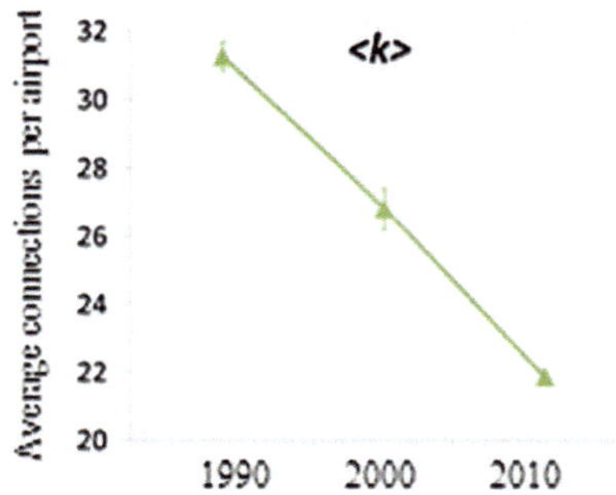

Fig. 10.7 Average connections per airport as a function of time

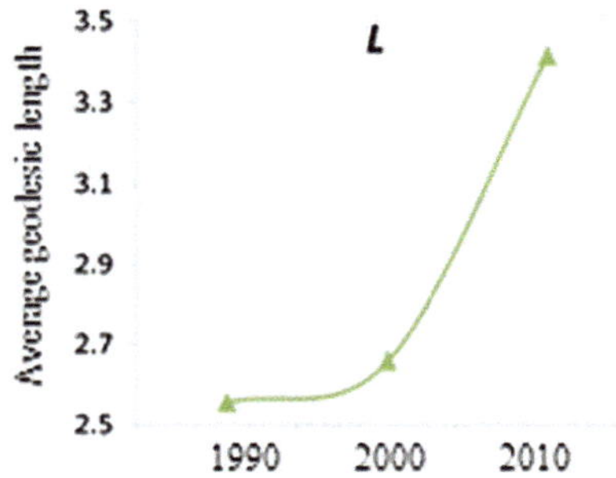

Fig. 10.8 Average geodesic length as a function of time

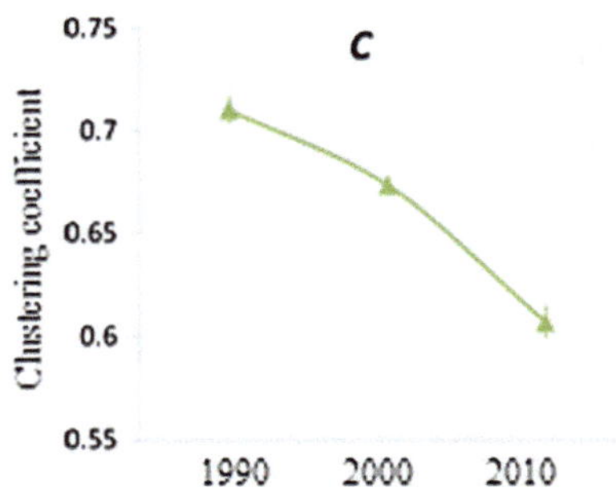

Fig. 10.9 Clustering coefficient as a function of time

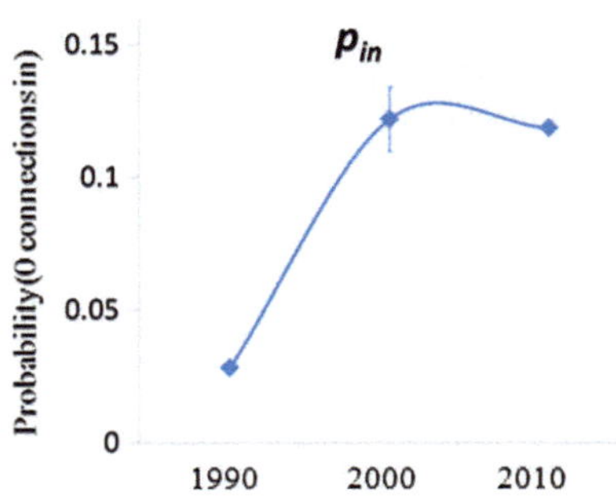

Fig. 10.10 Probability (0 connections in) as a function of time

Fig. 10.11 Probability (0 connections out) as a function of time

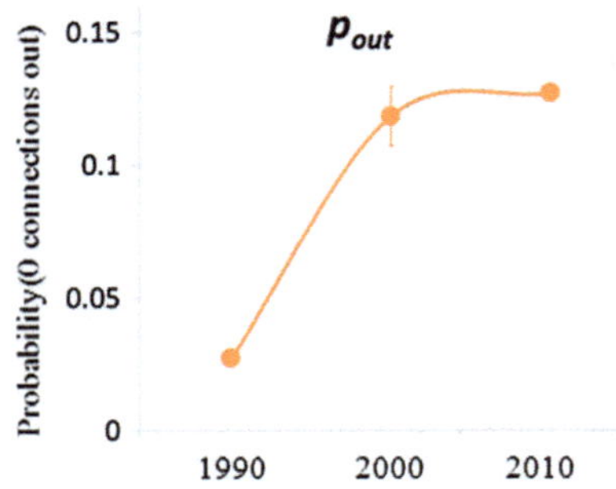

Fig. 10.12 Probability (1 connection in) as a function of time

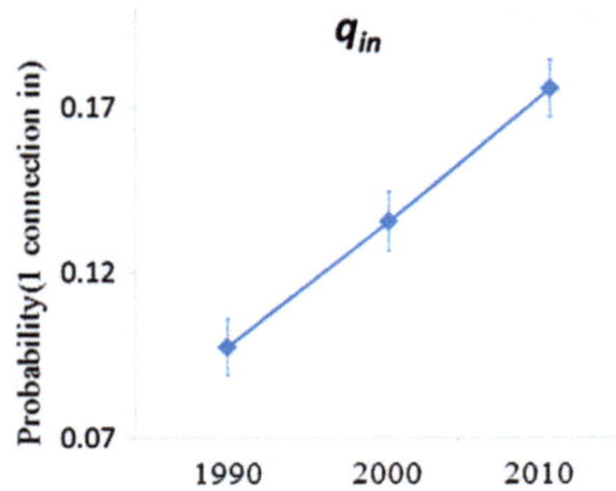

Fig. 10.13 Probability (1 connection out) as a function of time

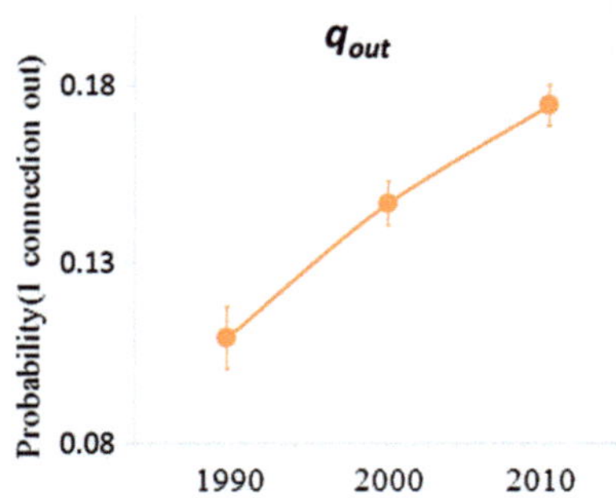

Fig. 10.14 Scaling factor a_{in} as a function of time

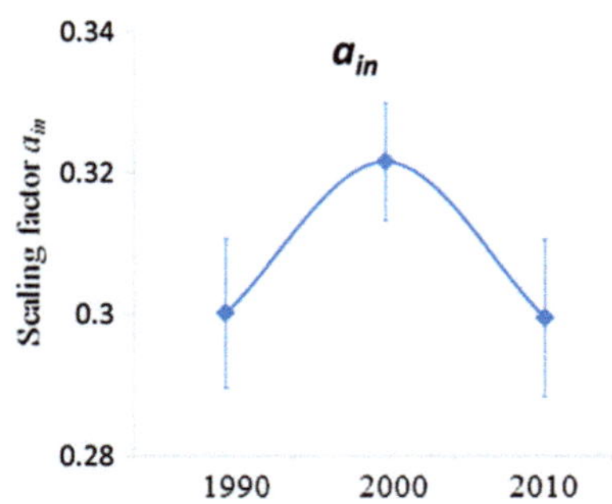

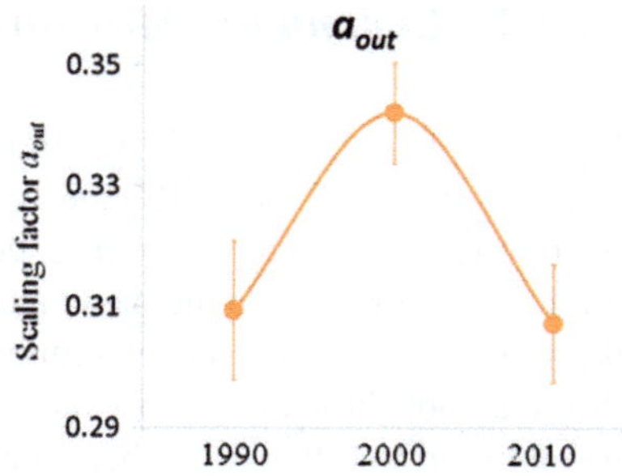

Fig. 10.15 Scaling factor a_{out} as a function of time

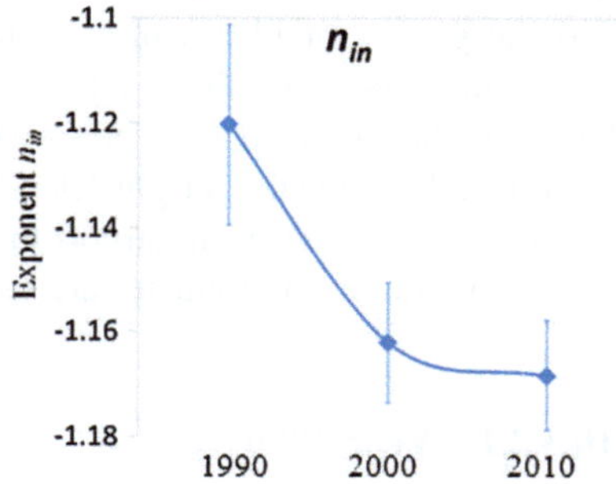

Fig. 10.16 Exponent n_{in} as a function of time

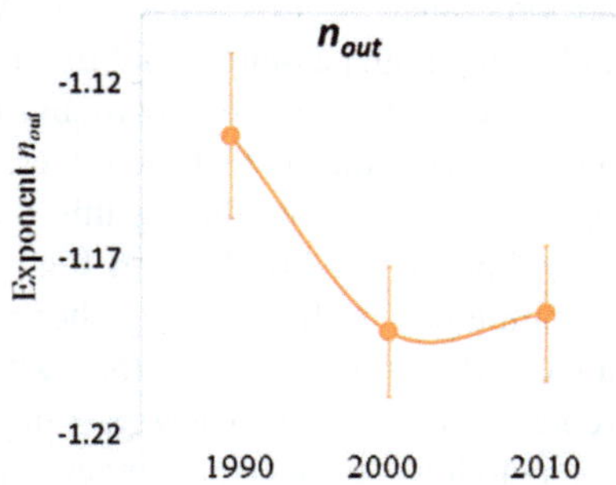

Fig. 10.17 Exponent n_{out} as a function of time

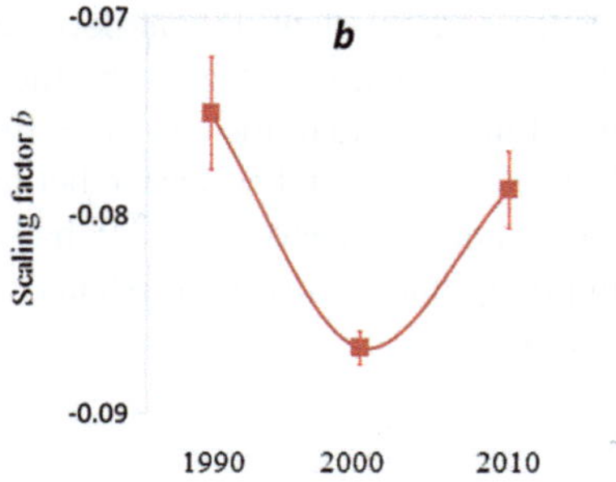

Fig. 10.18 Scaling factor b as a function of time

10.3.2 Community Structure

Figures A.1, A.2, A.3, A.4, A.5, A.6, A.7, A.8, A.9, A.10, A.11, A.12, A.13, A.14, A.15, A.16, A.17 and A.18 (see Appendix) represent the USAN at various stages over time, where each airport is denoted by a circle with a surface area that is directly proportional to the passenger flow (inbound and outbound passengers), and the colour represents the community. Airport connections and airport-to-airport flows are not shown for clarity, and colour is not consistent across the networks as it is only used to differentiate between different communities in a single network (the software used does not allow the user to consistently assign colours to communities). In other words, the figures depict the size of airports by passengers handled, and the groups of identically coloured airports that have particularly strong passenger flows between them. Alaska, Hawaii and the Mariana Islands are not shown here but they represent a very small fraction of the network. The airport in the bottom right is for the Virgin Islands. In the following analysis of results, the term hub is used to describe an airport that handles a high volume of passengers, and the terms community and cluster are used interchangeably.

10.3.2.1 Year 1990

In January–February (Fig. A.1) there is a well-defined cyan community of west-coast airports, such as Los Angeles (LA) and San Francisco, together with Chicago, indicating high passenger mobility between those locations. In Fig. A.2, the network for March–April implies a particularly large community (light-green) of the main US airports. This means that there were particularly active interactions between all the light-green locations during this time, in contrast to the previous image for January–February. May–June in Fig. A.3 displays a geographically clustered set of communities in the east, together with the largest community in red which spans almost the entire US. In other words, the geographically clustered communities represent the regions where passengers mainly flew locally, and the red community refers to long-distance passengers. July–August (Fig. A.4) shows a very inter-mixed network, with significant long-distance travel suggested by the spatial spanning of the communities. However, the cyan Dallas cluster is an exception, as it covers only Dallas and small nearby airports. September–October (Fig. A.5) sees an overall decline in air travel flagged by the noticeable reduction in general size of circles, matching the end of the tourist season, and two large communities in blue and green. In Fig. A.6, November–December has no major change in traffic patterns apart from the fact that Chicago (a key US hub) is taken over by the spanning blue community, implying that it was used extensively for air travel, particularly among these blue regions.

10.3.2.2 Year 2000

January–February in Fig. A.7 displays a prevailing cyan community of most major airports dominating the west and a large part of the rest of the US. In Fig. A.8, March–April displays a very similar pattern but the number of passengers has increased, which is reflected by the larger circles. In particular, yellow Atlanta (ATL) is clearly the leading US airport in terms of passengers handled during this period. May–June in Fig. A.9 suggests that Dallas and Chicago have separated from the largest community in the previous image, forming their own community (in blue) with a few more airports in the north–east. Again, Atlanta is nearly the only member of its yellow cluster, but its size implies that it plays the role of the main hub in the US, connecting many of the other regions. This is explored in more detail in the discussion section. July–August (Fig. A.10) appears similar to the networks for January–April, with a main green cluster covering most of the US and Atlanta still on its own. In Fig. A.11 September–October the number of passengers has predictably decreased. The east appears to be mixed while the west, Dallas and Chicago are all part of the same red cluster. November–December in Fig. A.12 is similar to the previous network for September–October.

10.3.2.3 Year 2010

Figure A.13 January–February has two large clusters in red and green covering the west and a large part of the US, respectively. Atlanta (blue) is still the largest hub but passenger demand is low due to the low season. March–April in Fig. A.14 shows an increase in passengers and a clearly dominating red community in the west. The south is covered by the pink Dallas cluster, and yellow Atlanta and light-green Chicago are the first and second largest hubs, respectively. May–June in Fig. A.15 is different in two respects. First, Chicago has formed a yellow cluster covering the south-west and the east, and second, orange Dallas has separated from the south cluster, so it has become more of a long-distance travel airport than in the previous 2 months. Atlanta is still the largest airport by far, providing the connections for many more passengers than any other airport in the US. July–August (Fig. A.16) is very similar to May–June. This means that there is a particularly high volume of travellers among the east coast, the west coast and Chicago. September–October (Fig. A.17) has a good mix of many clusters, suggesting that during these months there has been more long-distance travel within the US. The green, yellow and blue communities are particularly well spread out, highlighting the extent of long-range travel. November–December (Fig. A.18) is similar to the previous 2 months but now the Chicago and LA clusters have merged again (see May–June and July–August), forming one of the two largest clusters (red and green).

10.4 Discussion

10.4.1 Network Parameters

The results obtained from the parameter analysis are discussed in three parts. The first part addresses the individual parameters, which are based on the global structure of the entire network. The second part covers the degree distribution parameters, which describe the structure of the air transportation networks in terms of the airports' number of incoming and outgoing connections from/to other airports. The final part focuses on the weight distribution parameters, which highlight the high heterogeneity in the number of passengers on different connections.

10.4.1.1 Individual Network Parameters

Figures 10.4, 10.5 and 10.6 show the growth of the network in terms of airports, connections, and connected airports. Clearly, the expansion is much larger from 2000 to 2010, indicating a non-linear growth process. This observed behaviour is not unusual, as any transportation network is constantly affected by economic decisions, supply and demand, and many other factors. What is rather unusual is the fact that the average number of airport connections, Fig. 10.7, displays a linear decline in time, due to the faster increase in number of airports compared to the number of airport connections. This means that many (probably small size) airports were introduced but they were not inter-connected that well, unless already established airports lost some connections. Because of this rapid growth, the average geodesic length (Fig. 10.8) between any two airports in the US jumped from 2.5 to 3.5, within the past 10 years. However, this does not imply that the average journey would need more changes; to the contrary, the network was optimised over time to reduce the changes of the average passenger by interconnecting airports with higher passenger demands, and disconnecting those less profitable. This is evident from the recent boom in low-cost airlines, providing many point-to-point flights between poorly connected destinations. Based on these facts, it is natural to assume that the clustering in the network increases, but Fig. 10.9 contradicts this; again, this must be due to the huge number of new airports. All these parameters have confirmed the immense development of the USAN, particularly in the first decade of the twenty-first century, and the next section explains this phenomenon in more detail.

10.4.1.2 Degree Distribution Parameters

Figures 10.10 and 10.11 show the probability of an airport having zero incoming and outgoing connections, respectively. In other words, this parameter measures the proportion of very remote airports that only have some arrivals, or departures, per

2 months. Clearly, the fraction rises from 1990 to 2000, indicating a significant increase in such poorly connected airports, but more interesting is the 2000 to 2010 period, which experienced no major change. Figures 10.12 and 10.13 present the fraction of airports with just one incoming and outgoing connection, respectively. Again, these trends quantify the presence of minor airports, which increases linearly over the two decades. Figures 10.14 and 10.15 report the fitting functions' estimates for the parameters from the previous two figures. Basically, they confirm that the fits are not able to approximate (especially for the year 2000) the first two data points that were extracted as p and q, since they do not obey the power-law relationship that the rest of the data do. The key parameter in a power-law is the exponent, as it controls the skew of the distribution. Therefore, between 1990 and 2000, Figs. 10.16 and 10.17 suggest an increasing exponent in absolute terms, since the scale of the figures is negative. This implies stronger preferential attachment, which means that already highly connected airports obtained more connections, while poorly connected airports received few new, or even lost existing, connections. The fact that the change between 2000 and 2010 is small, suggests that although there was a lack of point-to-point flights in the 90s, it may have been resolved in the 00s.

10.4.1.3 Weight Distribution Parameters

The ranked passenger distribution is the only characteristic of the dynamics on the network that is considered in this work, and as such, cannot be taken as a complete description of the function of the network. Nevertheless, the results are interesting, and can be used as a basis for further analysis. Figure 10.19 depicts the two parameters of the logarithmic fit, and although further work is necessary to arrive at more precise conclusions, one thing is certain: the USAN exhibits considerable passenger variability over the course of a year. The results obtained are in line with this phenomenon—as also demonstrated by the error bars in the figure—because transport demand vary in time and space.

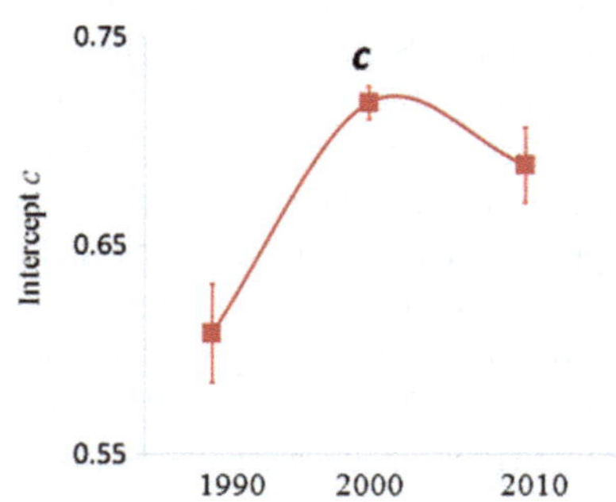

Fig. 10.19 Intercept c as a function of time

10.4.2 Community Structure

First of all, it is important to highlight the fact that some communities have airports that are very far apart, suggesting that spatial community detection discovers more meaningful communities that are not occupying a single region on the map. The seasonal variation within each of the 3 years and the long-term evolution of the network between those years are explored in the following two sections. In addition, the obtained space-independent community structure is validated through comparison to the standard space-dependent community structure.

10.4.2.1 Seasonal Variation

The seasonal variation in passenger flows within each year is investigated qualitatively by visually examining the obtained community structure, and quantitatively, using Normalised Mutual Information (NMI). In terms of qualitative analysis, there appear to be significant changes in the community structure of the USAN in 1990. In other words, there were considerable seasonal variations in the volume of passengers on network connections. Specifically, January–February had a very mixed structure, March–April had a large (green) super-cluster, and the rest of the year was mixed again, with some similarities between May–June and September–October. In the last 2 months of the year, Chicago joined the blue LA cluster, forming a similar structure to January–February, which indicates the presence of an annual cycle of passenger demand. Throughout 2000 (apart from May–June), the community structure remained fairly stable, implying low seasonal variation. In particular, the network had a large super-cluster covering most of the US, and Atlanta was the super-hub. May–June, however, was different as Dallas and Chicago were in a separate cluster of their own, so there was a particularly strong passenger flow between them and other smaller airports in the north-east during these months. In 2010, similar to 1990, there were notable fluctuations in the community structure of the network. January–February was mixed, March–April had a dominant red cluster, and in the rest of the year there were two dominant clusters (Denver and Chicago). LA and San Francisco formed their own community in green in September–October.

Quantitative analysis of network snapshots involves NMI, which measures the similarity between two network partitions (in this case two consecutive snapshots), returning 1 if they are identical and 0 if they are completely independent. It is typically used to quantify the stability of community structure over time, but it is also used in tests of community detection algorithms [21]. In order to calculate NMI, it was necessary to filter airports that do not appear in all snapshots for a given year. These few, small airports are rarely used and their traffic is very low, so their effect on the network is insignificant.

Figure 10.19 presents NMI over time. For example, January–April refers to the stability of the community structure in the period January to April, using the NMI

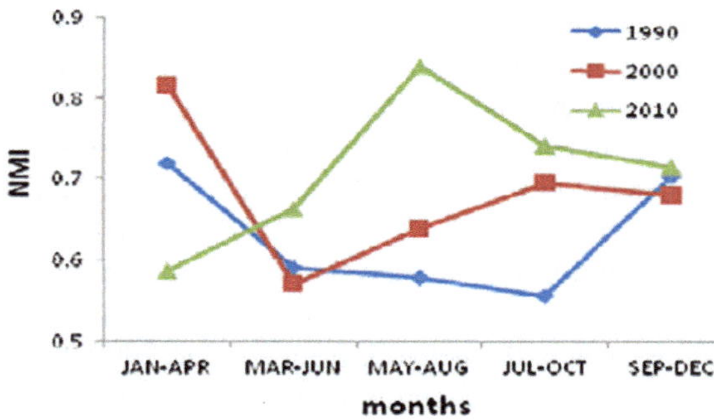

Fig. 10.20 Normalised Mutual Information (NMI) of consecutive network snapshots

of the partitions for January–February and March–April. The connecting lines do not indicate continuity, but are there to facilitate interpretation of the graph. Figure 10.20 suggests that in general, the community structure is fairly stable over the course of a year as the NMI is always above 0.5. In addition, annual stability has increased over the 3 years investigated as the average NMIs for 1990, 2000 and 2010 are 0.63, 0.68 and 0.71, respectively. Specifically, for May–August and July–October the network has shown consistent improvement in stability over its evolution, whereas for January–April it has become more unstable. The intervals March–June and September–December are virtually unchanged over the two decades. In particular, January–April 2000 and May–August 2010 were highly stable (NMI > 0.8), while 1990 was a relatively unstable year. The existence of an annual cycle is confirmed and quantified by calculating the NMI of the pair January–February and November–December, which is 0.69, 0.79 and 0.52 for the years 1990, 2000 and 2010, respectively. In other words, in terms of community structure, January–February resembles November–December (not so much in 2010), indicating the presence of an annual cycle of passenger demand.

10.4.2.2 Evolution

This part describes the evolution of the USAN from 1990 to 2010 by focusing on three key issues: volume of air travel, bi-monthly snapshots, and the main hub Atlanta. In addition, the migration levels between, and within, the four US macro-regions are discussed.

The quantity of domestic air traffic can be described by the total number of passengers carried across the USAN. Since the surface area of the airport nodes in Figs. A.1, A.2, A.3, A.4, A.5, A.6, A.7, A.8, A.9, A.10, A.11, A.12, A.13, A.14, A.15, A.16, A.17 and A.18 is proportional to the number of passengers; it is easy to determine the volume of air travel by observing the size of the nodes. The volume of air travel grew significantly from 1990 to 2000, with a particularly strong concentration of travellers via Atlanta. The first decade of the twenty-first century, however, did not see a significant increase in air travel, which, to a certain extent, may have been caused by key events, such as the September 11 terrorist attacks in

2001, and the start of the global economic recession in 2008. It is interesting that although most airports did not grow much from 2000 to 2010, there are some, such as Denver, that did experience a steady growth in terms of passengers. The specific changes in passenger distribution among airports are highlighted by the changes in the size of circles in Figs. A.1, A.2, A.3, A.4, A.5, A.6, A.7, A.8, A.9, A.10, A.11, A.12, A.13, A.14, A.15, A.16, A.17 and A.18.

In addition to the analysis of seasonal variation, it is also necessary to study long-term evolution, by focusing on individual bi-monthly snapshots and observing the changes in the network from 1990 to 2000, and from 2000 to 2010. Therefore, each of the six bi-monthly periods is analysed separately in order to illustrate the precise changes in passenger flows and community structure for the specified period, that have occurred in each of the two decades.

January–February: In terms of community structure, 1990 has a mixed pattern of clusters apart from the south (Fig. A.1), 2000 has a large cyan super-cluster covering all of the US (Fig. A.7), and 2010 again has a mixed structure (Fig. A.13). This indicates that in 1990 and 2010 there were numerous popular connections that saw a large number of air passengers, but in 2000 the passengers were more evenly distributed among the possible connections, resulting in a single super-community. In addition, Atlanta and some airports in the south and north–east had their own specific traffic patterns, as shown in Fig. A.7.

March–April: Generally, the community structure for this period is stable, but from 2000 to 2010 there is a clear transition of two hubs—Chicago and Dallas—from the main cluster to their own local-scale clusters. In other words, these two airports became regional hubs in the first decade of the twenty-first century, at least for the months of March and April.

May–June: Community structure changes significantly for the period 1990–2010, highlighting the specific changes in passenger trends over the years. In particular, 1990 is composed of one large red cluster covering all but the south, one medium-sized pink cluster in the south, and several regional clusters (Fig. A.3). This structure indicates that the red airports are the national long-range hubs, the south is somewhat more isolated, and the rest of the airports provide more local services. On the other hand, in 2000 Chicago and Dallas belong to the same cluster, and Atlanta is by far the top airport in the US (Fig. A.9). In 2010, there are two main clusters—the Chicago cluster in yellow, and the Denver cluster in blue—that cover the US together with Atlanta and Dallas, acting as national super-hubs (Fig. A.15).

July–August: Community structure in July–August suggests that in 1990 passengers preferred specific long-range connections (Fig. A.4). Most clusters cover large areas of the US, so many people travelled all over the US, specifically among airports of the same colour. On the other hand, in 2000 passengers were more evenly distributed within the green cluster, and more intricately concentrated on certain routes only in the north–east (Fig. A.10); while in 2010 the picture is, again, completely different, with two large clusters in red and pink, and two key hubs—Atlanta and Dallas—in blue and green, respectively (Fig. A.16).

September–October: The network in 1990 (Fig. A.5) is mainly composed of the blue LA cluster and the green Dallas cluster, with Chicago and Atlanta as hubs, and the usual mix of clusters in the densely populated north–east. In 2000, however, there is one red super-cluster, Atlanta is the main hub, and there is also a lot of activity in the Chicago region, as illustrated by the many colours that indicate the specific passenger trends in September–October (Fig. A.11). 2010 has a mix of multiple large clusters revealing new passenger flows (Fig. A.17). This is a sign of long-range travel among community members that are far apart.

November–December: In 1990 (Fig. A.6) the USAN is split into a large blue cluster and a yellow Dallas cluster in the south, but in 2000 (Fig. A.12) they have converged to a single yellow super-cluster, covering all but some regions in the north-east and the main hub Atlanta. In 2010 (Fig. A.18), the super-cluster has broken down, leaving Dallas as a national hub, and two red and green clusters spanning a large part of the US.

The role of Atlanta (ATL) as a leading US airport depends on factors, such as air services and their locations, as well as investments into growth and development. In 1967, the city of Atlanta and the airlines began to work on a master plan for the future development of the airport. Many investments were made in the following years, leading to new passenger terminals, runways, and facilities both inside (such as the people mover system linking parts of the terminal), and outside (such as the Red/Gold rail line, operated by the Metropolitan Atlanta Rapid Transit Authority, linking the airport to the counties of Fulton and DeKalb, in addition to Atlanta itself). ATL is also the primary base of many airlines, such as Delta Air Lines, who built one of the world's largest airline bases in 1930. Delta was an early adopter of the hub-and-spoke system, with Atlanta as its primary hub between the Midwest and Florida. This gave it an early competitive advantage, as Florida has been an attractive destination within the US for many decades. Although there is a decrease in the volume of migration in recent years, Florida and the South are still very popular destinations. In 1990, Atlanta was one of the three leading US airports for domestic flights. By 2000 it became the top airport (Figs. A.1, A.2, A.3, A.4, A.5, A.6, A.7, A.8, A.9, A.10, A.11, A.12, A.13, A.14, A.15, A.16, A.17 and A.18). Atlanta is also the only significant member in its community for all three years. This implies that it is equally well-connected to other airports, thereby possibly serving as a national hub. Since ATL handles so many passengers but there are no other major airports of the same colour, it follows that all ATL connections have relatively similar traffic loads, with longer connections having less traffic due to the effect of spatial separation. Therefore, ATL has no strong preferential attachment to any other major airport. To verify that Atlanta is a national hub, it is also necessary to check its number of direct connections. Table 10.1 summarises ATL's number of

Table 10.1 Atlanta's connections

	1990	2000	2010
Atlanta	101	142	167
Max	139	142	172

connections and the highest number of connections for the months January–February, in each of the years studied.

Clearly, Atlanta ranks very high in terms of connections, so it has a direct influence on a large part of the US territory. For example, in January–February 2010, ATL handled 10.7 million passengers (top in the US) on 167 connections, with an average of 64,000 passengers per connection, compared with the US highest figures of 172 and 73,000, respectively. In summary, ATL became the top US hub for domestic flights by the year 2000.

According to recent figures and US Census data, American people move many times during their adult lives, mainly in their 1920s. Preferred destinations of domestic migration were Southern states, mainly Florida, possibly because they are considered attractive places to live and work. Although US domestic migration has fallen noticeably since the 1980s, it is still higher than that within most other developed countries and during the period 2000–2004 it continued to redistribute the country's population (Perry, 2006). Nevertheless, the current slowdown in domestic migration due to the impact of the economic situation has changed the picture of movements within the US. In-migration towards states like Arizona, Florida and Nevada has slowed down, while Massachusetts, New York and California now have considerably less out-migration [20, 30]. In the years 2009 and 2010 mobility among states slowed nationwide and only a small percentage difference was observed during the 2-year period (Table 10.2). Migration is considered only for 2010 as this is the most recent year in the airport network model but a comprehensive investigation into the long-term relationships between migration and air travel is beyond the scope of this work.

Despite the current tendency to stagnancy, the role of the airport network in the context of US domestic migration is important. Since an airport network is continuously evolving depending on passenger demand, it is increasingly well-optimised for a number of functions, such as carrying more passengers, minimising flight changes for the average passenger, and making profit. As the USAN has evolved to attract passengers that are typically travelling to popular destinations, it is directly facilitating migration. Although most passengers fly for short-term business or leisure, there is evidence that a significant fraction of passengers are in fact migrating with a migration probability inversely proportional to the distance [23, 29]. According to Census data, Figs. 10.20, 10.21, 10.22 and 10.23 show the migration patterns for the years 1990, 2000 and 2010.

Data refer to people that are moving to a given macro-region or within it. The scale is relative to the maximum value and therefore not consistent across the 3 years, but they are comparable, in order to identify any potential variations in migration patterns over the two decades. Migration within the macro-regions is higher than that among them (decay of interaction as distance increases), and migration within the South is the highest, suggesting strong dynamics among the member States. Furthermore, the South region attracts the most people from outside for all 3 years. This is in line with the fact that Atlanta airport (located in the South) has the highest passenger flow, as discussed above, but it does not necessarily follow that the entire flow is related to the South, as many of the passengers change

Table 10.2 In-migration representing the number of people migrating to specific US states in 2009–2010

State	Year 2010	Year 2009	Diff %	State	Year 2010	Year 2009	Diff %
Alabama	108,951	124,658	−0.14	Montana	35,641	31,015	0.13
Alaska	36,345	40,474	−0.11	Nebraska	51,290	53,214	−0.04
Arizona	223,324	226,457	−0.01	Nevada	103,179	109,257	−0.06
Arkansas	79,214	85,857	−0.08	New Hampshire	39,423	37,940	0.04
California	445,972	460,161	−0.03	New Jersey	130,101	136,212	−0.05
Colorado	187,240	182,854	0.02	New Mexico	74,237	64,797	0.13
Connecticut	79,360	81,546	−0.03	New York	276,167	277,482	0.00
Delaware	31,713	35,085	−0.11	North Carolina	265,206	284,171	−0.07
District of Columbia	51,244	38,907	0.24	North Dakota	30,100	29,970	0.00
Florida	495,857	475,871	0.04	Ohio	174,773	171,894	0.02
Georgia	250,469	280,221	−0.12	Oklahoma	106,720	117,850	−0.10
Hawaii	53,581	53,270	0.01	Oregon	117,521	127,489	−0.08
Idaho	55,871	57,790	−0.03	Pennsylvania	241,855	232,316	0.04
Illinois	206,014	206,151	0.00	Rhode Island	32,335	32,108	0.01
Indiana	127,925	132,755	−0.04	South Carolina	152,710	33,616	0.78
Iowa	72,706	74,704	−0.03	South Dakota	25,777	145,873	−4.66
Kansas	95,127	102,695	−0.08	Tennessee	159,778	29,632	0.81
Kentucky	118,622	122,184	−0.03	Texas	490,738	168,174	0.66
Louisiana	98,291	90,957	0.07	Utah	78,163	511,166	−5.54
Maine	27,962	24,672	0.12	Vermont	22,529	90,375	−3.01
Maryland	165,096	174,958	−0.06	Virginia	260,813	19,390	0.93
Massachusetts	143,247	148,500	−0.04	Washington	191,784	271,600	−0.42
Michigan	117,581	118,054	0.00	West Virginia	39,791	192,654	−3.84
Minnesota	89,911	90,944	−0.01	Wisconsin	93,586	50,155	0.46
Mississippi	73,135	67,245	0.08	Wyoming	28,046	95,475	−2.40
Missouri	146,093	150,271	−0.03	Puerto Rico	31,732	30,889	0.03

flights in Atlanta en route to other regions. Nevertheless, the migration patterns do have a clear overlap with the community structure discovered in the USAN.

10.4.2.3 Comparative Validation

The purpose of this section is to demonstrate and to evaluate the effectiveness of Expert's (2011) space-independent community structure detection in comparison to [25] general community detection. To this end, Newman's method (referring to the null model proposed by Newman and Girvan) was applied to all eighteen network

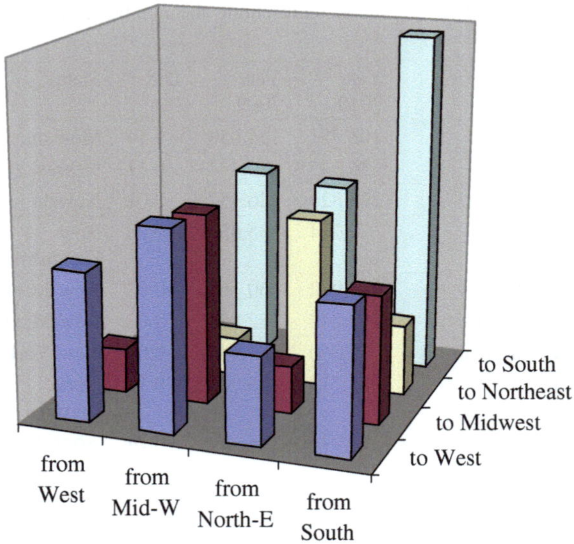

Fig. 10.21 1990 migration patterns among the four macro-regions: West, Midwest, North–east and South

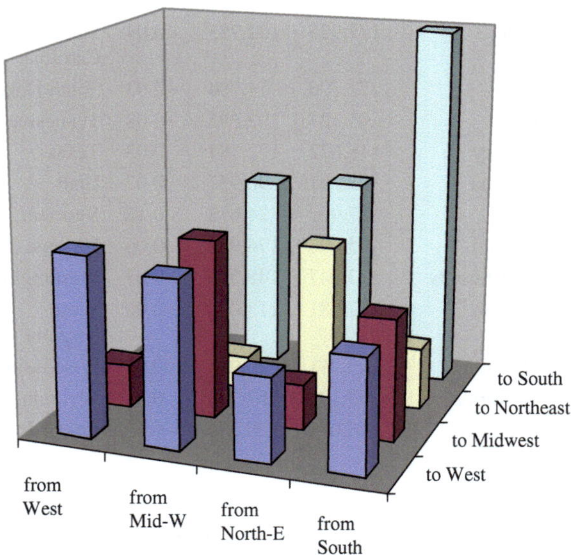

Fig. 10.22 2000 migration patterns among the four macro-regions: West, Midwest, North–east and South

snapshots of the USAN in order to compare the communities obtained. In addition, Ball, Karrer and Newman (2011) have identified an overlapping community structure in the USAN but the resolution of the communities is low since only two major communities are identified (splitting the US into east and west). This, however, does not provide any detailed information regarding particularly

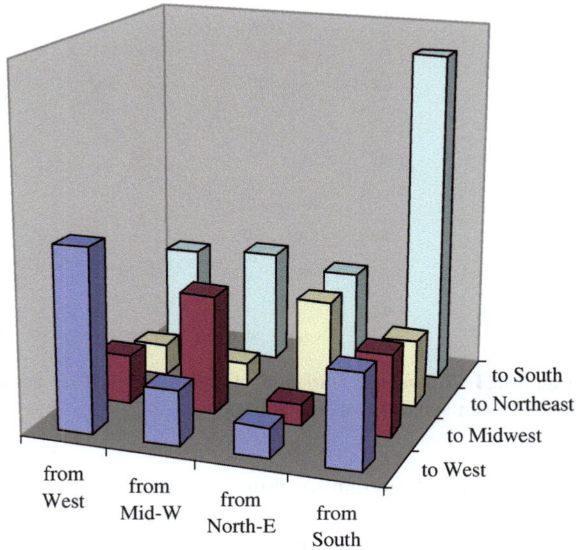

Fig. 10.23 2010 migration patterns among the four macro-regions: West, Midwest, North–east and South

high-traffic connections in the US, as identified by Expert's null model in this work. The application of Newman's method to the networks presented in this work is illustrated in Figs. B.1, B.2, B.3, B.4, B.5, B.6, B.7, B.8, B.9, B.10, B.11, B.12, B.13, B.14, B.15, B.16, B.17 and B.18 in Appendix B. In comparison, in the illustration of the application of Expert's method shown in Figs. A.1, A.2, A.3, A.4, A.5, A.6, A.7, A.8, A.9, A.10, A.11, A.12, A.13, A.14, A.15, A.16, A.17 and A.18 in Appendix A, the communities are not region-based but cover a large area of the US, exposing particularly high-traffic connections, given their distance.

All partitions identified using Newman's non-spatial NG null model reveal an identical and trivial community structure within the USAN. Specifically, there are four main regional communities that always cover the same region within the US: the east, west, north or south. These isolated communities of airports only provide a very low-resolution picture of the major flows within the US, based solely on the passenger volumes among airports, and disregard the non-linear spatial influence on passenger flows. In summary, Expert's spatial null model has revealed many particularly high flows among distant airports within the US, but Newman's general null model only reveals four regions of high internal traffic, which results from the spatial networks bias towards stronger short-range interactions in terms of passenger flows.

In order to provide a quantitative comparison of the results obtained using Expert's [17, 25] null models, it is possible to use a community structure comparison measure, such as Normalised Mutual Information (NMI) or Normalised Variation of Information (NVI). Since the purpose of this section on comparative validation is to highlight the contribution of Expert's spatial null model, NVI is a better candidate since it measures the difference between two partitions (in this case

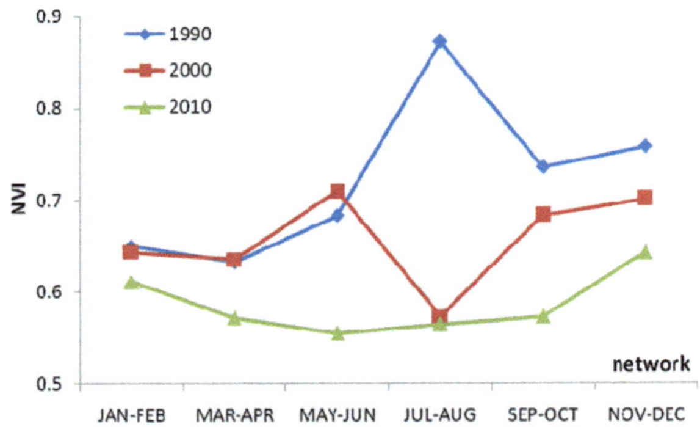

Fig. 10.24 Normalised Variation of Information (NVI) among community structure identified using Expert's and Newman's null models

Expert's and Newman's), thus quantifying the significance of using spatial information in the detection of communities.

NVI for Expert's (2011) and [25] community structure is shown in Fig. 10.24 where each trend represents the NVI over the course of a given year. Basically, the plot shows by how much Expert's and Newman's results differ over time, in each of the 3 years studied. High NVI means high variation (large difference) between the two partitions. Hence, since NVI ranges from 0 to 1, values above 0.5 are large and therefore the plot suggests that there are large differences in the community structure obtained by Newman's method and Expert's method. Specifically, 1990 generally has the highest NVI (especially July–August); 2000 is almost identical apart from a much lower NVI in July–August; and 2010 generally has the lowest and steadiest NVI. In summary, quantitative comparison of the communities obtained using Expert's and Newman's null models suggests that there is a significant difference between the partitions obtained. This means that there is indeed a need to use tailored spatial community detection techniques for spatial networks, as the results obtained would be very different. Assuming that Expert's model accurately captures the bias in the spatial networks, it follows that the higher the NVI, the better Expert's model performs in comparison to Newman's model.

10.5 Conclusion

The USAN is a complex system that is continuously evolving to meet the growing demands for air travel. Investigating the community structure within has illuminated important hidden characteristics of the network's topology and dynamics. Specifically, the findings reveal high heterogeneity in both space and time. In other words, the network is non-uniform (in space) and non-linear (in time) in terms of its connections and traffic. In addition, spatial community detection has identified a more realistic picture of the intricate structure within the network, which is invaluable for understanding this critical transportation system. Furthermore, this network model may be used for forecasting future trends in the USAN. For example, the identification of reliable communities can be the first step to study how external factors, such as natural disasters (e.g. tornados, which are common in large parts of the US), affect the function of the network. Moreover, the communities emerging from socio-economic interactions, as in the case of migration, reflect both the social influence radii and the activity system configuration (the distribution of activities in terms of location). Variations in the activity system will possibly modify such relationships and the resulting community structure. Finally, there is a clear relationship between domestic US air travel and migration. In particular, the identified community structures map well onto the migration patterns among the four macro-regions and within the region.

Future work in the general field of complex networks needs to propose new theoretical methods for the analysis and understanding of complex systems, and new simulation models that display the behaviour of empirical models of complex systems. Specifically, it is necessary to develop and evaluate alternative methods for the discovery of space-independent community structure. One important question regarding the application of Expert's (2011) method is exactly how much bias there is towards stronger short-range interactions in the network being studied. Also, it is important to identify new network parameters that describe sufficiently well both dynamics on and of the network. Finally, given the proposed evolution-based network models, it would be useful to forecast potential future trends in the networks.

Appendix A

The following figures depict the community structure found in the USAN using Expert's method. In each figure, all community members are assigned the same colour.

Year 1990

Figures A.1, A.2, A.3, A.4, A.5 and A.6 depict bi-monthly snapshots of the USAN for the year 1990

Fig. A.1 January–February 1990 community structure with Expert

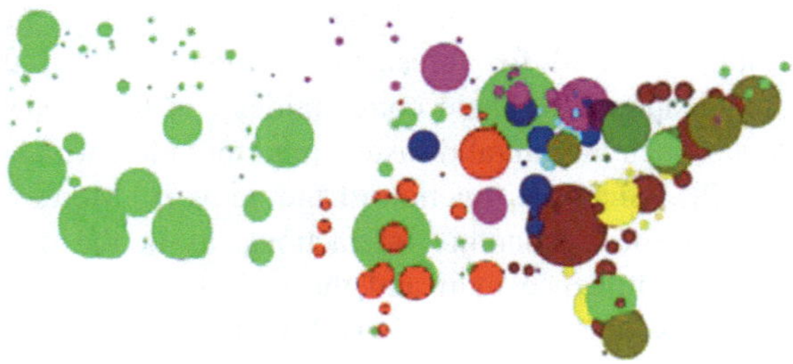

Fig. A.2 March–April 1990 community structure with Expert

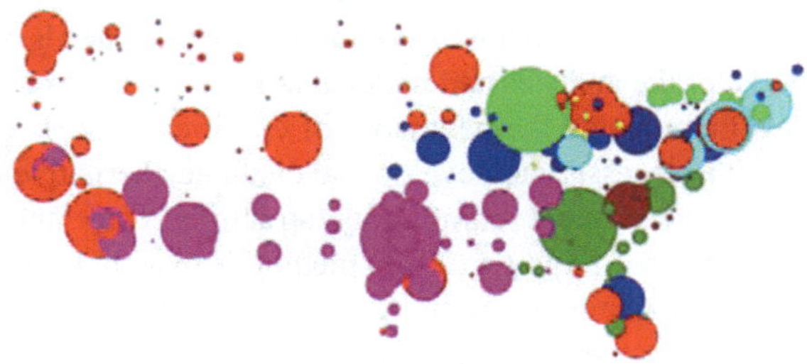

Fig. A.3 May–June 1990 community structure with Expert

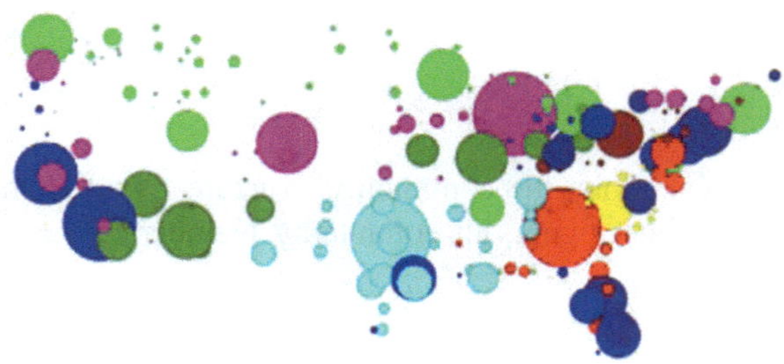

Fig. A.4 July–August 1990 community structure with Expert

Fig. A.5 September–October 1990 community structure with Expert

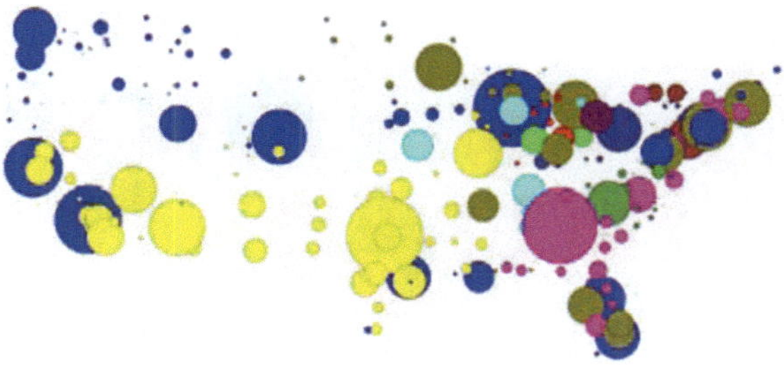

Fig. A.6 November–December 1990 community structure with Expert

Year 2000

Figures A.7, A.8, A.9, A.10, A.11 and A.12 depict bi-monthly snapshots of the USAN for the year 2000.

Year 2010

Figures A.13, A.14, A.15, A.16, A.17 and A.18 depict bi-monthly snapshots of the USAN for the year 2010.

Fig. A.7 January–February 2000 community structure with Expert

Fig. A.8 March–April 2000 community structure with Expert

Fig. A.9 May–June 2000 community structure with Expert

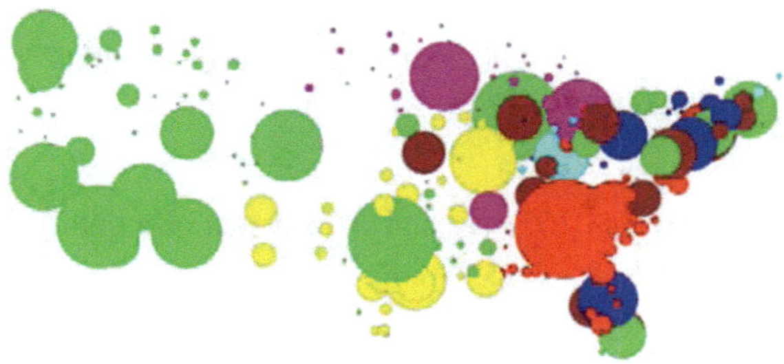

Fig. A.10 July–August 2000 community structure with Expert

Fig. A.11 September–October 2000 community structure with Expert

Fig. A.12 November–December 2000 community structure with Expert

Fig. A.13 January–February 2010 community structure with Expert

Fig. A.14 March–April 2010 community structure with Expert

Fig. A.15 May–June 2010 community structure with Expert

Fig. A.16 July–August 2010 community structure with Expert

Fig. A.17 September–October 2010 community structure with Expert

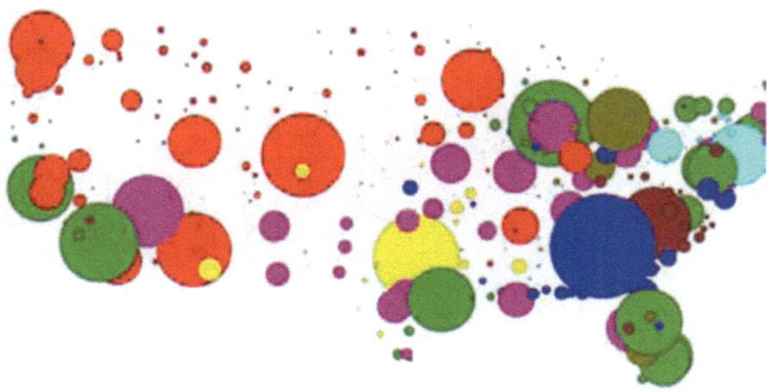

Fig. A.18 November–December 2010 community structure with Expert

Appendix B

The following figures depict the community structure found in the USAN using Newman's method. In each figure, all community members are assigned the same colour.

Year 1990

Figures B.1, B.2, B.3, B.4, B.5 and B.6 depict bi-monthly snapshots of the USAN for the year 1990.

Year 2000

Figures B.7, B.8, B.9, B.10, B.11 and B.12 depict bi-monthly snapshots of the USAN for the year 2000.

Year 2010

Figures B.13, B.14, B.15, B.16, B.17 and B.18 depict bi-monthly snapshots of the USAN for the year 2010.

Fig. B.1 January–February 1990 community structure with Newman

Fig. B.2 March–April 1990 community structure with Newman

Fig. B.3 May–June 1990 community structure with Newman

Fig. B.4 July–August 1990 community structure with Newman

Fig. B.5 September–October 1990 community structure with Newman

Fig. B.6 November–December 1990 community structure with Newman

Fig. B.7 January–February 2000 community structure with Newman

Fig. B.8 March–April 2000 community structure with Newman

Fig. B.9 May–June 2000 community structure with Newman

Fig. B.10 July–August 2000 community structure with Newman

Fig. B.11 September–October 2000 community structure with Newman

Fig. B.12 November–December 2000 community structure with Newman

Fig. B.13 January–February 2010 community structure with Newman

Fig. B.14 March–April 2010 community structure with Newman

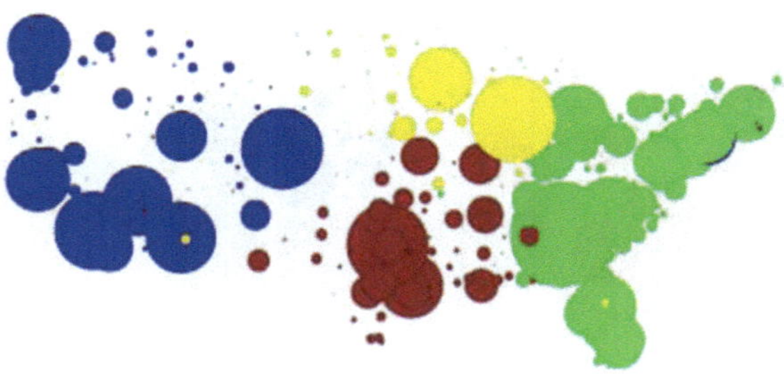

Fig. B.15 May–June 2010 community structure with Newman

Fig. B.16 July–August 2010 community structure with Newman

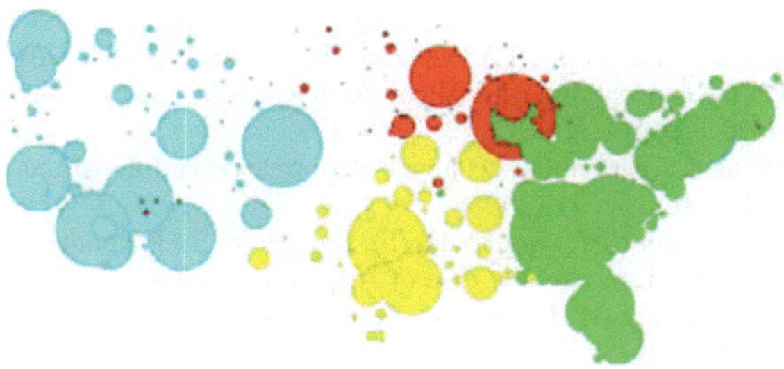

Fig. B.17 September–October 2010 community structure with Newman

Fig. B.18 November–December 2010 community structure with Newman

References

1. Almendral, J.A., Leyva, I., Li, D., Sendiña-Nadal, I., Havlin, S., Boccaletti, S.: Dynamics of overlapping structures in modular networks. Phys. Rev. E – Stat. Nonlinear Soft Matter Phys. **82**(1) (2010)
2. Amaral, L.A.N., Scala, A., Barthélémy, M., Stanley, H.E.: Classes of small-world networks. Proc. Natl. Acad. Sci. U.S.A. **97**(21), 11149–11152 (2000)
3. Ball, B., Karrer, B., Newman, M.E.J.: Efficient and principled method for detecting communities in networks. Phys. Rev. E – Stat. Nonlinear Soft Matter Phys. **84**, 036103 (2011)
4. Barrat, A., Barthélemy, M., Pastor-Satorras, R., Vespignani, A.: The architecture of complex weighted networks. Proc. Natl. Acad. Sci. U.S.A. **101**(11), 3747–3752 (2004)
5. Blondel, V.D., Guillaume, J., Lambiotte, R., Lefebvre, E.: Fast unfolding of communities in large networks. J. Stat. Mech.: Theory Exp. **2008**(10) (2008)
6. Bounova, G.: Topological Evolution of Networks: Case Studies in the US Airlines and Language Wikipedias. MIT (2009)
7. Bulu, M.: City Competitiveness and Improving Urban Subsystems: Technologies and Applications. IGI Global (2012)
8. Burghouwt, G., de Wit, J.: Temporal configurations of European airline networks. J. Air Transp. Manage. **11**(3), 185–198 (2005)
9. Bureau of Transportation Statistics: http://www.rita.dot.gov/bts/. Accessed Feb 2012
10. Button, K.J.: Economic aspects of regional airport development. In: Postorino, M.N. (ed.) Developments of Regional Airports. WIT Press, Southampton, UK (2010)
11. Calabrese, F., Dahlem, D., Gerber, A., Paul, D., Chen, X., Rowland, J., Rath, C., Ratti, C.: The connected states of America: quantifying social radii of influence. In: Proceedings—2011

IEEE International Conference on Privacy, Security, Risk and Trust and IEEE International Conference on Social Computing, PASSAT/SocialCom 2011, p. 223 (2011)
12. Camagni, R.: On the concept of territorial competitiveness: sound or misleading? Urban Stud. **39**(13), 2395–2411 (2002)
13. Cervero, R.: Efficient urbanisation: economic performance and the shape of the metropolis. Urban Stud. **38**(10), 1651–1671 (2001)
14. Chorianopoulos, I., Pagonis, T., Koukoulas, S., Drymoniti, S.: Planning, competitiveness and sprawl in the Mediterranean city: the case of Athens. Cities **27**(4), 249–259 (2010)
15. De Montis, A., Barthélemy, M., Chessa, A., Vespignani, A.: The structure of interurban traffic: a weighted network analysis. Environ. Plan. **34**(5), 905–924 (2007)
16. Estrada, E., Hatano, N.: Communicability graph and community structures in complex networks. Appl. Math. Comput. **214**(2), 500–511 (2009)
17. Expert, P., Evans, T.S., Blondel, V.D., Lambiotte, R.: Uncovering space-independent communities in spatial networks. Proc. Natl. Acad. Sci. U.S.A. **108**(19), 7663–7668 (2011)
18. Guimerà, R., Mossa, S., Turtschi, A., Amaral, L.A.N.: The worldwide air transportation network: anomalous centrality, community structure, and cities' global roles. Proc. Natl. Acad. Sci. U.S.A. **102**(22), 7794–7799 (2005)
19. Hsu, C., Wen, Y.: Determining flight frequencies on an airline network with demand-supply interactions. Transp. Res. Part E: Logist. Transp. Rev. **39**(6), 417–441 (2003)
20. Internal Revenue Service, Migration Data, 2010. http://www.irs.gov/
21. Lancichinetti, A., Fortunato, S.: Community detection algorithms: a comparative analysis. Phys. Rev. E – Stat. Nonlinear Soft Matter Phys. **80**(5) (2009)
22. Lancichinetti, A., Radicchi, F., Ramasco, J.J., Fortunato, S.: Finding statistically significant communities in networks. PLoS ONE **6**(4) (2011)
23. Levy, M.: Scale-free human migration and the geography of social networks. Phys. A **389**(21), 4913–4917 (2010)
24. Meunier, D., Lambiotte, R. and Bullmore, E. (2010) "Modular and hierarchically modular organization of brain networks", Frontiers in Neuroscience, vol. 4
25. Newman, M.E.J., Girvan, M.: Finding and evaluating community structure in networks. Phys. Rev. E – Stat. Nonlinear Soft Matter Phys. **69**(2 2), 026113-1–026113-15 (2004)
26. Postorino, M.N.: City Competitiveness and Airport: Information Science Perspective, in City Competitiveness and Improving Urban Subsystems: Technologies and Applications, pp. 61–83. IGI Global, Hershey Pa (USA) (2011)
27. Patuelli, R., Reggiani, A., Gorman, S.P., Nijkamp, P., Bade, F.: Network analysis of commuting flows: a comparative static approach to German data. Netw. Spat. Econ. **7**(4), 315–331 (2007)
28. Rouwendal, J.: Search theory and commuting behavior. Growth and Change **35**(3), 391–418 (2004)
29. Schwartz, A.: Interpreting the effect of distance on migration. J. Polit. Econ. **81** (1973)
30. United States Census Bureau, Census Data, 2011. http://www.census.gov/
31. Wuellner, D.R., Roy, S., D'Souza, R.M.: Resilience and rewiring of the passenger airline networks in the United States. Phys. Rev. E – Stat. Nonlinear Soft Matter Phys. **82**(5) (2010)

Chapter 11
Decentralized Control of Complex Dynamic Systems Employing Function Emulation by Neural Networks

Yuanwei Jing, Yanxin Zhang, Vesna M. Ojleska, Tatjana D. Kolemisevska-Gugulovska and Georgi M. Dimirovski

Abstract A novel robust adaptive control design synthesis, which employs both high-order neural networks and math-analytical results for a class of mechatronic nonlinear systems possessing similarity property has been derived. This approach makes adequate use of the structural feature of composite similarity systems and neural networks to solve the representation issue of uncertainty interconnections and subsystem gains by updating online the weight of the neural networks. Lyapunov stability theory and attraction domain analysis are used. This synthesis guarantees the real stability in closed loop but also requires skills to obtain larger attraction domains around the operating equilibrium. The benchmark example of elastically interconnected two inverted pendulums on carts, thus creating a complex nonlinear dynamic system possessing inherent uncertainties, is investigated. The decentralized control of this benchmark plant is solved and its simulation results are given to illustrate the proposed technique.

Y. Jing (✉)
College of Information Science and Engineering, Northeastern University, Shenyang 110004, Liaoning, People's Republic of China
e-mail: jingyuanwei@ise.neu.edu.cn

Y. Zhang
School of Electronic and information Engineering Haidian, Beijing Jiaotong University, No.3 Shangyuancun, Haidian, Beijing 100044, People's Republic of China
e-mail: yxzhang@bjtu.edu.cn

V.M. Ojleska · T.D. Kolemisevska-Gugulovska · G.M. Dimirovski
School FEEIT, St. Cyril and St. Methodius University, Karpos 2, Rujger Boskovic BB, Skopje MK-1000, Republic of Macedonia
e-mail: vojleska@feit.ukim.edu.mk

T.D. Kolemisevska-Gugulovska
e-mail: tanjakg@feit.ukim.edu.mk

G.M. Dimirovski
e-mail: dimir@feit.ukim.edu.mk; gdimirovski@dogus.edu.tr

G.M. Dimirovski
Faculty of Engineering, Dogus University, Acibadem, Kadikoy, 34722 Istanbul, Turkey

G.M. Dimirovski (ed.), *Complex Systems*, Studies in Systems, Decision and Control 55, DOI 10.1007/978-3-319-28860-4_11

11.1 Introduction

Control applications in advanced technological systems make use of various types of representation models of so-called complex systems [1, 27] that are developed or chosen according to certain specific properties of the plant process to be controlled. The control of complex mechatronic plant systems, consisting of several subsystems, has become a rather important research focus of the control community at large. In particular, the adaptive control of mechatronic systems that are composite interconnected, nonlinear, and uncertain is a problem still open to investigation, due to various phenomena simultaneously coexisting in the plant [1, 23, 27]. Therefore, there is indispensible need for a generalized conceptualization of representation models for dynamic plants (Fig. 11.1) and understanding the compatible usage of different mathematical formalisms.

Typically, known adaptive control approaches are based on the model adaptive control concepts (e.g., see [23, 24, 30–32]) given the fact that various uncertainties in plant subsystems may exist. Then the stability of the overall system is ensured by a certain condition imposed on the M matrix that is related to the boundary of the interconnection. Thus complete and precise knowledge of the parameters in interconnections and/or subsystems is no longer a necessity. A typical deficiency of these approaches stems from the fact that it is difficult to validate the positive definiteness of the M matrix beforehand, because elements of the M matrix depend on the uncertain subsystem parameters. Additional progress along these lines in adaptive control of strongly interconnected systems has also been reported (e.g., see [9, 24, 27, 28]).

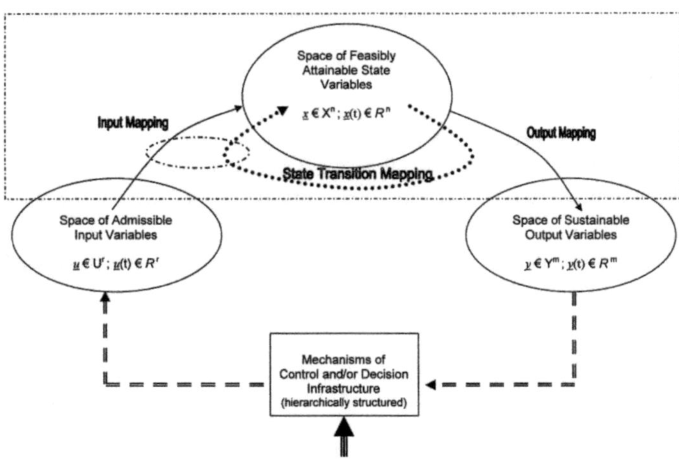

Fig. 11.1 Conceptualization of categories of general systems and control tasks in engineering terms that provide freedom in choosing representation formalisms, either maths-analytical or computational-intelligence based, and synthesis design methodologies [4]

Zhang [34] has made new discoveries of the structural impact on complex control systems when properties of similarity or symmetry among interconnected plant subsystems exist,called the Zhang's similarity–symmetry impact. To further clarify considerations about the uncertainty composite nonlinear systems in motion guidance and electromechanical plants controls have been contributed in [7, 8]. The comprehensive study by Gao [7] used time-varying linear composite uncertainty system to design a nonlinear controller that made a system real-stable in closed loop by exploiting Riccati equations. Studies [12, 31] first provided solutions to control nonlinear composite systems that exploit essentially the plant similarity property. In [7, 25], respectively, the guarantee for the overall system to be made real-stable was obtained using adaptive control. However, the case of subsystem uncertainty input gains is not included in these studies.

More recently, Liu and Jia [22], Ge and Wang [9], Zhou [38], and Zhu et al. [39] gave certain novel solutions to this problem by making use of neural networks. These studies investigate specific composite systems with non varying parameters, while the latter two studies investigate only linear interconnected systems. Besides, these results are far from ideal with regard to the uncertainty input gains (boundary of which was uncertain), although referring to common large-scale systems. Recently, Liu [21] derived a corrective solution to adaptive control for uncertain similarity composite systems. Alternative techniques [3, 13, 26] that exploit the universal approximation capacity of neural networks [10] and, in particular, the high-order neural networks [6, 18, 33] have emerged. For the purpose of emulating uncertainty interconnections in composite systems, it was first proposed by Zhang [36]. In addition, Zhang et al. [37] also proposed a similar approach for nonlinear similarity composite systems that possess uncertainties in the subsystem input gains and in the interconnections, which was further explored and extended in [14]. However, considerably profound debates on the relevance of approximating emulation of neural networks has been preceding these effective control applications, as it can be inferred from the works of Heht-Nielsen [11], Katsuura and Sprecher [15], Kurkova [19, 20] despite the generally powerful functional representation theorem [17].

On the other hand, Siljak and Zecevic [28] thoroughly explored state-of-the-art control of large-scale systems beyond the decentralized feedback strategy recently and made a critical appraisal pointing out to further innovations based on the sound background of dynamic graphs [29]. On these grounds and with regard to overview article [2] as well as the generalized conceptualization of categories of systems and control [4], this study revisits work of [14] and extends the results via modifying the latter technique for applications to complex mechatronic plants that possess uncertain couplings and may posses Zhang's similarity–symmetry impact.

The most recent work by Geun Bum Koo et al. [16] should be especially noted at this point for an alternative contribution to nonlinear large-scale systems beyond the decentralized feedback strategy that is based on fuzzy system observers and employing fuzzy models. This research has extended considerably the previously existing knowledge based on the previous successful results of decentralized control of composite and complex nonlinear plants, (e.g., see Chen et al. [5], and Yang and Zhou [32]). Also, recently some novel results by employing neural networks were reported in Liu et al. [22], Tong et al. [30], and Zhang et al. [35].

Further this paper is organized as follows. Section 11.2 presents the formal problem statement and the basic assumptions adopted. Section 11.3 presents the main new result along with the constructive control design. Section 11.4 presents the application to the case study of a real-world complex mechatronic system of axis-tray drive to illustrate the proposed design synthesis. Conclusions and references are given thereafter.

11.2 Problem Statement and Assumptions

Consider the class of complex nonlinear systems possessing similarity and uncertainties that are represented by the following set, $i = 1, 2, \ldots, N$, of equations:

$$\dot{x}_i = A_i x_i + B_i[u_i + f_i(x_i, t)] + h_i(x). \tag{11.1}$$

Here, the respective symbols denote: $x_i \in R^n$ and $u_i \in R^m$ are the state and input vectors of the ith subsystem, respectively; the overall system state vector is $x = (x_1^T, x_2^T, \ldots, x_N^T)^T$ by definition; $A_i \in R^{n \times n}$ and $B_i \in R^{n \times m}$ are subsystem state and input matrices, respectively. The basic assumption about this representation model includes further: all pairs (A_i, B_i) are controllable; all $f_i(x_i, t)$ are uncertain vector-valued functions; and all $h_i(x)$ are uncertain smooth vector-valued functions. The latter two at the zero equilibrium state also become zero vector-valued, which is a standard property of real-world physical systems.

Remark 11.1 In this paper, we suppose that system (11.1) belongs to the class of composite similarity dynamic systems in the sense of Zhang's similarity/symmetry impact, which is a subclass in the class of general complex dynamic systems.

Definition 11.1 ([34]) System (11.1) is said to be a similarity composite system if there is a parameter matrix T_i such that

$$T_i^{-1} A_i T_i = A,\ T_i^{-1} B_i = B. \tag{11.2}$$

Also, it is necessary to notice and observe the following property definitions and results, respectively, as given below.

Assumption 11.1 There exist uncertain functions $\rho_i(x_i, t) > 0$ such that

$$\|f_i(x_i, t)\| \le \rho_i(x_i, t), i = 1, 2, \ldots, N. \tag{11.3}$$

Lemma 11.1 ([39]) *If pair* (A, B) *is controllable, then for a positive-definite matrix,* Q*, and a positive number,* $\gamma > 0$*, there exists a unique solution* P *for the following matrix Riccati equation:*

$$A^T P + PA - \gamma^2 PBB^T P + Q = 0. \tag{11.4}$$

Definition 11.2 ([25]) Suppose $V = V(x)$ is a Lyapunov function of a given continuous system that satisfies

$$\gamma_1(\|x\|) \le V(x, t) \le \gamma_2(\|x\|), \tag{11.5}$$

$$\dot{V}(x, t) \le -\gamma_3(\|x\|) + \varphi(t), \tag{11.6}$$

where $\lim_{t \to \infty} \gamma_j(s) = \infty$, $1 \le j \le 3$, is a strictly continuous positive-definite function. If there exists a positive continuous function $\varphi(t)$ such that it satisfies $\varphi(t) \le r^2 < \infty$, then the system state converges to some neighbourhood of the origin, $\|x(t)\| \le r$, and the system is said to be *real-stable*.

11.3 ANN Emulation-Base Decentralized Control of the Similarity Class of Complex Dynamic Systems

It should be noted first, due to the function approximation [17] capacity of artificial neural networks [11, 20, 15], an arbitrary continuous function $g(x, t)$ can be represented by an ANN, which consists of a directed graph with ideal weights, W_g, summing junction and output activation function $\sigma(\cdot)$ [3, 13]. That is, $g(x, t) = W_g^T \sigma_g(x) + \varepsilon_g(t)$. Suppose the artificial neural network (ANN) estimation of $g(x, t)$ is given by means of the formula $\widehat{g}(x, t) = \widehat{W}_g^T \sigma_g(x)$. Then the function estimate error $\widetilde{g}(x, t)$ can be calculated by the following equation:

$$\widetilde{g}(x, t) = W_g^T \sigma_g(x) - \widehat{W}_g^T \sigma_g(x) + \varepsilon_g(t) \tag{11.7a}$$

or, putting $\widetilde{W}_g = W_g - \widehat{W}_g$ yields

$$\widetilde{g}(x, t) = \widetilde{W}_g^T \sigma_g(x) + \varepsilon_g(t). \tag{11.7b}$$

Second, a model of the nonlinear interconnections $h_i(x)$, $i = 1, 2, \ldots, N$, can be built by employing high-order neural nets [35, 36]. Namely, suppose x is the input of the high-order neural network and y_i is the output, then the respective representation model is given as

$$y_i = \widehat{W}_i s(x), \quad s(x) = [s_i(x), \ldots, s_L(x)]^T, \tag{11.8a}$$

$$s_i(x) = \prod_{k=1}^{N} \prod_{j \in I_i} [s(x_{kj})]^{d_j(i)}, \quad i = 1, 2, \ldots, N, \tag{11.8b}$$

$$s(x_{kj}) = \frac{\mu_0}{1 + e^{-l_0 x_{kj}}} + \lambda_0, \quad j = 1, 2, \ldots, n, \quad k = 1, 2, \ldots, N, \tag{11.8c}$$

where $\widehat{W}_i \in R^{n \times L}$ is the matrix of weights; $s_i(x)$ is an element of $s(x) \in R^{L \times 1}$; $\{I_i | i = 1, 2, \ldots, L\}$ is a collection of L not-ordered subsets of $\{1, 2, \ldots, n\}$; $d_j(i)$ is a nonnegative integer; and l_0, λ_0, μ_0 appropriate nonnegative constants. For these models of high-order neural networks, there exists an integer L, an integer $d_j(i)$ and an optimized matrix W_i^*, such that for any $\varepsilon > 0$, $|h_i(x) - W_i^* s(x)| \leq \varepsilon$ is satisfied. In other words, if the high-order network is large enough, there exists a matrix of weights such that $W_i^* s(x)$ can approximate $h_i(x)$ to any degree of accuracy; moreover, W_i^* is bounded, i.e., $\|W_i^*\| \leq M_W, M_W > 0$.

It is therefore that the system (11.1), $i = 1, 2, \ldots, N$, can be rewritten in the following form:

$$\dot{x}_i = Ax_i + B_i[u_i + f_i(x_i, t)] + W_i^* s(x) + \varepsilon_i(x), \tag{11.9}$$

where $\varepsilon_i(x) = h_i(x) - W_i^* s(x)$ is the weight error estimate. Furthermore, there exists $\varepsilon \geq 0$ such that $|\varepsilon_i(x)| \leq \varepsilon$, $i = 1, 2, \ldots, N$. Should W_i denote the estimation of the uncertain weights matrix W_i^*, $i = 1, 2, \ldots, N$, then one finds:

$$\dot{x}_i = A_i x_i + B_i[u_i + f_i(x_i, t)] - \widetilde{W}_i s(x) + W_i s(x) + \varepsilon_i(x), \tag{11.10}$$

where $\widetilde{W}_i = W_i - W_i^*$.

Theorem 11.1 *Consider system* (11.1) *and construct the set of controllers, $i = 1, 2, \ldots, N$, which are represented by*

$$u_i = u_i^a + u_i^b + u_i^c, i = 1, 2, \ldots, N, \tag{11.11}$$

$$u_i^a = -(\gamma + 1) B^T P T_i^{-1} x_i, \tag{11.12}$$

$$u_i^b = -\xi_i^{-1} \widehat{\rho}_i^2(x, t) B^T P T_i^{-1} x_i, \tag{11.13}$$

$$u_i^c = \frac{B_i^T W_i s(x)}{\lambda_i [1 + \|B_i\|^2]} + \frac{B_i^T \Theta_i}{\lambda_{1,i}[1 + \|B_i\|^2]}, \tag{11.14}$$

where $\widehat{\rho}_i^2$ *is the estimation of* ρ_i^2 *by means of a neural-net model, i.e.,* $\widehat{\rho}_i^2(x,t) = \widehat{Z}_i(t)\sigma_i(x_i)$ *with* $\widehat{Z}_i(t)$ *as the corresponding weight vector;* $\Theta_i \in R^{n\times L}$ *and* $\Theta_i = [\theta_i, 0, \ldots, 0]^T$*; and the quantities defined in the course of theorem proving* β_i, γ, $\gamma_{1,j}$, λ_i, $\lambda_{1,i}$ *are chosen as*

$$\lambda_i \geq \frac{k_{0,i}s}{\sqrt{2\bar{k}_2^i\beta_i - sk_{0,i}}}, \tag{11.15a}$$

$$\lambda_{1,i} \geq \frac{k_{0,i}}{\sqrt{2k_{0,i}\bar{k}_2^i\gamma_{1,i} - k_{0,i}}}, \tag{11.15b}$$

$$\beta_i > \frac{s^2k_{0,i}^2}{2\bar{k}_2^i}, \ \gamma_{1,i} > \frac{k_{0,i}}{2\bar{k}_2^i}; \tag{11.15c}$$

along with the adaptation law

$$\widehat{Z}_i = -\Gamma_{i1}\widehat{Z}_i + \Gamma_{i2}\sigma_i(x_i)\left\|B^TPT_i^{-1}x_i\right\|, \ \Gamma_{i1}, \Gamma_{i2} > 0, \tag{11.16}$$

$$\dot{W}_i = \begin{Bmatrix} 2k_{0,i}P^TT_i^{-1}x_iS(x)^T, \ \|W_i\| < M_W \\ -\beta_iW_i + 2k_{0,i}P^TT_i^{-1}x_iS(x)^T, \ \|W_i\| \geq M_W \end{Bmatrix}, \tag{11.17}$$

$$\theta_i = -\gamma_{1,i}\theta_i + 2k_{0,i}\left|P^TT_i^{-1}x_i\right|, \tag{11.18}$$

with Γ_{i1}, Γ_{i2}, $k_{0,i}$, β_i, *and* $\gamma_{1,i}$ *all positive constant design parameters, and* M_W *is a large design constant that confines* θ_i *within a ball of radius* M_W *to be chosen in the course of the design. Then assuming that the neural-network approximation error* $\varepsilon_{\rho_i}(t)$ *is time varying and bounded with an uncertain boundary, all the subsystem state vectors* x_i*, respectively, are consistent ultimately bounded on sets*

$$D_i = \left\{ x_i \in R^n \middle| v_{0i}(x) \leq \frac{\mu_i}{k_{0,i}\alpha_i}, \frac{\bar{k}_2^i}{\bar{k}_1^i} \leq k_{0,i} \leq 1 \right\}. \tag{11.19}$$

The proof of Theorem 11.1 is derived in two steps. In Step 1, it is proved first the existence of nominal controllers $u_i = u_i^a + u_i^b : R^n \to R^m$ and Lyapunov functions $V_{0i}(x_i)$ for the nominal subsystems

$$\dot{x}_i = A_ix_i + B_i[u_i + f_i(x_i, t)],$$

such that inequalities (11.5) and (11.6) are satisfied. Thus, the nominal subsystems proved can be guaranteed as real-stable. Then in Step 2, the result of the previous step is adopted as a priori assumption, and then the proof is derived to the full via

Lyapunov function approach by finding the domains of attraction that are represented by sets (11.19).

Step 1. Due to the condition that (A_i, B_i) is controllable by formula (11.2), it can be deduced that (A, B) is also controllable. Moreover, by virtue of Lemma 11.1, there exists P such that

$$A^T P + PA - \gamma^2 PBB^T P + Q = 0$$

is satisfied. Next, consider Lyapunov functions

$$v_{0i}(x_i) = x_i^T (T_i^{-1})^T P T_i^{-1} x_i + \widehat{Z}_i^T \Gamma_{i2}^{-1} \widehat{Z}_i. \tag{11.20}$$

Then by using (11.1), the time derivative of Lyapunov functions (11.20) can be expressed as follows:

$$\begin{aligned}
\dot{v}_{0i}(x_i) &= \{x_i^T A_i^T + [u_i^T + f_i^T(x_i,t)]B_i^T\}[(T_i^{-1})^T P T_i^{-1}]x_i \\
&\quad + x_i^T[(T_i^{-1})^T P T_i^{-1}]\{A_i x_i + B_i[u_i + f_i(x_i,t)]\} + 2\widehat{Z}_i^T \Gamma_{i2}^{-1} \widehat{Z}_i \\
&= x_i^T A_i^T (T_i^{-1})^T P T_i^{-1} x_i + x_i^T (T_i^{-1})^T P T_i^{-1} A_i x_i \\
&\quad + 2x_i^T (T_i^{-1})^T P T_i^{-1} B_i [u_i + f_i(x_i,t)] + 2\widehat{Z}_i^T \Gamma_{i2}^{-1} \widehat{Z}_i \\
&= x_i^T (T_i^{-1})^T [T_i^T A_i^T (T_i^{-1})^T] P T_i^{-1} x_i + x_i^T (T_i^{-1})^T P [T_i^{-1} A_i T_i] T_i^{-1} x_i \\
&\quad + 2x_i^T (T_i^{-1})^T P T_i^{-1} B_i [u_i + f_i(x_i,t)] + 2\widehat{Z}_i^T \Gamma_{i2}^{-1} \widehat{Z}_i \\
&= x_i^T (T_i^{-1})^T [A^T P + PA] T_i^{-1} x_i + 2x_i^T (T_i^{-1})^T PB[u_i + f_i(x_i,t)] + 2\widehat{Z}_i^T \Gamma_{i2}^{-1} \widehat{Z}_i.
\end{aligned}$$

Furthermore using (11.10) and (11.11), one can obtain

$$\begin{aligned}
\dot{v}_{0i}(x_i) &= -x_i^T (T_i^{-1})^T Q T_i^{-1} x_i - x_i^T (T_i^{-1})^T \gamma^2 PBB^T P T_i^{-1} x_i + 2x_i^T (T_i^{-1})^T PB \\
&\quad \times [-(\gamma+1)B^T P T_i^{-1} x_i - \xi_i^{-1} \widehat{\rho}_i^2(x,t) B^T P T_i^{-1} x_i + f_i(x_i,t)] + 2\widehat{Z}_i^T \Gamma_{i2}^{-1} \widehat{Z}_i \\
&= -x_i^T (T_i^{-1})^T Q T_i^{-1} x_i - x_i^T (T_i^{-1})^T \gamma^2 PBB^T P T_i^{-1} x_i + 2x_i^T (T_i^{-1})^T PB\gamma B^T P T_i^{-1} x_i \\
&\quad - 2x_i^T (T_i^{-1})^T PBB^T P T_i^{-1} x_i + 2x_i^T (T_i^{-1})^T PB \\
&\quad \times [-\xi_i^{-1} \widehat{\rho}_i^2(x,t) B^T P T_i^{-1} x_i + f_i(x_i,t)] + 2\widehat{Z}_i^T \Gamma_{i2}^{-1} \widehat{Z}_i \\
&= -x_i^T (T_i^{-1})^T Q T_i^{-1} x_i - x_i^T (T_i^{-1})^T \gamma^2 PBB^T P T_i^{-1} x_i - 2x_i^T (T_i^{-1})^T PB\gamma B^T P T_i^{-1} x_i \\
&\quad - 2x_i^T (T_i^{-1})^T PBB^T P T_i^{-1} x_i + 2[f_i(x_i,t) B^T P T_i^{-1} x_i - \xi_i^{-1}(\rho_i^2 - \varepsilon_{\rho_i})\|B^T P T_i^{-1} x_i\|^2] + 2\widehat{Z}_i^T \Gamma_{i2}^{-1} \widehat{Z}_i.
\end{aligned}$$

By making use of (11.16), the above formulas can be rewritten as

$$\begin{aligned}
\dot{v}_{0i}(x_i) &\le -x_i^T (T_i^{-1})^T Q T_i^{-1} x_i - x_i^T (T_i^{-1})^T PBB^T P T_i^{-1} x_i \\
&\quad + \varepsilon_{\rho_i} x_i^T (T_i^{-1})^T PBB^T P T_i^{-1} x_i - \widehat{Z}_i^T \Gamma_{i2}^{-1} \Gamma_{i1} \widehat{Z}_i + \xi_i \\
&\le -x_i^T (T_i^{-1})^T Q T_i^{-1} x_i + \varepsilon_{\rho_i}^2 - \widehat{Z}_i^T \Gamma_{i2}^{-1} \Gamma_{i1} \widehat{Z}_i + \xi_i \le -\lambda_i^* v_{0i} + \varepsilon_{\rho_i}^2 + \xi_i
\end{aligned} \tag{11.21}$$

with $\lambda_i^* = \min\left\{\frac{\lambda_{\min}(Q)}{\lambda_{\max}(P)}, \lambda_{\min}(\Gamma_{i1})\right\}$. Moreover, there can be found another form for the derivatives of Lyapunov functions:

$$\begin{aligned}\dot{v}_{0i}(x_i) &\le -x_i^T Q x_i + \varepsilon_{\rho_i}^2 - \widehat{Z}_i^T \Gamma_{i2}^{-1} \Gamma_{i1} \widehat{Z}_i + \xi_i \le -x_i^T Q x_i + \varepsilon_{\rho_i}^2 + \xi_i \\ &\le -\frac{\lambda_{\min}(Q)}{\lambda_{\max}(PP^T)} x_i^T (T_i^{-1})^T PP^T T_i^{-1} x_i + \varepsilon_{\rho_i}^2 + \xi_i \le -\bar{k}_i \left|\frac{\partial v_{0i}(x_i)}{\partial x_i}\right|^2 + \varepsilon_{\rho_i}^2 + \xi_i\end{aligned} \tag{11.22}$$

with $\bar{k}_i = \frac{\lambda_{\min}(Q)}{4\lambda_{\max}(PP^T)}$. From (11.21) and (11.22) along with Definition 11.1, it is readily inferred all the nominal subsystems are real-stable in closed loop.

Step 2: Construct Lyapunov function for system (11.1) in the following form:

$$V = V\left(x, \widetilde{W}_1, \widetilde{W}_2, \ldots, \widetilde{W}_N, \widetilde{\theta}_1, \widetilde{\theta}_2, \ldots, \widetilde{\theta}_N\right) = \sum_{i=1}^{N} V_i\left(x, \widetilde{W}_i, \widetilde{\theta}_i\right), \tag{11.23}$$

$$V_i\left(x, \widetilde{W}_i, \widetilde{\theta}_i\right) = k_{0,i} v_{0i}(x_i) + \frac{1}{2} tr\{\widetilde{W}_i^T \widetilde{W}_i\} + \frac{1}{2}\widetilde{\theta}_i^2, \tag{11.24}$$

with $\widetilde{\theta}_i = \theta_i - \varepsilon$. By virtue of (11.10), the derivatives of $V_i\left(x, \widetilde{W}_i, \widetilde{\theta}_i\right)$ are found to be

$$\begin{aligned}\dot{V}_i = {} & k_{0,i} \frac{\partial v_{0i}}{\partial x_i} \{A_i x_i + B_i [u_i^a + u_i^b + f_i(x_i, t)]\} \\ & + k_{0,i} \frac{\partial v_{0i}}{\partial x_i} B_i u_i^c - k_{0,i} \frac{\partial v_{0i}}{\partial x_i} \widetilde{W}_i S(x) \\ & + k_{0,i} \frac{\partial v_{0i}}{\partial x_i} W_i S(x) + k_{0,i} \frac{\partial v_{0i}}{\partial x_i} \varepsilon(x) + tr\left\{\dot{W}_i^T \widetilde{W}_i\right\} + \widetilde{\theta}_i \theta_i.\end{aligned} \tag{11.25}$$

By making use of (11.17), one obtains

$$\begin{aligned}\dot{V}_i = {} & k_{0,i} \dot{v}_{0i} + k_{0,i} \frac{\partial v_{0i}}{\partial x_i} B_i u_i^c + k_{0,i} \frac{\partial v_{0i}}{\partial x_i} W_i S(x) \\ & + k_{0,i} \frac{\partial v_{0i}}{\partial x_i} \varepsilon_i(x) - \beta_i I_W tr\{W_i^T \widetilde{W}_i\} + \widetilde{\theta}_i \theta_i,\end{aligned} \tag{11.26}$$

where I_W is the indicator function of W that satisfies

$$I_W = \left\{\begin{array}{ll} 1 & \text{if } \|W_i\| \ge M_W \\ 0 & \text{if } \|W_i\| < M_W \end{array}\right\}. \tag{11.27}$$

Since $tr\{W_i^T\widetilde{W}_i\} = \frac{1}{2}\|W_i\|^2 + \frac{1}{2}\|\widetilde{W}_i\|^2 - \frac{1}{2}\|W_i^*\|^2$, one can derive

$$\dot{V}_i = k_{0,i}\dot{V}_{0i} + k_{0,i}\frac{\partial v_{0i}}{\partial x_i}B_i u_i^c + k_{0,i}\frac{\partial v_{0i}}{\partial x_i}W_i S(x) + k_{0,i}\frac{\partial v_{0i}}{\partial x_i}\varepsilon_i(x)_i$$
$$-\frac{\beta_i}{2}tr\{\widetilde{W}_i^T\widetilde{W}_i\} + \frac{\beta_i}{2}(1-I_W)tr\{\widetilde{W}_i^T\widetilde{W}_i\} - \frac{\beta_i}{2}I_W\|W_i\|^2 + \frac{\beta_i}{2}I_W\|W_i^*\|^2 + \widetilde{\theta}_i\theta_i. \tag{11.28}$$

The u_i^c $(\lambda_i, \lambda_{1,i})$ (notice that these two parameters can be adjusted) are put into (11.24) and, due to Assumption 11.1, it is found in turn that the time derivatives of V_i satisfy the following inequality:

$$\dot{V}_i \le k_{0,i}\bar{k}_i\left|\frac{\partial v_{0i}}{\partial x_i}\right|^2 + k_{0,i}\left|\frac{\partial v_{0i}}{\partial x_i}\right|\frac{\|B_i\|^2\|W_i\||S(x)|}{\lambda_i\left[1+\|B_i\|^2\right]}$$
$$+ k_{0,i}\left|\frac{\partial v_{0i}}{\partial x_i}\right|\|W_i\||S(x)|\,k_{0,i}\left|\frac{\partial v_{0i}}{\partial x_i}\right|\frac{\|B_i\|^2\|W_i\||\theta_i|}{\lambda_i\left[1+\|B_i\|^2\right]}$$
$$+ \frac{\beta_i}{2}(1-I_W)tr\{\widetilde{W}_i^T\widetilde{W}_i\} - \frac{\beta_i}{2}I_W\|W_i\|^2 + \frac{\beta_i}{2}I_W\|W_i^*\|^2 + k_{0,i}(\xi_i + \varepsilon_{\rho_i}^2(t)). \tag{11.29}$$

Because of $\frac{\|B_i\|^2}{1+\|B_i\|^2} \le 1$, relationship (11.25) can be rewritten as follows:

$$\dot{V}_i \le k_{0,i}\bar{k}_i\left|\frac{\partial v_{0i}}{\partial x_i}\right|^2 + k_{0,i}\left|\frac{\partial v_{0i}}{\partial x_i}\right|\|W_i\||S(x)|\left(1+\frac{1}{\lambda_i}\right)$$
$$+ k_{0,i}\left|\frac{\partial v_{0i}}{\partial x_i}\right|\frac{|\theta_i|}{\lambda_{1,i}} + k_{0,i}\left|\frac{\partial v_{0i}}{\partial x_i}\right|\theta_i - k_{0,i}\left|\frac{\partial v_{0i}}{\partial x_i}\right|\theta_i$$
$$+ k_{0,i}\left|\frac{\partial v_{0i}}{\partial x_i}\right|\varepsilon_i + \widetilde{\theta}_i\theta_i - \frac{\beta_i}{2}tr\{\widetilde{W}_i^T\widetilde{W}_i\}$$
$$+ \frac{\beta_i}{2}(1-I_W)tr\{\widetilde{W}_i^T\widetilde{W}_i\} - \frac{\beta_i}{2}I_W\|W_i\|^2 + \frac{\beta_i}{2}I_W\|W_i^*\|^2 + k_{0,i}(\xi_i + \varepsilon_{\rho_i}^2(t)). \tag{11.30}$$

Now, should the following definition $\bar{k}_i = \bar{k}_1^i + \bar{k}_2^i + \bar{k}_3^i$ be introduced, then relationship (11.26) can be transformed into the following one:

$$
\begin{aligned}
\dot{V}_i \leq & -k_{0,i}\bar{k}_1^i \left|\frac{\partial v_{0i}}{\partial x_i}\right|^2 - k_{0,i}\bar{k}_2^i \left|\frac{\partial v_{0i}}{\partial x_i}\right|^2 \\
& - k_{0,i}\bar{k}_3^i \left|\frac{\partial v_{0i}}{\partial x_i}\right|^2 + k_{0,i}s\left|\frac{\partial v_{0i}}{\partial x_i}\right| \|W_i\| \left(1 + \frac{1}{\lambda_i}\right) \\
& + k_{0,i}\left|\frac{\partial v_{0i}}{\partial x_i}\right| \theta_i \left(1 + \frac{1}{\lambda_{1,i}}\right) - \frac{\gamma_{1,i}}{2}\tilde{\theta}_i^2 - \frac{\gamma_{1,i}}{2}\theta_i^2 + \frac{\gamma_{1,i}}{2}\varepsilon^2 - \frac{\beta_i}{2} tr\{\widetilde{W}_i^T \widetilde{W}_i\} \\
& + \frac{\beta_i}{2}(1 - I_W) tr\{\widetilde{W}_i^T \widetilde{W}_i\} - \frac{\beta_i}{2} I_W \|W_i\|^2 + \frac{\beta_i}{2} I_W \|W_i^*\|^2 + k_{0,i}(\xi_i + \varepsilon_{\rho_i}^2(t)).
\end{aligned} \tag{11.31}
$$

Via appropriate choice of $\lambda_i \geq \frac{k_{0,i}s}{\sqrt{2\bar{k}_2^i\beta_i - sk_{0,i}}}$, $\lambda_{1,i} \geq \frac{k_{0,i}}{\sqrt{2k_{0,i}\bar{k}_2^i\gamma_{1,i} - k_{0,i}}}$, $\beta_i > \frac{s^2k_{0,i}^2}{2\bar{k}_2^i}$, (see inequalities (11.15a–11.15c)) and $\gamma_{1,i} > \frac{k_{0,i}}{2\bar{k}_2^i}$, from (11.31) it can be found that

$$
\begin{aligned}
\dot{V}_i \leq & -k_{0,i}\bar{k}_1^i \left|\frac{\partial v_{0i}}{\partial x_i}\right|^2 - k_{0,i}\bar{k}_3^i \left|\frac{\partial v_{0i}}{\partial x_i}\right|^2 \\
& - \left[\bar{k}_2^i \left|\frac{\partial v_{0i}}{\partial x_i}\right|^2 - 2\sqrt{\frac{\bar{k}_2^i\beta_i}{2}} \left|\frac{\partial v_{0i}}{\partial x_i}\right| \|W_i\| + \frac{\beta_i}{2}\|W_i\|^2\right] \\
& + \left[\left(\sqrt{k_{0,i}\bar{k}_2^i}\left|\frac{\partial v_{0i}}{\partial x_i}\right|\right)^2 - 2\sqrt{\frac{k_{0,i}\bar{k}_2^i\gamma_{,i1}}{2}}\theta \left|\frac{\partial v_{0i}}{\partial x_i}\right| + \frac{\gamma_{1,i}}{2}\theta_i^2\right] \\
& + \frac{\beta_i}{2}\|W_i\|^2 + \bar{k}_2^i \left|\frac{\partial v_{0i}}{\partial x_i}\right|^2 - \frac{\gamma_{1,i}}{2}\tilde{\theta}_i^2 - \frac{\gamma_{1,i}}{2}\theta_i^2 \\
& + \frac{\gamma_{1,i}}{2}\varepsilon^2 - \frac{\beta_i}{2} tr\{\widetilde{W}_i^T \widetilde{W}_i\} + \frac{\beta_i}{2}(1 - I_W) tr\{\widetilde{W}_i^T \widetilde{W}_i\} \\
& - \frac{\beta_i}{2} I_W \|W_i\|^2 + \frac{\beta_i}{2} I_W \|W_i^*\|^2 + k_{0,i}(\xi_i + \varepsilon_{\rho_i}^2(t)).
\end{aligned} \tag{11.32}
$$

Notice that if $k_{0,i} \geq \frac{\bar{k}_2^i}{\bar{k}_1^i}$ is satisfied, then $-k_{0,i}\bar{k}_1^i + \bar{k}_2^i \leq 0$ $\left(\bar{k}_1^i \geq 0\right)$ is true; otherwise, if $k_{0,i} \leq 1$ is true, then $k_{0,i} \geq \frac{\bar{k}_2^i}{\bar{k}_1^i}$. Thus, if $\frac{\bar{k}_2^i}{\bar{k}_1^i} \leq k_{0,i} \leq 1$ is satisfied, then (11.32) can be modified to give

$$
\begin{aligned}
\dot{V}_i \leq & -k_{0,i}\bar{k}_3^i \left|\frac{\partial v_{0i}}{\partial x_i}\right|^2 - \frac{\beta_i}{2}\|\widetilde{W}_i\|^2 + \frac{\beta_i}{2}\|W_i^*\|^2 + \beta_i(1 - I_W) tr\{\widetilde{W}_i^T \widetilde{W}_i\} \\
& - \frac{\gamma_{1,i}}{2}\tilde{\theta}_i^2 + \frac{\gamma_{1,i}}{2}\varepsilon^2 + (1 - I_W)\frac{\beta_i}{2} M_W^2 + k_{0,i}(\xi_i + \varepsilon_{\rho_i}^2(t)).
\end{aligned} \tag{11.33}
$$

Since

$$\beta_i(1-I_W)tr\{\widetilde{W}_i^T\widetilde{W}_i\} = \begin{Bmatrix} \beta_i tr\{\widetilde{W}_i^T\widetilde{W}_i\}, \text{if } |W_i| < M_W \\ 0, \text{ if } |W_i| \geq M_W \end{Bmatrix}, \tag{11.34}$$

it follows that

$$\beta_i(1-I_W)\, tr\{\widetilde{W}_i^T\widetilde{W}_i\} \leq \beta_i M_W^2. \tag{11.35}$$

Moreover, because of

$$(1-I_W)\frac{\beta_i}{2}M_W^2 \leq \frac{\beta_i}{2}M_W^2, \tag{11.36}$$

inequality (11.33) can be transformed into the following form:

$$\begin{aligned} \dot{V}_i \leq & -k_{0,i}\bar{k}_3^i\left|\frac{\partial v_{0i}}{\partial x_i}\right|^2 - \frac{\beta_i}{2}\left\|\widetilde{W}_i\right\|^2 + \frac{\beta_i}{2}M_W^2 + \beta_i M_W^2 \\ & + \frac{\beta_i}{2}M_W^2 - \frac{\gamma_{1,i}}{2}\widetilde{\theta}_i^2 + \frac{\gamma_{1,i}}{2}\varepsilon^2 + k_{0,i}(\xi_i + \varepsilon_{\rho_i}^2(t)). \end{aligned} \tag{11.37}$$

Hence, making use of (11.21) and (11.22), one obtains

$$\begin{aligned} \dot{V}_i \leq & -\frac{k_{0,i}\bar{k}_3^i\lambda_i^*}{\bar{k}_i}v_{0i}(x_i) - \frac{\beta_i}{2}\left\|\widetilde{W}_i\right\|^2 - \frac{\gamma_{1,i}}{2}\widetilde{\theta}_i^2 \\ & + 2\beta_i M_W^2 + \frac{\gamma_{1,i}}{2}\varepsilon^2 + k_{0,i}(\xi_i + \varepsilon_{\rho_i}^2(t)), \end{aligned} \tag{11.38}$$

and therefore

$$\dot{V}_i \leq -\alpha_i V_i + \mu_i, \quad \dot{V} = \sum_{i=1}^{N}\dot{V}_i \tag{11.39}$$

with

$$\alpha_i = \min\left\{\frac{\bar{k}_3^i\lambda_i^*}{\bar{k}_i}, \beta_i, \gamma_{1,i}\right\}, \tag{11.40a}$$

$$\mu_i = 2\beta_i M_W^2 + \frac{\gamma_{1,i}}{2}\varepsilon^2 + k_{0,i}(\xi_i + \varepsilon_{\rho_i}^2(t)). \tag{11.40b}$$

Integration of both sides of inequality (11.39) yields

$$V_i(t) \leq \frac{\mu_i}{\alpha_i} + \left[V_i(0) - \frac{\mu_i}{\alpha_i}\right] e^{-\alpha_i t}, \forall t \geq 0. \tag{11.41}$$

Thus, it is readily deduced from (11.41) that x, $\theta_i(x)$, $W_i(x)$ are bounded consistently. On the other hand, from (11.24) it is seen that

$$k_{0,i} v_{0,i}(x_i) \leq V_i \tag{11.42}$$

hence for all $i = 1, 2, \ldots, N$ and $\forall t \geq 0$ it is valid that

$$v_{0,i}(x_i) \leq \frac{\mu_i}{k_{0,i}\alpha_i} + \frac{1}{k_{0,i}}\left[V_i(0) - \frac{\mu_i}{\alpha_i}\right] e^{-\alpha_i t}. \tag{11.43}$$

From inequality (11.43), it follows at once that all vector-valued state variables x_i, $i = 1, 2, \ldots, N$, are consistently ultimately bounded on the sets

$$D_i = \left\{ x_i \in R^n : v_{0i}(x) \leq \frac{\mu_i}{k_{0,i}\alpha_i}, \frac{\bar{k}_2^i}{\bar{k}_1^i} \leq k_{0,i} \leq 1 \right\} \tag{11.44}$$

thus defining the attraction domains for all the subsystems, respectively. It is therefore that the original class of composite similarity systems (11.1) under synthesized controls (11.11)–(11.18) are real-stable in the closed loop on the sets D_i defined by (11.44). This completes the proof.

Remark 11.2 The assumption on uncertain boundary on time-varying approximation error $\varepsilon_{\rho_i}(t)$ in this theorem is more appropriate and rational than the previous one proposed in [25]. Hence matrices Γ_{i1}, Γ_{i2}, $k_{0,i}$, β_i, $\gamma_{1,i}$, and M_W are well conceptualized.

11.4 Application to Complex Inverted Pendulums Mechatronic Plant System

In this section, we explore the application of the proposed decentralized control design method, which is essentially based on Theorem 11.1, to the complex mechatronic system constructed by two elastically coupled inverted pendulums on carts [27, 36]. The schematic diagram of this plant is given in Fig. 11.2, which is known to represent an important benchmark plant in the considered class of complex dynamic systems.

The basic differential equations describing the motion of this composite inverted pendulum on cart system are as follows:

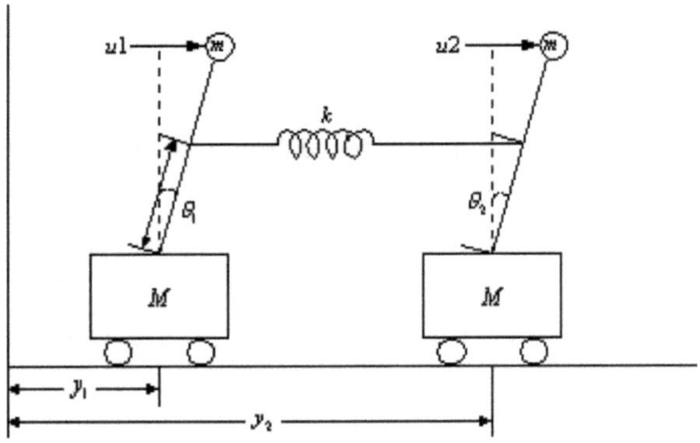

Fig. 11.2 The benchmark plant system of coupled inverted pendulums on carts [27]

$$\dot{\theta}_1 + \beta_1\theta_1^2 - f_1\theta_1 - f_3u_1 - f_2\theta_2 + f_4 = 0,$$
$$\dot{\theta}_2 + \beta_2\theta_2^2 - f_1\theta_2 - f_3u_2 - f_2\theta_1 - f_4 = 0.$$

These can be rewritten in the form of the considered class of complex dynamic systems, namely the uncertainty complex system possessing similarity property, hence Zhang's similarity–symmetry impact. For this purpose, apparently the following state variables can be defined:

$$x_{11} = \theta_1,\ x_{12} = \theta_1,\ x_{21} = \theta_2,\ x_{22} = \theta_2.$$

Then the system dynamics is represented by means of the following model:

$$\begin{aligned}
\dot{x}_{11} &= x_{12},\\
\dot{x}_{12} &= f_1x_{11} + f_3u_1 + f_2x_{21} - (\beta_1x_{12}^2 + f_4),\\
\dot{x}_{21} &= x_{22},\\
\dot{x}_{22} &= f_1x_{21} + f_3u_2 + f_2x_{11} - (\beta_2x_{22}^2 - f_4),
\end{aligned}$$

where the individual coefficients f_i and other coefficients are given by means of the following expressions:

$$\begin{aligned}
&f_1 = g/(c^*l) - f_2, f_3 = 1/(c^*m^*l^2),\\
&f_2 = k^*a(t)^*(a(t) - c^*l)^*f_3,\\
&f_4 = k^*(a(t) - c^*l)^*(z_1 - z_2)^*f_3,\\
&\beta_1 = m/M^*(\sin(x_1)), \beta_2 = m/M^*(\sin(x_3)).
\end{aligned}$$

In these expressions, g represents the gravity constant, M is the mass of the cart, m is the mass of the cycloid ball, L is the normal length of the spring, l is the length of the cycloid bar, z_1, z_2 are, respectively, the distances of the gravity centers of carts 1 and 2 to the origin with zero longitudinal coordinate. The initially perturbed subsystem state vectors were taken as $x_1(0) = [-1 \quad 1]^T$, $x_2(0) = [-1 \quad 1]^T$.

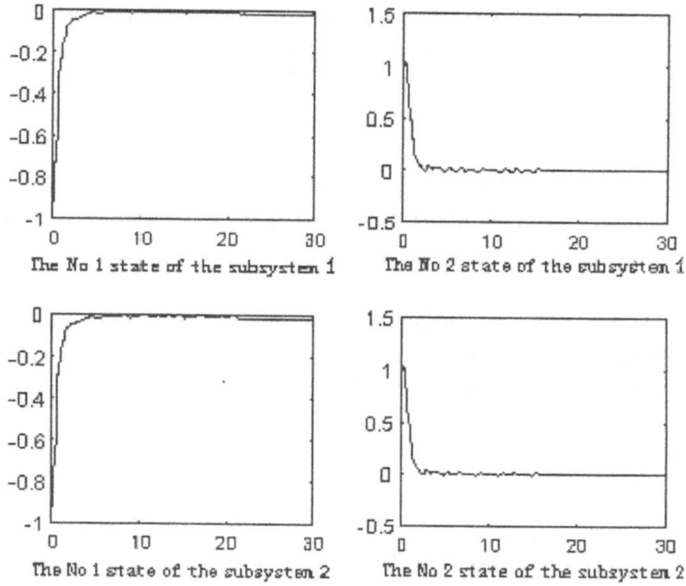

Fig. 11.3 Computer simulation results for the controlled system state variables of the mechatronic benchmark plant system of coupled inverted pendulums on carts

For the sake of comparison of the simulation results, we have chosen the same parameters as in the source reference, namely

$$a(t) = \sin(5t), z_1 = \sin(2t), z_2 = L + \sin(3t),$$
$$k = 1, c = 0.5, M = m = 10, l = 1, L = 2, g = 1.$$

The simulation results are shown as in Fig. 11.3.

Should one choose $\gamma = 40$, $Q = I$, then solving the Riccati equation yields

$$P = \begin{bmatrix} 1.6066 & 0.7906 \\ 0.7906 & 1.2701 \end{bmatrix}.$$

Using formulae of Theorem 11.1, the resulting decentralized controller design is given as follows:

$$u_i^a = -(\gamma + 1)B^T P x_i = -41*[0 \quad 0.2]*Px_i,$$
$$u_i^b = -\xi_i^{-1}\hat{\rho}^2 B^T P x_i = -50*\hat{\rho}^2*[0 \quad 0.2]*Px_i,$$
$$u_i^c = \frac{B^T W_i s(x)}{\lambda_i[1 + \|B\|^2]} = \frac{[0 \quad 0.2]\begin{bmatrix} W_{11} \\ W_{12} \end{bmatrix} s(x_{12})s(x_{22})}{0.08*[1 + (0.2)^2]},$$

where the values of crucial design parameters are $\xi_i = 0.02$, $\lambda_i = 0.08$.

For evaluating the approximated quantity, $\widehat{\rho}_i^2$, we made use of an ANN with three layers each one having six nodes. For emulating the function, $h_i(x)$, we made use of a network with three layers each having two nodes. In addition, the resulting learning law is

$$\begin{aligned}\hat{\widehat{z}}_i &= -\Gamma_1\widehat{z}_i + \Gamma_2\sigma(x_i)\left\|B^T P x_i\right\| \\ &= -0.02{*}\widehat{z}_i + 100{*}\sigma(x_i){*}\|[0 \quad 0.2]Px_i\|, \\ W_{i,j} &= 2k_{0.i}P_i^T x_i s(x) = 2{*}0.005{*}P_i^T x_i s(x_{12})s(x_{22}).\end{aligned}$$

Additional computer simulations were carried out using some academic examples of mathematically constructed composite systems representing the respective 'plants'. For instance, one such academic example [36] is

$$\dot{x}_1 = \begin{bmatrix} -1 & 1 \\ 0 & 0 \end{bmatrix} x_1 + \begin{bmatrix} 0 \\ 1 \end{bmatrix} \left[u_1 + x_{11}^2 e^{x_{11}x_{12}} \cos 10t + x_{12}e^{x_{11}} \sin 10t\right] + 5\begin{pmatrix} \cos x_{21} \\ \cos x_{22} \end{pmatrix},$$

$$\dot{x}_2 = \begin{bmatrix} -1 & 1 \\ 0 & 0 \end{bmatrix} x_2 + \begin{bmatrix} 0 \\ 1 \end{bmatrix} \left[u_2 + x_{21}^2 e^{x_{21}x_{22}} \cos 10t + x_{22}e^{x_{21}} \sin 10t\right] + 5\begin{pmatrix} \cos x_{11} \\ \cos x_{12} \end{pmatrix}.$$

All these demonstrated that the claims of the designed control system in Theorem 11.1 are satisfied to the full with no transient oscillations in the states. Though occasionally, some extremely constrained 'chattering' may occur locally for some finite period of time due to adaptive emulation of interconnecting terms before the system finally reaches its steady state.

It may well be seen from the example that the overall system, employing the designed adaptive controller, exhibits high-quality performance in terms of both fast response, no overshooting, and no transient oscillations. The latter seems to be an important intrinsic property of the control design derived, not guaranteed previously to the best of our knowledge. Theorem 11.1 guarantees the real stability and unifom ultimate boundedness of all state vectors x_i on the respective sets D_i, defined by Eq. (11.13), and also convergency of the on line learning ANN approximators.

11.5 Conclusions

A novel robust adaptive, neural-network based, control solution is presented for complex nonlinear mechatronic systems possessing uncertainties in both subsystem gains and interconnections provided the plant has a similarity structure. The low-order neural networks and the high-order one are banded together to resolve the uncertainty issues on both the input subsystem gains and the interconnections. Therefore, the controllers designed using our technique possess robustness and adaptability.

Two steps are adopted in solving the control design problem. The first step makes use of low-order artificial neural nets to design decentralized controllers in order to make the nominal subsystems real-stable. In this step, due to on line adaptation of the ANN weight, little a priori knowledge is needed and yet the transient performance of the system is considerably improved. The second step, in fact, makes use of the first step as an assumption and employs high-order artificial neural nets to handle the interconnections term of the system.

Acknowledgments Yuanwei Jing and Georgi M. Dimirovski express their gratitude to Prof. Si-Ying Zhang, Academician of Academia Sinica, and to late Prof. Miomir K. Vukobratovic, Academician of Serbian Academy or Arts & Sciences, respectively, for their long-lasting cooperation. This is dedicated to these great scientists and teachers.

References

1. Astrom, K.J., Albertos, P., Blanke, M., Isidori, A., Shaufelbreger, W., Sanz, R. (eds.): Control of Complex Systems. Springer, London (2001)
2. Bakule, M.: Decentralized control: an overview. Annu. Rev. Control **32**, 87–98 (2008)
3. Dimirovski, G.M., Andreeski, C.J.: A memory ANN computing structure for nonlinear system emulation identification. IU J. Electr. Electron. Eng. **3**(2), 905–915 (2003)
4. Dimirovski, G.M., Gough, N.E., Barnett, S.: Categories in systems and control theory. Int. J. Syst. Sci. **7**(9), 1081–1090 (1977)
5. Chen, B., Liu, X., Liu, K., Lin, C.: Direct adaptive fuzzy control for nonlinear systems with time-varying delays. Inf. Sci. **180**, 776–792 (2010)
6. Ferrari, S., Stengel, R.F.: Smooth function approximation using neural networks. IEEE Trans. Neural Netw. **16**, 24–38 (2005)
7. Gao, W.B.: Nonlinear Control System Guidance. Science Publication, Singapore (1988)
8. Gao, Z.S., Wang, Y.L.: Electromechanical Control Engineering. Qing-hua University Publication 9, Qing-hua (1994)
9. Ge, S.S., Wang, C.: Adaptive NN control of uncertain nonlinear pure-feedback systems. Automatica **38**, 671–682 (2002)
10. Haykin, S.: Neural Networks and Learning Machines, 3rd edn. Pearson Education Inc, Upper Saddle River (2009)
11. Heht-Nielsen, R.: Kolmogorov's mapping neural network existence theorem. In: Proceedings of IEEE Joint Conference on Neural Networks, vol. 3, pp. 11–14. The IEEE, New York, NY (1987)
12. Jiang, B., Liu, X.P., Zhang, S.Y.: The output tracking of nonlinear composite system with similar structure. Inf. Control **24**, 65–69 (1995)
13. Jing, Y.-W., Dimirovski, G.M.: Comparative analysis of Kolmogorov ANN and process characteristic input-output modes. IU J. Electr. Electron. Eng. **3**(1), 745–758 (2003)
14. Jing, Y., Zhang, Y. Kolemisevska-Gugulovska, T.D., Dimirovski, G.M., Vukobratovic, M.K.: Robust adaptive control of complex systems employing ANN emulation of the nonlinear plant models. Facta Univeristatis—Series Mechanics, Automatic Control & Robotics, Special Issue 2007, vol. 6, pp. 25–38 (2008)
15. Katsuura, H., Sprecher, D.A.: Computational aspects of Kolmogorov's theorem. Neural Netw. **7**, 455–461 (1994)
16. Koo, G.B., Park, J.B., Joo, Y.H.: Decentralized fuzzy-observer based output-feedback control for nonlinear large-scale systems: an LMI approach. IEEE Trans. Fuzzy Syst. **22**(2), 406–419 (2014)

17. Kolmogorov, A.N.: On representation of continuous functions of several variables by superposition of continuous functions of one variable and addition. Dokladi Akademii Nauk SSSR **114**, 953–956 (1957) (in Russian)
18. Kosmatopoulos, E.B., Polycarpou, M.M., Christodoulu, M.A., Ioannou, P.A.: High-order neural-network structures for identification of dynamical systems. IEEE Trans. Neural Netw. **6** (2), 422–431 (1995)
19. Kurkova, V.: Kolmogorov's theorem is relevant. Neural Comput. **3**, 617–622 (1991)
20. Kurkova, V.: Kolmogorov's theorem and multi-layer neural networks. Neural Netw. **5**, 501–506 (1992)
21. Liu, F.L.: Decentralized adaptive output control for uncertain similar composite system. Doctoral thesis, Northeastern University, Shenyang, CN (1999)
22. Liu, Y., Jia, Y.M.: Adaptive H_∞ control for a class of non-linear systems using neural networks. IET Control Theory Appl. **3**, 813–822 (2009)
23. Narendra, K.S., Annaswamy, A.M.: Stable Adaptive Systems. Prentice Hall, Englewood Cliffs (1989)
24. Ossman, K.A.: Indirect adaptive control for interconnected systems. IEEE Trans. Autom. Control **34**, 908–911 (1989)
25. Qu, Z.: Attenuation of nonlinearly state-dependent uncertainties: robust control design and its application to robotic manipulators. Int. J. Control **63**, 27–40 (1996)
26. Park, J., Sandberg, I.W.: Universal approximation using radial basis-function networks. Neural Comput. **3**(2), 246–257 (1991)
27. Siljak, D.D.: Decentralized Control of Complex Systems. Academic Press, Boston, MA (1991)
28. Siljak, D.D., Zecevic, A.I.: Control of large-scale systems: beyond decentralized feedback. Annu. Rev. Control **29**, 167–179 (2005)
29. Siljak, D.D.: Dynamic graphs. Nonlinear Anal. Hybrid Syst. **2**, 544–567 (2008)
30. Tong, S.C., Li, Y.M., Zhang, H.G.: Adaptive neural network decentralized backstepping output feedback control for nonlinear large-scale systems with time delays. IEEE Trans. Neural Netw. **22**, 1073–1086 (2005)
31. Yan, X.-G., Wang, J.-C., Lu, X.-Y., Zhang, S.-Y.: Decentralized output feedback robust stabilization for a class of nonlinear composite systems possessing similar subsystems. Presented at the 35th IEEE Conference on Decision and Control, Kobe, JP (1996), (private communication of the extended manuscript)
32. Yang, Y., Zhou, C.: Adaptive fuzzy H_∞ stabilization for strict-feedback canonical nonlinear systems via backstepping and small-gain approach. IEEE Trans. Autom. Control **13**, 104–114 (2005)
33. Zainuddin, Z., Pauline, O.: Function approximation using artificial neural networks. WSEAS Trans. Math. **7**, 333–338 (2008)
34. Zhang, S.-Y.: The structures of symmetry and similarity of complex control systems. Control Theory Appl. **11**, 231–237 (1994)
35. Zhang, T., Ge, S.S., Hang, C.C.: Adaptive neural network control for strict feedback nonlinear systems using backstepping design. Automatica **36**, 1835–1846 (2000)
36. Zhang, Y.-W.: Robust adaptive control of a class of composite systems using high-order neural networks. Doctoral thesis. Northeastern University, Shenyang, CN (2000)
37. Zhang, Y.-X., Jing, Y.-W., Zhang, S.-Y., Dimirovski, G.M.: Robust adaptive control for a class of uncertain nonlinear composite similarity systems using neural networks. In: Dimirovski, G.M. (ed.) Automatic Systems for Building the Infrastructure in Developing Countries, pp. 69–74. Pergamon Elsevier Science, Oxford (2001)
38. Zhou, D., Mu, C.D., Xu, W.L.: Composite control of space interceptor gesture. Qing-hua Univ. J. **39**, 28–33 (1999)
39. Zhu, R.-J., Chai, T.-Y., Fu, J.-D.: Robust control of uncertain nonlinear system using neural networks. Control Decis. **14**, 73–76 (1999)

Chapter 12
Neural Networks with Strong Anticipation and Some Related Problems of Complexity Theory

Oleksandr S. Makarenko

Abstract We proposed and realized one new type models of neural networks, which takes into account property of anticipation. As the base model, the Hopfield type models with anticipation have been explored. The basic new qualities, discovered at research there is that possible multi-valued solutions of given neural networks. Different types of behaviour of such systems have been explored depending on parameters of networks. Some problems of self-organized behaviour are proposed. The problems of complex solutions and stored information have been considered, including the measures of complexity in deterministic and non-deterministic cases. Presumable applications of such models for living and social systems are discussed.

12.1 Introduction

Recently artificial neural networks (ANN) are well known and recognized tools for data processing, mathematical modelling and artificial intelligence (see for example [1–4]). Since the 50 years of twentieth century many ways for ANN development had been proposed: back-propagation [1]; Hopfield type neuronets [1, 2, 5, 6]; different architectures of ANN [1]; networks with changing structures including networks with delay property [1, 2, 7, 8]; cellular neural networks [9]; neuronets with complex values [10]; neuronets with non-smooth activation functions [11, 12] and many others. Usually, the main sources of developments of ANN concepts were the real problems from different fields of investigations: physics, technique, economy, biology with different types of new systems and phenomena.

O.S. Makarenko (✉)
Institute for Applied System Analysis, National Technical University of Ukraine (KPI), 37 Prospect Peremogy, Kyiv 03056, Ukraine
e-mail: makalex@i.com.ua

G.M. Dimirovski (ed.), *Complex Systems*, Studies in Systems, Decision and Control 55, DOI 10.1007/978-3-319-28860-4_12

Now rather new phenomena are under intensive investigations—namely anticipatory property (see [13–18], papers at ten conferences on computing anticipatory systems CASYS 1999–2011; investigations on anticipation at social and economical systems [13, 14]).

It is known well, that the general systems with anticipating have large prospects both in theoretical, and in the applied plan. But for subsequent development of the topic large number of examples of the concrete dynamic systems is necessary. As it was indicated in previous works of author (O. Makarenko) very perspective are neuronets with anticipating elements, that correspond to the models of individual [13, 14].

Remember (because it is important for ANN with anticipation property) that earlier a discrete map as the model with anticipation had been used for money-economic processes) [13–15]. In this paper [15], two-step in time discrete model with anticipating had been proposed for research on the complex systems, especially related to economy. The transition function $f(x)$ in paper [15] has a piece-linear character and belongs to the class of neuron responses function. Piece-linear character of function allowed making the thorough numerical–analytical analysis. During the analysis of such maps the origin of multi-valued transitions had been found. Also a new type of significant behaviour in [15] had been found, when the region of multivaluedness is localized in phase space.

So the anticipatory phenomenon and systems with anticipation attracts more and more attention. One of the directions of such objects investigation is connected with using of neural networks models. Multivaluednes of solutions as the main new peculiarities had been found. It follows to the possibilities of familiar behaviour in ANN with anticipation property. Thus, such presumable properties (multivaluedness of solution) should be accounting in ANN design from the beginning.

In this paper, we propose to explore the models of neuron network with anticipation. First we propose the general structure of ANN with strong anticipation. Second we propose as example the counterpart with anticipation to common Hopfield and also some numerical illustration of presumable behaviour of such objects. And finally, we discuss some future research task especially taking into account presumable behaviour of such ANN solutions.

12.2 Strong Anticipation Property

The term ‘anticipation’ is introduced and used in biology and applied mathematics by Robert Rosen [16]. This name is well known for common researchers. Less known is contribution by Daniel Dubois to strong anticipation.

Since the beginning of 1990s in the works of Dubois—see [17, 18], the idea of strong anticipation had been introduced: “Definition of an incursive discrete strong anticipatory system …: an incursive discrete system is a system which computes its current state at time t, as a function of its states at past times , $\ldots, t-3, t-2, t-1$, present time, t, and even its states at future times $t+1, t+2, t+3, \ldots$

$$x(t+1) = A(\ldots, x(t-2), x(t-1), x(t), x(t+1), x(t+2), \ldots, p), \tag{12.1}$$

where the variable x at future times $t+1, t+2, t+3, \ldots$ is computed in using the equation itself.

Definition of an incursive discrete weak anticipatory system: an incursive discrete system is a system which computes its current state at time t, as a function of its states at past times , $\ldots, t-3, t-2, t-1$, present time, t, and even its predicted states at future times $t+1, t+2, t+3, \ldots$

$$x(t+1) = A(\ldots, x(t-2), x(t-1), x(t), x^*(t+1), x^*(t+2), \ldots, p), \tag{12.2}$$

where the variable x^* at future times $t+1, t+2, t+3, \ldots$ are computed in using the predictive model of the system" [18].

Following the description of anticipatory properties in this subsection, we can propose some new and important problems in the field of artificial neural network, especially in case of strong anticipation accounting. Here first we remember some well-known facts on common artificial networks [1–7]. For the goals of this paper, we will focus our description around the Hopfield type networks but the transition to other neuronets is more or less evident.

The description of general schemes for such ANN includes the next components: ANN has many elements (artificial neurons); each element has single state from some set of states; each pears of elements has bond between element with some single value from some set; the evolution of element's state takes part at continuous or discrete time; special dynamical rules for evolution of elements exist; special learning rules exist for establishing bonds between elements (Hebb type rules) from previous learning states of ANN.

Let us take that ANN system consisting of $N \gg 1$ elements and each element is characterized by state $s_i \in M, i = 1, 2, \ldots, N$, where M is a set of possible values for s_i. There are many possibilities to compose the elements in blocks and levels in such models. In sufficiently developed system, elements have many complex connections. Let us formalize this. We assume that there are connections J_{ij} between i and j elements. Thus, the set $Q = (\{s_i\}, \{J_{ij}\}, \ i,j = 1, \ldots, N)$ characterizes state of neural system. From the analysis of recent neuronet models, the rather general ANN with discrete time gave the form

$$s_i(t+1) = G(\{s_i(t)\}, \{J_{ij}\}, R), t = 0, 1, 2, \ldots, \tag{12.3}$$

where G is the so-called activation function, R–some parameters. Usually, [1–6] the activation function in ANN has sigmoid form and the dynamics of ANN elements is accepted as depending on mean field of influence

$$h_i(t) = \sum_j J_{ij} s_j(t), j = 1, 2, \ldots, N: \tag{12.4}$$

$$s_i(t+1) = G(h_i(t), R), t = 0, 1, 2, \ldots, \tag{12.5}$$

On of the main examples of such type models is Hopfield type neuronet. Other rather general type of neural networks are the networks of Hopfield type with continuous time case [1–6], see for illustration

$$\begin{cases} C_j \frac{\mathrm{d}v_j(t)}{Et} &= -\frac{v_j(t)}{R_j} + \sum_{i=1}^{N} w_{ji}(t)\varphi_i(v_i(t)) + I_j; \quad j = 1, \ldots, N, \\ \frac{\mathrm{d}w_{lk}(t)}{\mathrm{d}t} &= \lambda\varphi_l(v_l(t))\varphi_k(v_k(t)) - \gamma w_{lk}(t), k \neq l; \quad k, l = 1, \ldots, N. \end{cases} \tag{12.6}$$

Here for illustration we depict the Hopfield presumable dependence on time also for bonds w_{lk} between elements of neuronet.

As we had argued above one of the further research problem in the field of neural networks should be considered the neural system with strong anticipation. The first common example for neural networks investigations is the set of couple neurons with strong anticipation with one forward discrete step. In this case, the general form of ANN is the next:

$$s_i(t+1) = G(\{s_i(t)\}, \{s_i(t+1)\}, \{J_{ij}\}, R), t = 0, 1, 2, \ldots. \tag{12.7}$$

The counterpart to ANN (12.6) with continuous time then has the form

$$\begin{cases} C_j \frac{\mathrm{d}v_j(t)}{\mathrm{d}t} = -\frac{v_j(t)}{R_j} + (1-\alpha)\sum_{i=1}^{N} w_{ji}(t)\varphi_i(v_i(t)) \\ \qquad + \alpha \sum_{i=1}^{N} w_{ji}(t+\tau)\varphi_i(v_i(t+\tau)) + I_j; j = 1, \ldots, N, \\ \frac{\mathrm{d}w_{lk}(t)}{\mathrm{d}t} = (1-\alpha)(\lambda\varphi_l(v_l(t))\varphi_k(v_k(t)) - \gamma w_{lk}(t)) \\ \qquad + \alpha(\lambda\varphi_l(v_l(t+\tau))\varphi_k(v_k(t+\tau)) - \gamma w_{lk}(t+\tau)), \\ k \neq l; \quad k, l = 1, \ldots, N. \end{cases} \tag{12.8}$$

In Eq. (12.8) τ is time of anticipation and $\alpha \in [0, 1]$ is weight of anticipation ($\alpha = 0$ corresponds to the case of anticipation absence). The Eqs. (12.7) and (12.8) are rather new objects (especially in ANN science) and are difficult objects for investigation, especially because of presumable multivaluedness of the solutions. So first it is useful considering the simplest examples of such ANN with strong anticipation. So in next section of the paper, we propose the results of investigation of simplest but rather important ANN—Hopfield type model with anticipation.

12.3 Neural Network Example with Anticipation

Below we propose illustration of properties of ANN with strong anticipation and the description of counterpart of classical Hopfield models. Remark that other ANN with anticipation has the same behaviour. Let us consider first the recurrent network model proposed by Hopfield [1–6]. The 'pattern' of the states of the Hopfield's type networks at given model of time is presented at Fig. 12.1 below. It consists of a layer of elements, whose number is the number of units with inputs and outputs of the network. Each neuron has connections with all of N other neurons.

12.3.1 Classical Hopfield Model

In classical Hopfield model, each element of neuronet takes two values: {0, 1} and activation function has smooth sigmoid form [5, 6]. In modification of Hopfields model, the elements take values from unit interval $I \in [0, 1]$ ('grey' networks). In this paper we take the 'grey' case of element states. Also for the sake of easy solution of nonlinear equations of ANN we take the piecewise linear activation functions.

So in our case the state s_i of ith elements of network belong to the unit interval I. The connections between neurons set matrix J, with elements J_{ij}, where $i,j = 1, \ldots, N$. For each neuron, the mean field is determined by the size of the external field interaction h_i (Eq. 12.4 from Sect. 12.2):

The evolutionary process that occurs in such network is described by the following recurrence relation:

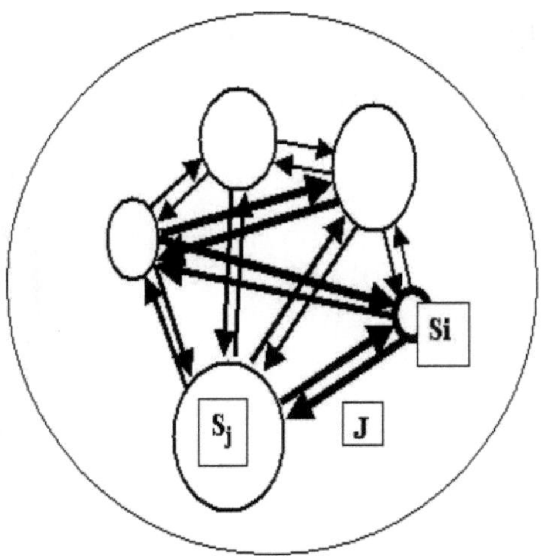

Fig. 12.1 'Pattern' of the classical Hopfield's type network

$$s_j(t+1) = f(h_j) = f\left(\sum J_{ji} s_i(n)\right), \tag{12.9}$$

where we use $y = f(x)$—piecewise linear activation function:

$$\begin{cases} f(x) = 0, x \leq 0; \\ f(x) = x, x \in (0, 1]; \\ f(x) = 1, x > 1; \end{cases} \tag{12.10}$$

12.3.2 Neural Network with Anticipation

In the proposed simple model with the anticipation, we took the influence function as the sum of current influence and virtual future influence from next future moment of time (Fig. 12.2).

As the first objects for investigation we can propose two presumable forms of such dependence:

$$h_j(t) = \sum_{i=1}^{N} J_{ji} s_i(t) + \alpha \sum_{i=1}^{N} J_{ji} s_i(t+1) \tag{12.11}$$

or

$$h_j(t) = (1 - \alpha) \sum_{i=1}^{N} J_{ji} s_i(t) + \alpha \sum_{i=1}^{N} J_{ji} s_i(t+1). \tag{12.12}$$

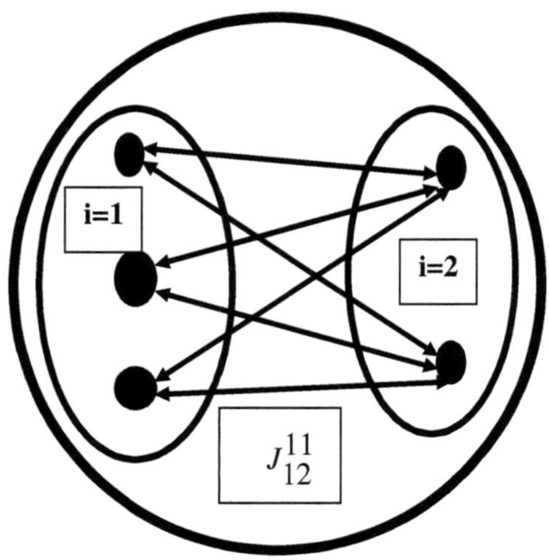

Fig. 12.2 'Pattern' for fixed moment of time for Hopfield's network with anticipation

The coefficient α corresponds to the anticipation accounting. When $\alpha = 0$ we have the case of usual Hopfield network. In case of Eqs. (12.11) and (12.12), the recurrence relations (12.5) of Hopfield's model transform to the next forms after the introducing formulas (12.7) and (12.8):

$$s_j(t+1) = f\left(\sum_{i=1}^{N} J_{ji} s_i(t) + \alpha \sum_{i=1}^{N} J_{ji} s_i(t+1)\right) \tag{12.13}$$

or

$$s_j(t+1) = f\left((1-\alpha)\sum_{i=1}^{N} J_{ji} s_i(t) + \alpha \sum_{i=1}^{N} J_{ji} s_i(t+1)\right). \tag{12.14}$$

Because of piecewise linear character of activation functions, we receive in our case the resulting system of linear equations for unknowns $\{s_i\}$ for each linear interval of activation function, and upshot of which are outputs of neural layer. Iteration process will be supplied by one separately with each output of the previous layer. So as the result we receive the expanding iterative process.

Thus as $f(x)$ is piecewise linear sigmoid function, we examined the case of three different possible values for solutions of Eqs. (12.13) and (12.14). After solving each of these equations it is checked whether the solution is suitable for implementation of this system.

12.4 Numerical Realization of Simple Model with Anticipation

12.4.1 Calculation Results

In this section, we give some results of computing investigations of the models from previous section with anticipation. The mail goal of this subsection is not the detailed investigation of such models (such investigations will be reported at further publications) but illustration presumable variety of ANN with anticipation behaviour. Below we proposed the pictures of solutions of ANN with for 2-, 6- and 8- with different conditions for cases one-valued and multi-valued solutions.

Two coupled neurons. When the coefficient of anticipation α is near the 0, then the behaviour of such system of two coupled elements is single-valued and remembers the behaviour without anticipation. But for large deviation of α from zero the types of regimes become reach; in a network with two neurons we have such modes, as stability is single fixed point, stable single-valued cycle,

Fig. 12.3 Example of multi-valued cycle in two-neuron system ($\alpha = -7$)

multi-valued stability and multi-valued cycle. In Fig. 12.3, we show the multi-valued cycle. These modes are set with a large negative coefficient of anticipation.

Here each of the horizontal plates correspond some future discrete moments of time (not zero), two points with arrows correspond to two neurons and each of the separate arrows correspond to separate state of neuron at the same time moment. The length of the arrow corresponds to the amplitude of given value of neuron state. We can see repeating states with many values after some discrete time steps.

In Figs. 12.4 and 12.5, the values of the dynamics of each neuron are shown. Horizontal axe corresponds to discrete moments of time. The values of neuron's state are represented at vertical axe.

We can see many-valued states in Figs. 12.4 and 12.5; branches of the solutions and termination of solution branches.

Case of six coupled neurons. When functioning network with many neurons, there are also modes such as increasing the number of solutions branches,

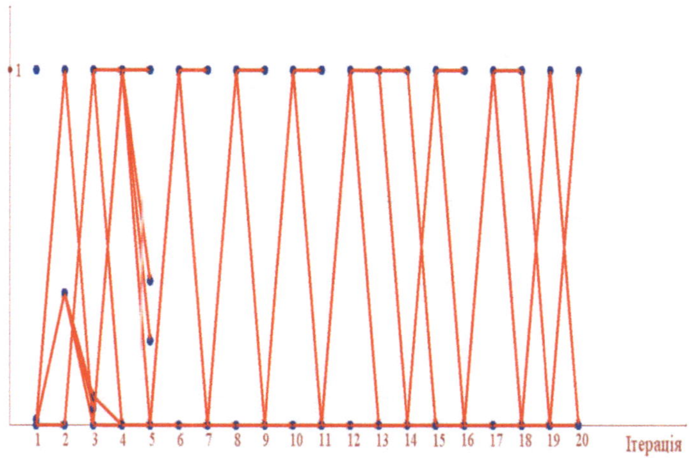

Fig. 12.4 First neuron dynamics

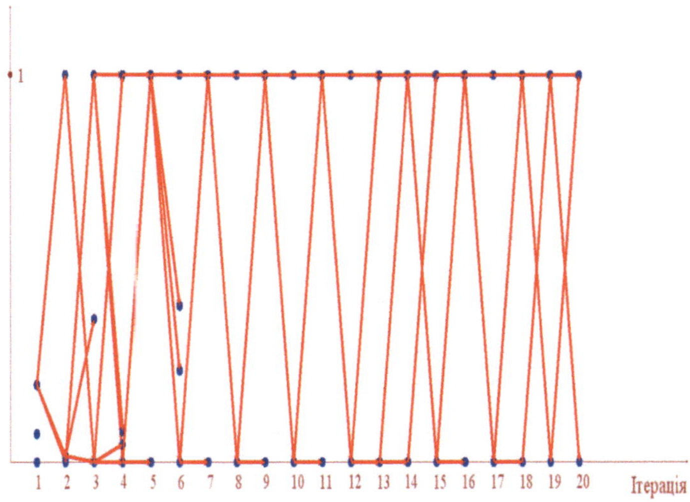

Fig. 12.5 Second neuron dynamics

single-valued cycles of solution, multi-valued cycles and increasing of multiplicity. Consider the following case (Fig. 12.6).

Let us represent the examples of the dynamic behaviour of some of neurons (Fig. 12.7):

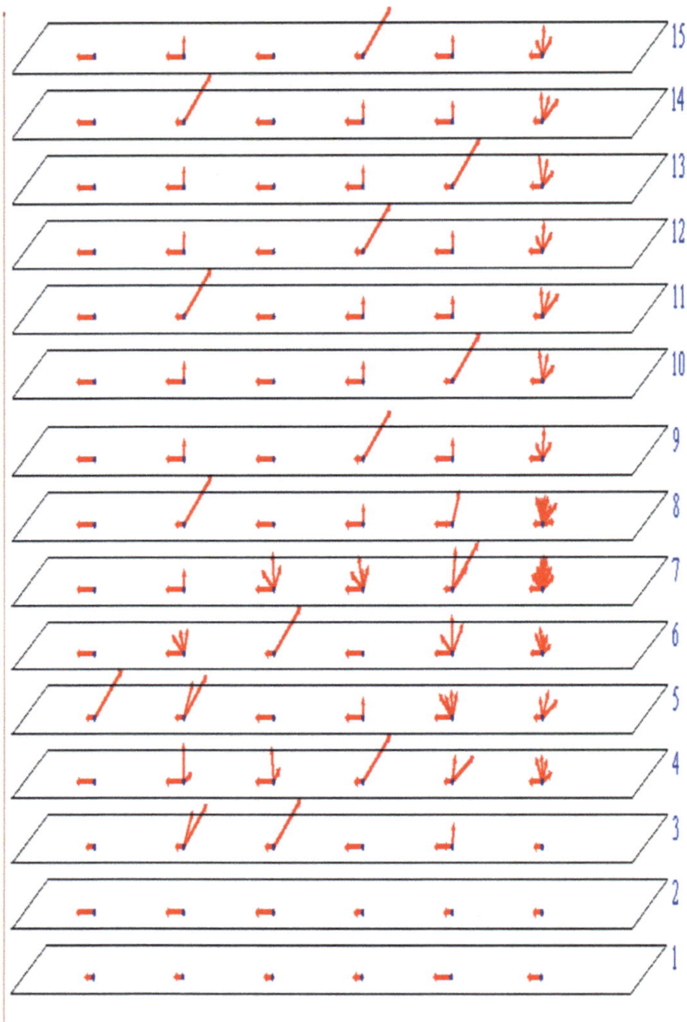

Fig. 12.6 Multi-valued cyclic behaviour and increasing of multiplicity ($\alpha = -0.7$)

in Fig. 12.8, we see the increase of multiplicity and accompanying termination of some branches. Also very important is the possibility of non-homogeneous behaviour for different neurons.

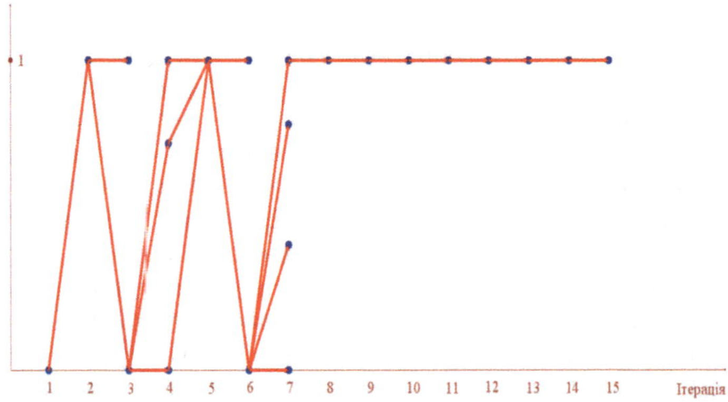

Fig. 12.7 Third neuron dynamic behaviour

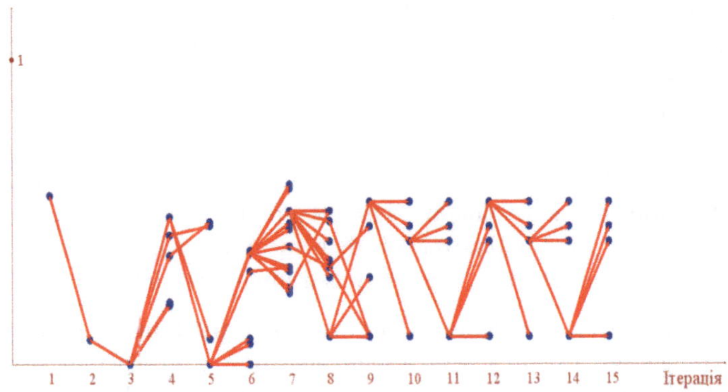

Fig. 12.8 Sixth neuron dynamic behaviour

12.5 Mathematical and Applied Problems for Future Investigations

Mathematical problems for consideration complex regimes. The detailed consideration of such multi-valued dynamical systems may constitute the subject of further mathematical investigations. Here we only pose some general comments. First, because the map acts on the multi-valued functions of [0, 1] then the dynamical system is infinite-dimensional.

Very important is the asymptotic solutions, when $t \to \infty$. For single-valued case (without anticipation) by A. Sharkovsky and other it has been found the existence of so-called relaxation oscillation regimes when the solution tends to the function with finite number of discontinuities at fix length time interval and so-called turbulent solutions when the number of discontinuities tends on fix length interval tends to infinity as $t \to \infty$. Some limit objects for such solution may have fractal structure. Moreover, it has been proved that the limiting regime is multi-valued function for which some probability distribution corresponds. And finally, very interesting for such problems is the structure of attractors in single-valued case. It had been found that attractors are infinite-dimensional and contained generalized functions [19]. The attractors are with simple dynamics: "… the movement on attractors is rather simple—periodical or almost periodical, also the elements of attractor are functions acting from $I = [0, 1]$ into I, may be very complex (for example with cantorian set of multivaluedness points" [19].

Based on our results, it may be proposed some generalizations for cases with accounting of anticipating. First as we described above, the solutions in such case may be multi-valued. So the fractal dimension of solutions may be bigger then that in single-valued case. Moreover, the behaviour of the solutions may be much more complex. Different regimes on the different branches of solutions may exist (some examples of such different branches see in [20]. So each branch may have different stochastic properties, different types of relaxation behaviour, etc. But may be the most interesting is the common behaviour of mixture of the branches in multi-valued case. Especially, interesting is the problem of limiting behaviour as $t \to \infty$: limit solutions, attractors of such solutions and probability properties of limiting objects. But also the new problems arrows: observability of solutions, selection the single-valued trajectories of the system, its limiting objects complexity and such complexity measures, searching adequate mathematical spaces for considering the problems and solutions. Just the problems of adequate definition of periodic behaviour (and moreover of dynamical chaos), Lemeray's staircase, Poincare's bifurcation diagram, ergodicity and mixing are interesting.

Specific mathematical problems for network investigations. Considering in this paper, investigations of ANN with anticipation pose a number of new research problems including mathematical. First such class of problems constitutes the correct mathematical description of such ANN. Because of presumable multivaluedness of the solutions in such case as multi-valued dynamical rules as investigations of corresponding functional—analogy to the 'energy' functional in classical neuronets should be considered. Remark that differential and difference inclusions may be useful. In such case set valued analysis should replace the classical calculus. The problems of attractors in ANN received new aspects because of multivaluedness. Also presumable case of multiple-valued bonds between elements brings the new possibilities. Remark that in such case the pictures which may correspond to the dynamics of such case looks like web on connections between 'shaggy' elements (elements covered by 'cloud' of virtual states). A very important research problem is learning problem. In given paper, we realized the variant of ANN with single-valued (classical) bonds between elements. Classical learning rules may correspond to such particular case. However, just such case may be interesting for applications because after such usual learning of single-valued bonds from single-valued time series measuring in practice we can use such single-valued bonds for evaluation of some part presumable branches of considered ANN with anticipation. But of course much more possibilities for investigations and applications open the case of multi-valued bonds between states including increasing of ANN storage capacity.

Some new research direction outlines. The accounting of anticipation in neural networks follows not only to the new properties of network solutions but also to new views on their interpretations. Here we briefly remark some possibilities (some papers with more detailed description are in preparation processes). Before in a series of the papers, we had considered the properties of social systems as the first examples of such interpretations. The examples are geopolitical problems, general properties of the society, pedestrian crowd elements, stock market behaviour, etc. But new networks models with anticipation property may also be interesting for other research fields where the neuronet models already have been used.

One of such fields is the investigation of the processes of neural activity in the brain. That is in this the artificial neural networks with the anticipation may revised to the original problems [21]. Note that recently the anticipatory property of intelligent organisms including human had been discussed in different contents (as in experimental as in theoretical). But a number of the publications are small and topic is very new for neuroscience. It is appeared that considering of neural networks with anticipation may help in understanding one of the most intriguing contemporary problems namely the problem of consciousness. Recently, many different concepts of conscious exist: computational neuroscience based on the classical neuronets; holographic theory; quantum ensembles theory; electromagnetic theory; extra-dimensions theory; microtubule orchestra; nonlocal quantum theory; p-adic theory, etc. According proposed models, the conscious may be related to the properties of the real brain networks. The neurons in the brain may have some amount of anticipatory property. The brain networks may produce multiple states. The act of consciousness correlate with the choice of single states for each involved neurons. External experiment allows receiving only the single-valued states. Note that the behaviour of such models closely remembers the behaviour of pure quantum systems where potentially many possibilities exist through the wave function and measurement reduces the system to definite single state. Also some analogies with causal nodes networks quantum description exist Thus the concepts of information processing in the brain and of consciousness should re-considered on the base of 'multiple states' considerations. As one of the consequences are the needs in considering non-turing machines in artificial intelligence theories.

12.6 Conclusions and Further Lines of Investigations

Thus, in this work suggested and implemented one of the models of neural network which takes into account some properties that are present in living systems. Feature of networks is that they put an additive way ahead by one step. The basic principle new system identified in the study is that possible multi-valued solution of the given neural network. This property opens up a whole new area of research of neural network models. Different types of behaviours such systems to analyze the dynamics of systems depending on parameters.

A study of similar models there are some problems: how to interpret the results of the online application mode network, optimization of the network, etc.

Possible areas of research are input for each layer of network decision-making system that would chose the one output layer, rejecting all others. Continuation of the given model may explore the possibility of increasing the memory at fixed network of neurons. Pattern recognition of images can give a result not one closest image, and a few with some probabilities to calculate the probability is also maybe a problem for research. Note that considered problem arose from considering of network models with [13, 14], where given the possible interpretation of these large models for socio-economical in the system.

However, more interesting and promising is a question of interpretation the solutions and their analogues in living systems. Thus in [20] a range of problems and their possible solutions was proposed, namely the problem of consciousness, new principles of computing, communications with the quantum-mechanical description, non-Turing computers, cellular automata and systems and many others.

Acknowledgements The author is gratitude for A. Yatsuk and V. Biluga for the help in numerical calculations.

References

1. Haykin, S.: Neural Networks. A Comprehensive Foundation, 2nd edn. Prentice Hall, New Jersey (1999)
2. Hertz, J., Krogh, A., Palmer, R.G.: Introduction to the Theory of Neural Computation. Addison-Wesley, Reading (1991)
3. Sipser, M.: Introduction to the Theory of Computation, 2nd edn. Thomson Course Technology, USA (2006)
4. Aubin, J.-P.: Neural Networks and Qualitative Physics. Cambridge University Press, Cambridge (1996)
5. Hopfield, J.J.: Neural networks and physical systems with emergent collective computational abilities. Proc. Nat. Acad. Sci. U.S.A **79**(8), 2554–2558 (1982)
6. Hopfield, J.J.: Neurons with graded response have collective computational properties like those of two-state neurons. Proc. Nat. Acad. Sci. U.S.A **79**(8), 2554–2558 (1984)
7. Sutton, J.P., Beis, J.S., Trainor, L.E.H.: Hierarchical model of memory and memory loss. J. Phys. A: Math. Gen. **21**, 4443–4454 (1988)
8. Guyon, I., Personnaz, L., Nadal, J.P., Dreyfus, G.: Storage and retrieval of complex sequences in neural networks. Phys. Rev. A **38**(12), 6365–6372 (1988)
9. Chua, L.O., Yang, L.: Cellular neural networks: Theory. IEEE Trans. Circuits Syst. I **35**, 1257 −1272 (1988)
10. Jankowski, S., Lozowski, A., Zarada, J.M.: Complex-valued multistate neural associative memory. IEEE Trans. Neural Netw. **6**(6), 1491–1496 (1996)
11. Forti, M., Grazzini, M., Nistri, P., Pancioni, L.: Generalized Lyapunov approach for convergence of neural networks with discontinuous or non–lipshitz activations. Physica D **214**, 88–99 (2006)

12. Forti, M., Nistri, P.: Global convergence of neural networks with discontinuous neuron activation. IEEE Trans. Circuits Syst.—I Fundam. Theory Appl. **50** (11), 1421—1435 (2003)
13. Makarenko, A.: Anticipating in modeling of large social systems—neuronets with internal structure and multivaluedness. Int. J Comput. Anticipatory Syst. **13**, 77–92 (2002)
14. Makarenko, A.: Anticipatory agents, scenarios approach in decision-making and some quantum–mechanical analogies. Int. J Comput. Anticipatory Syst. **15**, 217–225 (2004)
15. Makarenko, A., Stashenko, A.: Some two-steps discrete-time anticipatory models with 'boiling' multivaluedness. In: Dubois, D.M. (ed.) American Institute of Physics, AIP Conference Proceedings, vol. 839, pp. 265–272 (2006)
16. Rosen, R.: Anticipatory Systems. Pergamon Press, Oxford (1985)
17. Dubois, D.: Introduction to computing anticipatory systems. Int. J. Comput. Anticipatory Syst. **2**, 3–14 (1998)
18. Dubois, D.: Incursive and hyperincursive systems, fractal machine and anticipatory logic. In: Computing Anticipatory Systems: CASYS 2000—The Fourth International Conference. American Institute of Physics, AIP Conference Proceedings, vol. 573, pp. 437–451 (2001)
19. Scharkovski, A., Maystrenko, Yu., Romanenko, E.: Discrete Equations and Their Applications. Kluwer Academic, Dordrecht (1993)
20. Makarenko, A.: Cellular Automata with anticipation: some new research problems. Int. J. Comput. Anticipatory Syst. (Belgium) **20**, 230–242 (2008)
21. Makarenko, A.: On presumable role of anticipatory effects in neurophysiology and consciousness (short abstract in Russian). In: Proceedings of the XVI International Conference On Neurocybernetics (ICNC-12), 3 pp. Rostov-on-Don, Russia 24–28 Sept 2012

Part III
Control and Supervision of Complex Mechanical Structures and Robots

It may seem a paradox but all the exact science is dominated by the idea of approximation.

Bertrand Russell

Chapter 13
How to Cope with Disturbances in Biped Locomotion?

Miomir Vukobratović, Branislav Borovac, Mirko Raković and Milutin Nikolić

Abstract It is expected that the humanoid robots of the near future will 'live' and work in a common environment with humans. This imposes the requirement that their operational efficiency ought to be close to that of men. The main prerequisite to achieve this is to ensure the robot's efficient motion that is its ability to compensate for the ever-present disturbances. The work considers the strategies of how to compensate for the disturbances of different intensities: small which are permanently present and large that jeopardize the robot's dynamic balance instantly. It was illustrated that those two classes of disturbances require quite different compensation approaches.

13.1 Introduction

Although the problem of bipedal gait has been in the focus of researchers for almost 40 years, not all aspects of the gait synthesis and control have been solved yet in a satisfactory way. Besides, as the moment of regular involvement of humanoid robots in everyday human activities is getting closer, the quality of their behavior (where anthropomorphism is one of important aspects) is of growing importance. It is widely accepted that the future home or personal robots will become more and more human-like regarding motion, intelligence, and communication [1]. However,

M. Vukobratović
Mihajlo Pupin Institute, Volgina 15, 11000 Belgrade, Serbia
e-mail: vuk@robot.imp.bg.ac.rs

B. Borovac (✉) · M. Raković · M. Nikolić
Faculty of Technical Sciences, University of Novi Sad, Trg Dositeja Obradovića 6, 21000 Novi Sad, Serbia
e-mail: borovac@uns.ac.rs

M. Raković
e-mail: rakovicm@uns.ac.rs

M. Nikolić
e-mail: milutinn@uns.ac.rs

G.M. Dimirovski (ed.), *Complex Systems*, Studies in Systems, Decision and Control 55, DOI 10.1007/978-3-319-28860-4_13

basic requirement is to ensure motion, but it has to be clear that it is subjected to various but permanent disturbances This paper is concerned with humanoid's "reaction" to the disturbances, i.e., how to cope with them during the gait. Of course, the compensation mode will depend on the type of disturbance. Both, gait synthesis and its realization is based on use of Zero Moment Point (ZMP) [2–7].

Basic requirements to be fulfilled in a successful realization of the gait should be the simultaneous realization of a coordinated and functional motion of the joints (realization of the given gait type) and constant preservation of dynamic balance. All robot's movements are controlled in joint state space (*internal synergy*), whereas the verification of the realization efficiency is based on the robot's behavior in the external (Cartesian) coordinates (*external synergy*). In order to have the control task realized in one and verified in another state space it is necessary to have a unique relationship between these two spaces. During the regular gait this unique relationship is ensured by fulfilling the requirement for preserving dynamic balance, which is manifested as the requirement that at least terminal link of the kinematic chain of the supporting leg (or of both legs in the case of double-support phase) is immobile with respect to the ground.

13.2 Basic Issues

It is accustomed to compare the performances of humanoid robots with those of humans. This is especially related to the gait and compensation of disturbances arising during its realization. However, it should be borne in mind that despite of the apparent simplicity and ease of its realization, the gait is a complex activity. Man, at the beginning of its life, learns how to walk during several months [8] (this stage may also be called *basic learning*), whereas afterwards he practices walk all the time, constantly improving and refining it. This is related not only to the realization of basic walking patterns (say walking forward on a flat surface, climbing, and descending the staircases, etc.), but also to learning and perfecting the reactions to all kinds of disturbances that inevitably arise during the gait.

The main task of a locomotion system is to maintain the gait, i.e., to "keep on the feet," either standing or walking. For a system whose gait is not jeopardized we say that it is dynamically balanced. In the presence of disturbances (it should be emphasized that small disturbances are always present and cannot be avoided) the man reacts so as to remove the effect of the disturbance in the way that affects the least the realization of the activity that has been pursued before the occurrence of the disturbance. Of course, if the disturbance is of too high an intensity and man (humanoid) is under the threat of immediate fall, the whole attention has to be paid to its prevention and the realization of the former activity is to be abandoned.

The main indicator of dynamic balance of a bipedal robot is the ZMP. All the time the ZMP is within the support area, excluding the edges. (both in the single–and double-support phase), the system will be dynamically balanced. The action of the disturbance causes displacement of the ZMP from its reference position to the edge of the supporting foot, and the action of the compensating movements ought to be in the opposite direction, to return it close to the reference position (i.e., to bring it to the "safety zone").

Disturbances may differ in respect of both the type and action on the humanoid robot. As far as we know, their only categorization [9] which will be briefly repeated here, distinguishes small, medium and large disturbances. The categorization is based on the effect by which the disturbance jeopardizes dynamic balance and possibility of the continuation of the realization of the motion performed to the instant of the disturbance occurrence. In all three cases, it is assumed that the humanoid in the beginning of the considered period performs the reference motion, and then the disturbance begins to act.

(i) Because of the action of external disturbance force of **smaller intensity**, ideal tracking of joint trajectories is disturbed, as well as the ZMP position, but in such a manner that it still remains within a "safety zone." Appropriate control actions can return the system to the reference trajectory.
(ii) Assume now that a disturbance force of **medium intensity** is involved. In this case too, the force disturbs the ideal tracking of joint trajectories, and increases the deviation of the ZMP from its reference position. To preserve dynamic balance, the humanoid must undertake a 'more resolute' action (e.g., arms swinging), in order to ensure that the ZMP remains within the 'safety zone' and the system returns to the reference motion.
(iii) In the case of a **high-intensity** disturbance, as mentioned earlier, unpowered (passive) DOFs arise, and the system as a whole starts to rotate about the edge of the support area. The attempt to minimize deviations of joint trajectories and resume reference motion is senseless if the system has lost the dynamic balance. Hence, it is of highest interest to preserve (or reestablish) dynamic balance, taking no care at the moment of joints trajectories tracking. Because of that, the humanoid has to abandon the realization of the previous reference motion and, for example, step by one leg in the direction of falling, support on it, and—in the next several steps—return to the reference trajectory.

In all cases, control task is to minimize the deviations of the real humanoid state from the reference one for all the joints, while preserving dynamic balance whereby, especially in third case, the priority is to prevent the system from falling down, i.e., to preserve its dynamic balance [10].

In practice, small disturbances are most common, so that the problem of gait control reduces to case 1, which has been investigated in details in [11].

In the case of the compensation of large disturbances it is necessary to take care of two requirements that are to be simultaneously satisfied by the compensating movement:

(i) Forces induced by the compensating movement have to act so that the ZMP position, usurped by the action of the disturbance, is maintained within the support area.

(ii) Bearing in mind that the action of the forces induced by the compensating movement lasts short (it ends immediately after the ending of the compensating movement) *the humanoid's position at the end of the compensating movement should be such to ensure dynamic balance without the action of the forces induced by the motion*. This means that the projection of the system's CM at the end of the compensating movement has to be within the support area.

These two requirements impose serious constraints. Namely, compensating movements are often very complex (a number of joints move synchronously) and man learns them in the course of mastering the skill of walking by the trial-and-error procedure and then selects and adopts them. Hence, it is plausible to observe the characteristic ways of disturbance compensation used by the man, and then, by analyzing them, "discover" the reasons why man uses them as such.

13.3 Biped Mechanical Structure

The structure of the basic mechanism having 36 DOFs and one-link trunk, used in the present work, is shown in Fig. 13.1. The first kinematic chain represents the legs (links 1–27), the second chain extends from the pelvis and comprises the trunk and the right hand (links 28–33), and the third chain (links 34–36) forms the left shoulder and arm.

The multi-DOF joints were modeled as a set of "fictitious" links (massless links of zero-length) interconnected with the joints having one DOF. For example, the hip joints, which are in reality spherical joints with three DOFs, are modeled as sets of three one-DOF joints whose axes are mutually orthogonal. Thus, the right hip is modeled by a set of simple joints 13, 14, and 15 (with the unit vectors of rotation axes $\boldsymbol{e}_{13}$, $\boldsymbol{e}_{14}$, and $\boldsymbol{e}_{15}$), and the left hip by the set of joints 16, 17, and 18 (the unit vectors $\boldsymbol{e}_{16}$, $\boldsymbol{e}_{17}$, and $\boldsymbol{e}_{18}$). The links connecting these joints (for the right hip the links 13 and 14, and for the left links 16 and 17) were needed only to satisfy the mathematical formalism of modeling a kinematic chain. The other links (those that are not part of the joints with more DOF's) whose characteristics correspond to the links of an average human body (link 9 corresponds to the shank, link 12 to the thigh, link 30 to the trunk, etc.), are presented by solid lines in Fig. 13.1. In the same figure, the links that were needed only for modeling "complex" joints with more DOFs (having no mass and with the moment of inertia and length being equal zero) are presented by dashed lines, to indicate their "fictitious" nature.

Of special importance is the way of modeling the foot-ground contact in order to determine the exact position of the ZMP during the motion and observe the moment when the mechanism is out of dynamic balance. The loss of dynamic balance means that the mechanism collapses by rotating about one of the edges of the supporting foot, and this situation, obviously, has to be prevented. The contact of the mechanism with the ground is modeled by two rotational joints, determined with the unit vectors $\boldsymbol{e}_1$ and $\boldsymbol{e}_2$ (Fig. 13.1), mutually perpendicular. At the ZMP for dynamically balanced motion, it is constantly ensured that $\mathbf{M_Y} = 0$ (($\mathbf{M_X} \perp \mathbf{M_Y}$) $\wedge$ ($\mathbf{M_X}$, $\mathbf{M_Y} \in$

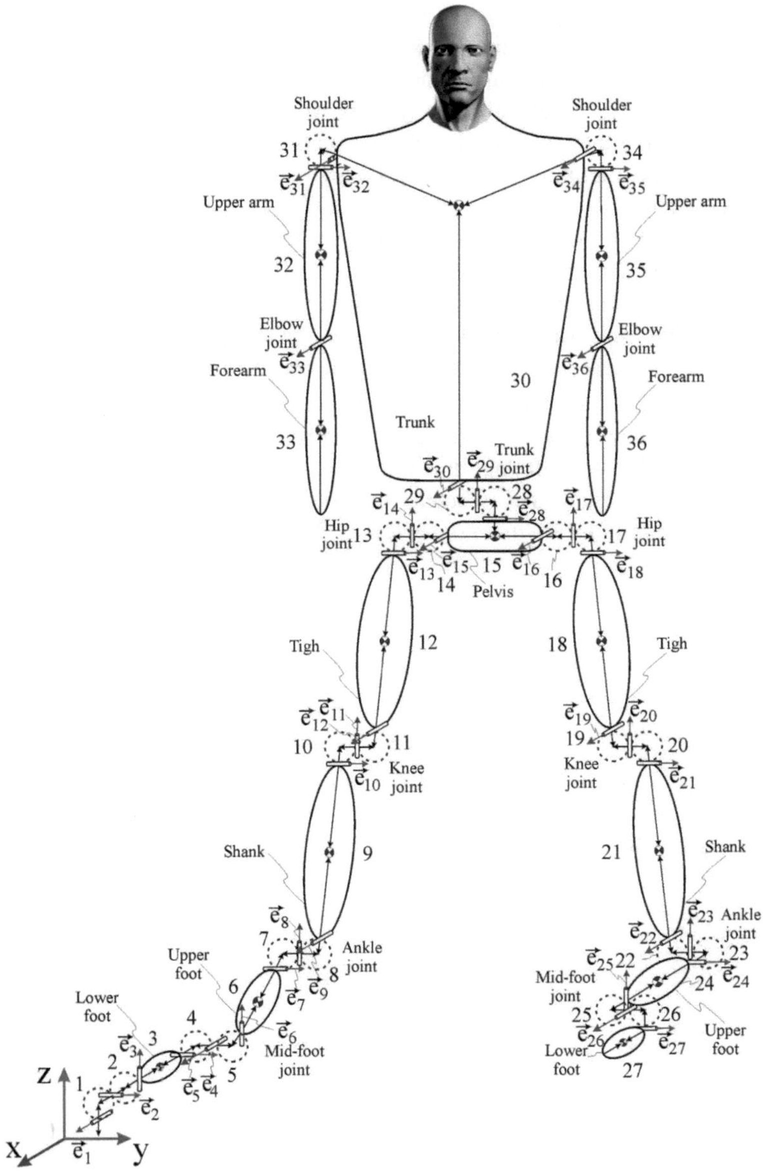

Fig. 13.1 Schematic of the robot's mechanical configuration

XoY)). It should be especially emphasized that the mechanism feet were modeled as the two-link ones. In Fig. 13.1 the anterior part of the right foot (toes) is presented by link 3, and its main part (foot body) by link 6. The toes of the left foot are presented by link 27 and the foot body by link 24. The trunk is customarily considered as one-link rigid body. However, since the trunk's base is the flexible spinal column, the assumption of one-link rigid trunk is not quite justified. To investigate the influence of the trunk bending on the synthesized motion of the mechanism and its role in control, the trunk was modeled as being multilink. The trunk was divided into several links (in this work it was modeled as being composed of 10 links) interconnected with the joints having two DOFs each (rotation about the y-axis (inclining forward-backward) and rotation about the x-axis (inclining left-right)). By this, the number of DOFs of the overall model was significantly increased, amounting to 54. The way of modeling the 10-link trunk and its "fitting into" the overall model of the humanoid is shown in Fig. 13.2.

In case of large disturbances system motion was simulated on our new software [12]. The pelvis link was chosen as the base one, Fig. 13.3, and its position and orientation in the space are presented by three translatory and angular coordinates $\mathbf{X} = [x, y, z, \theta, \varphi, \psi]^T$. To the pelvis are connected the other kinematic chains, of which the first chain (links 1–9) represents the right leg; the second chain (links 1, 10–17) stands for the left leg; the third chain (links 1, 18–42) represents the backbone and the right arm, whereas the fourth kinematic chain represents the 10-link trunk (the links are interconnected by 2-DOF joints) and the left arm (links

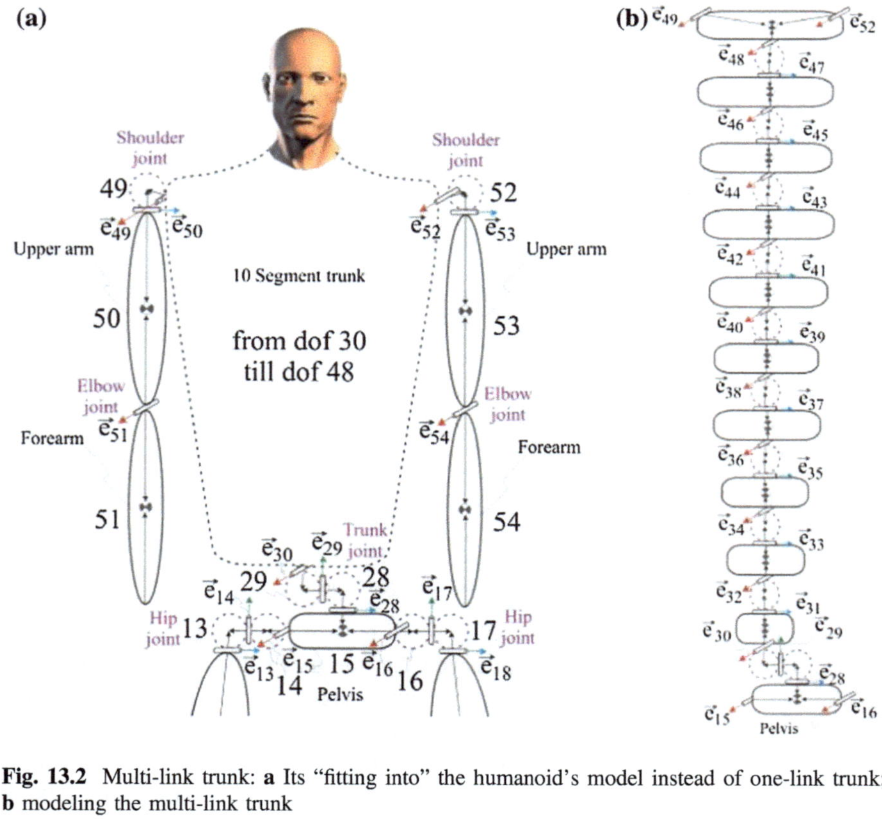

Fig. 13.2 Multi-link trunk: **a** Its "fitting into" the humanoid's model instead of one-link trunk; **b** modeling the multi-link trunk

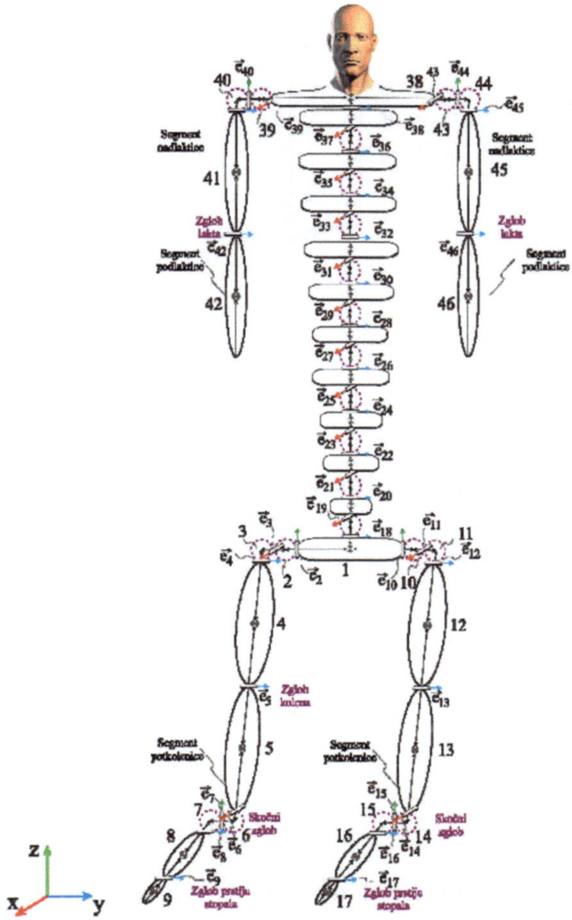

Fig. 13.3 Mechanical structure of the robot with 46 links and 45 joints

1, 18–38, 43–46). The head is included into the link 38. The feet are two-link ones, whereby between the foot body and toes there exists only one DOF (left foot: links 8 and 9; right foot: links 16 and 17).

As already described, the motion the ZMP trajectory is prescribed. Different ZMP trajectories may result in completely different motions of the trunk. In this work, the ZMP path at the foot-ground contact surface divided into six segments was adopted in advance. Then, the ZMP traveling time along each particular segment was varied in such a way that the overall step duration remained unchanged. For each of these cases, the trunk motion was synthesized. The motion synthesis was performed for the mechanism with a one-link trunk and two-link feet (the mechanism with 36 DOFs), shown in Fig. 13.1. The basic ZMP path (which includes both the single- and double-support phase) was adopted as shown in Figs. 13.4 and 13.5. Time dependence of the ZMP position during the half-step was realized in the following way. The path was divided into six segments, in each of

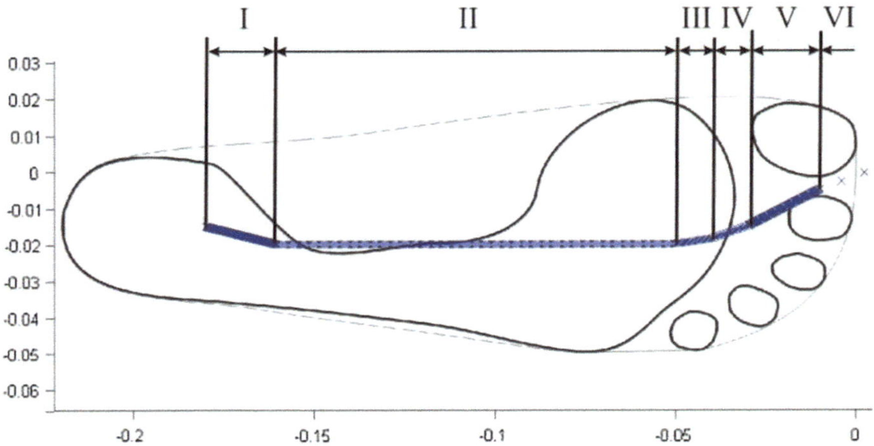

Fig. 13.4 Reference ZMP path under foot

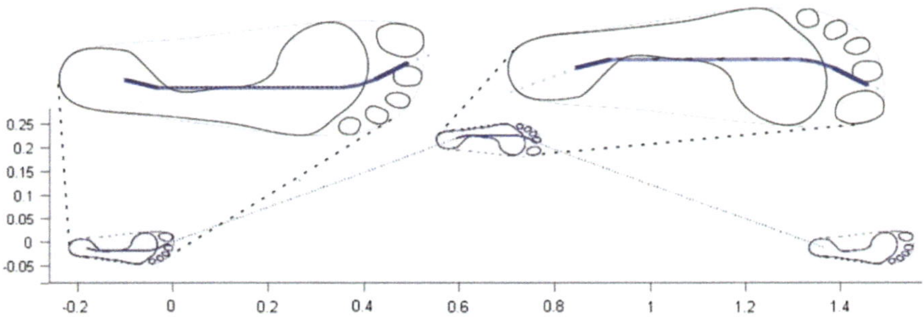

Fig. 13.5 Reference ZMP path for full step

them the ZMP, traveling from the starting to the final position, remained for a certain time. In segment I ZMP remained 10 % of the overall step duration, in II 42 %, in III 4 %, in IV 4 %, in V 19 % and in VI 21 %.

13.4 Compensation of Small Disturbances

Any deviation of internal synergy from the reference caused by the disturbance action, produces also deviation of the ZMP position from its reference, which can endanger dynamic balance. In this work, we consider disturbance in the form of the force action at the humanoid's right shoulder (Fig. 13.6). Figure 13.6a shows a perspective view of the action of the force (components F_x and F_y), Fig. 13.6b presents the change of the intensity of the force (components F_x and F_y) during a half-step. The half-step lasts 0.8 s, the force starts to act 0.15 s after the beginning of

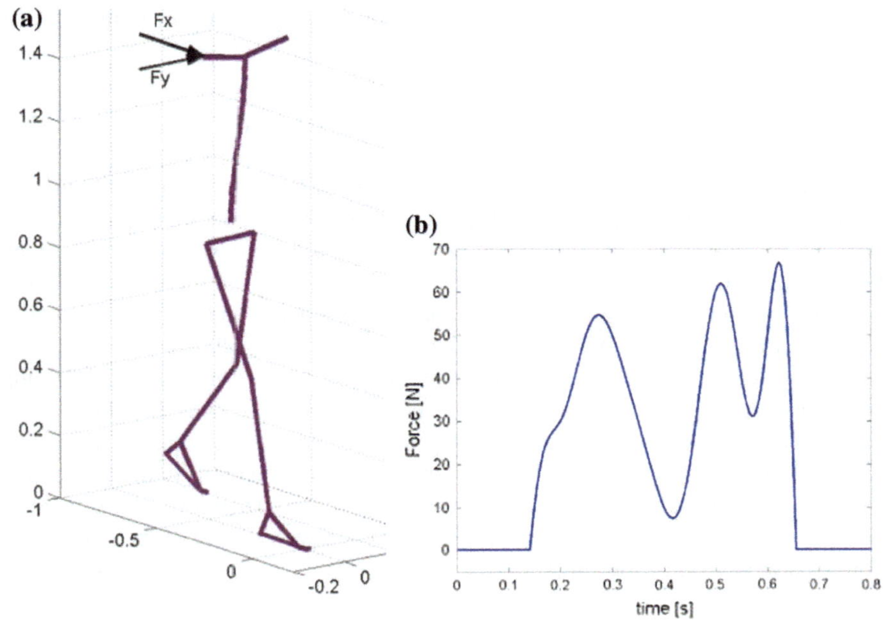

Fig. 13.6 **a** Force action at the mechanism's shoulder, **b** time profile of the force intensity

motion, and its action terminates after 0.65 s. During the second half-step the force does not act. However, in view of the disturbance of internal synergy in the beginning of the second half-step, there exists the disturbance of initial conditions type. The control system attempts to bring the internal synergy closer to the reference and preserve simultaneously dynamic balance. We consider the cases of acting of only component F_x, only component F_y, and both components simultaneously.

Generally speaking, in each time instant t_i, the total control signal at each of the joints consists of the reference control (calculated for the case of motion without disturbances) and corrections, which depend of the disturbance intensity. In other words:

$$u_{n_i}^{(t_i)} = u_{\text{reference } n_i}^{(t_i)} + \Delta u_{\text{at joint } n_i}^{(t_i)}.$$

The corrective part $\Delta u_{\text{at joint } n_i}^{(t_i)}$ consists also of two parts:

- One part of the corrective control serves to preserve dynamic balance, i.e., to minimize the ZMP position deviation from the reference $\Delta u_{\text{ZMP at joint } n_i}^{(t_i)}$ one. This task can be allocated to one or more mechanism's joints.
- The other part of the corrective control ($\Delta u_{\text{local at joint } n_i}^{(t_i)}$) should ensure minimization of the deviation of the actual synergy from the reference one at each joint. This part of the corrective control we call local control, as the regulators involved act locally, i.e., only at the individual joints.

Hence, the overall compensation (at each of the joints) can be written in the following way:

$$\Delta u_{n_i}^{(t_i)} = \Delta u_{\text{ZMP at joint } n_i}^{(t_i)} + \Delta u_{\text{local at joint } n_i}^{(t_i)}. \tag{13.1}$$

In the above expressions n_i refers to the ith joint of the mechanism configuration having in all n joints. Naturally, each joint tends to preserve its motion as close as possible to the desired one, so that the task of forming local corrective control may be imposed on all the joints where there exist deviations from the desired motion. The task of compensating for ZMP deviation can not be (and is not necessarily) realized at every joint at the same time, but the task may be allocated to only some joints. In this, it should be borne in mind that the axis of the joint performing compensation is perpendicular to the direction of ZMP deviation. In other words, the ZMP deviation in the x-direction may be compensated for only by the joints whose axes are parallel to the y-direction, and the ZMP deviations in the y-direction by the joints whose axes are parallel to the x-direction. Therefore, if we want to compensate for the ZMP deviations in the x-direction with the aid of the joint n_i, the control law will be of the form:

$$\Delta u_{\text{ZMP at joint } n_i}^{(t_i)} = k_{p\,\text{ZMP}\,n_i} \cdot \Delta \text{ZMP}_x^{(t_i)} + k_{i\,\text{ZMP}\,n_i} \cdot \sum_{i=1}^{t_i} \Delta \text{ZMP}_x^{(i)} + k_{d\,\text{ZMP}\,n_i} \cdot \left(\Delta \text{ZMP}_x^{(t_i)} - \Delta \text{ZMP}_x^{(t_i-1)}\right), \tag{13.2}$$

where $k_{p\ \text{ZMP}\ ni}$ represents the position feedback gain at the joint n_i, for the compensation of ZMP deviation; $k_{i\ \text{ZMP}\ ni}$ and $k_{d\ \text{ZMP}\ ni}$ represent integral and derivative feedback gains at the joint n_i, also for compensating the ZMP deviation; $\Delta \text{ZMP}_x^{(t_i)}$ stands for the deviation of ZMP in the direction of the x-axis at a time instant t_i.

The law of local control to compensate for the deviations from reference values at the joint n_i can be written as:

$$\Delta u_{\text{local } i \text{ at joint } n_i}^{(t_i)} = k_{p\,n_i\,\text{local}} \cdot \Delta q_{n_i}^{(t_i)} + k_{i\,n_i\,\text{local}} \cdot \sum_{i=1}^{t_i} \Delta q_{n_i}^{(i)} + k_{d\,n_i\,\text{local}} \cdot \left(\Delta q_{n_i}^{(t_i)} - \Delta q_{n_i}^{(t_i-1)}\right). \tag{13.3}$$

Analogously to the previous case, $k_{p\,n_i\,\text{local}}$ represents the position feedback gain at the joint n_i for the compensation of its deviation from the reference value, whereas $k_{i\,n_i\,\text{local}}$ and $k_{d\,n_i\,\text{local}}$ represent the integral and derivative feedback gains, also for the compensation of deviations at the joint n_i from the reference values.

Let us consider now in more detail how ZMP deviations are compensated for. The ZMP deviation in the x-direction may be compensated for by the ankle (joint 7), hip (joint 13), or the waist (joint 28) (see Fig. 13.1). Equation (13.2), applied to any of these cases, would be of the form:

$$\Delta u_{\text{ZMP at joint 7}}^{(t_i)} = k_{p\,\text{ZMP 7}} \cdot \Delta\text{ZMP}_x^{(t_i)} + k_{i\,\text{ZMP 7}} \cdot \sum_{i=1}^{t_i} \Delta\text{ZMP}_x^{(i)} + k_{d\,\text{ZMP 7}} \cdot \left(\Delta\text{ZMP}_x^{(t_i)} - \Delta\text{ZMP}_x^{(t_i-1)}\right), \tag{13.4}$$

$$\Delta u_{\text{ZMP at joint 13}}^{(t_i)} = k_{p\,\text{ZMP 13}} \cdot \Delta\text{ZMP}_x^{(t_i)} + k_{i\,\text{ZMP 13}} \cdot \sum_{i=1}^{t_i} \Delta\text{ZMP}_x^{(i)} + k_{d\,\text{ZMP 13}} \cdot \left(\Delta\text{ZMP}_x^{(t_i)} - \Delta\text{ZMP}_x^{(t_i-1)}\right), \tag{13.5}$$

$$\Delta u_{\text{ZMP at joint 28}}^{(t_i)} = k_{p\,\text{ZMP 28}} \cdot \Delta\text{ZMP}_x^{(t_i)} + k_{i\,\text{ZMP 28}} \cdot \sum_{i=1}^{t_i} \Delta\text{ZMP}_x^{(i)} + k_{d\,\text{ZMP 28}} \cdot \left(\Delta\text{ZMP}_x^{(t_i)} - \Delta\text{ZMP}_x^{(t_i-1)}\right). \tag{13.6}$$

In the case, we want to compensate for the ZMP deviation in the y-direction, the expressions to calculate compensational control become:

$$\Delta u_{\text{ZMP at joint 9}}^{(t_i)} = k_{p\,\text{ZMP 9}} \cdot \Delta\text{ZMP}_y^{(t_i)} + k_{i\,\text{ZMP 9}} \cdot \sum_{i=1}^{t_i} \Delta\text{ZMP}_y^{(i)} + k_{d\,\text{ZMP 9}} \cdot \left(\Delta\text{ZMP}_y^{(t_i)} - \Delta\text{ZMP}_y^{(t_i-1)}\right), \tag{13.7}$$

$$\Delta u_{\text{ZMP at joint 15}}^{(t_i)} = k_{p\,\text{ZMP 15}} \cdot \Delta\text{ZMP}_y^{(t_i)} + k_{i\,\text{ZMP 15}} \cdot \sum_{i=1}^{t_i} \Delta\text{ZMP}_y^{(i)} + k_{d\,\text{ZMP 15}} \cdot \left(\Delta\text{ZMP}_y^{(t_i)} - \Delta\text{ZMP}_y^{(t_i-1)}\right), \tag{13.8}$$

$$\Delta u_{\text{ZMP at joint 30}}^{(t_i)} = k_{p\,\text{ZMP 30}} \cdot \Delta\text{ZMP}_y^{(t_i)} + k_{i\,\text{ZMP 30}} \cdot \sum_{i=1}^{t_i} \Delta\text{ZMP}_y^{(i)} + k_{d\,\text{ZMP 30}} \cdot \left(\Delta\text{ZMP}_y^{(t_i)} - \Delta\text{ZMP}_y^{(t_i-1)}\right). \tag{13.9}$$

The local control to minimize deviations at each particular joint from the reference can be written in the following way:

$$\Delta u_{\text{local at joint}\,n_i}^{(t_i)} = k_{p\,n_i\,\text{local}} \cdot \Delta q_{n_i}^{(t_i)} + k_{i\,n_i\,\text{local}} \cdot \sum_{i=1}^{t_i} \Delta q_{n_i}^{(i)} + k_{d\,n_i\,\text{local}} \cdot \left(\Delta q_{n_i}^{(t_i)} - \Delta q_{n_i}^{(t_i-1)}\right). \tag{13.10}$$

The feedback gains of local regulators and for preserving dynamic balance, k_p, k_i, and k_d, are not constant but their values vary in a certain range in dependence of the instantaneous position of the ZMP. The values for k_p, k_i, and k_d are determined in a heuristic manner. The change in the values of the feedback gains must not be of "step function character" but a gradual one, taking place in the following way. Around the

ZMP trajectory, at a distance of 5 mm, a zone is formed. If the ZMP is within that zone, the feedback gains of the regulator for maintaining dynamic balance begin to decrease (up to the minimal values), and the coefficients of local regulators start to rise (up to the maximal values). After that they retain their (minimal and maximal) values till the next change. If the ZMP is out of the 5-mm zone, the feedback gains of the regulator for maintaining dynamic balance begin to increase (up to the maximal values), and the feedback gains of local regulators start to decrease (up to the minimal values). After that they retain their (minimal and maximal) values till the next change, which will happen when ZMP crosses over 5-mm border. Values of the increments and decrements for all the feedback gains k_p, k_i, and k_d were also determined based on heuristic.

Minimal and maximal values of the feedback gains in the case of compensating for ZMP deviation and the values of increments and decrements are presented in Table 13.1, whereas Table 13.2 gives the minimal and maximal values of feedback gains, along with the values of increments and decrements for the local regulators.

In Fig. 13.7 is illustrated the simulation of humanoid's motion when the disturbance force F_x acted at the shoulder (pushing the humanoid forward) and compensation was performed by the ankle joint. The trunk was a 10-link one. Reference control was applied at all joints. Additional control for correcting ZMP deviation, acting at the joints 7 and 9 was defined by Eqs. (13.4) and (13.7), while the local regulators were defined by Eq. (13.10). In view of the fact that the disturbance force was pushing the mechanism forward and the compensation was performed at the ankle of the supporting leg, the entire system inclined forward, so that the leg being in the swing phase might kick against the ground.

In order to eliminate the inclination of the overall system forward and allow regular pace, additional compensation was performed at the hip of the leg in the swing phase (joint 18) so that the same Δu that has been added to the ankle (joint 7) of the supporting leg was added to the hip joint of the swing leg in order to increase humanoid's step in the gait direction. Such correction of the internal synergy has been called preventive correction. Although essential stepping out was made in this half-step, it is evident from the figure that the ZMP trajectory deviated only a little from the reference one. The increased step length during the first half-step was fully

Table 13.1 Feedback gains for ZMP compensation and the corresponding increments and decrements

Gain	Joint	Min.	Max.	Inc.	Dec.
$k_{p\ \mathrm{ZMP}}$	4, 7, 10, 13	1	9	0.1	0.01
$k_{p\ \mathrm{ZMP}}$	9, 15	5	13	0.1	0.01
$k_{p\ \mathrm{ZMP}}$	28–48	2	10	0.1	0.01
$k_{i\ \mathrm{ZMP}}$	All	3	3.08	0.0001	0.00001
$k_{d\ \mathrm{ZMP}}$	All	2	2.8	0.001	0.0001

Table 13.2 Feedback gains for local regulators and the corresponding increments and decrements

Gain	Joint	Min.	Max.	Inc.	Dec.
$k_{p\ \mathrm{local}}$	All	10	200	0.1	10
$k_{i\ \mathrm{local}}$	All	2	3	0.0005	0.05
$k_{d\ \mathrm{local}}$	All	3	8	0.002	0.2

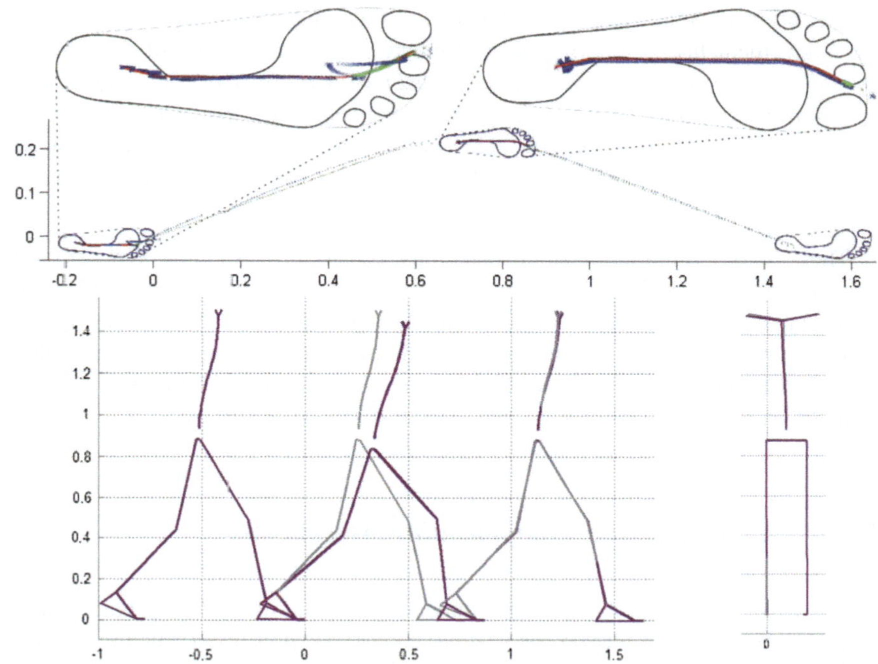

Fig. 13.7 Action of the force F_x on the mechanism shoulder. Compensation was performed by the ankle joint of the supporting leg. The *upper* picture shows the ZMP trajectory, the *lower left* a lateral view of the stick diagram, the *lower right* stick diagram in the sagittal plane at the end of the second half-step

cancelled out in the second half-step, and the rest of the humanoid mechanism too, came close to its reference position.

13.5 Compensation of Large Disturbances

The main task of any locomotion system is to "keep on the feet" (i.e., to preserve dynamic balance), either standing or walking. If the disturbance is of too high an intensity and humanoid is under the threat of immediate fall, the whole attention has to be paid to its prevention and the realization of the former activity is to be abandoned. In such a case two requirements have to be simultaneously satisfied by the compensating movement:

(i) Forces induced by the compensating movement have to act so that the ZMP position, usurped by the action of the disturbance, is maintained within the support area.

(ii) Bearing in mind that the action of the forces induced by the compensating movement lasts short (it ends immediately after the ending of the compensating movement) *the humanoid's position at the end of the compensating movement should be such to ensure dynamic balance without the action of the forces induced by the motion*. This means that the projection of the system's CM at the end of the compensating movement has to be within the support area.

These two requirements impose serious constraints. Namely, compensating movements are often very complex (a number of joints move synchronously) and man learns them in the course of mastering the skill of walking by the trial-and-error procedure and then selects and adopts them. Hence, it is plausible to observe the characteristic ways of disturbance compensation used by the man, and then, "discover" the reasons why man uses them as such [10].

13.5.1 Characteristic Compensating Movements

Let us consider, the case when the humanoid is standing in an upright position (Fig. 13.8a), with both feet on the ground, and on which a disturbance in the form of the impulse force is acting horizontally on its back at a height $h = 1.6$ m, whose profile is given in Fig. 13.8b. Depending on the direction and intensity of the disturbance force, there are two characteristic ways by which the humanoid may react. One is to attempt to maintain dynamic balance while not moving the feet from the ground and the other is to step in the direction of the force action. In the former case, there are two possible scenarios. The first is to generate compensating movements simultaneously at both ankles, while the whole system behaves as a unique link (Fig. 13.8c), i.e., as a single inverted pendulum. The other way (Fig. 13.8d) is to generate the compensating movements simultaneously at the ankle and the hip, but such that they act in the opposite directions. Figure 13.8d shows the

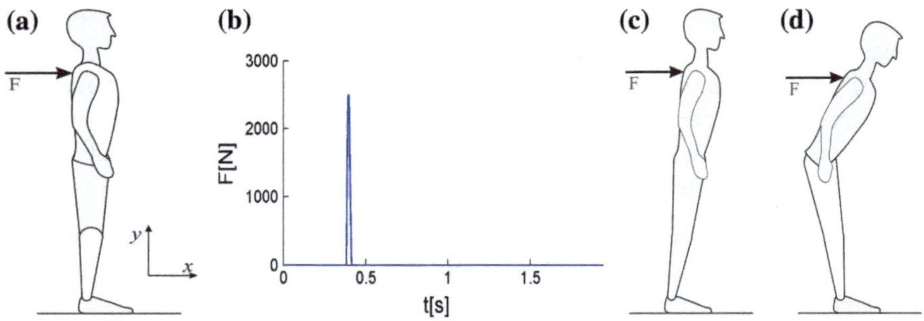

Fig. 13.8 **a** Initial posture of the humanoid **b** Profile of the disturbance force **c** Compensation by the moment at the ankle **d** Simultaneous compensation by the ankle and the hip

links of the legs (the lower leg and the thigh) and of the trunk (with the neck and head) as an unique link, so that the system can be approximated as a two-link one.

13.5.2 Analysis of Compensating Movements

Now, we will analyze two essentially different types of compensating movements: the case when the entire body behaves as a unique link (Fig. 13.8c) and as a two-link mechanism (Fig. 13.8d). In order to analyze both types of movements and establish their main characteristics we will form a planar model of the humanoid (Fig. 13.9) in which one-link stands for the legs and the other for the trunk together with the hands and the head. The figure shows also the foot, but it is considered to be immobile with respect to the support and not involved in the motion.

The motion of such a system is described by the following differential equation:

$$\begin{bmatrix} \tau_1 \\ \tau_2 \end{bmatrix} = \begin{bmatrix} H_{11} & H_{12} \\ H_{21} & H_{22} \end{bmatrix} \begin{bmatrix} \ddot{q}_1 \\ \ddot{q}_2 \end{bmatrix} + \begin{bmatrix} h_1 \\ h_2 \end{bmatrix}, \tag{13.11}$$

where

$$H_{11} = m_1 \cdot l_{C1}^2 + I_1 + m_2(l_1^2 + l_{C2}^2 + 2 \cdot l_1\, l_{C2} \cdot \cos q_2) + I_2, \tag{13.12}$$

$$H_{22} = m_2\, l_{C2}^2 + I_2, \tag{13.13}$$

$$H_{12} = H_{21} = m_2\, l_1\, l_{C2}\ \cos q_2 + m_2\, l_{C2}^2 + I_2, \tag{13.14}$$

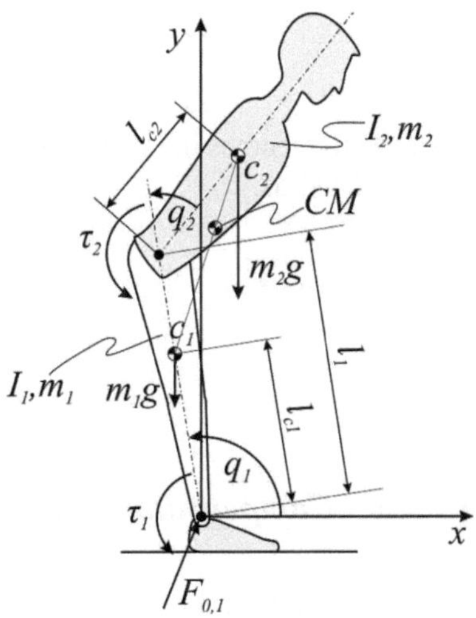

Fig. 13.9 Two-link planar mechanism

$$h_1 = -m_2\, l_1\, l_{C2} \cdot \sin q_2\left((\dot{q}_2)^2 + 2\dot{q}_1\dot{q}_2\right) + m_1\, l_{C1}\, g\ \cos q_1 \\ + m_2\, g\ (l_{C2}\ \cos\ (q_1 + q_2) + l_1\ \cos q_1), \tag{13.15}$$

$$h_2 = m_2\, l_1\, l_{C2} \cdot \sin q_2 (\dot{q}_1)^2 + m_2\, l_{C2}\, g \cdot \cos\ (q_1 + q_2) \tag{13.16}$$

with the designation being given in Fig. 13.9. It should be noticed that H_{11}, H_{12} and H_{22} in the case considered are greater than zero. The motion of the CM of the overall system can be described as:

$$\ddot{\vec{r}}_{\mathrm{CM}} = \frac{1}{m_1 + m_2}\left(\begin{bmatrix} A_{11} & A_{12} \\ A_{21} & A_{22} \end{bmatrix}\begin{bmatrix} \ddot{q}_1 \\ \ddot{q}_2 \end{bmatrix} + \begin{bmatrix} a_1 \\ a_2 \end{bmatrix}\right), \tag{13.17}$$

where

$$A_{11} = -m_1 \cdot l_{C1} \sin q_1 - m_2 l_1 \sin q_1 - m_2 l_{C2} \sin(q_1 + q_2), \tag{13.18}$$

$$A_{12} = -m_2 l_{C2} \sin(q_1 + q_2), \tag{13.19}$$

$$A_{21} = m_1 \cdot l_{C1} \cos q_1 + m_2 l_1 \cos q_1 + m_2 l_{C2} \cos(q_1 + q_2), \tag{13.20}$$

$$A_{22} = m_2 l_{C2} \cos(q_1 + q_2), \tag{13.21}$$

$$a_1 = -m_1 \cdot l_{C1}\dot{q}_1^2 \cos q_1 - m_2 l_1 \dot{q}_1^2 \cos q_1 - m_2 l_{C2}(\dot{q}_1 + \dot{q}_2)^2 \cos(q_1 + q_2), \tag{13.22}$$

$$a_2 = -m_1 \cdot l_{C1}\dot{q}_1^2 \sin q_1 - m_2 l_1 \dot{q}_1^2 \sin q_1 - m_2 l_{C2}(\dot{q}_1 + \dot{q}_2)^2 \sin(q_1 + q_2). \tag{13.23}$$

Since we deal with a planar model, the ZMP position can be calculated in the following way

$$\mathrm{ZMP}_X = \tau_1 / F_{0,1Y}, \tag{13.24}$$

where $F_{0,1Y}$ denotes the vertical component of the ground reaction force, which can be calculated as:

$$F_{0,1Y} = (m_1 + m_2)\, \ddot{r}_{\mathrm{CMY}} + (m_1 + m_2) g = [A_{21} \quad A_{22}]\begin{bmatrix} \ddot{q}_1 \\ \ddot{q}_2 \end{bmatrix} + a_2 + (m_1 + m_2) g. \tag{13.25}$$

13.5.3 Compensation Involving Only the Ankle Joint

When the compensation is performed only by the torque at the ankle joint, the hip joint remains stiff, and the values of joint-space coordinates, velocities and accelerations are zero ($q_2 = 0$, $\dot{q}_2 = 0$, $\ddot{q}_2 = 0$). In that case, the system moves as an inverted physical pendulum and Eq. (13.11) reduces to:

$$\ddot{q}_1 = \frac{\tau_1 - h_1}{H_{11}}, \tag{13.26}$$

The relationship between the ZMP position and driving torque in the case when the hip joint is stiff is given by the following relation:

$$\tau_1 = \mathrm{ZMP}_X(L_{CM}\,\ddot{q}_1 \cos q_1 - L_{CM}\dot{q}_1^2 \sin q_1 + (m_1 + m_2)g), \tag{13.27}$$

where

$$L_{CM} = m_1 \cdot l_{C1} + m_2 l_1 + m_2 l_{C2}. \tag{13.28}$$

By combining Eqs. (13.15), (13.26), and (13.27) one obtains the relationship between the angular acceleration at the ankle and ZMP position as:

$$\begin{aligned}(H_{11} - \mathrm{ZMP}_X L_{CM} \cos q_1)\,\ddot{q}_1 &= -\mathrm{ZMP}_X L_{CM}\dot{q}_1^2 \sin q_1 \\ &+ \mathrm{ZMP}_X(m_1 + m_2)g - g\,L_{CM} \cos q_1.\end{aligned} \tag{13.29}$$

Let us assume that in the initial moment the humanoid stands still and that this state is suddenly perturbed by a force as shown in Fig. 13.8a. Due to the momentum thus received, the humanoid will start to rotate around the ankle joint at an angular velocity[1] $\dot{q}_1$ in the direction indicated in Fig. 13.10 (this direction is designated as the negative one) and the CM will move in the direction of the action of the disturbance force (positive direction of the x-axis), causing the CM displacement to the toes, i.e., to the edge of the support area. In order to recover from the disturbance, the system has to generate a negative acceleration of the CM in the direction of the x-axis, which is achieved by a positive angular acceleration at the ankle. Based on (13.26), it can be seen that the increase of the torque at the ankle generates a positive acceleration that will stop further inclination of the humanoid and prevent the CM projection from going outside the support area. However, as is evident from (13.27), the torque at the ankle influences in a direct way the position of the ZMP, so that the increase in the torque yields the increase in the x-coordinate of the ZMP position, which means that the ZMP will shift toward the toes, as shown in Fig. 13.10. To preserve dynamic balance it is necessary to keep the ZMP within the support area (must not reach its edge), so that there will exist a maximum torque that can be applied at the ankle. If the torque applied is of higher intensity, the system will rotate around the toes and dynamic balance will be lost. On the other hand, if the torque is insufficient to prevent the CM motion, this can again lead to the fall, or some other strategy has to be applied, e.g., stepping.

[1]Its intensity depends on the momentum applied to the body and humanoid's mass.

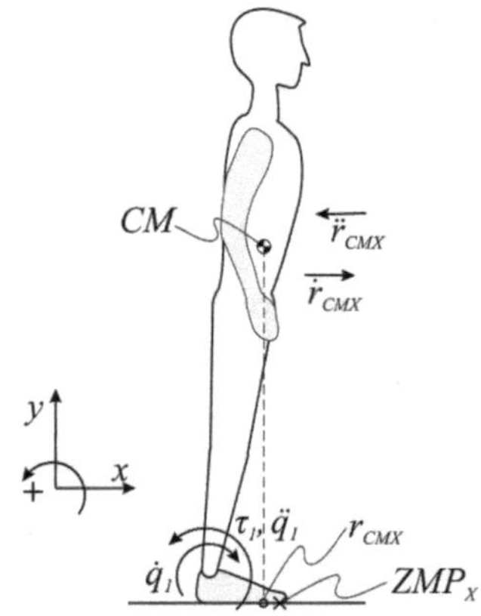

Fig. 13.10 System motion during the compensation by the ankle

To summarize, in order to preserve dynamic balance, the ZMP must remain within the support area, and the ZMP position is influenced by all the forces (gravitational and motion-induced) acting on the system. To preserve dynamic balance during the compensating movement (while the motion persists), the ZMP must be within the support area, but the CM projection need not to. At the end of the movement (when the motion stopped), the ground reaction is equal to the gravitational force, which can be seen from Eq. (13.25). Based on that and Eq. (13.24), it can be concluded that the ZMP position depends solely on gravitational forces. In that case, the ZMP and CM projection (CMP) coincide, so that the CMP will also be within the support area.

From this it follows that the driving torque at the ankle must not be greater than the value determined by the admissible ZMP displacement, so that the intensity of acceleration/deceleration of the system's CM will have its maximum allowed value. If the maximum permissible torque is not sufficient to stop the motion of the CM before its projection comes outside the support area, such strategy is not appropriate. In such situation, it is necessary to employ some other way to prevent the fall (to use different compensation strategy), for example, by the simultaneous motion at the ankle and at the hip. When the CMP comes outside the support area, then it holds that:

$$\mathrm{ZMP}_X < (m_1 + m_2)^{-1} L_{\mathrm{CM}} \cos q_1. \tag{13.30}$$

By inserting this inequality into expression (13.29) it can be seen that the angular acceleration at the ankle is negative, which causes additional acceleration of the CM. And this is just the reason why it is not possible to stop the motion of the CM if its projection is outside the support area. It should be noticed that the choice of the compensation strategy must be made immediately in the beginning of the disturbance action. Man is to learn how to do this, and it is believed that the decision is made on the basis of the simultaneous processing of more information on the intensity of the disturbance force, instantaneous posture, the state of the system, etc.

13.5.4 Compensation Simultaneous Motion of the Ankle and the Hip Joints

Let us have a look again at the humanoid standing still upright, with no disturbance force acting on it (Fig. 13.8a). The action of the disturbance will transfer to it a certain momentum, and the CMP will move in the direction of the disturbance (positive direction of the x-axis).

Let us suppose then that the humanoid performs the compensating movements simultaneously at the ankle and at the hip as is shown in Fig. 13.11. The motion at the ankle joint will directly contribute to the displacement of the system's CM in the direction that is opposite to the motion caused by the disturbance, which brings about the settling of the system.

In this case both joints of the model shown in Fig. 13.9 are active. Based on the equation of the system's motion (13.11) we get:

$$\ddot{q}_1 = \frac{1}{H_{11}H_{22} - H_{21}H_{12}}(H_{22}(\tau_1 - h_1) - H_{21}(\tau_2 - h_2)), \tag{13.31}$$

$$\ddot{q}_2 = \frac{1}{H_{11}H_{22} - H_{21}H_{12}}(-H_{12}(\tau_1 - h_1) + H_{11}(\tau_2 - h_2)). \tag{13.32}$$

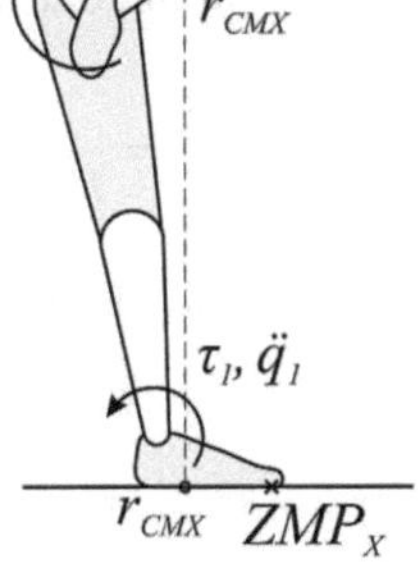

Fig. 13.11 Compensation performed by the ankle and hip joints

For a two-link planar mechanism it holds that $H_{11}H_{22} - H_{21}H_{12} > 0$, $H_{ii} > 0$, so that on the basis of (13.31) it can be concluded that if the torque at the ankle (τ_1) is greater than h_1 and the torque at the hip (τ_2) is smaller than h_2, the angular acceleration at the ankle will be positive. This means that this acceleration will tend to rotate the system backward. Similarly, based on Eq. (13.32), we conclude that if the torque at the ankle is greater than h_1, and that at the hip smaller than h_2, the angular acceleration at the hip will be negative. This means that in this case the trunk will rotate forward around the hip joint. It should be noticed that h_1 and h_2 are the moments generated by the Coriolis, centrifugal and gravitational accelerations at the ankle and hip joints. When the angular velocities are small, then, based on Eqs. (13.15) and (13.16), we can take that h_1 and h_2 are approximately equal to the moments generated by the gravitational forces at the ankle and hip, respectively. In order to stop the motion, the acceleration of the humanoid's CM in the x-direction must be negative, as shown in Fig. 13.11. Since A_{11} is by absolute value significantly greater than A_{12} (Eqs. 13.18 and 13.19), the effect of the ankle motion on the CM motion is much greater than the effect on the motion of the trunk (Eq. 13.17). By the same token, in order to stop the system it is necessary to generate a positive angular acceleration at the ankle. Also, the ZMP position is directly related to the driving torque at the ankle, which is given by the following equation:

$$\begin{aligned}\tau_1 = \ & \mathrm{ZMP}_X((m_1 C1 \cos q_1 + m_2 l_1 \cos q_1 + m_2 l_{C2} \cos(q_1 + q_2))\, \ddot{q}_1 \\ & + m_2 l_{C2} \cos(q_1 + q_2)\, \ddot{q}_2 \ - m_1 \cdot l_{C1} \dot{q}_1^2 \sin q_1 - m_2 l_1 \dot{q}_1^2 \sin q_1 \\ & - m_2 l_{C2} (\dot{q}_1 + \dot{q}_2)^2 \sin(q_1 + q_2) + (m_1 + m_2) g).\end{aligned} \tag{13.33}$$

As in the previous case, this equation provides a limit to the maximum τ_1 that may be applied at the ankle. However, the maximum torque τ_1 does not guarantee that the angular acceleration at the ankle will be positive and that CM will decelerate. The maximum angular acceleration of the CM for the case when only the ankle joint is active is:

$$\ddot{q}_1^{\mathrm{MAX}} = \frac{\tau_1^{\mathrm{MAX}} - h_1}{H_{11}}. \tag{13.34}$$

Maximum angular acceleration when both the ankle and the hip are active is:

$$\ddot{q}_1^{\mathrm{MAX}} = \frac{1}{H_{11}H_{22} - H_{21}H_{12}} \left(H_{22}(\tau_1^{\mathrm{MAX}} - h_1) - H_{21}(\tau_2^{\mathrm{MIN}} - h_2) \right). \tag{13.35}$$

It is easy to show that:

$$\frac{H_{22}}{H_{11}H_{22} - H_{21}H_{12}} > \frac{1}{H_{11}} \tag{13.36}$$

from which it follows that the same torque at the ankle joint in the case when both joints are active can generate a much greater angular acceleration at the ankle than when only this joint is active. Therefore, the simultaneous motion of the ankle and the hip may produce a significantly greater deceleration of the CM motion that is to compensate for a significantly larger disturbance. Besides, this way allows simultaneous deceleration of the CM motion and control of the ZMP position, so that if during the compensating movement the CM projection is found outside the support area, simultaneous motion at the ankle and the hip can "pull in" the CM projection into the support area. When the compensation is performed by the motion only at the ankle, there will be a direct relationship between the angular acceleration at the ankle (and thus the CM acceleration) and ZMP position (Eq. 13.29). Since the ZMP must be within the support area, this narrows the margin for the deceleration of the CM motion in the x-direction. And these are just the reasons why the compensation involving both the ankle and the hip can compensate for the disturbances of greater intensity compared to the case when compensation is performed by the ankle only.

However, this strategy has also its limitations. From (13.25) one can see that the angular accelerations at the ankle and hip are limited because the foot-ground contact is unilateral (the foot can push, but not pull, the support), so that the vertical component of the ground reaction force is always directed upwards. Another constraint comes out from the maximum amplitude of the torque that can be generated by the actuators at the hip.

13.5.5 Compensation Simulation—Only by Ankle and Using Simultaneously Ankle and the Hip Joints

In the first case we consider the realization of compensation using only the ankle. In the initial moment, the robot stands still with both feet on the ground. The torques at the joints (there exists only gravitational loads) are constant, so that the control inputs are also constant and we can write:

$$u_j^{t_i} = u_j^0, \quad j \in \{2, \ldots, 46\}, \tag{13.37}$$

where $u_j^{t_i}$ is the motor voltage activating the joint j at the moment t_i, whereas u_j^0 is the reference control voltage, which in this example corresponds to the gravitational load of the joints prior to the disturbance. At a certain moment, the disturbance force of an amplitude of 510 N and profile shown in Fig. 13.8a, starts to act in the direction of the x-axis. The force transfers a momentum of 18 kgm/s, which brings about the CM motion at a speed of 0.21 m/s.

Immediately after the occurrence of the disturbance, the compensation by the action at the ankle is to start, to stop the CM motion in the x-direction, but so that dynamic balance is preserved (the ZMP has to remain within the support area). The deviation of the CM from its initial position, desired ZMP position,[2] and the ZMP position error are calculated in the following way:

$$\Delta \mathrm{CM}_X^{t_i} = \mathrm{CM}_X^0 - \mathrm{CM}_X^{t_i}, \tag{13.38}$$

$$\begin{aligned} {}^{\mathrm{ref}}\mathrm{ZMP}_X^{t_i} = {}& {}^{\mathrm{ref}}\mathrm{ZMP}_X^{t_{i-1}} + \Delta\mathrm{CM}_X^{t_i}\left(K_{\mathrm{pcz}} + K_{\mathrm{icz}} + K_{\mathrm{dcz}}\right) \\ & - \Delta\mathrm{CM}_X^{t_{i-1}}\left(K_{\mathrm{pcz}} - 2K_{\mathrm{dcz}}\right) + \Delta\mathrm{CM}_X^{t_{i-2}} K_{\mathrm{dcz}}, \end{aligned} \tag{13.39}$$

$$\Delta\mathrm{ZMP}_X^{t_i} = {}^{\mathrm{ref}}\mathrm{ZMP}_X^{t_i} - \mathrm{ZMP}_X^{t_i}. \tag{13.40}$$

K_{pcz} represents the proportional feedback gain aiming to control the CM position via the ZMP position, while K_{dcz} and K_{icz} are the derivative and integral gains used in (13.39). In view of the fact that the system is symmetric at both ankles (joints 8 and 16 in Fig. 13.3), the same control voltage is applied, calculated in the following way:

$$\begin{aligned} \Delta u_{8,16}^{t_i} = {}& \Delta u_{8,16}^{t_{i-1}} + \Delta\mathrm{ZMP}_X^{t_i}\left(K_{\mathrm{pz}} + K_{\mathrm{dz}} + K_{\mathrm{iz}}\right) \\ & - \Delta\mathrm{ZMP}_X^{t_{i-1}}\left(K_{\mathrm{pz}} + 2K_{\mathrm{dz}}\right) + \Delta\mathrm{ZMP}_X^{t_{i-2}} K_{\mathrm{dz}}, \end{aligned} \tag{13.41}$$

$$u_{8,16}^{t_i} = u_{8,16}^{t_{i-1}} + \Delta u_{8,16}^{t_i}, \tag{13.42}$$

where $\Delta u_{8,16}^{t_i}$ designates the control (voltage) increment at the moment t_i (joints 8 and 16) calculated in the form of a PID regulator minimizing the deviation of the ZMP position. Such obtained control increment is added to the control from the preceding time instant. Expression (13.41) is obtained in a similar way as (13.39). K_{pz}, K_{dz} and K_{iz} represents proportional, derivative and integral gains with respect to the deviation of ZMP position, used in (13.41).

All the joints apart from two ankles should be stiff. Because of that, the actuators at these joints are velocity-controlled, and the desired velocity is zero. The control law is:

[2]The ZMP position when the compensating action is ended coincides with the CM position. Hence, it is important that the ZMP position to which the system returns (denoted here as the reference one) is defined so as to minimize the deviation of the actual CM position from its position prior to the action of the disturbance. This can be done in the form of the PID regulator in the following way:

$${}^{\mathrm{ref}}\mathrm{ZMP}_X^{t_i} = K_{\mathrm{pcz}}\Delta\mathrm{CM}_X^{t_i} + K_{\mathrm{icz}} \sum_{j=1}^{i} \Delta\mathrm{CM}_X^{t_j} + K_{\mathrm{dcz}}\left(\Delta\mathrm{CM}_X^{t_i} - \Delta\mathrm{CM}_X^{t_{i-1}}\right)$$

Expression (13.39) is obtained by subtracting the above expression, written for the time instant t_{i-1}, from the expression written for the time instant t_i.

$$\Delta \dot{q}_j^{t_i} = -\dot{q}_j^{t_i}, \tag{13.43}$$

$$\begin{aligned} u_j^{t_i} = u_j^{t_i-1} &+ \Delta \dot{q}_j^{t_i}\left(K_{\text{pvj}} + K_{\text{ivj}} + K_{\text{dvj}}\right) - \Delta \dot{q}_j^{t_i-1}\left(K_{\text{pvj}} - 2K_{\text{dvj}}\right) \\ &+ \Delta \dot{q}_j^{t_i-2} K_{\text{dvj}}, \quad j \in \{2..46\} \backslash \{8, 16\}. \end{aligned} \tag{13.44}$$

K_{pvj} represents proportional, K_{dvj} derivative and K_{ivj} integral gain to control the velocity at the joint j. It should be noted that these gains are different for each of the joints (Table 13.3).

Figure 13.12a shows the resulting humanoid motion obtained by the simulation. The CM position prior to the disturbance coincides with the reference one. Since only the ankle joint is active, the angle at this joint prior to the disturbance is equal to the angle at the instant when the CM position attains its reference value. From this, it comes out that the initial posture is identical to the final one, as shown in Fig. 13.12a. The figure shows also the posture when the robot is inclined most, and this corresponds to the moment when the CM velocity in the x-direction drops to

Table 13.3 Gains used in the simulation

	CM position feedback gains (Eq. 13.39)				ZMP position feedback gains (Eq. 13.41) applied at ankle joints		
	K_{pcz}	K_{dcz}	K_{icz}	Joint	K_{pz}	K_{dz}	K_{iz}
	15.64	7750	0.0022	8 and 16	15.64	39.72	0.4443
				Gains for the regulation of the angular velocity at the joint			
Joint	K_{pvj}	K_{dvj}	K_{ivj}	Joint	K_{pvj}	K_{dvj}	K_{ivj}
2 and 10	112.3	142.7	6.38	27	541.7	688	30.8
3 and 11	178.3	170.1	13.48	28	545.2	692.3	31.0
4 and 12	140.2	133.6	10.61	29	543.3	689.9	30.9
5 and 13	174.8	222	9.93	30	544.3	691.2	30.9
6 and 14	60.89	77.35	3.45	31	544.9	692.0	31.0
7 and 15	63.5	80.67	3.607	32	546.9	694.6	31.1
8 and 16	64.1	81.42	3.64	33	546.6	694.2	31.1
9 and 17	59.34	75.38	3.37	34	532.4	676.1	30.2
18	144.4	137.6	10.9	35	534.9	679.3	30.4
19	180.3	171.8	13.7	36	599.2	761.1	34.0
20	597.6	758.9	33.9	37	604.3	767.6	34.3
21	592.6	752.7	33.7	38	82.0	104.2	4.7
22	533.2	677.2	30.3	39 and 43	217.8	276.6	12.4
23	535	679.6	30.4	40 and 44	62.4	79.2	3.5
24	547.2	695	31.1	41 and 45	91.6	116.4	5.2
25	547.7	695.7	31.1	42 and 46	69.6	88.5	4.0
26	544.4	691.4	30.9				

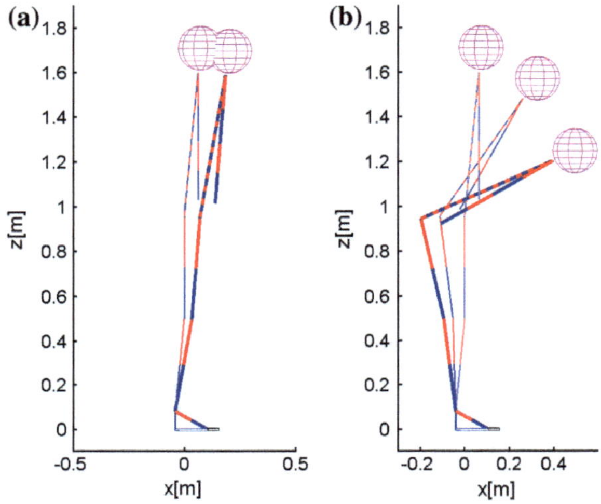

Fig. 13.12 **a** Compensating movement obtained by the ankle joint, **b** by the ankle and hip joints simultaneously

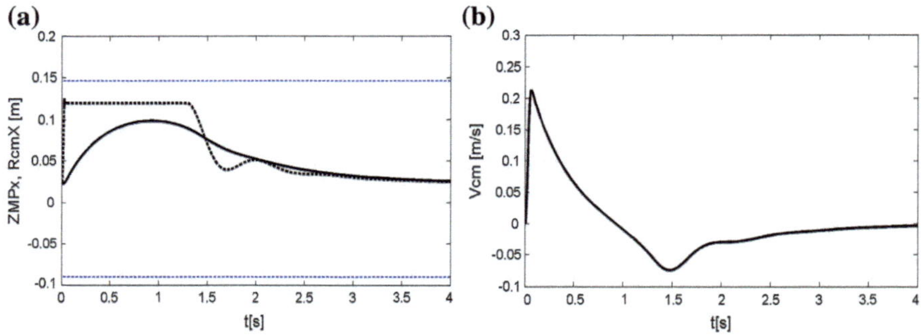

Fig. 13.13 **a** ZMP position (*thick dashed line*) and CM position (*thick full line*), foot borders (*thin dashed line*) and **b** CM velocity

zero, after which the robot gradually returns to the initial posture. Figure 13.13a shows the motion of the ZMP and CM in the x-direction, whereas Fig. 13.13b shows the velocity of the CM motion in the same direction.

As can be seen from Fig. 13.13b, there is at the beginning an abrupt jump of the CM velocity, which is a consequence of the disturbance force influence on the trunk. The maximum reference ZMP position (${}^{ref}ZMP_X^{MAX}$) is selected so that its distance to the foot front edge is 25 mm. This ensures that in the case of an overshoot or disturbance, control system can keep the ZMP within "safe area" and the system remains dynamically balanced. After the occurrence of the disturbance, the reference

ZMP position becomes equal to maximum allowed one. The actual ZMP attains quickly the reference position (Fig. 13.13a). After approximately 1 s the CM velocity drops to zero (Fig. 13.13b), after which the CM returns gradually to its initial position, and after about 4 s the robot settles down, and the CM position coincides with the position that existed prior to the action of the disturbance. If the disturbance force increases, and the CM moves in the x-direction, further from the maximum allowed ZMP position, the angular acceleration will become positive and the robot will fall down. As is evident from Fig. 13.13a, the CM position is close to the ZMP position, which means that the intensity of the applied disturbance force was close to the maximum that can be compensated for by applying this strategy. In the second case, the disturbance compensation is performed by simultaneous movement at the ankle and hip joints. In the initial moment, the robot stands still with both feet on the ground. Control applied before disturbance start is the same as in the previous case (Eq. 13.37).

The second case, we consider is compensation by ankle and hip joints simultaneously. The disturbance force acting on the robot produces a momentum of 43 kgm/s, which will cause the movement of the CM of the overall system. The disturbance is immediately followed by the compensating movement, simultaneously at both the ankle and the hip. Thus, with the aid of the ankle it will be possible again to regulate the ZMP position (Eqs. 13.41 and 13.42), but in this case we will adopt the reference ZMP position to be the maximum one that is allowed by the foot dimension:

$$ {}^{\mathrm{ref}}\mathrm{ZMP}_X^{t_i} = {}^{\mathrm{ref}}\mathrm{ZMP}_X^{\mathrm{MAX}}. \tag{13.45}$$

Instead of controlling the motion of the hip (joints 4 and 12), the inclination will be performed by the first joint of trunk (joint 18). By the trunk motion is regulated the position of the CM projection, and itis reference position is at the same place as prior to the occurrence of the disturbance. The control for this joint can be written in the form:

$$ u_{18}^{t_i} = u_{18}^{t_{i-1}} + \Delta\mathrm{CM}_X^{t_i}\left(K_{\mathrm{pc}} + K_{\mathrm{ic}} + K_{\mathrm{dc}}\right) - \Delta\mathrm{CM}_X^{t_{i-1}}\left(K_{pc} - 2\mathrm{Kdc}\right) + \Delta\mathrm{CM}_X^{t_{i-2}} K_{\mathrm{dc}}. \tag{13.46}$$

In the case considered, the $K_{\mathrm{pc}} = 3636$ is proportional, $K_{\mathrm{ic}} = 1.652$ integral and $K_{\mathrm{dc}} = 5.77 \cdot 10^5$ derivative gaint.

As in the previous case, all the joints apart from the two ankle joints (8 and 16) and the first trunk joint (18), are immobile. The motion of the ankle joints is controlled according to Eqs. (13.41) and (13.42), the control of the trunk joint (18) is given by Eq. (13.46), whereas all remaining joints are stiff and their control is performed according to Eqs. (13.43) and (13.44).

When the velocity of the CM becomes sufficiently small, the compensation is (say at the instant T) switched to the action of the ankle joint only. The motion of the first link of the trunk (joint 18) should be stopped, and this has to be done gradually, to avoid abrupt acceleration change and undesired effects on the ZMP position. Hence the control of this joint will be velocity-based:

$$\Delta\dot{q}_{18}^{t_i} = \left(1 - f\left(\frac{t_i - T}{0.8s}\right)\right)\dot{q}_{18}^{T} - \dot{q}_{18}^{t_i}, \tag{13.47}$$

$$u_{18}^{t_i} = u_{18}^{t_{i-1}} + \Delta\dot{q}_{18}^{t_i}\left(K_{\mathrm{pv}18} + K_{\mathrm{iv}18} + K_{\mathrm{dv}18}\right) - \Delta\dot{q}_{18}^{t_{i-1}}\left(K_{\mathrm{pv}18} - 2K_{\mathrm{dv}18}\right) + \Delta\dot{q}_{18}^{t_{i-2}}K_{\mathrm{dv}18}. \tag{13.48}$$

The gains for the regulation of angular velocity at the joint 18 are given in Table 13.3. In this case, the reference velocity changes with time. The function f ensures a smooth convergence of the instantaneous velocity to zero. Since this function in the initial moment is zero, the reference velocity is equal to the instantaneous one. After 0.8 s, (predefined time for the trunk deceleration), the value of the function f becomes 1, so that the reference velocity at the joint 18 will be reduced to zero. Of course, to drive the joint 18 to such state appropriate control (Eq. 13.48) is needed.

Control by the ankle and calculation of the reference ZMP position are the same as in the case of compensation by the ankle only (Eqs. 13.38–13.42). All the joints apart from the two ankles joints (8 and 16) and first joint of the trunk (18) should remain stiff, and the control law for these joints will not change (Eqs. 13.43 and 13.44).

Figure 13.12b depicts the movement obtained in three characteristic moments. It shows the posture when the disturbance force is starting, the moment when the compensation by the ankle and hip switches to the compensation by the ankle only, and the posture when the system stopped.

Figure 13.14a shows the positions of the ZMP and CM in the x-direction, whereas Fig. 13.14b shows the CM velocity in the same direction. In both figures is marked the moment when the simultaneous compensation by the ankle and hip switches to the compensation by the ankle only.

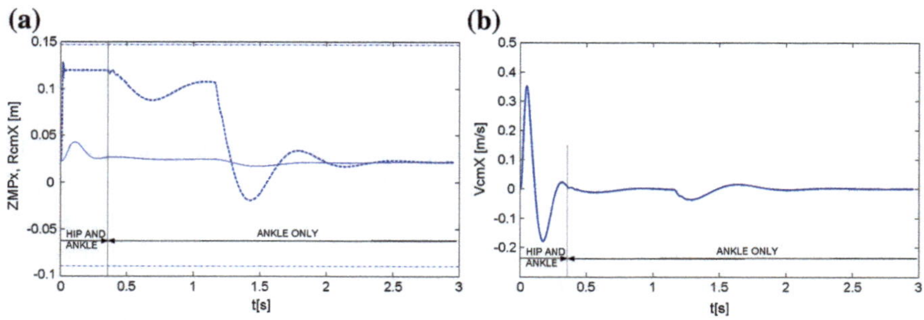

Fig. 13.14 **a** ZMP position is shown by the *thick dashed line* and CM position by the *thin full line*. *Thin dashed lines* show the foot borders **b** CM velocity. Also, it is shown the moment when the compensation by the ankle and hip switches to the compensation by the ankle only

Like in the previous case, an abrupt jump of the CM velocity can be seen (Fig. 13.13b) in the beginning of the disturbance action. By comparing Figs. 13.13b and 13.14b one can see that in the case of the compensation by the simultaneous movement at the ankle and the hip, the CM velocity drops much faster to zero, although the disturbance intensity was more than twice higher than in the previous case. As can be seen from Fig. 13.13a, the CMP shifts very little during the whole movement and in none of the instants there exists a danger that the CM comes out the support area. However, Fig. 13.11b shows large inclination of the trunk (almost 90°) what does not correspond to behavior of humans we may often observe. However, what is the disturbance intensity that can be compensated for this way, and when it is necessary to switch to the compensation by stepping, it is still an open issue. The answer to this question should be obtained by the future research.

13.6 Conclusions

In this paper, we showed that the two types of disturbances may act on humanoid robot: small and large. Each of them should be compensated for in a different way.

When considering small disturbances the compensation should be distributed onto the compensation of internal synergy (preservation of a desired form of motion) and compensation aimed at preserving dynamic balance. When there is no danger of falling, it is necessary to intensify the activity on compensation of internal synergy; when dynamic balance is endangered an urgent action is needed to prevent fall, while all other actions can be postponed. An indicator of the necessity of intensifying one or the other compensational action is the ZMP. The work demonstrated the application of a simple and efficient way of changing the "engagement degree" of corrective control for compensating internal synergy and preservation of dynamic balance.

In case of the compensation for large disturbances, it is of absolute priority to prevent the fall, i.e., to ensure that the ZMP position during the walk remains within the support area. Since the disturbances are of higher intensity, the compensating movements have to be appropriate, and the main role in them play an inertial forces. However, their action ends after the realization of the "energetic part" of the movement (which is always very short), and the moments of gravitational forces take over the dominant influence on the ZMP position. As a consequence, the compensating movement has to be such that at its end the humanoid remains in the position that ensures its static balance.

Acknowledgements This work was in part supported by the Ministry of Science and Technological Development of the Republic of Serbia under contract III44008, and in part by the Autonomous Province of Vojvodina, Provincial Secretariat for Science and Technological Development, Serbia, under contract 114-451-2116/2011-03.

References

1. Fukuda, T., Michelini, R., Potkonjak, V., Tzafestas, S., Valavanis, K., Vukobratović, M.: How far away is "artificial man". IEEE Robot. Autom. Mag. 66–73 (2001)
2. Juričić, D., Vukobratović, M.: Mathematical Modeling of Biped Walking Systems. ASME Publlications. 72-WA/BHF-13 (1972)
3. Vukobratović, M.: How to control the artificial anthropomorphic systems. IEEE Trans. Syst. Man Cybern. SMC-3, 497–507 (1973)
4. Vukobratović, M.: Legged Locomotion Systems and Anthropomorphic Mechanisms, Mihajlo Pupin Institute, Belgrade, (1975); also published in Japanese by Nikkan Shimbun Ltd. Tokyo (1976), in Russian by "MIR", Moscow (1976), in Chinese, Beijing (1983)
5. Vukobratović, M., Borovac, B., Surla, D., Stokić, D.: Biped Locomotion—Dynamics, Stability Control and Application. Springer, Berlin (1990)
6. Vukobratović, M., Borovac, B.: Zero-moment point—thirty five years of its life. Int. J. Humanoid Rob. **1**(1), 157–173 (2004)
7. Vukobratović, M., Borovac, B.: Note on the article, zero-moment point—thirty five years of its Life. Int. J. Humanoid Rob. **2**(2), 225–227 (2005)
8. Breniere, Y., Bril, B.: Development of postural control of gravity forces in children during the first 5 years of walking. Exp. Brain Res. **121**(3), 255–262 (1998)
9. Vukobratović, M., Borovac, B., Potkonjak, V.: Towards a unified understanding of basic notions and terms in humanoid robotics. Robotica **25**, 87–101 (2007)
10. Vukobratović, M., Borovac, B., Raković, M., Potkonjak, V., Milinović, M.: On some aspects of humanoid robots gait synthesis and control at small disturbances. Int. J. Humanoid Rob. **5** (1), 225–227 (2008)
11. Borovac, B., Nikolić, M., Raković, M.: How to compensate for the disturbances that jeopardize dynamic balance of a humanoid robot? Int. J. Humanoid Rob. **8**(3), 533–578 (2011)
12. Vukobratović, M., Potkonjak, V., Borovac, B., Babković, K.: Simulation model of general human and humanoid motion. Multibody Sys. Dyn. **17**(1), 71–96 (2007)

Chapter 14
New Adaptive Algorithm of Flexible Spacecraft Control

Vladislav Yu. Rutkovsky, Victor M. Glumov and Victor M. Sukhanov

Abstract A new adaptative algorithm for the system with reference model is proposed. Its operation does not depend on the intensity and spectral composition of the input actions and its realization does not require external disturbances estimation. The proposed algorithm functioning is illustrated in the example of large space structure (LSS) control. It is a new type of large-size spacecraft (space energy stations and large orbiting reflectors). Such an object cannot be inserted into orbit in assembled condition because of its big size. LSS, therefore, is assembled in orbit and it is a discretely evolving structure. As the control object, it is a multi-frequency oscillating system with discretely time-varying parameters and a number of freedom degrees. In the case of large space structure control proposed algorithm was simplified. Nevertheless, the control system of LSS realizes high precision.

14.1 Introduction

Since the emergence of adaptive systems with a reference model (ASRM) ([1–7] and others), they have been addressed in hundreds of monographs and thousands of papers. The interest in ASRM is motivated by the constructive statement of the problem and by the beauty of the mathematical tools involved. Indeed, the existence of a reference model provides the goal to be achieved. This implies the possibility of applying analogs of the feedback principle to the designing of the system. This demonstrates the constructive character of ASRM. The beauty of the mathematical tools is manifested in that the Lyapunov's direct method is effectively used not only to analyze motion but also to seek algorithms for functioning ASRM that are nonlinear, nonstationary multiply connected dynamical system.

V.Yu. Rutkovsky (✉) · V.M. Glumov · V.M. Sukhanov
Institute of Control Sciences, Russian Academy of Science, Profsoyuznaya 65, 117997, Moscow, Russia
e-mail: rutkov@ipu.ru

G.M. Dimirovski (ed.), *Complex Systems*, Studies in Systems, Decision and Control 55, DOI 10.1007/978-3-319-28860-4_14

However, despite the intensive theoretical development of ASRM, the application of this class of systems in practice is rather modest, which is caused by a number of substantial shortcomings of ASRM. Specifically, the dynamic accuracy of ASRM motion depends on the intensity and spectral composition of input actions, on the rate of variations of parametric and coordinate disturbances, and on the rather detailed information on the state vector of the system. All these factors make the dynamics of ASRM motion less predictable, which discourages the developers of actual systems. Along with this, very high dynamic accuracy of motion (precision dynamics) is required in modern promising control system operating under uncertainty with regard to coordinate and parametric disturbances.

It is well known that the class of variable structure systems (VSS) was under active development for the control of nonstationary objects [8]. In the VSS class, the dependence of the system operation on the intensity and spectral composition of input actions and on the rate of variations in parametric and coordinate disturbances decreases to a substantial degree. However, the VSS class has its own shortcomings. For example, the range of parametric disturbances must be bounded. Another factor is that the coefficients of regulators vary in jump, which can adversely activate the neglected system dynamics.

In the paper by [9] a new ASRM operation algorithm is proposed that combines the ASRM and VSS construction principles in the control of nonstationary systems so as to converge the positive features of each of these principles and, at the same time, to ameliorate their negative features and to achieve their smooth synergy. As a net result, the proposed class of ASRM acquires new properties. These properties include the independence of the ASRM motion on the intensity and spectral composition of input actions, its independence of the range and the rate of variations of parametric and coordinate disturbances, and high dynamic accuracy of ASRM motion.

The proposed algorithm functioning is illustrated by application to the example of controlling a large space structure (LSS). This example case is of particular interest because of the ongoing exploration endeavors in astronautics and interplanetary space explorations.

Developments of some global projects for the next stage of mastering the space started in the mid-1980s in several countries worldwide. The required design of a new type of spacecraft was christened the large space structures (LSS) [10]. The scale of these projects may be illustrated by two well-known examples. For instance, to replace the decreasing resources of energy carriers, it is planned to construct the large solar power stations provided with solar cell panels, with a size running up to that of a football ground, in the near-Earth space for convenience of future explorations. There also exists a project of using the large orbiting reflectors to illuminate by solar light the northern regions during the polar nights in order to promote development of these underpopulated territories.

It is clear that realization of such projects requires insertion into orbits or in-orbit assembly of large structures and control of their angular position. At the same time, the specific properties of this new class of objects poses new challenges because of: the infra-low (of the order of $0.01 \div 0.1$ Hz) structural frequencies approaching the control frequencies of the LSS "rigid" motion; the impossibility of carrying out the ground-based checkout tests and, as a result, very poor definiteness not only of the object model coefficients but of its structure as well; almost complete lack of external and internal damping of the vibrations; and the large number of the LSS degrees of freedom. These facts have suggested the following conclusion, which appeared in the USA Government-owned journal [10]: "... All solutions of the problems of control of the elastic satellites that were by now established by the international scientific community only superficially touched upon the specificity that properly manifests itself only in the problems of controlling the large space structures, a new class of the space objects of the near future. Therefore, new efforts of the experts are required to solve these problems which are important for the entire mankind."

Such object as LSS cannot be inserted into orbit in assembled condition because of its big size. Hence, LSS in the course of its in-orbit assembly is a discretely evolving structure. As the control object it is a multi-frequency oscillating system with discretely time-varying parameters and a number of freedom degrees. In the case of the LSS control, the proposed algorithm was simplified.

Further this manuscript is organized as follows. Section 14.2 is focused on the synthesis of the new adaptation algorithm on the grounds of an adequate dynamics model adopted. In Sect. 14.3, the exposition continues by investigating the model of such an LSS planet. The solution to the actual problem of controlling such an LSS plant is then presented in Sect. 14.4. The paper ends with a certain relevant conclusion and references used.

14.2 Adaptation Algorithm Synthesis

Let us consider a control system obeying the following mathematical model (MM)

$$\mathrm{d}\varphi/\mathrm{d}t + A(t)\varphi = B(t)g(t) + f(t) + S, \tag{14.1}$$

where t is time and t_0 is the initial point of considering $t \geq t_0$; $\varphi^T \in R^n$, $g \in R^m$, $f \in R^n$ $(m \leq n)$—are the vectors of state, control action, and coordinate disturbance, respectively; $A(t)$ and $B(t)$ are matrices with unknown variable components; $S \in R^n$ is the vector of additional feedback, its components are used to compensate for the coordinate and parametric disturbances. The vectors $\varphi(t)$ and $g(t)$ are assumed to be measurable and their components are continuously differentiable functions.

The matrices $A(t)$ and $B(t)$ are represented in the form

$$A(t) = A_0 + \Delta A(t), \quad B(t) = B_0 + \Delta B(t). \tag{14.2}$$

Here A_0 and B_0 are the matrices that are chosen from the condition of the system desired reaction on the control action at $f(t) \equiv 0$ and $S \equiv 0$, $(-A_0)$ is a Hurwitz matrix, $\Delta A(t)$ and $\Delta B(t)$ are unknown parametric disturbances and their components are continuously differentiable functions.

Equation (14.1) with regard to the correlation (14.2) are rewritten as

$$d\varphi/dt + A_0\varphi = B_0 g(t) + d(t) + S, \tag{14.3}$$

where $d(t) = -\Delta A(t)\varphi(t) + \Delta B(t)g(t) + f(t)$ is a vector function with nonmeasurable components.

The goal is to find algorithms for varying the components of S so as to compensate for the influence of the disturbing vector $d(t)$ on the motion of the system described by model (14.3). The problem is solved assuming that the components $d_i(t),\ i = (\overline{1,n})$ of the vector $d(t)$ are not bounded, whereas the rate of their variation is bounded, that is, $|d(d(t))/dt| \leq \mu_{i0},\ \mu_{i0} = \text{const}$, and numerical values of μ_{i0} are given and can be rather large. So the object of our control system is essentially a nonstationary one.

Notice further that in the considered case the equation of reference model will be as

$$d\varphi_m/dt + A_0\varphi_m(t) = B_0 g(t), \tag{14.4}$$

where $\varphi_m \in R^n$.

Algorithm of adaptation (algorithm for varying the components of the vector S) is sought depending on the measurable information so that $\varphi(t)$ converges asymptotically to the motion $\varphi_m(t)$ irrespective of the intensity and the spectral composition of the input disturbances and the ranges of parametric and coordinate disturbances with predictability and high dynamic accuracy of ASRM motion.

Introducing the notation

$$\varepsilon = \varphi - \varphi_m,\ y = d(t) + S,\ d(d(t))/dt = \mu,\ dS/dt = \psi, \tag{14.5}$$

we derive from (14.3) and (14.4) the system of equations

$$d\varepsilon/dt + A_0\varepsilon = y,\ d/dt y = \mu(t) + \psi. \tag{14.6}$$

Here, $\psi = \psi(t, \varepsilon)$ is the desired algorithm of adaptation.

Consider now the equilibrium motion

$$\varepsilon = 0,\ y \equiv 0 \tag{14.7}$$

of system (14.6). The algorithm for varying ψ is chosen from the condition that the motion $\varepsilon(t)$ and $y(t)$ of system (14.6) converges asymptotically to (14.7). For this purpose, we define the Lyapunov function

$$V(\varepsilon, y) = \chi(\varepsilon^T P\varepsilon) + y^T y, \tag{14.8}$$

where $\chi = \text{const} > 0$, P is a symmetric positive definite matrix defined by the equality $-(A_0^T P + PA_0) = Q$ and Q is a prescribed negative definite matrix following the Lyapunov matrix equation.

In view of system (14.6), the time derivative of function (14.8) is

$$\mathrm{d}V(\varepsilon, y)/\mathrm{d}t = \chi\left(\varepsilon^T Q\varepsilon\right) + 2y^T[\chi\sigma + \mu(t) + \psi], \tag{14.9}$$

where $\sigma = P\varepsilon$.

The desired algorithm ψ is chosen from the condition $\mathrm{d}V(\varepsilon, y)/\mathrm{d}t < 0$ for $\varepsilon \neq 0$ and $y \neq 0$. For this purpose, we consider the algorithm [9]

$$\psi = -\chi\sigma - K\,\mathrm{sign}(y), \tag{14.10}$$

where K is diagonal matrix $K = \mathrm{diag}(k_1, \ldots, k_n)$ and $\mathrm{sign}(y)$ is a vector of the form $[\mathrm{sign}(y)]^T = [\mathrm{sign}(y_1), \ldots, \mathrm{sign}(y_n)]$.

With regard to (14.10) equality (14.9) is rearranged in

$$\mathrm{d}V(\varepsilon, y)/\mathrm{d}t = \chi\left(\varepsilon^T Q\varepsilon\right) - 2|y|^T \rho(t), \tag{14.11}$$

where $|y|^T = (|y_1|, \ldots, |y_n|)$, $[\rho(t)]^T = [\rho_1, \ldots, \rho_n]$ and $\rho_i(t) = k_i - \mu_i \mathrm{sign}(y_i)$, $i = \overline{1, n}$.

Under the condition

$$k_i > |\mu_i(t)|, \tag{14.12}$$

we have $\rho_i(t) > 0$, and it follows from (14.8) and (14.11) that

$$V(\varepsilon, y) > 0, \ \mathrm{d}V(\varepsilon, y)/\mathrm{d}t < 0, \tag{14.13}$$

inequalities (14.13), which were obtained irrespective of the form of the components of $g(t)$ and $\mathrm{d}(t)$ under condition (14.12), ensure the asymptotic stability of the motion (14.7) of the system (14.6).

We represent the manifold-based motion (MM) of the system (14.6) with algorithm (14.10) for ψ as

$$\mathrm{d}^2\varepsilon/\mathrm{d}t^2 + A_0\,\mathrm{d}\varepsilon/\mathrm{d}t + \chi P\varepsilon = -R(t)\,\mathrm{sign}(y). \tag{14.14}$$

Here $R(t) = \mathrm{diag}(\rho_1(t), \ldots, \rho_n(t))$ is a diagonal matrix. We can state that motion

$$\varepsilon \equiv 0, \ \ \mathrm{d}\varepsilon/\mathrm{d}t \equiv 0 \tag{14.15}$$

is asymptotically stable. In conjunction with this statement there should be noted two peculiarities of algorithm (14.10).

First, system (14.14) goes over to the class of systems with mathematical models described by differential equations with discontinuous right-hand side. In such system, a sliding mode may occur on one or more surfaces of discontinuity. This problem requires a special consideration. But it is very important that sliding mode does not violate inequalities (14.13).

Second, algorithm (14.10) was obtained under the condition of measurability of the vector $\mathrm{sign}(y)$. However, it was not indicated as to how to acquire this information. Indeed, if one assumes that the vector y is measurable, then the problem of adaptation vanishes. This can be seen from the correlation $y = \mathrm{d}(t) + S$ where the vector $\mathrm{d}(t)$ is not measurable by problem formulation. It is obvious that the vector can be calculated from the first equation of MM (14.6), but the derivative $\mathrm{d}\varepsilon/\mathrm{d}t$ is not measurable.

To solve this contradiction an additional link is introduced in the mathematical model (14.6). This link corresponds to the small parameters that was not taken into consideration earlier

$$\tau\,\mathrm{d}x/\mathrm{d}t + x = \varepsilon, \tag{14.16}$$

where $x \in R^n$, $\tau = \mathrm{const} > 0$ is a known small value.

Taking into consideration (14.16) the system motion is described as

$$\tau \mathrm{d}^2x/\mathrm{d}t^2 + (E + \tau A_0)\mathrm{d}x/\mathrm{d}t + A_0 x = y, \quad \mathrm{d}y/\mathrm{d}t = \mu(t) + \psi. \tag{14.17}$$

Here $E \in R^{n \times n}$ is identity matrix. Now the vector $\mathrm{d}x/\mathrm{d}t$ in the second order Eq. (14.17) is measurable.

Further it is required to solve the next problems: (14.1) what is the motion of system (14.17) with the adaptation algorithm (14.10) at small but final values of τ or is there an interval of values of τ for which the motion $x \equiv 0,\ y \equiv 0$ asymptotically stable; (14.2) if such interval of τ does not exist then is there a domain into which the motion of the system (14.16) enters after a finite time interval from any initial point and remains there, at that the domain includes motion $x \equiv 0,\ y \equiv 0$? If such domain does exist, then what are its sizes? And we are to estimate these sizes in order to draw a conclusion about practical acceptability of the system.

Let us consider these problems to investigate the relevant solving answers.

Before proceeding further we introduce the notations $z = (z_1, z_2)$, where $z_1 = x,\ z_2 = \mathrm{d}x/\mathrm{d}t$ and $y = (y_1, y_2),\ y_1 = 0$, and rewrite system (14.17) as

$$\mathrm{d}z/\mathrm{d}t = A_1 z + Dy, \quad \mathrm{d}y_2/\mathrm{d}t = \mu + \psi, \tag{14.18}$$

where block matrices $A_1 \in R^{2n \times 2n}$, $D \in R^{2n \times 2n}$ are:

$$A_1 = \begin{vmatrix} 0 & E \\ -\tau^{-1}A_0 & -\tau^{-1}(E + \tau A_0) \end{vmatrix}, \quad D = \begin{vmatrix} 0 & 0 \\ 0 & \tau^{-1}E \end{vmatrix}.$$

Let us consider non-perturbed motion

$$z \equiv 0, \quad y_2 \equiv 0. \tag{14.19}$$

Again Lyapunov function is chosen as

$$V(z, y_2) = \chi(z^T P_1 z) + (y_2^T y_2), \tag{14.20}$$

where $\chi = \text{const} > 0$, $P_1 \in R^{2n \times 2n}$ is a symmetric positive definite matrix defined by the equality $(A_1^T P_1 + P_1 A_1) = Q_1$, Q_1 is a prescribed negative definite matrix, A_1 is the Hurwitz matrix.

In view of system (14.19) the time derivative of function (14.20) at adaptation algorithm that is the analog of algorithm (14.10)

$$\psi = -\chi\tau^{-1} P_{21} z_1 - K\,\text{sign}(N) \tag{14.21}$$

it follows that

$$\mathrm{d}V/\mathrm{d}t = \chi\left(z^T Q_1 z\right) + 2\chi\tau^{-1}\left(y_2^T P_{22} z_2\right) - 2[y_2 \bullet \text{sign}(N)]^T \rho(t). \tag{14.22}$$

Here $N = z_2 + A_0 z_1$, $P_{ij} \in R^{n \times n}$, $\left(i, j = \overline{1, n}\right)$ are block matrices of the matrix $P_1 = \begin{vmatrix} P_{11} & P_{12} \\ P_{21} & P_{22} \end{vmatrix}$. The other involved quantities are as follows: $K = \text{diag}(k_1, \ldots, k_n)$, $[\text{sign}(N)]^T = [\text{sign}(N_1), \ldots, \text{sign}(N_n)]$, $[y_2 \bullet \text{sign}(N)]^T = [y_{21}\text{sign}(N_1), \ldots, y_{2n}\text{sign}(N_n)]$, $[\rho(t)]^T = [\rho_1(t), \ldots, \rho_n(t)]$, $\rho_i(t) = k_i - \mu_i \text{sign}(N_i)$, $k_i > |\mu_1|$, $\rho_i^{\min} \le \rho_i \le \rho_i^{\max}$, $\rho_i^{\min}, \rho_i^{\max} > 0$.

Nonetheless, it is worthwhile to consider a modified algorithm such as

$$\psi = -\chi\tau^{-1} P_{21} z_1 - \chi\tau^{-1} P_{22} z_2 - K\,\text{sign}(N). \tag{14.23}$$

With this algorithm the time derivative of Lyapunov function is

$$\mathrm{d}V/\mathrm{d}t = \chi(z^T Q_1 z) - 2[y_2 \bullet \text{sign}(N)]^T \rho(t). \tag{14.24}$$

Expression (14.24) can be rewritten as

$$\mathrm{d}V/\mathrm{d}t = \chi(z^T Q_1 z) - 2|N|^T \rho(t) - 2\tau[(z_3 + A_0 z_2) \bullet \text{sign}(N)]^T \rho(t), \tag{14.25}$$

where $z_3 = \mathrm{d}z_2/\mathrm{d}t$.

From the equality (14.25) it is obvious that this function is not negative definite one. Thus, there is no interval of τ at which motion

$$x \equiv 0, \quad dx/dt \equiv 0, \quad y_2 \equiv 0 \tag{14.26}$$

or

$$x \equiv 0, \quad dx/dt \equiv 0, \quad d^2x/dt^2 \equiv 0 \tag{14.27}$$

is asymptotically stable.

However, the derivative (14.24) in the domain

$$-\chi(z^T Q_1 z) > 2|y_2|^T \rho(t) \tag{14.28}$$

takes only negative values. It is, therefore, possible to guarantee: all phase trajectories of phase space of $3n$-dimension from any initial position at $t = t_0$ into domain $V(z_0, y_{20}) > 0$, $dV(z_0, y_{20})/dt < 0$ after a finite time interval enters into the domain that contains the origin of coordinates and, moreover, it is limited from above by the surface

$$V(z, y_2) = C_b, \quad C_b = V(z_b, y_{2b}). \tag{14.29}$$

Here z_b, y_{2b} is the solution of Eq. (14.28) at replacement of symbol of inequality on the symbol of equality. Domain (14.29) determines the guaranteed accuracy of the control system. But it is important to notice that this estimation of the system guaranteed accuracy is considerably understated. Real accuracy appears much higher.

From the expression (14.25) it follows that one can choose a very small value of τ at which the guaranteed accuracy can be considered as a precision one. Notice that, at $\tau = 0$, the derivative (14.25) coincides with the expression (14.11).

14.3 Mathematical Model of the Large Space Structure

It is supposed that the LSS plant has a framed (trussed) construction and it is assembled on orbit of light rod elements by means of free-flying space manipulation robots. During mounting and subsequent operation LSS must be controllable in order to maintain the desired orientation and stabilization. To this end, the LSS needs a control unit which plays the role of the carrying (main) body to which the other elements eventually making up the desired framed surface are attached discretely in a certain sequence.

Further, we consider for definiteness a model of LSS of an "umbrella" type (see Fig. 14.1), which can be suited for description of the regular structures such as large solar power stations or large orbiting reflectors [11].

At low-angular velocities of the LSS and to a first approximation its three-dimensional motion can be considered as three independent plane motions

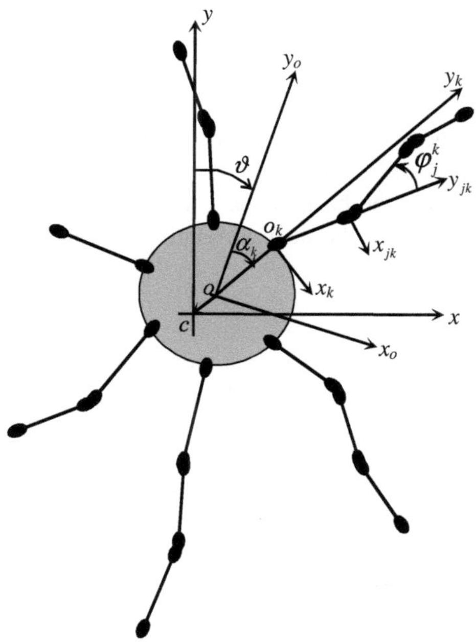

Fig. 14.1 Current structure of the DES

(i.e., yaw, pitch, and roll). In this case the total mathematical model can be separated on three independent modal-physical models [12]:

$$\begin{aligned}
&\ddot{\bar{x}} = m_n(u), m_n(u) = M_r(u)I_{\Sigma n}^{-1}, \quad I_{\Sigma n} = (I_n + J);\\
&\ddot{\tilde{x}}_i + \omega_{in}^2\tilde{x}_i = \tilde{k}_{in}m_n(u);\\
&x_n = \bar{x}_n + \tilde{x}_n, \quad \tilde{x}_n = \sum \tilde{x}_{in}, \qquad i = \overline{1, n}.
\end{aligned} \tag{14.30}$$

Here x is a measurable controlled angular coordinate, $\bar{x}$ is the object coordinate if it would be rigid, $\tilde{x}$ is an additional deviation of the main body due to the object elasticity, and parameters $\tilde{\omega}_{in}, \ \tilde{k}_{in}$ are the fundamental frequencies and the vibration modes coefficients, $I_{\Sigma n}$ is the summary moment of inertia of the LSS together with motor-flywheel (*J*); $M_r(u)$ is the control action; u is the control law, n is the number of the assembling stage, $n = \overline{1, N}$.

Control actions $M_r(u)$ in Eq. (14.30) are the moments of dynamic response forces that appear at flywheel acceleration or deceleration. These control moments are applied to the LSS main body

$$M_r = -M_d + M_f, \tag{14.31}$$

where M_d is the moment created by the motor-flywheel and M_f is the drag torque moment on the engine shaft.

For the ball-barring DC and AC motors, as employed actuator drivers, we get

$$\Delta M_d = k_u \Delta u - k_\Omega \Delta \Omega, \tag{14.32}$$

where $k_u,\ k_\Omega > 0$ are constant electromechanical coefficients of the barring motor and $\Delta\Omega$ is the increment of flywheel speed at changes of the control voltage Δu.

On the basis of the law of the momentum conversation for mechanical system FS-motor the next correlation is correct

$$\Delta\Omega = -I_n J^{-1} \dot{x}, \tag{14.33}$$

where I_n, J are inertia moments of the LSS and the motor respectively.

Substitution of (14.33) into (14.32) we provide the expression for the barring motor moment

$$\Delta M_{dn}(u) = k_{\omega n}\dot{x} + k_u \Delta u, \quad k_{\omega n} = I_n J^{-1} k_\Omega > 0. \tag{14.34}$$

Now, on the assumption that $\Delta M_f \approx 0$ the control moment M_r acting on the flexible structure (FS) from the motor-flywheel is given by

$$M_{rn}(u) = -(k_{\omega n}\dot{x} + k_u u). \tag{14.35}$$

By substituting (14.35) into Eq. (14.30) we obtain modal-physical model of the LSS plane motion with gyro-force stabilizer like motor-flywheel as the actuator

$$\begin{aligned}
&\ddot{\bar{x}} + k_{\omega n}\dot{\bar{x}} = k_{un}u - k_{\omega n}\dot{\tilde{x}},\\
&\ddot{\tilde{x}}_i + \tilde{k}_{in}k_{\omega n}\dot{\tilde{x}}_i + \tilde{\omega}_{in}^2\tilde{x}_i = \tilde{k}_{in}k_{un}u - \tilde{k}_{in}k_{\omega n}\Big(\dot{\bar{x}} + \sum_{j=1,\, j\neq i}^{n} \dot{\tilde{x}}_j\Big),\\
&x = \bar{x} + \tilde{x}, \tilde{x} = \textstyle\sum_1^n \tilde{x}_i, \quad i = \overline{1,n}, \ n = \overline{1,N}, \ i \le n.
\end{aligned} \tag{14.36}$$

where $k_{\omega n} = k_\omega I_{\Sigma n}^{-1} > 0, \quad k_{un} = -k_u I_{\Sigma n}^{-1}$; $k_{\omega n}\dot{\tilde{x}} \doteq \tilde{f}_{1n}$ is a perturbation caused by the elastic vibrations and acting on the LSS "rigid" motion via the motor-flywheel.

As it can be seen from (14.36), use of the motor-flywheel as the actuator of the LSS orientation system gave rise to damping in the LSS motion equations that is defined by the moment dependence M_d from flywheel angular velocity. Nevertheless, since the coefficient $k_{\omega n} = k_\omega I_\Sigma^{-1}$ is sufficiently small the aforementioned damping is insignificant and the system (14.9) motion stability can be lost because of the perturbations on the right-hand sides $\tilde{f}_{1n} = k_{\omega n}\dot{\tilde{x}}$ and $\tilde{f}_{2n} = \tilde{k}_{in}k_{\omega n}(\dot{\bar{x}} + \sum_{j=1,\, j\neq i}^{n} \dot{\tilde{x}}_j)$.

14.4 Design of the LSS Control

The LSS as a control object is a nonstationary one. Under terrestrial conditions the coefficients of its mathematical model cannot be calculated precisely. All frequencies $\tilde{\omega}_{in}$ and the exciting coefficients $\tilde{k}_{in}$ of vibration modes are changed at the very attaching of the LSS new element to the previous construction. In this connection the vibration damping is a rather complicated problem.

Let us consider the possibility of suggested algorithm used for this problem's solution. We shall assume that our LSS has a main body and eight attached to it flexible elements, so $N=8$.

The control law is realized on the computer (therefore the control system is a discrete one)

$$u_k = k_0(-k_1x_k - k_2\frac{\Delta x_k}{T_0} + S_k),$$
$$\Delta x_k = x_k - x_{k-1}, \quad k = 1, 2, \ldots. \tag{14.37}$$

Here S_k is the signal of a second loop of the regulator which is destined for decreasing the vibration modes influenced on the attitude system accuracy. The period of discreteness was chosen $T_0 = 0.01$ s, other parameters were: $k_0 = 0.03$, $k_1 = 7.5$, $k_2 = 275$.

As the reference model was chosen the system (14.36), at $\tilde{x}_i \equiv 0$, that is,

$$\dddot{x}_M + k_{\omega M}\dot{x}_M = k_{uM}u, \quad k_{\omega M} = k_{\omega n}, \quad k_{uM} = k_{un}, \quad n = 3. \tag{14.38}$$

For the considered task, adaptive algorithm (14.21) was simplified to the next one

$$\psi = -\chi(p_{21}\varepsilon + p_{22}\frac{d\varepsilon}{dt}) - K\,\mathrm{sign}(c_1\varepsilon + c_2\frac{d\varepsilon}{dt} + \frac{d^2\varepsilon}{dt^2}), \tag{14.39}$$

where $p_{21}, p_{22}, c_1, c_2 = \mathrm{const}$.

Therefore at $n = \overline{1, 8}$ in our system which has the total order $n_1 = 2(n+1)$, that is, $n_1 = \overline{2, 18}$. Only the first and second derivatives in the adaptive algorithm (14.39) are used. In discrete form (using Euler method) we shall have

$$S_k = S_{k-1} + T_0[-\chi(p_{21}\varepsilon_k + p_{22}\Delta\varepsilon_k)$$
$$- K\,\mathrm{sign}(c_1\varepsilon_k + c_2\Delta\varepsilon_k + \Delta^2\varepsilon_k)], \tag{14.40}$$

where $\varepsilon_k = x_k - x_{kM}$, $\Delta\varepsilon_k = T_0^{-1}(\Delta x_k - \Delta x_{kM})$, $\Delta^2\varepsilon_k = \Delta\varepsilon_k - \Delta\varepsilon_{k-1}$.

The coefficients of algorithm (14.40) were as follows:

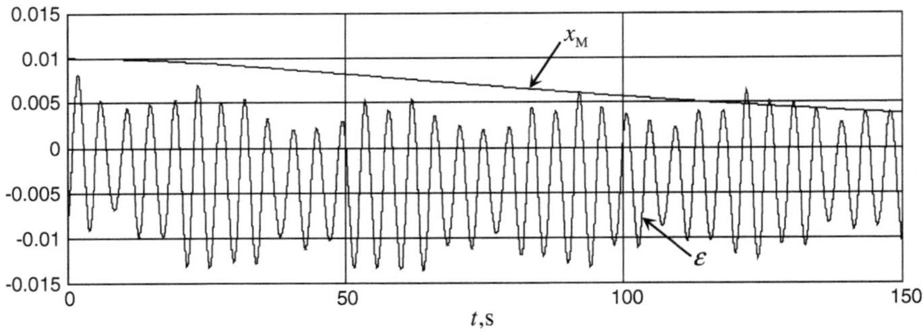

Fig. 14.2 The processes $x_M(t)$, $\varepsilon(t)$ at $S_k \equiv 0$, $n=3$

$$\chi = 0.2;\ p_{21} = 267;\ p_{22} = 887;\ K = 350;\ c_1 = 0.0075;\ c_2 = 0.3.$$

The results of the LSS dynamics computer simulations for the case $n=3$ without the adaptive control ($S_k = 0$) are shown in Fig. 14.2.

Parameters of the LSS for $n=3$ were: $k_{\omega 3} = 0.03$; $k_{u3} = 0.002$. These small coefficient values are explained by very big values of the LSS moment of inertia $I_{\Sigma n}$. And parameters of vibration modes were: $\tilde{\omega}_{13}^2 = 2.15\,\mathrm{s}^{-2}$, $\tilde{k}_{13} = 0.04$; $\tilde{\omega}_{23}^2 = 2.75\,\mathrm{s}^{-2}$, $\tilde{k}_{23} = 0.11$; $\tilde{\omega}_{33}^2 = 4.4\,\mathrm{s}^{-2}$, $\tilde{k}_{33} = 0.12$. One can see that undamped vibrations occur in the system.

With the suggested additional control S_k that corresponds to the algorithm (14.40) the vibrations are damped very effectively (see Fig. 14.3). At $t = 75$ s the error ε_k (or $\varepsilon(t)$) is practically equal to zero.

The results of the LSS dynamics computer simulations for the assembling stage $n=6$ are shown in Figs. 14.4 and 14.5. The LSS parameters on this stage were chosen as follows:

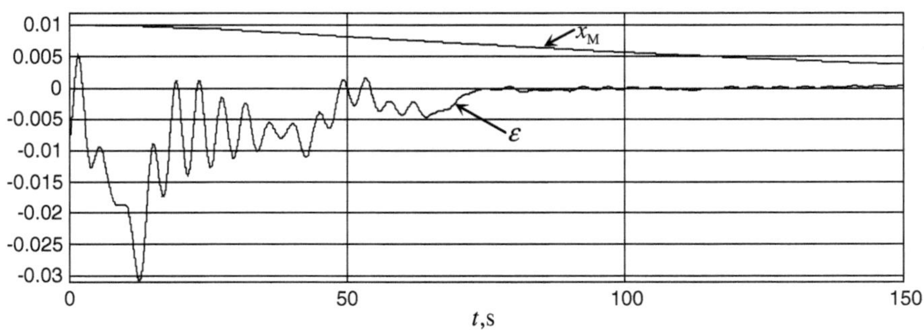

Fig. 14.3 The processes $x_M(t)$, $\varepsilon(t)$ at $S_k \neq 0$, $n=3$

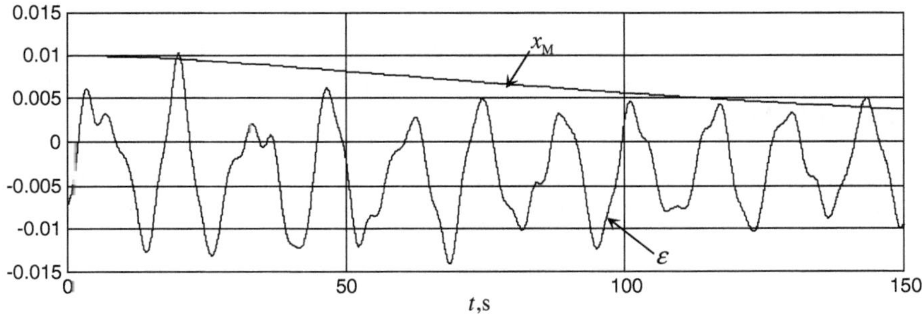

Fig. 14.4 The processes $x_M(t)$, $\varepsilon(t)$ at $S_k \equiv 0$, $n=6$

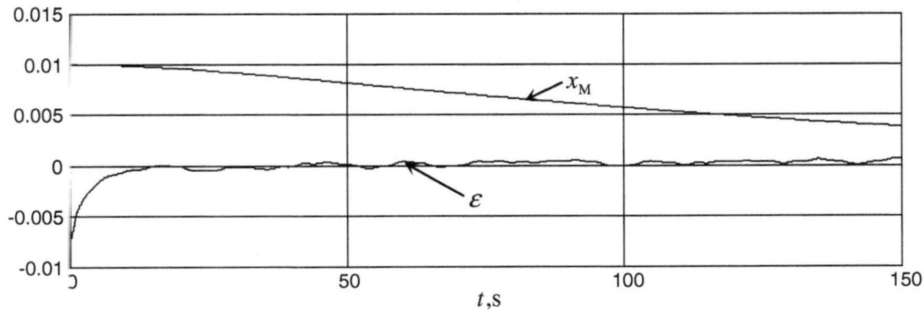

Fig. 14.5 The processes $x_M(t)$, $\varepsilon(t)$ at $S_k \neq 0$, $n=6$

$k_{\omega 6}=0.025$; $k_{u6}=0.0015$; $\tilde{\omega}_{16}^2=0.21\,\mathrm{s}^{-2}$, $\tilde{k}_{16}=0.12$; $\tilde{\omega}_{26}^2=0.5\,\mathrm{s}^{-2}$, $\tilde{k}_{26}=0.55$;
$\tilde{\omega}_{26}^2=0.5\,\mathrm{s}^{-2}$, $\tilde{k}_{26}=0.55$; $\tilde{\omega}_{36}^2=2.2\,\mathrm{s}^{-2}$, $\tilde{k}_{36}=0.14$; $\tilde{\omega}_{46}^2=3.5\,\mathrm{s}^{-2}$, $\tilde{k}_{46}=0.28$;
$\tilde{\omega}_{56}^2=9.5\,\mathrm{s}^{-2}$, $\tilde{k}_{56}=0.01$; $\tilde{\omega}_{66}^2=15.2\,\mathrm{s}^{-2}$, $\tilde{k}_{66}=0.04$.

In Fig. 14.4 is shown the processes of the case when the adaptive control contour was cut out. The vibrations were not damped.

When the adaptive control was switched on the vibrations were damped to the instant $t=25$ s (see Fig. 14.5).

14.5 Conclusion

The suggested novel adaptation algorithm (14.39) for LSS was obtained as the result of a very significant simplification of algorithm (14.21). For LSS, it can be considered only as a heuristic algorithm, but the quality of control with this algorithm appeared rather high.

The presented numerical examples and simulation results have confirmed the validity of the proposed control design method by simultaneously achieving two objectives: (a) the minimal feasible steady-state vector norm and (b) the prescribed operational SM dynamics.

Acknowledgments The work reported in this paper is a contribution to Project 14-08-01091 funded by the Russian Foundation for Basic Research, for which the authors remain grateful.

References

1. Donalson, D.D., Leondes, C.T.: A model referenced parameter tracking technique for adaptive control systems. IEEE Trans. Appl. Ind. **82**, 241–262 (1963)
2. Hiza, J.G., Li, C.C.: Analytical synthesis of class model reference time-varying control systems. IEEE Trans. Ind. Appl. **9**, 432–441 (1963)
3. Whitaker, H.P., Jarom, J., Kezer, A.: Design of Model Reference Adaptive Control Systems for Aircraft. Massachusetts Institute of Technology, Lab. Report R-164, Sept 1958
4. Parks, P.C.: Lyapunov redesign of model reference adaptive control systems. IEEE Trans. Automat. Control AC **11**(3), 362–367 (1966)
5. Zemlyakov, S.D., Rutkovsky, V.Y.: On the synthesis of self-adjusting control systems with reference model. Avtomatika i Telemechanika **3**, 70–77 (1966)
6. Morgan, A.P., Narendra, K.S.: On the stability of non-autonomous differential equations dx/dt = [A + B(t)]x with skew symmetric matrix. SIAM J. Control Optim. **15**(1), 163–176 (1977)
7. Yuan, J.S.C., Wonham, W.M.: Probing Signals for Model Reference Identification. University of Toronto, Systems Control Report No 7613 (1976)
8. Emel'yanov, S.V.: Automatic Control Systems with Variable Structure. Nauka, Moscow (1967)
9. Zemlyakov, S.D., Rutkovsky, V.Y.: Precision control algorithm for a dynamic system under uncertainty based on an adaptive system with reference model. Dokladi Rossiyskoy Akademii Nauk **429**(2), 176–179 (2009)
10. Nurre, G.S., Ryan, R.S., Skofield, H.N., Sims, J.I.: Dynamics and control of large space structures. J. Guidance **77**(5), 514–526 (1984)
11. Rutkovsky, V.Y., Sukhanov, V.M.: Large space structure: models, methods of study and control, parts I and II. Autom. Remote Control Part 1 in **57**(7), 953–963; Part 2 in **57**(8), 1108–1117 (1996)
12. Glumov, V.M., Zemlyakov, S.D., Rutkovsky, V.Y., Sukhanov, V.M.: Spatial angular motion of flexible spacecraft. the modal physical model and its characteristics part 1. Autom. Remote Control **59**(12), 1728–1738 (1998)

Chapter 15
State-Dependent Riccati Equation-Based Tracking Control of a Hydraulic Seismic Isolator Test Rig

Stefano Pagano, Riccardo Russo, Salvatore Strano and Mario Terzo

Abstract This paper presents a nonlinear control for a hydraulic seismic isolator test rig with significant nonlinearities. These are the key factors causing delay and error in the hydraulic actuation response and highly limit the performances of the classical linear control. Hence, a nonlinear control design based on the mathematical model of the hydraulic actuator is employed. The proposed approach consists of a nonlinear feedback control based on the state-dependent Riccati equation (SDRE). The control performance is demonstrated by means of real-time experiments.

15.1 Introduction

Structural responses under strong external excitations, such as earthquakes and wind storms, can be mitigated by means of base isolation strategies [1, 2]. The base isolation is typically effected using passive, semi-active or active systems [3, 4]. Passive and semi-active isolators have to be opportunely tested in order to evaluate their characteristics in terms of restoring force and energy loss. To this end, these devices are tested in order to obtain the force-displacement cycle that allows an analytical description of their dynamic characteristics to be deduced [5, 6]. The test rigs typically use hydraulic cylinders to impose periodic deformations on the isolator and a control system guarantees that the desired displacement law is tracked. The electrohydraulic actuation system is characterized by a high power/mass ratio and a fast response [7]. At the same time, it exhibits significant nonlinear behaviour due to the pressure-flow rate relationship, the dead zone of the control valve and frictions [8]; these nonlinearities make the mathematical model more complex and,

S. Pagano · R. Russo · S. Strano · M. Terzo (✉)
Department of Industrial Engineering, University of Naples Federico II, Naples, Italy
e-mail: m.terzo@unina.it

G.M. Dimirovski (ed.), *Complex Systems*, Studies in Systems, Decision and Control 55, DOI 10.1007/978-3-319-28860-4_15

at the same time, highly limit the performance achieved by the classical linear controller [9].

The traditional and widely used approach to the control of electrohydraulic systems is based on a linear model or on the local linearization of the nonlinear dynamics about the nominal operating point. The test rig mainly consists of a sliding table, driven by a hydraulic cylinder, on which the vibration isolator under test is connected; the other end of the isolator is connected to a reaction structure. In this paper, a model-based hydraulic actuator control system, able to follow the desired displacement law is presented and its robustness properties are checked. Indeed in these applications it is important to design control systems that are insensitive to the unknown forces caused by devices under test. The main step to achieve this target is to design the controller based on a mathematical model that is very close to the system to be controlled.

To this end, a nonlinear mathematical model is derived and employed for a feedback control design able to enhance the performances in terms of table positioning. The proposed approach consists of a nonlinear optimal control based on the state-dependent Riccati equation (SDRE). This strategy provides a very effective algorithm for synthesizing nonlinear feedback controls. The controller design procedure is carried out with no isolator employed as specimen under test. On the contrary, in order to evaluate controller sensitiveness, tests have been executed in presence of an isolator model. The results highlight the controller effectiveness both in terms of tracking error and robustness.

15.2 Test Rig Description

The experimental test rig is a machine utilized to perform shear tests on seismic isolators. The actuation system consists of a double-ended hydraulic actuator placed between a fixed base and a sliding Table (1.8 m × 1.59 m). The hydraulic actuator is constituted by a mobile barrel, integral with the sliding table (A), and two piston rods linked to the fixed base (B) (Fig. 15.1).

The isolator under test (C) is located between the sliding table and a slide that can move vertically with respect to the horizontal reaction structure (D).

A hydraulic jack (E) is positioned between the vertical reaction structure (F) and the slide in order to make the isolator under test vertically loaded.

The supply circuit of the hydraulic actuator is mainly constituted by an axial piston pump, powered by a 75 kW AC electric motor, a pressure relief valve and a four-way-three-position proportional valve. The pump is characterized by a variable displacement and it is able to generate a maximum pressure of 210 bar and a maximum flow rate equal to 313 l/min.

The maximum horizontal force is 190 kN, the maximum speed is 2.2 m/s and the maximum stroke is 0.4 m (±0.2 m).

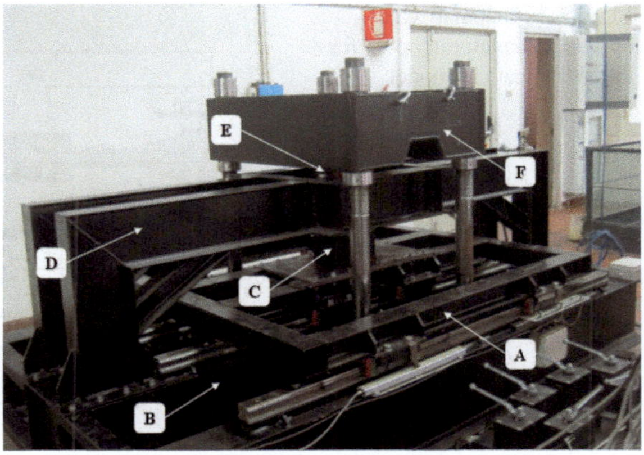

Fig. 15.1 The experimental test rig

The removal of the reaction structures allows the testing machine to be used as a shaking table.

The full-state feedback has been obtained with the following measurements:

- table position by means of magnetostrictive position sensor (FS = 0.4 m – estimated uncertainty = ±1.2e−4 m);
- pressure in the two chambers of the hydraulic cylinder by means of strain gauge sensor (FS = 400 bar – estimated uncertainty = ±1 bar);
- valve spool position by means of built-in LVDT sensor.

15.3 Mathematical Model

The mathematical model regards the test rig without the isolator installed: the hydraulic cylinder has to move only the sliding table mass (Fig. 15.2).

In order to write the mathematical model of the hydraulic circuit, the following hypothesis has been proposed: (a) fluid properties not depending on the temperature; (b) equal piston areas; (c) equal oil volume for each side (with the barrel in a central position); (d) negligible internal and external fluid leakages.

Furthermore, the table has been modelled as a single Degrees of Freedom (DOF) system on which the actuation force and the friction forces act. In particular, the actuator can be modelled as a double-ended hydraulic cylinder driven by a four-way spool valve.

In the following, the differential equations governing the hydraulic actuation system dynamics are given [7]. The pressure dynamics is given by:

$$\frac{V_0}{2\beta}\dot{P}_L = -A_p\dot{y} + Q_L, \tag{15.1}$$

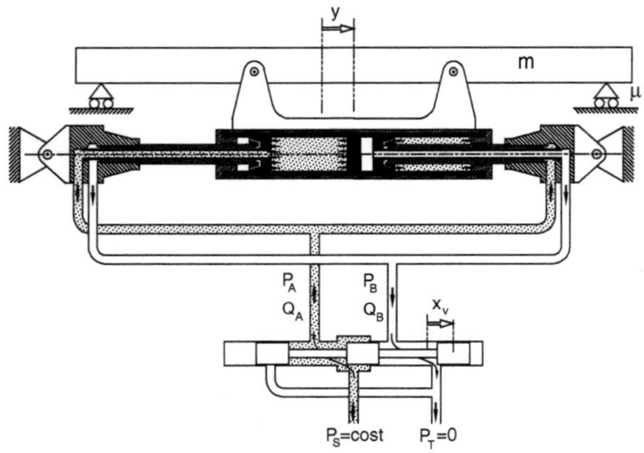

Fig. 15.2 Schematic of hydraulic actuator adopted for the mathematical model

where $P_L = P_A - P_B$ is the load pressure, P_A and P_B are the pressures in the cylinder chambers, $V_O = V_A = V_B$ is the oil volume between the piston and the valve in each side for the centred barrel position, A_p is the piston area, $Q_L = (Q_A + Q_B)/2$, commonly called load flow, β is the effective Bulk modulus and y is the table displacement.

The motion equation of the sliding table is:

$$m\ddot{y} + \mu N\,\mathrm{sgn}(\dot{y}) + F_f = A_p P_L, \tag{15.2}$$

where m is the movable mass, N the vertical load on the linear guides, μ the Coulombian friction coefficient and F_f the hydraulic friction force.

In the following, the hydraulic friction is modelled as the sum of a viscous and a Coulombian term:

$$F_f = \sigma\dot{y} + F_c\,\mathrm{sgn}(\dot{y}), \tag{15.3}$$

where σ is the viscous friction coefficient and F_c the Coulomb friction force.

The valve under consideration is an overlapped four-way valve; typically, this kind of valve is characterized by the lands of the spool that are greater than the annular parts of the valve body. Consequently, when the spool is in the neighbourhood of its central position, the flow rate is zero (dead zone).

Considering the tank pressure P_T equal to zero, the analytical expression of the load flow depends on the supply pressure P_s, load pressure and valve spool position and is given by:

$$Q_L = DZ(v_e)\sqrt{P_s - \mathrm{sgn}(v_e)P_L}, \tag{15.4}$$

where v_e is the voltage signal proportional to the valve spool position x_v and $DZ(v_e)$ is the dead zone function. In general, neither the break-points nor the slopes of the dead band are equal and the analytical expression for $DZ(v_e)$ is therefore:

$$DZ(v_e) = \begin{cases} k_{qn}(v_e - v_{en}) & \text{if} \quad v_e < v_{en} \\ 0 & \text{if} \quad v_{en} \le v_e \le v_{ep}\,, \\ k_{qp}(v_e - v_{ep}) & \text{if} \quad v_e > v_{ep} \end{cases} \tag{15.5}$$

where v_{en} and v_{ep} are the limits of the dead zone, k_{qn} and k_{qp} are the adopted gains if v_e is negative or positive, respectively.

As regards the valve dynamics, the following second-order differential equation is adopted:

$$\frac{\ddot{v}_e}{\omega_{nv}^2} + \frac{2\xi_v}{\omega_{nv}}\dot{v}_e + v_e = v_c, \tag{15.6}$$

where parameters ω_{nv} and ξ_v are the natural frequency and the damping ratio of the valve, respectively; v_c is the valve command voltage due to its electronic driver. The relationship between v_c and the valve input voltage u is:

$$v_c = \begin{cases} k_{ep}u + v_{e0} & \text{if} \quad u > 0 \\ v_{e0} & \text{if} \quad u = 0\,, \\ k_{en}u + v_{e0} & \text{if} \quad u < 0 \end{cases} \tag{15.7}$$

where v_{e0} is the bias, and k_{ep}, k_{en} the gains.

Finally, the equations governing the dynamics of the whole system (sliding table + electrohydraulic system) are:

$$\begin{cases} m\ddot{y} + (\mu N + F_c)\mathrm{sgn}(\dot{y}) + \sigma\dot{y} = A_p P_L \\ \frac{V_0}{2\beta}\dot{P}_L = -A_p\dot{y} + Q_L \\ Q_L = DZ(v_e)\sqrt{P_s - \mathrm{sgn}(v_e)P_L} \\ \ddot{v}_e = -\omega_{nv}^2 v_e - 2\xi_v\omega_{nv}\dot{v}_e + \omega_{nv}^2 v_c \end{cases}. \tag{15.8}$$

The developed fifth-order model fully describes the nonlinear dynamical behaviour of the electrohydraulic actuation system and takes the nonlinear friction forces and the nonlinear flow rate distribution into account.

15.4 Controller Design

The isolator test rig needs a suitable position control system that enhances tracking performance. Taking into account the highly nonlinear behaviour of the system, a nonlinear feedback controller has been developed. The proposed approach is based on the state-dependent Riccati equation (SDRE) technique [10]. The SDRE algorithm fully preserves system nonlinearities, bringing the nonlinear system to a linear structure with state-dependent coefficient (SDC) matrices. The SDC form is obtained by factorization of the nonlinear dynamics into the state vector and matrices that depend on the state itself. The Linear Quadratic Regulator (LQR) synthesis method can be applied to this state-dependent state-space equation characterized by the SDC matrices. The last ones constitute the coefficients of the algebraic Riccati equation that is solved online to give the optimal control law.

The nonlinear system (15.8) can be written in the following form that brings the system itself towards SDC factorization:

$$\begin{cases} \ddot{y} = \left[-\frac{\sigma}{m} - \frac{(\mu N + F_c)\text{sgn}(\dot{y})}{\dot{y}m} \right] \dot{y} + \frac{A_p P_L}{m}, \\ \dot{P}_L = -\frac{2\beta A_p}{V_0} \dot{y} + \frac{2\beta DZ(v_e)_{ss} \sqrt{P_s - \text{sgn}(v_e) P_L}}{V_0} v_e, \\ \ddot{v}_e = -2\xi_v \omega_{nv} \dot{v}_e - \left(1 - \frac{v_{e0}}{v_e} \right) \omega_{nv}^2 v_e + \omega_{nv}^2 k_e u, \end{cases} \tag{15.9}$$

where:

$$DZ(v_e)_{ss} = \begin{cases} k_{qn}(1 - \frac{v_{en}}{v_e}) & \text{if} \quad v_e < v_{en} \\ 0 & \text{if} \quad v_{en} \le v_e \le v_{ep} \\ k_{qp}(1 - \frac{v_{ep}}{v_e}) & \text{if} \quad v_e > v_{ep} \end{cases}. \tag{15.10}$$

The nonlinear system (15.9) can be expressed as a linear-like state-space equation using SDC factorization. The generic expression of the state-dependent state-space equation is:

$$\dot{\mathbf{x}} = \mathbf{A}(\mathbf{x})\mathbf{x} + \mathbf{B}(\mathbf{x})u \tag{15.11}$$

in which $\mathbf{x}$ is the state vector, $\mathbf{A}(\mathbf{x})$ the state matrix and $\mathbf{B}(\mathbf{x})$ the input matrix. Taking into account the nonlinear system (15.9), it follows:

$$\mathbf{x} = \begin{bmatrix} \dot{y} \\ y \\ P_L \\ \dot{v}_e \\ v_e \end{bmatrix}, \tag{15.12}$$

$$\mathbf{A}(\mathbf{x}) = \begin{bmatrix} -\frac{\sigma}{m} - \frac{(\mu N + F_c)\text{sgn}(\dot{y})}{\dot{y}m} & 0 & \frac{A_p}{m} & 0 & 0 \\ 1 & 0 & 0 & 0 & 0 \\ -\frac{2\beta A_p}{V_0} & 0 & 0 & 0 & \frac{2\beta DZ(v_e)_{ss} \sqrt{P_s - \text{sgn}(v_e) P_L}}{V_0} \\ 0 & 0 & 0 & -2\xi_v \omega_{nv} & -\left(1 - \frac{v_{e0}}{v_e} \right) \omega_{nv}^2 \\ 0 & 0 & 0 & 1 & 0 \end{bmatrix}, \tag{15.13}$$

$$\mathbf{B} = \begin{bmatrix} 0 \\ 0 \\ 0 \\ \omega_{nv}^2 k_e \\ 0 \end{bmatrix} \tag{15.14}$$

in which the state transition through zero is handled by means of numerical constraints.

The system (15.9), represented in the state-space form, is nonlinear in the state, autonomous and characterized by a fully known state by means of measurements.

The position control system of the sliding table, based on the SDRE technique, finds an input u that minimizes the following performance index:

$$J = \frac{1}{2}\int_0^\infty \left(\mathbf{x}^T\mathbf{Q}\mathbf{x} + u^T R u\right)dt. \tag{15.15}$$

$\mathbf{Q} = diag(0, q, 0, 0, 0)$ is the weight matrix and R the weight coefficient. On the basis of the LQR synthesis method, the input voltage u is:

$$u = -\mathbf{k}(\mathbf{x})\mathbf{x}, \tag{15.16}$$

where

$$\mathbf{k}(\mathbf{x}) = R^{-1}\mathbf{B}^T\mathbf{P}(\mathbf{x}) \tag{15.17}$$

and $\mathbf{P}(\tilde{\mathbf{x}})$ is the solution of the following state-dependent Riccati equation:

$$\mathbf{A}^T(\mathbf{x})\mathbf{P}(\mathbf{x}) + \mathbf{P}(x)\mathbf{A}(\mathbf{x}) - \mathbf{P}(\mathbf{x})\mathbf{B}R^{-1}\mathbf{B}^T\mathbf{P}(\mathbf{x}) + \mathbf{Q} = \mathbf{0}. \tag{15.18}$$

15.5 Real-Time Experiments

In order to test the effectiveness of the proposed nonlinear tracking controller, experimental studies have been conducted on the seismic isolator test rig.

A dSPACE DS1103 controller board, equipped with a 16-bit A/D and D/A converter, has been used for the real-time experiments. All experiments have been conducted with a sample frequency of 1 kHz. All measured signals have been processed through a low-pass filter to attenuate the influence of the noise. The supplied pressure has been fixed to 6e6 Pa.

The experiments on the seismic isolator testing machine have been conducted in presence of a typical seismic isolator [8, 9]. Hence, the hydraulic actuation system has been utilized only for the sliding table positioning.

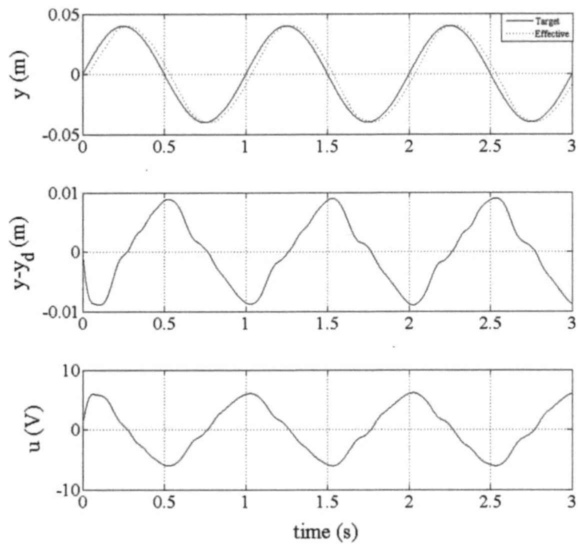

Fig. 15.3 Experimental results of the nonlinear SDRE-based control

The problem of computing the SDRE feedback gains reduces to solving (15.5). The proposed approach is based on finding the eigenvalues of the associated Hamiltonian matrix

$$H(x) = \begin{bmatrix} A(x) & -B(x)R^{-1}B^T(x) \\ -Q & -A^T(x) \end{bmatrix}. \tag{15.20}$$

The SDRE-based feedback gains have been obtained with a pole placement algorithm in terms of the stable eigenvalues of *H(x)*.

The SDRE feedback action has been implemented as a C-code function downloaded to the controller board and implemented in real-time.

The experimental results are reported in Fig. 15.3 in terms of the target and the experimental effective table displacement, the tracking error and the control action.

Analyzing the experimental results, it is possible to assert that the maximum amplitude error is equal to 3e−5 m and the actuator phase lag is equal to 0.035 s. Concerning the tracking error, its maximum value is equal to 0.009 m.

It has been experimentally demonstrated that the proposed nonlinear controller can achieve very good performance in terms of tracking and stability.

15.6 Conclusions

An optimal nonlinear control has been proposed in this paper for a seismic isolator test rig. A fifth-order dynamic model of the system has been derived taking into account the typical nonlinearities of the hydraulic actuators. The parameterization of the SDRE feedback control has been obtained directly from the fifth-order model. The real-time implementation of the SDRE nonlinear optimal problem has been performed with a pole placement algorithm in terms of the stable eigenvalues of the Hamiltonian matrix.

Experiments have been conducted on the hydraulically actuated system, validating the proposed approach and theoretically obtained results.

References

1. Skinner, R.I., Robinson, W.H., McVerry, G.H.: An Introduction to Seismic Isolation. Wiley, Chichester (1993)
2. Naeim, F., Kelly, J.M.: Design of Seismic Isolated Structures: From Theory to Practice. Wiley, Chichester (1999)
3. Yoshioka, H., Ramallo, J.C., Spencer Jr, B.F.: "Smart" base isolation strategies employing magnetorheological dampers. J. Eng. Mech. **128**(11), 540–551 (2002)
4. Chang, C.M., Spencer Jr, B.F.: Active base isolation of buildings subjected to seismic excitations. Earthq. Eng. Struct. Dynam. **39**, 1493–1512 (2010)
5. Hwang, J.S., Wu, J.D., Pan, T.C., Yang, G.: A mathematical hysteretic model for elastomeric isolation bearings. Earthq. Eng. Struct. Dynam. **31**, 771–789 (2002)
6. Kikuchi, M., Aiken, I.D.: An analytical hysteresis model for elastomeric seismic isolation bearings. Earthq. Eng. Struct. Dynam. **26**(2), 215–231 (1997)
7. Merritt, H.E.: Hydraulic Control Systems. Wiley (1967)
8. Pagano, S., Russo, R., Strano, S., Terzo, M.: Non-linear modelling and optimal control of a hydraulically actuated seismic isolator test rig. Mech. Syst. Signal Process. **35**(1–2), 255–278 (2013)
9. Pagano, S., Russo, M., Strano, S., Terzo, M.: A mixed approach for the control of a testing equipment employed for earthquake isolation systems. In: Proceedings IMechE, Part C: Journal of Mechanical Engineering Science. doi:10.1177/0954406213484424
10. Çimen, T.: Survey of state-dependent riccati equation in nonlinear optimal feedback control synthesis. AIAA J. Guidance Control Dyn. **35**(4), 1025–1047 (2012)

Chapter 16
Multi-Robot Navigation Using Market-Based Optimization

Rainer Palm, Abdelbaki Bouguerra and Muhammad Abdullah

Abstract This paper deals with artificial force potential fields for obstacle avoidance and their optimization by a market-based approach in scenarios where several robots are acting in a shared area. Specifically, the potential field method is enhanced by fuzzy logic, traffic rules, and market-based optimization (MBO). Fuzzy rules are used to deform repulsive potential fields in the vicinity of obstacles to produce smoother motions around them. Traffic rules are used to deal with situations where robots are crossing each other. MBO, on the other hand, is used to strengthen or weaken repulsive potential fields generated due to the presence of other robots. For testing and verification, the navigation strategy is implemented and tested in simulation of more realistic vehicles. Issues while implementing this method and limitations of this navigation strategy are also discussed. Extensive simulation experiments are performed to examine the improvement of the traditional potential field (PF) method by the MBO strategy.

16.1 Introduction

In the last decades, several methods of robot navigation and obstacle avoidance have been discussed. A prominent method for obstacle avoidance is the artificial potential field method [1]. A review on this method given by Borenstein and Koren addresses the advantages and disadvantages of this approach with respect to stability and deadlocks [2]. To deal with these problems the so-called Vector Field Histogram approach was presented, being able to cope with fast moving mobile robots in the presence of obstacles [3]. Aicardi et al. addressed team building among mobile robots sharing the same robot task and discussed the appropriate decentralized control [4]. In

R. Palm (✉) · A. Bouguerra · M. Abdullah
Örebro University, Örebro, Sweden
e-mail: rub.palm@t-online.de

A. Bouguerra
e-mail: abaki.bouguerra@gmail.com

M. Abdullah
e-mail: m_abdullah87@gmail.com

G.M. Dimirovski (ed.), *Complex Systems*, Studies in Systems, Decision and Control 55, DOI 10.1007/978-3-319-28860-4_16

approaching situations robots act as moving obstacles where coordination is done by online local path planning using the so-called *via points*. Another approach can be found in [5] where local groups of robots share information on common potential field regions for navigation among static and dynamic obstacles. Werling et al. deal with tracking of nonholonomic autonomous vehicles for inner-city scenarios dealing with precalculated state trajectories and their stabilization [6]. Further research results regarding navigation of nonholonomic mobile robots can be found in [7, 8].

These few examples show the variety of methods for performing different subtasks like

- reaching a target
- avoiding obstacles
- following traffic rules

under the assumption of stable trajectories.

A heuristic method to cope with obstacle avoidance is the *fuzzy logic approach* which has been widely used for mobile robots since the early 1990s. Martinez et al. described a system of heuristic rules based on interaction of mobile robots and traffic rules [9]. A fuzzy obstacle controller using so-called negative fuzzy rules is reported by Lilly [10] where a negative rule is a rule like "IF A THEN DO NOT B" in contrast to a positive rule "IF A THEN DO B". Stingu and Lewis combined a motion control fuzzy rule base using an occupancy map of the environment similar to an artificial potential field within which the robots interact [11].

Different tasks to perform at the same time makes a decentralized optimization necessary in which context different weights for the tasks are generated. Decentralized methods like multiagent control can handle optimization tasks for a large number of complex local systems more efficiently than centralized approaches. Mobile robot navigation is an important application for agent-based control. One popular example is the flow control of mobile platforms in a manufacturing plant using intelligent agents [12]. Other decentral navigation strategies are the so-called utility approach [13] and the behavioral approach [14]. Barrett et al. developed a behavioral control strategy dealing with communicating decentralized controllers [15]. One of the most interesting and promising approaches to cope with large decentralized systems is the market-based optimization (MBO). MB algorithms imitate economical systems where producer and consumer agents both compete and cooperate on a market of commodities.

Dias et al. give an overview on MB multi-robot coordination, which is based on bidding processes [16]. The method deals with motion planning, task allocation, and team cooperation, whereas obstacles are not considered. Gerber et al. describe a MB recource allocation method for vehicle routing applications [17]. This method is based on auction mechanisms where the trucks and the auctioneer are modeled as local agents with planning and bidding capabilities.

In order to improve the performance of safe navigation of multiple robots based on artificial potential fields the present paper adopts many ideas from [9, 18–23] in order to combine fuzzy methods and MBO methods.

In the context of MB navigation, combinations of competing tasks, that should be optimized, can be manifold, for example, the presence of a traffic rule and the necessity for avoiding an obstacle at the same time. Another case is the accidental meeting of more than two robots within a small area. This requires a certain minimum distance between the robots and appropriate (smooth) maneuvers to keep stability of trajectories to be tracked. This paper addresses exactly this point where optimization takes place between "competing" potential fields of mobile robots: Some potential fields are strengthened and some are weakened by a combination of MBO and fuzzy methods depending on the local situation. Repulsive forces both between robots and between robots and obstacles are computed under the assumption of central symmetrical force fields meaning that forces are computed between the centers of mass of the objects considered. The work presented here extends the work reported in [22] through more details of the MBO approach including discussions of its limitations. Another contribution with respect to [22] is the extensive experimental setup aimed at looking at the performance of multiple robots in a wider variety of random scenarios.

Section 16.2.1 addresses the navigation principles applied to the task. In Sect. 16.2 navigation and obstacle avoidance using potential fields and fuzzy rules in the framework of a multi-robot system are outlined. Section 16.3 gives an introduction to the MB optimization used in this paper. The connection between the MB approach and the system to be controlled is outlined in Sect. 16.4. Section 16.5 shows extensive simulation experiments [24] and Sect. 16.6 draws conclusions and highlights possible future work.

16.2 Navigation, Modeling, and Obstacle Avoidance

16.2.1 Navigation Principles

A multi-robot system is constituted of individual mobile robots whose functions can be arranged with the help of a hierarchical control architecture which adopts the idea of a control hierarchy for industrial robots introduced by [25].

The navigation of a mobile robot is more or less located in the control levels "High level control" and "Trajectory Planner" receiving information from higher and lower control levels, and from the environment that consists of targets, obstacles, moving objects (e.g., other robots), and possible team members. To illustrate the navigation problems, let n mobile platforms (autonomous mobile robots) perform special tasks in a working area like loading materials from a starting station, bringing them to a target station and unloading the materials there. The task of the platforms is to reach their targets while avoiding obstacles and other platforms.

Navigation principles for a mobile robot (platform) P_i are meant to be *heuristic rules* to perform a specific task under certain restrictions originating from the environment, obstacles O_j, and other robots P_j. Each platform P_i is supposed to have an estimation about position/orientation of itself and the target T_i. The position of another platform P_j relative to P_i can be measured if it lies within the sensor cone of P_i. Five navigation principles are used here.

1. Move in direction of target T_i.
2. Avoid an obstacle O_j (static or dynamic) if it appears in the sensor cone at a certain distance. Always orient platform in the direction of motion.
3. Decrease speed if a dynamic (moving) obstacle O_j comes from the right.
4. Move to the right if the obstacle angles β (see [26]) of two approaching platforms are small (e.g., $\beta < 10$) (see Fig. 16.3).
5. Brake in the presence of an imminent collision.

Apart from the *heading-to-target movement* all other navigation calculations and actions take place in the local coordinate system of platform P_i. The positions of obstacles (static or dynamic) O_j or other platforms P_j are also formulated in the local frame of platform P_i.

16.2.2 Modeling of the System

In the following two nonholonomic models are presented—a car-like system which is already discussed in [22] and a differential-drive system on the basis of which the simulations presented in this paper are done.

16.2.2.1 Kinematic Model of a Car-Like Robot

The kinematic model of rear-driven car-like nonholonomic vehicle (see also Fig. 16.1) is described by

$$\begin{aligned} \dot{q}_i &= R_i(q_i) \cdot u_i, \\ q_i &= (x_i, y_i, \theta_i, \phi_i), \\ R_i(q_i) &= \begin{pmatrix} \cos\theta_i & 0 \\ \sin\theta_i & 0 \\ \frac{1}{l_i} \cdot \tan\phi_i & 0 \\ 0 & 1 \end{pmatrix}, \end{aligned} \tag{16.1}$$

where
$q_i \in \Re^2$—state vector.
$u_i = (u_{1i}, u_{2i}) \in \Re^2$—control vector, pushing/steering speed.
$x_{ip} = (x_i, y_i)^T \in \Re^2$—position vector of platform P_i.
θ_i—orientation angle.
ϕ_i—steering angle.
l_i—length of vehicle.

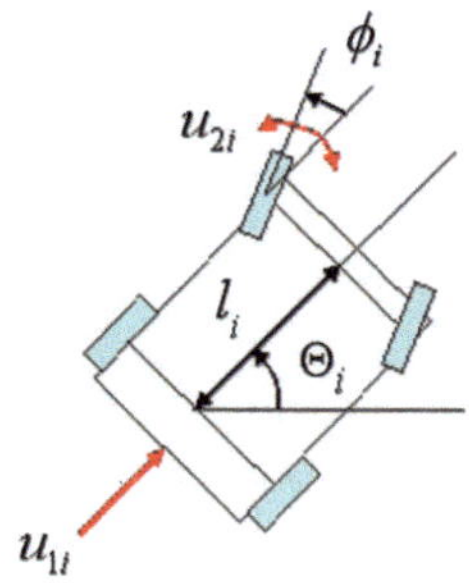

Fig. 16.1 Nonholonomic vehicle

16.2.2.2 Kinematic Model of a Differential-Drive Robot

We consider differential-drive robots that have two driving wheels, each with diameter r. Point P is the center point between the driving wheels. The distance between each wheel and P is l, see Fig. 16.2.

Let the robot's target (goal) position be (x_g, y_g) in a global frame as shown in Fig. 16.2b, where

d_l = distance traveled by left wheel
d_r = distance traveled by right wheel
$d = (d_r + d_l)/2$ travel distance covered by robot
$\Delta = (d_r - d_l)/D$, central angle, where $D = 2l$
$R = d/\Delta$, turning radius

The error in position is E_{pos} and the error in orientation is E_θ. The aim is to reduce E_{pos}, which can be done by controlling the linear and angular speeds. In our implementation, a proportional control strategy is used.

$$\begin{bmatrix} v_t \\ w \end{bmatrix} = \begin{pmatrix} k_p & 0 \\ 0 & k_\theta \end{pmatrix} \cdot \begin{bmatrix} E_{pos} \\ E_\theta \end{bmatrix}, \tag{16.2}$$

where
$E_{pos} = \sqrt{D_x{}^2 + D_y{}^2}$ and $D_x = x_g - x_r$, $D_y = y_g - y_r$
$E_\theta = atan2(D_y, D_x) - \theta_r$.

Here k_p and k_θ are proportional terms.

To move the robot the velocity vector (v_t, w) is converted to (v_r, v_l) by Eq. (16.3) (see [27] and Fig. 16.2b).

$$\begin{aligned} v_r &= v_t + (D/2)w, \\ v_l &= v_t - (D/2)w, \end{aligned} \tag{16.3}$$

where
v_t = linear velocity
w = angular velocity
v_r = linear velocity of right wheel
v_l = linear velocity of left wheel

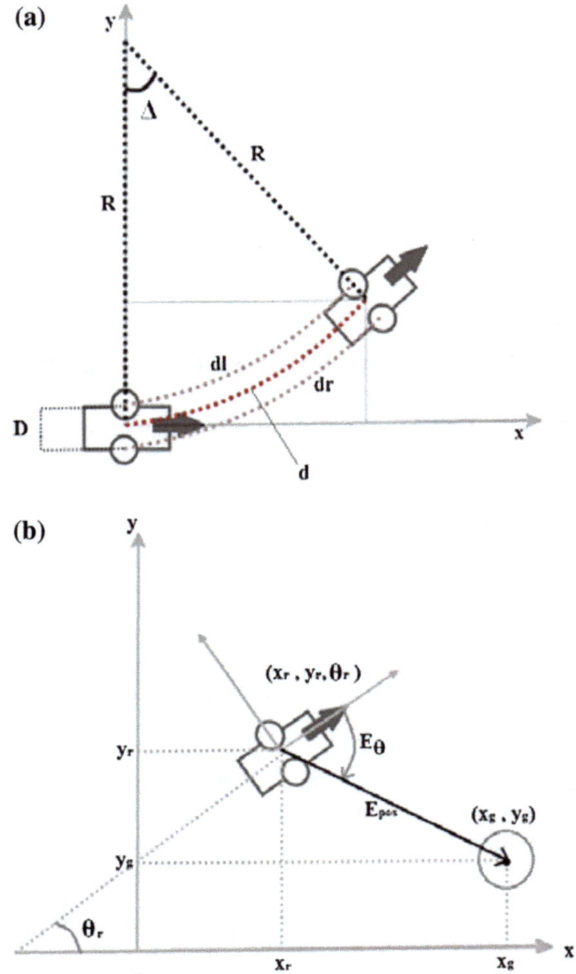

Fig. 16.2 Differential-drive robot **a** Schematic diagram of a differential-drive robot. **b** Position error from a target position

16.2.3 Obstacle Avoidance Using Artificial Force Fields

The tracking velocity in the direction of a target is designed as a control term which works as an *attractive force*

$$v_{ti} = k_{ti}(x_{i_p} - x_{ti}), \tag{16.4}$$

$x_{ti} \in \Re^2$—position vector of target T_i
$k_{ti} \in \Re^{2\times2}$—gain matrix (diagonal).

In addition to that, *repulsive forces* may appear between platforms P_i and obstacles O_j leading to repulsive velocities

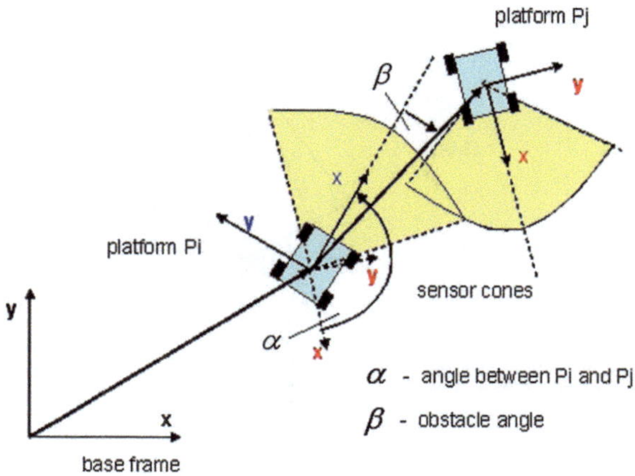

Fig. 16.3 Geometrical relationship between platforms

$$v_{ij_{ob}} = -c_{ij_{ob}}(x_{j_p} - x_{j_{ob}})d_{ij_{ob}}^{-2}, \tag{16.5}$$

$v_{ij_{ob}} \in \Re^2$—repulsive velocity vector between platform P_i and obstacle O_j
$x_{j_{ob}} \in \Re^2$—position vector of obstacle O_j
$d_{ij_{ob}} \in \Re$—Euclidian distance between platform P_i and obstacle O_j
$c_{ij_{ob}} \in \Re^{2\times2}$—gain matrix (diagonal).

Repulsive forces may also appear between platforms P_i and P_j from which we get the repulsive velocities

$$v_{ij_p} = -c_{ij_p}(x_{i_p} - x_{j_p})d_{ij_p}^{-2} \tag{16.6}$$

$v_{ij_p} \in \Re^2$—repulsive velocity between platforms P_i and P_j
$d_{ij_p} \in \Re$—Euclidian distance between platforms P_i and P_j
$c_{ij_p} \in \Re^{2\times2}$—gain matrix (diagonal)

The resulting velocity v_{di} is the sum

$$v_{di} = v_{ti} + \sum_{j=1}^{m_{ob}} v_{ij_{ob}} + \sum_{j=1}^{m_p} v_{ij_p}, \tag{16.7}$$

where m_{ob} and m_p are the numbers of contributing obstacles and platforms. It should be emphasized that the force fields are switched on/off according to the actual scenario: distance between interacting systems, state of activation according to the sensor cones of the platforms, positions and velocities of platforms w.r.t. to targets, obstacles, and other platforms. All calculations of the velocity components (16.4)–(16.7), angles, and sensor cones are formulated in the local coordinate systems of the platforms (see Fig. 16.3).

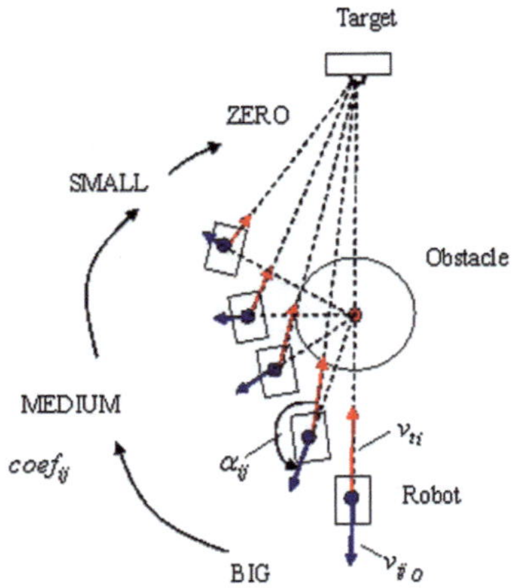

Fig. 16.4 Deformation of potential fields around obstacles using fuzzy rules

16.2.4 "Deformation" of Potential Fields Using Fuzzy Rules

Potential fields of obstacles (static and dynamic) act normally independently of the attractive force of the target. This may cause unnecessary repelling forces especially in the case when the platform can "see" the target.

Another situation occurs when the tracking velocity $|v_{ti}|$ becomes zero for some reason. In this case a platform would be pushed away from an obstacle even if it should keep its position. The goal is therefore to "deform" the repulsive potential field so that it is strong if the obstacle hides the target and weak if the target "can be seen" from the platform. In addition, the potential field should also be strong for a high tracking velocity and weak for a small one (see Fig. 16.4). These requirements can be achieved by introducing a coefficient $coef_{ij} \in [0, 1]$ that is multiplied to $v_{ij_{ob}}$ to obtain a new $v_{ij_{ob}}$ as follows:

$$v_{ij_{ob}} = -coef_{ij} \cdot c_{ij_{ob}} \cdot (x_{i_p} - x_{j_{ob}})d_{ij_{ob}}^{-2}. \tag{16.8}$$

The coefficients $coef_{ij}$ can be calculated by a set of 16 fuzzy rules like

$$\begin{aligned} IF \quad |v_{ti}| = B \quad AND \quad \alpha_{ij} = M,& \\ THEN \quad coef_{ij} = M,& \end{aligned} \tag{16.9}$$

where α_{ij} is the angle between $v_{ij_{ob}}$ and v_{ti}. The set of 16 rules can be summarized in a table shown in Fig. 16.5a. *Z*—ZERO, *S*—SMALL, *M*—MEDIUM, *B*—BIG are fuzzy sets (see [28]). The corresponding membership functions μ_α, μ_{vt}, and μ_{coef} are triangular and shown in Fig. 16.5b. Finally, (16.7) can be rewritten into

$$v_{di} = v_{ti} + \sum_{j=1}^{m_{ob}} w_{ij_{fuzz}} v_{ij_{ob}} + \sum_{j=1}^{m_p} v_{ij_p}, \tag{16.10}$$

(a)

v_{ti} \ α_{ij}	Z	S	M	B
B	Z	S	M	B
M	Z	S	M	M
S	Z	M	M	S
Z	Z	Z	Z	Z

(b)

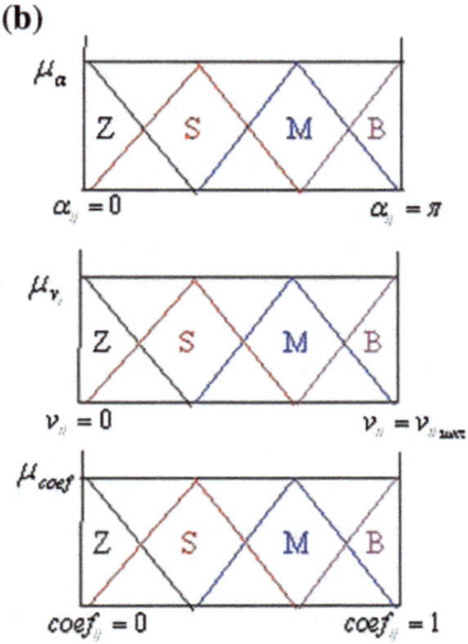

Fig. 16.5 Fuzzy rules for deformation of potential fields. **a** Fuzzy table. **b** Membership functions

where

$$w_{ij\,fuzz} = coef_{ij} = \frac{\sum_{l=1}^{s} \mu^l(|v_{ti}|, \alpha_{ij}) \cdot coef_{ij}^l}{\sum_{l=1}^{s} \mu^l(|v_{ti}|, \alpha_{ij})}, \tag{16.11}$$

$$\mu^l = min(\mu_{vt}, \mu_\alpha),$$

s—number of rules.

16.3 MB Approach

Market-based optimization is used to improve the motion behavior of robots when other robots are present in their vicinity. The improvement aims at producing smooth motions that avoid abrupt changes of motion due to the other moving robots. The improvement is sought through appropriate weighting of the repulsive forces/ velocities $v_{ij_{ob}}$ and v_{ij_p} using MBO methods.

The desired motion of platform P_i is then described by

$$v_{d_i} = vo_i + \sum_{j=1, i \neq j}^{m_p} w_{ij} v_{ij_p} + \sum_{j=1, i \neq j}^{m_{ob}} w_{ij_{ob}} v_{ij_{ob}}, \tag{16.12}$$

where vo_i is a combination of

- tracking velocity depending on distance between platforms i and goal i
- Traffic rules

w_{ij}—weighting factors for repelling forces where $\sum_{j=1, i \neq j}^{m_p} w_{ij} = 1$
$w_{ij_{ob}}$—weighting factors for repelling forces between platform i and obstacle j.

The first objective is to change the weights w_{ij} so that all contributing platforms show a smooth dynamical behavior during avoiding each other. One possible option for tuning the weights w_{ij} is to find a global optimum over all contributing platforms. This, however, is rather difficult especially in the case of many interacting platforms. Therefore, a multiagent approach has been preferred. The determination of the weights is done by producer–consumer agent pairs in a MB scenario that is presented in the following.

Assume that to every local system S_i (platform) belongs a set of m producer agents Pag_{ij} and m consumer agents Cag_{ij}. Producer and consumer agents sell and buy, respectively, the weights w_{ij} on the basis of a common price p_i. Producer agents Pag_{ij} supply weights w_{ij_p} and try to maximize specific local profit functions ρ_{ij} where "local" means "belonging to system S_i". On the other hand, consumer agents Cag_{ij} demand for weights w_{ij_c} from the producer agents and try to maximize specific local utility functions U_{ij}. The whole "economy" is in equilibrium as the sum over all supplied weights w_{ij_p} is equal to the sum over all utilized weights w_{ij_c}.

$$\sum_{j=1}^{m} w_{ij_p}(p_i) = \sum_{j=1}^{m} w_{ij_c}(p_i). \tag{16.13}$$

A "trade" between a producer and consumer agent is managed by cost functions for both types of agents. We define a local utility function for the consumer agent Cag_{ij}

$$\begin{aligned} Utility &= benefit - expenditure \\ U_{ij} &= \tilde{b}_{ij} w_{ij_c} - \tilde{c}_{ij} p_i (w_{ij_c})^2, \end{aligned} \tag{16.14}$$

where $\tilde{b}_{ij}, \tilde{c}_{ij} \geq 0$, $p_i \geq 0$. Furthermore, a local profit function is defined for the producer agent Pag_{ij}

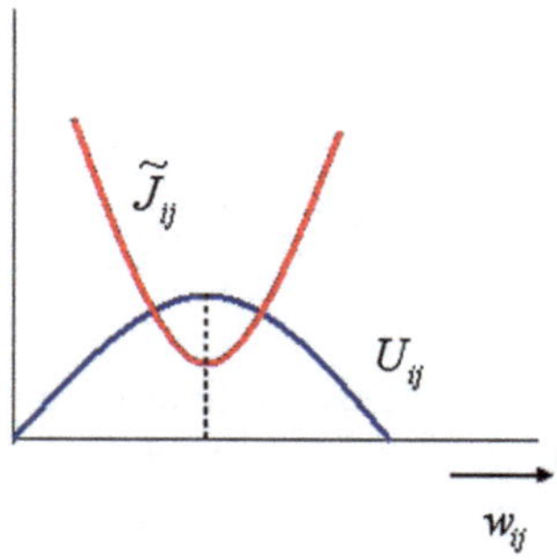

Fig. 16.6 Energy and utility function

$$\begin{aligned} profit &= income - costs, \\ \rho_{ij} &= g_{ij}p_i(w_{ij_p}) - e_{ij}(w_{ij_p})^2, \end{aligned} \tag{16.15}$$

where $g_{ij}, e_{ij} \geq 0$ are free parameters which determine the average price level. It has to be emphasized that both cost functions (16.14) and (16.15) use the same price p_i on the basis of which the weights w_{ij} are calculated. From the system Eq. (16.12) we define further a local energy function to be minimized

$$\begin{aligned} \tilde{J}_{ij} &= v_{d_i}^T v_{d_i}, \\ &= a_{ij} + b_{ij}w_{ij} + c_{ij}(w_{ij})^2 \rightarrow min, \end{aligned} \tag{16.16}$$

where $\tilde{J}_{ij} \geq 0, a_{ij}, c_{ij} > 0$. The question is how to combine the local energy function (16.16) and the utility function (16.14), and how are the parameters in (16.14) to be chosen? An intuitive choice

$$\tilde{b}_{ij} = |b_{ij}|, \quad \tilde{c}_{ij} = c_{ij} \tag{16.17}$$

guarantees $w_{ij} \geq 0$. It can also be shown that, independently of a_{ij}, near the equilibrium $v_{d_i} = 0$, and for $p_i = 1$, the energy function (16.16) reaches its minimum, and the utility function (16.14) its maximum, respectively, (see Fig. 16.6). With (16.17) the utility function (16.14) becomes

$$U_{ij} = |b_{ij}|w_{ij_c} - c_{ij}p_i(w_{ij_c})^2. \tag{16.18}$$

Maximization of the local utility function (16.18) leads to

$$w_{ij_c} = \frac{|b_{ij}|}{2c_{ij}} \cdot \frac{1}{p_i}. \tag{16.19}$$

Maximization of the local profit function (16.15) yields

$$w_{ij_p} = \frac{p_i}{2\eta_{ij}} \quad where \quad \eta_{ij} = \frac{e_{ij}}{g_{ij}}. \tag{16.20}$$

Substituting (16.19) and (16.20) into (16.13) gives the prices p_i for the weights w_{ij_p}

$$p_i = \sqrt{\frac{\sum_{j=1}^{m} |b_{ij}|/c_{ij}}{\sum_{j=1}^{m} 1/\eta_{ij}}}. \tag{16.21}$$

Substituting (16.21) into (16.19) yields the final weights w_{ij} to be implemented in each local system. Once the new weights w_{ij} are calculated, each of them has to be normalized with respect to $\sum_{j=1}^{m} w_{ij}$ which guarantees the above requirement $\sum_{j=1}^{m} w_{ij} = 1$.

16.4 MB Optimization of Obstacle Avoidance

16.4.1 MBO Between Active Mobile Platforms

In the following the optimization of obstacle avoidance between moving platforms by MB methods will be addressed. Coming back to the equation of the system of mobile robots (16.12)

$$v_{d_i} = vo_i + \sum_{j=1,i\neq j}^{m_p} w_{ij} v_{ij_p}, \tag{16.22}$$

where vo_i is a subset of the RHS of (16.7)—a combination of different velocities (tracking velocity, control terms, etc.), v_{ij_p} reflects the repelling forces between platforms P_i and P_j. The *global* energy function (16.16) reads

$$\begin{aligned}\tilde{J}_i &= vo_i^T vo_i + 2vo_i^T \sum_{j=1,i\neq j}^{m_p} w_{ij} v_{ij_p}, \\ &+ (\sum_{j=1,i\neq j}^{m_p} w_{ij} v_{ij_p})^T (\sum_{j=1,i\neq j}^{m_p} w_{ij} v_{ip}).\end{aligned} \tag{16.23}$$

The *local* energy function reflects only the energy of a pair of two interacting platforms P_i and P_j

$$\begin{aligned}\tilde{J}_{ij} &= vo_i^T vo_i + (\sum_{k=1,k\neq i,j}^{m_p} w_{ik} v_{ikp})^T (\sum_{k=1,k\neq i,j}^{m_p} w_{ik} v_{ikp}) \\ &+ 2 \sum_{k=1,k\neq i,j}^{m_p} w_{ik} vo_i^T v_{ikp}\end{aligned}$$

$$+ 2w_{ij}(vo_i^T + \sum_{k=1,k\neq i,j}^{m_p} w_{ik}v_{ik_p}^T)v_{ij_p}$$

$$+ w_{ij}^2(v_{ij_p}^T v_{ij_p}). \tag{16.24}$$

Comparison of (16.24) and (16.16) yields

$$b_{ij} = 2(vo_i^T + \sum_{k=1,k\neq i,j}^{m_p} w_{ik}v_{ik_p}^T)v_{ij_p},$$

$$c_{ij} = (v_{ij_p}^T v_{ij_p}), \tag{16.25}$$

while neglecting a_{ij} because a_{ij} does not contribute to the MBO process.

16.5 Simulation Results

In order to validate the MBO navigation strategy, extensive experiments were performed. Limitations of the MBO navigation strategy are also discussed by showing the scenarios where mobile robots run into problems like deadlocks or emergency stops.

1. Some experiments were performed to test the effect of fuzzy rules around obstacles.
2. Experiments were performed to test MBO for different numbers of mobile robots moving in a small area. In these experiments, simple scenarios were considered and detailed results are described. Experiments were performed for 3, 4, and 5 robots.
3. For extensive experimentation to check the validation of MBO, an experimental setup is used where random scenarios were generated. Experiments were performed with and without obstacles; for 3, 4, and 5 robots, 20 random scenarios were generated for each experiment.

16.5.1 Software Framework and Packages

The implementation is based on the Robot operating system (ROS) [29] using the 3D robot simulation environment Gazebo [30]. The simulated robots were all of differential-drive type, where the erratic robot packages in ROS were extended to incorporate the proposed navigation strategy.

An erratic robot is a differential-drive robot. It is selected for testing, as the "*erratic_robot*" stack is available in ROS. This stack has a basic configuration, hardware drivers, sensor drivers, and a simulation for erratic robots in Gazebo. A differential-drive plug-in for the Gazebo simulator is available in the "*erratic_robot*" stack.

Table 16.1 Possible options combination of algorithm

	MBO	TF	Algorithm behavior
1	OFF	OFF	Traditional PF
2	OFF	ON	Traditional PF with TF
3	ON	OFF	Traditional PF with MBO
4	ON	ON	Traditional PF with TF and MBO

16.5.2 Experimental Setup

The traditional potential field algorithm (PF)—by which we mean the pure PF algorithm without any enhancement—is implemented in the ROS package "*potentialfields*." A ROS node can be started on any robot with different options. Traffic rules (TF) and MBO are options and can be selected with the following four combinations given in Table 16.1.

The navigation strategy is tested by the Gazebo simulator in ROS, using different scenarios with different numbers of erratic robots. Every "*erratic robot*" is equipped with a laser range finder for obstacle detection. Every robot uses its local frame to compute its velocity vectors and tries to avoid other robots to reach its goal position.

Experiments are performed in different scenarios for each of which the data (linear and angular velocities, position in the map at time t) are logged. In this work, five parameters are used for the performance evaluation of each robot: "accuracy," "completion time," "path length," "curvature change," and "lateral stress."

16.5.2.1 Evaluation Parameters

In [31] an evaluation framework is given for the testing of motion methods of mobile robots. This framework can be used to evaluate and compare different motion methods but only on a single robot.

On the other hand, the MBO navigation strategy can be evaluated only for multiple robots navigating in a shared area where each robot uses the same method for navigation. This makes it difficult to evaluate multi-robot navigation methods. In the following an evaluation framework proposed by [31] is used where the evaluation parameters are computed for every robot taking part in multi-robot navigation in a shared area. The evaluation parameters used are as follows.

Accuracy

"Accuracy" is the distance between a robot's actual final position $x_{finalpos}$ and its target position x_{ti}. This parameter tells us whether the robot was able to reach its goal position and how far it was from its goal? "Accuracy" is calculated by

$$Accur_i = ||x_{ti} - x_{finalpos}||. \tag{16.26}$$

Completion Time

"Completion time" is the time in seconds that a robot will take to complete its navigation task. For a good performance, a robot should have low completion (execution) time. (Here, the ROS simulation time is considered because we are working in simulation)

Path Length

The "path length" is the total travel distance covered by the robot to reach its target position. Shorter path lengths are desirable for a better performance.

Curvature Change (CC)

"Curvature change" is associated with car-like robots and is helpful to find unwanted turns in the trajectory [31]. Since we have information about linear and angular velocities of robots, the curvature can be found using Eq. (16.27) [31]

$$k(i) = \frac{w(i)}{v(i)}, \tag{16.27}$$

where $w(i)$ and $v(i)$ are angular and linear velocities, respectively. The parameter "curvature change" is computed as

$$CC = \frac{\sum_{i=1}^{N}(|k(i)'| \ . \ \Delta t)}{N}, \tag{16.28}$$

where N is total number of time steps. A small value of CC is desirable for smooth robot motion.

Lateral Stress (LS)

The parameter "Lateral stress" helps to find the behavior of robot while turning. It is computed by Eq. (16.29) [31]

$$LS = \sum_{i=1}^{N}(\frac{v(i)^2}{|r(i)|} \ . \ \Delta t), \tag{16.29}$$

where $r(i) = \frac{1}{k(i)}$ and $v(i)$ and $r(i)$ are the linear velocity and the curvature radius, respectively. If the robot is turning at a high speed it means that *LS* will be high. So a small value of *LS* is desirable for safe turning.

It is highly very difficult to achieve good performance in all of these parameters at the same time. It is possible that, for example, a robot with less completion time has a higher curvature change. The same may also be valid for other parameters. Therefore, a trade-off between these parameters is necessary.

16.5.3 Results

16.5.3.1 Obstacle Avoidance (Comparison with and Without Fuzzy Rules)

Experiments were run to check how fuzzy rules can optimize obstacle fields. Three different tests were run in this context. In the first test, 3 obstacles were placed between robot initial and target positions. The robot's movement was tested with and without fuzzy rules as shown in Fig. 16.7a.

It can be seen that without fuzzy rules, due to an unwanted rotation of the robot near obstacles, the robot took more time to reach its target while in the case of using fuzzy rules, the robot can pass between two obstacles without being strongly disturbed by them during the passage.

In the second test, three obstacles were placed with different configurations (non-collinear). In Fig. 16.7b, it can be seen that without fuzzy rules, the robot path is strongly effected by obstacle potential fields.

In the third test, the goal point was situated in the effective range of an obstacle. In Fig. 16.7c, it can be seen that without fuzzy rules the robot was unable to reach its target position, while in the case of using fuzzy rules the robot reached its target position with success.

16.5.3.2 Results—Simple Scenarios

The MBO navigation approach was tested on several simple scenarios. In these scenarios, robots interact in a small area while heading to their goal positions. Experiments are performed with and without obstacles.

Movement of Three Robots Without Obstacles

In the case of the 3 robots "α," "β," and "γ," initial and target positions were selected in such a way that—while moving—the robots are forced to interact with each other. Three robots reached their target positions under the four combinations given in Table 16.1. Navigation results can be seen in Fig. 16.8. Detailed results of each robot can be seen in Table 16.2.

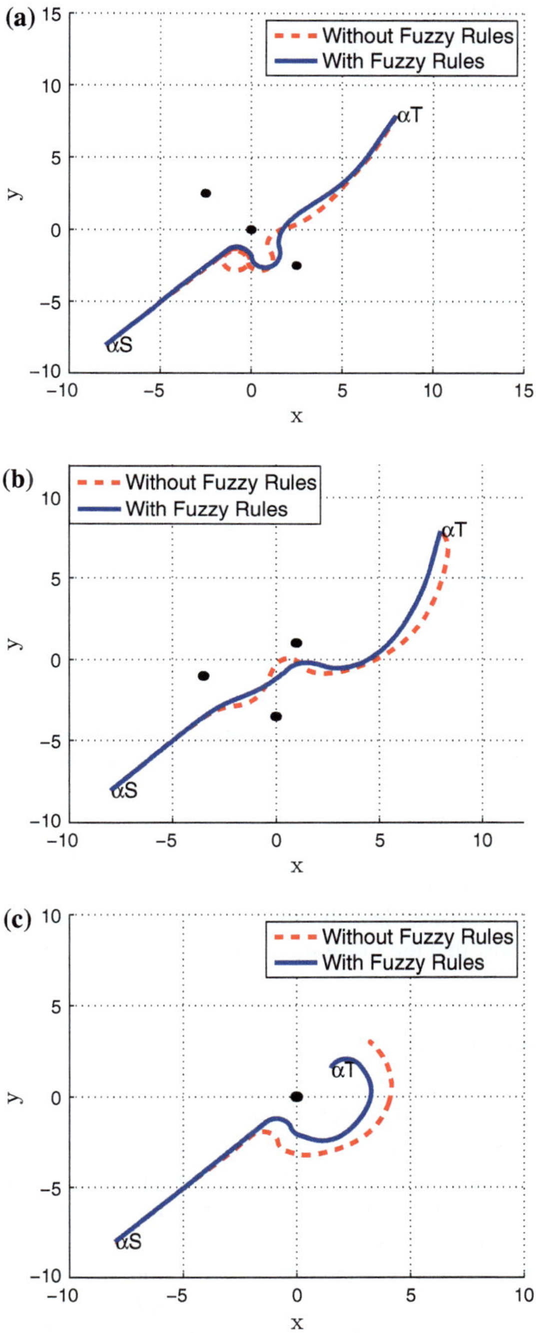

Fig. 16.7 Performance comparison with and without fuzzy rules. α_S (start position) and α_T (target position), **a** Three obstacles placed in a line, **b** Three obstacles placed at different places, **c** Goal in obstacle range

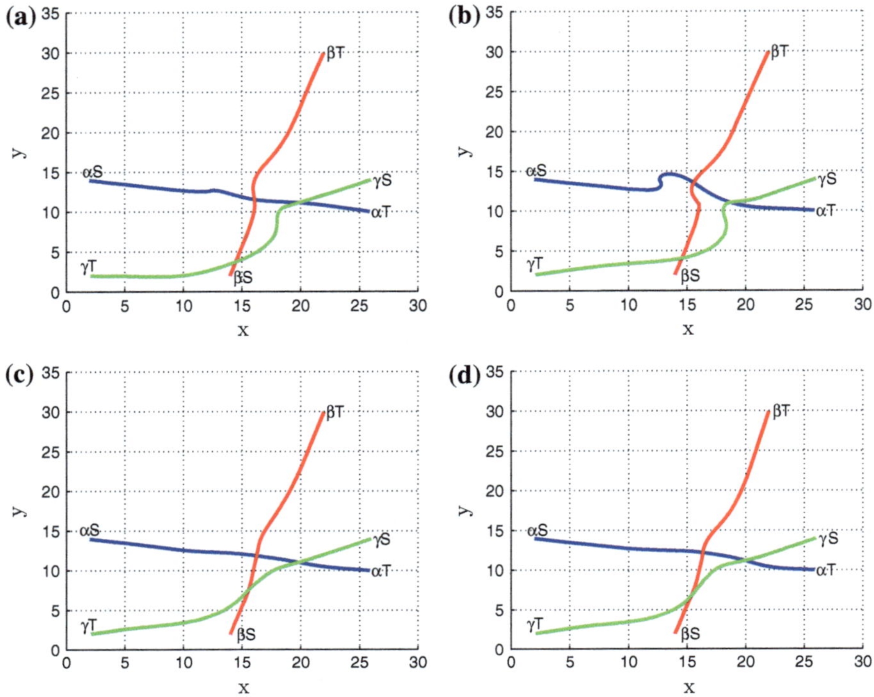

Fig. 16.8 Movement of three robots without obstacles: Navigation results of different options of the navigation algorithm, **a** mboOff–tfOff, **b** mboOff–tfOn, **c** mboOn–tfOff, **d** mboOn–tfOn

Table 16.2(*d*) shows the combined (averaged) result of 3 robots. It can be seen from Table 16.2(*d*) that in case of MBO navigation strategy, the robots took less time and covered less distance to complete the navigation task. *CC* and *LS* parameters show that the movement was smooth while using this algorithm.

Movement of Three Robots with Obstacles

In this experiment, navigation of three robots was tested in a shared area where static obstacles were also placed. Starting and target positions were selected for robots "α," "β," and 'γ'. Three obstacles were placed in the area. These positions were selected in such a way that while moving, these robots were interacting with each other and sensing static obstacles at the same time. With different options (from Table 16.1) of the navigation approach, all the robots were able to reach their target positions.

Navigation results can be seen in Fig. 16.9.

Detailed results of each robot can be seen in Table 16.3. The accuracy parameter shows that all robots succeeded in reaching their target positions. "α," "β," and 'γ' robots took less time in the case of option "mboON and tfON." Distance covered by

Table 16.2 Movement of three robots without obstacles: Performance comparison of different options of the navigation algorithm

a	mboOFF–tfOFF	mboOFF–tfON	mboON–tfOFF	mboON–tfON
Accur	0.100	0.100	0.100	0.099
Time	68.100	71.400	63.000	63.900
Length	24.370	26.861	24.269	24.350
CC	0.317	0.992	0.355	0.425
LS	2.270	3.345	2.085	2.185
b				
Accur	0.100	0.100	0.099	0.099
Time	67.000	69.100	65.600	67.000
Length	29.311	29.624	29.125	29.192
CC	1.681	2.375	1.459	1.558
LS	1.106	1.293	1.073	1.031
c				
Accur	0.099	0.100	0.099	0.099
Time	68.900	70.200	65.800	66.500
Length	29.380	29.757	27.668	27.926
CC	0.735	0.731	0.557	0.480
LS	2.873	3.201	2.553	2.683
d				
Avg Accur	0.100	0.100	0.099	0.099
Avg Time	68.000	70.233	64.800	65.800
Avg Length	27.687	28.748	27.021	27.156
Avg CC	0.911	1.366	0.790	0.821
Avg LS	2.083	2.613	1.904	1.966

each robot was also less when MBO was used. In the case of using the MBO method the CC value was less for "α" and "β" robots but for "γ" robot it was almost same as compared with traditional PF. The value of LS parameter was also less when the MBO method was used showing the safe turning of robots. A combined (averaged) result of this scenario is shown in Table 16.3(*d*), due to MBO a better performance can be clearly observed from the different values of the metrics.

Movement of Five Robots Without Obstacles

In this experiment, the algorithm was tested with five robots, "α," "β," "γ," "δ," and "ϵ." Navigation results with different options (from Table 16.1) in Fig. 16.10 show that the MBO navigation strategy is far better than the traditional PF, because it reduced the unwanted effects of superimposing potential fields.

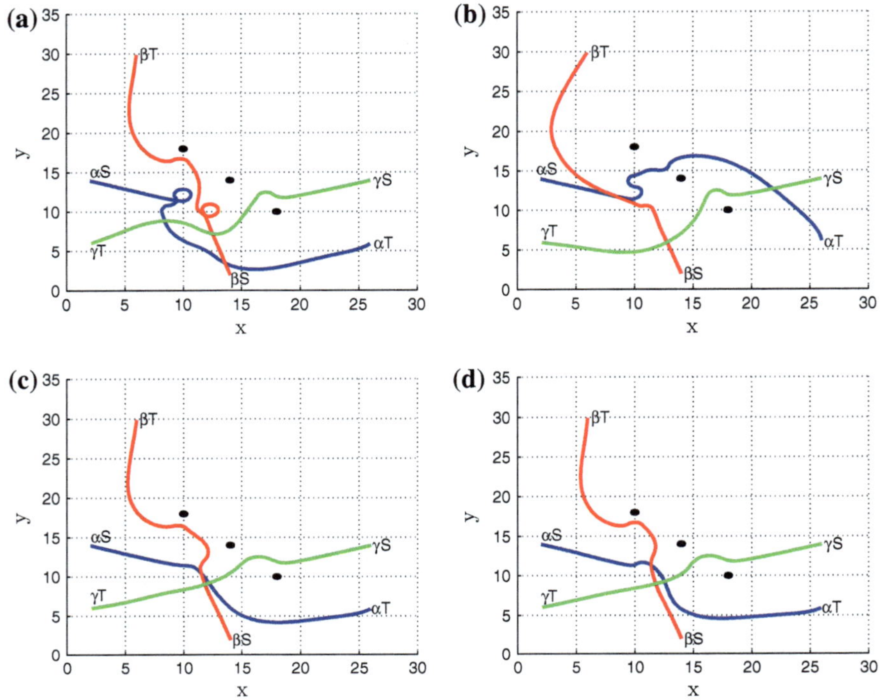

Fig. 16.9 Movement of three robots with obstacles: Navigation results of different options of the navigation algorithm, **a** mboOff–tfOff, **b** mboOff–tfOn, **c** mboOn–tfOff, **d** mboOn–tfOn

When "mboOFF and tfOFF" option was used all the robots safely reached their target positions but showed unwanted movements in Fig. 16.10a. In the case of "mboOFF and tfON," traffic rules option was turned on (Fig. 16.10b). In this case conditions for applying traffic rules were fulfilled by "β" and "γ" robots. When "mboOn and tfOFF" option was used all the robots safely reached at their target positions but with a far better result than in Fig. 16.10a, b.

Detailed results for each robot can be seen from Table 16.4. It can be analyzed from combined (averaged) results (Table 16.4(*f*)) of this scenario that the MBO navigation strategy performs really well compared with traditional PF.

Movement of Five Robots with Obstacles

In this test, five robots were moving in a shared area in the presence of static obstacles. Placement of starting and target positions of robots were selected in such a way that while movement, robots were sensing static obstacles and other robots at the same time. In this experiment the movement of the robots was tested with different options (from Table 16.1) of algorithm. Navigation results can be seen in Fig. 16.11.

Table 16.3 Movement of three robots with obstacles: Performance comparison of different options of the navigation algorithm

a	mboOFF–tfOFF	mboOFF–tfON	mboON–tfOFF	mboON–tfON
Accur	0.099	0.100	0.100	0.099
Time	93.600	97.600	70.000	71.300
Length	36.626	34.636	28.171	28.690
CC	1.762	1.547	0.703	0.513
LS	5.378	4.880	2.573	3.228
b				
Accur	0.100	0.099	0.100	0.099
Time	93.300	81.900	77.300	76.999
Length	36.548	33.017	32.249	32.314
CC	3.459	1.092	2.153	2.374
LS	4.406	1.711	2.188	2.481
c				
Accur	0.100	0.100	0.100	0.099
Time	90.700	77.100	70.100	70.900
Length	28.402	28.646	26.213	26.254
CC	0.413	0.434	0.381	0.428
LS	3.003	2.874	2.770	2.748
d				
Avg Accur	0.100	0.099	0.100	0.099
Avg Time	92.533	85.533	72.467	73.066
Avg Length	33.859	32.099	28.878	29.086
Avg CC	1.878	1.024	1.079	1.105
Avg LS	4.263	3.155	2.511	2.819

In the case of "mboOFF and tfOFF" option, traditional PF was used that helped every robot to reach its target position safely but with unwanted movements of robots Fig. 15.11a.

Detailed results of each robot are shown in Table 16.5. Table 16.5(*f*) shows the combined (averaged) result of this scenario, and the performance of the MBO method can be compared with traditional PF.

16.5.3.3 Results of Random Scenarios

Navigation algorithms can behave differently in different environments and situations. So, there is a possibility that a navigation algorithm showing good results in one scenario can give unwanted results in other scenarios. We designed a set of random scenarios to test the performance of the proposed approach under different crossing situations. To generate each random scenario, we subdivide the environ-

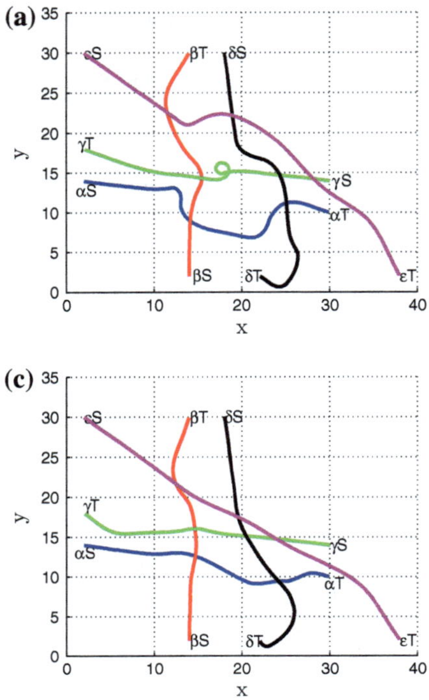

Fig. 16.10 Movement of five robots without obstacles: Navigation results of different options of the navigation algorithm, **a** mboOff–tfOff, **b** mboOff–tfOn, **c** mboOn–tfOff, **d** mboOn–tfOn

ment using a rectangular grid of cells, where each cell has a size of 4 by 4 meters. The goal position of each robot is generated by randomly picking up a cell among a set of candidate target cells. The initial position of the robot is however fixed.

A specific time period was given as an upper bound for each robot to reach its target. In this respect, we set the upper bound of completion time to be 7, 10, and 12 min for 3, 4, and 5 robots, respectively. Failures can occur when one/many robots are not able to reach their goal positions within the given time. For each number of robots, we generated 20 random scenarios. The aggregated result of each navigation option in every scenario is normalized with respect to the nonoptimized option (mboOFF and tfOFF). The normalization is computed as the ratio of the result of each option and the result of the nonoptimized option. In other words,

normalized result of option i = result of option i / result of option 1

By doing this normalization, the results of every scenario are equally scaled. Median, mean, and standard deviation of normalized results of 20 random scenarios are calculated for each number of robots (3, 4, and 5).

The MBO navigation strategy was tested with three robots "α," "β," and "γ" in 20 random scenarios. Figure 16.12 shows the fixed starting positions "α_S," "β_S," and "γ_S," and the possible target positions "α_T," "β_T," and "γ_T" to be selected randomly for the three robots in an environment without obstacles. The starting and target positions in an environmnet with obstacles are shown in Fig. 16.13.

Table 16.4 Movement of five robots without obstacles: Performance comparison of different options of the navigation algorithm

a	mboOFF–tfOFF	mboOFF–tfON	mboON–tfOFF	mboON–tfON
Accur	0.100	0.100	0.099	0.100
Time	87.000	86.800	81.600	80.900
Length	33.851	33.378	29.201	29.097
CC	0.605	0.394	0.335	0.354
LS	4.154	3.834	2.810	2.817
b				
Accur	0.099	0.099	0.099	0.099
Time	72.400	74.900	67.300	70.400
Length	29.751	32.625	28.797	29.706
CC	2.165	3.513	2.674	2.814
LS	0.920	1.974	0.850	0.842
c				
Accur	0.099	0.099	0.099	0.100
Time	77.399	78.200	67.400	72.100
Length	33.555	30.457	28.891	29.280
CC	0.757	0.484	0.458	0.625
LS	5.725	3.525	3.303	3.154
d				
Accur	0.099	0.099	0.099	0.100
Time	82.000	76.600	74.099	67.900
Length	35.772	34.036	33.121	31.092
CC	1.466	2.008	1.318	1.220
LS	2.228	2.055	1.627	1.407
e				
Accur	0.099	0.099	0.100	0.099
Time	89.000	89.500	84.700	84.200
Length	48.636	49.071	46.237	46.180
CC	0.616	0.730	0.694	0.708
LS	5.305	5.205	4.579	4.779
f				
Avg Accur	0.099	0.099	0.099	0.099
Avg Time	81.560	81.200	75.020	75.100
Avg Length	36.313	35.913	33.250	33.071
Avg CC	1.122	1.426	1.096	1.144
Avg LS	3.667	3.319	2.634	2.600

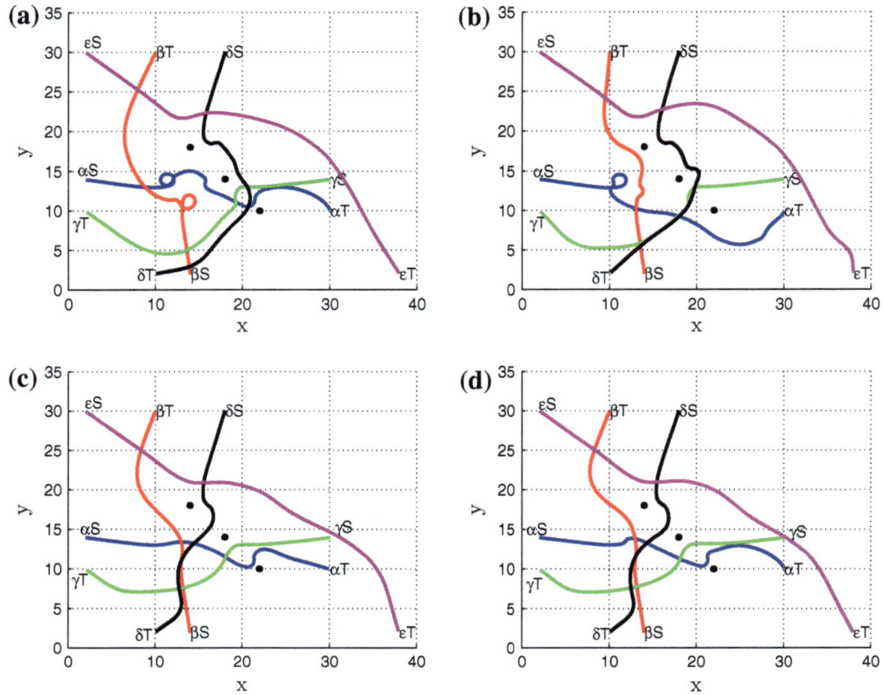

Fig. 16.11 Movement of five robots with obstacles: Navigation results of different options of the navigation algorithm, **a** mboOff–tfOff, **b** mboOff–tfOn, **c** mboOn–tfOff, **d** mboOn–tfOn

The combined results for the three-robot test case can be seen in Fig. 16.14. The results show that the MBO navigation method is better compared to traditional PF. We got small standard deviation values for time and length parameters but large standard deviation for CC and LS parameters. This can be mainly attributed to the maneuvers aiming at avoiding possible collision with other nearby robots.

A similar behavior of the different options of the navigation approach was noticed in the case of scenarios involving four and five robots. The results of random scenarios involving four robots with and without obstacles are shown in Fig. 16.15, while the results of scenarios with five robots are displayed in Fig. 16.16.

16.5.4 Results Summary

The proposed algorithm presented in this paper is a combination of multiple methods. Potential fields are used for a safe movement of robots from starting positions to target positions. Fuzzy rules are used to reduce the effect of repulsive potential fields in the vicinity of obstacles. Traffic rules are used to handle situations where robot paths are crossing each other. The MBO navigation strategy is used to optimize repul-

Table 16.5 Movement of five robots with obstacles: Performance comparison of different options of the navigation algorithm

a	mboOFF–tfOFF	mboOFF–tfON	mboON–tfOFF	mboON–tfON
Accur	0.129	0.099	0.100	0.099
Time	106.500	93.700	83.000	85.300
Length	37.963	38.011	30.623	31.547
CC	2.182	1.749	0.629	1.451
LS	6.477	5.547	3.575	3.919
b				
Accur	0.099	0.099	0.100	0.099
Time	88.700	79.400	72.600	71.800
Length	37.010	30.110	29.915	30.142
CC	2.505	4.349	2.124	2.119
LS	4.049	1.725	1.021	1.137
c				
Accur	0.099	0.100	0.100	0.099
Time	82.400	78.400	71.200	71.200
Length	34.008	33.524	31.185	31.283
CC	1.294	0.838	0.427	0.543
LS	3.290	3.588	3.655	3.760
d				
Accur	0.099	0.125	0.099	0.100
Time	91.100	98.500	79.700	80.300
Length	36.522	34.891	31.202	31.502
CC	1.239	0.982	1.662	1.480
LS	2.905	2.435	1.321	1.444
e				
Accur	0.100	0.100	0.191	0.099
Time	93.900	93.300	83.200	89.200
Length	49.098	50.403	48.249	47.893
CC	0.543	1.237	0.731	0.677
LS	4.309	4.744	4.816	4.556
f				
Avg Accur	0.105	0.104	0.118	0.099
Avg Time	92.520	88.660	77.940	79.560
Avg Length	38.920	37.388	34.235	34.473
Avg CC	1.553	1.831	1.115	1.254
Avg LS	4.206	3.608	2.877	2.963

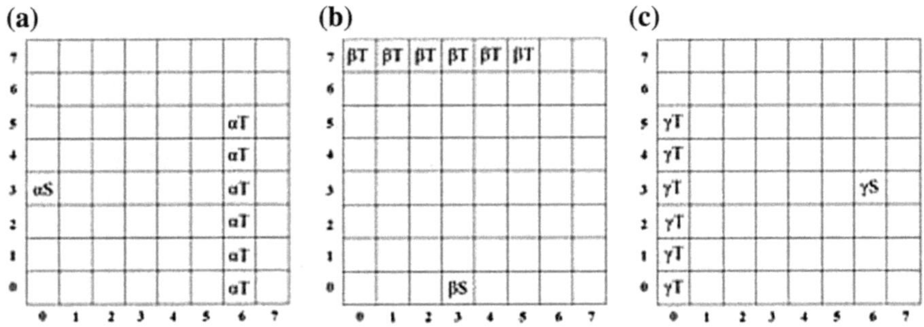

Fig. 16.12 Placement of three robots in random scenarios without obstacles: possible grid map positions, **a** α robot positions, **b** β robot positions, **c** γ robot positions

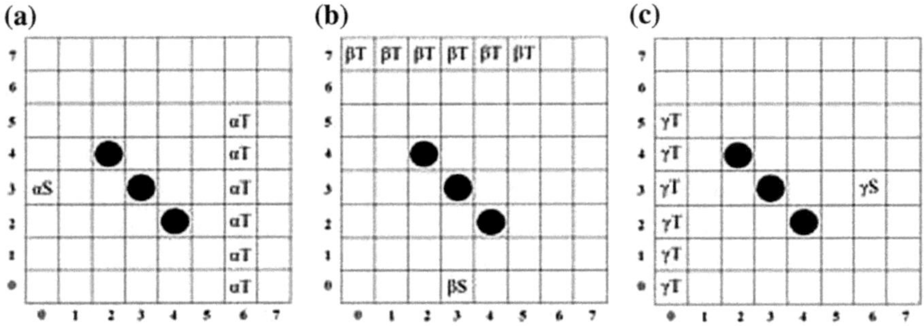

Fig. 16.13 Placement of three robots in random scenarios with obstacles: possible grid map positions, **a** α robot positions, **b** β robot positions, **c** γ robot positions

sive potential fields of robots acting in a shared area. After using all these methods a final velocity vector is computed and provided to the robot. It has been observed that the navigation of mobile robots can be effectively implemented using a combination of these methods. First, fuzzy rules were tested for obstacle avoidance. It could be observed that fuzzy rules can reduce unnecessary potential fields in the effective region of obstacles.

In further experiments the validation of the MBO method was tested, and results were compared with traditional PF. It has been observed that while using the traditional PF method repulsive potential fields caused unwanted motions of robots, but after activation of the MBO option smoother results were obtained. From the experiments it has been examined that robots using the MBO method take less time and cover less travel distance, while values of LS parameters in different scenarios showed that the MBO method helps in safe turnings of robots while moving. During experiments it has been observed that the MBO method has more effect on time and length parameters, while CC and LS are more related with low-level robot control. In the results of random scenarios it is observed that MBO method showed better results in most of the scenarios. However, in some scenarios MBO leads to failures due to the limitations of MBO method.

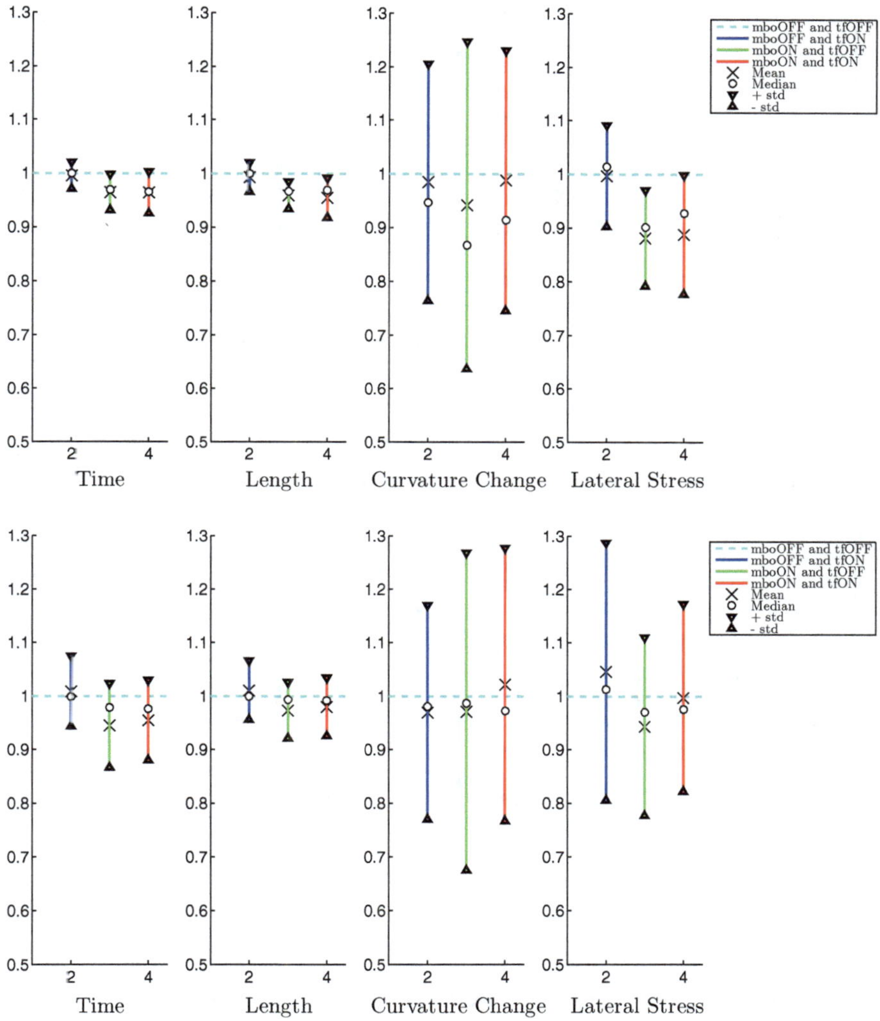

Fig. 16.14 Combined results of random scenarios involving three robots with and without obstacles in the environment. The results are the median, mean, and standard deviation of 20 random scenarios after normalizing w.r.t. nonoptimized case

Another point is that an increased number of robots affects the performance of the MBO strategy. On the one hand the performance becomes better compared with traditional PF. On the other hand some risk of collision remains due to limitations of all methods used here. Since the proposed algorithm is a combination of different methods, the final movement of robots will be—more or less—effected by all these methods. This is also valid for the limitations of every method, e.g., potential fields have a local minimum problem, traffic rules may add additional velocity vectors causing unwanted movements of the robots. The MBO method has two limitations: One is due to the normalization of weights to 1 that can lead to too small weights to

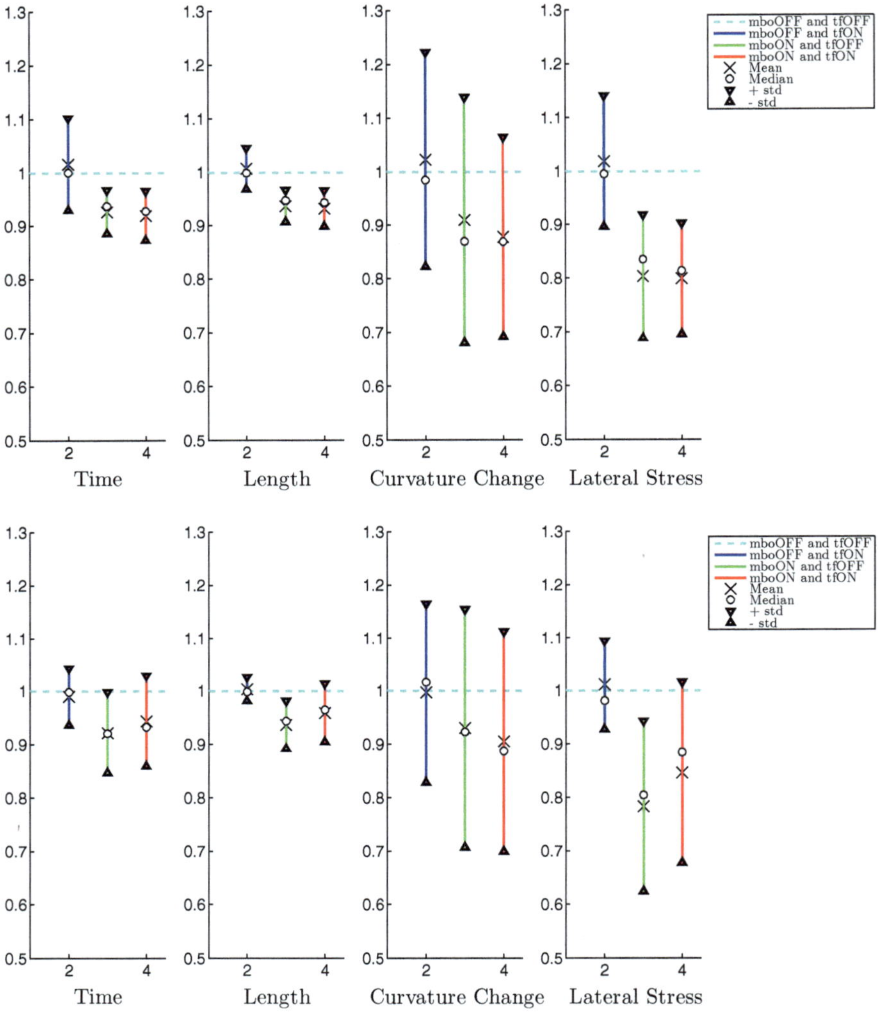

Fig. 16.15 Combined results of random scenarios involving four robots with and without obstacles in the environment. The results are the median, mean, and standard deviation of 20 random scenarios after normalizing w.r.t. nonoptimized case

some robots if the number of robots in a small area grows. The second limitation of the MBO method is that while computing the weights it only considers the positions and does not consider orientations of other robots. These issues can be solved by changing the normalization of weights accordingly and by considering the directions of motion of other robots in the MBO method. These issues will be considered in future developments.In future work, some features will be added to enable a robot to differentiate between static and dynamic obstacles using laser range finder and to localize itself in the global map.

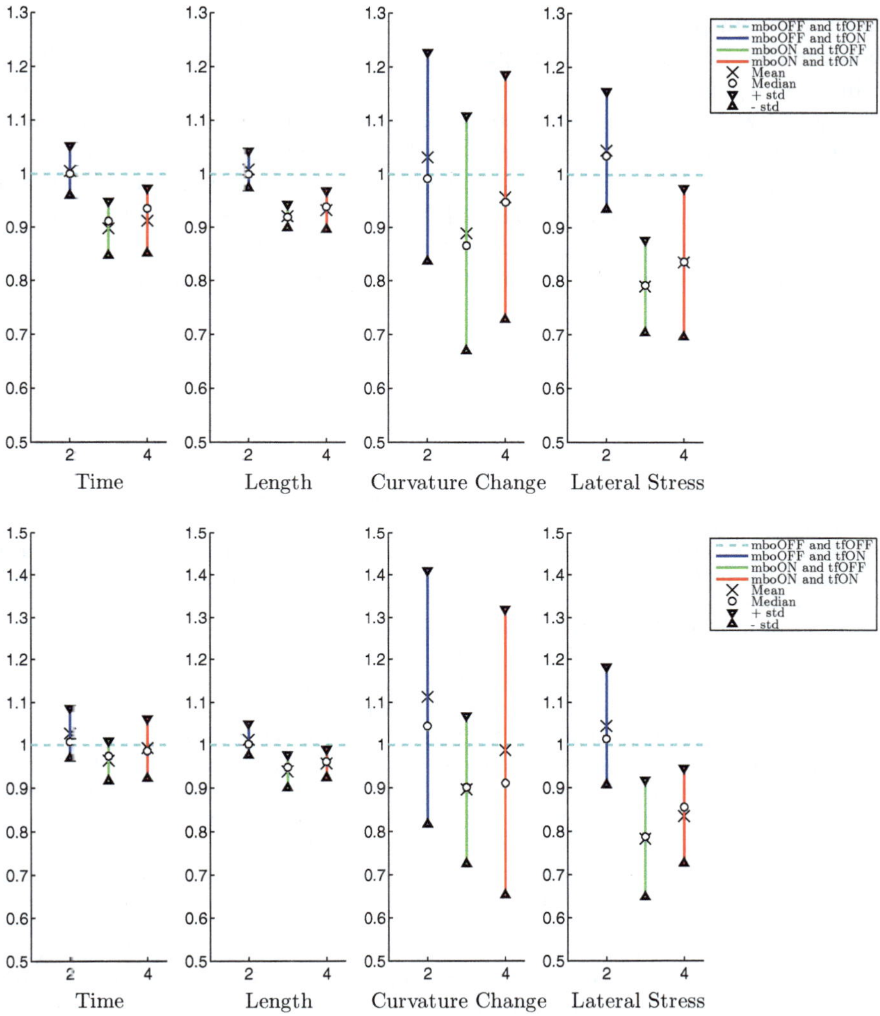

Fig. 16.16 Combined results of random scenarios involving five robots with and without obstacles in the environment. The results are the median, mean, and standard deviation of 20 random scenarios after normalizing w.r.t. nonoptimized case

16.6 Conclusion

Navigation and obstacle avoidance of mobile robots can be performed by a variety of principles like artificial potential fields, traffic rules, and control methods. It has also been shown that a "deformation" of central symmetry using fuzzy rules may be helpful because it takes the robot-object scenario better into account. An important aspect is the market-based optimization (MBO) of competing potential fields of mobile platforms. MBO imitates economical behavior and the competition between consumer and producer agents. By means of MBO some potential fields will be strengthened and some weakened depending on the actual scenario. This is required when more than two robots compete within a small area which makes a certain minimum distance between the robots and appropriate maneuvers necessary. Therefore, MB navigation allows smooth motions in such situations. Simulation experiments have shown the feasibility of the presented method. One main aspect of this paper was to implement and test extensively the MBO navigation strategy on the realistic simulation environment ROS and Gasebo to be able to implement the results to real robots.

References

1. Khatib, O.: Real-time obstacle avoidance for manipulators and mobile robots. In: IEEE International Conference on Robotics and Automation, pp. 500–505. St. Loius, Missouri (1985)
2. Koren, Y., Borenstein, J.: Potential field methods and their inherent limitations for mobile robot navigation. In: Proceedings of the IEEE Conference on Robotics and Automation, pp. 1398–1404. Sacramento, California, 7–12 April 1991
3. Borenstein, J., Koren, Y.: The vector field histogram—fast obstacle avoidance for mobile robots. IEEE Trans. Robot. Autom. **7**(3), 278–288 (1991)
4. Aicardi, M., Baglietto, M.: Decentralized supervisory control of a set of mobile robots. In: Proceedings of the European Control Conference 2001, 557–561 (2001)
5. Baxter, J.L., Burke, E.K., Garibaldi, J.M., Norman, M.: Multi-robot search and rescue: a potential field based approach. In: Mukhopadhyay, S., Gupta, G.S. (eds.) Autonomous Robots and Agents Series. Studies in Computational Intelligence, vol. 76, pp. 9–16. Springer, Berlin (2007)
6. Werling, M., Grœll, L., Bretthauer, G.: Invariant trajectory tracking with a full-size autonomous road vehicle. IEEE Trans. Robot. **26**(4), 758–765 (2010). August
7. Sorensen, M.J.: Artificial potential field approach to path tracking for a nonholonomic mobile robot. In: Proceedings of the 11th Mediteranean Conference on Control And Automation. Rhodes, Greece (2003)
8. Alonso-Mora, J., Breitenmoser, A., Rufli, M., Beardsley, P., Siegwart, R.: Optimal reciprocal collision avoidance for multiple non-holonomic robots. In: Proceedings of the 10th International Symposium on Distributed Autonomous Robotic Systems (DARS), Switzerland, November 2010
9. Martinez, A., Tunstel, E., Jamshidi, M.: Fuzzy logic based collision avoidance for a mobile robot. Robotica **12**(6), 521–527 (1994). November–December
10. Lilly, John H.: Evolution of a negative-rule fuzzy obstacle avoidance controller for an autonomous vehicle. IEEE TFS **15**(4), 719–839 (2007)
11. Stingu, P.E., Lewis, F.L.: Motion path planning for mobile robots. Report: University of Texas at Arlington, Automation and Robotics Research Institute, 1–7 April 2007. http://arri.uta.edu/acs/ee5322/lectures/Motion/planning.pdf
12. Wallace, A.: Flow control of mobile robots using agents. In: 29th International Symposium on Robotics Birmingham, UK, pp. 273–276 (1998)
13. Gold, T.B., Archibald, J.K., Frost, R.L.: A utility approach to multi-agent coordination. In: Proceedings of the 2000 IEEE International Conference on Robotics and Automation, pp. 2052–2057, San Francisco (1996)
14. Large, E.W., Christensen, H.I., Bajcsy, R.: Scaling the dynamic approach to path planning and control: competition among behavioral constraints. **18**(1), 37–58 (1999)

15. Barrett, G., Lafortune, S.: Some issues concerning decentralized supervisory control with communication. In: Proceedings of the 38th CDC, pp. 2230–2236, Phoenix, Arizona USA (1999)
16. Dias, M.B., Zlot, R., Kalra, N., Stentz, A.: Market-based multirobot coordination: a survey and analysis. Proc. IEEE **94**(7), 1257–1270 (2006)
17. Gerber, C., Russ, C., Vierke, G.: On the suitability of market-based mechanisms for telematics applications. In: Proceedings of the 3rd International Conference on Autonomous Agents, pp. 409–409. Seattle, USA, 1–5 May 1999
18. Clearwater, S.H. (ed.): Market-based control: a paradigm for distributed resource allocation. In: Proceedings of the 38th CDC, Phoenix, Arizona USA., World Scientific, Singapore (1996)
19. Guenther, O., Hogg, T., Huberman, B.A.: Controls for unstable structures. Proceedings of the SPIE, pp. 754–763. San Diego, CA, USA (1997)
20. Voos, H., Litz, L.: A new approach for optimal control using market-based algorithms. In: Proceedings of the European Control Conference ECC99, Karlsruhe (1999)
21. Palm, R.: Synchronization of decentralized multiple-model systems by market-based optimization. IEEE Trans. Syst. Man Cybern. B **34**, 665–672 (2004)
22. Palm, R., Bouguerra, A.: Navigation of mobile robots by potential field methods and market-based optimization. ECMR 2011, Oerebro, Sweden, 7–9 September 2011
23. Palm, R., Bouguerra, A.: Market-based optimization for the navigation of mobile robots. COSY 2011, Ohrid, Macedonia, 16–20 September 2011
24. Abdullah, M.: Mobile Robot Navigation using potential fields and market based optimization. Master thesis, Oerebro University, Oerebro, Sweden (2013)
25. Albus, J., McLean, Ch., Barbera, A., Fitzgerald, M.: An architecture for real-time sensory-interactive control of robots in a manufacturing facility, IFAC Information Control Problems in Manufacturing Technology, pp. 81–90. USA (1982)
26. Fajen, B.R., Warren, W.H.: Behavioral dynamics of steering, obstacle avoidance, and route selection. J. Exp. Psychol. **29**(2), 343–362 (2003). (Copyright by the American Psychological Association, Inc., Human Perception and Performance)
27. Cook, G.: Mobile robots: Navigation, control and remote sensing (Ch.1), pp. 1–9. Wiley-IEEE Press (2011)
28. Palm, R., Driankov, D., Hellendoorn, H.: Model based fuzzy control. Springer, Berlin (1997)
29. Willow Garage. Ros. http://ros.org/wiki/ROS/Introduction
30. Gazebo. Gazebo user guide. http://gazebosim.org/userguide/overviewintro.html
31. Calisi, D., Nardi, D.: Performance evaluation of pure-motion tasks for mobile robots with respect to world models. Auton. Robots **27**(4), 465–481 (2009)

Chapter 17
Fault Tolerant Estimation of UAV Dynamics via Robust Adaptive Kalman Filter

Chingiz Hajiyev and Halil Ersin Soken

Abstract A covariance scaling based robust adaptive Kalman filter (RAKF) algorithm is developed for the case of sensor/actuator faults. The proposed RAKF uses variable scale factors for scaling the process and measurement noise covariances and eliminating the effect of the faults on the estimation procedure. At first, the existing covariance estimation based adaptation techniques are reviewed. Then the covariance scaling methods with single and multiple factors are discussed. After choosing the efficient adaptation method an overall concept for the RAKF is proposed. In this concept, the filter initially isolates the fault, either in the sensors or in the actuators, and then it applies the required adaptation process such that the estimation characteristic is not deteriorated. The performance of the proposed filters is investigated via simulations for the UAV state estimation problem. The results of the presented algorithms are compared for different types of sensor/actuator faults and recommendations about their application are given within this scope.

17.1 Introduction

Since 1990s the unmanned aerial vehicles (UAVs) has taken place in our life. There is an overwhelming interest in them, especially in the military field, due to their advantages such as reducing human life loss and cost of the performed mission. They have been preferred as a part of wide range of missions from disaster control to missile decoy [1] and can perform all these missions with low failure risk.

C. Hajiyev (✉)
Aeronautics and Astronautics Faculty, Istanbul Technical University, 34469 Maslak, Istanbul, Turkey
e-mail: cingiz@itu.edu.tr

H.E. Soken
Institute of Space and Astronautical Science (ISAS), Japan Aerospace Exploration Agency (JAXA), Yoshinodai, 3-1-1, Sagamihara, Kanagawa, Japan
e-mail: ersin_soken@ac.jaxa.jp

G.M. Dimirovski (ed.), *Complex Systems*, Studies in Systems, Decision and Control 55, DOI 10.1007/978-3-319-28860-4_17

Nonetheless, as an obligation for the successful mission, they must be excellent in terms of autonomy because of the absence of a human pilot who can take the initiative. In this sense, navigation system design for the UAVs is an open discussion [2] and fault-tolerant UAV control system constitutes an important part of these discussions [3].

The preliminary phase for the UAV control system design is building an observer to accurately estimate the states that will be controlled. That is a desired procedure since it is important to know the parameters like velocity, altitude, attitude, etc., precisely. Only if these states are estimated properly, the aircraft can be controlled successfully. However, for such accurate estimation procedure both the sensors and actuators should be fault free. In contrary, the estimator gives inaccurate results and diverges by time unless it is built robust in order to overrun these issues.

The Kalman filter (KF) approach to the state estimation is quite sensitive to any malfunctions. If the condition of the real system does not correspond to the models, used in the synthesis of the filter, then these changes, resulting from some possible failures at the sensors or actuators, significantly decrease the performance of the estimation system. In such cases, the KF can be adapted and adaptive or robust Kalman filters can be used to recover the possible malfunctions.

The KF can be adapted and hence made insensitive to the sensor or actuator failures by using various different techniques. The basic approaches to the adaptive Kalman filtering problem are the multiple-model-based adaptive estimation (MMAE) [4–6], innovation-based adaptive estimation (IAE) [7–9] and residual-based adaptive estimation (RAE) [10, 11]. While in the first approach, a bank of Kalman filters run in parallel under different models for the filter's statistical information, in the rest the adaptation is performed by directly estimating the covariance matrices of the measurement and/or system noises based on the changes in the innovation or residual sequences.

The MMAE methods described in [4–6] assume the faults are known. The parallel running KFs are designed for the known sensor/actuator faults and the algorithm works for the faults within these limits. Besides the MMAE approach requires high computational load due to several KFs running parallel. As a result of these drawbacks the MMAE method can be used only for a limited number of applications.

Estimation of the covariance matrices by the IAE and RAE [7–11] requires using the innovation or residual vectors for N epoch. This increases the storage burden and presents the determination of the moving window width, N, as another problem. Furthermore, the IAE and RAE estimators require that the number, type, and distribution of the measurements for all epochs within a window should be consistent. If they are not, the covariance matrices cannot be estimated based on the innovation or residual vectors.

In [12] an approach to detect and isolate the aircraft sensor/actuator faults affecting the mean of the KF innovation is presented. The effects of the sensor and actuator faults in the innovation of the channels are investigated, and a decision approach for isolating the sensor and actuator faults is proposed. When a KF is used, the decision statistics changes regardless the fault is in the sensors or in the actuators. On the other hand, in [12] Hajiyev and Caliskan present a robust Kalman filter (RKF) based on the Doyle–Stein condition and show that distinguishing the sensor and actuator faults is easy by using this RKF. Although the paper proposes an effective method for the fault isolation, this work examines the sensor and actuator faults affecting only the mean of the innovation, and, as another drawback, the given measurement noise covariance adaptation procedure is hard to apply.

Another concept for the KF adaptation is to scale the noise covariance matrix by multiplying it with a time-dependent variable. One of the methods for constructing such algorithm is to use a scale factor as a multiplier to the process or measurement noise covariance matrices [13]. These types of algorithms are also named as fading memory algorithms and the adaptive Kalman filter (AKF) is referred as the adaptive fading Kalman filter (AFKF) [14]. When the process noise covariance matrix is scaled (Q-adaptation) there is an actuator fault to be detected. Otherwise, when the measurement noise covariance matrix is scaled (R-adaptation), a sensor fault is dealt with. In [14, 15], an adaptation procedure based on single-scale factor and in [16] a scheme based on multiple-scale factors is presented. The essence of using multiple-scale factors is the performance of the KF that varies for each variable. Unlike the adaptation with a single-scale factor, the multiple factor based method scales only the required terms of the process/measurement noise covariances, so any unnecessary information loss is prevented by disregarding only the data of the faulty sensor/actuator. Nonetheless, these papers [14–16] take only the Q-adaptation procedure into consideration and do not examine the R-adaptation methods. Per contra, the R-adaptation is investigated for possible sensor faults in [17] with a considerably simpler technique, but the Q-adaptation is not touched upon. It is shown that if there is a malfunction in the measurement system, the RKF algorithm can be utilized, and by correcting the measurement noise covariance matrix with the measurement noise scale factor, insensitiveness of the filter to the current sensor faults can be ensured.

Moreover, there are also published works in the literature which are interested in both the R- and Q-adaptations [18–20]. However, in these papers an isolation scheme for the fault is absent. These filters can detect whether there is a fault or not, but the type of the fault (e.g., sensor or actuator) cannot be found. Besides, they are based on single-scale factor and for complex multivariable systems they may not be sufficiently accurate.

In this chapter, first we review the existing estimation based adaptation techniques for the process and measurement noise covariance matrices to build a robust adaptive Kalman filter (RAKF) algorithm. Then we propose novel noise covariance scaling based methods for the filter adaptation and compare their performance with the existing methods. We discuss the simultaneous R- and Q-adaptations by using different combinations of these adaptation methods and investigate their applicability for the UAV state estimation in case of sensor/actuator fault. In the last part, we introduce a RAKF algorithm, which incorporates a two-stage adaptive filter procedure for the detection and isolation of the sensor or actuator faults, and apply for the UAV state estimation problem.

The chapter proceeds as follow; in Sect. 17.2 flight dynamics model of the UAV is given. In Sect. 17.3 the KF is introduced and in Sect. 17.4 a brief description about the KF adaptation is given. In Sect. 17.5, the noise covariance estimation based R- and Q-adaptation techniques are reviewed. Section 17.6 presents the novel noise covariance scaling based R- and Q-adaptation methods for building the RAKF. In Sect. 17.7 the proposed methods are compared with the existing ones and the simultaneous Q- and R-adaptation methodology is discussed. In Sect. 17.8 an overall scheme for the RAKF is presented in the light of the results obtained in advance. In Sect. 17.9 the paper is concluded.

17.2 Mathematical Model of the UAV Flight Dynamics

The dynamic characteristic of an aircraft must be known in order to build a KF for the state estimation. In general, equation derivation process for an aircraft may be examined in two steps; derivation of the rigid body equations of motion and the linearization.

In the first step, six rigid body equations which consist of three force and three moment equations, are obtained for the UAV. With the assumptions of

- the aircraft is a rigid body,
- the mass of the aircraft remains constant for a relatively short duration of time,
- the xz plane of the aircraft is the plane of the symmetry,

following force and moment equations can be written for an aircraft [21]:

$$F_x = m(\dot{u} + qw - rv), \tag{17.1}$$

$$F_y = m(\dot{v} + ru - pw), \tag{17.2}$$

$$F_z = m(\dot{w} + pv - qu), \tag{17.3}$$

$$L = \dot{p}I_{xx} + qr(I_{zz} - I_{yy}) - (\dot{r} + pq)I_{xz}, \tag{17.4}$$

$$M = \dot{q}I_{yy} - pr(I_{zz} - I_{xx}) + (p^2 - r^2)I_{xz}, \tag{17.5}$$

$$N = \dot{r}I_{zz} + pq(I_{yy} - I_{xx}) + (qr + \dot{p})I_{xz}. \tag{17.6}$$

Here, m is the mass of the aircraft; u, v, and w are the velocity components and F_x, F_y, and F_z are the forces acting on the plane in the body frame along x, y, and z directions; p, q, *and* r are the angular rates and L, M, *and* N are the moments about x, y, and z axes in the body frame; I_{xx}, I_{yy}, I_{zz} and I_{xz} are the moments of inertia and the product of inertia terms for the related axes.

These nonlinear equations can be linearized by using the small perturbation theory [22]. Hereafter, $\Delta(\cdot)$ term is used for representing the perturbed state.

In order to apply the method, these assumptions are also made:

- The motion of the aircraft consists of small perturbations about the steady flight condition.
- Reference flight condition is symmetric.
- The propulsive forces remain constant.
- x axis is considered as it coincides with the direction of the aircraft velocity vector.
- If $order\ of\ (\Delta q), \ldots, (\Delta v) = \varepsilon$ neglect the terms with the order of ε^2.

After that, linearized equations of motion for the aircraft can be derived. In general, the equations are examined in two phases; longitudinal and lateral. Consequently, the linearized longitudinal equations of motion for the UAV in the state space form is,

$$\begin{bmatrix} \Delta\dot{u} \\ \Delta\dot{w} \\ \Delta\dot{q} \\ \Delta\dot{\theta} \\ \Delta\dot{h} \end{bmatrix} = \begin{bmatrix} X_u & X_w & 0 & -g & 0 \\ Z_u & Z_w & u_0 & 0 & 0 \\ M_u + M_{\dot{w}} Z_u & M_w + M_{\dot{w}} Z_w & M_q + M_w u_0 & 0 & 0 \\ 0 & 0 & 1 & 0 & 0 \\ 0 & -1 & 0 & u_0 & 0 \end{bmatrix} \begin{bmatrix} \Delta u \\ \Delta w \\ \Delta q \\ \Delta\theta \\ \Delta h \end{bmatrix} + \begin{bmatrix} X_{\delta e} & X_{\delta T} \\ Z_{\delta e} & Z_{\delta T} \\ M_{\delta e} + M_{\dot{w}} Z_{\delta e} & M_{\delta T} + M_{\dot{w}} Z_{\delta T} \\ 0 & 0 \\ 0 & 0 \end{bmatrix} \begin{bmatrix} \Delta\delta_e \\ \Delta\delta_T \end{bmatrix} \tag{17.7}$$

and the linearized lateral equations of motion for the UAV in the state space form is,

$$\begin{bmatrix} \Delta\dot{\beta} \\ \Delta\dot{p} \\ \Delta\dot{r} \\ \Delta\dot{\phi} \end{bmatrix} = \begin{bmatrix} \frac{Y_\beta}{u_0} & \frac{Y_p}{u_0} & -\frac{u_0 - Y_r}{u_0} & \frac{g\cos(\theta_0)}{u_0} \\ L_\beta & L_p & L_r & 0 \\ N_\beta & N_p & N_r & 0 \\ 0 & 1 & 0 & 0 \end{bmatrix} \begin{bmatrix} \Delta\beta \\ \Delta p \\ \Delta r \\ \Delta\phi \end{bmatrix} + \begin{bmatrix} 0 & \frac{Y_{\delta r}}{u_0} \\ L_{\delta a} & L_{\delta r} \\ N_{\delta a} & N_{\delta r} \\ 0 & 0 \end{bmatrix} \begin{bmatrix} \Delta\delta_a \\ \Delta\delta_r \end{bmatrix}. \tag{17.8}$$

Here, $\Delta\delta_e$, $\Delta\delta_a$ and $\Delta\delta_r$ are the elevator, aileron and the rudder deflections, $\Delta\delta_T$ is the change in the thrust, $\Delta\theta$ is the pitch angle about y axis, $\Delta\phi$ is the roll angle about x axis, $\Delta\beta$ is the sideslip angle, Δh is the height, θ_0 and u_0 are the values of related terms in the steady state flight, g is the gravity constant and X_u, X_w, $X_{\delta e}, X_{\delta T}, Z_u$, $Z_w, Z_{\delta e}, Z_{\delta T}$, M_u, M_w, M_q, $M_{\dot{w}}$, Y_r, Y_p, Y_β, $Y_{\delta r}$, L_β, L_p, $L_r, L_{\delta a}$, $L_{\delta r}$, $M_{\delta e}$, $M_{\delta T}, N_{\delta a}$, $N_{\delta r}$, N_β, N_p, N_r are the stability derivatives.

17.3 Optimal Kalman Filter for UAV State Estimation

When we built the KF algorithm for the UAV state estimation, we worked on the combined longitudinal and lateral dynamics of the aircraft. Hence, the state vector to be estimated is formed of $n = 9$ states as

$$x = [\,\Delta u \quad \Delta w \quad \Delta q \quad \Delta \theta \quad \Delta h \quad \Delta \beta \quad \Delta p \quad \Delta r \quad \Delta \phi\,]^T, \tag{17.9}$$

while the control input vector is

$$u = [\,\Delta \delta_e \quad \Delta \delta_T \quad \Delta \delta_a \quad \Delta \delta_r\,]^T. \tag{17.10}$$

Then, let us introduce the UAV process and observation models for the combined dynamics in state space form as follows:

$$x_k = F_k x_{k-1} + B_k u_{k-1} + G_k w_k, \tag{17.11}$$

$$y_k = H_k x_k + v_k, \tag{17.12}$$

where, F_k is the system dynamics matrix, B_k is the control distribution matrix, y_k is the measurement vector, G_k is the transition matrix of system noises, H_k is the measurement matrix, which is 9×9 identity matrix in case, and w_k and v_k are white Gaussian system process and measurement noises, respectively;

$$E\left[w_k w_j^T\right] = Q_k \delta_{kj}, \tag{17.13}$$

$$E\left[v_k v_j^T\right] = R_k \delta_{kj}, \tag{17.14}$$

$$E\left[w_k v_j^T\right] = 0. \tag{17.15}$$

Here, Q_k is the process noise covariance matrix, R_k is the measurement noise covariance matrix, and δ_{kj} is the Kronecker delta function as

$$\delta_{kj} = \begin{cases} 1, & k = j \\ 0, & k \neq j \end{cases}. \tag{17.16}$$

After that, the KF for such combined UAV model can be given by those following steps [23]:

State prediction:

$$\tilde{x}_{k/k-1} = F_k \hat{x}_{k-1/k-1} + B_k u_{k-1}. \tag{17.17}$$

Covariance prediction:

$$P_{k/k-1} = F_k P_{k-1/k-1} F_k^T + G_k Q_k G_k^T. \tag{17.18}$$

Innovation:

$$\tilde{e}_k = y_k - H_k \tilde{x}_{k/k-1}. \tag{17.19}$$

Optimal Kalman Gain:

$$K_k = P_{k/k-1} H_k^T \left(H_k P_{k/k-1} H_k^T + R_k\right)^{-1}. \tag{17.20}$$

State estimation:

$$\hat{x}_{k/k} = \tilde{x}_{k/k-1} + K_k \tilde{e}_k. \tag{17.21}$$

Covariance estimation:

$$P_{k/k} = (I - K_k H_k) P_{k/k-1}. \tag{17.22}$$

Here $\tilde{x}_{k/k-1}$ is the predicted state vector, $P_{k/k-1}$ is the predicted covariance matrix, $\tilde{e}_k$ is the innovation sequence, K_k is the optimal Kalman gain, $\hat{x}_{k/k}$ is the estimated state vector, and $P_{k/k}$ is the estimated covariance matrix for discrete step k. Since the Kalman gain for the KF presented here is optimal without any adaptation, the filter may be also referred as the optimal Kalman filter (OKF).

17.4 Adaptive Kalman Filtering

17.4.1 A Priori Uncertainty

The KF, which is examined in the previous section, is valid only if all a priori statistical data is known. Practically, we may encounter with situations where a priori data is partially known or known with predictions, not the exact information. In this case, the developed estimators are not optimal and the estimated values via these algorithms may not converge. As a result, another approach might be required to build the estimation algorithms.

A priori uncertainty degree may vary as follows.

a. Complete a priori *statistical uncertainty*

In this case, neither shape nor parameters of the probability distribution rule for the components of the measured and estimated random processes are known. However, the limited variation areas for the components of the random processes are given. In such a case, the estimation algorithms can be formed only on the basis of a guaranteed approach [24].

b. Partial a priori *statistical uncertainty*

In this case, probability distribution rule is known for some of the components of the measured and estimated random processes. The number for the parameters with unknown probability characteristics should not be high. Increase in this number deteriorates the quality of the solution for the problem. In this sense, when there is a priori uncertainty in the parameters, distribution set is given rather than the probability distribution rule for the random processes. Estimation algorithm should be chosen such that it satisfies the optimality criteria for the given distribution set. It means that the estimation algorithm must be adaptive.

17.4.2 Adaptation

There are three approaches for the adaptation problem: the parametric approach, invariant principle based approach, and structural approach. The parametric approach is the most common one. In this approach, the algorithm uses the measurement data for both estimating the required components of the random processes and restoring a priori statistical characteristics of the dynamical system and measurements.

For most of the parametric adaptive estimation methods, a self-tuning circuit is added to the regular KF algorithm. When the KF is designed, it is assumed that the statistical characteristics of the system model, measurements, and noises are known accurately. Yet this assumption is not often ensured and the filter becomes suboptimal. Suboptimality might be also caused by the simplified process calculations in the filter algorithm. As a result the adaptation of the noise covariances becomes a necessity. Such adaptation procedure should be also followed for determining the type of the fault caused by the suboptimality of the filter. These types of KFs with adapted noise covariances are called as the AKF.

Possible techniques for designing the AKF are presented below:

- The MMAE is used. In the MMAE approach, a bank of KFs run in parallel under different models for the statistical filter information matrices, i.e., the process noise covariance matrix Q and/or the measurement noise covariance matrix R.
- Unknown noise covariances of the KF are determined by the statistical analysis of the innovation or residual series. This might be either as direct estimation of the covariance matrices or as an adaptation performed on the basis of covariance scaling (Fig. 17.1).
- Probability methods such as the maximum likelihood are used for the noise estimation. Estimated values for the unknown noise covariances are periodically used to renew the noise information in the filtering algorithm.

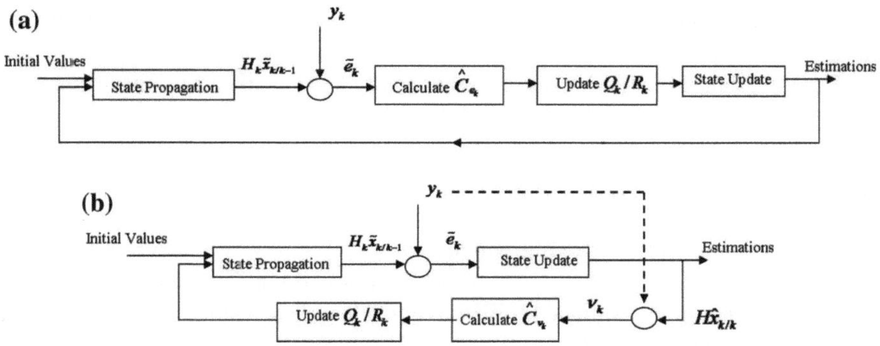

Fig. 17.1 The KF adaptation by the statistical analysis of the innovation and residual series: **a** Innovation-based KF adaptation; **b** Residual-based KF adaptation. For both adaptation techniques update for the process and measurement noise covariances might be either as estimation or scaling

- An iterative procedure is followed for determining the Q and R matrices. The unknown covariances are estimated by analyzing the residual error that is a result of using the same covariance values at the previous iteration step. This method is generally used when the computational load is not the primary concern.

In the next section, we review some of the common techniques for the KF adaptation.

17.5 Adaptive Kalman Filtering with Noise Covariance Estimation

In this section, the existing methods for the adaptive Kalman filtering, which are based on noise covariance estimation, are reviewed and their drawbacks are discussed.

17.5.1 Innovation-Based Adaptive Estimation (IAE)

Among the methods mentioned in the previous section the IAE is the mostly used method for the filter adaptation [7–9]. The essence of this approach is to investigate the behavior of the innovation sequence and determine whether the real characteristics of the noise match with its a priori characteristics. When the real innovation sequence is different than the white noise, adaptation of the process noise covariance (Q) or measurement noise covariance (R) is necessary. Investigation of the innovation sequence is achieved by checking if it has zero mean expected value and given covariance.

The covariance of the innovation sequence ($\tilde{e}_j$) is calculated by the formula

$$\hat{C}_{e_k} = \frac{1}{N} \sum_{j=k-N+1}^{k} \tilde{e}_j \tilde{e}_j^T \tag{17.23}$$

Here N is the size of the moving window.

If a priori and real characteristics of the noise do not match, the R and Q covariances should be estimated, respectively. Then these estimations are included in the filtering algorithm. Hence for the AKF with noise covariance estimation the filter is used for both estimating the states and characteristics of the noise covariances.

In the IAE, the R and Q matrices should be updated at each iteration in accordance with the estimations. The equations for estimating the matrices in the IAE based AKF are

$$\hat{R}_k = \hat{C}_{e_k} - H_k P_{k/k-1} H_k^T, \tag{17.24}$$

$$\hat{Q}_k = K_k \hat{C}_{e_k} K_k^T. \tag{17.25}$$

17.5.2 Residual Based Adaptive Estimation (RAE)

In the RAE approach, the adaptation is applied on the measurement and process noise covariances based on the changes in the residual sequence [10, 11]. The residual sequence is formulated as

$$\nu_k = y_k - H\hat{x}_{k/k}, \tag{17.26}$$

where $\hat{x}_{k/k}$ is the estimated state vector. The residual covariance is given as

$$\hat{C}_{\nu_k} = \frac{1}{N} \sum_{j=k-N+1}^{k} \nu_j \nu_j^T. \tag{17.27}$$

In order to estimate the R and Q matrices the following equations are used:

$$\hat{R}_k = \hat{C}_{\nu_k} + H_k P_{k/k} H_k^T \tag{17.28}$$

$$\hat{Q}_k = \frac{1}{N} \sum_{j=k-N+1}^{k} \Delta x_j \Delta x_j^T + P_{k/k} - F P_{k-1/k-1} F^T, \tag{17.29}$$

Here $P_{k/k}$ is the estimated covariance for the recent step (k), $P_{k-1/k-1}$ is the estimated covariance for the previous step $(k-1)$ and Δx_k is the state correction sequence, which is the difference between the estimated and predicted states as

$$\Delta x_k = \hat{x}_{k/k} - \hat{x}_{k/k-1}. \tag{17.30}$$

In the steady state we may consider just the first term of (17.29). Then regarding $\Delta x_k = K_k \nu_k$ equality, (17.29) may be approximated with (17.25) [10].

For some application areas such as the integrated INS/GPS system it is experienced that the RAE is more appropriate than the IAE for adaptive estimation.

17.5.3 Drawbacks of Adaptive Estimation Methods

Both the IAE and RAE have some certain drawbacks when they are used for the KF adaptation. These are:

a. We need to use the innovation/residual sequences for N epoch. In this case the computational load increases and determination of the moving window size N appears as another problem.
b. The IAE and RAE estimators require that the number, type, and distribution of the measurements for all epochs within a window should be consistent. If they do not, the covariance matrices of the measurement noises cannot be estimated based on the innovation or the residual vectors.
c. For the R estimation in the IAE, if $H_k P_{k/k-1} H_k^T > \hat{C}_{e_k}$ then $\hat{R}_k$ becomes negative definite and the KF may collapse.

17.6 Adaptive Kalman Filtering with Noise Covariance Scaling

In this section, we present the novel KF adaptation techniques based on the noise covariance scaling.

17.6.1 Innovation-Based Adaptive Scaling

R-Adaptation. The essence of the innovation-based R-adaptation is the comparison of real and theoretical values of the innovation covariance [17]. When the operational condition of the measurement system mismatches with the model used in the synthesis of the filter, then the KF gain changes related to the differentiation in the innovation covariance matrix. Under these circumstances, the innovation covariance differs as

$$\hat{C}_{e_k} = H_k P_{k/k-1} H_k^T + S_k R_k \tag{17.31}$$

and so the Kalman gain becomes

$$K_k = P_{k/k-1} H_k^T \left(H_k P_{k/k-1} H_k^T + S_k R_k \right)^{-1}, \tag{17.32}$$

Here S_k is the measurement noise scale factor (SF).

In this approach, the Kalman gain is changed when the predicted observation $H_k \hat{x}_{k/k-1}$ is considerably different from the actual observation, y_k, because of the significant changes in the operational condition of the measurement system. In other words, if the real value of the filtration error exceeds the theoretical error (or if a priori and real characteristics of the noise do not match) as

$$tr\left\{ \tilde{e}_k \tilde{e}_k^T \right\} \geq tr\left\{ H_k P_{k/k-1} H_k^T + R_k \right\} \tag{17.33}$$

the filter must be run adaptively. Here $tr\{\cdot\}$ denotes the trace of the related matrix.

There are two possible techniques for scaling the R matrix. The first one is to use just a single-scale factor (SSF) and the second one is to scale with a matrix formed of multiple-scale factors (MSF).

For obtaining the SSF let us substitute (17.31) into (17.33) and regard that the adaptation begins at the point where the equality condition for (17.33) is satisfied,

$$tr\left\{ \tilde{e}_k \tilde{e}_k^T \right\} = tr\left\{ H_k P_{k/k-1} H_k^T \right\} + S_k tr\{ R_k \}. \tag{17.34}$$

Then, in the light of $tr\left\{ \tilde{e}_k \tilde{e}_k^T \right\} = \tilde{e}_k^T \tilde{e}_k$ equality, the SSF, S_k, can be written as

$$S_k = \frac{\tilde{e}_k^T \tilde{e}_k - tr\left\{ H_k P_{k/k-1} H_k^T \right\}}{tr\{ R_k \}}. \tag{17.35}$$

In case of malfunction in the measurement system, the adaptation of the KF (17.17)–(17.22) is performed via automatically correcting the Kalman gain. If the inequality condition for (17.33) is met, then the scale factor S_k increases. Bigger S_k causes a smaller Kalman gain (17.32), because of the covariance of the innovation sequence (17.31) which also increases. Consequently, small Kalman gain value reduces the effect of the faulty innovation sequence on the state estimation process (17.21). In all other cases, where measurement system operates normally, the SSF takes the value of $S_k = 1$ and so the filter runs optimally.

On the other hand, as clearly demonstrated with simulations in [25, 26], using the MSF is a healthier procedure for the adaptation since the filter disregards only the measurement of the faulty sensor in this case rather than disregarding all as for the SSF. When we scale the R matrix using the MSF, the relevant term of the Kalman gain matrix, which corresponds to the innovation channel of the faulty sensors, is fixed individually.

In order to determine the scale matrix formed of the MSF, the real and theoretical values of the innovation covariance matrix must be compared like we did for the SSF. When there is a measurement malfunction in the estimation system, the real error will exceed the theoretical one. Hence, if a scale matrix, S_k, is added into the algorithm as

$$\hat{C}_{e_k} = H_k P_{k/k-1} H_k^T + S_k R_k, \tag{17.36}$$

then it can be determined by the formula

$$S_k = \left(\hat{C}_{e_k} - H_k P_{k/k-1} H_k^T\right) R_k^{-1}. \tag{17.37}$$

In case of normal operation, the scale matrix will be a unit matrix as $S_k = I$. Here I represents the unit matrix. Nonetheless, S_k matrix, found by the use of (17.37) may be non-diagonal and have diagonal elements which are "negative" or less than "one." This is mainly because of the limited N number and the approximation errors. S_k matrix should be diagonal because only its diagonal terms have significance on the adaptation since each diagonal term corresponds to the noise covariance of each measurement (for the adaptation procedure S_k matrix is multiplied with the diagonal R matrix). Besides the measurement noise covariance matrix must be positive definite (that is why the multiplier S_k matrix cannot have negative terms) and also any term of this matrix cannot decrease in time for this specific problem since there is no possibility for increasing the performance of the onboard sensor (that is why the multiplier S_k matrix cannot have terms less than one). Therefore, in order to avoid such situation, composing the scale matrix by the following rule is suggested:

$$S^* = diag\left(s_1^*, s_2^*, \ldots, s_n^*\right), \tag{17.38}$$

where,

$$s_i^* = \max\{1, S_{ii}\} \qquad i = 1, n. \tag{17.39}$$

Here, S_{ii} represents the ith diagonal element of the matrix S. Apart from that point, if the measurements are faulty, S_k^* will change and so will affect the Kalman gain matrix;

$$K_k = P_{k/k-1} H_k^T \left(H_k P_{k/k-1} H_k^T + S_k^* R_k\right)^{-1}. \tag{17.40}$$

In case of any kind of malfunction, the element of the scale matrix, which corresponds to the faulty component of the innovation vector, increases and so the terms in the related column of the Kalman gain decreases. As a consequence, the effect of the faulty innovation term on the state update process reduces and accurate estimation results can be obtained even in case of measurement malfunctions.

We name the R-adapted KF as the RKF since it is robust against sensor faults.

Q-Adaptation. When there is an actuator fault in the system that ends up with changes in the control distribution matrix, the real characteristics of the noise does not match with its a priori characteristics. As a solution the Q-adaptation should be performed. The essence of the method is again comparing the real and theoretical values of the innovation covariance matrix. When there is an actuator fault, the real error will exceed the theoretical one.

Besides, it is known that using multiple factors for adaptation satisfies better estimation accuracy [25, 26]. Hence, a fading matrix is used for the adaptation rather than a fading factor. We name the matrix used for the Q-adaptation as fading matrix in order to distinguish from the scale matrix used for the R-adaptation, otherwise the function of scale and fading matrices are same. The fading matrix,Λ_k, built of multiple fading factors (MFF), is added into the algorithm as

$$\hat{C}_{e_k} = H_k(F_k P_{k-1/k-1} F_k^T + \Lambda_k G_k Q_k G_k^T) H_k^T + R_k. \tag{17.41}$$

Then the fading matrix can be determined as

$$\Lambda_k = \left(\hat{C}_{e_k} - H_k F_k P_{k-1/k-1} F_k^T H_k^T - R_k\right) \times \left(H_k G_k Q_k G_k^T H_k^T\right)^{-1}. \tag{17.42}$$

In a similar manner with the R-adaptation, the fading matrix should be diagonalized since the Q matrix must be a diagonal, positive definite matrix.

$$\Lambda^* = diag\left(\lambda_1^*, \lambda_2^*, \ldots, \lambda_n^*\right), \tag{17.43}$$

$$\lambda_i^* = \max\{1, \Lambda_{ii}\} \qquad i = 1, n. \tag{17.44}$$

Here, Λ_{ii} represents the ith diagonal element of the matrix Λ. Apart from that point, if there is a fault in the system, Λ_k^* must be put into process as

$$P_{k/k-1} = F_k P_{k-1/k-1} F_k^T + \Lambda_k^* G_k Q_k G_k^T. \tag{17.45}$$

17.6.2 Residual-Based Adaptive Scaling

We may also perform a residual-based scaling for the adaptation of the R matrix. In this case, the scale matrix formed of the MSF is

$$\tilde{S}_k = \left(\hat{C}_{\nu_k} + H_k P_{k/k} H_k^T\right) R_k^{-1}. \tag{17.46}$$

Same as the proposed method for the innovation-based R scaling, $\tilde{S}_k$ should be diogonalized using the following rule:

$$\tilde{S}^* = diag\left(\tilde{s}_1^*, \tilde{s}_2^*, \ldots, \tilde{s}_n^*\right), \tag{17.47}$$

$$\tilde{s}_i^* = \max\left\{1, \tilde{S}_{ii}\right\} \qquad i = 1, n. \tag{17.48}$$

Here, $\tilde{S}_{ii}$ represents the ith diagonal element of the matrix $\tilde{S}$.

On the other hand, a residual-based scaling for the process noise covariance adaptation is rather complicated because of the inverse matrix calculations. The analytical solution for obtaining the residual-based MFF, Λ_k, is difficult. Therefore, it is not presented and discussed in this chapter.

17.7 Adaptive Kalman Filtering for the UAV State Estimation

17.7.1 Comparison of the Noise Covariance Estimation and Scaling Techniques for the UAV State Estimation

In this section we compare the KF adaptation methods based on the noise covariance estimation and scaling. Comparison is made regarding their performance for the UAV state estimation in case of simulated sensor/actuator faults. Fault cases are examined individually.

In order to compare two different adaptation procedures, with the noise covariance estimation and scaling, a set of simulations have been realized.

Simulations are run for a period of 100 s with 0.1 s of sampling time, Δt. As an experimental platform the ZAGI UAV is chosen and KFs are run for estimating the states of this UAV [27]. For the KF adaptation the size of the moving window is selected as $N = 30$. For the Q and R scaling multiple-scale and fading factors are used.

The R-adaptation algorithms are tested by implementing a sensor fault to the filter in between 30 and 40th s. For this period of 10 s a constant term is added to the pitch angle measurement $\Delta\theta$.

In Table 17.1, means of the absolute values of error for estimations in between 30 and 40th s are given. The results are the mean for five different runs for each filter.

The table clearly shows that, as expected, the OKF cannot provide accurate estimation results in case of sensor fault. Moreover, the RKF with innovation-based R estimation diverges as a result of the negative definite R matrix which appears when $H_k P_{k/k-1} H_k^T > \hat{C}_{\tilde{e}_k}$. A negative definite R estimation even for a single step affects the filter's estimation characteristic for a long time and if it lasts for few steps the filter diverges. When the size of the moving window for estimation, N, is

Table 17.1 Means of the absolute values of error for estimations in between 30 and 40th s in case of sensor fault scenario

Parameter	OKF	RKF with IRE[a]	RKF with RRE	RKF with IRS	RKF with RRS
Δu (m/s)	1.9272	N/A	0.2479	0.0992	0.1153
Δw (m/s)	0.5627	N/A	0.0958	0.0603	0.0805
Δq (rad/s)	0.0822	N/A	0.0661	0.0575	0.0588
$\Delta \theta$ (rad)	0.5118	N/A	0.0474	0.0251	0.0378
Δh (m)	1.0979	N/A	0.7587	0.0839	0.1053
$\Delta \beta$ (rad)	0.0214	N/A	0.0389	0.0231	0.0229
Δp (rad/s)	0.0335	N/A	0.0484	0.0288	0.0321
Δr (rad/s)	0.0208	N/A	0.0458	0.0257	0.0227
$\Delta \phi$ (rad)	0.0383	N/A	0.0533	0.0410	0.0439

IRE: innovation-based R estimation, RRE: residual-based R estimation, IRS: innovation-based R scaling, RRS: residual-based R scaling. ([a])Results for the RKF with IRE are not available since the RKF diverges with negative definite R estimation

smaller the risk for negative definite R estimation increases. Possible method to overcome this problem is to force the R components to be higher than a minimum threshold or evaluate the estimations within a low pass filter but in this case the sensor fault may not be tolerated.

When we compare the performance of the RKFs, which are working properly, the RKF with the innovation-based R scaling gives slightly better estimation results than the others. The main advantage of the innovation-based R scaling is updating the R matrix in the actual time step when the filter detects the sensor fault. Nevertheless, the residual-based update (both for scaling and estimation) is performed with one step of delay and that brings about slightly poorer estimation accuracy especially at the initial phase of the fault (at around 30th s).

The Q-adaptation algorithms are tested for the actuator failure case. The UAV actuator fault is simulated by taking the first column elements of longitudinal control distribution matrix nearly zero in between 70 and 80th s.

Table 17.2 gives the means of the absolute values of error for estimations in between 70 and 80th s. The results are the mean for five different runs for each filter. As seen the OKF fails about providing accurate estimations in case of actuator fault. As for the AKFs, the AKF with the innovation-based Q estimation cannot tolerate the fault since it is an approximation for (17.29). We understand that the steady state assumption is not valid for this case. The AKF with the residual-based Q estimation and the AKF with the innovation-based Q scaling are not affected by the fault and can satisfy good estimation performance. The estimation performance for these two filters is not too much different. The only drawback of the AKF with the residual-based Q estimation is the higher computational load mainly caused by the extra calculation for the residual sequence. Our examinations show that when the Q is estimated using the residual sequence the filter requires almost 1.4 times more computation than it does for the case the Q is scaled by the innovation sequence.

Table 17.2 Means of the absolute values of error for estimations in between 70 and 80th s in case of actuator failure scenario

Parameter	OKF	AKF with IQE	AKF with RQE	AKF with IQS
Δu (m/s)	0.8164	1.4045	0.3474	0.1803
Δw (m/s)	1.3785	0.9983	0.2532	0.2416
Δq (rad/s)	0.3030	0.0778	0.0892	0.0654
$\Delta \theta$ (rad)	0.1784	0.1805	0.0344	0.0407
Δh (m)	0.0911	0.1091	0.0973	0.0891
$\Delta \beta$ (rad)	0.0208	0.0301	0.0240	0.0204
Δp (rad/s)	0.0304	0.0331	0.0173	0.0332
Δr (rad/s)	0.0230	0.0276	0.0271	0.0203
$\Delta \phi$ (rad)	0.0277	0.1135	0.0435	0.0461

IQE: innovation-based Q estimation, RQE: residual-based Q estimation, IQS: innovation-based Q scaling

As a consequence we may state that the novel noise covariance scaling based KF adaptation techniques are more advantageous than the existing noise covariance estimation based adaptation techniques. The innovation-based R scaling method should be preferred for fault tolerance against sensor faults because of its accuracy which is better than other adaptation methods. The innovation-based Q scaling should be used for fault tolerance against actuator faults because of both its performance and lower computational load than the estimation based techniques.

17.7.2 Simultaneous Q- and R-Adaptation

After deciding the KF adaptation techniques for the UAV state estimation in case of sensor/actuator faults, the next step should be discussing how to integrate two different noise covariance scaling procedures, the Q- and R-adaptations.

First of all in [8, 10] the R estimation algorithm is derived assuming full knowledge for the Q matrix and vice versa. Therefore, when we use the R (or Q) estimation method (both innovation- and residual-based) first we should assume that the Q (or R) matrix is fully known. Using both estimation algorithms at the same time cannot be in question.

Moreover all the adaptation procedures presented in this chapter are based on the covariance (either innovation or residual) matching method. The adaptation methods rely on each other when the estimation or scaling is performed. As also discussed in [28] full estimation of Q and R based on the covariance matching is questionable.

One possibility might be using the innovation sequence for adapting one of the matrices and residual sequence for the other. However, even in this case the adaptations are sensitive against each other and both the sensor and actuator faults cannot be tolerated at the same time. In Fig. 17.2 the estimation results for the pitch angle of the UAV is given. The innovation-based Q scaling and residual-based R

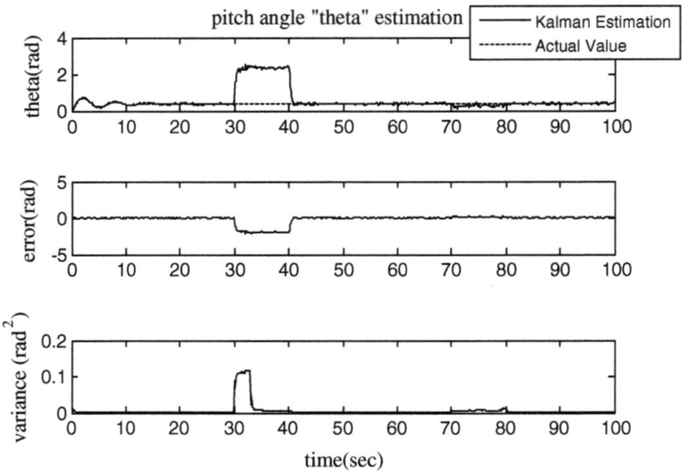

Fig. 17.2 Estimation of the pitch angle in case of sensor fault (between 30 and 40th s) and actuator fault (between 70 and 80th s). Innovation-based Q scaling and residual-based R scaling methods are run at the same time for the KF adaptation

scaling methods are run at the same time. There is a sensor fault in between 30 and 40th s and actuator faults in between 70 and 80th s. The top plot gives the KF estimation and actual value. The middle plot presents the estimation error and the lower one shows the estimation variance. As seen although the actuator fault is tolerated the effects of the sensor fault cannot be removed. The results are similar if we try to integrate two adaptation methods with different combinations such as the innovation-based R scaling and residual-based Q estimation. That proves the necessity for a fault isolation scheme for using the Q- and R-adaptation methods at the same time. In the next section, we propose the overall RAKF algorithm including the fault isolation procedure.

17.8 The RAKF Algorithm for the UAV State Estimation

17.8.1 The Overall RAKF Algorithm

In this section, an integration scheme for the R- and Q-adaptation methods is proposed such that a RAKF, which is insensitive against both the sensor and actuator faults, is built at the end. Multiple factors based adaptation is performed in both methods. Moreover because of the proved superiority, the noise covariance scaling technique is used for the adaptation.

First of all, remark that, due to the scale matrix or fading matrix (depending on the fault type) the covariance of the estimation error of the filter increases compared to the OKF (as seen in Fig. 17.2). Therefore, the RAKF algorithm is operated only when the fault is detected and in all other cases the procedure is run optimally with the OKF (17.17)–(17.22). The process is controlled by a kind of statistical information. At that point, following two hypotheses may be introduced:

- γ_0; the system is normally operating,
- γ_1; there is a malfunction in the estimation system.

To detect the failures a statistical function may be defined as

$$\beta_k = \tilde{e}_k^T \left[H_k P_{k/k-1} H_k^T + R_k\right]^{-1} \tilde{e}_k. \tag{17.49}$$

This statistical function has χ^2 distribution with s degree of freedom where s is the dimension of the innovation vector. If the level of significance, α, is selected as

$$P\{\chi^2 > \chi^2_{\alpha,s}\} = \alpha;; \qquad 0 < \alpha < 1, \tag{17.50}$$

then the threshold value, $\chi^2_{\alpha,s}$ can be found. Hence, when the hypothesis γ_1 is correct, the statistical value of β_k will be greater than the threshold value $\chi^2_{\alpha,s}$, i.e.,

$$\begin{array}{ll} \gamma_0 : \beta_k \le \chi^2_{\alpha,s} & \forall k \\ \gamma_1 : \beta_k > \chi^2_{\alpha,s} & \exists k. \end{array} \tag{17.51}$$

On the other hand, after detecting the fault in the system, the key point is detecting the type of the fault (either a sensor or actuator fault). After that the appropriate adaptation (the R- or Q-adaptation) may be applied. The fault isolation can be realized by an algorithm similar to the one proposed for the aircrafts in [12, 29]. When a KF is used, the decision statistics changes regardless the fault in the sensors or in the actuators. If a RKF based on the Doyle–Stein condition is used, it is easy to distinguish the sensor and actuator faults. The KF that satisfies the Doyle–Stein condition is referred in [29] as the RKF insensitive to the actuator failures and may be used for fault isolation as well in our study. If the sensor fault occurs, the R-adaptation is realized; or the procedure is continued with the Q-adaptation. The Doyle–Stein condition is

$$K(I + H\phi_s K)^{-1} = B_s (H\phi_s B_s)^{-1}. \tag{17.52}$$

Here K is the filter gain, I is unit matrix, H is the system measurement matrix, B_s is the control distribution matrix in continuous time, and $\phi_s = (sI - F_s)^{-1}$, where F_s is the system matrix in continuous time. The RKF is very useful for isolating the sensor and control surface failures as it is insensitive to the latter failures. If the KF process noise covariance matrix is chosen as

$$Q(q_r) = Q + q_r^2 B_s V B_s^T, \tag{17.53}$$

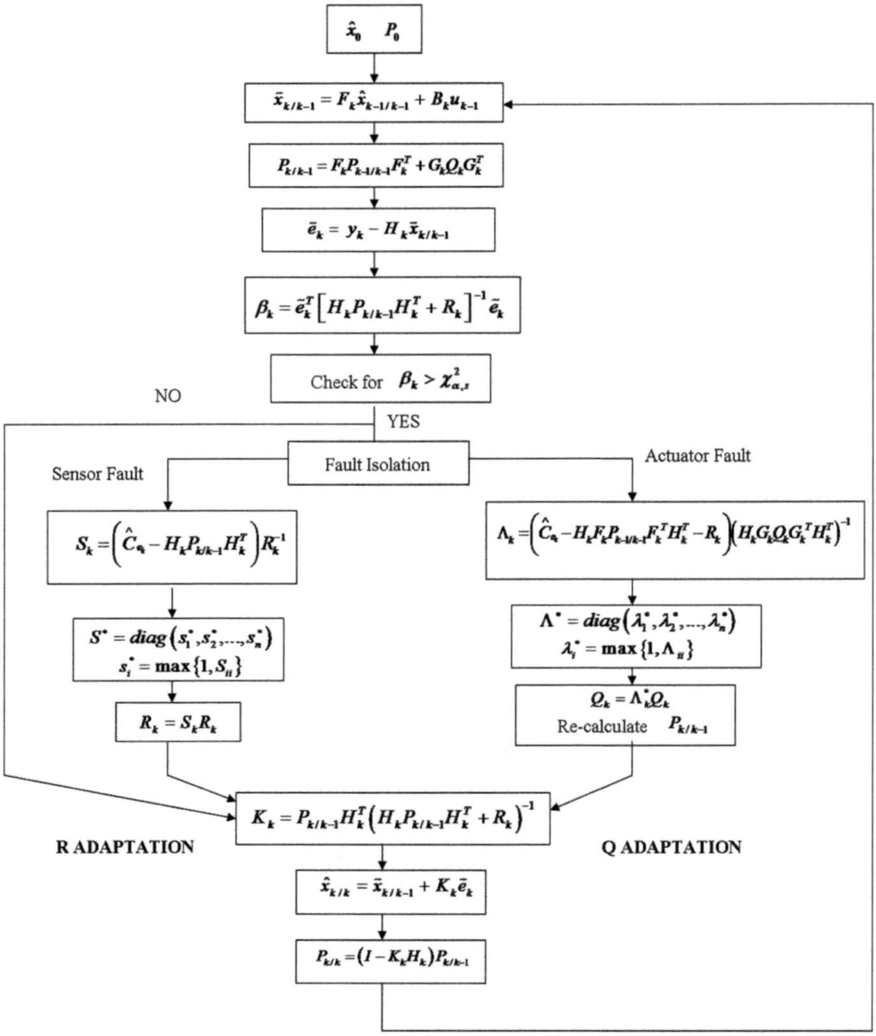

Fig. 17.3 System architecture for the proposed RAKF algorithm

then the RKF is obtained. Here, Q is the process noise covariance matrix for the nominal plant, q_r is a parameter that approaches to infinity and V is any positive definite symmetric matrix.

The algorithm architecture given in Fig. 17.3 summarizes the integrated R- and Q-adaptation procedures [30].

17.8.2 Numerical Example

In order to examine the effectiveness of the proposed RAKF, it has been applied for the state estimation of the ZAGI UAV. Simulations are run for 100 s and at some specific period sensor/actuator faults are implemented to the system. Simulations are also repeated for the OKF so as to compare results with the RAKF algorithm.

In simulations, $\chi^2_{\alpha,s}$ is taken as 21.7 and this value comes from the chi-square distribution when the degree of freedom is 9 and the reliability level is 99 %.

The graph for the statistical values of β_k in case of using the OKF is shown in Fig. 17.4. As seen, except the periods, where the sensor and actuator failures occur, β_k is lower than the threshold. On contrary, when one of these failures occurs β_k grows rapidly and exceeds the threshold value. Hence γ_1 hypothesis is judged to be true and it is shown that by using such procedure the fault can be detected.

The results for the fault isolation process are given in Fig. 17.5. As mentioned, the RKF insensitive to actuator failures is used for this isolation process. The behavior of the statistics β_k in the case of sensor/actuator failures shows that detecting the actuator failure is not possible when the RKF is used since it is insensitive against such faults; on the other hand, the sensor failure is detected immediately. Hence the fault isolation can be realized by using such procedure. The RKF, which is insensitive to actuator failures and sensitive to sensor failures, is employed to isolate the sensor and actuator failures in the proposed RAKF algorithm (See Fig. 17.3).

In order to test the RAKF algorithm, first, sensor fault is formed by adding a constant term to the measurement of pitch angle, $\Delta\theta$ in between 30 and 40th s. The constant term is nearly at same magnitude with the measurement, so that the measurement's magnitude is doubled for those instants of time. Then the actuator fault is simulated by taking the first column elements of longitudinal control distribution matrix nearly zero in between 70 and 80th s. In the presented figures, the

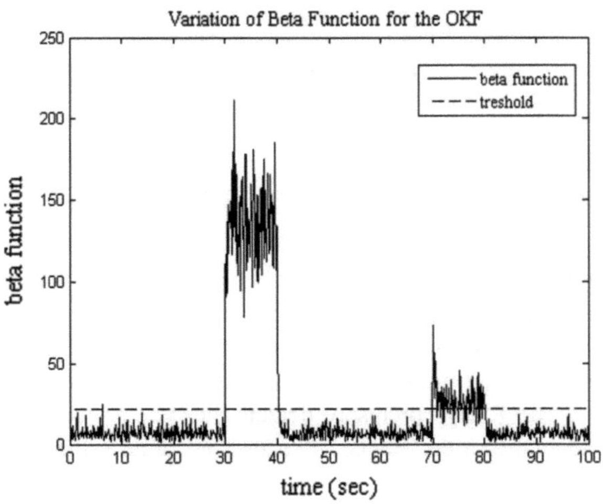

Fig. 17.4 Behavior of the statistics β_k when the OKF is used: Measurement failure in between 30 and 40th s and actuator failure in between 70 and 80th s

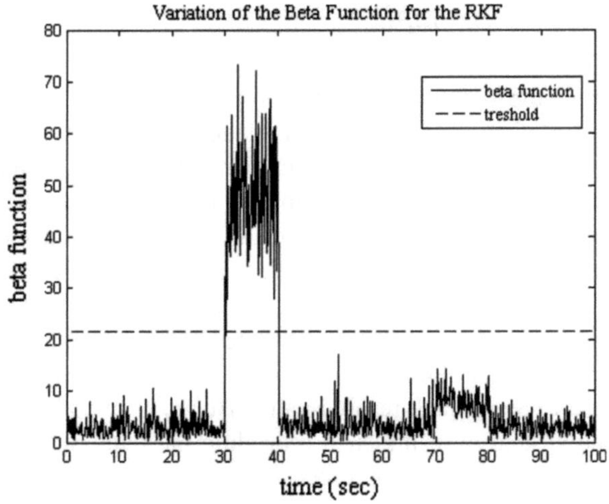

Fig. 17.5 Behavior of the statistics β_k when the RKF is used: Measurement failure in between 30 and 40th s and actuator failure in between 70 and 80th s

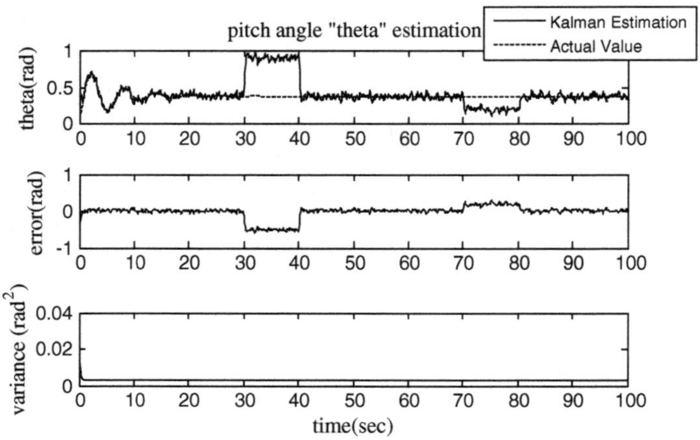

Fig. 17.6 Estimation of the pitch angle by the OKF in case of bias sensor fault (between 30 and 40th s) and actuator fault (between 70 and 80th s)

top plot gives the KF estimation and actual value, the middle plot presents the estimation error and the lower one shows the estimation variance (17.22).

As it is clear from Figs. 17.6 and 17.7, the RAKF gives superior estimation results than the OKF in both cases.

The RAKF precisely detects the type of fault and tunes itself according to the adaptation rule. Hence, it secures accurate estimation results for the whole process

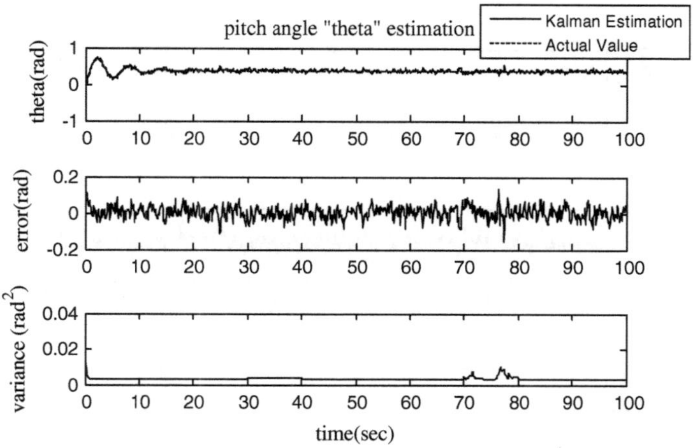

Fig. 17.7 Estimation of the pitch angle by the RAKF in case of bias sensor fault (between 30 and 40th s) and actuator fault (between 70 and 80th s)

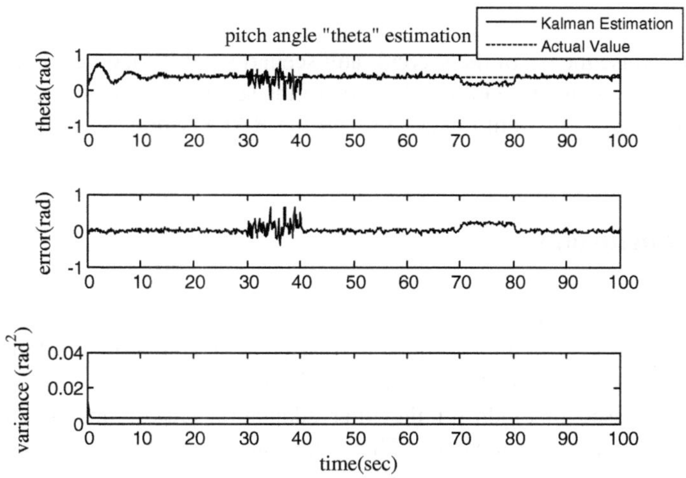

Fig. 17.8 Estimation of the pitch angle by the OKF in case of noise increment sensor fault (between 30 and 40th s) and actuator fault (between 70 and 80th s)

while the OKF fails in both sensor and actuator fault conditions. Also, note that similar results are obtained for all other estimated longitudinal parameters.

As the second testing scenario for the RAKF, this time sensor fault is characterized by multiplying the variance of the noise of the velocity component $\Delta\theta$ measurement with a constant term in between 30 and 40th s (measurement noise increment) whereas the actuator fault is not changed. As Figs. 17.8 and 17.9 show, again the OKF outputs contain error while the RAKF algorithm achieves estimation of the states accurately.

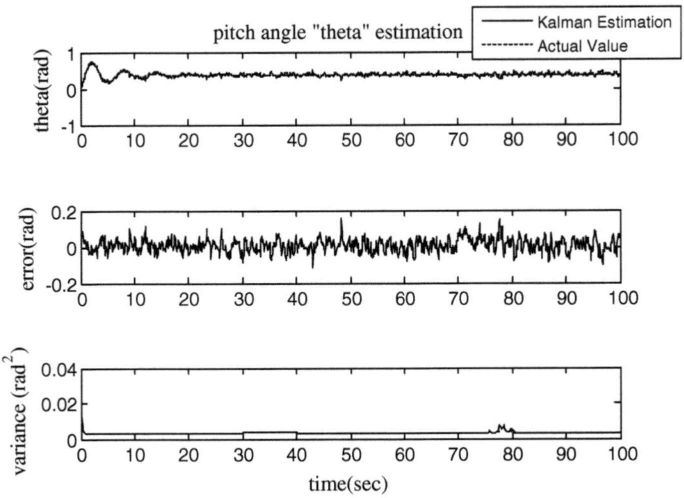

Fig. 17.9 Estimation of the pitch angle by the RAKF in case of noise increment sensor fault (between 30 and 40th s) and actuator fault (between 70 and 80th s)

Another consequence of the second fault scenario is the good performance of the fault isolation scheme. Regardless the type of the sensor fault, isolation scheme can distinguish type of the fault effectively and the RAKF works properly.

17.9 Conclusions

In this chapter, first we reviewed the existing estimation-based adaptation techniques for the process and measurement noise covariance matrices to build a robust adaptive Kalman filter (RAKF) algorithm. Then we proposed novel noise covariance scaling based methods for the filter adaptation and compared their performance with the existing methods. These comparisons showed that the proposed noise covariance scaling based adaptation methods satisfy better performance. After that we discussed simultaneous R- and Q-adaptations by using different combinations of these adaptation methods and investigated their applicability for the UAV state estimation in case of sensor/actuator fault. In the last part, we introduced a RAKF algorithm, which incorporates a two-stage adaptive filter procedure for the detection and isolation of the sensor or actuator faults, and applied it for the UAV state estimation problem. The results prove that the RAKF algorithm is capable of isolating the sensor/actuator faults and tolerating their effects on the estimation procedure.

References

1. Austin, R.: Unmanned Aircraft Systems: UAVs Design, Development and Deployment. Wiley, West Sussex (2010)
2. Parra, S.L., Angel, J.: Low Cost Navigation System for UAVs. Aerosp. Sci. Technol. **9**, 504–516 (2005)
3. Kobayashi, Y., Takahashi, M.: Design of intelligent fault-tolerant flight control system for unmanned aerial vehicles. In: Agneta, B. (ed.) Advances in Flight Control Systems. InTech, Rijeka, Croatia (2011)
4. Zhang, Y., Li, X.R.: Detection and diagnosis of sensor and actuator failures using interacting multiple-model estimator. In: Proceedings of 36th IEEE Conference on Decision and Control, **5**, 4475-4480 (1997)
5. Maybeck, P.S.: Multiple model adaptive algorithms for detecting and compensating sensor and actuator/surface failures in aircraft flight control systems. Int. J. Robust Nonlinear Control **9**, 1051–1070 (1999)
6. White, N.A., Maybeck, P.S., DeVilbiss, S.L.: MMAE detection of interference/jamming and spoofing in a DGPS-aided inertial system. IEEE Trans. Aerosp. Electron. Syst. **34**, 1208–1217 (1998)
7. Mehra, R.K.: On the identification of variance and adaptive kalman filtering. IEEE Trans. Autom. Control **15**, 175–184 (1970)
8. Maybeck, P.S.: Stochastic Models, Estimation, and Control, vols. I and II. Academic Press, New York (1982)
9. Salychev, O.S.: Special studies in dynamic estimation procedures with case studies in inertial surveying, ENGO 699. In: 26 Lecture Notes, Department of Geomatics Engineering, University of Calgary. Calgary, Canada (1994)
10. Mohamed, A.H., Schwarz, K.P.: Adaptive falman filter for INS/GPS. J. Geodesy **73**, 193–203 (1999)
11. Wang, J., Stewart, M.P., Tsakiri, M.: Adaptive kalman filtering for integration of GPS with GLONASS and INS. In: Proceedings of the 1999 International Geodesy Symposia (IUGG 99). Birmingham, UK (1999)
12. Hajiyev, C., Caliskan, F.: Sensor/actuator fault diagnosis based on statistical analysis of innovation sequence and robust kalman filtering. Aerosp. Sci. Technol. **4**, 415–422 (2000)
13. Hide, C., Moore, T., Smith, M.: Adaptive kalman filtering algorithms for integrating GPS and low cost INS. In: Proceedings on the Position Location and Navigation Symposium, Monterey, USA. pp. 227-233. Monterey (2004)
14. Hu, C., Chen, W., Chen, Y., Liu, D.: Adaptive kalman filtering for vehicle navigation. J. Global Positioning Syst. **2**, 42–47 (2003)
15. Ding, W., Wang, J., Rizos, C.: Improving adaptive kalman estimation in GPS/INS integration. J. Navig. **60**, 517–529 (2007)
16. Geng, Y., Wang, J.: Adaptive estimation of multiple fading factors in kalman filter for navigation applications. GPS Solutions **12**, 273–279 (2008)
17. Hajiyev, C.: Adaptive filtration algorithm with the filter gain correction applied to integrated INS/Radar altimeter. Proc. Inst. Mech. Eng. Part G: J. Aerosp. Eng. **221**, 847–885 (2007)
18. Kim, K.H., Lee, J.G., Park, C.G.: Adaptive two-stage kalman filter in the presence of unknown random bias. Int. J. Adapt. Control Signal Process. **20**, 305–319 (2006)
19. Jwo, D.J., Weng, T.P.: An adaptive sensor fusion method with applications in integrated navigation. J. Navig. **61**, 705–721 (2008)
20. Jwo, D.J., Chang, S.C.: Particle swam optimization for GPS navigation kalman Filter. Aircr. Eng. Aerosp. Technol.: Int. J. **81**, 343–352 (2009)
21. Yechout, T.R., Morris, S.L., Bossert, D.E., Hallgren, W.F.: Introduction to Aircraft Flight Mechanics: Performance, Static Stability, Dynamic Stability, and Classical Feedback Control. AIAA Education Series, Virginia (2003)
22. Nelson, R.C.: Flight Stability and Automatic Control. McGraw-Hill, Reading (1998)

23. Kalman, R.E.: A new approach to linear filtering and prediction problems. ASME J. Basic Eng. **82** Series D, 35–45 (1960)
24. Ogarkov, M.A.: Methods for Statistical Estimation of the Random Process' Parameters (in Russian). Energoatomizdat, Moscow (1990)
25. Hajiyev, C., Soken, H.E.: Fault tolerant estimation of UAV dynamics via robust adaptive kalman filter. In: Proceedings of COSY 2011—Special International Conference on Complex Systems: Synergy of Control, Computing and Communications, Sept 16–20 2011, Ohrid, Macedonia, pp. 311–320. The ETAI Society, Skopje, Macedonia (2011)
26. Hajiyev, C., Soken, H.E.: Robust estimation of UAV dynamics in presence of measurements faults. ASCE J. Aerosp. Eng. **25–1**, 1–10 (2012)
27. Matthews, J.S.: Adaptive Control of Micro Air Vehicles, M.Sc. Thesis, Department of Electrical and Computer Engineering, Brigham Young University, Utah, USA (2006)
28. Almagbile, A., Wang, J., Ding, W.: Evaluating the performance of adaptive kalman filter methods in GPS/INS integration. J. Global Positioning Syst. **9–1**, 33–40 (2010)
29. Hajiyev, C., Caliskan, F.: Sensor and control surface/actuator failure detection and isolation applied to F-16 flight dynamics. Aircr. Eng. Aerosp. Technol. Int. J. **77–2**, 152–160 (2005)
30. Hajiyev, C., Soken, H.E.: Robust Adaptive Kalman Filter for Estimation of UAV Dynamics in the Presence of Sensor/Actuator Faults. Aerosp. Sci. Technol. **28**, 376–383 (2013)

Chapter 18
Guidance Laws and Navigation Systems for Quadrotor UAV: Theoretical and Practical Findings

Stojche Deskovski, Vasko Sazdovski and Zoran Gacovski

Abstract There is an increasing interest in the unmanned aerial vehicles (UAVs) technologies in the recent years. UAVs are essential for many applications where human presence is considered unnecessary, dangerous, or impossible. These applications include variety of scientific, civilian, and military applications. This paper reflects the efforts that we are taking over the years toward a deeper understanding of these technologies. Presentation of low-cost, small-size quadrotor UAV that we are using is given. Our research activities in the Guidance Navigation and Control (GN&C) algorithms for quadrotor UAV are discussed here.

18.1 Introduction

Remotely operated unmanned aerial vehicles (UAVs), semiautonomous or autonomous aerial vehicles (AAVs) are finding increasing use in a variety of civilian, scientific, and military applications. Civilian applications include traffic monitoring, pollution monitoring, urban planning, crop yield prediction, livestock inventory, inspection of man-made structures (power lines and pipelines), and many others. Scientific applications include the use of UAVs for weather/atmospheric monitoring, climate research and monitoring, agricultural monitoring aerial imagery and

S. Deskovski (✉)
Faculty of Technical Sciences, University St. Kliment Ohridski, Bitola, Republic of Macedonia
e-mail: stojce.deskovski@gmail.com

V. Sazdovski
Faculty of Electrical Engineering, University Goce Delchev, Shtip, Republic of Macedonia
e-mail: vasko.sazdovski@ugd.edu.mk

Z. Gacovski
Faculty for Information & Communication Technologies, FON University, Skopje, Republic of Macedonia
e-mail: zgacovski@yahoo.com

G.M. Dimirovski (ed.), *Complex Systems*, Studies in Systems, Decision and Control 55, DOI 10.1007/978-3-319-28860-4_18

mapping, ore detection, and many others. Military applications include intelligence gathering and surveillance, airborne reconnaissance, electronic countermeasures and electronic warfare, search and rescue (SAR) support, weapons delivery systems, airborne drones for weapons training, etc.

Although modeling and control of aerial vehicles are well-understood problems, today's operational UAVs are still limited in terms of their autonomy. They are typically remotely operated during takeoff and landing, while inflight navigation is done by controlling the aerial vehicle along a programmed (way point) trajectory using global positioning system (GPS) and inertial navigation system (INS). Planning of the trajectory usually is done by human operators on the ground, prior the flight. Today's UAVs are equipped with state of the art vision, infrared and laser sensors. But still most of the analysis of the collected data is done after the flight and the detected failures are normally remotely handled.

Future UAVs have a need of high levels of autonomy and independence. These requirements are critically connected to the sensing, computational, and interpretational competence of the vehicles. Future autonomous navigation systems for UAVs need to provide reliable navigation parameters (position, velocity, and attitude) in environments where there is limited knowledge or there is not any knowledge of the environment. Usually in these applications the vehicles encounter obstacles. This brings a need to change the planned path, i.e., revise the motion in order to achieve collision-free path. The environment uncertainty and complexity is a critical issue here. The capability of trajectory planning in dynamic (changing) and unknown environments is to be very important for the future autonomous vehicles.

This paper is organized as follows. Section 18.2 presents a concept of low-cost, small-size quadrotor UAV which is very attractive for research and development. Section 18.3 is reserved for the navigation system of the quadrotor UAV. The current trends are discussed as well as some advanced ideas are presented. Section 18.4 discusses the guidance and control laws for quadrotor UAV. Simulation results are given in Sect. 18.5. The conclusions and future work are part of Sect. 18.6.

18.2 Concept of Low-Cost Unmanned Aerial Vehicle–Quadrotor

Quadrotor is a type of UAV, which consists of four independent propellers attached at each corner of a cross-frame, see Fig. 18.1. The propellers 1 and 3 rotate anti clockwise, whereas propellers 2 and 4 rotate clockwise with angular speeds $\vec{\Omega}_1, \vec{\Omega}_3$ and $\vec{\Omega}_2, \vec{\Omega}_4$, respectively.

Because of this kind of orientation of angular speeds, the resultant angular speed becomes $\vec{\Omega} = -\vec{\Omega}_1 + \vec{\Omega}_2 - \vec{\Omega}_3 + \vec{\Omega}_4$, see Fig. 18.1, where its direction is along the z axis in the body frame $B(P; x, y, z)$. On this way the yaw angular speed $\dot{\psi}\psi$ becomes smaller and can be maintained to zero value. The propellers thrust have

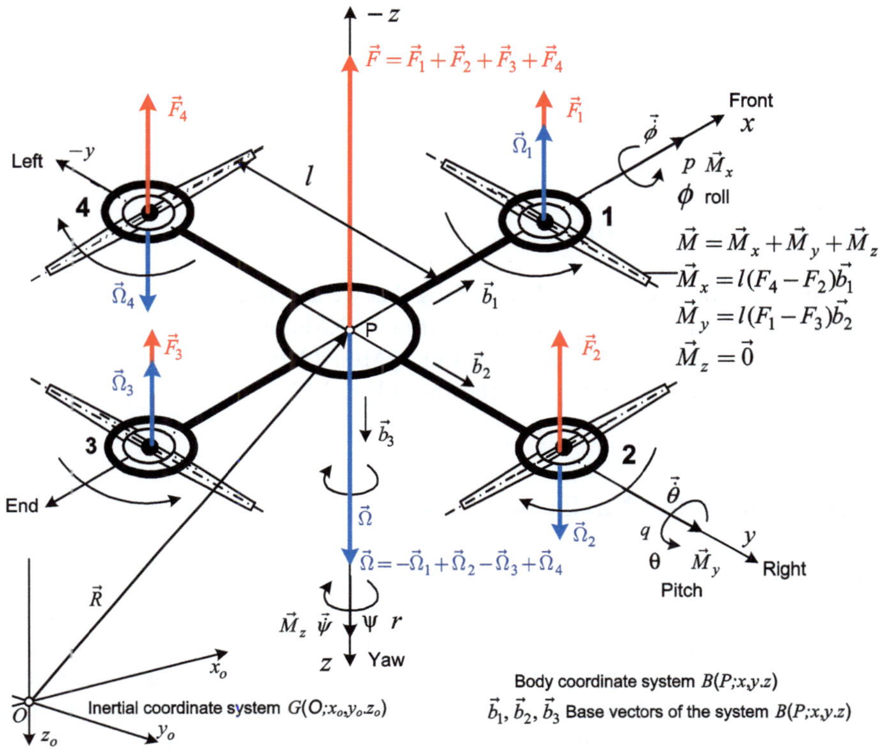

Fig. 18.1 Quadrotor helicopter, geometry, acting forces and moments, coordinate frames, Euler's angles and angular rates

same direction because of their different attack angles under which the attack edges are set. The total thrust force is $\vec{F} = \vec{F}_1 + \vec{F}_2 + \vec{F}_3 + \vec{F}_4$.

This configuration also offers the advantage of lateral motion without changing the pitch of the propeller blades. Fixed pitch simplifies rotor mechanics and reduces the gyroscopic effects. The control of the quadrotor is achieved by commanding different speeds to different propellers, which in turn produces differential aerodynamic forces and moments. For hovering, all four propellers rotate at same speed. For vertical motion, the speed of all four propellers is increased or decreased by the same amount, simultaneously [5].

In order to pitch and move laterally in that direction, speed of propellers 3 and 1 is changed conversely. Similarly, for roll and corresponding lateral motion, speed of propellers 2 and 4 is changed conversely. To produce yaw, the speed of one pair of two oppositely placed propellers is increased while the speed of the other pair is decreased by the same amount. This way, overall thrust produced is same, but differential drag moment creates yawing motion. In spite of four actuators, the quadrotor is still an under-actuated system.

Fig. 18.2 Quadrotor Flying Frog

The quadrotor UAV named Flying Frog that we are using in our research work presents a custom upgrade of flying set from KKMultiCopter. The quadrotor is upgraded with Inertial Measurement Unit (IMU) (Atomic IMU from Sparkfun Electronics) and a smart phone, as shown on Fig. 18.2.

18.3 Quadrotor Navigation System

Many of today's operational UAV navigation systems rely on inertial sensors as a primary measurement source. Inertial navigation (IN) is the process of calculating position by integration of velocity and computing velocity by integration of the total acceleration, known as specific force acceleration [10, 11, 15]. IN as a stand-alone navigation system suffers many drawbacks that complicate its usage. Errors from instrumentation, computational, alignment, and environmental nature cause the INto diverge slowly with time. From the 1990s and still today the GPS is the dominant navigation aid for IN. This GPS/IN integration was addressed in the research work of [12].

GPS-aided IN today can be considered as a well-understood and mature integration solution. A large number of articles and papers that discuss about GPS-aided IN are available today. Specialized books that treat IN, GPS, and their integration in a unified framework, are available [3, 7, 8]. Current research activities in this area are focused on implementation of different filter techniques then the classical Kalman Filter techniques [14]. The apparent goal is "to improve'' the Kalman filter results and implement nonlinear filtering techniques.

In a number of scenarios GPS measurements may be completely unavailable or simply they may be not precise enough in order to update the INS. Tunnels, canyons, forests, and urban areas are typical examples. Also the GPS signal can be jammed. Driven primarily by this reasons alternative methods of aiding IN have seen great attention. Vision-aided IN apparently has been the favored solution in the past several years. This is a case particularly because of its low cost, long range and high resolution and most importantly because of its passive property [14].

The single video camera is favorite hardware setup for aerial vehicles. It has small weight and can be easily mounted on the vehicle. The lack of depth information presents a serious drawback of the single camera. Motion of the vehicle is required to estimate the depth information. Large uncertainty may be reflected in the depth estimate because of the correlation of the camera motion with the depth information. Laser scanners and laser-range finders as well as ultrasound range finders may provide the needing depth information and provide aiding information to the single video camera. But their big size and greater weight limits it use on quadrotor UAVs. Quadrotor UAVs will become small, highly maneuverable, and very agile. Quadrotor UAV equipped with single video camera with an ability to perceive and understand the three-dimensional world is to be fundamental for autonomous navigation systems. Motion, i.e., maneuvers may provide the necessary aiding information to the single video camera. Therefore, there will be a need of intelligent maneuvering techniques so that reliable and accurate navigation parameters are provided. This will require coordination with the guidance and control measurements and the vehicle task itself for performing the needing vehicle maneuvers (movements) and achieving better navigation accuracy.

18.4 Quadrotor Guidance and Control System

Structure of the developed quadrotor guidance and control system is presented in Fig. 18.3. Quadrotor control system is designed using *two-degree-of-freedom controller design* concept [4, 9] which consists of *feedforward controller* and *linear feedback controller*.

The quadrotor feedforward controller consists of *trajectory generator* and *inverse model of quadrotor dynamics*. The trajectory generator and the dynamic inverse model together generate the *feasible-feedforward reference trajectory*, or *nominal state vector* $\mathbf{x}^o$ and *nominal input vector* $\mathbf{u}^o$ required to track a given reference trajectory. In this work, the inverse model derivation is based on six-degree-of-freedom model of quadrotor.

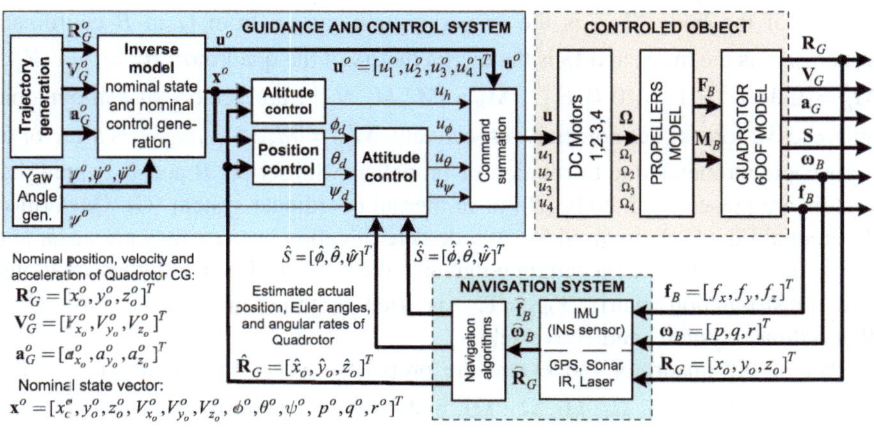

Fig. 18.3 Block diagram of quadrotor guidance and control system

The linear feedback controller is used to correct errors between the desired (reference) and actual trajectories. In case of quadrotor, linear feedback controller has three parts: *attitude controller* for control and stabilization of quadrotor rotational motion which is represented with Euler's angles, ϕ, θ, ψ, and angular rates p, q, r, or $\dot{\phi}$, $\dot{\theta}$, $\dot{\psi}$; *altitude controller* which corrects error between desired (nominal, reference) high $h^o = -z_o^o$ and actual high $h = -z_o$; and *position controller* correcting errors between nominal and actual position of CG in plane parallel to horizontal plane Ox_oy_o. The design of the linear controllers can be made using different design techniques known in linear control theory.

Block *Quadrotor 6DOF model* contains quadrotor motion equations which are derived from equations of motion for a six-degree-of-freedom rigid body, neglecting earth's rotation and slopes. Therefore, earth's fixed axis system $G(O; x_o, y_o, z_o)$ (see Fig. 18.1) is assumed to be the inertial reference. Quadrotor body-fixed system $B(P; x, y, z)$, has its origin in the quadrotor's center of mass (CM), point P, Fig. 18.1.

Rigid body dynamic and kinematic equations of motion in vector-matrix form [5], are:

$$\dot{\mathbf{V}}_B = -\tilde{\boldsymbol{\omega}}_B \mathbf{V}_B + m^{-1}(\mathbf{R}_B^A + \mathbf{F}_B) + \mathbf{T}_{GB}\mathbf{g}_{G,} \tag{18.1}$$

$$\dot{\omega}_B = -\mathbf{I}_B^{-1}\tilde{\omega}_B \mathbf{I}_B \boldsymbol{\omega}_B + \mathbf{I}_B^{-1}(\mathbf{M}_B^A + \mathbf{M}_B^F + \mathbf{M}_{gB}), \tag{18.2}$$

$$\dot{\mathbf{q}} = \mathbf{f}(\boldsymbol{\omega}_B, \mathbf{q}), \tag{18.3}$$

$$\dot{\mathbf{R}}_G = \mathbf{T}_{BG}^T \mathbf{V}_B, \tag{18.4}$$

where: $\mathbf{V}_B = [V_x, V_y, V_z]^T$ is the velocity vector of CM; $\boldsymbol{\omega}_B = [p, q, r]^T$ is angular rate vector; $\mathbf{R}_G = [x_o, y_o, z_o]^T$ is the position vector of CM; $\mathbf{g}_G = [0, 0, g]^T$ is the acceleration of gravity; $\tilde{\omega}_B$ is a skew-symmetric matrix constituted by components of the vector $\boldsymbol{\omega}_B$; $\mathbf{q} = [q_0, q_1, q_2, q_3]^T$ is quaternion vector which uniquely determines the attitude of the body; $\mathbf{T}_{BG}$ is the transformation matrix from G to B coordinate systems; m is the mass, and $\mathbf{I}_B$ is the inertia matrix of the quadrotor; $\mathbf{R}_B^A = [X, Y, Z]^T$, $\mathbf{M}_B^A = [L, M, N]^T$, $\mathbf{F}_B = [0, 0, F_z]^T$, $\mathbf{M}_B^F = [M_x, M_y, N_z]^T$, are vectors of aerodynamic and reactive forces and moments, respectively. $\mathbf{M}_{gB}^F = [M_{gx}, M_{gy}, N_{gz}]^T$ is a vector of gyroscopic moments due to rotation of propellers. Subscripts B and G denote that vectors are presented in body (B), or in inertial coordinate system (G). Quaternion differential Eq. (18.3) are used for attitude determination because they are linear and do not have singularities at certain points like the standard Euler's angles equations. The transformation matrix $\mathbf{T}_{BG} = \mathbf{T}_{GB}^T$ is used to transform vectors from G to B coordinate systems and conversely.

Part of the quadrotor model are the propellers where the inputs are the DC motor's angular rates, Ω_1, Ω_2, Ω_3, Ω_4, and outputs are the forces $\vec{F}_1$, $\vec{F}_2$, $\vec{F}_3$, $\vec{F}_4$

which generate thrust and control moments (see Fig. 18.3). This model is nonlinear and has the following form [1, 2]:

$$
\begin{aligned}
F_z &= b(\Omega_1^2 + \Omega_2^2 + \Omega_3^2 + \Omega_4^2), \qquad \Omega = -\Omega_1 + \Omega_2 - \Omega_3 + \Omega_4, \\
[M_x, M_y, M_z]^T &= [lb(-\Omega_2^2 + \Omega_4^2),\ lb(-\Omega_1^2 + \Omega_3^2),\ d(-\Omega_1^2 + \Omega_2^2 - \Omega_3^2 + \Omega_4^2)]^T,
\end{aligned} \tag{18.5}
$$

where l is the length from CM to the propellers, b [Ns^2] is the thrust factor, d [Nms^2] is the drag factor. DC motors are represented with first-order transfer function $G_{mi}(s) = \Omega_i(s)/u_i(s) = K_{mi}/(T_{mi}s + 1)$, $i = 1, 2, 3, 4$ which is sufficient to reproduce the dynamics between the input (command) signals u_i, $i = 1, 2, 3, 4$, and propeller's true speeds Ω_i, $i = 1, 2, 3, 4$.

Quadrotor dynamic inversion is based on the six-degree-of-freedom mathematical model of the quadrotor. This approach is different than the model (18.1)–(18.2) used for quadrotor dynamic simulation. In this case the force Eq. (18.1) is described in earth-fixed (inertial) coordinate $G(O; x_o, y_o, z_o)$. Instead the quaternion Eq. (18.3) for rotational kinematic equation, the Euler kinematic equations are used and in the force and moment equations, the aerodynamic and gyroscopic effects are neglected. This model is presented in [6].

Block diagram of the inverse model implemented in quadrotor guidance and control system is given in Fig. 18.4. The inverse model consists of three modules: *quadrotor dynamic inversion*, *inversion of propeller equation*, and *inversion of DC motors equation*.

The desired output of the quadrotor system is $\mathbf{y}^o = [\mathbf{R}_G^{oT}, \psi^o]^T$ where $\mathbf{R}_G^o = [x_o^o, y_o^o, z_o^o]^T$ is three-dimensional trajectory given in earth-fixed coordinate $G(O; x_o, y_o, z_o)$, and ψ^o is given yaw angle, and both are results from the *path planning process*. Output of the module *Trajectory generation* are the position

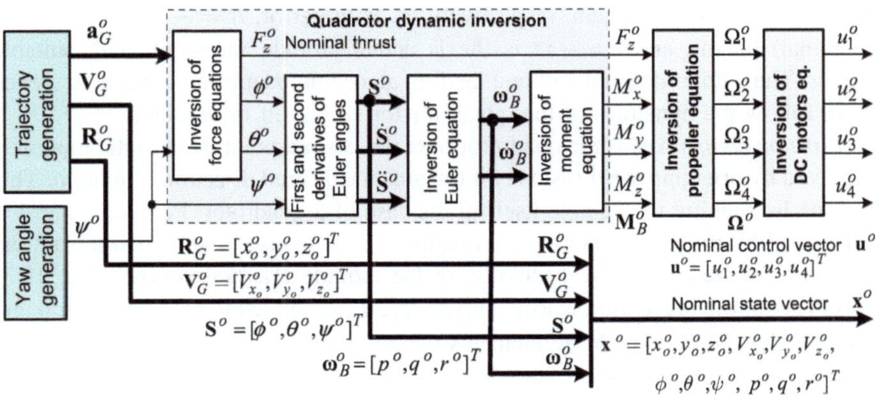

Fig. 18.4 Inverse model block diagram

$\mathbf{R}_G^o = [x_o^o, y_o^o, z_o^o]^T$, velocity $\mathbf{V}_G^o = [V_{x_o}^o, V_{y_o}^o, V_{z_o}^o]^T$, and acceleration $\mathbf{a}_G^o = [a_{x_o}^o, a_{y_o}^o, a_{z_o}^o]^T$ vectors of desired quadrotor CG motion, Fig. 18.3.

Inversion procedure and equations in the blocks in quadrotor dynamic inversion module are described in more detail in [6]. Outputs of the quadrotor dynamic inversion module are nominal force and moments, $F_z^o, M_x^o, M_y^o, M_z^o$.

Now by inverting the propeller Eq. (18.5), we get the nominal angular rates of the propeller's rotors, $\Omega_1^o, \Omega_2^o, \Omega_3^o, \Omega_4^o$. This nominal angular rates we use for determination of nominal inputs in DC motors, $u_1^o, u_2^o, u_3^o, u_4^o$. For our inversion we use only the gains of DC motors.

Next we find the nominal control vector $\mathbf{u}^o = [u_1^o, u_2^o, u_3^o, u_4^o]^T$. At the end of the process of forming the inverse model, we form the nominal state vector, $\mathbf{x}^\mathbf{o}$, (Fig. 18.4).

The nominal control vector $\mathbf{u}^\mathbf{o} = [u_1^o, u_2^o, u_3^o, u_4^o]^T$ will keep quadrotor near the nominal trajectory with error which is dependent on the present disturbances then on the parameters of the nominal trajectory, quadrotor dynamics, and other factors. For stabilization and for reducing the error linear feedback controllers are designed.

Design of these controllers is based on linear model of quadrotor which describes the dynamics with respect to nominal trajectory/state of quadrotor. In Fig. 18.3, the three modules for *attitude and position control are presented.*

Attitude control keeps the three-dimensional orientation of the quadrotor to desired (nominal) values. For roll, pitch and yaw control, and stabilization we used PD controllers which are determined from

$$u_v = K_{pv}\varepsilon_v + K_{dv}\dot{\varepsilon}_v, \quad \varepsilon_v = v_d - v, \;\; v = \phi, \theta, \psi, \;\; v_d = \phi_d, \phi_d, \psi^o, \tag{18.6}$$

where K_p is the proportional gain, K_d is the derivative gain, and ε is the error (difference) between nominal (desired or referent one) and the measured variables.

The altitude controller keeps the distance of the quadrotor from the ground at the desired value. For altitude control we use Proportional Integral Derivative (PID) controller which is determined by the following equation: $u_h = K_{ph}\varepsilon_h + K_{dh}\dot{\varepsilon}_h + K_{ih}\int_0^t \varepsilon_h dt, \quad \varepsilon_h = h^o - h$, where K_{ph},K_{dh},K_{ih} are proportional, derivative, and integral gain, respectively, ε_h is the error, $h^o = -z_o^o$ is the desired (nominal) height, and $h = -z_o$ is the is actual altitude measured with suitable sensor. The autonomous takeoff and landing algorithm adapts the altitude reference h^o to follow the dynamics of the quadrotor for taking off or landing.

Position control keeps the quadrotor over the desired point. Horizontal motion is achieved by orienting the thrust vector towards the desired direction of motion. This is done by rotating the vehicle itself in the case of a quadrotor. Position control is performed by rolling or pitching the quadrotor in response to a deviation from the $y_o^o(t)$ or $x_o^o(t)$ references, respectively. In the module *Position control* (Fig. 18.3) two PD controllers are used whose outputs, ϕ_y, θ_x, are added to the nominal roll and pitch angles, ϕ^o,θ^o, which are elements of the nominal state vector $\mathbf{x}^o(t)$, and so, they are part of the inputs of attitude control: $\phi_d = \phi^o + \phi_y, \;\; \theta_d = \theta^o + \theta_x$.

Inputs in the *Command summation* module are the *nominal control* $\mathbf{u}^o$ and the commands from the linear controllers: $u_h, u_\phi, u_\theta, u_\psi$ summarized in appropriate way to get control signal $\mathbf{u}(t) = \mathbf{u}^o(t) + \Delta\mathbf{u}(t)$, $\Delta\mathbf{u}(t) = [\Delta u_1,\ \Delta u_2,\ \Delta u_3,\ \Delta u_4]^T$, $\Delta u_1 = u_h + u_\theta - u_\psi$, $\Delta u_2 = u_h - u_\phi + u_\psi$, $\Delta u_3 = u_h - u_\theta - u_\psi$, $\Delta u_4 = u_h + u_\phi + u_\psi$.

18.5 Simulation and Results

In accordance with block diagram in Fig. 18.3 and the equations that describe each module we developed the simulation model of the quadrotor in MATLAB/Simulink. The parameters of the quadrotor which are used in the simulation are: $m = 1$ kg-mass, $I_x = I_y = 8.1 \times 10^{-3}$, $I_z = 14.2 \times 10^{-3}$ kgm^2—moments of inertia, $d = 1.1 \times 10^{-6}$ Nms^2-drag factor, $l = 0.24$ m–length from CG to propellers, $b = 54.2 \times 10^{-6}$ Ns^2 trust factor, $J_{tp} = 104 \times 10^{-6}$ Nms^2—propeller total moment of inertia [13].

Several experiments were carried out for testing the mathematical model of the quadrotor and its guidance and control system. Here we present results from two simulation experiments, named Experiments 18.1 and 18.2.

Experiment 18.1 Input here is the nominal trajectory in spiral form which is generated using three-dimensional kinematic model of moving point. Figure 18.5 presents how the quadrotor follows this trajectory. Figure 18.6 shows the time history of the nominal and actual coordinates of the vehicle.

Experiment 18.2 The generated referent trajectory has closed form and has three parts: takeoff phase, flight on constant altitude, and landing at the start position. Figure 18.7 shows how the quadrotor tracks given reference trajectory. Figure 18.8 shows the time history of the nominal and actual altitude and we can see how well

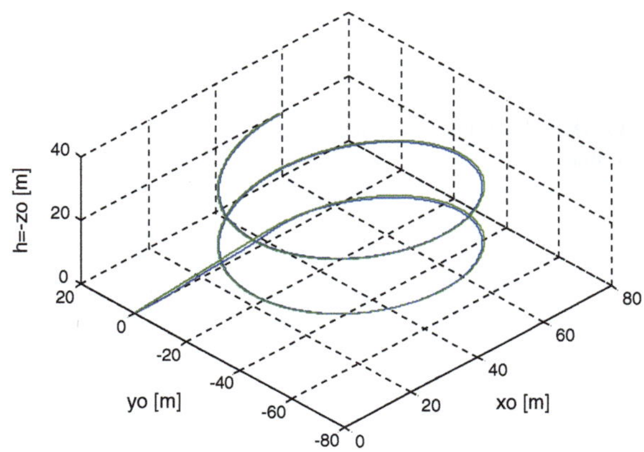

Fig. 18.5 Nominal and actual trajectories

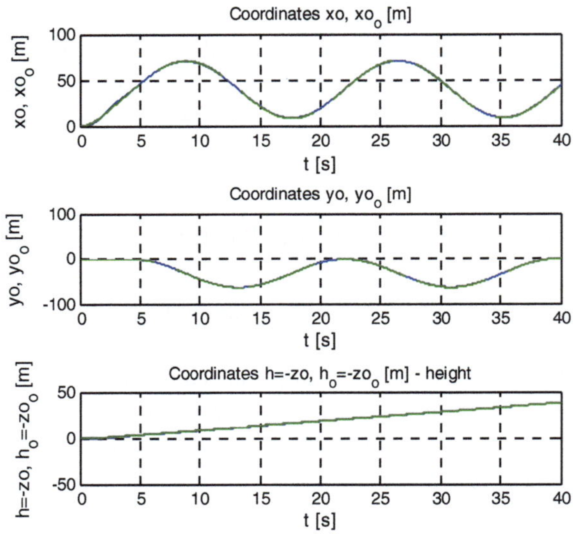

Fig. 18.6 Nominal and actual coordinates

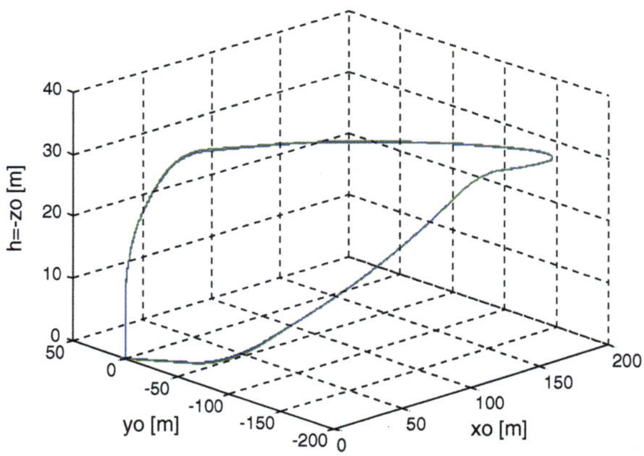

Fig. 18.7 Nominal and actual trajectory

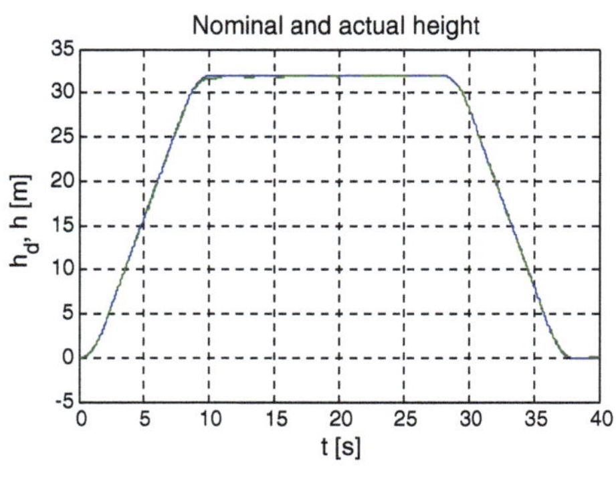

Fig. 18.8 Nominal and actual altitude

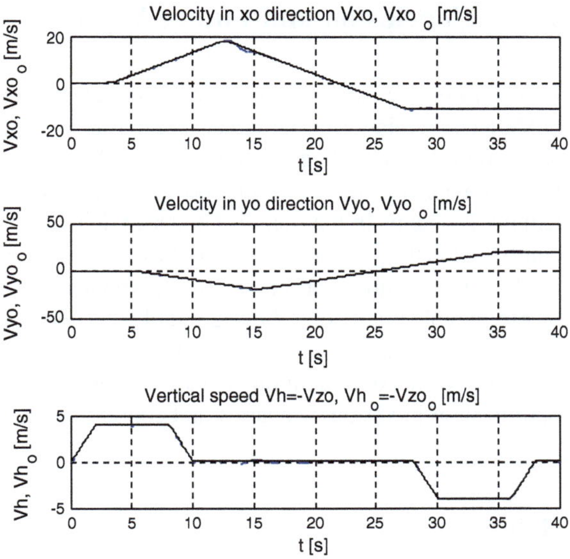

Fig. 18.9 Nominal and actual velocities

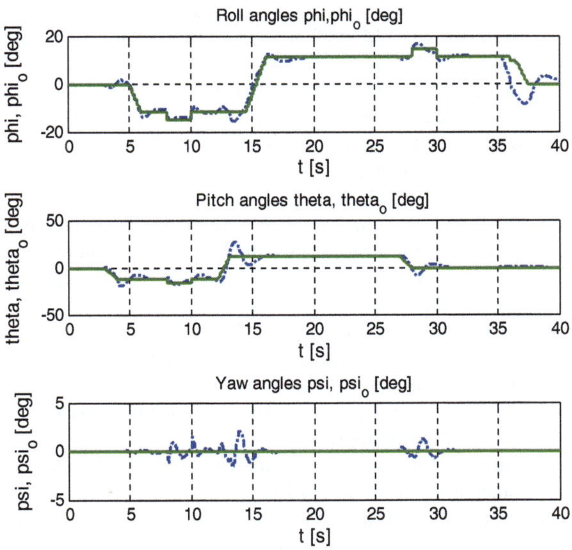

Fig. 18.10 Nominal and actual Euler's angles

quadrotor tracks the desired altitude. In Fig. 18.9 we can see the profile of the nominal and actual velocity components. Figure 18.10 presents the time history of the nominal and actual Euler's angles.

18.6 Conclusion and Future Work

In this paper, we gave an overview of our research work throughout the years in the UAV technologies. Our particular research interests are in the Guidance Navigation and Control (GN&C) algorithms for UAVs. We are using low-cost, small-size quadrotor UAV. This UAV platform is helping us and our students to tackle challenging problems in the field of Guidance, Navigation and Control, where advanced control, global optimization approaches, state and parameter estimation techniques, modeling techniques can be practiced. Applications will focus on fault tolerant and reconfigurable flight control design, flights envelop clearance and protection, optimal terminal area energy management, GPS/Inertial/Air data integrated navigation system. Further, we are continuing on developing algorithms for trajectory generation and its optimization in various scenarios and conditions (fast maneuvers, fast winds, etc.).

References

1. Bouabdallah, S.: Design and Control of quadrotors with application to autonomous flying. Ph. D. Thesis, Ecole Polytechnique Federale de Lausanne, CH (2007)
2. Bresciani, T.: Modelling, Identification and Control of a Quadrotor Helicopter, Lund University Department of Automatic Control, Box 118, SE-221 00 Lund, SW (2006)
3. Biezad, D.J.: Integrated Navigation and Guidance Systems. AIAA Education Series, Washington (1999)
4. Deskovski, S.: Synthesis of a Missile Guidance System Using Inverse Model of Motion (Ph.D. Thesis; in Croatian). Faculty of Electrical Engineering, University of Zagreb, Zagreb, SR Croatia, SFRY (1990)
5. Deskovski, S., Sazdovski, V., Dedinec, A.: Quaternion based modelling and control of quadrotor. In: Proceedings of X Triennial International SAUM Conference on Systems Automatic Control and Measurements, Niš, Serbia, Nov 10–12th, pp. 60–64. The SAUM Society, Belgrade (2010)
6. Deskovski, S., Sazdovski, V., Gacovski, Z.: Advanced guidance laws and navigation systems for UAV: theoretical and practical findings in a developing country. In: Proceedings of COSY 2011–Special International Conference on Complex Systems: Synergy of Control, Computing and Communications, 16–20 Sept 2011, Ohrid, Macedonia, pp. 333–344, The ETAI Society, Skopje, Macedonia (2011)
7. Grewal, M.S., Weill, L.R., Andrews, A.P.: Global Positioning System Inertial Navigation and Integration. Wiley Interscience, New York (2001)
8. Farrell, J.A., Barth, M.: The Global Positioning System and Inertial Navigation. The McGraw-Hill Inc, New York (1999)
9. Murray, R.M.: Optimization-Based Control: Control and Dynamical Systems. California Institute of Technology, Pasadena (2010)
10. Savage, P.G.: Strapdown inertial navigation integration algorithm design, Part 1: Attitude algorithms. J. Guidance Control Dyn. **21**(1), 19–28 (1998a)
11. Savage, P.G.: Strapdown inertial navigation integration algorithm design, Part 2: Velocity and position algorithm. J. Guidance Control Dyn. **21**(2), 208–221 (1998b)
12. Sazdovski, V., Gugulovska, T.K., Stankovski, M.: Kalman filter implementation for unmanned aerial vehicles navigation developed within a graduate course. In: Proceedings of

the 16th *FAC World Congress*, Prague, Czech Republic. The IFAC and CVUT, Prague, CZ, in vol. 6, Part. 1 (2005a)
13. Sazdovski, V., Deskovski, S., Gacovski, Z.: Simulation model of navigation guidance and control systems for unmanned aerial vehicles. In: Proceedings of the ETAI Conference, Ohrid, Macedonia. The Society for ETAI, Skopje, MK, Paper, A1–5 (2005b)
14. Sazdovski, V.: Inertial Navigation Aided by Simultaneous Localization and Mapping, Ph.D. Thesis, Centre for Autonomous systems, Department for Informatics and Systems Engineering, Cranfield Defence and Security, Cranfield University, UK (2012)
15. Titterton, D.: Strapdown Inertial Navigation Technology. Peter Peregrinus Ltd, London (1997)

Part IV
Control and Supervision in Multi-Agent and Industrial Systems

Logic habitually assumes numerals are being employed, and therefore it is not applicable to this life but rather to some celestial existence.

Bertrand Russell

Chapter 19
Distributed Supervisory Strategies for Multi-agent Networked Systems

Alessandro Casavola, Emanuele Garone and Francesco Tedesco

Abstract Novel distributed supervisory strategies for multi-agent linear systems connected via data networks and subject to coordination constraints are presented in this paper. Such a *coordination-by-constraint* paradigm is based on the online management of the prescribed set points and it is characterized by a set of spatially distributed dynamic systems, connected via communication channels, with possibly dynamical coupling amongst them which need to be supervised and coordinated in order to accomplish their overall objective. Two distributed strategies will be fully described and analysed. First, a "sequential" distributed strategy will be presented where only one agent per decision time is allowed to manipulate its own reference signal. Such a strategy will be instrumental to introduce a more effective "parallel" distributed strategy, in which all agents are allowed, under certain conditions, to modify their own reference signals simultaneously. Finally, some cases of study will be presented to show the effectiveness of the proposed methods.

19.1 Introduction

The problem of interest here is the design of distributed supervision strategies based on Command Governor (CG) ideas for multi-agent systems in situations where the use of a centralized coordination unit is impracticable because requiring unrealistic or unavailable communication infrastructures. A centralized solution to this problem has been recently proposed in [2] in the quite general context depicted in Fig. 19.1. There, the master station is in charge of supervising and coordinating the slave systems via a data network. In particular, r_i, g_i, x_i, y_i, and c_i represent, respectively,

A. Casavola (✉) · F. Tedesco
Università della Calabria, Via Pietro Bucci 42c, Rende, Italy
e-mail: casavola@dimes.unical.it

F. Tedesco
e-mail: ftedesco@dimes.unical.it

E. Garone
Université Libre de Bruxelles, Avenue F.D. Roosvelt, Brussel, Belgium
e-mail: egarone@ulb.ac.be

G.M. Dimirovski (ed.), *Complex Systems*, Studies in Systems, Decision and Control 55, DOI 10.1007/978-3-319-28860-4_19

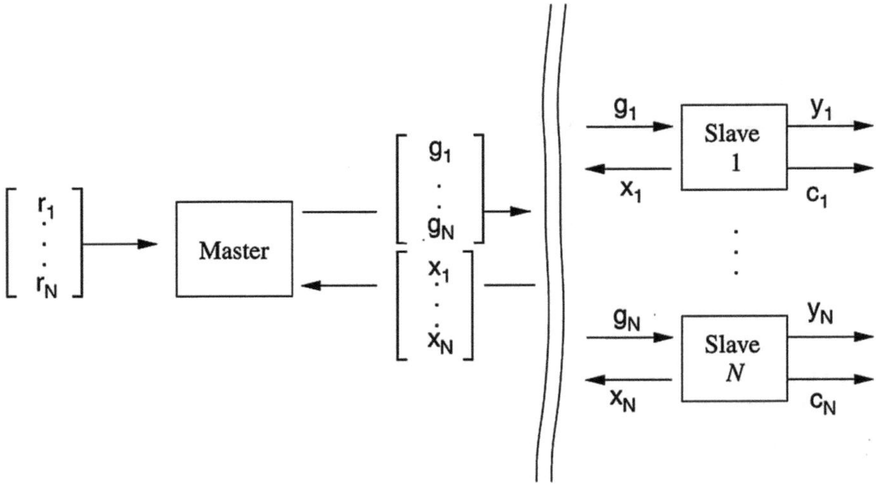

Fig. 19.1 Multi-agent master/slave architectures

the nominal references, the feasible references, the states, performance-related, and coordination-related outputs of the slave systems. In such a context, the supervision task can be expressed as the requirement of satisfying some tracking performance, viz. $y_i \approx r_i$, whereas the coordination task consists of enforcing some pointwise-in-time constraints $c_i \in \mathscr{C}_i$ and/or $f(c_1, c_2, \ldots, c_N) \in \mathscr{C}$ on each slave system and/or on the overall network evolutions. To this end, the supervisor is in charge of modifying the nominal references into the feasible ones, when the tracking of the nominal references would produce constraint violations and hence loss of coordination. Examples of such a situation include groups of vehicles cooperatively converging to a desired formation [11], large-scale chemical processes [20], supply chain management systems [9], and coordination of generators in networked power systems [15] to mention a few.

In this paper we move toward distributed strategies for solving the above task in large-scale applications based on novel CG ideas recently proposed in [13]. The novel distributed context is depicted in Fig. 19.2, where the supervisory task is now distributed amongst several agents which are assumed to be able to communicate amongst them and with the regulated plants as well. See also [10, 11] for recent results on distributed Model Predictive Control (MPC) schemes of relevance here.

The CG approach (see [2, 8, 14]) is a well-known and established methodology that provides a simple and effective way to enforce pointwise-in-time constraints along the trajectories of a closed-loop system. The CG is a nonlinear device which is added to a precompensated control system. Whenever necessary, the CG modifies the reference to the closed-loop system so as to avoid constraint violations.

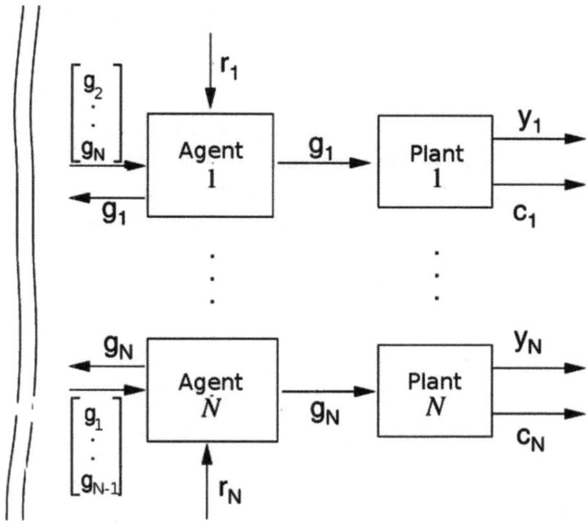

Fig. 19.2 Multi-agent architectures

In the above "traditional" contexts, the CG action is determined on the basis of the knowledge of the actual measured state. In this paper, on the contrary, we will make use of a recently proposed *Feed-Forward* CG solution [6, 13] to the CG design problem, which, at the price of some additional conservativeness, is able to accomplish the CG task in the absence of an explicit measure or estimate of the state.

It is important to remark that the use of such a novel FF-CG scheme introduces many technical challenges for the development of distributed schemes which have to be carefully managed. In this respect, in [5] and [3] several theoretical aspects of this novel sequential distributed scheme have been clarified, in particular those related to the *liveness* of the method, which is the analysis of conditions avoiding the occurrence of deadlock situations.

The main advantages of such a scheme are in its simplicity and in the low communication rates required for its implementation, remarkably lower than other distributed approaches—e.g., those based on consensus mechanisms. The basic idea is that only one agent per time is allowed to modify its own reference signal. This approach, although behaving increasingly slower for an increasing number of agents, is anyway of interest in all situations whereby the coordination problem consists of few and slow set point adjustments, e.g., in all small-/medium-scale situations where the set points change infrequently or slower than the system dynamics. Its derivation is also instrumental to build up faster "parallel" version of the scheme in which, whenever possible, all agents are allowed to modify their own reference signals simultaneously. See [4, 12] for a preliminary study on parallel distributed CG schemes and [7, 18] for more complete results. For distributed CG schemes tailored to solve the collision avoidance problem for unmanned vehicles please refer to [17, 19].

19.2 System Description and Problem Formulation

Consider a set of N subsystems $\mathscr{A} = \{1, \ldots, N\}$, each one being a Linear Time Invariant (LTI) closed-loop dynamical system regulated by a local controller which ensures stability and good closed-loop properties when the constraints are not active (small-signal regimes when the coordination is effective). Let the ith closed-loop subsystem be described by the following discrete-time model:

$$\begin{cases} x_i(t+1) = \Phi_{ii} x_i(t) + G_i g_i(t) + \sum\limits_{j \in \mathscr{A}-\{i\}} \Phi_{ij} x_j(t), \\ y_i(t) \quad = H_i^y x_i(t) \\ c_i(t) \quad = H_i^c x(t) + L_i g(t), \end{cases} \tag{19.1}$$

where $t \in \mathbb{Z}_+$, $x_i \in \mathbb{R}^{n_i}$ is the state vector (which includes the controller states under dynamic regulation), $g_i \in \mathbb{R}^{m_i}$ is the manipulable reference vector which, if no constraints (and no CG) were present, would coincide with the desired reference $r_i \in \mathbb{R}^m$ and $y_i \in \mathbb{R}^{m_i}$ is the output vector which is required to track r_i. Finally, $c_i \in \mathbb{R}^{n_i^c}$ represents the local constrained vector which has to fulfil the set-membership constraint

$$c_i(t) \in \mathscr{C}_i, \ \forall t \in \mathbb{Z}_+, \tag{19.2}$$

$\mathscr{C}_i$ being a convex and compact polytopic sets. It is worth pointing out that, in order to possibly characterize global (coupling) constraints amongst states of different subsystems, the vector c_i in (19.1) is allowed to depend on the aggregate state and manipulable reference vectors $x = [x_1^T, \ldots, x_N^T]^T \in \mathbb{R}^n$, with $n = \sum_{i=1}^N n_i$, and $g = [g_1^T, \ldots, g_N^T]^T \in \mathbb{R}^m$, with $m = \sum_{i=1}^N m_i$. Moreover, we denote by $r = [r_1^T, \ldots, r_N^T]^T \in \mathbb{R}^m$, $y = [y_1^T, \ldots, y_N^T]^T \in \mathbb{R}^m$ and $c = [c_1^T, \ldots, c_N^T]^T \in \mathbb{R}^{n^c}$, with $n^c = \sum_{i=1}^N n_i^c$, the other relevant aggregate vectors. The overall system arising by the composition of the above N subsystems can be described as

$$\begin{cases} x(t+1) = \Phi x(t) + G g(t) \\ y(t) \quad = H^y x(t) \\ c(t) \quad = H^c x(t) + L g(t), \end{cases} \tag{19.3}$$

where

$$\Phi = \begin{pmatrix} \Phi_{11} & \ldots & \Phi_{1N} \\ \vdots & \ddots & \vdots \\ \Phi_{N1} & \ldots & \Phi_{NN} \end{pmatrix}, G = \begin{pmatrix} G_1 & \ldots & 0 \\ \vdots & \ddots & \vdots \\ 0 & \ldots & G_N \end{pmatrix}.$$

$$H^y = \begin{pmatrix} H_1^y & \ldots & 0 \\ \vdots & \ddots & \vdots \\ 0 & \ldots & H_N^y \end{pmatrix}, H^c = \begin{pmatrix} H_1^c \\ \ldots \\ H_N^c \end{pmatrix}, L = \begin{pmatrix} L_1 \\ \ldots \\ L_N \end{pmatrix}$$

It is further assumed that

A1. The overall system (19.3) is asymptotically stable.
A2. System (19.3) is off-set free, i.e., $H^y(I_n - \Phi)^{-1}G = I_m$.

Roughly speaking, the CG design problem we want to solve is that of locally determine, at each time step t and for each agent $i \in \mathscr{A}$, a suitable reference signal $g_i(t)$ which is the best approximation of $r_i(t)$ such that its application never produces constraints violation, i.e., $c_i(t) \in \mathscr{C}_i, \forall t \in \mathbb{Z}_+, i \in \mathscr{A}$.

Classical centralized solutions to the above-stated CG design problem (see [2, 8]) have been achieved by finding, at each time t, a CG action $g(t)$ as a function of the current reference $r(t)$ and measured state $x(t)$:

$$g(t) := \underline{g}(r(t), x(t)) \tag{19.4}$$

such that $g(t)$ is the best approximation of $r(t)$ under the condition $c(t) \in \mathscr{C}$, where $\mathscr{C} \subseteq \{\mathscr{C}_1 \times \cdots \times \mathscr{C}_N\}$ is the global admissible region. In [13], the *Feed-Forward* CG (FF-CG) approach has been proposed, where a CG action having the following structure

$$g(t) = \underline{g}(r(t), g(t-\tau)) \tag{19.5}$$

was proved to have similar properties of the standard CG state-based approach when computed every τ steps and kept constant between two subsequent updatings, without hinging upon on the explicit knowledge of the state vector.

Here we recall the algorithm and notation of the FF-CG approach proposed in [13] which will be relevant for the forthcoming discussion. To this end, consider, for a given $\delta > 0$, the following sets:

$$\begin{array}{l} \mathscr{C}_\delta := \mathscr{C} \sim \mathscr{B}_\delta \\ \mathscr{W}_\delta := \{g \in \mathbb{R}^m : \ c_g \in \mathscr{C}^\delta\}, \end{array} \tag{19.6}$$

where $\mathscr{B}_\delta$ is the ball of radius δ centred at the origin and $\mathscr{A} \sim \mathscr{E}$ is the Pontryagin set difference defined as $\{a : \ a + e \in \mathscr{A}, \forall e \in \mathscr{E}\}$. In particular, $\mathscr{W}_\delta$, which we assume non-empty, is the convex and closed sets of all constant commands g whose corresponding equilibrium points $c_g := H^c(I_n - \Phi)^{-1}Gg + Lg$ satisfy the constraints with margin δ. By taking the definitions of $\Delta g(t) := g(t) - g(t-\tau)$ and $x_{\Delta g} := x_{g(t)-x_g(t-\tau)}$ into account, we can present the Feed-Forward CG selection algorithm as follows:

The FF-CG Algorithm

REPEAT AT EACH TIME t

1.1 IF $(t == \kappa\tau, \kappa = 1, 2 \ldots)$

1.1.1 SOLVE
$$g(t) = \arg\min_g \| g - r(t) \|_\Psi^2 \tag{19.7}$$

$$\text{SUBJECT TO}: \begin{cases} g \in \mathscr{W}_\delta \\ (g - g(t-\tau)) \in \Delta\mathscr{G}(g(t-\tau)) \end{cases} \tag{19.8}$$

1.2 ELSE $g(t) = g(t-1)$
2.1 APPLY $g(t)$
3.1 UPDATE $\rho(t) = \gamma\rho(t-\tau) + max_{k\geq 0} \parallel H_c\Phi^k(I-\Phi)^{-1}G\Delta g(t) \parallel$.

where $\Psi > 0$ is a weighting matrix and $\gamma \in (0,1)$ suitably chosen; moreover,

$$\Delta\mathscr{G}(g) := \left\{\Delta g : \parallel H^c\Phi^k(I-\Phi)^{-1}G\Delta g \parallel \leq \rho_{g+\Delta g} - \gamma\rho_g, \forall k \geq 0\right\} \quad (19.9)$$

is the set of all admissible τ-step incremental commands Δg which ensure constraint satisfaction with ρ_g representing the minimum distance between c_g and the border of $\mathscr{C}$. Details on the computation of ρ_g can be found in [13]. It is worth noticing that the sets $\mathscr{W}_\delta$, $\Delta\mathscr{G}(g)$ and the generalized settling time τ can be computed from the outset.

19.3 Distributed CG

Here we introduce two distributed CG schemes based on the above FF-CG approach assuming that the agents are connected via a communication network. Such a network is modelled with a *communication graph*: an undirected graph $\mathscr{G} = (\mathscr{A}, \mathscr{B})$, where $\mathscr{A}$ denotes the set of the N subsystems and $\mathscr{B} \subset \mathscr{A} \times \mathscr{A}$ the set of edges representing the communication links amongst agents. More precisely, the edge (i,j) belongs to $\mathscr{B}$ if and only if the agents governing the ith and the jth subsystems are able to directly share information within τ sampling times. The communication graph is assumed to be connected, i.e., for each couple of agents $i \in \mathscr{A}, j \in \mathscr{A}$, there exists at least one sequence of edges connecting i and j, with the minimum number of edges connecting the two agents denoted by $d_{i,j}$. The set of all agents with a direct connection with the ith agent will be referred to as *Neighbourhood of the ith agent* $\mathscr{N}_i = \{j \in \mathscr{A} : d_{i,j} = 1\}$.

19.3.1 Sequential Procedure (S-FFCG)

The following procedure is quite similar to the serial method presented in [20]. Let $\mathscr{G}$ be a Hamiltonian graph and assume, without loss of generality, that the sequence $\mathscr{H} = \{1, 2, \ldots, N-1, N\}$ is a Hamiltonian cycle. The idea behind the approach is that only one agent per decision time is allowed to manipulate its local command signal $g_i(t)$, while all others are instructed to hold their previous values. After each decision, the agent in charge will update the global command received from the previous updating agent and will forward this new value to the next updating agent in the cycle. Such a polling policy implies that, eventually after a preliminary initialization cycle, at each time instant the "agent in charge" always knows the whole aggregate vector $g(t-\tau)$.

By exploiting this observation we can define the following distributed SSCG algorithm:

Sequential-SSCG Algorithm (S-SSCG)—Agent i

REPEAT AT EACH TIME t

1.1 IF$(t == \kappa\tau, \kappa = 0, 1, \ldots)\&\&(\kappa \bmod N) == i$

1.1.1 RECEIVE $g(t-\tau)$ from the previous agent in the cycle $\mathscr{H}$

1.1.2 SOLVE

$$g_i(t) = \arg\min_{g_i} \| g_i - r_i(t) \|^2_{\Psi_i}$$
$$\text{SUBJECT TO}: \quad \begin{cases} g(t) = [g_1^T(t-\tau), \ldots, g_i^T, \ldots, g_N^T(t-\tau)]^T \in \mathscr{W}^\delta \\ (g_i - g_i(t-\tau)) \in \Delta\mathscr{G}^0_{g,i} \end{cases} \tag{19.10}$$

1.1.3 APPLY $g_i(t)$

1.1.4 UPDATE $g(t) = [g_1^T(t-\tau), \ldots, g_i(t), \ldots, g_N^T(t-\tau)]^T$

1.1.5 TRANSMIT $g(t)$ to the next agent in $\mathscr{H}$

1.2 ELSE

1.2.1 APPLY $g_i(t) = g_i(t-1)$

where $\Psi_i > 0$ is a weighting matrix, $\kappa \bmod N$ is the remainder of the integer division κ/N and

$$\Delta\mathscr{G}^0_{g,i} = \left\{ \Delta g_i \,\middle|\, [0, 0, \ldots, \Delta g_i^T, \ldots, 0]^T \in \Delta\mathscr{G}_g \right\}$$

is the set of all possible command variations for g_i in the case that the commands of all other agents are frozen.

In [16] the following proposition has been proved for the S-FFCG under the assumption:

A3—Viability Property. Each point belonging to $\partial(\mathscr{W}_\delta)$ in (19.6) is viable, $\partial(\mathscr{W}_\delta)$ denoting the border of $\mathscr{W}_\delta$.

For additional details about liveness and viability please refer to [3].

Proposition 19.1 *Let assumptions* **A1-A2-A3** *be fulfilled. Consider system (19.3) as the composition of N subsystems in form (19.1) along with the distributed* **S-FFCG** *selection rule and let an admissible aggregate command signal* $g(0) = [g_1^T(0), \ldots, g_N^T(0)]^T$ *be applied at* $t = 0$. *Then*

1. *for each agent* $i \in \mathscr{A}$ *the algorithm produces at each time step a feasible local command* $g_i(t)$*;*
2. *constraints are fulfilled for all* $t \in \mathbb{Z}_+$*;*
3. *the overall system is asymptotically stable. In particular, whenever* $r(t) \equiv r$*, the sequence of the aggregate vectors* $g(t) = [g_1^T(t), \ldots, g_N^T(t)]^T$ *converges in finite time either to r or to an admissible approximation in* $\mathscr{W}^\delta$ *which is Pareto-optimal w.r.t. the local objective functionals* $||g_i - r_i||^2_\Psi, i = 1, \ldots, N$.

19.3.2 Parallel FFCG (P-FFCG)

The main drawback of the S-FFCG algorithm is that, every τ time instants, only one agent per time is allowed to modify its local command. This yields to a reduced capability to track the desired reference $r(t)$. In order to overcome such a limitation, any agent should be enabled to select its local command every τ time instants. The two key points to be considered in building up such a kind of strategy are

(a) the definition of the information set available to each agent
(b) the determination of a set of decentralized "selection rules" such that the composition of all feasible local commands satisfies global constraints (19.8).

With regards to the information available to each agent, we will assume that each agent acts as a gateway in redistributing data amongst the other no directly connected agents. Then, at each time instant t, the ith agent has knowledge of the following vector:

$$\xi_i(t-\tau) = [g_1^T(t-d_{i,1}\tau), \dots, g_i^T(t-\tau), \dots, g_N^T(t-d_{i,N}\tau)]^T.$$

It results that the most recent information on the applied commands shared by all agents at each decision time t is given by the vector

$$\xi(t-\tau) = [g_1^T(t-d_{max,1}\tau), \dots, g_N^T(t-d_{max,N}\tau)]^T,$$

where $d_{max,i} = max_{j\in\mathscr{A}} d_{i,j}$.

The main idea behind the proposed selection rule is that of generating, every τ steps and on the basis of the information shared by all the agents in the network, a set of decoupled alternative constraints (one for each agent) such that their local fulfilment implies the fulfilment of global constraints (19.8). In other words, at each computation step, we substitute the admissible region (19.8) for Δg by its *set Cartesian decomposition* [1]. If such decomposition is opportunely performed, the problem decouples and each agent will have simply to fulfil the inclusion into a local set in the form

$$\Delta g_i(t) \in \Delta G_{v,i}(t), i = 1, \dots, N \tag{19.11}$$

with $\Delta G_{v,i}(t) \subseteq \mathbf{R}^{n_i^c}, i = 1, \dots, N$ convex and compact sets containing $0_{n_i^c}$ for all $t \geq 0$.

It remains now to understand how to generate local decoupled constraints that ensure global constraints satisfaction. The first step is to observe that if constraints (19.11) were satisfied at each time step, then we could define the set of all feasible values for the actual command $g(t)$, to be computed on the basis of the common information vector $\xi(t-\tau)$, as follows:

$$\begin{aligned}
&\Xi(t-\tau) = \{\xi(t-\tau)\} + \\
&+ \Big(\big(\Delta G_{v,1}(t-\tau) \times \cdots + \Delta G_{v,1}(t-d_{max,1}\tau)\big) \times \cdots \\
&\cdots \times \big(\Delta G_{v,N}(t-\tau) + \cdots + \Delta G_{v,N}(t-d_{max,1}\tau)\big) \Big),
\end{aligned} \tag{19.12}$$

where $+$ denotes the Pontryagin set sum defined as $X+Y=\{z=x+y : \forall x \in X, \forall y \in Y\}$. On the basis of the above feasible values for $g(t-\tau)$, the set of admissible aggregate command variations can be easily computed as follows:

$$\begin{aligned} \Delta\mathcal{G}^{\mathcal{W}_\delta}_{\Xi(t-\tau)} &= \{\Delta g | \Delta g \in \Delta\mathcal{G}(g), \forall g \in \Xi(t-\tau)\} \\ &\cap \{\Delta g | (g+\Delta g) \in \mathcal{W}_\delta, \forall g \in \Xi(t-\tau)\}. \end{aligned} \tag{19.13}$$

Finally, the (approximated) Cartesian decomposition generating the agent-wise decoupled constraints (19.11) should satisfy the following set inclusion condition:

$$\Delta G_{v,1}(t) \times \cdots \times \Delta G_{v,N}(t) \subseteq \Delta\mathcal{G}^{\mathcal{W}_\delta}_{\Xi(t-\tau)}. \tag{19.14}$$

Note also that, because $\Xi(t-\tau)$ is the common information shared by all agents in the network, then, once an opportune objective function is defined, all agents will be able to independently determine the same collection of sets $\Delta G_{v,i}(t), i=1,\ldots,N$. We can finally describe the Parallel FF-CG procedure to be performed every τ steps:

1.1 Each agent determines the collection of sets $\Delta G_{v,i}(t)$, $i=1,\ldots,N$ as the solution of own instance the following optimization problem

$$\begin{aligned} \max_{\Delta G_{v,i}(t), i=1,\ldots,N} \quad & V(\Delta G_{v,1}(t) \times \cdots \times \Delta G_{v,N}(t)) \\ \text{subject to} \quad & (1.14), \end{aligned} \tag{19.15}$$

where $V(\cdot)$ denotes a possible measure of the volume of a set (to achieve good dynamical properties we want $(\Delta G_{v,1}(t) \times \cdots \times \Delta G_{v,N}(t))$ to be as large as possible)

2.1 if the distance between $\xi(t-\tau)$ and the boundaries of $\mathcal{W}_\delta$ is larger, then a pre-determined threshold ϵ_{thr}; each agent can choose its own reference by solving the following convex optimization problem:

$$\begin{aligned} g_i(t) &= arg \min_{g_i} \|g_i - r_i(t)\|^2_{\Psi_i} \\ \text{subject to} \ & (g_i - g_i(t-\tau)) \in \Delta G_{v,i}(t) \end{aligned} \tag{19.16}$$

2.2 otherwise, after an opportune initialization phase, all agents simultaneously change the strategy and *go sequentially* under the S-FFCG strategy, where the agent in charge is constrained inside a pre-determined virtual constraint $\Delta G_{seq,i}$, $i=1,\ldots,N$.

The threshold ϵ_{thr} and the pre-determined sets $\Delta G_{seq,i} i=1,\ldots,N$ are design parameters and have to be chosen accurately in order to ensure good dynamical performance during the switching phases.

Remark 19.1 It has been proved in [18] that the same properties pertaining to the S-FFCG scheme and reported in Proposition 19.1 still apply for the P-FFCG algorithm here presented. Details are here omitted for space reasons. It is finally worth remarking that low communication efforts are necessary for implementing the distributed strategies, which anyway do not affect the online computational burdens because data are exchanged only after the computation and application of local commands.

19.4 Illustrative Example: Coordination of Autonomous Vehicles

In Fig. 19.3, a system consisting of two dynamically coupled masses is presented. Such a system, we will refer to as the *m2-system*, represents in essence a pair of unmanned vehicles transporting an elastic membrane. The system is described by the following equations:

$$\begin{aligned} m_i\ddot{x}_i &= -k(x_i - \textstyle\sum_{j\in\mathscr{N}_i} x_j) - \beta(\dot{x}_1 - \sum_{j\in\mathscr{N}_i} \dot{x}_j) + F_i^x \\ m_i\ddot{y}_i &= -k(y_i - \textstyle\sum_{j\in\mathscr{N}_i} y_j) - \beta(\dot{y}_1 - \sum_{j\in\mathscr{N}_i} \dot{y}_j) + F_i^y, \end{aligned} \tag{19.17}$$

where (x_i, y_i), $i \in \mathscr{A} = \{1, 2\}$ are the coordinates of the ith mass position w.r.t a Cartesian reference frame and (F_i^x, F_i^y), $i \in \mathscr{A}$, the components along the same reference frame of the forces acting as inputs for the two slave systems. $\mathscr{N}_i$ denotes the neighbourhood of the ith agent, in this case: $\mathscr{N}_1 = \{2\}$ and $\mathscr{N}_2 = \{1\}$. The following system parameters are assumed: $\beta = 1 \left[\frac{\text{N s}}{\text{m}}\right], k = 1 \left[\frac{\text{N}}{\text{m}}\right], m_i = 1$ [kg], $\forall i \in \mathscr{A}$. Discrete-time simulations with a sampling time of $T_c = 0.1$ [s] have been undertaken. Each subsystem has been precompensated by an optimal LQ state-feedback local controller.

The problem we consider here is the coordination of the planar motions of those two masses along two continuously parameterized paths $r_1(\alpha_1) \in \mathbb{R}^2, r_2(\alpha_2) \in \mathbb{R}^2$ where $\alpha_1 \in [0, \bar{\alpha}], \alpha_2 \in [0, \bar{\alpha}]$ are real parameters. It is required that each agent tracks its own path by progressively increasing the value of $\alpha_i(t)$ at the maximum speed compatible with the following constraints:

$$\begin{aligned} \left|F_i^j(t)\right| &\le 0.5 \text{ [N]} \quad j = x, y,\ i \in \mathscr{A}, \\ |v_i(t) - r_i(\alpha_i(t))| &\le 0.05 \text{ [m]} \qquad i \in \mathscr{A}, \\ \left|\alpha_i(t) - \alpha_j(t)\right| &\le 0.04, \quad i \in \mathscr{A}, j \in \mathscr{N}_i. \end{aligned} \tag{19.18}$$

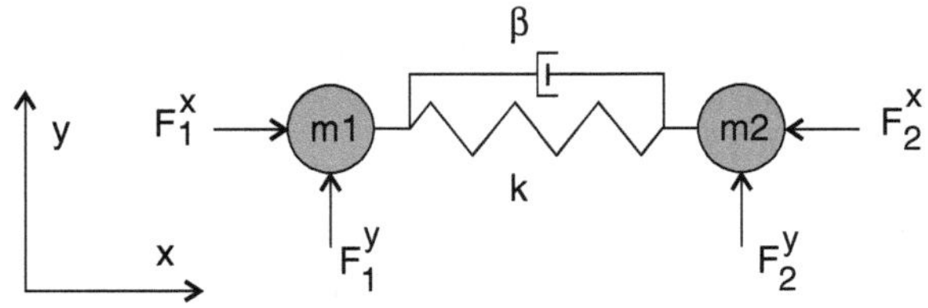

Fig. 19.3 Planar system of two dynamically coupled masses

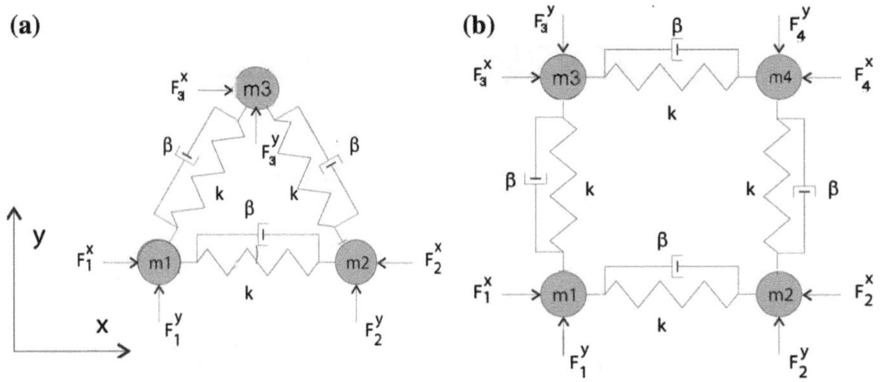

Fig. 19.4 **a** Planar system of three dynamically coupled masses (*m3-system*). **b** Planar system of four dynamically coupled masses (*m4-system*)

The first set of inequalities represents input-saturation constraints on the four forces F_i^x and F_i^y, $i \in \mathscr{A}$, acting as inputs to the vehicles. They have to be taken into account in order to avoid the generation of trajectories out of the actuator ranges. The second set of constraints represents the component-wise accuracy of the vehicle positions $v_i(t) = [x_i(t), y_i(t)]$ with respect to the target rigid motion $r_i(\alpha_i(t))$. Finally, the third group of constraints represents the coordination constraints between the agents: the agents have never to be "too far apart" in the parameter progression in order to maintain the formation shape (Fig. 19.4).

In this example, we will consider the rigid motions $r_1(\alpha), r_2(\alpha)$ for $\alpha \in [0, 4]$ reported in Fig. 19.5a, consisting of a combination of circular and translational uniform motions generated by assuming the two masses to be rigidly connected 0.2-m-long way apart and the rigid system rotating around its centre of mass which translates uniformly along a quadratic path of side $L = 1$ m. For comparisons, the system trajectories achieved by the use of the centralized Standard CG strategy [2] and the two presented distributed approaches are depicted in Fig. 19.5b–d for the first 1000 simulation steps. Those trajectories correspond to the evolution of $\alpha_i(t), i = 1, 2$ shown in Fig. 19.6. As expected, the use of distributed schemes introduces a certain level of conservativeness w.r.t. centralized CG approaches. However, with regards to the P-FFCG method it is worth pointing out here that, even if the Cartesian decomposition undertaken in this example was based on the very conservative computation of the maximal volume inscribed hypercube, the performance degradation is very moderate.

In order to evaluate the scalability of the proposed algorithms, formations of three and four masses, hereafter denoted as *m3-system* and *m4-system*, have been considered (see Fig. 19.4a). The same constraints (19.18) and parameterized paths used for the two-mass model (shown in Figs. 19.7a and 19.9a) have been considered. The simulative results are depicted in Figs. 19.7, 19.8 and Figs. 19.9, 19.10, respectively.

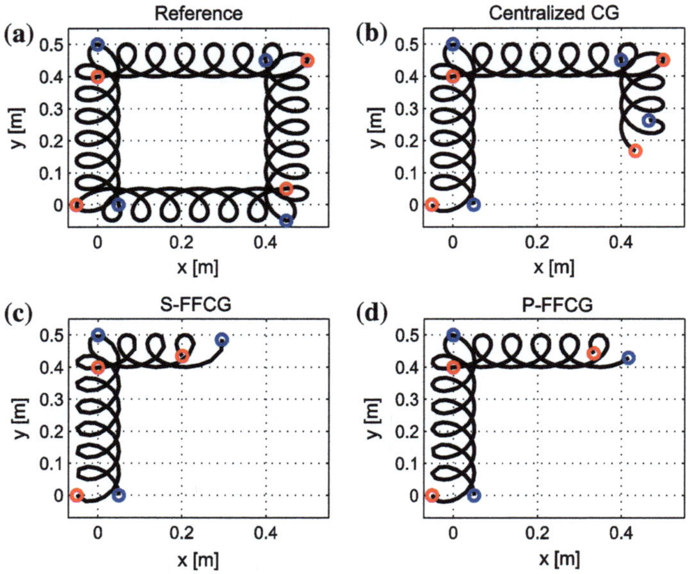

Fig. 19.5 Tracking trajectories for the *m2-system*: **a** Reference Trajectories, **b** Standard CG, **c** S-FFCG, **d** P-FFCG

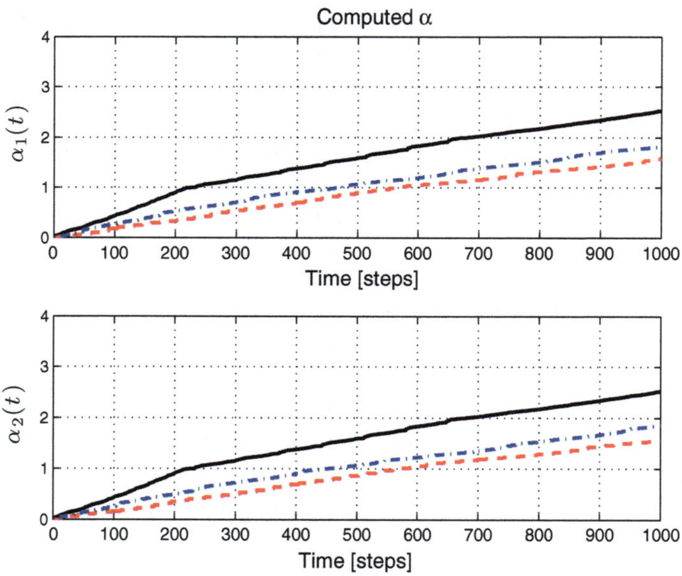

Fig. 19.6 Computed $\alpha_1(t)$, $\alpha_2(t)$ for the *m2-system*: Standard CG (-), P-FFCG (-.-), S-FFCG(- -)

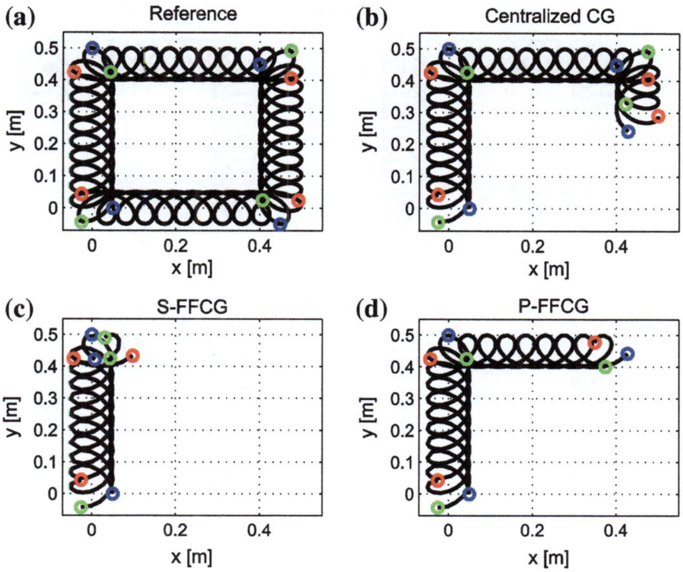

Fig. 19.7 Tracking trajectories for the *m3-system*: **a** Reference Trajectories, **b** Standard CG, **c** S-FFCG, **d** P-FFCG

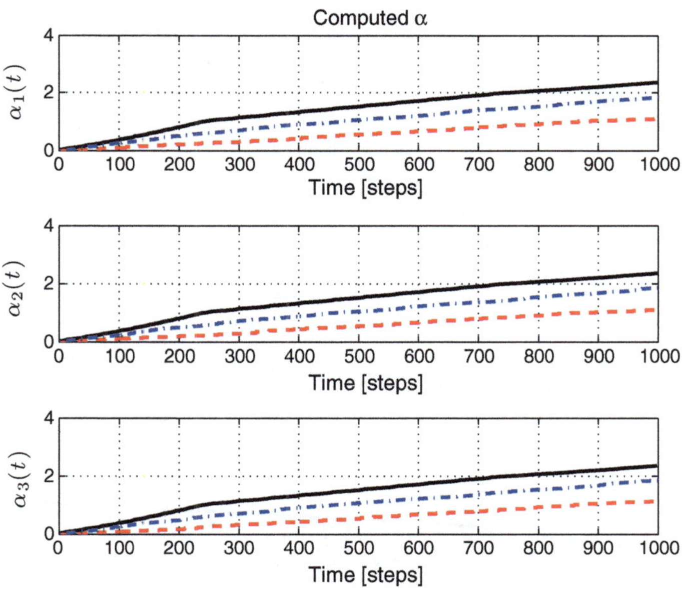

Fig. 19.8 Computed $\alpha_1(t), \alpha_2(t), \alpha_3(t)$ for the *m3-system*: Standard CG (-), P-FFCG (-.-), S-FFCG(- -)

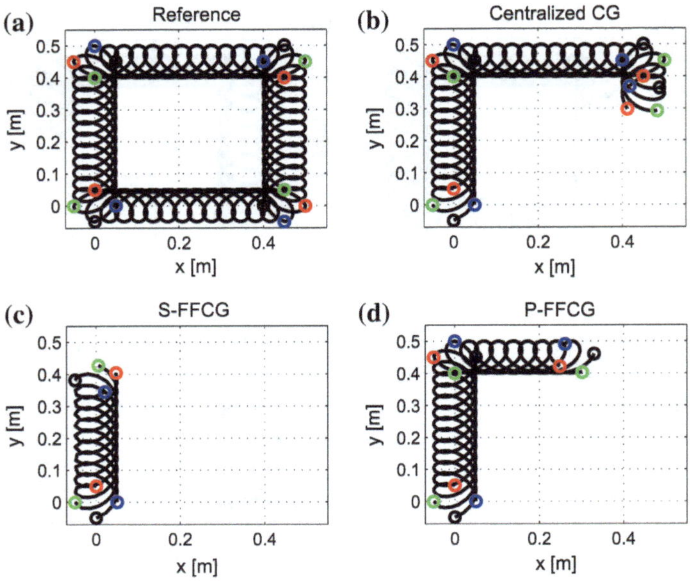

Fig. 19.9 Tracking trajectories for the *m4-system*: **a** Reference Trajectories, **b** Standard CG, **c** S-FFCG, **d** P-FFCG

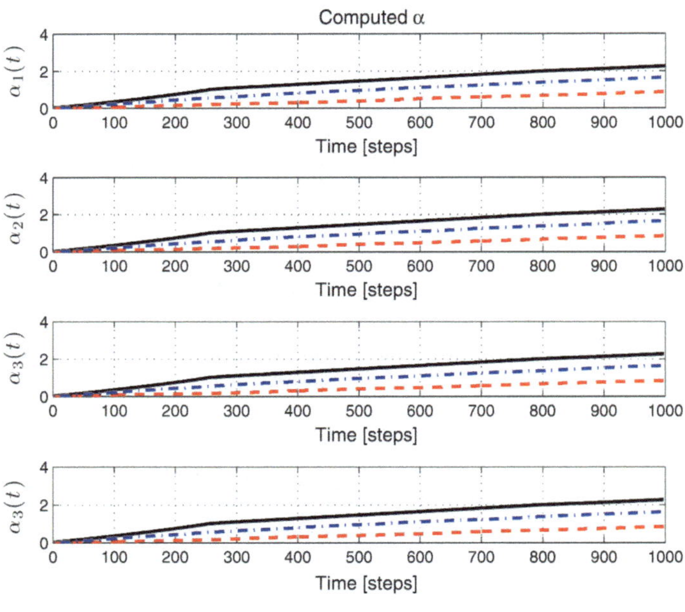

Fig. 19.10 Computed $\alpha_1(t), \alpha_2(t), \alpha_3(t), \alpha_4(t)$ for the *m4-system*: Standard CG (-), P-FFCG (-.-), S-FFCG(- -)

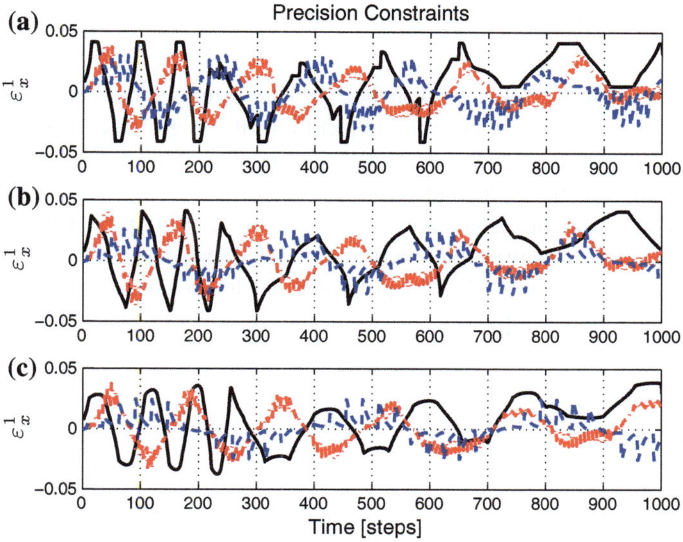

Fig. 19.11 Formation accuracy constraints of agent 1. **a** *m2-system*. **b** *m3-system*. **c** *m4-system*): Standard CG (-), P-FFCG (-.-), S-FFCG(- -)

In Fig. 19.11, the accuracy constraints for a single agent of each formation are depicted for all the simulated systems. The results of the centralized CG scheme are contrasted with those pertaining to both distributed algorithms. It is worth noticing that the centralized solution is more active and the constrained signal is often near to the constraints' boundaries. On the contrary, the distributed methods present a more conservative behaviour. It is also worth noticing the particular "pulsing" constrained signal behaviour for the S-FFCG algorithm: a remarkable variation is produced due to the action of the "agent in charge". After this time and by the time this agent is "in charge" again, no set point modifications are produced for this agent. A much more smoother behaviour can be noticed on the contrary for the P-FFCG policy.

Finally, in Table 19.1 the online computational burdens and network usage are reported for all schemes. The CPU times are related, in the distributed case to a single agent and in the centralized case to the unique supervisor in charge. The exchanged information amounts: in the distributed case to data exchanged by an agent with the rest of the network (it is the same for each agent in this example), in the centralized case the data received and transmitted by the unique CG device.

Table 19.1 CPU time—exchanged information

	CPU time (ms)			TX Data (bit/agent)		
	m2	m3	m4	m2	m3	m4
CG	5.0	5.5	6.1	384	576	768
P-FFCG	3.8	4.7	4.8	128	328	512
S-FFCG	0.47	0.38	0.35	128	192	256

19.5 Conclusions

In this paper, novel classes of distributed CG schemes have been discussed for dynamically coupled linear systems subject to local and global constraints. The key point was to resort to a novel FF-CG scheme that does not require an explicit measure of the state to be implemented. Two distributed coordination algorithms have been singled out and their constraints fulfilment and stability properties highlighted. Comparisons with a centralized solution have been also presented and commented in the final illustrative example. The presented results are encouraging and stimulate further research on the topic.

References

1. Bertsekas, D.P.: Separable dynamic programming and approximate decomposition methods. IEEE Trans. Autom. Control **52**, 911–916 (2007)
2. Casavola, A., Papini, M., Franzè, G.: Supervision of networked dynamical systems under coordination constraints. IEEE Trans. Autom. Control **51**(3), 421–437 (2006)
3. Casavola A., Garone E., Tedesco F.: A liveliness analysis of a distributed constrained coordination strategy for multi-agent linear systems. In: CDC 2011, Orlando, USA (2011)
4. Casavola A., Garone E., Tedesco F.: Distributed reference management strategies for a networked water distribution system. In: IFAC 2011, Milan, Italy (2011)
5. Casavola A., Garone E., Tedesco F.: Distributed coordination-by-constraint strategies for multi-agent networked systems. In: CDC 2011, Orlando, USA (2011)
6. Casavola, A., Garone, E., Tedesco, F.: Improved feed-forward command governor strategies for constrained discrete-time linear systems. IEEE Trans. Autom. Control **59**(1), 832–852 (2014)
7. Casavola, A., Garone, E., Tedesco, F.: A distributed multi-agent command governor strategy for the coordination of networked interconnected systems. IEEE Trans. Autom. Control **59**(8), 2099–2112 (2014)
8. Casavola, A., Mosca, E., Angeli, D.: Robust command governors for constrained linear systems. IEEE Trans. Autom. Control **45**, 2071–2077 (2000)
9. Dunbar, W.B., Desa, S.: Distributed model predictive control for dynamic supply chain management, pp. 26–30. International Workshop on NMPC, Freudenstadt-Lauterbad, Germany (2005)
10. Dunbar, W.B.: Distributed receding horizon of dynamically coupled nonlinear systems. IEEE Trans. Autom. Control **52**(7), 1249–1263 (2007)
11. Keviczky T.: Decentralized receding horizon control of large scale dynamically decoupled systems. Ph.D. Thesis, University of Minnesota, USA (2005)
12. Garone E., Tedesco F., Casavola A.: Distributed coordination strategies for interconnected multi-agent systems. In: 8th IFAC NOLCOS, 403–408, Bologna, Italy (2010)
13. Garone, E., Tedesco, F., Casavola, A.: Sensorless supervision of LTI dynamical sytems: the feed forward command governor. Automatica **47**(7), 1294–1303 (2011)
14. Gilbert, E.G., Kolmanovsky, I., Tan, K.T.: Discrete-time reference governors and the nonlinear control of systems with state and control constraints. Int. J. Robust Nonlinear Control **5**, 487–504 (1995)
15. Ibraheem, K., Kothari, D.P.: Recent philosophies of automatic generation control strategies in power systems. IEEE Trans. Power Syst. **20**(1) (2005)
16. Tedesco, F., Casavola, A.: Fault-tolerant distributed load/frequency supervisory strategies for networked multiarea microgrids. Int. J. Robust Nonlinear Control **24**(8–9), 1380–1402 (2014)

17. Tedesco F., Raimondo D.M., Casavola A., Lygeros J.: Distributed collision avoidance for interacting vehicles: a command governor approach. In: 2nd IFAC Workshop on Distributed Estimation and Control in Networked Systems (NecSys 10), Annecy, France (2010)
18. Tedesco F., Casavola A., Garone E: A parallel distributed coordination-by-constraint strategy for multi-agent networked systems. In: CDC 2012, Maui Hawaii, USA, Dec 2012
19. Tedesco, F., Raimondo, D.M., Casavola, A.: Collision avoidance command governor for multivehicle unmanned systems. Int. J. Robust Nonlinear Control **24**(16), 2309–2330 (2004)
20. Venkat A.N.: Distributed model predictive control: theory and applications. Ph.D. Thesis, University of Wisconsin, USA (2006)

Chapter 20
Petri Net-Based Synthesis of Agent Cooperation by Means of Modularity and Supervision Principles

František Čapkovič

Abstract A possibility how the principles of modularity and supervision can be utilized at synthesis of the agent cooperation is pointed out in this paper. Subsystems modelling agents of different kinds are understood to be discrete-event systems modules. They are modelled by means of place/transition Petri nets (P/T PN). A desired strategy of the mutual behaviour of agents during their cooperation is expressed by conditions for the discrete-event systems (DES) based supervisor synthesis. Then, the synthesized supervisor obtrudes the cooperation strategy on the agents at the realization of a common job. Such a procedure can be utilized step by step also at the synthesis of more complicated structures of cooperating modules (e.g. groups of agents) in order to achieve a prescribed behaviour of the global structure altogether. The supervisor synthesis is realized either by means of the P/T PN place invariants (P-invariants) or in virtue of the extended method where P-invariants are complemented by conditions imposed on P/T PN transitions and/or on the Parikh's vector (especially in order to express priorities). The soundness of the approach is illustrated by simple examples.

20.1 Introduction

While not long ago the term agents was exclusively used for abstract software entities (modules), at present we can speak also about the agents with a material substance (i.e., material agents) especially in manufacturing systems (e.g. robots, different intelligent machines and/or automatically guided vehicles, even whole production lines cooperating each other), transport systems (e.g. trains in railways, local tracks), communication systems (e.g. sender, receiver, communication channel), etc., or persons, e.g. in business processes (vendor, customer, manager, buyer, contractor) and in other social systems. The behaviour exhibited by such devices, companies,

F. Čapkovič (✉)
Institute of Informatics, Slovak Academy of Sciences, Dúbravská cesta 9, 845 07
Bratislava, Slovakia
e-mail: Frantisek.Capkovic@savba.sk
URL: http://www.ui.sav.sk/home/capkovic/capkhome.htm

G.M. Dimirovski (ed.), *Complex Systems*, Studies in Systems,
Decision and Control 55, DOI 10.1007/978-3-319-28860-4_20

persons have the character of discrete-event systems (DES). DES are systems discrete in nature. Their behaviour is driven by means of occurrence of discrete events. To describe DES different kinds of Petri nets (PN) are often used. Namely, PN yield both the graphical symbolization and the mathematical description of DES. In this paper the place/transition PN (P/T PN) will be applied to model the behaviour of autonomous DES modules as well as the behaviour of groups of cooperating modules. To ensure the desired cooperation of modules in a prescribed fashion the methods of DES supervising (beside others) can be utilized. Understand the structure of P/T PN to be the bipartite directed graph

$$\langle P, T, F, G\rangle, \; P \cap T = \emptyset, \; F \cap G = \emptyset, \; F \subseteq P \times T, \; G \subseteq T \times P, \tag{20.1}$$

where $P = \{p_1, p_2, \dots, p_n\}$ is the set of places p_i, $i = 1, \dots, n$, representing states of elementary processes, $T = \{t_1, t_2, \dots, t_m\}$ is the set of transitions $t_j, j = 1, \dots, m$, representing states of elementary discrete events, F is the set of directed arcs (the graph edges) oriented form places to transitions, and G is the set of directed arcs (the graph edges) oriented form transitions to places. Thus, F, G express, respectively, causal relations places $\rightarrow$ transitions and transitions $\rightarrow$ places and can be replaced by incidence matrices $\mathbf{F}$, $\mathbf{G}$. Formally, the dynamics of P/T PN can be expressed as

$$\langle X, U, \delta, \mathbf{x}_0\rangle, \; X \cap U = \emptyset, \tag{20.2}$$

where $X = \{\mathbf{x}_0, \mathbf{x}_1, \dots, \mathbf{x}_N\}$ is the set of state vectors; $\mathbf{x}_k \in X$, $k = 0, 1, \dots, N$ represents the state vectors of elementary places in the step k of the system dynamics development; $U = \{\mathbf{u}_0, \mathbf{u}_1, \dots, \mathbf{u}_N\}$ is the set of control vectors; $\mathbf{u}_k \in U$, $k = 0, 1, \dots, N$ represents the state vectors of elementary transitions in the step k of the system dynamics development; $\delta : X \times U \rightarrow X$ is the transition function; $\mathbf{x}_0$ is the initial state vector. The transition function δ can be transformed into the system form.

The system form of the P/T PN-based model of a DES module dynamics is given as follows:

$$\mathbf{x}_{k+1} = \mathbf{x}_k + \mathbf{B}.\mathbf{u}_k \quad , \quad k = 0, \dots, N, \tag{20.3}$$

$$\mathbf{B} = \mathbf{G}^T - \mathbf{F}, \tag{20.4}$$

$$\mathbf{F}.\mathbf{u}_k \leq \mathbf{x}_k, \tag{20.5}$$

where k is the discrete step of the dynamics development; $\mathbf{x}_k = (\sigma^k_{p_1}, \dots, \sigma^k_{p_n})^T$ is the n-dimensional state vector; $\sigma^k_{p_i} \in \{0, 1, \dots, c_{p_i}\}$, $i = 1, \dots, n$ express the states of atomic activities by 0 (passivity) or by $0 < \sigma_{p_i} \leq c_{p_i}$ (activity); c_{p_i} is the capacity of p_i; $\mathbf{u}_k = (\gamma^k_{t_1}, \dots, \gamma^k_{t_m})^T$ is the m-dimensional control vector (the state vector of transitions); its components $\gamma^k_{t_j} \in \{0, 1\}$, $j = 1, \dots, m$ represent occurring of elementary discrete events (e.g. starting or ending the activities, failures, etc.) by 1 (presence of the corresponding discrete event) or by 0 (absence of the event); $\mathbf{B}$, $\mathbf{F}$ (the incidence matrix of the graph edges corresponding to F), $\mathbf{G}$ (the incidence matrix of the graph edges corresponding to G) are the matrices of integers; $\mathbf{F} = \{f_{ij}\}_{n \times m}$, $f_{ij} \in \{0, M_{f_{ij}}\}$, expresses the causal relations among the states (as causes) and the discrete events occurring during the DES operation (as consequences) by 0 (nonexistence of the relation) or by $M_{f_{ij}} > 0$ (existence and multiplicity of the relation); $\mathbf{G} = \{g_{ij}\}_{m \times n}$, $g_{ij} \in \{0, M_{g_{ij}}\}$, expresses analogically the causal relations among the discrete events (as causes) and the DES states (as consequences); $\mathbf{B}$ is given according to (20.4); $(.)^T$ symbolizes the matrix or vector transposition. Just such an exact mathematical expression of P/T PN, in contrast to high-level PN, yield the possibility to deal with the cooperation synthesis in analytical terms.

Developing (20.3) at simultaneous respecting (20.5) it can be written that

$$\mathbf{x}_1 = \mathbf{x}_0 + \mathbf{B}.\mathbf{u}_0, \tag{20.6}$$

$$\mathbf{x}_2 = \mathbf{x}_1 + \mathbf{B}.\mathbf{u}_1 = \mathbf{x}_0 + \mathbf{B}.(\mathbf{u}_0 + \mathbf{u}_1), \tag{20.7}$$

$$\ldots \quad \ldots \ldots \ldots \ldots \ldots \ldots \ldots \ldots$$

$$\mathbf{x}_k = \mathbf{x}_0 + \mathbf{B}.(\mathbf{u}_0 + \mathbf{u}_1 + \cdots + \mathbf{u}_{k-1}) = \mathbf{x}_0 + \mathbf{B}.\mathbf{v}, \tag{20.8}$$

where $\mathbf{v} = \sum_{i=0}^{k-1} \mathbf{u}_i$ in (20.8) is named as the Parikh's vector. It gives us information about how many times the particular transitions are fired during the development of the system dynamics from the initial state $\mathbf{x}_0$ to the state $\mathbf{x}_k$.

Because not only the behaviour of agents but also the cooperation of agents exhibit attributes of DES we can utilize PN model for cooperating modules too.

This paper represents the extension of the conference COSY 2011 paper [12].

20.2 Modularity at Building des Models

Having the PN models of DES modules we can build different DES structures. Consider a group of N_A agents $\{A_1, \ldots, A_{N_A}\}$. When the agents A_i, $i = 1, \ldots, N_A$, are autonomous, the incidence matrices are block diagonal as follows:

$$\mathbf{F} = \begin{pmatrix} \mathbf{F}_1 & \mathbf{0} & \ldots & \mathbf{0} & \mathbf{0} \\ \mathbf{0} & \mathbf{F}_2 & \ldots & \mathbf{0} & \mathbf{0} \\ \vdots & \vdots & \ddots & \vdots & \vdots \\ \mathbf{0} & \mathbf{0} & \ldots & \mathbf{F}_{N_A-1} & \mathbf{0} \\ \mathbf{0} & \mathbf{0} & \ldots & \mathbf{0} & \mathbf{F}_{N_A} \end{pmatrix}; \quad \mathbf{G} = \begin{pmatrix} \mathbf{G}_1 & \mathbf{0} & \ldots & \mathbf{0} & \mathbf{0} \\ \mathbf{0} & \mathbf{G}_2 & \ldots & \mathbf{0} & \mathbf{0} \\ \vdots & \vdots & \ddots & \vdots & \vdots \\ \mathbf{0} & \mathbf{0} & \ldots & \mathbf{G}_{N_A-1} & \mathbf{0} \\ \mathbf{0} & \mathbf{0} & \ldots & \mathbf{0} & \mathbf{G}_{N_A} \end{pmatrix}.$$

Namely, the diagonal blocks $\mathbf{F}_i$, $\mathbf{G}_i$, $i = 1, \ldots, N_A$ represent the incident matrices of the agent A_i. However, when the agents are mutually connected the incidence matrices of the DES model can have three principal forms as follows. Each form depends on the fact how the modules are interconnected. The P/T PN models of DES modules can be connected

1. either by means of PN transitions
2. or by means of PN places
3. or, finally, by means of both the PN transitions and PN places.

However, the interconnections

- can be created by a qualified mathematical method of synthesis;
- can result from a real situation—e.g. mutual interconnections of production lines is prescribed by the production technology;
- can be created intuitively—i.e., ad hoc in a self-contained process of manual synthesis—see Example 20.5.

While the first possibility is supported by scientific methods, the second and third ones are based especially on human experience. However, sometimes they can be supported by the first one in order to improve the quality of cooperation or a degree of automation—e.g. see and compar the Examples 20.1 and 20.3.

20.2.1 Interconnections by PN Transitions

When interconnections consist (exclusively) of additional PN transitions the structure of the block diagonal matrices will be augmented for submatrices expressing the interconnections among the modules. Thus, the matrices **F**, **G** are the following:

$$\mathbf{F} = \left(\text{blockdiag}(\mathbf{F}_i)_{i=1,N_A} \,|\, \mathbf{F}_c\right); \quad \mathbf{G} = \left(\frac{\text{blockdiag}(\mathbf{G}_i)_{i=1,N_A}}{\mathbf{G}_c}\right),$$

$$\mathbf{B} = \left(\text{blockdiag}(\mathbf{B}_i)_{i=1,N_A} \,|\, \mathbf{B}_c\right),$$

where $\mathbf{B}_i = \mathbf{G}_i^T - \mathbf{F}_i$; $\mathbf{B}_{c_i} = \mathbf{G}_{c_i}^T - \mathbf{F}_{c_i}$; $i = 1, \ldots, N_A$; $\mathbf{F}_c = (\mathbf{F}_{c_1}^T, \mathbf{F}_{c_2}^T, \ldots, \mathbf{F}_{c_{N_A}}^T)^T$ $\mathbf{G}_c = (\mathbf{G}_{c_1}, \mathbf{G}_{c_2}, \ldots, \mathbf{G}_{c_{N_A}})$; $\mathbf{B}_c = (\mathbf{B}_{c_1}^T, \mathbf{B}_{c_2}^T, \ldots, \mathbf{B}_{c_{N_A}}^T)^T$. Here, $\mathbf{F}_i$, $\mathbf{G}_i$, $\mathbf{B}_i$ represent the parameters of the PN-based model of the module (agent) A_i. $\mathbf{F}_c$, $\mathbf{G}_c$, $\mathbf{B}_c$ represent the structure of the interface between the cooperating agents.

20.2.2 Interconnections by PN Places

Alike, when the interconnections consist (exclusively) of additional PN places, the shape of the matrices is as follows:

$$\mathbf{F} = \left(\frac{\text{blockdiag}(\mathbf{F}_i)_{i=1,\ldots,N_A}}{\mathbf{F}_d}\right); \quad \mathbf{G} = \left(\text{blockdiag}(\mathbf{G}_i)_{i=1,N_A} \,|\, \mathbf{G}_d\right),$$

$$\mathbf{B} = \left(\frac{\text{blockdiag}(\mathbf{B}_i)_{i=1,\dots,N_A}}{\mathbf{B}_d} \right),$$

where $\mathbf{B}_i = \mathbf{G}_i^T - \mathbf{F}_i$; $\mathbf{B}_{d_i} = \mathbf{G}_{d_i}^T - \mathbf{F}_{d_i}$; $i = 1, \dots, N_A$; $\mathbf{F}_d = (\mathbf{F}_{d_1}, \mathbf{F}_{d_2}, \dots, \mathbf{F}_{d_{N_A}})$; $\mathbf{G}_d = (\mathbf{G}_{d_1}^T, \mathbf{G}_{d_2}^T, \dots, \mathbf{G}_{d_{N_A}}^T)^T$; $\mathbf{B}_d = (\mathbf{B}_{d_1}, \mathbf{B}_{d_2}, \dots, \mathbf{B}_{d_{N_A}})$. $\mathbf{F}_i$, $\mathbf{G}_i$, $\mathbf{B}_i$, $i = 1, \dots, N_A$, represent the parameters of the PN-based model of the agent A_i, and $\mathbf{F}_d$, $\mathbf{G}_d$, $\mathbf{B}_d$ represent the structure of the interface between the cooperating agents.

20.2.3 Interconnections by both PN Places and PN Transitions

Finally, when the interconnections consist of both the additional PN places and the additional PN transitions the matrices acquire the following form:

$$\mathbf{F} = \begin{pmatrix} \text{blockdiag}(\mathbf{F}_i)_{i=1,N_A} & \mathbf{F}_c \\ \mathbf{F}_d & \mathbf{F}_{d\leftrightarrow c} \end{pmatrix}; \quad \mathbf{G} = \begin{pmatrix} \text{blockdiag}(\mathbf{G}_i)_{i=1,N_A} & \mathbf{G}_d \\ \mathbf{G}_c & \mathbf{G}_{c\leftrightarrow d} \end{pmatrix},$$

$$\mathbf{B} = \begin{pmatrix} \text{blockdiag}(\mathbf{B}_i)_{i=1,N_A} & \mathbf{B}_c \\ \mathbf{B}_d & \mathbf{B}_{d\leftrightarrow c} \end{pmatrix},$$

where $\mathbf{B}_i = \mathbf{G}_i^T - \mathbf{F}_i$; $\mathbf{B}_{d_i} = \mathbf{G}_{d_i}^T - \mathbf{F}_{d_i}$; $\mathbf{B}_{c_i} = \mathbf{G}_{c_i}^T - \mathbf{F}_{c_i}$; $i = 1, \dots, N_A$; $\mathbf{B}_{d\leftrightarrow c} = \mathbf{G}_{c\leftrightarrow d}^T - \mathbf{F}_{d\leftrightarrow c}$. Here $\mathbf{F}$, $\mathbf{G}$, $\mathbf{B}$ have a special structure. Each of them has the big diagonal block describing the structure of autonomous agents and the specific part in the form of the letter $\mathbf{L}$ turned over the vertical axe. $\mathbf{F}_{d\leftrightarrow c}$, $\mathbf{G}_{c\leftrightarrow d}$, $\mathbf{B}_{d\leftrightarrow c}$ are situated, respectively, on their diagonals just in the breakage of the turned L. Here, in effect the interconnection has a form of the cooperation with a new PN subnet (quasi another agent) containing n_d additional places and m_c additional transitions. Its structure is given by the $(n_d \times m_c)$-dimensional matrix $\mathbf{F}_{d\leftrightarrow c}$ and $(m_c \times n_d)$-dimensional matrix $\mathbf{G}_{c\leftrightarrow d}$. The row and the column consisting of corresponding blocks model the contacts of the kernel with the autonomous agents.

20.2.4 Connecting PN Modules of Agents by PN Transitions

Any system (even the continuous one) has minimally two discrete states—idle and working. Discrete events execute switching between them. The idle system can be switched on by a discrete event (e.g. start) while the running system can be switched off (stopped) by another discrete event. However, besides such events concerning the single agents, other discrete events can start or stop interconnections among the agents. Consequently, to illustrate the first kind of interconnecting the PN modules, i.e., by means of the PN transitions, let us introduce the following example.

Example 20.1 Consider six production lines in a factory recycling the collected waste plastic given in Fig. 20.1.

Four upper lines produce the plastic double foil from the granulate prepared from the waste plastic. The model of such a production line based on first-order hybrid Petri nets (FOHPN) is presented in [11]. Here, only the cooperation of the lines will be discussed.

Two lower lines produce rolls of plastic bags from the double foil. We can encounter a kind of such rolls of bags everyday in supermarkets when we buy fruits, bread, pastry, etc., or at home we insert bags from another kind of rolls into the bin in our kitchen or into the dust basket at cleaning our flat or into the waste paper basket in our office.

In Fig. 20.1 a rough schema of the cooperation of two groups of lines is displayed, namely, the group of four autonomous foil producing lines and the group of two autonomous lines producing bags.

The PN places $\{p_1, p_4, p_7, p_{10}\}$ represent the continuous production of the foil, $\{p_2, p_5, p_8, p_{11}\}$ represent the cutting a bales of the foil with a determined weight

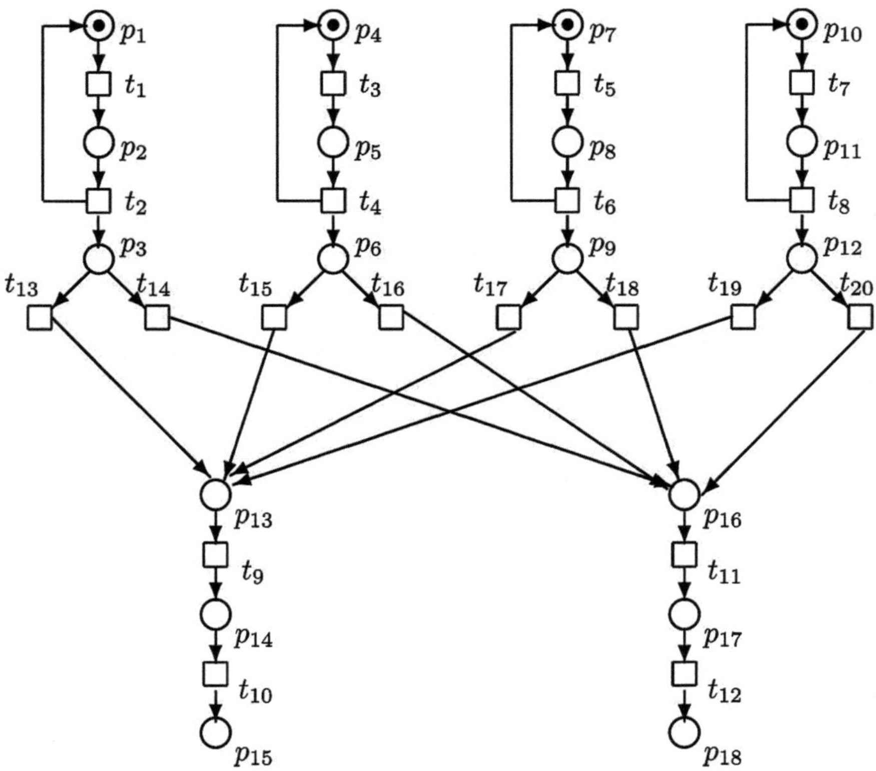

Fig. 20.1 The Petri net-based model of the rough conception of the supposed cooperation of the production lines

and delivering the bale to the buffers $\{p_3, p_6, p_9, p_{12}\}$. The PN places $\{p_{13}, p_{16}\}$ represent the continuous rolling lines processing the double foil into the form of the belt of bags, $\{p_{14}, p_{17}\}$ represent the rolling the belt into rolls of a prescribed length (prescribed number of bags) and $\{p_{15}, p_{18}\}$ represent buffers of the rolls.

The transitions $\{t_{13}, t_{14}\}$ make possible the cooperation of the first foil production line either with the first or with the second rolling lines. Likewise, $\{t_{15}, t_{16}\}$ makes possible the cooperation of the second foil production line with the rolling lines, $\{t_{17}, t_{18}\}$ the cooperation of the third foil production line with the rolling lines, and finally $\{t_{19}, t_{20}\}$ the cooperation of the fourth foil production line with the rolling lines.

We can built the system structure as follows:

$$\mathbf{F}_{i=1,\cdots,6} = \begin{pmatrix} 1\,0 \\ 0\,1 \\ 0\,0 \end{pmatrix}; \ \mathbf{G}_{i=1,\cdots,4} = \begin{pmatrix} 0\,1\,0 \\ 1\,0\,1 \end{pmatrix}; \ \mathbf{G}_{i=5,6} = \begin{pmatrix} 0\,1\,0 \\ 0\,0\,1 \end{pmatrix},$$

$$\mathbf{F}_c = \begin{pmatrix} 0\,0\,0\,0\,0\,0\,0\,0 \\ 0\,0\,0\,0\,0\,0\,0\,0 \\ 1\,1\,0\,0\,0\,0\,0\,0 \\ 0\,0\,0\,0\,0\,0\,0\,0 \\ 0\,0\,0\,0\,0\,0\,0\,0 \\ 0\,0\,1\,1\,0\,0\,0\,0 \\ 0\,0\,0\,0\,0\,0\,0\,0 \\ 0\,0\,0\,0\,0\,0\,0\,0 \\ 0\,0\,0\,0\,1\,1\,0\,0 \\ 0\,0\,0\,0\,0\,0\,0\,0 \\ 0\,0\,0\,0\,0\,0\,0\,0 \\ 0\,0\,0\,0\,0\,0\,1\,1 \\ 0\,0\,0\,0\,0\,0\,0\,0 \\ 0\,0\,0\,0\,0\,0\,0\,0 \\ 0\,0\,0\,0\,0\,0\,0\,0 \\ 0\,0\,0\,0\,0\,0\,0\,0 \\ 0\,0\,0\,0\,0\,0\,0\,0 \\ 0\,0\,0\,0\,0\,0\,0\,0 \end{pmatrix}; \mathbf{G}_c^T = \begin{pmatrix} 0\,0\,0\,0\,0\,0\,0\,0 \\ 0\,0\,0\,0\,0\,0\,0\,0 \\ 0\,0\,0\,0\,0\,0\,0\,0 \\ 0\,0\,0\,0\,0\,0\,0\,0 \\ 0\,0\,0\,0\,0\,0\,0\,0 \\ 0\,0\,0\,0\,0\,0\,0\,0 \\ 0\,0\,0\,0\,0\,0\,0\,0 \\ 0\,0\,0\,0\,0\,0\,0\,0 \\ 0\,0\,0\,0\,0\,0\,0\,0 \\ 0\,0\,0\,0\,0\,0\,0\,0 \\ 0\,0\,0\,0\,0\,0\,0\,0 \\ 0\,0\,0\,0\,0\,0\,0\,0 \\ 1\,0\,1\,0\,1\,0\,1\,0 \\ 0\,0\,0\,0\,0\,0\,0\,0 \\ 0\,0\,0\,0\,0\,0\,0\,0 \\ 0\,1\,0\,1\,0\,1\,0\,1 \\ 0\,0\,0\,0\,0\,0\,0\,0 \\ 0\,0\,0\,0\,0\,0\,0\,0 \end{pmatrix}.$$

However, it has to be said that in Fig. 20.1 is only a rough structure. Namely, the behaviour of such a structure does not determine any order or priorities as to the cooperation of the lines from the first group with the lines from the second group. In such a form it is not suitable either from the idle time point of view or from the technological point of view. Namely, the dimensionality of the double foil in bales from different lines is not the same—it may be different as to the width of the foil and as to the thickness of the foil itself. Therefore, the displayed structure is incomplete. Thus, it is necessary to define rules for cooperation, i.e., when and how the lines can cooperate. The supervision can help us on this way. The extended correct schema will arise after imbedding the supervisor synthesized by means of supervision—see Example 20.3 introduced below in the part Sect. 20.3.2.

20.3 Supervision and the Agent Cooperation

A supervisor for DES subsystems can be suitable in general for avoiding the egoistic effort of autonomous agents, especially in case of limited sources (working space, raw materials or semiproducts, energy, etc.). Namely, by means of prohibition some states of the global system describing the multi-agent system (MAS), e.g. like the so-called mutex (mutual exclusion), a useless 'haggle' of agents each other for a priority can be removed on behalf of the global system purposes. However, on the contrary, the supervision process can be understood to be a carrier of the cooperation in the sense of the global system politics. In such a way the conditions for the supervisor synthesis represent the desired cooperation of agents in a group of agents or in MAS. In both cases some constraints have to be satisfied in order to achieve the desired behaviour (i.e., to synthesize the supervisor). Here, two kinds of constraints known from supervising methodology in DES control theory will be considered. Namely, (i) the constraints based on the P-invariants—[3, 5, 7–10, 13]; (ii) the generalized constraints based not only on the PN places but also on the PN Parikh's vector and/or on the PN transitions—[2–4].

20.3.1 Using P/T PN P-Invariants at Agent Cooperation Synthesis

The principle of the method is based on the PN P-invariants [3, 4, 6] being the vectors, $\mathbf{v}$, with the property that multiplication of these vectors with any state vector $\mathbf{x}_k \in X_{reach}$ (i.e., reachable from a given initial state vector $\mathbf{x}_0 \in X_{reach}$) yields the same result (the relation of the state conservation): $\mathbf{v}^T.\mathbf{x}_k = \mathbf{v}^T.\mathbf{x}_0$. Because of (20.3) $\mathbf{v}^T.\mathbf{x}_k = \mathbf{v}^T.\mathbf{x}_0 + \mathbf{v}^T.\mathbf{B}.\mathbf{u}_k$. Hence, to satisfy the previous definition of P-invariants, the condition $\mathbf{v}^T.\mathbf{B} = \mathbf{0}$ has to be met. P-invariants are useful in checking the property of mutual exclusion. To eliminate a selfish behaviour of autonomous agents at exploitation of limited joint resources it is necessary to allocate the sources to individual agents rightly, with respect to the global goal of MAS. Such a constraint of the agents behaviour and violation of their autonomy is rather in favour of MAS than in disfavour. In case of the existence of several (e.g. n_x) invariants in a PN, the set of the P-invariants is created by the columns of the $(n \times n_x)$-dimensional matrix $\mathbf{V}$ being the solution of the homogeneous system of equations

$$\mathbf{V}^T.\mathbf{B} = \boldsymbol{\emptyset}. \tag{20.9}$$

This equation represents the base for the supervisor synthesis method. Some additional PN places (slacks) can be added to the PN model in question. The slacks create the places of the supervisor. Hence (20.9) can be rewritten as

$$[\mathbf{L}, \mathbf{I}_s].\begin{bmatrix}\mathbf{B}\\ \mathbf{B}_s\end{bmatrix} = \emptyset; \quad \mathbf{L}.\mathbf{B} + \mathbf{B}_s = \emptyset; \quad \mathbf{B}_s = -\mathbf{L}.\mathbf{B}; \tag{20.10}$$

where $\mathbf{I}_s$ is $(n_s \times n_s)$-dimensional identity matrix with $n_s \leq n_x$ being the number of slacks, $(n_s \times n)$-dimensional matrix $\mathbf{L}$ of integers represents (in a suitable form) the conditions $\mathbf{L}.\mathbf{x} \leq \mathbf{b}$ ($\mathbf{b}$ is the vector of integers), imposed on marking of the original PN and the $(n_s \times m)$-dimensional matrix $\mathbf{B}_s = \mathbf{G}_s^T - \mathbf{F}_s$ yields (after its finding by computing) the structure of the PN-based model of the supervisor. Thus, the supervisor structure respects the actual structure of the matrix $\mathbf{L}$. The supervised system (the group of autonomous agents augmented for the supervisor) is characterized by the augmented state vector and the augmented structural matrices as follows:

$$\mathbf{x}_a = \begin{bmatrix}\mathbf{x}\\ \mathbf{x}_s\end{bmatrix}; \quad \mathbf{F}_a = \begin{pmatrix}\mathbf{F}\\ \mathbf{F}_s\end{pmatrix}; \quad \mathbf{G}_a^T = \begin{pmatrix}\mathbf{G}^T\\ \mathbf{G}_s^T\end{pmatrix}, \tag{20.11}$$

where $\mathbf{F}_s$ and $\mathbf{G}_s^T$ represent the supervisor structure. They correspond to the interconnections of the supervisor and autonomous agents. Analogically, the initial state of the supervisor ${}^s\mathbf{x}_0$ can be computed as

$$[\mathbf{L} \mid \mathbf{I}_s].\begin{bmatrix}\mathbf{x}_0\\ {}^s\mathbf{x}_0\end{bmatrix} = \mathbf{b}\,; \quad {}^s\mathbf{x}_0 = \mathbf{b} - \mathbf{L}.\mathbf{x}_0, \tag{20.12}$$

where $\mathbf{b}$ is the vector of the corresponding dimensionality (i.e., n_s) with integer entries representing the limits for number of common tokens—i.e., the maximum numbers of tokens that the corresponding places can possess altogether (i.e., to share them).

Example 20.2 Let us show how easy the dining philosophers problem defined by [1] can be solved by means of the supervisor synthesis. In computer science, this problem is an illustrative example of a common computing problem in concurrency. It is a classic multi-process synchronization problem where five computers competed for access to five shared tape drive peripherals. It was named as dining philosophers problem. Namely, five philosophers are sitting at a circular table with a large bowl of spaghetti in the centre doing one of two things—eating or thinking. While eating, they are not thinking, and while thinking, they are not eating. A chopstick is placed in between each philosopher. Each philosopher has one chopstick to his left and one chopstick to his right. It is assumed that a philosopher must eat with two chopsticks. The philosopher can only use the chopstick on his immediate left or right.

The PN-based model of the situation for one philosopher is given in Fig. 20.2. In case of five philosophers the thinking is modelled by the PN places p_1, p_3, p_5, p_7, p_9 and eating is represented by the places p_2, p_4, p_6, p_8, p_{10}. In this situation all

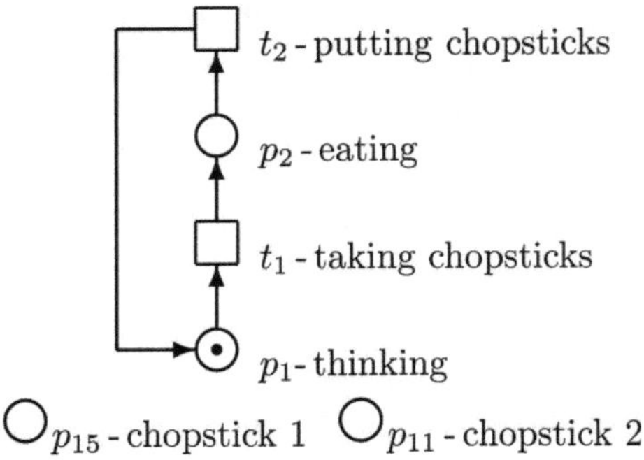

Fig. 20.2 The PN-based model of one philosopher activities

of the philosophers are thinking—p_1, p_3, p_5, p_7, p_9 are active—i.e., no chopsticks are necessary. However, formally they are expressed by means of the PN places p_{11}, $p_{12}, p_{13}, p_{14}, p_{15}$—see Fig. 20.3, apart from interconnections. The defined problem can be solved by the supervisor synthesis method. The incidence matrices of the PN models of the autonomous agents A_i, $i = 1, \ldots, 5$ are

$$\mathbf{F}_i = \begin{pmatrix} 1 & 0 \\ 0 & 1 \end{pmatrix}; \; \mathbf{G}_i^T = \begin{pmatrix} 0 & 1 \\ 1 & 0 \end{pmatrix}; \; \mathbf{B}_i = \begin{pmatrix} -1 & 1 \\ 1 & -1 \end{pmatrix}.$$

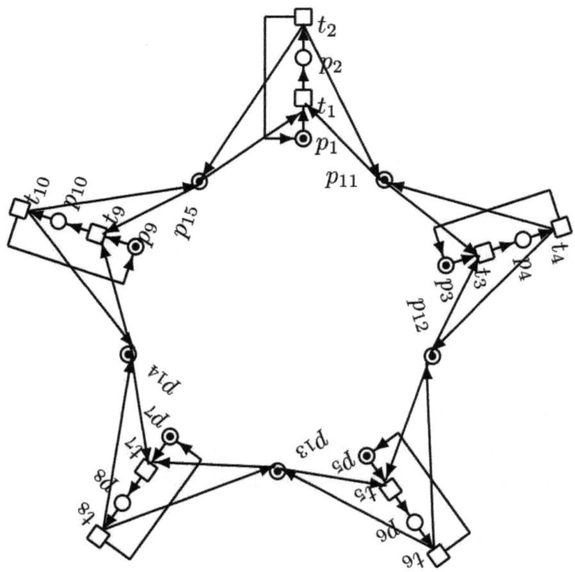

Fig. 20.3 The PN-based model of the supervised dining philosophers

Consider that the initial states are the same

$$^{i}\mathbf{x}_0 = (1, 0)^T; \quad i = 1, \dots, 5.$$

The parameters of the PN model of the group of autonomous agents can be expressed as follows:

$$\mathbf{F} = \text{blockdiag}(\mathbf{F}_i)_{i=1,5}; \ \mathbf{G} = \text{blockdiag}(\mathbf{G}_i)_{i=1,5},$$

$$\mathbf{x}_0 = ({}^{1}\mathbf{x}_0^T, {}^{2}\mathbf{x}_0^T, {}^{3}\mathbf{x}_0^T, {}^{4}\mathbf{x}_0^T, {}^{5}\mathbf{x}_0^T)^T.$$

The conditions imposed on the autonomous agents are

$$\begin{array}{lr}
\sigma_{p_2} + \sigma_{p_4} & \leq 1 \\
\sigma_{p_4} + \sigma_{p_6} & \leq 1 \\
\sigma_{p_6} + \sigma_{p_8} & \leq 1 \\
\sigma_{p_8} + \sigma_{p_{10}} & \leq 1 \\
\sigma_{p_{10}} + \sigma_{p_2} & \leq 1.
\end{array}$$

Verbally it means that two adjacent agents (neighbours) must not eat simultaneously. These conditions yield the matrix $\mathbf{L}$ and the vector $\mathbf{b}$ as follows

$$\mathbf{L} = \begin{pmatrix} 0\,1\,0\,1\,0\,0\,0\,0\,0\,0 \\ 0\,0\,0\,1\,0\,1\,0\,0\,0\,0 \\ 0\,0\,0\,0\,0\,1\,0\,1\,0\,0 \\ 0\,0\,0\,0\,0\,0\,0\,1\,0\,1 \\ 0\,1\,0\,0\,0\,0\,0\,0\,0\,1 \end{pmatrix}; \quad \mathbf{b} = (1,\ 1,\ 1,\ 1,\ 1)^T.$$

Hence,

$$\mathbf{B}_s = -\mathbf{L}.\mathbf{B}; \ {}^{s}\mathbf{x}_0 = \mathbf{b} - \mathbf{L}.\mathbf{x}_0,$$

$$\mathbf{B}_s = \begin{pmatrix} -1\,1 & -1\,1 & 0\,0 & 0\,0 & 0\,0 \\ 0\,0 & -1\,1 & -1\,1 & 0\,0 & 0\,0 \\ 0\,0 & 0\,0 & -1\,1 & -1\,1 & 0\,0 \\ 0\,0 & 0\,0 & 0\,0 & -1\,1 & -1\,1 \\ -1\,1 & 0\,0 & 0\,0 & 0\,0 & -1\,1 \end{pmatrix}, ; \quad {}^{s}\mathbf{x}_0 = \begin{pmatrix} 1 \\ 1 \\ 1 \\ 1 \\ 1 \end{pmatrix}$$

$$\mathbf{F}_s = \begin{pmatrix} 1010000000 \\ 0010100000 \\ 0000101000 \\ 0000001010 \\ 1000000010 \end{pmatrix}; \ \mathbf{G}_s^T = \begin{pmatrix} 0101000000 \\ 0001010000 \\ 0000010100 \\ 0000000101 \\ 0100000001 \end{pmatrix}.$$

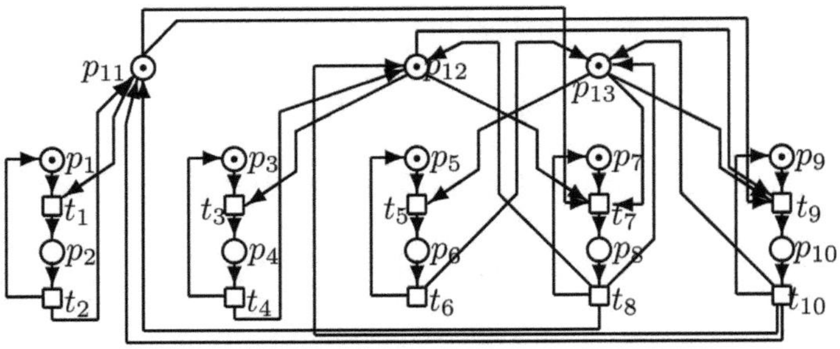

Fig. 20.4 The PN-based model of the three groups cooperation

The structural matrices $\mathbf{F}_s$, $\mathbf{G}_s$ of the supervisor give us the structural interconnections between the philosophers and the chopsticks. Using the supervisor synthesis the problem was easily resolved. The PN-based model of the solution—the cooperating agents—is given in Fig. 20.3. Of course, the conditions for cooperation can be more complicated. Consider, e.g. the group of five simple autonomous agents $Gr_A = \{A_1, A_2, A_3, A_4, A_5\}$ with the same structure like those handled above. Let us solve the situation when it is necessary to ensure that only one agent from the subgroup $Sgr_1 = \{A_1, A_4, A_5\}$ and only one agent from the subgroup $Sgr_2 = \{A_2, A_4, A_5\}$ and only one agent from the subgroup $Sgr_3 = \{A_3, A_4, A_5\}$ can simultaneously cooperate with other agents from Gr_A. In other words, the agents inside the designated subgroups must not work simultaneously. Even, the agents A_4 and A_5 can work only individually (any cooperation with other agents is excluded). However, the agents A_1, A_2, A_3 can work simultaneously. Now, the conditions prescribing the cooperation of agents are

$$\sigma_{p_2} + \sigma_{p_8} + \sigma_{p_{10}} \leq 1,$$
$$\sigma_{p_4} + \sigma_{p_8} + \sigma_{p_{10}} \leq 1,$$
$$\sigma_{p_6} + \sigma_{p_8} + \sigma_{p_{10}} \leq 1.$$

After the supervisor synthesis the PN model of the cooperating agents is displayed in Fig. 20.4.

20.3.2 A General Approach to Agent Cooperation Synthesis

Of course, the previous approach does not cover all cases occurring in practice. To widen a class of cooperation fashions the more general approach has to be used. One of possibilities how to do this is the extended method. The method gives us the possibility to impose some conditions also on the P/T PN transitions as well as on the components of the P/T PN Parikh's vector. On this way especially the Parikh's vector is very important and useful because it allows us to force conditions concerning the priorities of elementary agents in case of utilizing common sources (e.g. access to a critical space, power consumption, etc.). The general linear constraints for supervisor synthesis were described in [2] in detail. Concentrated form of them can be expressed as follows

$$\mathbf{L}_p.\mathbf{x} + \mathbf{L}_t.\mathbf{u} + \mathbf{L}_v.\mathbf{v} \leq \mathbf{b}, \tag{20.13}$$

where $\mathbf{L}_p$, $\mathbf{L}_t$, $\mathbf{L}_v$ are, respectively, $(n_s \times n)-$, $(n_s \times m)-$, $(n_s \times m)-$dimensional matrices. The matrix $\mathbf{L}_p$ corresponds to the matrix $\mathbf{L}$ introduced in the previous method. The matrix $\mathbf{L}_t$ expresses the conditions imposed on the linear combinations of transitions, while the matrix $\mathbf{L}_v$ expresses the conditions imposed on the linear combinations of the components of Parikh's vector. When $\mathbf{b} - \mathbf{L}_p.\mathbf{x} \geq \mathbf{0}$ is valid—see e.g. [2]—the supervisor with the following structure and initial state

$$\mathbf{F}_s = \max(\mathbf{0}, \mathbf{L}_p.\mathbf{B} + \mathbf{L}_v, \mathbf{L}_t); \ \mathbf{L}_{pv} = \mathbf{L}_p.\mathbf{B} + \mathbf{L}_v, \tag{20.14}$$

$$\mathbf{G}_s^T = \max(\mathbf{0}, \mathbf{L}_t - \max(\mathbf{0}, \mathbf{L}_{pv})) - \min(\mathbf{0}, \mathbf{L}_{pv}), \tag{20.15}$$

$${}^s\mathbf{x}_0 = \mathbf{b} - \mathbf{L}_p.\mathbf{x}_0 - \mathbf{L}_v.\mathbf{v}_0, \tag{20.16}$$

guarantees that constraints are verified for the states resulting from the initial state. Here, the max(.) is the maximum operator for matrices. However, the maximum is taken element by element. Namely, in general, for the matrices $\mathbf{X}$, $\mathbf{Y}$, $\mathbf{Z}$ of the same dimensionality ($n \times m$, the relation $\mathbf{Z} = \max(\mathbf{X}, \mathbf{Y})$ it holds that $z_{ij} = \max(x_{ij}, y_{ij})$, $i = 1, \ldots, n, j = 1, \ldots, m$. The same holds for min(.).

The following examples illustrate usage of the extended methods of the supervisor synthesis:

Example 20.3 Consider the production system mentioned above in Example 20.1 introduced in Sect. 20.2.4. At forming the rules defining the mutual cooperation of the lines we have to respect the facts as follows: (i) any bale of the foil from output buffers of the four foil production lines can enter only one of the two rolling machines; (ii) only one bale can enter any rolling machine; (iii) next bale can enter the rolling machines after finishing the rolling process, i.e., when all the rolls are put into their output buffers. Hence, the verbal conditions (i), (ii) mean that the transition functions of the PN transitions $t_{13} - t_{20}$ have to satisfy the conditions

$$\gamma_{t_{13}} + \gamma_{t_{15}} + \gamma_{t_{17}} + \gamma_{t_{19}} \leq 1, \tag{20.17}$$

$$\gamma_{t_{14}} + \gamma_{t_{16}} + \gamma_{t_{18}} + \gamma_{t_{20}} \leq 1, \tag{20.18}$$

while the verbal condition (iii) means that the places $p_{13} - p_{16}$ has to meet the conditions

$$\sigma_{p_{13}} + \sigma_{p_{14}} \leq 1, \tag{20.19}$$

$$\sigma_{p_{16}} + \sigma_{p_{17}} \leq 1. \tag{20.20}$$

Consequently, we have the matrices $\mathbf{L}_t$ and $\mathbf{L}_p$ as follows:

$$\mathbf{L}_t = \begin{pmatrix} 0\,0\,0\,0\,0\,0\,0\,0\,0\,0\,0\,0\,1\,0\,1\,0\,1\,0\,1\,0 \\ 0\,0\,0\,0\,0\,0\,0\,0\,0\,0\,0\,0\,0\,1\,0\,1\,0\,1\,0\,1 \end{pmatrix},$$

$$\mathbf{L}_p = \begin{pmatrix} 0\,0\,0\,0\,0\,0\,0\,0\,0\,0\,0\,0\,1\,1\,0\,0\,0\,0 \\ 0\,0\,0\,0\,0\,0\,0\,0\,0\,0\,0\,0\,0\,0\,0\,1\,1\,0 \end{pmatrix}.$$

Utilizing the general procedure for supervisor synthesis the lines cooperation can be synthesized. Thus, we obtain the supervisor with the structure described by the matrices

$$\mathbf{F}_s = \begin{pmatrix} 0\,0\,0\,0\,0\,0\,0\,0\,0\,0\,0\,0\,1\,0\,1\,0\,1\,0\,1\,0 \\ 0\,0\,0\,0\,0\,0\,0\,0\,0\,0\,0\,0\,0\,1\,0\,1\,0\,1\,0\,1 \end{pmatrix},$$

$$\mathbf{G}_s^T = \begin{pmatrix} 0\,0\,0\,0\,0\,0\,0\,0\,0\,1\,0\,0\,0\,0\,0\,0\,0\,0\,0\,0 \\ 0\,0\,0\,0\,0\,0\,0\,0\,0\,0\,0\,1\,0\,0\,0\,0\,0\,0\,0\,0 \end{pmatrix}$$

and the initial state

$${}^s\mathbf{x}_0 = \begin{pmatrix} 1 \\ 1 \end{pmatrix}.$$

Then, the PN model of cooperating lines is given in Fig. 20.5.

Example 20.4 Consider an agent with the PN model given in Fig. 20.6. It can be seen that although the PN-based model of the agent is relatively simple, it contains double oriented arcs. They induce more complicated behaviour, especially as to the dimension of the feasible states space represented by the PN reachability tree (RT). Number of places of the agent PN model is $n = 4$, while the number of transitions is $m = 3$. The agent is in the initial state ${}^1\mathbf{x}_0 = (2, 0, 1, 0)^T$. The model structural matrices are the following:

$$\mathbf{F} = \begin{pmatrix} 2\,0\,0 \\ 0\,1\,0 \\ 0\,0\,1 \\ 0\,2\,0 \end{pmatrix}; \ \mathbf{G} = \begin{pmatrix} 0\,1\,1\,0 \\ 1\,0\,0\,0 \\ 1\,0\,0\,2 \end{pmatrix}.$$

Have another agent with the same structure and with the same initial state ${}^2\mathbf{x}_0 = {}^1\mathbf{x}_0$. Impose the following constraints for the agents cooperation, exclusively on the entries v_1 and v_4 of the Parikh's vector $\mathbf{v} = (v_1, \ldots, v_6)^T$, namely

$$v_1 - v_4 \leq 2, \tag{20.21}$$
$$v_4 - v_1 \leq 2. \tag{20.22}$$

Let us analyse the influence of the vector $\mathbf{b}$ on the number of the nodes of the P/T PN reachability tree which represents the dimension of the module (agent) state space.

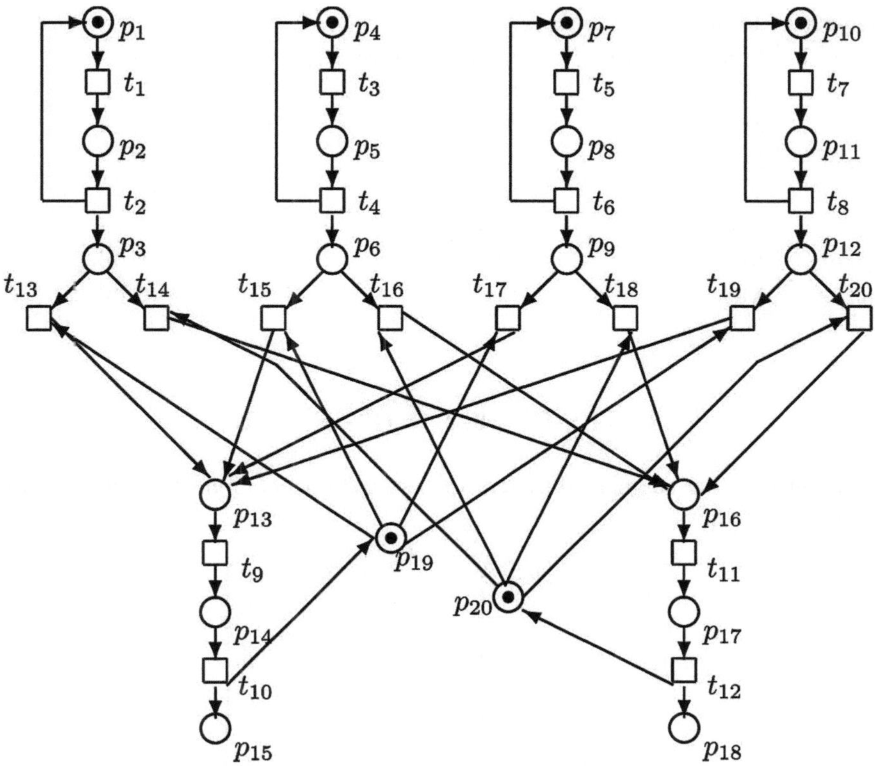

Fig. 20.5 The Petri net-based model of the supervised cooperation of the production lines

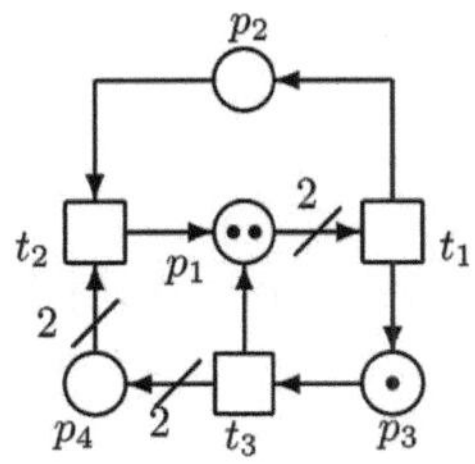

Fig. 20.6 The PN-based model of the DEDS modul (subsystem)

It means that the number of firing the transitions t_1, t_4 is limited in both directions (i.e., when either t_1 is fired before t_4 or vice versa). In such a case $\mathbf{L}_p = \mathbf{0}$, $\mathbf{L}_t = \mathbf{0}$. Because of the constraints (20.13)

$$\mathbf{L}_v.\mathbf{v} \leq \mathbf{b};\ \mathbf{L}_v = \begin{pmatrix} 1\,0\,0\,-1\,0\,0 \\ -1\,0\,0\quad 1\,0\,0 \end{pmatrix};\ \mathbf{b} = \begin{pmatrix} 2 \\ 2 \end{pmatrix}$$

and when e.g. $\mathbf{v}_0 = (1,\ 0,\ 1,\ 1,\ 0,\ 1)^T$ because of the relations (20.14)–(20.15) concerning the supervisor parameters we have

Table 20.1 The influence of the vector **b**

Vector **b**	Symmetric structure	Asymmetric structure	n_{RT}	${}^s\mathbf{x}_0$
$(2, 2)^T$	YES	NO	245	$(2, 2)^T$
$(2, 1)^T$	NO	YES	196	$(2, 1)^T$
$(1, 2)^T$	NO	YES	196	$(1, 2)^T$
$(1, 1)^T$	YES	NO	147	$(1, 1)^T$
$(0, 1)^T$	NO	YES	98	$(0, 1)^T$
$(1, 0)^T$	NO	YES	98	$(1, 0)^T$
$(0, 0)^T$	YES	NO	4	$(0, 0)^T$

$$\mathbf{F}_s = \mathbf{L}_v; \ \mathbf{G}_s^T = \emptyset; \ \mathbf{B}_s = \mathbf{G}_s^T - \mathbf{F}_s = -\mathbf{L}_v.$$

Consequently, the augmented system (supervised cooperation of agents) has the structure as follows

$$\mathbf{B}_a = \begin{pmatrix} \mathbf{B} \\ \mathbf{B}_s \end{pmatrix} = \begin{pmatrix} -2 & 1 & 1 & 0 & 0 & 0 \\ 1 & -1 & 0 & 0 & 0 & 0 \\ 1 & 0 & -1 & 0 & 0 & 0 \\ 0 & -2 & 2 & 0 & 0 & 0 \\ - & - & - & - & - & - \\ 0 & 0 & 0 & -2 & 1 & 1 \\ 0 & 0 & 0 & 1 & -1 & 0 \\ 0 & 0 & 0 & 1 & 0 & -1 \\ 0 & 0 & 0 & 0 & -2 & 2 \\ - & - & - & - & - & - \\ -1 & 0 & 0 & 1 & 0 & 0 \\ 1 & 0 & 0 & -1 & 0 & 0 \end{pmatrix}$$

Because of (20.16) the initial state of the supervisor is as follows

$${}^s\mathbf{x}_0 = \mathbf{b} - \mathbf{L}_v.\mathbf{v}_0 = \begin{pmatrix} 2 \\ 2 \end{pmatrix} - \begin{pmatrix} 0 \\ 0 \end{pmatrix} = \begin{pmatrix} 2 \\ 2 \end{pmatrix}.$$

Finally, the initial state of the augmented system is

$$\mathbf{x}_0 a = \begin{pmatrix} \mathbf{x}_0 \\ {}^s\mathbf{x}_0 \end{pmatrix} = (2\,0\,1\,0 \mid 2\,0\,1\,0 \mid 2\,2)^T.$$

The structure of the cooperating agents is given in Fig. 20.7 where p_9, p_{10} (i.e., the principle part of the supervisor) mediate (or interfere with) the agents cooperation. Besides this, the agents autonomy is fully conserved. Here, ${}^s\mathbf{x}_0 = (2, 2)^T$ and the number of the corresponding RT nodes is $n_{RT} = 245$.

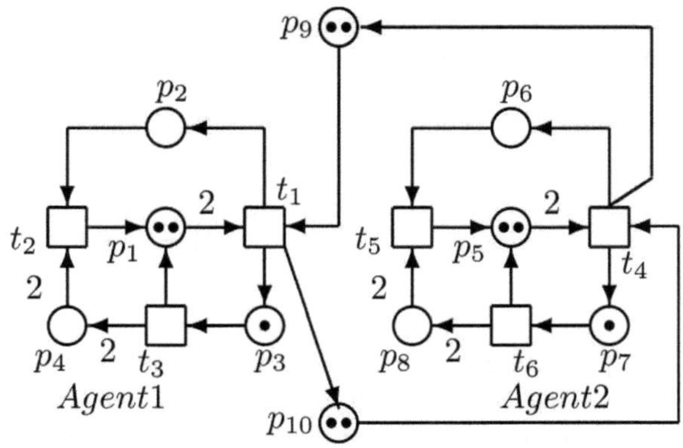

Fig. 20.7 The PN-based model of the cooperating DEDS subsystems

Let us test the influence of **b** on structural properties of the RT of the PN modelling the cooperating agents. When $\mathbf{b} = (2, 1)^T$ or $\mathbf{b} = (1, 2)^T$ $n_{RT} = 196$. When $\mathbf{b} = (1, 1)^T$ $n_{RT} = 147$, when $\mathbf{b} = (0, 1)^T$ or $\mathbf{b} = (1, 0)^T$ $n_{RT} = 98$ and $\mathbf{b} = (0, 0)^T$ $n_{RT} = 4$. As we can see in Table 20.1, the approach yields the big variability of the system dynamics (measured by the number of the nodes of the RT (n_{RT}) corresponding to the PN model of cooperating agents). In the last case (the last row of the Table 20.1) the transitions t_1, t_4 are absolutely blocked (i.e., excluded from the activities—they cannot be fired). The RT adjacency matrix $\mathbf{A}_{RT}$ of such a structure and corresponding reachable states (the RT nodes) being the columns of the matrix $\mathbf{X}_{reach}$ are:

$$\mathbf{A}_{RT} = \begin{pmatrix} 0&3&6&0 \\ 0&0&0&6 \\ 0&0&0&3 \\ 0&0&0&0 \end{pmatrix}; \quad \mathbf{X}_{reach} = \begin{pmatrix} 2&3&2&3 \\ 0&0&0&0 \\ 1&0&1&0 \\ 0&2&0&2 \\ 2&2&3&3 \\ 0&0&0&0 \\ 1&1&0&0 \\ 0&0&2&2 \\ 0&0&0&0 \\ 0&0&0&0 \end{pmatrix}.$$

In this case only events modelled by t_3, t_6 can be activated (i.e., fired). Thus, also events modelled by t_2, t_5 cannot occur because of passive p_2, p_6, respectively. Of course, we can change also the entries of $\mathbf{v}_0$ in order to change the ${}^s\mathbf{x}_0$, but the given structure of $\mathbf{L}_\nu$ does not give us many possibilities in this case (note the product $\mathbf{L}_\nu.\mathbf{v}_0$).

When we choose e.g. $\mathbf{v}_0 = (1, 0, 0, 0, 0, 0)^T$, we indeed obtain another initial state of the supervisor, namely, ${}^s\mathbf{x}_0 = (1, 3)^T$, but at the same $\mathbf{b} = (2, 2)^T$ we have the same number of RG nodes (namely, 245).

Thus, we can find that (i) on the one hand the approach is robust, but (ii) on the other hand it conserves a big autonomy of the agents. In such a way the global aim of a group of agents (or MAS), expressed by means of demands on the synthesized supervisor, is joined with local aims of the individual agents.

Example 20.5 When we consider other conditions for the supervisor synthesis of the agents in the precious example, namely, in the form

$$2.t_1 + p_5 \leq 4, \tag{20.23}$$

$$2.t_4 + p_1 \leq 4, \tag{20.24}$$

i.e., t_1 affecting A_1 can be fired even two times consecutively, but in dependency on the marking of the place p_5 belonging to A_2 and conversely, t_4 can be fired even two times consecutively, but in dependency on the marking of p_1. This yields (starting with the initial states of the agents ${}^{A_1}\mathbf{x}_0 = (2,\ 0,\ 1,\ 0)^T$ and ${}^{A_2}\mathbf{x}_0 = (0,\ 1,\ 1,\ 0)^T$) the cooperation given in Fig. 20.8. Namely, considering

$$\mathbf{u}_0 = (1,\ 0,\ 1,\ 1,\ 0,\ 1)^T; \quad \mathbf{v}_0 = \mathbf{0}; \quad \mathbf{b} = (4,\ 4)^T$$

and expressing the constraints in the following form

$$\mathbf{L}_t = \begin{pmatrix} 2\,0\,0\,0\,0\,0 \\ 0\,0\,0\,2\,0\,0 \end{pmatrix}; \quad \mathbf{L}_p = \begin{pmatrix} 0\,0\,0\,0\,1\,0\,0\,0 \\ 1\,0\,0\,0\,0\,0\,0\,0 \end{pmatrix},$$

we obtain the supervisor structure and its initial state as follows

$$\mathbf{F}_s = \begin{pmatrix} 2\,0\,0\,0\,1\,1 \\ 0\,1\,1\,2\,0\,0 \end{pmatrix}; \ \mathbf{G}_s^T = \begin{pmatrix} 2\,0\,0\,2\,0\,0 \\ 2\,0\,0\,2\,0\,0 \end{pmatrix}; \ {}^s\mathbf{x}_0 = (2,\ 0)^T.$$

It follows from this example that in general we can express practically arbitrary constraint to affect the behaviour of agents.

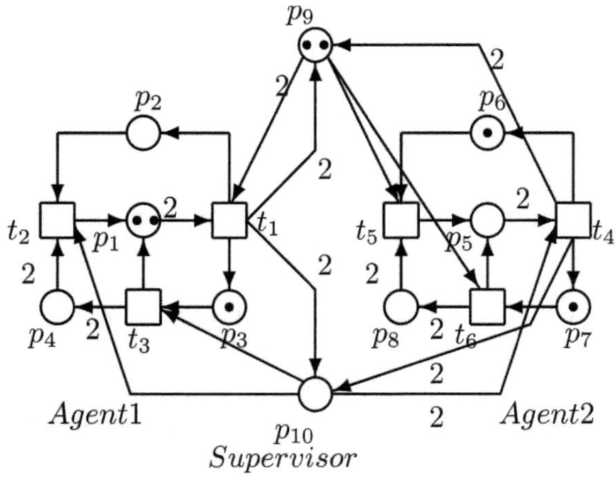

Fig. 20.8 Another PN-based model of the two agents cooperation

20.4 Self-Contained Synthesis of the Supervisor

Many times the cooperation can be synthesized manually, based on the experience or the structure results directly from a prescribed technology of production (e.g. in case of cooperation of production lines in a factory). No scientific method for synthesizing the agent cooperation is necessary there.

To illustrate such an approach of synthesizing the cooperation in virtue of experience, let us introduce the example of the agents' negotiation in general. Consider two subjects (robots, model of human managers, etc.) and the communication channel between them. The PN models of the subjects can be created manually based on the creator experience. Let us illustrate such a situation in detail by means of the following example.

Example 20.6 Consider two agents A_1, A_2 and the interface (the communication channel) between them given in Fig. 20.9. The agents can represent different kinds of technical devices (e.g. intelligent robots), model of the behaviour of human managers of factories, etc. The communication channel can be understood to be the third agent which makes possible to realize the negotiation process between the agents.

The interpretation of PN places is the following: p_1—A_1 does not want to communicate; p_2—A_1 is available; p_3—A_1 wants to communicate; p_4—A_2 does not want to communicate; p_5—A_2 is available; p_6—A_2 wants to communicate; p_7—communication and p_8—availability of the communication by means of the interface (a communication channel).

The interpretation of the PN transitions is clear, but let us emphasize: t_9—fires the communication when A_1 is available and A_2 wants t_{10}—fires the communication when A_2 is available and A_1 wants t_{12}—fires the communication when both A_1 and A_2 want to communicate each other. From Fig. 20.9 it is clear, the interface realizing the negotiation process has the form of the PN module (PN subnet). The P/T PN-based model of the agents communication has the following parameters:

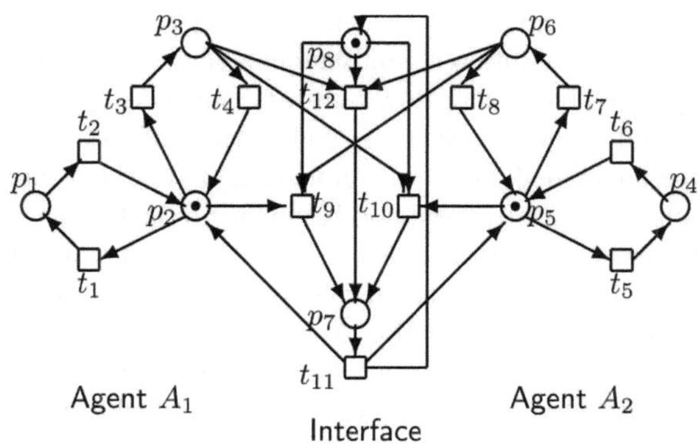

Fig. 20.9 The negotiation of two agents

$$
\mathbf{F} = \begin{pmatrix} \mathbf{F}_{A_1} & \mathbf{0} & \mathbf{F}_{A_1-A_2} \\ \mathbf{0} & \mathbf{F}_{A_2} & \mathbf{F}_{A_2-A_1} \\ \mathbf{0} & \mathbf{0} & \mathbf{F}_{neg} \end{pmatrix} = \left(\begin{array}{cccc|cccc|cccc}
0 & 1 & 0 & 0 & 0 & 0 & 0 & 0 & 0 & 0 & 0 & 0 \\
1 & 0 & 1 & 0 & 0 & 0 & 0 & 0 & 1 & 0 & 0 & 0 \\
0 & 0 & 0 & 1 & 0 & 0 & 0 & 0 & 0 & 1 & 0 & 1 \\
\hline
0 & 0 & 0 & 0 & 0 & 1 & 0 & 0 & 0 & 0 & 0 & 0 \\
0 & 0 & 0 & 0 & 1 & 0 & 1 & 0 & 0 & 1 & 0 & 0 \\
0 & 0 & 0 & 0 & 0 & 0 & 0 & 1 & 1 & 0 & 0 & 1 \\
\hline
0 & 0 & 0 & 0 & 0 & 0 & 0 & 0 & 0 & 0 & 1 & 0 \\
0 & 0 & 0 & 0 & 0 & 0 & 0 & 0 & 1 & 1 & 0 & 1
\end{array}\right),
$$

$$
\mathbf{G}^T = \begin{pmatrix} \mathbf{G}^T_{A_1} & \mathbf{0} & \mathbf{G}^T_{A_1-A_2} \\ \mathbf{0} & \mathbf{G}^T_{A_2} & \mathbf{G}^T_{A_2-A_1} \\ \mathbf{0} & \mathbf{0} & \mathbf{G}^T_{neg} \end{pmatrix} = \left(\begin{array}{cccc|cccc|cccc}
1 & 0 & 0 & 0 & 0 & 0 & 0 & 0 & 0 & 0 & 0 & 0 \\
0 & 1 & 0 & 1 & 0 & 0 & 0 & 0 & 0 & 0 & 1 & 0 \\
0 & 0 & 1 & 0 & 0 & 0 & 0 & 0 & 0 & 0 & 0 & 0 \\
\hline
0 & 0 & 0 & 0 & 1 & 0 & 0 & 0 & 0 & 0 & 0 & 0 \\
0 & 0 & 0 & 0 & 0 & 1 & 0 & 1 & 0 & 0 & 1 & 0 \\
0 & 0 & 0 & 0 & 0 & 0 & 1 & 0 & 0 & 0 & 0 & 0 \\
\hline
0 & 0 & 0 & 0 & 0 & 0 & 0 & 0 & 1 & 1 & 0 & 1 \\
0 & 0 & 0 & 0 & 0 & 0 & 0 & 0 & 0 & 0 & 1 & 0
\end{array}\right).
$$

To use the parameters at simulation it is necessary to choose an initial state—e.g. $\mathbf{x}_0 = (0, 0, 1, 0, 0, 1, 0, 1)^T$ or $\mathbf{x}_0 = (0, 0, 1, 0, 1, 0, 0, 1)^T$, etc.

In such a way, modelling of more cooperating agents is possible too.

20.5 Conclusion

The modular synthesis of the agent cooperation, especially that based on supervision, was presented and illustrated in this paper.

Agents of different kinds are modelled by modules (subsystems of discrete-event systems) based on place/transition Petri nets. The agent cooperation strategy to be forced to agents is synthesized by means of DES supervision methods. It represents the prescribed goal of the global agent system or multi-agent system (MAS). After this manner the supervisor, being the product of the synthesis, is obtained. The synthesized supervisor is the carrier of the cooperation strategy and its realizer. During the agents activity the supervisor ensures the fulfilling the prescribed goal represented by the cooperation strategy. Namely, at the realization of a common job of agents the supervisor obtrudes the cooperation strategy on the agents and in such a way it makes possible to reach the prescribed goal of the global system.

The supervisor synthesis is realized here by means of two methods: (i) the method based on the P/T PN place invariants (P-invariants) and (ii) in virtue of the extended method where P-invariants are complemented by different kinds of conditions imposed on P/T PN transitions and/or on the Parikh's vector.

The presented procedure can be utilized also for cooperation of more complicated structures of modules (represented by groups of agents) in order to achieve a prescribed behaviour of the global structure altogether. Consequently, the idea intrudes, that we can utilize this approach for organizing the MAS behaviour in a very big measure, namely, within of an exactly predefined global strategy.

Such an understanding the agent cooperation is very important especially for "material" agents, e.g. those working in manufacturing systems (like robots, automatically guided vehicles, etc.), in transport systems, etc. However it can be utilized also in other kind of cooperating agents comprehended in business processes, social systems, etc.

The application of the approach was illustrated by several simple examples. The approach seems to be sound and hopeful. It makes possible to examine very deep and detailed mutual interferences of agents in a group to achieve a global aim.

Moreover, there exist two advantages of the presented approach: (i) the approach in question is linear; (ii) it can be computed in analytical terms. In such a way it represents a contribution to computational intelligence.

Acknowledgments The author thanks the Slovak Grant Agency for Science VEGA for supporting the research performed in the project # 2/0075/09 (solved in the years 2009–2012) and in the consecutive project # 2/0039/13 (being solved in the years 2013–2016).

References

1. Dijkstra, E.: Hierarchical ordering of sequential processes. Acta Informatica **1**(2), 115–138 (1971)
2. Iordache, M.: Methods for the supervisory control of concurrent systems based on Petri nets abstraction. Ph.D. thesis, University of Notre Dame, USA (2003)
3. Iordache, M., Antsaklis, P.: Supervision based on place invariants: a survey. Discret. Event Dyn. Syst. **16**(4), 451–492 (2006)
4. Iordache, M., Antsaklis, P.: Supervisory Control of Concurrent Systems: A Petri Net Structural Approach. Birkhäuser, Boston (2006)
5. Moody, J., Antsaklis, P.: Supervisory Control of Discrete Event Systems Using Petri Nets. Kluwer Academic Publishers, Norwell (1998)
6. Murata, T.: Petri nets: properties, analysis and applications. Proc. IEEE **77**(4), 541–580 (1989)
7. Čapkovič, F.: Des modelling and control vs. problem solving methods. Int. J. Intell. Inform. Database Syst. **1**(1), 53–78 (2007)
8. Čapkovič, F.: Modelling, analysing and control of interactions among agents in mas. Comput. Inform. **26**(5), 507–541 (2007)
9. Čapkovič, F.: Des control synthesis and cooperation of agents. In: Nguyen, N., Kowalczyk, R., Chen, S. (eds.) ICCCI 2009: Computational Collective Intelligence, Wroclaw, Poland, October 5–7, 2009. Proceedings. Lecture Notes in Artificial Intelligence, vol. 5796, pp. 596–607. Springer (2009)
10. Čapkovič, F.: Automatic control synthesis for agents and their cooperation in mas. Comput. Inform. **29**(6+), 1045–1071 (2010)

11. Čapkovič, F.: Cooperation of hybrid agents in models of manufacturing systems. In: O'Shea, J., Nguyen, N., Crockett, K., Howlett, R., Lakhmi, C. (eds.) Proceedings of KES-AMSTA 2011: Agent and Multi-Agent Systems: Technologies and Applications, Manchester, UK, June 29-July 1, 2011. Lecture Notes in Artificial Intelligence, vol. 6682, pp. 221–230. Springer (2011)
12. Čapkovič, F.: Modularity and supervision in petri net based synthesis of agents cooperation. In: Proceedings of ETAI/COSY 2011, pp. 195–200. IEEE (2011)
13. Yamalidou, E., Moody, J., Lemmon, M., Antsaklis, P.: Feedback control of petri nets based on place invariants. Automatica **32**(1), 15–28 (1996)

Chapter 21
Adaptive Internal Model-Based Distributed Output Agreement in a Class of Multi-Agent Dynamic Systems

Esma Gül and Veysel Gazi

Abstract In this study we consider the distributed output agreement problem in a class of multi-agent dynamic systems with uncertainties. The problem is formulated as a nonlinear servomechanism problem and an adaptive internal model-based controller is used to achieve agreement of the agent outputs using local information. Various agent neighborhood topologies are considered and the performance of the system is verified using simple numerical simulations.

21.1 Introduction

The problem of distributed agreement or distributed consensus is one of the important problems in the recently popular multi-agent dynamic systems literature [6, 11]. Inspired by the behavior of swarms in nature, in which global agreement (also referred to as consensus or synchronization) is achieved based only on limited local information and local interactions, engineering studies have aimed to uncover the principles and mechanisms of such behavior.

One of the studies which analytically consider the distributed state agreement problem is [18], where the authors consider an averaging model with time-dependent bidirectional interaction links. A more general model as well as unidirectional time-dependent communication/interaction topology was considered by Moreau in [25]. Similarly, Ren and Beard considered a weighted averaging model with time-dependent unidirectional communication in [28]. Furthermore, a time-dependent unidirectional weighted averaging model with asynchronous operation and time delays in the information flow was considered in [9]. These studies are mostly in discrete time setting, although [18] has some parts in continuous time as well. Another

E. Gül
Turkish Aerospace Industries (TAI), Fethiye Mahallesi, Havacılık Bulvarı No: 17, 06980 Kazan, Ankara, Turkey

V. Gazi (✉)
Department of Electrical and Electronics Engineering, Istanbul Kemerburgaz University, Mahmutbey Dilmenler Caddesi No: 26, 34217 Bağcılar, Istanbul, Turkey
e-mail: veysel.gazi@kemerburgaz.edu.tr

G.M. Dimirovski (ed.), *Complex Systems*, Studies in Systems, Decision and Control 55, DOI 10.1007/978-3-319-28860-4_21

work which considers the problem mostly in continuous-time setting is [27] where the authors consider a model with time delays and investigate the problem for various communication/interaction topologies using tools from algebraic graph theory, matrix theory, and control theory. Another related work is [21] where the authors derive necessary and sufficient conditions for reaching agreement in a group of unicycles.

In this article we consider the distributed output agreement problem within the framework of output regulation. The output regulation problem has been studied extensively in the literature including output regulation of linear systems [3–5], nonlinear systems [16, 17], robust and structurally stable as well as semi-global and global regulation [1, 15, 19, 22, 23, 31], output regulation using adaptive internal models [24, 30], output regulation in systems with switched exosystems [8], and decentralized output regulation [10, 36]. Output regulation techniques have been also applied to formation control of multi-agent systems [2, 7, 13, 14].

In [32, 33] the authors considered output synchronization (i.e., distributed output agreement) in multi-agents systems and derived necessary conditions for achieving synchronization. In a follow-up study in [34] the same authors developed also synchronizing controllers for multi-agent systems composed of linear agents. Similar studies can be found in [20, 35], where the authors approach the distributed output agreement problem in multi-agent systems composed of agents with linear dynamics from the perspective of output regulation. In this article we develop a controller to achieve distributed output agreement in a class of multi-agent systems composed of agents with nonlinear dynamics. Moreover, we assume that there are unknown parameters and external disturbances in the agent dynamics. Furthermore, the exogeneous signals are also assumed to be generated by systems with unknown parameters. To overcome these difficulties we use an adaptive internal model-based procedure from [30] to achieve agreement. The output agreement considered here can also be viewed as aggregation of the multi-agent dynamic system and cohesive collective motion afterward. Various agent neighborhood topologies are considered. Numerical simulations are performed to illustrate the effectiveness of the control strategy. This work was presented in [12].

21.2 Agent Model and Problem Definition

Consider a multi-agent system consisting of N agents with motion dynamics

$$\begin{aligned}
\dot{\chi}_i &= f_{i0}(\chi_i, x_i, \mu_i, \omega_i) \\
\dot{x}_{i1} &= x_{i2} \\
&\dots \\
\dot{x}_{ir_i} &= f_{ir_i}(\chi_i, x_i, \mu_i, \omega_i) + b_i(\omega_i)u_i \\
y_i &= x_{i1},
\end{aligned} \tag{21.1}$$

where $y_i \in \mathbb{R}^m$ is the output, $x_i^{\top} = [x_{i1}^{\top}, \ldots, x_{ir_i}^{\top}]$ is the vector of the output and its derivatives, and $z_i = [\chi_i^{\top}, x_i^{\top}]^{\top} \in \mathbb{R}^n$ is the local state of agent i. Similarly, $u_i \in \mathbb{R}^m$ is its local control input, $\mu_i \in \mathbb{R}^t$ is its local exogenous input, and ω_i represents the unknown parameters of agent i. The functions f_{i0}, f_{ir_i}, and b_i are assumed to be smooth and known. Moreover, it is assumed that $b_i(\omega_i)$ are positive and bounded away from zero for all values of the unknown parameters ω_i within a set of interest.

The local exogenous inputs μ_i to the agent dynamics represent the local disturbances affecting the agent dynamics. These disturbances can adversely affect the agreement efforts by the agents and are to be rejected by the agent control inputs. It is assumed that μ_i are generated by parameterized neutrally stable systems whose poles lie on the imaginary axis and are simple poles for each value of the corresponding parameters. We assume that the local exogenous systems are of the form

$$\dot{\mu}_i = g_{\mu_i}(\eta_i)\mu_i, 1 \leq i \leq N, \tag{21.2}$$

where η_i represents the unknown parameters of the local exosystem i. These parameters are assumed to belong to a bounded set of appropriate dimensions.

Consider the problem in which the agents are required to reach consensus or agreement on a variable of interest using only local information from their neighbors without knowledge of the global picture. The variable of interest can be the whole state, part of the state, or some function of the state of the agents. The distributed state agreement problem (i.e., the case in which the variable of interest consists of the whole state of the agents) has been considered extensively in the literature. In this article we consider the problem of output agreement which is a more general and considerably more difficult problem. In particular, the output agreement problem allows for more heterogeneous systems. This is because in the state agreement problem the dimensions of the states (i.e., state spaces) of the agents must be the same. In contrast, in the output agreement problem the dimensions of only the outputs (i.e., the output spaces) of the agents must be the same allowing for multi-agent systems composed of agents with different internal dynamics (i.e., number of internal states). Moreover, some neighborhood definitions may bring extra restrictions for the analysis of the system. For example, there might be cases in which the neighborhood definition between agents is based not on the output itself (i.e., the agreement variable) but on different portions of the state. In such a case the controller should also keep connectivity in addition to achieving agreement.

If the outputs of the agents represent the agent positions, achieving agreement results in aggregation in the multi-agent system and in cohesive motion thereafter. The case considered in this paper can be interpreted in this manner. Note that the aggregation point and the resultant trajectory of motion are emergent entities. In other words, these variables depend on the initial positions/states, the neighborhood structure or the dynamic change in the neighborhood structure of the agents, and the effect of the disturbances none of which are known a priori.

Let us denote with $N_i, i = 1, \ldots, N$ the set of neighbors of agent i. This is the set of agents from which agent i can obtain information. Then, the distributed output agreement problem can be stated as follows.

Probelm 21.1 For every agent $i, i = 1, \ldots, N$, design the control inputs u_i based on information from only their neighbors N_i such that the outputs y_i of the agents satisfy

$$\lim_{t\to\infty} \|y_i(t) - y_j(t)\| = 0 \tag{21.3}$$

for all agents i and j $(i,j = 1, \ldots, N)$.

Note once more that in the definition in Problem 21.1 the agent control inputs u_i are required to use information obtained from only the neighbors N_i of agent i implying only local interactions and local information exchange between agents. The output agreement, on the other hand, is required to be achieved on a global scale. For this to be feasible the agent neighborhood topology plays an important role. In the following section we will briefly discuss the connectivity requirements and some neighborhood topologies considered in the simulations.

21.3 Neighborhood Topologies

Different topologies can be used to define the neighborhood relationship between agents. Moreover, the information flow can be unidirectional or bidirectional. The neighborhood topology of the multi-agent system can be represented using a directed graph. In order to guarantee continuation of the information flow in the multi-agent system and to facilitate achieving agreement it is required that the neighborhood graph satisfies some connectivity conditions. These conditions can be expressed in different forms and can also vary depending on the interconnection schemes such as unidirectional (non-reciprocal) or bidirectional (reciprocal) interaction, existence of cycles, etc. (see for example [9, 18, 21, 25, 27, 28] and the references therein). We assume that the necessary connectivity properties are satisfied and the information flow is not a real concern. For example, for unidirectional static neighborhood topology, a necessary and sufficient condition is that the interaction/interconnection graph has a spanning tree. This can be also stated as a requirement imposed on the Laplacian matrix of the interaction/interconnection graph in the form of multiplicity of the zero eigenvalue. In other words, the zero eigenvalue of the Laplacian matrix must be of multiplicity one. Another alternative statement of the same connectivity requirement is to have an agent which is connected to all agents in the multi-agent system. The reader should see the cited references for the related discussions. In this study we consider static neighborhood topologies only. I particular we consider fully connected, ring, and random neighborhood topologies which are briefly discussed below.

Ring Neighborhood Topology: The ring neighborhood topology is a static topology in which the overall neighborhood of the system constitutes a ring. If the information flow is bidirectional every agent has two neighbors—one on its virtual right and one on its virtual left. In other words, the neighborhood sets of the agents can be expressed as $N_i = \{i-1, i+1\}$ for $i = 2, \ldots, N-1$, $N_1 = \{N, 2\}$, and $N_N = \{N-1, 1\}$. If the information flow is unidirectional, the neighborhood sets are given by $N_i = \{i+1\}$ for $i = 1, \ldots, N-1$ and $N_N = \{1\}$. Note that in the ring topology there is a path from every agent to every other agent in the multi-agent system and information from any agent can flow/propagate to all other agents. The case when agents interact based on unidirectional ring topology is known as cyclic pursuit in the literature. There is a relevant literature on the problem of cyclic pursuit. One can see for example [29] and the references therein.

Fully Connected Neighborhood Topology: Fully connected neighborhood topology is another static topology. In this topology, every agent is a neighbor of every other agent and we have $N_i = \{1, \ldots, N\} \setminus i$ for all i. In this topology information from any agent reaches all the other agents in one step.

Random Neighborhood Topology: Random neighborhood is a neighborhood in which the agent neighbors are determined non-deterministically. Such a neighborhood can be defined to evolve in time as a dynamic neighborhood topology. However, in this study we consider a static implementation. In particular, in order to achieve random neighborhood at the beginning of the simulation for every agent i and for every corresponding pair (i,j) a random number ε_{ij} is generated. Then this number is compared to a predefined threshold ε and if $\varepsilon_{ij} < \varepsilon$ agent j is assigned a neighbor of agent i. Otherwise, i.e., if $\varepsilon_{ij} \geq \varepsilon$, they are considered not to be neighbors. In other words, the neighborhood of agent i can be expressed as

$$N_i = \{j, j \neq i : \varepsilon_{ij} < \varepsilon\}. \tag{21.4}$$

Note that the agent neighbors do not change during the simulation. Moreover, since ε_{ij} are independent from ε_{ji}, the topology is unidirectional in general. In case bidirectional (reciprocal) interactions are desired one can generate only one number (say ε_{ij}) for every pair (i,j) and simply assign $\varepsilon_{ji} = \varepsilon_{ij}$.

Note that as mentioned the above topologies are static topologies. In other words, the agent neighborhoods do not vary with time. We have not considered dynamic neighborhood in this study for two main reasons. First of all dynamic change or switching in the topology may result in instantaneous jump and discontinuities in the local error variables. Second, although some uniform connectivity requirements are still imposed, the agents may get disconnected and remain isolated for some period in a dynamic neighborhood topology. Given agents with nonlinear dynamics these issues need further attention and can be topics of further research.

The fully connected and ring neighborhood topologies satisfy the connectivity requirement for achieving consensus. The random neighborhood topology, on the other hand, may not satisfy the connectivity requirement. In other words, in the random topology it might be the case that there is no agent which is connected to all other agent in the multi-agent system. In any case, whether the connectivity requirement is satisfied or not is determined at the beginning of the simulation and remains unchanged throughout the run.

21.4 Controller Design

In this section we will present the development of the agreement controller. First of all note that when output agreement occurs, i.e., the condition (21.3) in Problem 21.1 is satisfied, the outputs of all agents satisfy

$$\lim_{t\to\infty} y_i(t) = \phi(t), 1 \leq i \leq N, \tag{21.5}$$

where the common output trajectory $\phi(t)$ is called synchronous or consensus trajectory [32, 33].

It was argued in [32, 33] that in order for the output synchronization (i.e., agreement) problem to be solvable, the agents need to have internal models which can generate the consensus trajectory $\phi(t)$ in (21.5). In other words, for some system of the form

$$\begin{aligned} \dot{s} &= g(s), \\ \phi &= q(s), \end{aligned} \tag{21.6}$$

whose output is $\phi(t)$ in (21.5), the agents need to have a copy (or an appropriate equivalent) of the system in (21.6) in their internal dynamics or local controllers in order to achieve consensus. If the consensus trajectory $\phi(t)$ is known one can define the output error for agent i as

$$\bar{e}_i(t) = y_i(t) - \phi(t) \tag{21.7}$$

and the problem reduces to individual output tracking problem. In other words, if $\phi(t)$ is known one can design the controllers $u_i(t)$ of the agents such that to track the $\phi(t)$ mapping in order to perform the desired behavior. However, since $\phi(t)$ is emergent (it depends on the initial conditions and the possibly time-varying interactions), requiring that the agents know it a priori is not realistic. Moreover, assuming that it is known by the agents effectively means introduction of a global variable and is in conflict with the properties of multi-agent dynamic systems. In particular, if $\phi(t)$ is known there is no need even for interagent interactions to achieve agreement. Therefore, since the agents do not know $\phi(t)$ and can obtain information from only their neighbors in the set N_i, the average output error for agent i with respect to its neighbors can be defined as

$$e_i(t) = \frac{1}{|N_i|} \sum_{j \in N_i} (y_i(t) - y_j(t)), \tag{21.8}$$

where $|N_i|$ denotes the number of agents in the set N_i. This is an unweighted average. It is also possible to define a weighted average in which the individual output errors $(y_i(t) - y_j(t))$ are scaled by weighting coefficients based on various means such as priority, reliability of information from the corresponding agents, and others.

Note also that the above average error does not require knowledge of absolute information. Instead, it can be calculated using only relative information/measurements. In other words, to calculate the error in (21.8) the agents need to know only the difference between their own output and the outputs of their neighbors and they do not need the exact values of neither their own output nor the outputs of their neighbors.

Note that although $\bar{e}_i(t)$ in (21.7) and $e_i(t)$ in (21.8) are different variables they are related. In particular, having $\bar{e}_i(t) = 0$ for all i implies $e_i(t) = 0$ for all i. The reverse, on the other hand, holds only when the interagent interaction graph is connected.[1] In other words, when the interagent interaction graph is connected $e_i(t) = 0$ for all i also implies $\bar{e}_i(t) = 0$ for all i. This implies that, given that the connectivity requirement is satisfied, one can design the local agent controllers $u_i(t)$ such that to drive the local errors $e_i(t)$ to zero in order to achieve output agreement.

Given the fact that it is a necessary condition to have local internal models but not realistic for the agents to know them or the consensus trajectory, one can introduce local internal models with unknown parameters and update their parameters until agreement is achieved. This is the approach we take here. We use a procedure from [30] to update the parameters of the local internal models. As inputs to the local internal models and in general to the agent (error feedback) controllers we use the local errors $e_i(t)$ in (21.8). In order to parameterize the internal model we need to assume some structure and parameterize the system in (21.6). In order to be able to approach the problem from the output regulation context we assume that (21.6) are linear of the form

$$\begin{aligned} \dot{s} &= G(\sigma)s \\ \phi &= Qs, \end{aligned} \tag{21.9}$$

where σ denotes the unknown parameters which belong to a bounded set of appropriate dimensions. We also assume that for all values of the unknown parameters σ the matrix $G(\sigma)$ has all its eigenvalues on the imaginary axis. Moreover, the eigenvalues are assumed to be simple excluding possibly one repeated eigenvalue at the origin. This type of system can model/generate wide range of signals including constant, periodic, and ramp signals or a combination of these. Stable eigenvalues are excluded since they lead to modes which decay to zero. Similarly, since the unstable eigenvalues lead to modes which can quickly diverge they are also outside of the scope of interest. From here one can deduce that it is expected that the agent outputs agree on an unknown trajectory which is possibly a combination of constant, ramp, and periodic (i.e., sinusoidal) signals. Note also that the local internal models (the local controllers) of the agents must also compensate for the local external disturbances in (21.2). Therefore, one can define the local unknown exosystems for the agents in the form

[1]Here we use the term connected loosely meaning that at least minimum connectivity conditions are satisfied and there are no agents or groups of agents which are disconnected from the rest of the multi-agent system. For example, as mentioned earlier for static unidirectional topology it is sufficient that the interaction topology has a spanning tree.

$$\dot{s}_i = g_i(\sigma_i)s_i = \begin{bmatrix} G(\sigma) & 0 \\ 0 & g_{\mu_i}(\eta_i) \end{bmatrix} s_i, \tag{21.10}$$

where $s_i = [s^\top, \mu_i^\top]^\top$ is the state of the local exosystem, and $\sigma_i = [\sigma^\top, \eta_i^\top]^\top$ represents its unknown parameters. The local controller of every agent will possess a parameterized adaptive local internal model which is updated based on the local error $e_i(t)$ in (21.8). The discussion below closely follows the controller development procedure in [30] with appropriate modifications and discussions wherever necessary.

The dynamic equations of the local error feedback controllers can be written in the form

$$\begin{aligned} \dot{\xi}_i &= v_i(\xi_i, e_i) \\ u_i &= \theta_i(\xi_i), 1 \leq i \leq N, \end{aligned} \tag{21.11}$$

where ξ_i represent the local controller states of appropriate dimensions. These control laws should be designed such that for all initial conditions in a neighborhood around the origin

$$\lim_{t\to\infty} e_i = 0, 1 \leq i \leq N \tag{21.12}$$

are satisfied for every σ_i. As mentioned earlier, provided that the connectivity requirement is satisfied, this will lead to satisfaction of (21.3) in Problem 21.1.

The problem is solvable if there exist

$$z_i = \pi_i(s_i, \omega_i) = \begin{bmatrix} \zeta_i(s_i, \omega_i) \\ \vartheta_i(s_i, \omega_i)) \end{bmatrix}$$

and

$$u_i = c_i(s_i, \omega_i)$$

with $\pi_i(0, \omega_i) = 0$ and $c_i(0, \omega_i) = 0, 1 \leq i \leq N$ mappings such that the equations

$$\begin{aligned} \vartheta_{i1} &= \phi \\ \vartheta_{ij} &= \frac{\partial \vartheta_{i(j-1)}}{\partial s_i} g_i(\sigma_i)s_i, j = 2, \ldots, r_i \\ \frac{\partial \vartheta_{ir_i}}{\partial s_i} g_i(\sigma_i)s_i &= f_{ir_i}(\zeta_i, \vartheta_i, s_i, \omega_i) + b_i(\omega_i)c_i \\ \frac{\partial \zeta_i}{\partial s_i} g_i(\sigma_i)s_i &= f_{i0}(\zeta_i, \vartheta_i, s_i, \omega_i) \end{aligned} \tag{21.13}$$

are satisfied. Note that in (21.13) the variables $\vartheta_{ij} = \vartheta_{ij}(s_i, \omega_i)$, $\phi = \phi(s_i, \omega_i)$, $\zeta_i = \zeta_i(s_i, \omega_i)$, and $c_i = c_i(s_i, \omega_i)$ are functions of the external inputs s_i and the unknown parameters ω_i. In this equation the first equality follows from the fact that when the error is zero we have $0 = y_i - \phi(s_i, \omega_i)$ and $x_{i1} = y_i$. In the above equation, the expression

$$\frac{\partial \vartheta_{i(j-1)}(s_i, \omega_i)}{\partial s_i} g_i(\sigma_i) s_i = L_{\dot{s}_i} \vartheta_{i(j-1)}(s_i, \omega_i)$$

represents the Lie derivative applied to the elements of the vector $\vartheta_{i(j-1)}(s_i, \omega_i)$. In this formulation in order for the problem to be solvable we assume that there exist mappings $\zeta_i(s_i, \omega_i)$, $1 \le i \le N$ such that the equation

$$\frac{\partial \zeta_i(s_i, \omega_i)}{\partial s_i} g_i(\sigma_i) s_i = f_{i0}(\zeta_i(s_i, \omega_i), \vartheta_i(s_i, \omega_i), s_i, \omega_i)$$

is satisfied for all i. In addition, we assume that for all i the equalities

$$\begin{aligned} L_{\dot{s}_i}^{\beta_i} c_i(s_i, \omega_i) &= a_{i0}(\sigma_i) c_i(s_i, \omega_i) + a_{i1}(\sigma_i) L_{\dot{s}_i} c_i(s_i, \omega_i) \\ &\quad + \cdots + a_{i(\beta_i - 1)}(\sigma_i) L_{\dot{s}_i}^{\beta_i - 1} c_i(s_i, \omega_i) \end{aligned} \tag{21.14}$$

are satisfied for corresponding integers β_i and real numbers $a_{i0}(\sigma_i)$, $a_{i1}(\sigma_i), \ldots,$ $a_{i(\beta_i-1)}(\sigma_i)$ such that for all σ_i the roots of the polynomial

$$p_i(\lambda) = \lambda^{\beta_i} + a_{i(\beta_i-1)}(\sigma_i)\lambda^{\beta_i-1} + \cdots + a_{i1}(\sigma_i)\lambda + a_{i0}(\sigma_i)$$

are simple roots on the imaginary axis. In the above equation, the expression

$$L_{\dot{s}_i}^k \phi(s_i, \omega_i) = \frac{\partial}{\partial s_i}(L_{\dot{s}_i}^{k-1} \phi(s_i, \omega_i)) g_i(\sigma_i) s_i$$

represents the k'th order Lie derivative for $k > 1$.

Under the above assumptions there exists solution of the equations in (21.13) for $z_i = \pi_i(s_i, \omega_i)$ and $u_i = c_i(s_i, \omega_i)$ with $\pi_i = \left[\zeta_i^{\mathsf{T}}, \vartheta_i^{\mathsf{T}}\right]^{\mathsf{T}}, \vartheta_i = \left[\phi^{\mathsf{T}}, L_{\dot{s}_i}\phi^{\mathsf{T}}, \cdots L_{\dot{s}_i}^{r_i-1}\phi^{\mathsf{T}}\right]^{\mathsf{T}}$ and

$$c_i = [b_i(\omega_i)]^{-1} \left(L_{\dot{s}_i}^{r_i} \phi(s_i, \omega_i) - f_{ir_i}(\zeta_i, \vartheta_i, s_i, \omega_i) \right). \tag{21.15}$$

Note that this vector controller linearizes and decouples the input–output channels and applies the corresponding regulating controller. However, it is not implementable since it contains unknown entities. Nevertheless, it can be estimated. In particular, the above assumptions imply the existence of

$$\tau_{\sigma_i}(s_i, \omega_i) = \begin{bmatrix} c_i(s_i, \omega_i) \\ L_{\dot{s}_i} c_i(s_i, \omega_i) \\ \cdots \\ L_{\dot{s}_i}^{\beta_i - 1} c_i(s_i, \omega_i) \end{bmatrix}$$

transformations which satisfy

$$\begin{aligned} L_{\dot{s}_i}\tau_{\sigma_i}(s_i,\omega_i) &= \Phi_i(\sigma_i)\tau_{\sigma_i}(s_i,\omega_i) \\ c_i(s_i,\omega_i) &= \Gamma_i\tau_{\sigma_i}(s_i,\omega_i) \end{aligned} \tag{21.16}$$

Using these transformations, the systems

$$\begin{aligned} \dot{s}_i &= g_i(\sigma_i)s_i, \\ \dot{\omega}_i &= 0, \\ u_i &= c_i(s_i,\omega_i),\, 1 \le i \le N \end{aligned} \tag{21.17}$$

which generate the controllers in (21.15) are immersed into linear and observable systems (internal models)

$$\begin{aligned} \dot{\xi}_i &= \Phi_i(\sigma_i)\xi_i, \\ u_i &= \Gamma_i\xi_i,\, 1 \le i \le N \end{aligned} \tag{21.18}$$

which can generate the same outputs as the systems in (21.17) (i.e., the controllers $c_i(s_i,\omega_i)$). The matrices $\Phi_i(\sigma_i)$ and Γ_i in this equation are in block diagonal form given by $\Phi_i(\sigma_i) = diag(\overline{\Phi}_i(\sigma_i),\ldots,\overline{\Phi}_i(\sigma_i))$ and $\Gamma_i = diag(\overline{\Gamma}_i,\ldots,\overline{\Gamma}_i)$ where $\overline{\Gamma}_i = \begin{bmatrix}1 & 0 & 0 & \cdots & 0\end{bmatrix}$ and $\overline{\Phi}_i(\sigma_i)$ are in controllable canonical form with the last row composed of the coefficients in (21.14) or basically $\{a_{i0}(\sigma_i), a_{i1}(\sigma_i), a_{i2}(\sigma_i), \cdots, a_{i(\beta_i-1)}(\sigma_i)\}$, respectively. Note that this is a completely decoupled system composed of m subsystems which can be treated independently. In addition, the parametrization of the internal model (21.18) has a very important role for the solution of the problem. Based on a result in [26] it was stated in [30] that the internal model $(\Phi_i(\sigma_i),\Gamma_i)$ in (21.18) can be parameterized as

$$M_\sigma\Phi(\sigma)M_\sigma^{-1} = F + G\Psi(\sigma), \quad \Psi(\sigma) = \Gamma M_\sigma^{-1}$$

for any Hurwitz matrix F and vector G such that (F,G) is controllable. Note that $\Phi(\sigma)$ and $F + G\Psi(\sigma)$ are similar matrices and have the same eigenvalues. Using this parametrization for each dimension j (i.e., for each input–output channel j) the internal models can be represented in a parameterized form with the equations

$$\begin{aligned} \dot{\xi}_{ij} &= (F_i + G_i\Psi_{ij}(\sigma_i))\xi_{ij}, \\ u_{ij} &= \Psi_{ij}(\sigma_i)\xi_{ij},\, 1 \le i \le N. \end{aligned} \tag{21.19}$$

As discussed above the values of the σ_i parameters are unknown. Therefore, for every agent $i = 1,\ldots,N$ and for each dimension $j = 1,\ldots,m$ the initial conditions $\Psi_{i0}(\sigma_i)$ of $\Psi_{ij}(\sigma_i)$ are obtained using nominal values and the actual $\Psi_{ij}(\sigma_i)$ are estimated. The estimation of the $\Psi_{ij}(\sigma_i)$ is represented with $\hat{\Psi}_{ij}$.

As discussed above the errors $e_i(t)$ are measurable. In case their derivatives are also known, following the strategy in [30], for the j'th input–output channel ($j = 1, \ldots, m$) the corresponding parameter vectors $\hat{\Psi}_{ij}$ can be updated by

$$\dot{\hat{\Psi}}_{ij} = -\gamma_i \left(e_{ij}^{(r_i-1)} + k_i^{r_i-1} b_{i0} e_{ij} + \cdots + k_i b_{i(r_i-2)} e_{ij}^{(r_i-2)} \right), \tag{21.20}$$

where $\gamma_i > 0$ and $k_i > 0$ are positive constants and the coefficients $b_{i0}, b_{i1}, \ldots, b_{i(r_i-2)}$ are chosen such that the polynomials

$$p_i(\lambda) = \lambda^{r_i-1} + b_{i(r_i-2)} \lambda^{r_i-2} + \cdots + b_{i1} \lambda + b_{i0}$$

are Hurwitz. Then defining

$$\theta_{ij} = e_{ij}^{(r_i-1)} + k_i^{r_i-1} b_{i0} e_{ij} + \cdots + k_i b_{i(r_i-2)} e_{ij}^{(r_i-2)}$$

as was shown in [30] for agent i and for each input–output channel $j = 1, \ldots, m$ the corresponding controller

$$\begin{aligned} \dot{\xi}_{ij} &= (F_i + G_i \hat{\Psi}_{ij}) \xi_i - K_i G_i \theta_{ij} \\ \dot{\hat{\Psi}}_{ij} &= -\gamma_i \theta_{ij} \\ u_{ij} &= \hat{\Psi}_{ij} \xi_{ij} - K_i \theta_{ij}, \end{aligned} \tag{21.21}$$

where the gains $K_i > 0$ are sufficiently large, can be used as local controller for regulating the output error to zero. In the formulation here, this controller leads to output agreement. As can be seen from Eq. (21.21) the control inputs $u_{ij}, j = 1, \ldots, m; i = 1, \ldots, N$ contain two components. The first components (i.e., $\hat{\Psi}_{ij} \xi_{ij}$ together with the dynamics of ξ_{ij} and the adaptation of $\hat{\Psi}_{ij}$) constitute the adaptive internal models, whereas the second components (i.e., $K_i \theta_{ij}$) are stabilizing controllers. The final controllers applied to the agents are $u_i = [u_{i1}, \ldots, u_{im}]$.

Although it is normal to know the output errors $e_i(t)$, in some applications we may not know their derivatives. In such a situation the controller given above cannot be implemented and needs to be modified. In case in which the derivatives of the error are unknown, they can be estimated using a high gain observer. This observer is a linear system with dynamic equation given by

$$\dot{\hat{x}}_{ij} = M_{gi} \hat{x}_{ij} + L_{gi} e_{ij}, \tag{21.22}$$

where the M_{gi} and L_{gi} matrices are of the form

$$M_{gi} = \begin{bmatrix} -\bar{g}_i\bar{c}_{i(r_i-1)} & 1 & 0 & \cdots & 0 \\ -\bar{g}_i^2\bar{c}_{i(r_i-2)} & 0 & 1 & \ddots & \vdots \\ \vdots & \vdots & \ddots & \ddots & 0 \\ -\bar{g}_i^{r_i-1}\bar{c}_{i1} & \vdots & & \ddots & 1 \\ -\bar{g}_i^{r_i}\bar{c}_{i0} & 0 & \cdots & \cdots & 0 \end{bmatrix}, L_{gi} = \begin{bmatrix} \bar{g}_i\bar{c}_{i(r_i-1)} \\ \bar{g}_i^2\bar{c}_{i(r_i-2)} \\ \vdots \\ \bar{g}_i^{r_i-1}\bar{c}_{i1} \\ \bar{g}_i^{r_i}\bar{c}_{i0}. \end{bmatrix}$$

For every agent i the $\bar{c}_{i0}, \bar{c}_{i1}, \ldots, \bar{c}_{i(r_i-1)}$ coefficients in its observer are chosen such that the corresponding polynomial

$$p_{ic}(\lambda) = \lambda^{r_i} + \bar{c}_{i(r_i-1)}\lambda^{r_i-1} + \cdots + \bar{c}_{i1}\lambda + \bar{c}_{i0}$$

is Hurwitz and $\bar{g}_i > 0$.

The states of the observer $\hat{x}_{ij}$ correspond to the estimates of the derivatives of the error e_{ij}. Therefore, the θ_{ij} variables in the above controller can be redefined as

$$\hat{\theta}_{ij} = \hat{x}_{ijr_i} + k_i^{r_i-1}b_{i0}\hat{x}_{ij1} + \cdots + k_i b_{i(r_i-2)}\hat{x}_{ij(r_i-1)},$$

where k_i and the coefficients $b_{i0}, b_{i1}, \ldots, b_{i(r_i-2)}$ are chosen as discussed above. Then for sufficiently large $K_i > 0$ the controllers defined as

$$\begin{aligned} \dot{\hat{x}}_{ij} &= M_{gi}\hat{x}_{ij} + L_{gi}e_{ij} \\ \dot{\xi}_{ij} &= (F_i + G_i\hat{\Psi}_{ij})\xi_{ij} - K_i G_i \mathrm{sat}(l_i, \hat{\theta}_{ij}) \\ \dot{\hat{\Psi}}_{ij} &= -\gamma_i \mathrm{sat}(l_i, \hat{\theta}_{ij}) \\ u_{ij} &= \hat{\Psi}_{ij}\xi_{ij} - K_i \mathrm{sat}(l_i, \hat{\theta}_{ij}) \end{aligned} \tag{21.23}$$

given in [30] can be used as local error feedback controllers to regulate the output errors to zero. In the setting in this study they lead to output agreement. The sat function used in (21.23) is a saturation function defined as

$$\mathrm{sat}(l, \alpha) = \begin{cases} \alpha & \text{if } |\alpha| \leq l \\ l\frac{\alpha}{|\alpha|} & \text{if } |\alpha| > l. \end{cases}$$

The high gain observer may lead to initial large approximation errors and therefore result in large control signals in the system, which, on the other hand, may lead to instabilities. This is the reason for using the saturation function since it limits the effect of the approximation errors and prevents the system from such instabilities.

21.5 Simulation Results

In this section, we present the simulation results obtained by applying the adaptive controller discussed in the preceding section. The system we consider consists of $N = 6$ unity-mass agents moving in $\mathbb{R}^2$ with motion dynamics

$$\begin{aligned} \dot{x}_i &= v_i \\ \dot{v}_i &= \mu_i + \omega_i u_i, \\ y_i &= x_i \end{aligned} \tag{21.24}$$

where $x_i \in \mathbb{R}^2$ is the position, v_i is the velocity ($z_i^\top = [x_i^\top, v_i^\top]$ is the state), $u_i \in \mathbb{R}^2$ is the control input of the i'th agent/robot, $\mu_i \in \mathbb{R}^2$ represents a periodic disturbance acting on the agents, and $\omega_i \in \mathbb{R}^{2\times 2}$ represents the coefficient matrix of unknown parameters which we choose as a diagonal matrix. The diagonal elements of ω_i have constant values which are randomly chosen from the set $\{0 < \omega_i < 5\}$ for each agent and the $\mu_i = [\mu_{i1}, \mu_{i2}]^\top$ disturbances are generated by neutrally stable systems of the form

$$\begin{bmatrix} \dot{\bar{\mu}}_{ij1} \\ \dot{\bar{\mu}}_{ij2} \end{bmatrix} = \begin{bmatrix} 0 & -\alpha_{ij} \\ \alpha_{ij} & 0 \end{bmatrix} \begin{bmatrix} \bar{\mu}_{ij1} \\ \bar{\mu}_{ij2} \end{bmatrix}$$
$$\mu_{i1} = \bar{\mu}_{ij1}$$

with random initial values within the interval $[3, 8]$ for $j = 1, 2$ and the actual values of the α_{ij} parameters are not known. The values of α_{ij} are also chosen randomly within the set $\{0 < \alpha_{ij} < 1\}$.

For each dimension $j = 1, 2$ of the agent dynamics we use a six-dimensional internal model which contains two pairs of eigenvalues on the imaginary axis and an eigenvalue at the origin with multiplicity two. One of the eigenvalue pairs on the imaginary axis is for the unknown disturbance μ_{ij}, whereas the remaining eigenvalues are for the unknown consensus trajectory. The nominal (initial) values of the eigenvalues and the off-diagonal coefficient of the repeated eigenvalue at the origin are chosen randomly such that one pair of the imaginary poles is within $[-j2, j2]$, another pair is within $[-j3, j3]$, and the off-diagonal coefficient related to the repeated eigenvalue is within $[0, 1]$. The F_i matrix is calculated such that the roots of its characteristic polynomial are at $\{-1, -2, -3, -4, -5, -6\}$. Based on this choice and the randomly chosen initial values the F_i, G_i, and $\Psi_{i0}(\sigma_i)$ matrices are calculated. They are typically in the form

$$\begin{bmatrix} -15.2814 & 8.6390 & 2.9589 & -1.0662 & 0.3143 & 0.1233 \\ -8.6390 & -1.4293 & -2.0711 & 0.5233 & -0.1638 & -0.0629 \\ 2.9589 & 2.0711 & -2.2021 & 1.7301 & -0.4145 & -0.1743 \\ 1.0662 & 0.5233 & -1.7301 & -1.5384 & 1.4055 & 0.3148 \\ 0.3143 & 0.1638 & -0.4145 & -1.4055 & -0.4010 & -0.5328 \\ -0.1233 & -0.0629 & 0.1743 & 0.3148 & 0.5328 & -0.1478 \end{bmatrix}$$

$$\left[-4.6374\ -0.8452\ 0.4827\ 0.1599\ 0.0479\ -0.0187\right]^{\top}$$

and

$$\left[-4.6374\ 0.8452\ 0.4827\ -0.1599\ 0.0479\ 0.0187\right].$$

Note that the entries of these matrices can be different for every agent i and for each dimension j. The entries also change from simulation to simulation based on the random initial parameters such that the eigenvalues of $F_i + G_i\Psi_{ij0}(\sigma_i)$ match the eigenvalues of $\Phi_i(\sigma_i)$ for the "nominal" values of the parameters σ_i (which are also chosen randomly).

We assume that only the output errors are measurable and use a high gain observer. The values of the $\bar{g}_i$ parameter of the high gain observer, and k_i, b_{i0}, K_i, γ_i, and l_i parameters of the controller are chosen as 100, 5, 0.5, 100, 1, and 10, respectively. The values of the $\bar{c}_{i0}$ and $\bar{c}_{i1}$ parameters of the high gain observer are chosen such that the roots of the $p_{ic}(\lambda)$ polynomials are $\{-12, -13\}$.

Simulations are performed for the ring, fully connected, and random neighborhood topologies.

Ring Neighborhood Topology: We consider the bidirectional ring neighborhood topology for which the output errors of the agents can be expressed as

$$\begin{aligned}
e_1 &= \tfrac{1}{2}\left((y_1 - y_N) + (y_1 - y_2)\right),\\
e_i &= \tfrac{1}{2}\left((y_i - y_{i-1}) + (y_i - y_{i+1})\right), i = 2, \ldots, N-1,\\
e_N &= \tfrac{1}{2}\left((y_N - y_{N-1}) + (y_N - y_1)\right).
\end{aligned}$$

As mentioned earlier, it is a static neighborhood and the neighbors of the agents do not change during motion. The agent interactions are reciprocal in this topology. Agents start their motion from random initial states with components within the set [0, 100]. Figure 21.1 shows example output trajectories for the agents. As can be seen from the figure the agent outputs agree in a short period of time and continue aggregated motion after that. The resultant trajectory varies from simulation to

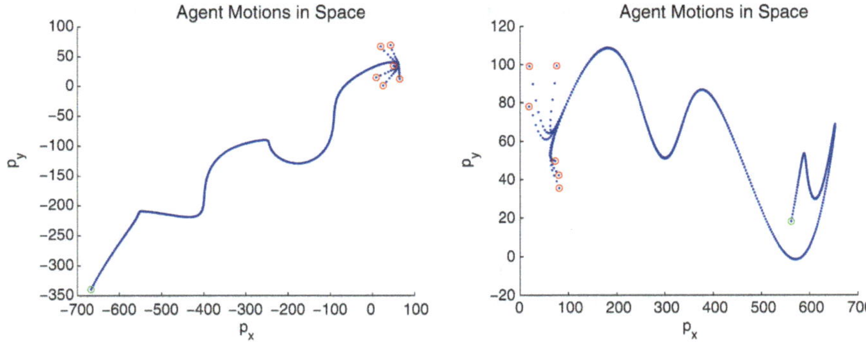

Fig. 21.1 Example output trajectories for bidirectional ring neighborhood topology

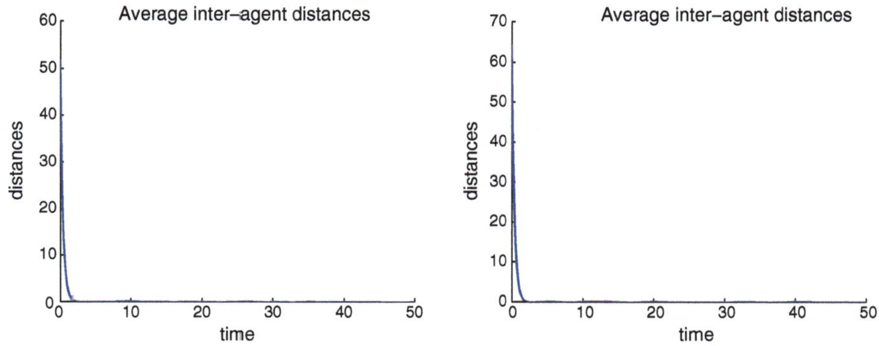

Fig. 21.2 Output errors for bidirectional ring neighborhood topology

simulation. It is emergent and depends on the initial states, initial and actual values of the unknown parameters, and the disturbances affecting the agent dynamics. The plots of the corresponding average interagent distances are shown in Fig. 21.2. One can observe from the figure that the average interagent distances quickly decay to the vicinity of zero and agreement is achieved.[2] Similar results are obtained for the unidirectional (non-reciprocal) ring neighborhood topology. The output errors in the unidirectional neighborhood topology can be expressed as

$$e_i = (y_i - y_{i+1}), i = 1, \dots, N-1,$$
$$e_N = (y_N - y_1).$$

The results obtained for the unidirectional neighborhood topology are in general qualitatively similar to the results obtained for the bidirectional topology and are not presented here. As mentioned earlier, the case when agents interact based on unidirectional ring topology is known as cyclic pursuit. For a relevant literature on the problem of cyclic pursuit one can consult for example [29] and the references therein.

The values of the updated parameters $\hat{\Psi}_{ij}$ (a six-dimensional vector for each $j = 1, 2$ or basically the x-coordinate and the y-coordinate) for one of the agents (agent 1) for the above two cases are plotted in Fig. 21.3. The plots on the left-hand side show the parameters on the x and y axes (i.e., for $j = 1$ and $j = 2$) for the first case, whereas the plots on the right-hand side show the parameters for the second case. Initial change in the parameter values is large. However, they are quickly updated and the rate of change decreases. The update continues until the error and its derivatives (or the states of the high gain observer) become zero.

Fully Connected Neighborhood Topology The fully connected neighborhood topology is also a static topology in which every agent is connected to every other agent in the group. The output error for this case for agent i can be expressed as

[2] At the time the simulation is stopped, the average interagent distance is very close to zero and continues to decrease.

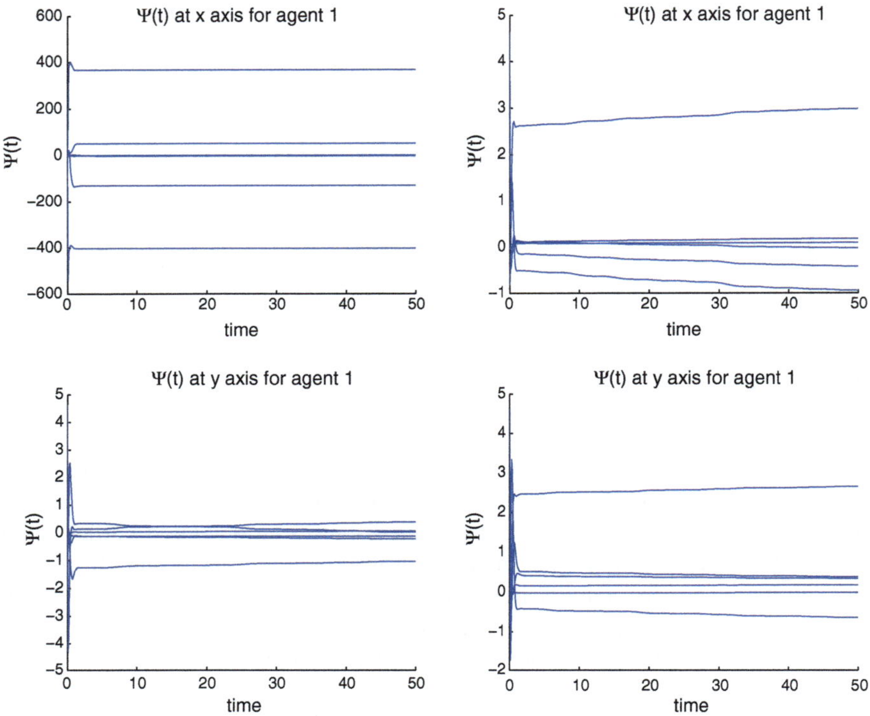

Fig. 21.3 Parameters for agent one for bidirectional ring neighborhood topology

$$e_i(t) = \frac{1}{N-1} \sum_{j=1, j \neq i}^{N} (y_i(t) - y_j(t)),$$

where N is the number of agents. We use the same values and procedures discussed above for determining the initial conditions and controller parameters.

Figure 21.4 shows example agent output trajectories for the case of the fully connected neighborhood topology. Both simulations shown in the figure last for 50 s. The results are in general qualitatively similar to the case of ring neighborhood topology. As mentioned, every simulation run results in emergent common output trajectory (due to the different initial conditions and disturbances). The possibilities of distinct agreement trajectories are infinite.

The plots of the average interagent distances and the $\hat{\Psi}_{ij}$ parameter values for the above cases are shown in Figs. 21.5 and 21.6, respectively. As expected the average interagent distances converge to zero in both cases and the parameters converge to relatively constant or slowly varying values (which are also expected to converge with time).

The nature of the agreement trajectory can be affected by the (size of the regions for the) initial conditions (both state and parameter) as well as the controller parame-

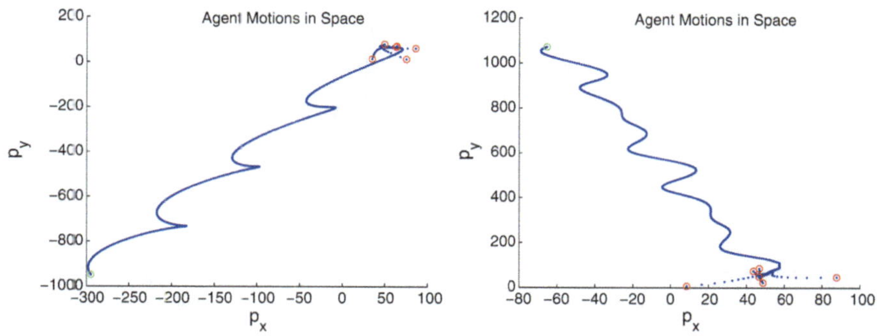

Fig. 21.4 Example output trajectories for fully connected neighborhood topology

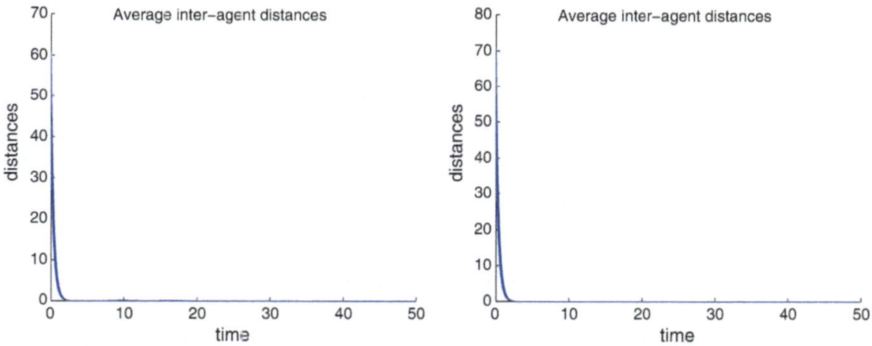

Fig. 21.5 Output errors for fully connected neighborhood topology

ters. Another important factor which affects the agreement trajectory is the external disturbances acting on the agent dynamics. Example plots of the disturbances acting on the agent dynamics for the above two cases (i.e., the simulations shown for the fully connected neighborhood topology) are shown in Fig. 21.7. Each of the plots in Fig. 21.7 shows the disturbance signals for all agents $i = 1, \dots, 6$ and for all dimensions $j = 1, 2$ (a total of 12 signals on each plot). In other words, different agents and different components of the state of the agents are affected by different unknown disturbances. However, agreement is still achieved despite these adverse effects.

Random Neighborhood Topology The last considered neighborhood is the random neighborhood topology in which the neighborhood relations between the agents are determined at the beginning of the simulation based on (21.4). They are kept constant throughout the simulation run. Note that since the neighborhood is random sometimes the resultant neighborhood graph may not satisfy the required connectivity property. We perform simulations for two cases. In the first case the necessary connectivity requirement is satisfied, whereas in the second case it is not.

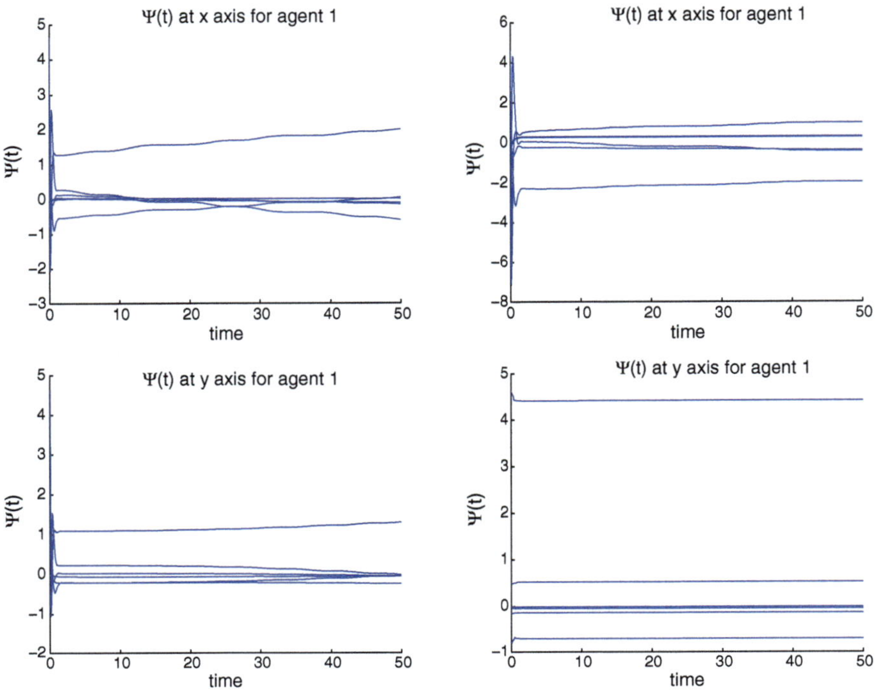

Fig. 21.6 Parameters for agent one for fully connected neighborhood topology

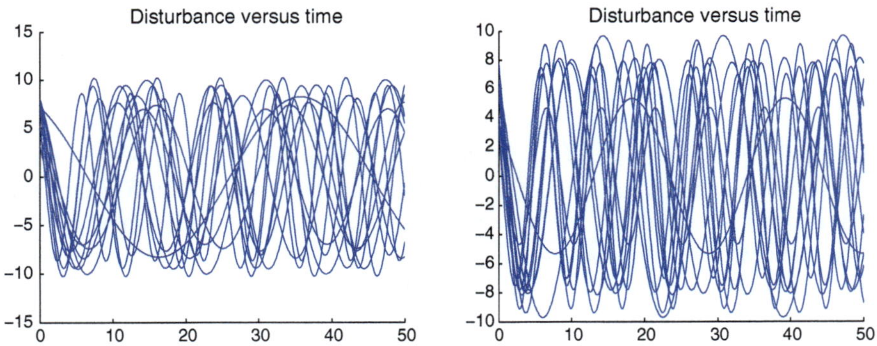

Fig. 21.7 Disturbances for fully connected neighborhood topology

For the first simulation, the value of the ε parameter in (21.4) is chosen as 0.5. This means that while the agent neighbors are assigned at the beginning of the simulation the chance of an agent to become a neighbor of another agent is 50 %. For the second simulation, on the other hand, we use $\varepsilon = 0.3$ which results in 30 % chance of an agent to become a neighbor of another agent. Figure 21.8 shows the agent trajectories and the output errors for these simulations. The plots on the left-hand side of

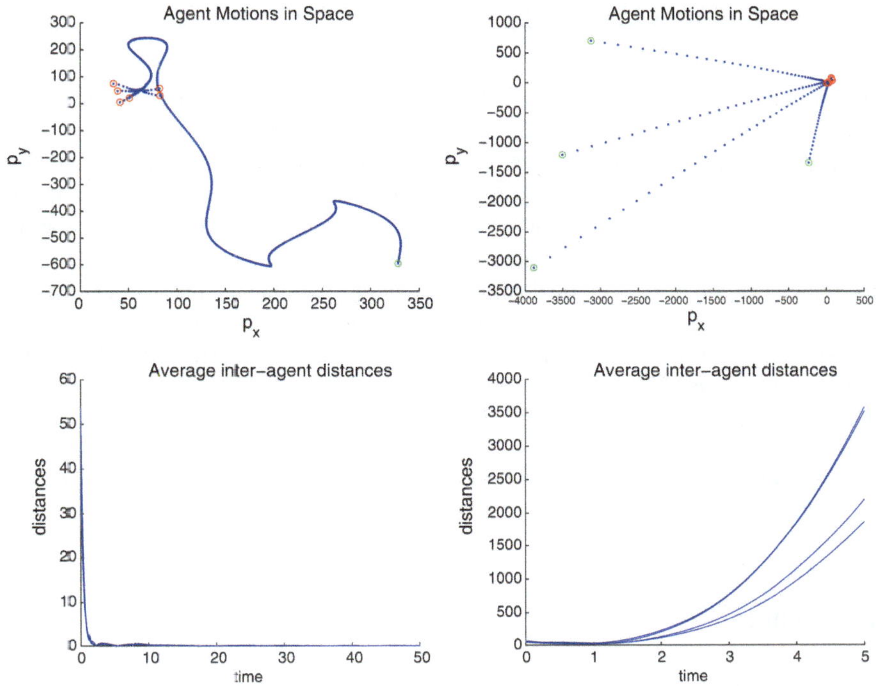

Fig. 21.8 Paths and output errors for random neighborhood topology

the figure correspond to the first simulation in which the connectivity requirement is satisfied, whereas those on the right-hand side correspond to the second simulation in which the connectivity requirement is not satisfied. As one can easily see from the figure the result for the first case is qualitatively similar to the results presented above for the other considered topologies. In particular, in the first simulation the agent outputs quickly converge to a common trajectory and the output errors converge to zero. In contrast, in the second simulation agent outputs diverge in different directions and the output errors continue to increase as time progresses and output agreement is not achieved. This result shows the importance of the connectivity for the information flow and therefore for achieving agreement in the multi-agent dynamic system. In particular, for the case of unidirectional topology if there is no spanning tree in the connectivity graph (implying that there are one or more agents which are not connected to the rest of the group and cannot obtain information from them), it is not possible to guarantee achieving of agreement. Therefore, for the random neighborhood topology with the neighborhood determination strategy in (21.4) one needs to choose larger value of ε in order to increase the probability of agent pairs becoming neighbors and therefore the probability of satisfaction of the connectivity requirement.

21.6 Concluding Remarks

In this study we considered the problem of distributed output agreement (output synchronization) in a class of multi-agent dynamic systems composed of agents with nonlinear dynamics with unknown parameters and external disturbances. The local disturbances are generated by neutrally stable exosystems with unknown parameters. We approach the problem within the context of output regulation using an adaptive internal model-based control strategy based on [30]. An adaptive internal model is introduced to counter effect the disturbances and to estimate the unknown consensus trajectory. The agents achieve agreement using only local information. A high gain observer is used to estimate the derivatives of the local output errors. Numerical simulations are performed in order to illustrate the performance of the controller. The obtained results verify the effectiveness of the proposed controller.

Acknowledgments This work was supported by the Scientific and Technological Research Council of Turkey (TÜBİTAK) under grant no 109E175.

References

1. Byrnes, C.I., Priscoli, F.D., Isidori, A., Kang, W.: Structurally stable output regulation of nonlinear systems. Automatica **33**(3), 369–385 (1997)
2. Chen, X., Serrani, A., Özbay, H.: Control of leader-follower formations of terrestrial uavs. In: Proceedings of the 42nd IEEE Conference on Decision and Control, pp. 498–503. Hawaii, USA, Dec 2003
3. Davison, E.J.: The robust control of a servomechanism problem for linear time-invariant multivariable systems. IEEE Trans. Autom. Control **21**(1), 25–34 (1976)
4. Francis, B.A.: The linear multivariable regulator problem. In: 1976 IEEE Conference on Decision and Control including the 15th Symposium on Adaptive Processes, vol. 15, pp. 873–878 (1976)
5. Francis, B.A., Wonham, W.M.: The internal model principle of control theory. Automatica **12**, 457–465 (1976)
6. Gazi, V., Fidan, B.: Coordination and control of multi-agent dynamic systems: models and approaches. In: Sahin, E., Spears, W.M., Winfield, A.F.T. (eds.) Proceedings of the SAB06 Workshop on Swarm Robotics, pp. 71–102. Lecture Notes in Computer Science (LNCS) vol. 4433, Springer, Berlin (2007)
7. Gazi, V.: Formation control of a multi-agent system using nonlinear servomechanism. Int. J. Control 78(8), 554–565 (2005)
8. Gazi, V.: Output regulation of a class of linear systems with switched exosystems. Int. J. Control **80**(10), 1665–1675 (2007)
9. Gazi, V.: Stability of a discrete-time asynchronous swarm with time-dependent communication links. IEEE Trans. Syst. Man Cybern: Part B **38**(1), 267–274 (2008)
10. Gazi, V., Passino, K.M.: Decentralized output regulation of a class of nonlinear systems. Int. J. Control **79**(12), 1512–1522 (2006)
11. Gazi, V., Passino, K.M.: Swarm Stability and Optimization. Springer, Heidelberg (2011)
12. Gül, E., Gazi, V.: Adaptive internal model based distributed output agreement in a class of multi-agent dynamic systems. In: Kolemisevska-Gugulovska, T.D., Stankovski, M.J. (eds.) Proceedings of the Special International Conference on Complex Systems: Synergy of Control, Computing, and Communications (COSY 2011), pp. 215–222. The Society for Electronics,

Telecommunications, Automation, and Informatics of Macedonia, Skopje, MK, Ohrid, Macedonia, Sept 2011
13. Gül, E., Gazi, V.: Adaptive internal model based formation control of a class of multi-agent systems with switched exosystems. In: Chinese Control and Decision Conference, pp. 6–13. Taiyuan, China, May 2012
14. Gül, E., Gazi, V.: Adaptive internal model based formation control of a class of multi-agent systems. In: American Control Conference, pp. 4800–4805. Baltimore, MD, USA, June–July 2010
15. Huang, J., Lin, C.: On a robust nonlinear servomechanism problem. IEEE Trans. Autom. Control **39**(7), 1510–1513 (1994)
16. Huang, J., Rugh, W.J.: On a nonlinear multivariable servomechanism problem. Automatica **26**(6), 963–972 (1990)
17. Isidori, A., Byrnes, C.I.: Output regulation of nonlinear systems. IEEE Trans. Autom. Control **35**(2), 131–140 (1990)
18. Jadbabaie, A., Lin, J., Morse, A.S.: Coordination of groups of mobile autonomous agents using nearest neighbor rules. IEEE Trans. Autom. Control **48**(6), 988–1001 (2003)
19. Khalil, H.K.: Robust servomechanism output feedback controllers for feedback linearizable systems. Automatica **30**(10), 1587–1599 (1994)
20. Kim, H., Shim, H., Seo, J.H.: Output consensus of heterogeneous uncertain linear multi-agent systems. IEEE Trans. Autom. Control **56**(1), 200–206 (2011)
21. Lin, Z., Francis, B., Maggiore, M.: Necessary and sufficient graphial conditions for formation control of unicycles. IEEE Trans. Autom. Control **50**(1), 121–127 (2005)
22. Mahmoud, N.A., Khalil, H.K.: Asymptotic regulation of minimum phase nonlinear systems using output feedback. IEEE Trans. Autom. Control **40**(10), 1402–1412 (1996)
23. Mahmoud, N.A., Khalil, H.K.: Robust control for a nonlinear servomechanism problem. Int. J. Control **66**(6), 779–802 (1997)
24. Marino, R., Tomei, P.: Output regulation for linear systems via adaptive internal model. IEEE Trans. Autom. Control **48**(12), 2199–2202 (2003)
25. Moreau, L.: Stability of multiagent systems with time-dependent communication links. IEEE Tran. Autom. Control **50**(2), 169–182 (2005)
26. Nikiforov, V.O.: Adaptive nonlinear tracking with complete compensation of unknown disturbances. Euro. J. Control **4**, 132–139 (1998)
27. Olfati-Saber, R., Murray, R.M.: Consensus problems in networks of agents with switching topology and time-delays. IEEE Trans. Autom. Control **49**(9), 1520–1533 (2004)
28. Ren, W., Beard, R.W.: Consensus seeking in multi-agent systems under dynamically changing interaction topologies. IEEE Trans. Autom. Control **50**(5), 655–661 (2005)
29. Şamiloğlu, A.T., Gazi, V., Koku, A.B.: Asynchronous cyclic pursuit. In: S. Nolfi (ed.) Proceedings of 9'th Conference on Simulation of Adaptive Behavior (SAB06), pp. 667–678. Lecture Notes in Artificial Intelligence (LNAI) 4095, Springer, Berlin (2006)
30. Serrani, A., Isidori, A., Marconi, L.: Semiglobal nonlinear output regulation with adaptive internal model. IEEE Trans. Autom. Control **46**(8), 1178–1194 (2001)
31. Serrani, A., Isidori, A.: Global robust output regulation for a class of nonlinear systems. Syst./Control Lett. **39**, 133–139 (1999)
32. Wieland, P., Allgöwer, F.: An internal model principle for consensus in heterogeneous linear multi-agent systems. In: 1st IFAC Workshop on Estimation and Control in Networked Systems, pp. 7–12. Venice, Italy, Sept 2009
33. Wieland, P., Allgöwer, F.: An internal model principle for synchronization. In: IEEE International Conference on Control and Automation (ICCA), pp. 285–290. Christchurch, New Zealand, Dec 2009
34. Wieland, P., Sepulchre, R., Allgöwer, F.: An internal model principle is necessary and sufficient for linear output synchronization. Automatica **47**(5), 1068–1074 (2011)
35. Xiang, J., Wei, W., Li, Y.: Synchronized output regulation of linear networked systems. IEEE Trans. Autom. Control **54**(6), 1336–1341 (2009)
36. Ye, X., Huang, J.: Decentralized adaptive output regulation for a class of large-scale nonlinear systems. IEEE Tran. Autom. Control **48**(2), 276–281 (2003)

Chapter 22
An Example of Fault Detection and Reconfiguration-Based Tolerance Within Distributed Embedded Control Systems

Matjaž Colnarič, Domen Verber and Matej Šprogar

Abstract This chapter introduces some devised solutions for fault detection within embedded control systems. These are a follow-on to the successful IST FW5 project IFATIS at the Laboratory for Real-Time Systems of the Faculty of Electrical Engineering and Computer Science. The topic will first be elaborated on and the overall results of the original project are presented. Then in continuation some later enhancements and improvements will be shown together with original implementations of specific parts, e.g. discrete field-programmable gate array (FPGA)- and programmable system on chip (PSoC)-based fault monitoring cells.

22.1 Introduction

Embedded computer control systems are the 'invisible minds' behind most modern industrial and other technical appliances. Their major spread can be observed even within critical safety environments, where failures can have serious or even fatal consequences. However, highly dependable programmable embedded systems for critical safety applications still lack proper scientific treatment. In addition to being dependable controllers must be flexible in order to cut production costs. Flexibility, which is achieved mainly by programmability is, however, in conflict with dependability because it requires a large number of components and features that

M. Colnarič (✉) · D. Verber · M. Šprogar
Faculty of Electrical Engineering and Computer Science,
University of Maribor, Smetanova 17, 2000, Maribor, Slovenia
e-mail: colnaric@uni-mb.si

D. Verber
e-mail: domen.verber@uni-mb.si

M. Šprogar
e-mail: matej.sprogar@uni-mb.si

G.M. Dimirovski (ed.), *Complex Systems*, Studies in Systems, Decision and Control 55, DOI 10.1007/978-3-319-28860-4_22

increase their complexities and thus decrease their abilities for predicting their safe functioning.

Faults within programmable control systems are unavoidable. Fault management as a discipline embraces four types of approaches [11]:

(a) F*ault avoidance* prevents faults during the design phase;
(b) F*ault removal* attempts to find faults before the system enters service (testing);
(c) F*ault detection* finds faults during system service and minimises their effects, and
(d) F*ault tolerance* allows the system to operate correctly even in the presence of faults.

A number of excellent elaborations of this domain can be found, for example [2, 9] and others.

Failure within a system can be handled, for example by redundancy, diversity, reconfiguration, etc. First, however, it must be detected. In regard to detection some sort of dependable monitoring subsystem must be used that detects abnormalities and triggers appropriate corrective actions. It is because of the complexities it necessarily introduces to the system that safety-related issues for the system should be designed, evaluated and implemented independently, and in parallel with the functional part within which it is finally integrated.

Fault detection, isolation and tolerance were the significant points of interest during the 3-year EC IST 5th Framework Program project IFATIS (Intelligent Fault-Tolerant Control in Integrated Systems), which started in 2002 and was concluded in 2005. Its major goals were to establish a framework for exploring intelligent fault-tolerant control technology, to devise a novel methodology, and to develop a corresponding software tool.

Based on the outcome and knowledge obtained during IFATIS, our work has continued towards enhancing and improving the techniques and measures. In this chapter we first provide a brief overview of the IFATIS project, together with some suggestions for improving the original architecture. In the second part we introduce a monitoring cell as a non-intrusive fault detecting device. Finally, two alternative implementations are compared in the discussion section.

This chapter is improved and partly enhanced version of the contribution to the COSY 2011 Special International Conference on Complex Systems, which was held in Ohrid, Republic of Macedonia, on September 16–20, 2011 in honour of Prof. Georgi M. Dimirovski [5].

22.2 Brief IFATIS Project Overview

Failures within the systems are usually dealt with by redundancy and diversity [1, 11]. The most obvious model is n-modular redundancy, where the system as a whole is replicated *n*-times (e.g. TMR—Triple Modular Redundancy). This model is used where very high dependability is required (i.e., control systems in avionics).

Nevertheless, such a scenario is often too expensive to be used in less demanding cases.

Instead of n-modular redundancy, the main idea of the IFATIS project was based on reconfiguration. In this case, the system normally works with full configuration in the absence of failures. In the cases of failures by any components, the remaining parts of the system are checked and the best possible configuration is sought, consisting of a sound subset of the parts. The performance of the system may be degraded but there is a chance that the system may survive the failure.

The architecture supporting the idea is hierarchically designed (for the details please refer to [10]. It is based on a logical structure called the Fault-Tolerant Control Cell (FTC) that controls a partial process. FTCs have access to process inputs and outputs, perform a control function on them and communicate with each other, see Fig. 22.1. In the data path from the process inputs via the control function to the outputs, there are a number of integrity checks and other measures enhancing the fault tolerance such as functionality checking, consistency checking, checking pairs, information redundancy, loop-back testing, watchdog timers, etc. Their detailed descriptions can be found in [11].

An important part of a system is its Function Monitor for supervising the integrity of the data and the functioning of the control function. If any abnormality is detected the Local Reconfiguration and Mode Control Block (LRMC) is activated in order to find an alternative solution to the function that has failed. Alternative or redundant resources and/or functions are employed if available or the system's performance is degraded gracefully in order to survive the situation.

If this is impossible, the problem is reported to the next hierarchically higher level, the Group Resource Reconfiguration Manager (GRRM), which is in charge of finding a solution from amongst the FTCs within the group. If that also fails, at the highest hierarchical level there is the Global Resource Reconfiguration Manager

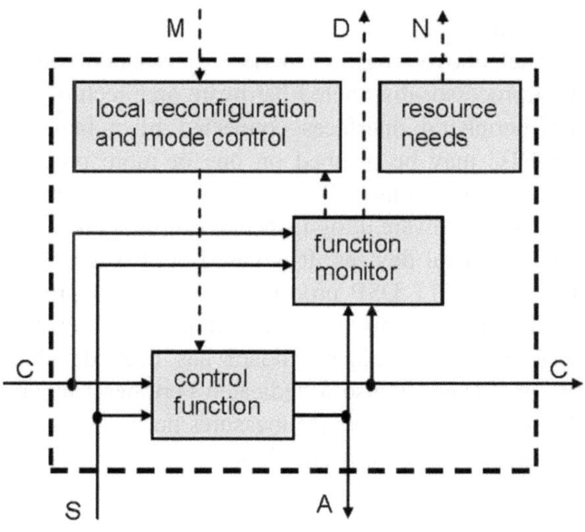

Fig. 22.1 Fault-tolerant control cell FTC. For a description of the signals, please see Fig. 22.2

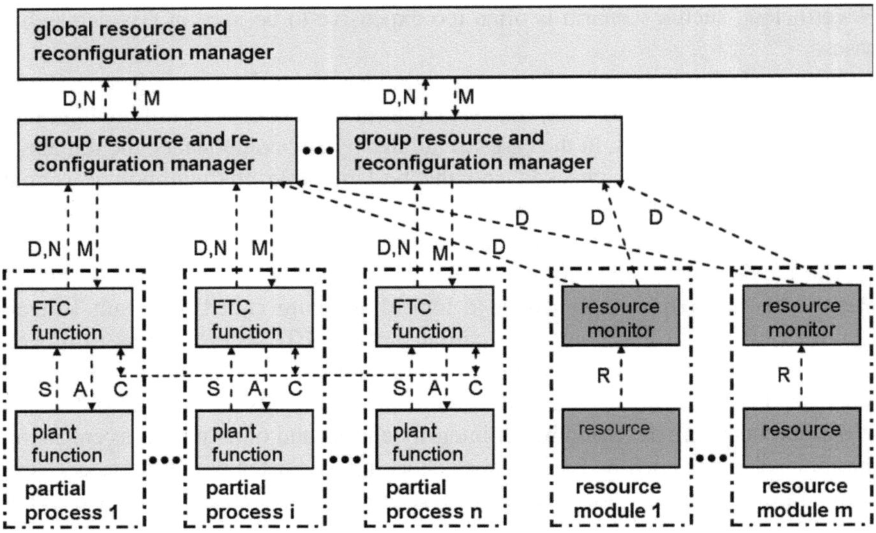

- A actuator information (mode dependent)
- C cross-communication between FTC functions
- D diagnosis results
- M mode decisions and resource allocations
- N resource needs and urgencies for all possible modes
- R resource-specific information
- S sensor information (mode dependent)
- FTC fault-tolerant control (function, implemented by software, running on controller modules)

Fig. 22.2 Fault detection architecture of the IFATIS approach

that tries to reconfigure the whole system. The architecture of the IFATIS model is given in Fig. 22.2.

The above logical model is mapped onto an essentially symmetrical multiprocessor-distributed hardware architecture consisting of multiple processors and peripheral interfaces connected by a fault-tolerant communication system. An FTC may be mapped on one or more processors or it may share part of a processor with other FTCs, as shown in the example in Fig. 22.3.

In the hardware design, special care was taken when selecting proper processing and peripheral devices. It is based on Texas C6711DSK processor boards with a TMS320C6711 DSP processor. In order to provide for the necessary low-level functionality (communication, enhanced fault tolerance, modularity, increased performance) a daughter board has been developed using XILINX Spartan II FPGAs. There is also a dedicated peripheral module implemented for supporting the specific fault tolerance measures during peripheral data transfer. A photo of the prototype is shown in Fig. 22.4.

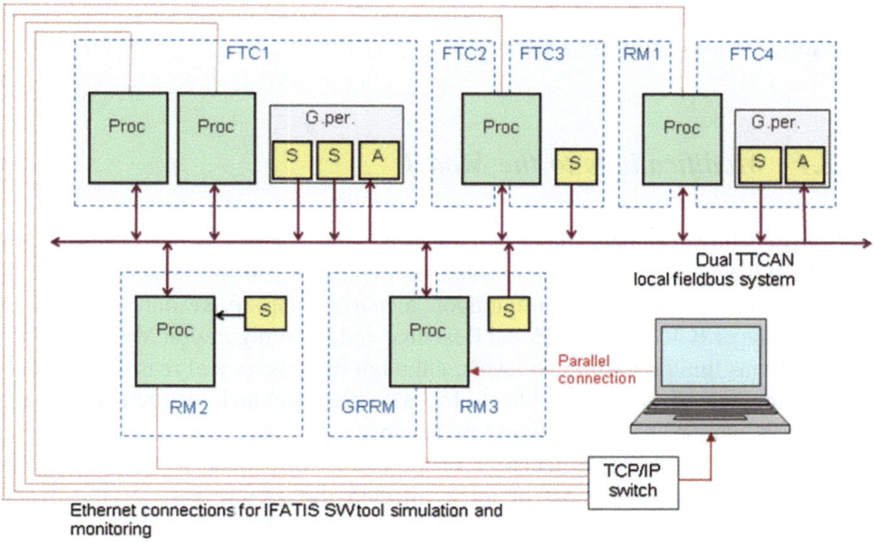

Fig. 22.3 Mapping of FTCs onto the architecture

Fig. 22.4 Implementation of the IFATIS architecture

Although it was not followed in the laboratory prototype for simplicity reasons, in order to prevent common mode failures the redundant resources should be designed diversely, thus offering different levels of service quality. In general, simpler systems should provide higher levels of integrity. In regard to ultimate safety, simple automata (even falling into the SIL4 [7]) could also be available as a last resort for providing fail-safe behaviour in those cases when the system encounters a non-recoverable failures.

22.2.1 *Modifications to the Model*

The architecture has been modified following the successful conclusion of the IFATIS project and based on its global concept. First, in the original model the hierarchy of higher level reconfiguration managers (Group Resource Reconfiguration Manager (GRRM) and Global Resource and Reconfiguration Manager), from Fig. 22.2 was limited to the two levels, although there is no real reason for this. In the new implementation we decided to leave it open. Any architecture from a single to arbitrary number of levels should be feasible.

Next, the software implementation of the blocks was found to be complex and resource-consuming and most of all it affected the performance of the overall control application. It was for this reason that we considered the possibility of moving the monitoring functions to autonomous, custom-designed modules, the functions of which could be generalised or even learned from the behaviour of the control system.

Further elaboration of the proposed model in more detail is given in [4].

22.3 Fault Detection by Monitoring Cells

In order to decouple fault detection from the control functions, a special component called a *monitoring cell* (MC) has been introduced; its function being to supervise the control function. This section of the chapter introduces the MC concept for embedded control systems. It will be shown how it can be implemented within hardware using either discrete components with FPGA or the PSoC.

In comparison to the IFATIS model of the fault-tolerant control cell (Fig. 22.1), the monitoring cell can be considered as that part of the function monitor that looks at the relationship between the sensory data (S) and the actuation signals (A). The function monitor itself is more application-specific and may include more knowledge about the expected behaviour of the control function.

The task of the MC is to detect those run-time faults that are difficult to discover because they are a consequence of an unpredictable event or chain of events. One way to detect them is to observe whether the system's states are within reasonable limits at all times. In order to recognise what is 'normal' we propose observing the system during normal operations by recording all inputs, outputs and internal states that are believed to affect a system's future behaviour. Based on these recordings a machine-learning technique can be used to 'learn' about the normal behaviour. Evolutionary algorithms (EAs), for example are one suitable paradigm for this task; as an early reference please consider [6].

The causes of faults within such systems are either hardware or software-related; and additionally the input and output signals may not comply with the functional specifications of the system thus also causing faulty behaviour. In hard real time a system's improper temporal behaviour is also considered as a fault. Another less obvious reason for errors is the temporal inconsistencies of the signals; e.g. the signals' changes can be too fast, the frequency of events can be too high, etc.

Figure 22.5 shows how the control function is monitored by the monitoring cell. The outputs $\boldsymbol{y}$ and the internal states $\boldsymbol{m}$ at the instant $t + 1$ are determined by the inputs x and internal states at the instant t.

$$(y(t+1), m(t+1)) = F(x(t), m(t)); \quad t \in \mathrm{N}.$$

The basic task of a monitoring cell is to monitor the validities of the input and output values and (possibly) the internal states of the control process. As a result, the correctness $\boldsymbol{c}$ is reported to a higher fault management layer that will handle the detected fault. If feasible, an additional diagnostics parameter d can also be provided.

The control cell under surveillance must have physically accessible input and output signals. Since it can implement any (sub) process it is important to identify those control functions with clear and explicit relationships between the input and output signals. The monitored control cell is considered a *grey* box with defined external functional behaviour and with at least a partly known internal structure that is observable through its obtainable internal states. There are several reasons for this decision:

- The monitored component is not necessarily a black box and this knowledge can reveal additional and more accurate information about any faulty behaviour.
- The white box (similar to the one chosen in IFATIS model) can be too complex. If chosen, it is necessary to implement the monitor cell physically inside the original software and hardware in order to limit any communication problems. This, however, is very intrusive and would increase the complexity and reduce the performance of the control part, which would be counterproductive.

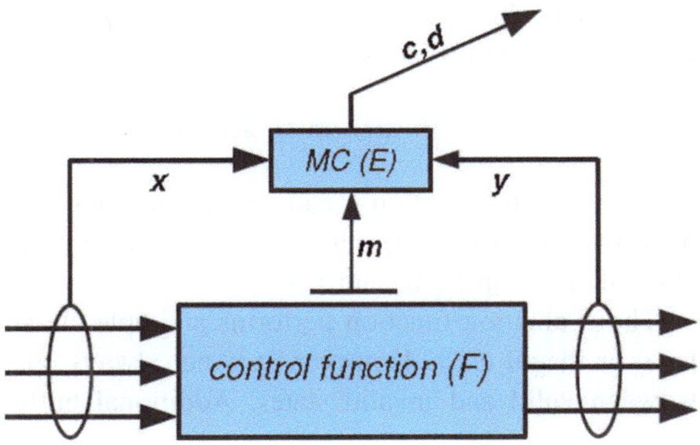

Fig. 22.5 Concept of the monitoring cell

- White box implementation on the same resources would introduce another central point of failure.
- Clear separation between the control and monitoring functions simplifies the design; both functions can be carried out separately and by different designers with diverse competences, thus enhancing dependability to some extent.

If the control function is a component with well-defined behaviour but unknown internal structure, it can also be taken as a black box. We propose physically separating the control part and the monitoring cell, i.e., to employ separate hardware. Whilst it may be more expensive, it provides much more competent implementation of the above guidelines. Also, complexity is kept lower by partitioning of the functions.

In order to allow for the grey box implementation the input and output digital and analogue signals need to be observable, as well as the internal states. There are a number of possibilities for the latter: either they are made accessible via standard parallel or serial interfaces at the control cell, or another feature of contemporary processors is made use of, such as JTAG boundary scan testing and in-system programming [8] or a concept similar to the 'background debugging mode' high-speed clocked serial debugger interface of the CPU32 Motorola family.

The monitoring cell should be in regard to order of magnitude less complex than the control cell and should have as little interference as possible with the control environment. It should be built from simple and robust components with low probabilities of failure. The consequence could be that the complexities of monitoring functions may be limited (e.g. no floating-point arithmetic, etc.). This limitation, however, could be advantageous because the simplicity such a system would allow for formal verification and possible certification by a certification authority.

When an abnormality is detected by a monitoring cell, the diagnostic mechanism will attempt to acquire as many details as possible. An error or a failure, together with a possible description, will be coded within the diagnostic signal and allotted to the upper layers of the fault management system where appropriate actions will be taken by, e.g. the reconfiguration manager. This, however, is beyond the scope of this chapter.

The monitoring cell operates in a PLC fashion: it collects all the signals from the process' inputs and outputs, as well as any accessible system states. It then elaborates the evaluation function E (here only the general outlook will be given; for a detailed description please refer to [13]:

$$E(\mathbf{x}(t), \mathbf{m}(t), \mathbf{x}(t+1), \mathbf{m}(t+1), \mathbf{y}(t+1)).$$

Note that the function E can operate on both new and old instances of $\boldsymbol{x}$ and $\boldsymbol{m}$, thus allowing for early detection of any discrepancies regarding the inputs and internal states at the instant $t + 1$.

The evaluation function performs a simple classification of input signals into a legal or illegal class. However, it is not always possible to find a sharp boundary between valid and invalid states. Additional buffer zones should be introduced where the validity of the signals cannot be determined. In addition, to simplify the analysis and implementation, a system under observation can be analysed at its subsystem's level. The partial evaluation function E_i, which evaluates the *i*th subsystem, determines the correctness $\boldsymbol{c}_i$ of the subset signal s_i. Let $\boldsymbol{s}_i$ represent the inputs $\boldsymbol{x}$, $\boldsymbol{m}$, $\boldsymbol{y}$ at instants t and $t + 1$. Function E_i is then evaluated to express correctness c_i for the signals $\boldsymbol{s}_i$:

$$c_i = E_i\left(s_i\right),$$

$$c_i \in C = \{\text{valid, invalid, undetermined}\}.$$

This enables a simple validation of each signal. For example, any invalid input value can be detected (possibly from a faulty sensor) and/or illegal output can be detected and, consequently, isolated.

In [13], the space of the possible signals $\boldsymbol{s}$ is roughly divided by orthogonal hypercubes, areas in which input signals $\boldsymbol{x}$, $\boldsymbol{m}$, $\boldsymbol{y}$ result in valid, invalid and undetermined states. As an illustration, please consider the next Fig. 22.6.

The round-shaped grey area represents the space regarding the instances of correct signals. This space is embraced by the three hypercubes H_1, H_2 and H_3. The MC classifies signals according to their positions: signals inside any hypercube are *valid*, signals outside *invalid*. The hatched area represents the model's error space, where the signals are considered valid, although they should be recognised as invalid. The *undetermined* tag can be given to signals either because they are (i) close to the hypercube's borders or, if feasible, (ii) have inconsistent classifications by two or more independent hyperspace models used simultaneously.

In simpler deterministic cases, the hypercubes can be drawn in advance at the design time. If the behaviour of the system is not fully known, however, some automated process of hypercube creation must be applied. In both cases the machine-learning techniques can be used. In any case, the task of drawing bounding hypercubes is a multi-objective optimisation task [3] with two conflicting goals: high accuracy and high generalisation. In addition, the more hypercubes employed

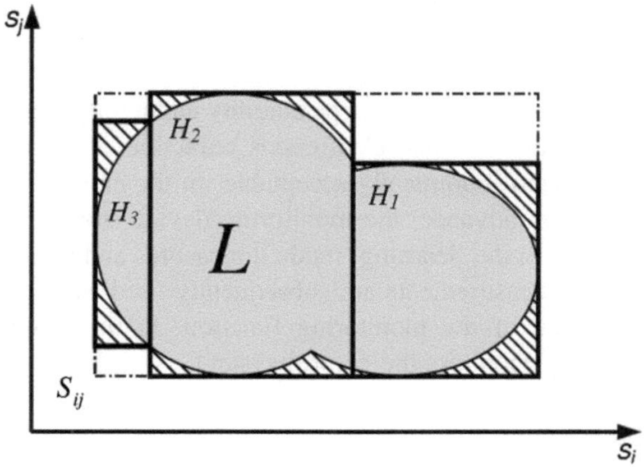

Fig. 22.6 A possible partitioning of the signal space

the more resources are needed for their implementation. Again, the details are given in [13].

22.4 Implementations of the MC

The MC can be implemented in several ways. If the original application code is available and can be modified, the MC can be implemented as a set of monitoring routines that run together with the control software on the same processing resources, as it was in the case of the IFATIS model. After the input is acquired, the pre-evaluation routine is called. It evaluates the individual signals, quantifies them and stores them for later assessment. Similarly, post-evaluation routine is called before output is produced. Additionally, it also checks for proper mappings between inputs and outputs.

The next example shows how these routines can be integrated into the code. Each input, output and other parts of the algorithm are associated with a constant that identifies them.

```
functon PerformTask:
AcquireInputs(i1,i2)
if not MCEvaluateInputInt(INPUT1_ID,i1) then
  HandleException
if not MCEvaluateInputFloat(INPUT2_ID,i2) then
  HandleException
MCSetDeadline(FUNCTION1_ID)
Compute(i1,i2,o1)
if not MCEvaluateOutputInt(OUTPUT1_ID,o1) then
  HandleException
if not MCEvaluateResults(FUNCTION1_ID) then
  HandleException
ProduceOutputs(o1)
```

The main disadvantage of this approach is that modifying the original code necessarily distorts the temporal characteristic of the system. Also, changing of the original application is not always possible. Therefore, because of this, this approach was unconsidered for further research.

A higher degree of dependability and agility can be achieved by using dedicated hardware solutions. Continuous reduction of prices for the hardware makes this approach economically acceptable. In the case that the evaluation function cannot be set in advance, the monitoring device should run in two different operational modes. In the 'learning' mode it measures and records the input and output signals. These measurements are subsequently used as a learning base for the offline construction of the monitoring function. In the 'monitoring' mode of operation it actually monitors the control system.

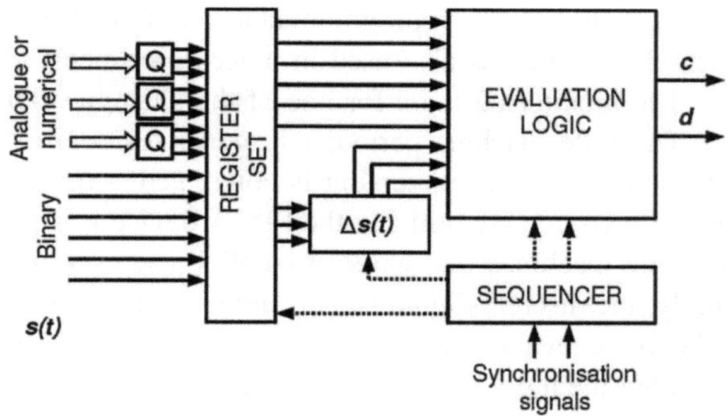

Fig. 22.7 Implementation of the monitoring cell

The conceptual diagram of the MC monitoring hardware implementation is shown in Fig. 22.7. Signals from the control system ***s*** are connected to a set of registers that implement the two-stage FIFO buffers. They hold their current value ***s***(***t***) and the previous values ***s***(***t*** − 1). The analogue and numeric values are processed by the quantisation blocks (labelled Q in Fig. 22.7). Each quantisation block transforms the input into a corresponding discrete value in order to simplify the evaluation. Signals ***s*** may also contain some internal states ***m*** of the control system. In order to observe the dynamics of the signal, the block $\Delta s_i(t)$ compares new and previous values and calculates the difference by producing additional information for the evaluation. The operation of the MC's components is synchronised using a simple sequencer according to those external synchronisation signals that indicate the beginnings of each execution cycle. The outputs c and *d* are generated by evaluation logic based on the rules set by the function *E*.

22.4.1 Implementation with Discrete Components

In the first case study, a solution was performed using discrete analogue devices, an FPGA unit, and a simple microcontroller. 12 bit A/D converters ADS7841 with serial communication interfaces were used for the A/D conversion. The FPGA was

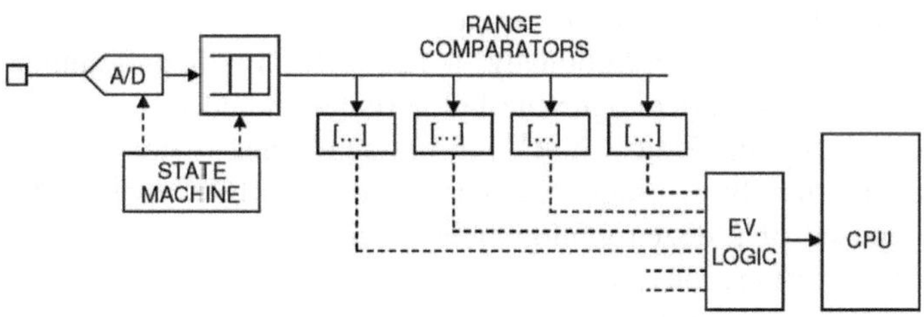

Fig. 22.8 Discrete monitoring logic for a single analogue signal

Spartan-IIe with 3072 programmable slices and 8 KB of dedicated memory blocks. This hardware was also used in other experiments [12].

The monitoring logic for one of the analogue signals is shown in Fig. 22.8.

The state machine periodically generates control signals that trigger A/D conversion and, after conversion is completed, stores the result within the two-stage FIFO registers. By utilising the FPGA device it is possible to implement a number of state machines that work in parallel, each serving a single A/D interface. Similarly, the discrete digital signals are acquired periodically by other state machines. The basic evaluation is performed by a set of range comparators, each testing whether the input value is within the predetermined range. The results from this and other evaluation channels are then combined with simple Boolean evaluation logic, which is implemented as a truth table inside the dedicated memory blocks of the FPGA—the status signals are interpreted as a memory address containing the appropriate output.

The output $\boldsymbol{c}$ states whether all the signals are within the valid ranges; whilst output $\boldsymbol{d}$ is a vector designating the validities of individual inputs. All the constants for the range comparators and the contents of the memory blocks are generated offline and may only be changed by full re-initialisation of the FPGA device. Approximately 150 slices are used for implementing the evaluation logic for a single evaluation channel with four range comparators. A simple 8-bit microcontroller is used for communication with the fault management system. It is also needed for initialisation of the MC at start-up, for initial data acquisition, and later for the debugging and diagnostic. It is possible to implement the microcontroller within the FPGA; however, some functionality (e.g. enhanced debugging) is lost.

The evaluation function $\boldsymbol{E}$, based on the hypercubes, is in the form of truth tables and range pairs loaded into the memory blocks that compose the evaluation logic. Unfortunately, the data consume a lot of sparse on-chip memory that limits the applicability of this approach.

22.4.2 Implementation Using Programmable SoC

The former solution requires separate analogue circuits for analogue signal manipulation. Nowadays, very compact and low-cost technology is available with programmable system on chip. These chips integrate a microcontroller and a number of configurable blocks of analogue and digital logic, as well as programmable interconnects. Additionally, a fast CPU, Flash programme memory, SRAM data memory and configurable I/O ports are included.

In the case study, the PSoC CY8C29466 Mixed Signal Array is used (PSoC is a trademark for a family of programmable SoC devices by Cypres). The analogue part is composed of a dozen configurable blocks, each allowing the creation of complex analogue functions such as A/D and D/A converters, comparators, filters, amplifiers, etc. More complex functions can be implemented by combining several primitive cells. The digital part is composed of several digital blocks. Each block is

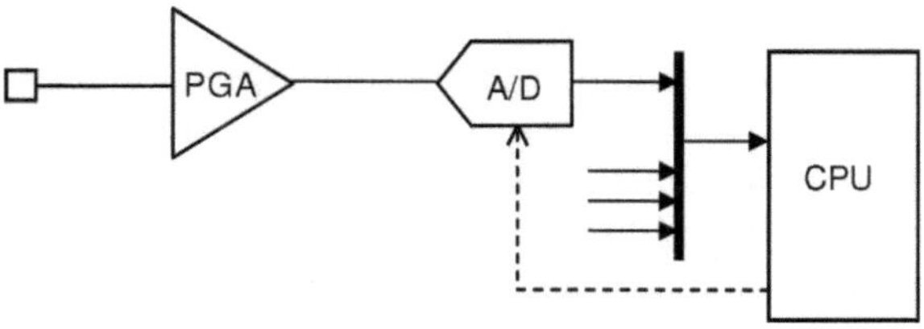

Fig. 22.9 Logical organisation of a single analogue acquisition channel with PSoC

an 8-bit resource that can be used alone or combined with other blocks to form 8, 16, 24 and 32-bit peripherals. The capabilities of these blocks are greater than their counterparts in a typical FPGA device and may be configured as counters and timers, PWMs, different serial communication interfaces, etc. The CPU has full control over the configuration of the analogue and digital blocks.

A conceptual diagram of the MC evaluation logic with PSoC is shown in Fig. 22.9. Each channel consists of a Programmable Gain Amplifier (PGA) and a 6-bit Successive Approximation Register (SAR) A/D converter. The PGA allows for adaptation to different signal levels. In this way low-voltage signals can be measured with a lower loss of precision. An off-chip voltage divider is required for measuring signals greater than 5 V. The hardware provides four analogue data acquisition channels that can be expanded to eight by using the also provided two-way analogue multiplexers. The digital part of the device allows for up to 16 bits of discrete data acquisition. Although it is possible to implement buffering and preliminary signal evaluation with digital cells on this device these features were implemented in software because of various limitations regarding the digital part.

The sequencing and the evaluation logics are executed by the microprocessor. It is implemented as a series of tests that check the inclusion of variables in the hypercubes. The hypercube parameters are loaded into the device during the initialisation phase.

22.4.3 Discussion About Different Implementations of MC

The solution using discrete analogue devices is somewhat more robust and adaptable to various situations. It can use different kinds of A/D converters for accommodating various kinds of signals. In addition, the external A/D converters are usually much less sensitive to voltage overloads. Another important benefit of using this approach is its speed because multiple monitoring channels and evaluation logic can be constructed within the same device. The sampling, quantisation and evaluation of different signals occur in parallel. If the evaluation logic is simple enough, the execution cycle time is within a range of several microseconds. This is much shorter than the time needed for A/D conversion. Therefore, it is possible to evaluate the signals from one execution cycle whilst acquiring the signals from the next one. The drawback of this solution is its inflexibility (it has to be done for each application), its limited scalability to more complex evaluation functions and its price. The estimated price for the parts used in the experiments is more than 30 € and increases with each additional A/D converter. In contrast the programmable SoC solution can cost (for up to four analogue channels) less than 5 €.

The solution with programmable SoC devices is much more compact and requires less supporting components than the previous one. However, apart from the A/D conversion, all the data processing is performed by the microprocessor. This impacts on the execution time because all evaluations must be done sequentially. For example, for a configuration where two analogue signals were observed and evaluated within four regions (as in Fig. 22.7), the execution time of the evaluation was 83 μs (using a 24 MHz system clock). In contrast, a typical A/D conversion takes only 25 μs. Therefore the evaluation cannot be performed in parallel with the conversion. If more dimensions are needed, the execution time increases accordingly. This is in contrast to the first approach where an additional variable has almost no impact on the execution time due to the parallel nature of the execution.

The case studied use only simple evaluation functions, although more complex regions than the hypercubes could be used, possibly also requiring higher mathematical operations such as multiplications. Modern FPGA and PSoC devices are capable of implementing instant multiplications by means of dedicated multiplication circuits. However, only integer or fixed-point arithmetic is feasible.

22.5 Future Work

Although both implementations of the MC were successfully done, the integration of (possibly only one of) them still needs to be employed. Based on this, it will be tested as to whether the approach is practically feasible for a typical control application. In the meantime, much faster and more capable programmable hardware devices have become available and they will be used for the experiments. Furthermore, the design of the system, its configuration and composition of the function $\boldsymbol{E}$ should be automated and included within the application design process.

There are some other possible usages for the MCs. When mapping functions between inputs and outputs are simple enough, MC can be implemented before the actual process is built, thus emulating its behaviour. Consequently, it could substitute for the process or serve as an additional test bed.

MC could also be used as a last resort for fault tolerance in situations of primary process failure. Based on the simplified knowledge when mapping inputs onto outputs, it could be expected to generate rough output results.

Acknowledgments The authors would like to thank the technical staff of the Real-Time Systems Laboratory of the Faculty of Electrical Engineering and Computer Science, University of Maribor, Slovenia, for implementing the prototypes.

References

1. Bowen, J.P., Hinchey, M.G.: High-Integrity System Specification and Design. Springer, Berlin, New York (1999)
2. Chiang, L., Russell, E., Braatz, R.D.: Fault Detection and Diagnosis in Industrial Systems Springer, Berlin, New York (2001)
3. Coello, C.A.C.: A comprehensive survey of evolutionary-based multiobjective optimization techniques. Int. J. Knowl. Inf. Syst. **3**(1), 169–308 (1999)
4. Colnarič, M., Verber, D., Halang, W.A.: Distributed embedded control systems: improving dependability with coherent design. In: Advances in Industrial Control. Springer, Berlin (2008)
5. Colnarič, M.: An example of fault detection and reconfiguration-based tolerance in distributed embedded control systems. In: Kolemisevska-Gugulovska, T.D., Stankovski, M.J. (eds.) Special International Conference on Complex Systems: Synergy of Control, Communications and Computing—Proceedings of COSY 2011 Papers, 16–20 Sept 2011, Ohrid, Macedonian Society for Electronics, Telecommunications, Automation, and Informatics, Skopje, pp. 281–286 (2011)
6. Holland, J.: Adaptation in Natural and Artificial Systems. The University of Michigan Press, Ann Arbor (1975)
7. IEC 61508: Functional safety of electrical/electronic programmable electronic systems. Generic aspects. Part 1: General Requirements. International Electrotechnical Commission, Geneva (1992)
8. IEEE: Test Access Port & Boundary Scan Architecture. IEEE, New York (1990)
9. Isermann, R.: Fault Diagnosis Systems: An Introduction from Fault Detection to Fault Tolerance. Springer, Berlin, New York (2005)
10. Maier, U., Colnarič, M.: Some basic ideas for intelligent fault tolerant control systems design. In: Proceedings of 15th IFAC World Congress. Barcelona, Spain (2002)
11. Storey, N.: Safety-Critical Computer Systems. Addison-Wesley, Longman (1996)
12. Verber, D., Łent, B., Halang, W.A.: Firmware support for disjunctive dataflow driven distributed control applications. In: Bradač, Z., Zezulka, F., Polanski, M., Jirsik, V. (eds.) Proceedings of IFAC Workshop on Programmable Devices and Embedded Systems PDeS, pp. 84—89. IFAC, Brno (2006)
13. Verber, D., Šprogar, M., Colnarič, M.: Implementation of non-intrusive fault detection in embedded control systems. Inf. MIDEM **37**(1), 23–30 (2007)

Chapter 23
Predictive Control of Complex Industrial Thermal Processes

Goran Stojanovski and Mile Stankovski

Abstract This paper is oriented toward presenting the advanced predictive control methods in thermal processes which have been developed and/or practically implemented at the ASE Institute in FEEIT Skopje. The thermal processes usually have high fuel consumption and therefore the optimization of the fuel costs is extremely important. With reducing of these costs we dramatically reduce the costs of the final product. In such direction, the advanced control methods for thermal processes usually lean toward faster response and increased robustness, in this case, using predictive techniques.

23.1 Introduction

Control and supervision as well as identification and simulation modeling of hybrid complex systems have been a subject of extensive research for a longer period. The overall control of heat treatment plants, that involves both regulation and task-oriented controls, is a typical problem of this kind, therefore very attractive and interesting to the scientist.

The real-world dynamical processes can be described as a non-separable interplay of the three fundamental natural quantities of energy, matter, and information. The processes in which energy and matter are also simultaneously carriers of information, which is governing the former ones, largely coincide with the dynamics and the feasible thermodynamics equilibrium of the thermal systems [1, 2].

G. Stojanovski (✉) · M. Stankovski
Department of Automatics and System Engineering, Faculty of Electrical Engineering and Information Technology, SS Cyril and Methodius University Skopje, Ruger Boskovic bb, P.O. Box 574, 1000 Skopje, Macedonia
e-mail: goranst@feit.ukim.edu.mk

M. Stankovski
e-mail: milestk@feit.ukim.edu.mk

G.M. Dimirovski (ed.), *Complex Systems*, Studies in Systems, Decision and Control 55, DOI 10.1007/978-3-319-28860-4_23

In heat treatment plants, the accuracy of temperature control has a major impact on the quality of the product and the used energy [3–7]. One of the early reviews of control practice, [8], showed there have been surprisingly few attempts to use advanced types of control. This is due, in part, to the difficulty of obtaining representative models of furnaces [3, 4, 9–12]. Moreover, on the grounds of the results in [13–15] and of an innovated deeper understanding, [9], now it can be argued that a good and well-validated identified process model could lead to a rather poor controller design which may cause serious stability difficulties at the real-world plant [14]. Thus, facing the issue of matching criteria for control and identification [16] the necessity of iterative identification and control redesign is inevitable [13, 14].

It should be noted that this research has been instrumental to derive model predictive controllers for high consumption industrial furnaces of different types as an addition to the standard decoupling and Multiple Input Multiple Output (MIMO) controllers for multivariable processes. These controllers based on a well-identified model aim to increase the productivity of the plant.

23.2 System Identification of the Industrial Furnace in FZC 11-Oktomvri

Thermal systems such as high-power, multi-zone furnaces usually consist of very complex processes of energy conversion and transfer processes which seem to be ideally suited for new designs of improved control and supervision strategies.

Typical steel mill slab pusher furnaces as in Skopje Steelworks [1, 2, 9] have high installed power and are operated with combined gas/oil fuel in a heavy-duty field environment. Also, they do require considerable maintenance support. In order to control the thermal processes properly, we must make proper process identification. In this paper we have used the MIMO model derived with identification of the furnace in the factory "FZC 11 Oktomvri" defined in [10].

This model was derived with structural and parameter identification using step response and Pseudorandom Binary Sequence (PRBS) techniques. For the dynamical heating regulation, the furnace process is represented by its 3×3 system model (transfer function for the heating zones only, and not the equalizing maintenance zone); the family of 3×3 models are used throughout this paper all having nine transfer paths. In this section we will explain only the key points from the process of the system identification and we will present the complete matrix of transfer functions.

23.2.1 *Steady-State Gain Matrix*

Steady-state gain matrix is given as in (23.1) where values are in [°C/MJ/min]. Here, we see a symmetrical matrix K, which is also clear from recorded input–output characteristics; for, these results out of the furnace construction. Taking into consideration furnace load changes, i.e., heating pipes/tubes with different dimensions, the steady-state gain values change is inverse proportional. That is, the bigger is the pipe mass the smaller is the gain, because for the same temperature much larger heat quantity is needed, that require much more combustion fuel, e.g. bigger fuel flow. On the other hand, smaller mass pipes require lower heat quantity, and steady-state gain gets bigger. For this reason, direct path steady-state gains are bigger for one-third than those in the first-neighbor paths, and bigger for one-tenth from gains in the direct paths in the non-neighbor paths. Note that these gain values demonstrate that interactions between transfer paths are very strong to be neglected, and do independent process control design for each zone separately.

$$K = \begin{bmatrix} 12 & 8 & 1.25 \\ 8 & 12 & 8 \\ 1.25 & 8 & 12 \end{bmatrix}. \tag{23.1}$$

23.2.2 Time-Delay Matrix

Taking into consideration everything what has been given above, time delays can be approximated with matrix with time delays values which are not average values between maximal and minimal time delay for different loads, but values nearer to the maximal ones. This is for reasons that time delays presence in systems are much unfavorable case for control. On the grounds of recorded step response transients with thermocouples and the stochastic identification experiments, a family set of approximate values for path time delays has been found. A good set is given (23.2) where matrix values are expressed in minutes.

$$\tau = \begin{bmatrix} 1.2 & 2.5 & 5 \\ 2.5 & 1.2 & 2.5 \\ 5 & 2.5 & 1.2 \end{bmatrix}. \tag{23.2}$$

Dynamic of system transfer paths is aperiodic. However via model identification using PRBS signal, and then data processed through SYSTEMS IDENTIFICATION TOOLBOX, we concluded that the best results give the set of second-order models

$$G_{ij}(s) = K_{ij} \exp(-\tau_{dij} s) / (T_1 s + 1)(T_2 + 1), \tag{23.3}$$

where matrix of process gains is given by

$$K = \Delta\Theta / \Delta Q. \tag{23.4}$$

By estimating the transfer function, we concluded that all transfer paths have almost the same dynamics; this is obvious from poles within the unit circle. For this reason, and for the fact that the model has to correspond for a family of pipe dimensions, we take all path transfer dynamics be given by an average of models for maximal and minimal loading. Therefore transfer function matrix for MIMO system gas-fired furnace system is given by

$$G(S) = \begin{bmatrix} \frac{12e^{-1.2s}}{(6.22s+1)(0.7s+1)} & \frac{8e^{-2.5s}}{(6.22s+1)(0.7s+1)} & \frac{1.25e^{-5s}}{(6.22s+1)(0.7s+1)} \\ \frac{8e^{-2.5s}}{(6.22s+1)(0.7s+1)} & \frac{12e^{-1.2s}}{(6.22s+1)(0.7s+1)} & \frac{8e^{-2.5s}}{(6.22s+1)(0.7s+1)} \\ \frac{1.25e^{-5s}}{(6.22s+1)(0.7s+1)} & \frac{8e^{-2.5s}}{(6.22s+1)(0.7s+1)} & \frac{12e^{-1.2s}}{(6.22s+1)(0.7s+1)} \end{bmatrix}. \tag{23.5}$$

This transfer function matrix (Eq. 23.5) is a result of identification of gas-fired furnace system in operation, and thus used as plant representation in appropriate control design.

23.3 Predictive Control of Industrial Thermal Processes

23.3.1 Predictive Control

Process industries need an easy to setup predictive controller that costs low, and maintains an adaptive behavior which accounts for time-varying dynamics as well as potential plant mismodeling. Accounting these requirements, the *Model Predictive Control (MPC)* has evolved into one of the most popular techniques for control of complex processes.

As presented in [17] the essence of model-based predictive control (MBPC or MPC) lies in optimization of the future process behavior with respect to the future values of the executive (or manipulated) process variables. The use of linear, nonlinear, hybrid, and time-delay models in MPC is motivated by the drive to improve the quality of the prediction of inputs and outputs [1, 2, 17, 18].

23.3.2 Switched Predictive Control

For control, the process is approximated with p linear affine models (23.6) that build a hybrid PWA state space model, as presented in [19]

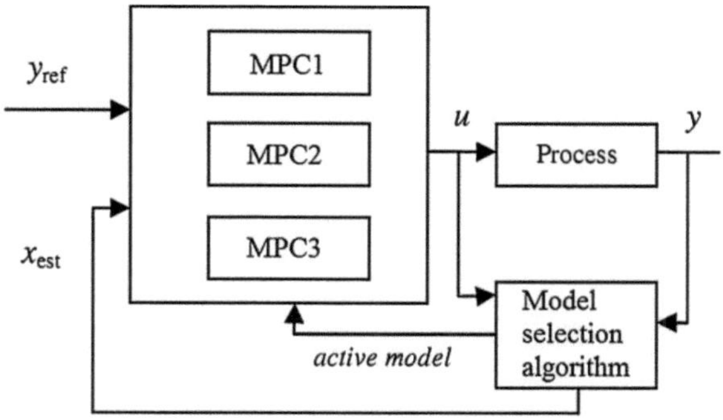

Fig. 23.1 Multiple model predictive control scheme

$$\begin{aligned} x(k+1) &= A^i x(k) + B^i u(k) + f^i \\ y(k) &= C^i x(k) + D^i u(k) + g^i \end{aligned} \quad \text{if} \begin{bmatrix} x(k) \\ u(k) \end{bmatrix} \in P^i, i \in \{1, \ldots, p\}, \tag{23.6}$$

where k is the discrete time index, A^i, B^i, C^i, D^i state space matrices, f^i, g^i the affine vectors, $u \in R^m$ input, $x \in R^n$ state, and P^i valid region of the state + input space in R^{m+n}. The system is subject to input and state constraints. For each region P^i a model exists and for it the corresponding mp-MPC controller is designed. The active model is determined by model selection algorithm from estimated state values. Each step time, the active controller computes the control signal. The control scheme of such controller is presented in Fig. 23.1.

This MPCs use linear models of the nonlinear system to predict the future behavior. The models are linearized around different working points of the plant.

The model selection algorithm is the most important part of the multiple model MPC. Usually, it is a function depending on the inputs and outputs of the system which results with appropriate model of the system. In more complicated systems Kalman filter is used to estimate the system states, and afterward the algorithm selects the appropriate model. The stability of the proposed controller will depend mainly on the process of switching with this algorithm.

23.3.3 Predictive Control with Genetic Algorithms Optimization

The genetic algorithm for optimization of MPC was proposed by [20, 21]. Since then there are few more applications papers and algorithm improvements like the ones presented in [22, 23]. Nevertheless the potential for implementing GA-MPC applications is widely unused.

Model predictive control algorithms are usually implemented on models with linear or fixed constraints of the process and control variables. Although sufficient

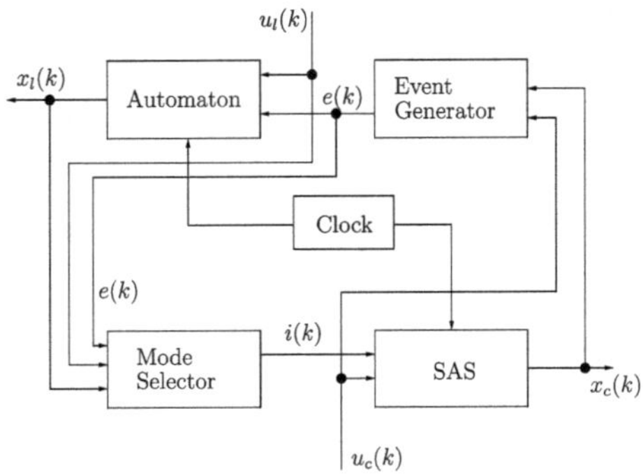

Fig. 23.2 Discrete time hybrid automata

for most of the controllers and processes, in some particular cases, nonstandard constraints cannot be neglected. In this work we present an easy to go method for incorporating nonstandard constraints in model predictive controllers using genetic algorithm optimization.

23.3.4 Hybrid Model Predictive Control

A hybrid system denotes in general a system composed of two unlike components. A hybrid control system is a control system with both continuous and binary/integer signals. Such a system generates a mixture of continuous and discrete signals, which take values in a continuum (such as the real numbers **R**) and a finite set (such as *a*, *b*, *c*), respectively.

Here we present a hybrid model for a high consumption industrial furnace that should represent the real plant more accurately. On the basis of this model, we will design controller(s) that will lead to increasing of the control system performance.

The basic idea behind the discrete hybrid automaton which is essential part of the model predictive control systems is presented in Fig. 23.2.

23.3.5 Discrete Time Hybrid Model

In order to implement hybrid control we must introduce the hybrid model based on the furnace dynamics. The hybrid model of the furnace was previously defined in [24]. The model for the temperature is discrete and it is represented with Eqs. (23.7)–(23.10).

$$T_i[k+1] = T_{out} - (T_i[k] - T_{out})\{0.5/(T_{max} - T_{out}) - \alpha_i\} + 0.05(T_i[k-1] - T_{out}) + \theta_i F_i(u)(T_{max} - T_i[k])/\in, \quad (23.7)$$

$$\begin{aligned} \alpha_1 &= 0.945 - hc_F - hc_B/5 \\ \alpha_2 &= 0.945 - hc_F/3 - hc_B/3 \\ \alpha_3 &= 0.945 - hc_F/5 - hc_B, \end{aligned} \quad (23.8)$$

$$\begin{aligned} F_1(u) &= n_{S1} \cdot U1_{k-3} + n_{S2} \cdot U1_{k-4} + n_{F1} \cdot U2_{k-4} \\ &\quad + n_{F2} \cdot U2_{k-5} + n_{D1} \cdot U3_{k-6} + n_{D2} \cdot U3_{k-7} \\ F_2(u) &= n_{S1} \cdot U2_{k-3} + n_{S2} \cdot U2_{k-4} + n_{F1} \cdot U1_{k-4} \\ &\quad + n_{F2} \cdot U1_{k-5} + n_{F1} \cdot U3_{k-4} + n_{F2} \cdot U3_{k-5} \\ F_3(u) &= n_{S1} \cdot U3_{k-3} + n_{S2} \cdot U3_{k-4} + n_{F1} \cdot U2_{k-4} \\ &\quad + n_{F2} \cdot U2_{k-5} + n_{D1} \cdot U1_{k-6} + n_{D2} \cdot U1_{k-7}, \end{aligned} \quad (23.9)$$

$$\theta_i = \begin{cases} 1 & \text{if pipe}_i = 0 \\ 0.95 & \text{if pipe}_i = 1, \end{cases} \quad (23.10)$$

$$\begin{aligned} hc_F &= \begin{cases} 0 & \text{if Frontdoor is closed``} = 0\text{"} \\ 0.005 & \text{if Frontdoor is open``} = 1\text{"} \end{cases} \\ hc_B &= \begin{cases} 0 & \text{if Backdoor is closed``} = 0\text{"} \\ 0.005 & \text{if Backdoor is open``} = 1\text{"}, \end{cases} \end{aligned} \quad (23.11)$$

where $n_{S1} = 1.195$; $n_{S2} = 0.6232$; $n_{F1} = 0.07968$; $n_{F2} = 0.04155$; $n_{D1} = 0.01245$; $n_{D2} = 0.006492$. T_{max} represents the maximum temperature that can be achieved in this furnace and is equal to 1300. Signals hc_F, hc_B (23.11), and θ_i (23.10) are logic signals that can change their value according to the process dynamics and represent disturbances of the system. The outdoor temperature around the furnace T_{out} is continuous state disturbance to this system.

23.4 Simulation Results

23.4.1 Model Predictive Control

In order to discuss the quality of the MPC algorithm we will compare the results with standard industrial PID decoupling control as we have described in [25]. The results for the temperature in the first furnace zone are presented on Fig. 23.3. The solid (red) line represents the control with decoupling PID, and the intercepted (blue) line represents the results obtained with MPC. It is obvious that MPC drives the system faster to the steady state, and they sustain the direct output disturbances in proper manner.

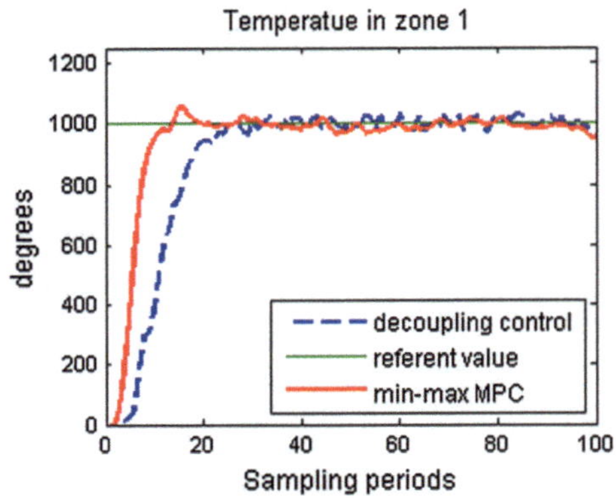

Fig. 23.3 The temperature in the first zone of the furnace

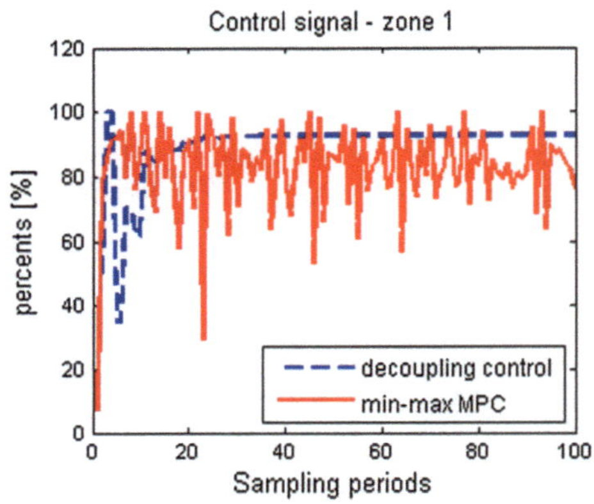

Fig. 23.4 The control signal applied on the first control valve

The control signal in zone 1 is shown in Fig. 23.4. The values of the control signals (which represent the fuel consumption from 0 to 100 %) vary a lot. While the decoupling control has very slow reaction to the direct output disturbances, the MPC tries to eliminate their influence to the system all the time. Furthermore, if we calculate the fuel consumption norms during the time of the experiment we obtain that the norm for the consumed fuel for control with MPC is 18,817, and the norm for fuel consumption with the decoupling control algorithm is 20,699. As we can see there is a difference, and it is in favor of MPC. This means that with using the MPC algorithm we managed to obtain faster response and more robust control (regarding the direct output disturbances) compared to the decoupling PID control, for a smaller fuel consumption norm.

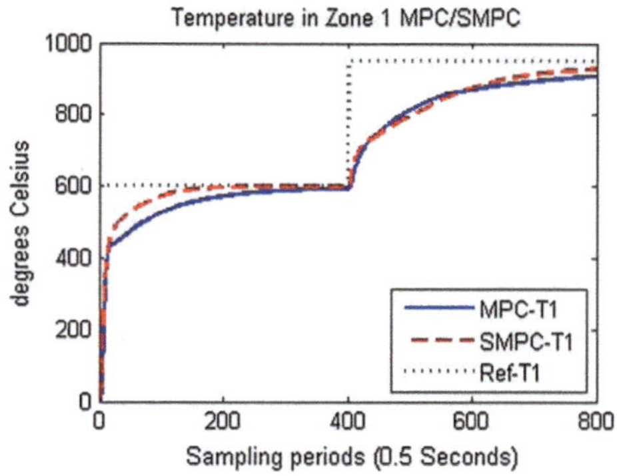

Fig. 23.5 Temperature in zone 1 MPC versus SMPC

23.4.2 Switched MPC

As another advanced control method for thermal processes we will present the multiple model MPC algorithm for control of high consumption industrial furnace. The executed simulation [25] compares the results from the control of the plant with regular MPC and multiple model MPC, which can also be known as switched MPC. The results of the simulation for the temperature in zone 1 are presented in Fig. 23.5. Additionally with the MMMPC we slightly reduce the fuel consumption norm for the furnace.

23.4.3 Predictive Control with Genetic Algorithm Optimization

In this subsection we will present an effective algorithm for GA-MPC for dealing with nonstandard and/or nonlinear constraints. Genetic Algorithms (GAS) inspired by Darwinian theory, represents powerful nondeterministic iterative search heuristic [21]. Genetic algorithms operate on a population which consist of encoded strings, each string represents a solution. This algorithm uses the crossover operator in order

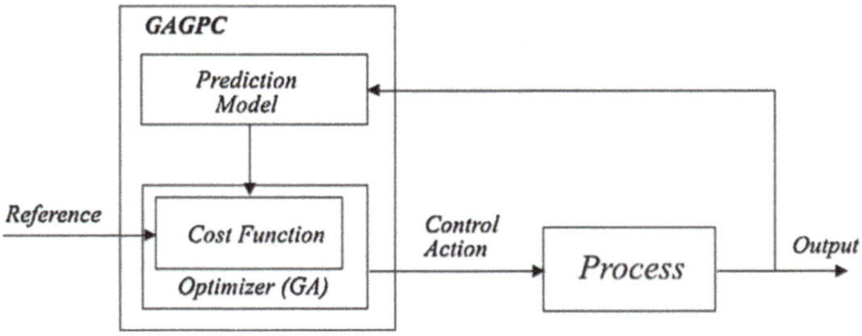

Fig. 23.6 Predictive control loop with GA

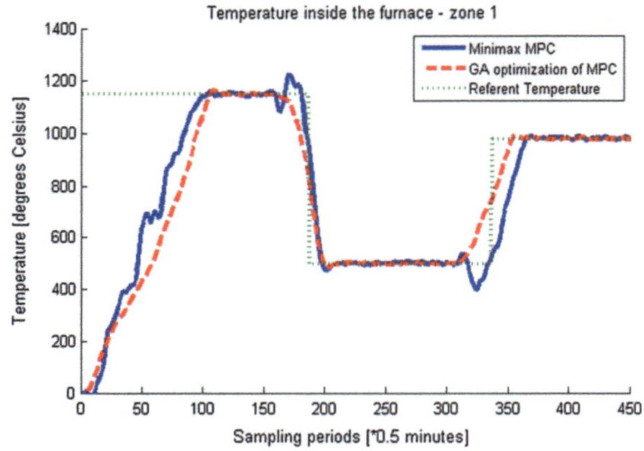

Fig. 23.7 Temperature in the first zone of the furnace

to obtain the new solutions. We use this algorithm in order to make an optimization of the cost function of the MPC. The basic structure of a GA-MPC algorithm is presented in Fig. 23.6.

The model of the high consumption industrial furnace has nontypical constraint. The maximum power generation from pump supplying the three control valves is 155 units, while each from the valves can use up to 100 units (if available). Although it can implement standard constraints as in (23.7), the MPC algorithm cannot implement the constraint as presented with (23.8).

$$0 \leq u_i \leq 100,\ i=1,2,3, \tag{23.7}$$

$$u_1+u_2+u_3 \leq 155. \tag{23.8}$$

The proposed GA-based MPC algorithm was simulated on a MIMO system and the results are found to be reasonably good. We applied the algorithm on a model of a high consumption industrial furnace [24] and compared the result with conventional industry control methods such as Min–Max MPC. The control goal is to keep the temperature in the three zones of the furnace at the referent temperature, with

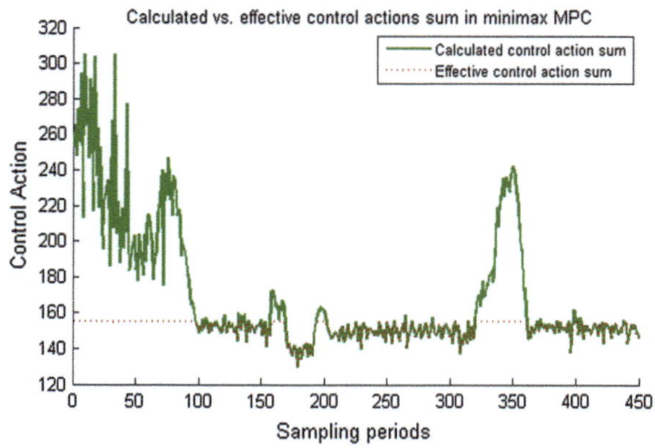

Fig. 23.8 Comparison of the calculated and effective control actions regarding the energy consumption constraint

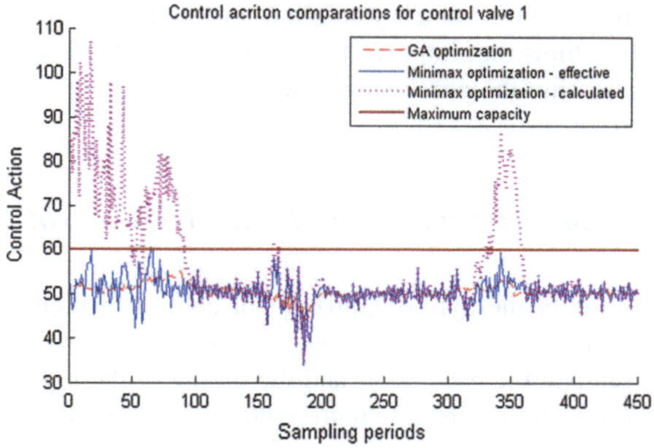

Fig. 23.9 Sum of the control signals in the furnace

minimum possible fuel consumption. The simulation experiment will be conducted in conditions of malfunctioning of one of the power lines. The temperature in the first zone of the industrial furnace is presented in Fig. 23.7.

The sum of the control signals that is subject to the nonstandard constraint is presented in Fig. 23.8. We can notice that the control signal calculated using the genetic optimization-based MPC algorithm never violate the constraints. On the other hand, when using the min–max optimization-based MPC algorithm, we have two different values for the control signals. The first one is the calculated value by the algorithm that has no maximum fuel constraint, and the second one is the effective value of the fuel that has been used in the respective sampling period. The values of the control signals, for the first zone of the furnace, both for min–max and GA-based MPC are depicted in Fig. 23.9.

The simulations of the proposed MPC with genetic algorithm optimization can track the referent temperature reasonably good under the defined constraints. We

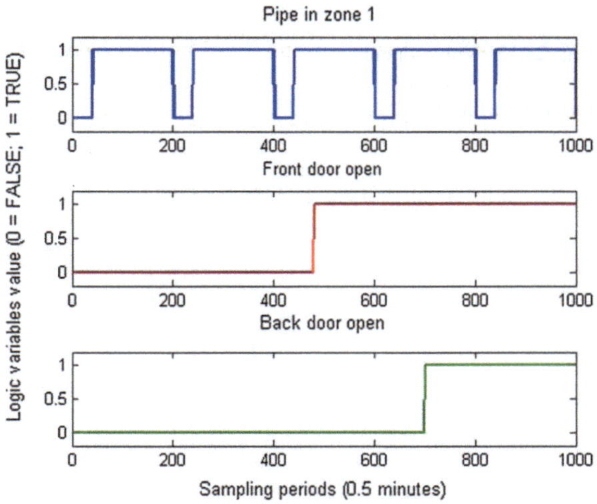

Fig. 23.10 Timing of the logic variable disturbances of the furnace during the simulation

have presented an easy solution for implementing nonstandard constraints in MPC algorithms that improve the results of the standard MPC algorithms while slightly increasing the processing power.

23.4.4 Hybrid Model Predictive Control

To verify the hybrid approach for control of high consumption industrial furnace the authors have conducted series of simulations. The Disturbance signals from the front and the back door, and the timing of the pipe entering in the first zone of the furnace are graphically represented in Fig. 23.10. In Fig. 23.10 the logic variable for pipes entering zone 1 is presented. The logic variables for zone 2 and 3 have deterministic dependence on this value with fixed delay. In reality this delay is

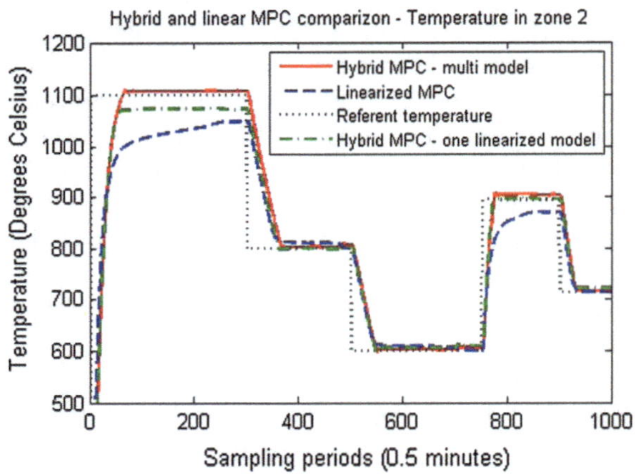

Fig. 23.11 Temperature in the second zone of the furnace during the simulation

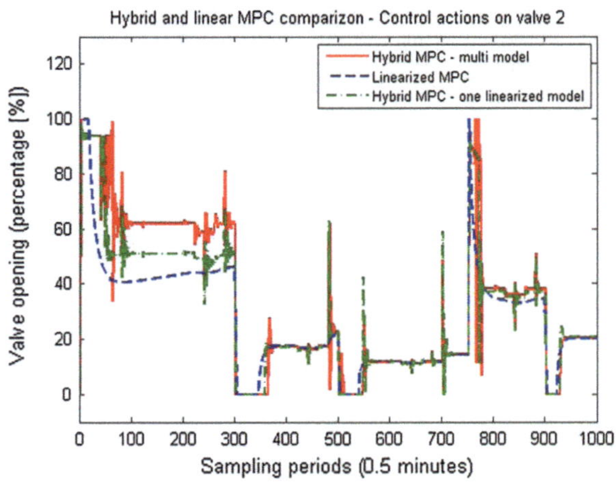

Fig. 23.12 Valve openings on the second valve of the furnace during the simulation

represented through the line speed of the conveyor driving the pipes in the furnace, but this is to be done in near future. During this simulation a fixed delay time of 10 min between zones is adopted. During the simulation the continuous disturbance signal T_{out} has value of 15.

The main results for the second zone are presented in Fig. 23.11 where the temperatures in the respective zones of the furnace are presented along with the reference signal. The control signals applied to the three control valves respectively are presented in Fig. 23.12.

From the presented results it is obvious that introducing the hybrid control approach for high consumption industrial furnace improves the quality of the control. The controller leads the system faster to the referent set point and the steady-state error is acceptable. The hybrid MPC—one linearized model method, has also satisfactory results. Nevertheless we must point out that the tracking of the referent trajectory is best when it is near the linearization point(s), and as the referent trajectory moves from this point we have bigger error in the control algorithm. This is more expressed in the hybrid controller with only linearization point, which is linearized near 800°. In this case it is obvious that output tracks the reference without any problem near this region, but if we have work plans that require a lot of temperature changes throughout the temperature domain of the furnace, the multimodel hybrid approach is to be considered. The previous remark regarding the performance of the controller near the linearization point also stands for the multimodel hybrid approach. The difference here is that we have several models and the difference between the set point and the active model cannot be very big. Logically, if we introduce more models linearized in different operating points we will increase the performance of the controller, but also we will increase the complexity and the time necessary to perform the optimization.

Regarding the control signals, from the presented control signal in the second zone (Fig. 23.12) we can note that the hybrid controllers have fast reaction time to the disturbances. When there is new pipe entering in the one of the zones of the furnace, the control signal in the respective zone, acts toward stabilization of the temperature. Also we can note that when the furnace is operating near 800°, the controller generates the same control value, but if we move far from this central linearization point, the calculated values for the control action differ a lot.

The presented new model incorporates the logic signals that act as disturbances to the furnace (new pipe entering in the zone, opening of the back and the front cooling door). Also in order to improve the performance of the furnace, multipoint linearization was implemented on five characteristic points in the temperature domain. These results are confirmed with the presented simulation results.

23.5 Further Control of Industrial Thermal Processes

The future of thermal processes control is definitely the new intelligent control methods. These processes usually have high consumption and reducing this consumption even slightly can significantly contribute to the cause of less expensive production.

We have presented some of the modern control methods which although effective in testing and simulation still are not largely used in the industry. Nowadays we have overrun the problem of computer burden which these algorithms introduce into the system and a path is open for their massive use.

On another side using these algorithms (especially two or multilevel algorithms) we can dramatically increase the quality of the products. Adding a supervising level in the system allows the operator or the management team to compromise between low-cost products on one side and extra quality products on the other.

23.6 Conclusions

In this paper we have presented predictive control methods for high consumption thermal processes. Sophisticated MIMO solutions based on CPA/CVE approach are more difficult for effective functional maintenance. Moreover, operating personnel is reluctant to accept them, which rules out the idea of on-site in-process redesign. The same conclusion may be drawn for the intelligent control methods such as two-level fuzzy control, model predictive control, and switched model predictive control.

References

1. Dimirovski, G.M., Dourado, A., Ikonen, E., Kortela, U., Pico, J., Ribeiro, B., Stankovski, M. J., Tulunay, E.: Learning control of thermal systems. In: Astrorm, K.J., Albertos, P., Blanke, M., Isidori, A., Schaufelberger, W., Sanz, R. (eds.) Control of Complex Systems, pp. 317–337. Springer, London (2001)
2. Dimirovski, G.M., Dinibutun, A.T., Vukobratovic, M., Zhao, J.: Optimizing supervision and control for industrial furnaces: predictive control based design. In: Sgurev, V., Dimirovski, G., Hadjiski, M. (eds.) Automatic Systems for Building the Infrastructure in Developing Countries —Global and Regional Aspects DECOM-04. Union for Automation & Informatics of Bulgaria and the IFAC, Sofia, BG 17–28 (2004)
3. Dimirovski, G.M., Angelovski, R.S., Stankovski, M.J.: Enhanced control of induction heating processes of welded tubes. In: De Carli, A. (ed.) Low Cost Automation: Techniques, Components and Instrument Applications, pp. 461–469. The IFAC and Pergamon Press, Oxford (1990)
4. Dimirovski, G.M., Hadzi-Nicev, N., Ivanoski, M., Gyorsoski, J.: Design and implementation of an automated industrial electrical reheating furnace. In: Goodwin, G.C., Evans, R.J., (eds.)

Automatic Control—World Congress 1993, Applications II, vol. 4, 451–454. Pergamon Press, Oxford, UK (1994)
5. Sadaoui, N., Gough, N.E., Ting, I.H., Dimirovski, G.M.: Computer-aided study of non-linear systems using input-output approach. In: Atherton D.P., (General Chair) Proceedings of the 30th IEEE Conference on Decision and Control, Brighton, UK, vol. 3, pp. 2799–2804. The IEEE, New York, NY (1991)
6. Ting, I.H., Gough, N.E., Dimirovski, G.M., Deskov, V.P.: Application of the characteristic pattern methodology to the design of a reheat furnace multi-variable control system. In: Proceedings of the Institution of Mechanical Engineers Pt. I Systems & Control Engineering, **206** (I2), pp. 29–34 (1992)
7. Zhu, Y.C., Backx, A.C.P.M.: MIMO process identification for controller design: test signals, nominal model and error bounds. In: Identification and System Parameter Estimation 1991, Proceedings of Selected Papers from the 9th IFAC/IFORS Symposium, vol. 1, pp. 127–132. The IFAC and Pergamon Press, Oxford, UK (1992)
8. Stankovski, M.J., Gough, N.E., Dimirovski, G.M.: Temperature Control of Large Furnaces—A Survey. Technical report IDP290, School of Computing and Information Technology, University of Wolverhampton, Wolverhampton, UK (1992)
9. Rhine, J.M., Tucker, R.J.: Modeling of Gas-Fired Furnaces and Boilers. McGraw-Hill, New York (1991)
10. Stankovski, M.J.: Non-conventional Control of Industrial Energy Processes in Large Heating Furnaces. Ph.D. thesis; in Macedonian, SS Cyril and Methodius University, Faculty of EE, Skopje, MK (1997)
11. Stankovski, M.J., Dimirovski, G.M., Gough, N.E., Hanus, R.: Industrial furnace control: an experiment in iterative identification and design. In: Patton, R.J., Goodall, R., Fleming, P. (eds.) Proceedings of the UKACC International Conference on Control, IEE Publication No. 455, vol. II, pp. 946–951. The Institution of Electrical Engineers, London, UK (1998)
12. Stankovski, M.J., Dimirovski, G.M., Gough, N.E., Ting, I.H., Kolemishevska-Gugulovska, T. D.: Identification and digital control system design for 20 MW gas-fired pipe heating furnace. In: Dourado, A., et al. (eds.) Controlo 98 Proceedings, vol. 1, pp. 87–92. The APCA and Universidade de Coimbra, Coimbra, PT (1998)
13. Albertos, P.: Iterative controller design. In: Proceedings ESF-COSY'97 Workshop on Learning Control Systems, Sao Silvestre (PT). The ESF & CIS – UC, Coimbra. Paper 11, pp. 1–15 (1997)
14. Gevers, M.: Learning from identification and control design (Plenary Lecture). In: Albertos, P. (ed.) Proceedings Joint ESF-COSY'96 Workshop, Valencia (E): The ESF & DISCA-UPV, Valencia, ES, Theme 3, PL.3, pp. 1–60 (1996)
15. Keviczky, L., Banyasz, Cs.: On the dialectics of identification and control in iterative learning schemes. In: Albertos, P. (ed.) Proceedings of Joint ESF-COSY'96 Workshop, Valencia: The ESF & DISCA-UPV, Valencia ES, Theme 3, Paper 1, pp. 1–24 (1996)
16. Astrom, K.J.: Matching criteria for control and identification. In: Proceedings of the 2nd European Control Conference, Groningen, The Netherlands, pp. 248–251. The EUCA, Groningen, NL (1993)
17. Allgöwer, F., Badgwell, T., Qin, J., Rawlings, J., Wright, S.: Nonlinear predictive control and moving horizon estimation—An introductory overview. In: Frank, P.M. (ed.) Advances in Control: Highlights of ECC'99, pp. 391–449. Springer, Berlin (1999)
18. Jing, Y., Dimirovski, G.M.: Decentralized stabilization control for composite systems with time delays and uncertainties. In: Solaiman, B., (ed.) Proceedings of the 3rd IEEE Conference Information and Communication Technologies, Damascus, Syria, vol. 1, pp. 1404–1409. The IEEE, Piscataway, NJ, USA (2006)
19. Pregelj, B., Gerkšič, S.: Multiple model approach to multi-parametric model predictive control of a nonlinear process—A simulation case study. http://as.utia.cas.cz/files/113.pdf (2006)
20. Onnen, C., Babuska, R., Kaymak, U., Sousa, J.M., Verbruggen, H.B., Isermann, R.: Genetic algorithms for optimization in predictive control. Control Eng. Pract. **5**(10), 1363–1372 (1997)

21. Blasco, X., Martinez, M., Senent, J., Sanchis, J.: Generalized predictive control using genetic algorithms (GAGPC). An application to control of a non-linear process with model uncertainty. Eng. Appl. Artif. Intell. **11**, 355–367 (1998)
22. Potocnik, P., Grabec, I.: Model predictive control using neural networks and genetic algorithms. In: Dville, M., Owens, R. (eds.) Proceedings of the 16th IMACS World Congress 2000 on Scientific Computation, Applied Mathematics and Simulation, pp. 1–6. Brussels Paris, The IMACS (2000)
23. Al-Duwaish, H., Naeem, W.: Nonlinear model predictive control of Hammerstein and Wiener models using genetic algorithms. In: Control Applications 2001 (CCA '01), Proceedings of the 2001 IEEE International Conference, the IEEE, Piscataway, NJ, USA, pp. 465–469 (2001)
24. Stojanovski, G.S., Stankovski, M.J.: Model predictive controller employing genetic algorithm optimization of thermal processes with non-convex constraints. In: Popa, R. (ed) Genetic Algorithms in Applications, pp. 19–34 (2012)
25. Stojanovski, G.S., Stankovski, M.J., Dimirovski, G.M.: Multiple-model model predictive control for high consumption industrial furnaces. FACTA UNIVERSITATIS Ser. Autom. Control Robot. **9**(1), 131–139 (2010)

Chapter 24
Closed-Loop Control with Evolving Gaussian Process Models

Jus Kocijan and Dejan Petelin

Abstract This contribution presents a new development in the design of control system based on evolving Gaussian process (GP) models. GP models provide a probabilistic, nonparametric modelling approach for black-box identification of nonlinear dynamic systems. They can highlight areas of the input space where prediction quality is poor, due to the lack of data or its complexity, by indicating the higher variance around the predicted mean. GP models contain noticeably less coefficients to be optimised than commonly used parametric models. While GP models are Bayesian models, their output is normal distribution, expressed in terms of mean and variance. Latter can be interpreted as a confidence in prediction and used in many fields, especially in control system. Evolving GP model is the concept approach within which various ways of model adaptations can be used. Successful control system needs as much as possible data about process to be controlled. If the prior knowledge about the system to be controlled is scarce or the system varies with time or operating region, this control problem can be solved with an iterative method which adapts model with information obtained with streaming data and concurrently optimises hyperparameter values. This contribution provides: a survey of adaptive control algorithms for dynamic systems described in publications where GP models have been used for control design, a novel and improved closed-loop controller with evolving GP models and an example for the illustration of proposed control algorithm.

J. Kocijan (✉) · D. Petelin
Jožef Stefan Institute, Jamova 39, 1000 Ljubljana, Slovenia
e-mail: jus.kocijan@ijs.si

D. Petelin
e-mail: dejan.petelin@ijs.si

J. Kocijan
University of Nova Gorica, Vipavska 13, 5000 Nova Gorica, Slovenia

G.M. Dimirovski (ed.), *Complex Systems*, Studies in Systems,
Decision and Control 55, DOI 10.1007/978-3-319-28860-4_24

24.1 Introduction

Increasingly complex systems are expected to be handled with new technologies among them with control system technologies. There exist a range of control design methods depending on the sort of system model and amount of information that is available. Various adaptive, self-learning or other learning approaches are frequent to tackle the problem of low initial prior knowledge about the system to be controlled or in the case of time-variant or nonlinear systems. Often various kinds of computational intelligence methods for model development that result in so-called black-box models are used for these kinds of control problems. This paper deals with control system design based on Gaussian process (GP) models.

GP models provide a probabilistic, nonparametric modelling approach for black-box identification of nonlinear dynamic systems. They can highlight areas of the input space where model prediction quality is poor, due to the lack of data or its complexity, by indicating the higher variance around the predicted mean. This property can be incorporated in the closed-loop control design. GP models contain noticeably less coefficients to be optimised than parametric models that are frequently used in control design.

The aim of this chapter is to present an improved closed-loop controller with evolving GP models and place it within contemporary research on adaptive control methods based on GP models and to demonstrate a proposed control algorithm based on GP model.

The chapter is structured as follows. First the modelling with Gaussian processes in general and the modelling of dynamic systems with GP models is explained. Then a short review of adaptive control methods based on GP models is given. Adaptive control with evolving GP model is introduced next. The control method is demonstrated with an illustrative example. This example demonstrates the performance and the adaptation of closed-loop tracking control in different operating regions with a computer simulation study.

24.2 Systems Modelling with Gaussian Processes

A GP model is a flexible, probabilistic, nonparametric model that enables the prediction of output-variable distributions. Contrary to parametric modelling methods where a structure is usually presumed and the parameters are optimised with regression, the GP-based modelling is different in the sense that the structure of the mapping function is not presumed, but the data themselves are used to describe the mapping function. The modelled system is, therefore, not approximated by fitting the parameters of the selected basis functions, but rather with the relationship among the measured data. The model of the nonlinear mapping function is called the GP model as the output of the GP model is by prior belief considered to be a GP. GP model's properties and application potentials are reviewed in [29].

A GP is a collection of random variables that have a joint multivariate Gaussian distribution. Assuming a relationship of the form $y = f(\mathbf{z})$ between the input $\mathbf{z}$ and the output y, we have $y_1, \ldots, y_N \sim \mathcal{N}(0, \Sigma)$, where $\Sigma_{pq} = \mathrm{Cov}(y_p, y_q) = C(\mathbf{z}_p, \mathbf{z}_q)$ gives the covariance between the output points corresponding to the input points described by vectors $\mathbf{z}_p$ and $\mathbf{z}_q$. Thus, the mean $\mu(\mathbf{z})$ and the covariance function $C(\mathbf{z}_p, \mathbf{z}_q)$ fully specify the Gaussian process.

The value of the covariance function $C(\mathbf{z}_p, \mathbf{z}_q)$ expresses the correlation between the individual outputs $f(\mathbf{z}_p)$ and $f(\mathbf{z}_q)$ with respect to the inputs $\mathbf{z}_p$ and $\mathbf{z}_q$. Note that the covariance function $C(\cdot, \cdot)$ can be any function that generates a positive semi-definite covariance matrix. It is usually composed of two parts,

$$C(\mathbf{z}_p, \mathbf{z}_q) = C_{\mathrm{f}}(\mathbf{z}_p, \mathbf{z}_q) + C_{\mathrm{n}}(\mathbf{z}_p, \mathbf{z}_q), \tag{24.1}$$

where C_{f} represents the functional part and describes the unknown system we are modelling, and C_{n} represents the noise part and describes the model of the noise.

For the noise part it is most common to use the constant covariance function, presuming white noise. The choice of the covariance function for the functional part also depends on the stationarity of the data used for modelling. Assuming stationary data the most commonly used covariance function is the square exponential covariance function. The composite covariance function is therefore

$$C(\mathbf{z}_p, \mathbf{z}_q) = v_1 \exp\left[-\frac{1}{2}\sum_{d=1}^{D} w_d (z_{pd} - z_{qd})^2\right] + \delta_{pq} v_0, \tag{24.2}$$

where w_d, v_1 and v_0 are the 'hyperparameters' of the covariance function, D is the input dimension and $\delta_{pq} = 1$ if $p = q$ and 0 otherwise. In contrast, assuming non-stationary data the polynomial or its special case, the linear covariance function, can be used. Other forms and combinations of covariance functions suitable for various applications can be found in [29]. The hyperparameters can be written as a vector $\theta = [w_1, \ldots, w_D, v_1, v_0]^{\mathrm{T}}$. The parameters w_d indicate the importance of the individual inputs: if w_d is zero or near zero, it means the inputs in dimension d contain little information and could possibly be neglected.

To accurately reflect the correlations present in the training data, the hyperparameters of the covariance function need to be optimised. Due to the probabilistic nature of the GP models, the common model optimisation approach, where model parameters and possibly also the model structure are optimised through the minimisation of a cost function defined in terms of model error (e.g. mean square error), is not readily applicable. A probabilistic approach to the optimisation of the model is more appropriate. Actually, instead of minimising the model error, the probability of the model is maximised.

GP models can be easily utilised for a regression calculation. Consider a matrix $\mathbf{Z}$ of N D-dimensional input vectors where $\mathbf{Z} = [\mathbf{z}_1, \mathbf{z}_2, \ldots, \mathbf{z}_N]^{\mathrm{T}}$ and a vector of the output data $\mathbf{y} = [y_1, y_2, \ldots, y_N]$. Based on the data $(\mathbf{y}, \mathbf{Z})$, and given a new input vector $\mathbf{z}^*$, we wish to find the predictive distribution of the corresponding output y^*.

Based on the training input data $\mathbf{Z}$, a covariance matrix $\mathbf{K}$ of size $N \times N$ is determined. The overall problem of learning unknown parameters from data corresponds to the predictive distribution $p(y^*|\mathbf{y}, \mathbf{Z}, \mathbf{z}^*)$ of the new target y^*, given the training data $(\mathbf{y}, \mathbf{Z})$ and a new input $\mathbf{z}^*$. In order to calculate this posterior distribution, a prior distribution over the hyperparameters $p(\theta|\mathbf{y}, \mathbf{Z})$ can first be defined, followed by the integration of the model over the hyperparameters

$$p(y^*|\mathbf{y}, \mathbf{Z}, \mathbf{z}^*) = \int p(y^*|\theta, \mathbf{y}, \mathbf{Z}, \mathbf{z}^*)p(\theta|\mathbf{y}, \mathbf{Z})d\theta. \qquad (24.3)$$

The computation of such integrals can be difficult due to the intractable nature of the nonlinear functions. A solution to the problem of intractable integrals is to adopt numerical integration methods such as the Monte Carlo approach. Unfortunately, significant computational efforts may be required to achieve a sufficiently accurate approximation.

In addition to the Monte Carlo approach, another standard and general practice for estimating hyperparameters is the maximum marginal-likelihood estimation, i.e., to minimise the following negative log marginal-likelihood function [29]:

$$\ell(\theta) = -\frac{1}{2}\ln(|\mathbf{K}|) - \frac{1}{2}\mathbf{y}^{\mathrm{T}}\mathbf{K}^{-1}\mathbf{y} - \frac{N}{2}\ln(2\pi). \qquad (24.4)$$

As the likelihood is, in general, nonlinear and multi-modal, efficient optimisation routines usually entail the gradient information. The computation of the derivative of ℓ with respect to each of the parameters is as follows:

$$\frac{\partial \ell(\theta)}{\partial \theta_i} = -\frac{1}{2}\mathrm{trace}\left(\mathbf{K}^{-1}\frac{\partial \mathbf{K}}{\partial \theta_i}\right) + \frac{1}{2}\mathbf{y}^{\mathrm{T}}\mathbf{K}^{-1}\frac{\partial \mathbf{K}}{\partial \theta_i}\mathbf{K}^{-1}\mathbf{y}. \qquad (24.5)$$

For performing a regression, the availability of the training input data described with matrix $\mathbf{Z}$ and the corresponding output data described with vector $\mathbf{y}$ is assumed. As already mentioned, the aim is to find the distribution of the corresponding output y^* for some new input vector $\mathbf{z}^* = [z_1(N+1), z_2(N+1), \ldots, z_D(N+1)]^{\mathrm{T}}$.

For the collection of random variables $[y_1, \ldots, y_N, y^*]$ we can write:

$$[\mathbf{y}, y^*] \sim \mathcal{N}(0, \mathbf{K}^*) \qquad (24.6)$$

with the covariance matrix

$$\mathbf{K}^* = \left[\begin{array}{c|c} \mathbf{K} & \mathbf{k}(\mathbf{z}^*) \\ \hline \mathbf{k}^{\mathrm{T}}(\mathbf{z}^*) & \kappa(\mathbf{z}^*) \end{array}\right], \qquad (24.7)$$

where $\mathbf{y} = [y_1, \ldots, y_N]$ is a $1 \times N$ vector of training targets. The predictive distribution of the output for a new test input has a normal probability distribution with a mean and variance [29]

$$\mu = \mathbf{k}(\mathbf{z}^*)^{\mathrm{T}} \mathbf{K}^{-1} \mathbf{y}, \tag{24.8}$$
$$\sigma^2 = \kappa(\mathbf{z}^*) - \mathbf{k}(\mathbf{z}^*)^{\mathrm{T}} \mathbf{K}^{-1} \mathbf{k}(\mathbf{z}^*), \tag{24.9}$$

where $\mathbf{k}(\mathbf{z}^*) = [C(\mathbf{z}_1, \mathbf{z}^*), \ldots, C(\mathbf{z}_N, \mathbf{z}^*)]^{\mathrm{T}}$ is the $N \times 1$ vector of covariances between the test and training cases, and $\kappa(z^*) = C(\mathbf{z}^*, \mathbf{z}^*)$ is the covariance between the test input itself.

The obtained model, in addition to the mean value, provides information about the confidence in the prediction by the variance of the predictive distribution. Usually, the confidence of the prediction is depicted with a 2σ interval, which is about 95 % confidence interval. This confidence region can be seen in the example in Fig. 24.1 as a grey band. It highlights the areas of the input space where the prediction quality is poor, due to the lack of data or noisy data, by indicating a wider confidence band around the predicted mean.

GP models can, like neural networks, be used to model static nonlinearities and can therefore be used for the modelling of dynamic systems [1, 16, 17] as well as time series, if lagged samples of the output signals are fed back and used as regressors. A review of recent developments in the modelling of dynamic systems using GP models and its applications can be found in [15] and [14]. It is important to stress that the model prediction in the form of GP is just an approximation when the Gaussian assumption is not fulfilled, which is in line with common engineering practice.

A dynamic GP model is trained as the nonlinear autoregressive model with an exogenous input (NARX) representation, where the output at time instant k depends on the delayed output y and the exogenous control input u:

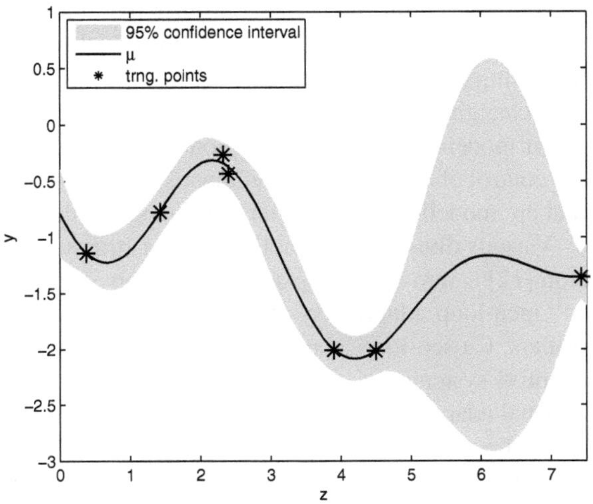

Fig. 24.1 Using GP models: in addition to the prediction mean value (*full line*), we obtain a 95 % confidence interval (*grey band*) for the underlying function y

$$y(k) = f_S(y(k-1), \dots, y(k-n), u(k-1), \dots, u(k-n)) + \nu(k), \qquad (24.10)$$

where f_S denotes a function, $\nu(k)$ is white noise disturbance with normal distribution and the output $y(k)$ depends on the regression vector $\mathbf{z}(k) = [y(k-1), y(k-2), \dots, y(k-n), u(k-1), u(k-2), \dots, u(k-n)]$ at time instant k. This model notation can be generalised to mutivariable cases, i.e., cases with multiple inputs and outputs.

For the validation of obtained dynamic GP model the nonlinear output error (NOE), also called parallel, model is used. This means that the NARX model is used to predict a further step ahead by replacing the data at instant k with the data at instant $k+1$ and using the prediction $\hat{y}(k)$ from the previous prediction step instead of the measured $y(k)$. This is then repeated indefinitely. The latter possibility is equivalent to simulation. Simulation, therefore, means that only on the basis of previous samples of a process input signal $u(k-i)$ can the model simulate future outputs. Frequently, the mean value of prediction $\hat{y}(k)$ is used to replace $y(k)$, which is called 'naive' simulation. Other possibilities, where the entire distribution is used, are described in, e.g. [17].

24.3 Adaptive Control Algorithms Based on Gaussian Process Models

Control is the activity that makes a system behave in the desired way. There are many reference books available describing a variety of control methods, their design procedures and their applications. This section provides only a review of some of the adaptive control methods that are based on GP model and were published in literature. Reader is referred to [15] and [13] for more comprehensive review of control methods based on GP model.

Adaptive controller is the controller that continuously adapts to some changing process. Adaptive controllers emerged in early 60s of the previous century. At the beginning these controllers were mainly adapting themselves based on linear models with changing parameters. Since then several authors have proposed the use of nonlinear models as a base to build nonlinear adaptive controllers. These are meant for the control of time-varying nonlinear systems or of time-invariant nonlinear systems that are modelled as parameter-varying simplified nonlinear models.

Various divisions of adaptive control structures are possible. One possible division [12] is into open-loop and closed-loop adaptive systems.

Open-loop adaptive systems are gain-scheduling or parameter-scheduling controllers. Closed-loop adaptive systems can be further divided to dual and non-dual adaptive systems.

Dual adaptive systems [11, 36] are those where the optimisation of the information collection and the control action are pursued at the same time. The control signal should ensure that the system output cautiously tracks the desired set-point value and at the same time excites the plant sufficiently to accelerate the identification process. The solution to the dual-control problem is based on dynamic programming and the resulting functional equation is often the Bellman equation. Not a large number of such controllers have been developed.

The difficulties in finding the optimal solution for adaptive dual control lead to suboptimal dual adaptive controllers [11, 36] obtained by either various approximations or by reformulating the problem. Such a reformulated adaptive dual control problem is when a special cost function is considered, which consists of two added parts: control losses and an uncertainty measure. This is appealing for application with the GP model that provides measures of uncertainty.

Many adaptive controllers in general are based on the separation principle [36] that implies separate estimation of system model, i.e., system parameters, and the application of this model for control design. When the identified model used for control design and adaptation is presumed to be the same as the true system then the adaptive controller of this kind is said to be based on certainty equivalence principle and such an adaptive controller is named non-dual adaptive controller. The control actions of non-dual adaptive controller do not take any active actions that will influence the uncertainty.

When using the GP model for the adaptive control, different from gain-scheduling control, the GP model is identified online and this model is used in the control algorithm. The block scheme showing the general principle of adaptive control with the GP model identification is given in Fig. 24.2. It is sensible that advantages of GP models are considered in the control design, which relates the GP model-based adaptive control at least to suboptimal dual adaptive control principles. The uncertainty of model predictions obtained with the GP model prediction are dependent, among others, on local learning-data density, and the model complexity is automatically related to the amount and the distribution of the available data—more complex models need more evidence to make them likely. Both aspects are very useful in sparsely populated transient regimes. Moreover, since weaker prior assumptions are typically applied in a nonparametric model, the bias is typically lower than in parametric models.

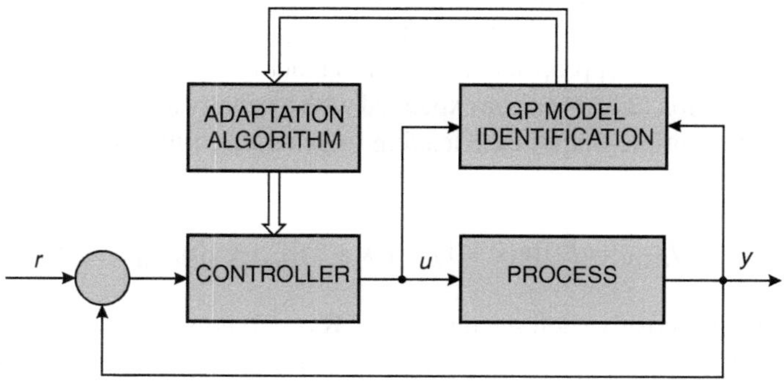

Fig. 24.2 General *block* scheme of the closed-loop system with adaptive controller

The above ideas are indeed related to the work done on the adaptive dual control, where the main effort has been concentrated on the analysis and design of adaptive controllers based on the use of the uncertainty associated with parameters of models with a fixed structure [11, 32].

The major differences in up-to-now published adaptive systems based on GP models are in the way how the online model identification is pursued.

Increasing the size of the covariance matrix, i.e., ‘blow-up model’, with the in-streaming data and repeating model optimisation is used in papers [19, 20, 30–32], where more attention is devoted to control algorithms and their benefits based on information gained from the GP model and not on the model identification itself.

Another adaptive control algorithm implementation is control with feedback for cancelling nonlinearities with the online learning of the inverse model. This sort of adaptive control with the increasing covariance matrix with the in-streaming data is described in [22]. Two sorts of online learning for the mentioned feedforward contained control are described in [23]. The first sort is with moving window strategy, where the old data are dropped from the online learned model, while the new data is accommodated, the second one accommodates only new data with sufficient information gain. These applications of referenced inverse GP models do not use entire information from the prediction distribution, but they focus on the mean value of prediction.

A lot of GP model-based adaptive control algorithms from the referenced publications are based on the minimum-variance controller. One of the reasons is that the minimum-variance controller explores the variance that is readily available with the GP model prediction.

The minimum-variance controller in general [12] looks for a control signal $u(k)$ in time instant k that will minimise the following cost function:

$$J_{\text{MV}} = E(\|\mathbf{r}(k+p) - \mathbf{y}(k+p)\|^2). \tag{24.11}$$

In this case, J_{MV} refers to the covariance of the error between the vector of reference values $\mathbf{r}(k+p)$ and the controlled outputs p-time steps in the future, $\mathbf{y}(k+p)$. The desired controller is thus the one that minimises these variances, hence the name minimum-variance control. The optimal control signal $\mathbf{u}_{\text{opt}}$ can be obtained by minimising selected cost function. The minimisation can be done analytically, but also numerically, using any appropriate optimisation method.

The cost function (24.11) can be expanded with a penalty terms and generalised to multiple-input multiple-output case leading to generalised minimum-variance control [31].

$$J_{\text{GMV}} = E(||\mathbf{r}(k+1) - \mathbf{y}(k+1)||^2_{\mathbf{Q}}) + ||\mathbf{u}_k||^2_{\mathbf{R}}, \tag{24.12}$$

where matrix $\mathbf{Q}$ is positive definite matrix and $\mathbf{R}$ is polynomial matrix with the backward shift operator q^{-1}. The matrix $\mathbf{Q}$ elements and matrix $\mathbf{R}$ polynomial coefficients can be used as tuning parameters.

The method named Gaussian Process Dynamic Programming (GPDP) is a Gaussian-process-model-based adaptive control algorithm with a proximity to adaptive dual control. The details of the method are described in [8]. The following description is summarised from [8, 10]. The evolution of the method can be followed through time with publications [8–10, 27, 28].

GPDP is an approximate dynamic programming method, where cost functions, so-called value functions in the dynamic programming recursion are modelled by GPs.

The reader is referred to [8] for details and a demonstration of the method. Unfortunately, according to the method's authors [8], GPDP cannot be directly applied to a dynamic system because it is often not possible to experience arbitrary state transitions. Moreover GPDP method does not scale that well to high dimensions.

More promising for engineering control applications is Probabilistic Inference and Learning for COntrol (PILCO) method, described in [5–7].

PILCO is a policy search method and an explicit value function model is not required as in GPDP method. The general idea of the method is to learn the system model with reinforcement learning and control the closed-loop system, taking into account the probabilistic model of the process. The algorithm can be divided into three layers: a top level for the controller adaptation, an intermediate layer for the approximate inference for long-term predictions and a bottom layer for learning the model dynamics.

The PILCO method was applied to real systems, e.g. robotic systems [7].

24.4 Evolving GP-Model-Based Control

An adaptive minimum-variance controller based on evolving Gaussian process model [24] is presented in this section. The basic idea of control based on evolving system model is that system GP model evolves with in-streaming data and the information about system from the model is used for its control. One option is that the information can be in the form of GP model prediction for one or several steps ahead which is then used to calculate optimal control input in the controlled system. The other option would be, for example prediction of some particular part of the model, e.g. parameters, like in [2], and online calculation of controller.

The proposed control is actually a variation of the control proposed in [25]. The main difference is an online learning method used for adapting GP model. It should be noted that an efficient adaptation of the GP model is crucial, as the calculation of the optimal control input is based on the GP model's prediction. However, a noticeable drawback of GP model identification is the computation load that increases with the third power of the amount of input data due to the calculation of the inverse of the covariance matrix. This computational complexity restricts the amount of training data to, at most, a few thousand cases. To overcome the computational limitation issues, only a subset of the most informative data is to be used. In the literature, such a subset is called the active set or the basis vectors set [4] and its elements are called basis vectors [4], inducing variables [26] or basis functions [18]. The basic idea is to retain the bulk of the information contained in the full training dataset, but reduce the size of the covariance matrix so as to facilitate a less computationally demanding implementation of the GP model. As the data is in-streaming the GP model should be adapted continuously. In other words, the online learning method should process every new piece of streaming data sequentially.

In the previous version of the controller [25] we used Csato's method named sparse online Gaussian processes [4]. Its main disadvantage is twofold. The first one is a lack of ability to update hyperparameter values in an online fashion. In other words, for adequately learning optimised hyperparameters are needed in advance, so their values should be optimised before we apply controller to the system. Usually, there is some available data that can be used for optimisation of hyperparameter values, but obtained values are optimal only for the current data presenting the system's dynamics. But, if the data presents only one region of the system's dynamics or if the system is time variant, obtained hyperparameter values most likely will not be optimal enough in other regions or time spans. The second disadvantage is potential computational instabilities [34] or unguaranteed convergence of the algorithm for nonlinear systems [3]. Therefore, we propose evolving GP models [24] to be used for adapting controller's GP model.

The basic idea of evolving GP models is that all influential parts of the GP model should be adapted online. The GP model is fully determined by the training data and the covariance function. In the case of sparse approximation which we use, the training data is actually represented by the active set. Usually, the training data, especially in case of dynamic systems, has more than one input dimensions, so-called regressors, which have various influence on output data. Moreover, some regressors may be fallacious and can present additional noise to model. Thus, selection of regressors is important part of modelling as well. As described in Sect. 24.2, with selection of appropriate type or a combination of various types of covariance function a prior knowledge of the system is included in the model. Nevertheless, with optimisation of hyperparameter values the model is even more adjusted to real system. So, there are four parts that can evolve:

- the regressors,
- the active set,
- the type of covariance function and
- the hyperparameter values.

Although the proposed concept considers all four parts that can evolve, our implementation is somewhat more facile. In dynamic nonlinear systems, where the nonlinear mapping between input and output can not be easily formulated, frequently the squared exponential covariance function is used presuming smoothness and stationarity of the system. That means the covariance function is fixed and does not need to evolve. Furthermore, the squared exponential covariance function can be used with automatic relevance determination (ARD) which is able to find influential regressors [29]. With the optimisation of the hyperparameter values, uninfluential regressors have smaller values and as a consequence have smaller influence to the result. Therefore, all available regressors can be used. Consequently, only the active set and hyperparameter values have left to be adapted sequentially.

With every incoming data, first its information gain regarding the current GP model is estimated. This is done in two phases. In the first phase a prediction of the current input data is calculated based on the current GP model. If the difference between mean value of prediction and current data output is small enough, it means that the current GP model can accurately predict the output of the incoming data. Furthermore, if also the variance of the prediction is small, the current data most probably does not contain any new information regarding the current GP model, thus there is no need to include current data into the GP model. Otherwise, we include current data into the GP model. If this inclusion causes the excess of the preset maximal active set size, the less informative data in the active set should be removed. The less informative data is found by calculating the negative log marginal likelihood from Eq. (24.4) for all subsets of the active set of length $m = n - 1$, where m is the preset maximal size of the active set and n is the size of the exceeded active set.[1] The subset with the lowest negative log marginal likelihood is preserved and the remaining data is removed. After every update of the active set, the hyperparameter values are optimised by minimising negative log marginal likelihood. This can be done with any suitable optimisation method. In our case the conjugate gradients optimisation method is used.

The described online learning method is, besides the selected control algorithm, the main difference between the control we propose and PILCO method [5]. Due to the nature of the policy search methods, PILCO is implemented as batch learning method. That means the model is updated at the end of every cycle, during which new measurements are collected. As the number of collected measurements is usually high enough that a sequential learning is not convenient, but rather offline learning is used, since it can consider all recently collected data. By default, a full GP model is used, but in case the number of data points exceeds a particular threshold a sparse pseudo-input Gaussian process [35] can be used.

An illustrative example that shows operation of minimum-variance control based on evolving GP model is given in the following section.

24.5 Illustrative Example

To assess the proposed controller we used the nonlinear dynamic system [21] described by

$$y(k+1) = \frac{y(k)}{1 + y^2(k)} + u^3(k) + v, \tag{24.13}$$

[1]It should be noted that for calculation of the negative log marginal likelihood the inverse of the covariance matrix is needed. In our case the inverse of the covariance matrix should be calculated for every subset. But, this computational demanding task can be speeded up by calculating the Cholesky decomposition of the exceeded active set of length n only once and then downdates it for every data in it by using low-rank updates [33].

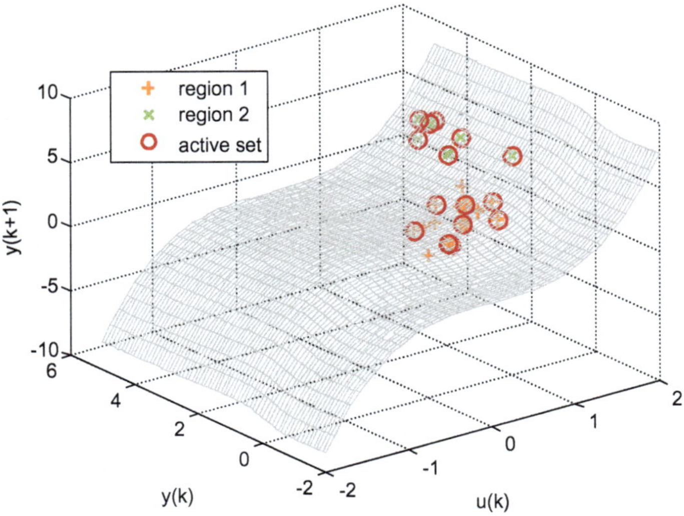

Fig. 24.3 Observed data and the most informative data—active set showed on the surface of selected nonlinear system. The *grey* mesh denotes the nonlinear mapping of the system to be controlled, *orange* pluses and *green* crosses denote first and second regions, respectively, while *red* circles denote the active set

where u is system's input signal, y is system's output signal, ν is white noise of normal distribution with standard deviation 0.005 that contaminates the system response and the sampling time is one second. The nonlinearity in the region of interest for the benchmark system is depicted by a grey mesh in Fig. 24.3. The illustrative example is adapted from [15].

The requirement of the closed-loop control is to follow the set-point, depicted in Fig. 24.4, as close as possible. We start off with an empty GP model's active set and with some default hyperparameter values $\log \theta = [0; 0; 1; -1]$, which are quite different comparing to the optimal ones. The set-point signal is a combination of periodic pulses in two different regions. The first region is between 0.5 and 1.5 and the second one between 3 and 4. The priority of such a signal is to show that the proposed approach for control system based on evolving GP models is able to learn from scratch, without any prior model, and to update regarding the dynamics changes. The data stream contains only 250 data points (shown in Fig. 24.4), which serve the demonstration requirements. We preset the maximal active size to 50 data points. The used control cost function is variation of minimum variance cost function from Eq. (24.11)

$$J(k) = [r(k) - [y(k-1) - E(\hat{y}(k-1))] - E(\hat{y}(k))]^2. \qquad (24.14)$$

The term $y(k-1) - E(\hat{y}(k-1))$ is to make the control algorithm insensitive to errors in the steady-state gain with subtracting the discrepancy between the latest plant output and the latest model most likely output.

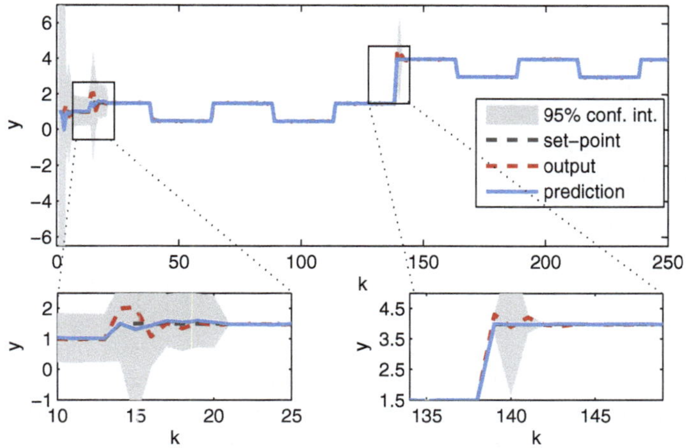

Fig. 24.4 Simulation of controller based on evolving GP model. The *black dashed line* denotes the set-point signal and the *red dashed line* denotes the output of the system, while the *blue solid line* denotes a mean value of the prediction and a *grey* band denotes the 95 % prediction confidence interval based on the current GP model

Fig. 24.5 Optimal control signal $u(k)$ based on the minimum variance controller using GP model's predictions

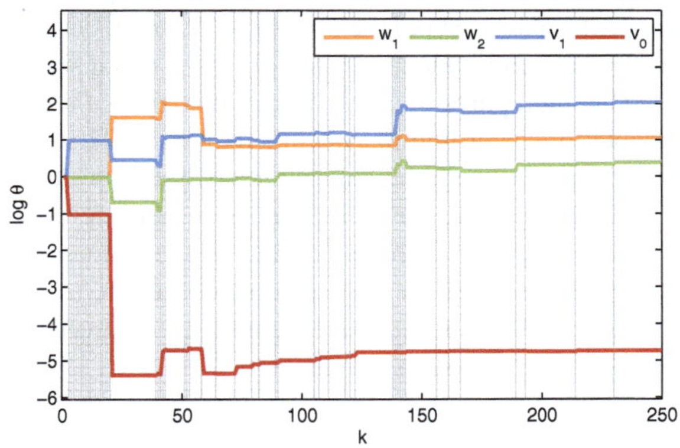

As the controller has no prior knowledge about the system, the system's output oscillates at the beginning, Fig. 24.4. Figure 24.5 shows the corresponding optimal control signal. Nevertheless, the controller observes enough data to successfully, but with some overshoot, follow the first step. Afterwards, the controller easily follows the set-point signal, even in the second region, where the nonlinearity is locally different. The complete nonlinearity, including both regions, can be seen in Fig. 24.3, where orange pluses and green crosses denote first and second regions, respectively. However, at the beginning of these regions also some overshoots appear, but the output quickly gets settled.

The final active set is depicted in Fig. 24.3. The most informative data points, selected from in-streaming data with the proposed evolving method, are denoted with

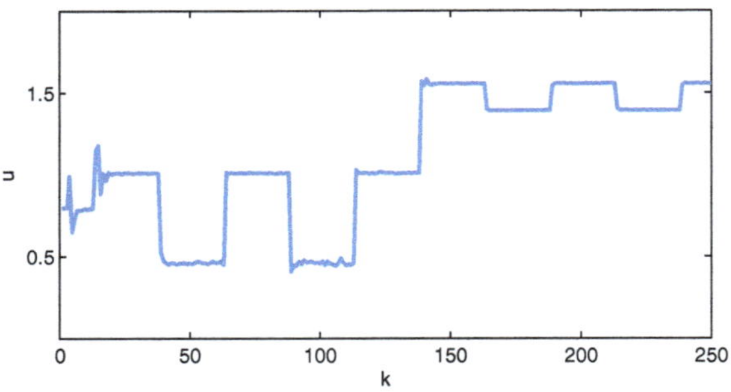

Fig. 24.6 Traces of the hyperparameter values changing through time and the active set updates. Coloured *solid lines* denote hyperparameter values and *vertical grey lines* denote time instants when the GP model was updated with new data

red circles. It is clear that the selected data points are evenly distributed through all nonlinear space, which indicates that the proposed method successfully adapts the GP model according to operating regions. The times when the GP model is updated are depicted in Fig. 24.6 as light grey lines. It can be seen that most updates occur, as expected, at changes of the set-point signal, especially at changes in dynamics. Similarly, holds for hyperparameter values, whose traces through the process are also shown in Fig. 24.6, denoted as coloured solid lines. It can be seen that hyperparameter values are mostly changing in a region of the first three steps, when most new information about the system is obtained. Once near-optimal values are reached, hyperparameter values are changing in a much smaller scale.

The main purpose of this implementation of the closed-loop control is the model adaptation according to system's dynamics. Therefore, once enough data about the system is obtained, the controller easily follows the set-point signal and adapts the GP model. But the controller can be further improved to somehow explore unknown space, especially at the beginning of the process or in any other cases when it is not possible to follow the set-point signal due to the lack of information about the dynamics. With such an improvement the controller is able to follow almost arbitrary changes in the process dynamics.

24.6 Conclusions

In this chapter an adaptive closed-loop control based on evolving GP models is described. Evolving GP model is the model where not only hyperparameters, but also the content of covariance matrix is changing with new in-streaming data that have enough information content in comparison with the previously contained data. This way the model keeps up with the dynamics which changes with the change of operating region.

The evolving GP model is upgraded with the minimum-variance controller that completes the proposed adaptive closed-loop system.

The illustrative example shows the satisfactory performance of the adaptive control with no initial knowledge and with a rapid change of operating region during its operation.

The proposed control algorithm which, we believe, is an improvement over similar and previously proposed algorithms is meant as a step forward in the control of nonlinear and time-variable dynamic systems with possible presence of uncertainties and disturbances. Potential applications of the proposed control are foreseen in fields of rehabilitation engineering and biotechnology, robotics, process and power engineering and elsewhere where nonlinear and time-variable dynamic systems can be found.

Acknowledgments This work has been supported by the Slovenian Research Agency, grant No. P2-0001

References

1. Ažman, K., Kocijan, J.: Application of Gaussian processes for black-box modelling of biosystems. ISA Trans. **46**, 443–457 (2007)
2. Ažman, K., Kocijan, J.: Fixed-structure Gaussian process model. Int. J. Syst. Sci. **40**(12), 1253–1262 (2009)
3. Cornford, D., Csato, L., Opper, M.: Sequential, sparse learning in Gaussian processes. In: Proceedings of the 7th International Conference on GeoComputation, vol. 44. Southampton, UK (2003)
4. Csató, L., Opper, M.: Sparse online Gaussian processes. Neural Comput. **14**(3), 641–668 (2002)
5. Deisenroth, M.P.: Efficient Reinforcement Learning using Gaussian Processes. Ph.D. thesis, Karlsruhe Institute of Technology, Karlsruhe (2010)
6. Deisenroth, M.P., Rasmussen, C.E.: PILCO: a model-based and data-efficient approach to policy search. In: Proceedings of the 28th International Conference on Machine Learning (ICML 2011). Bellevue, WA (2011)
7. Deisenroth, M.P., Rasmussen, C.E., Fox, D.: Learning to control a low-cost manipulator using data-efficient reinforcement learning. In: Proceedings of the International Conference on Robotics: Science & Systems (R:SS 2011). Los Angeles, CA (2011)
8. Deisenroth, M.P., Rasmussen, C.E., Peters, J.: Gaussian process dynamic programming. Neurocomputing **72**(7–9), 1508–1524 (2009)
9. Deisenroth, M., Peters, J., Rasmussen, C.: Approximate dynamic programming with Gaussian processes. In: Proceedings of American Control Conference (ACC), pp. 4480–4485. Seattle, WA (2008)
10. Deisenroth, M., Rasmussen, C.: Bayesian inference for efficient learning in control. In: Proceedings of Multidisciplinary Symposium on Reinforcement Learning (MSRL). Montreal, Canada (2009)
11. Filatov, N., Unbehauen, H.: Survey of adaptive dual control methods. IEE Proc.— Control Theory Appl. **147**(1), 119–128 (2000)
12. Isermann, R., Lachman, K.H., Matko, D.: Adaptive Control Systems. Systems and Control Engineering. Prentice Hall International, New York (1992)
13. Kocijan, J.: Control algorithms based on Gaussian process models: a state-of-the-art survey. In: Kolemisevska-Gugulovska, T.D., Stankovski, M.J. (eds.) Special International Conference on Complex systems: Synergy of Control, Communications and Computing—Proceedings of COSY 2011 Papers, September 16–20, 2011, Ohrid, Macedonia. The Society for Electronics,

Telecommunications, Automation, and Informatics of Macedonia, pp. 69–80. Skopje, Macedonia, Sept 2011
14. Kocijan, J.: Dynamic GP models: an overview and recent developments. In: Recent Researches in Applied Mathematics and Economics: proceedings of the 6th International Conference on Applied Mathematics. Simulation, Modelling, (ASM'12), pp. 38–43. Vougliameni, Greece (2012)
15. Kocijan, J.: Modelling and Control of Dynamic Systems Using Gaussian Process Models. Springer International Publishing, Cham (2016)
16. Kocijan, J., Girard, A., Banko, B., Murray-Smith, R.: Dynamic systems identification with Gaussian processes. Math. Comput. Model. Dyn. Syst. **11**(4), 411–424 (2005)
17. Kocijan, J., Likar, B.: Gas-liquid separator modelling and simulation with Gaussian-process models. Simul. Model. Pract. Theory **16**(8), 910–922 (2008)
18. Lázaro-Gredilla, M., Quiñonero Candela, J., Rasmussen, C.E., Figueiras-Vidal, A.R.: Sparse spectrum Gaussian process regression. J. Mach. Learn. Res. **11**, 1865–1881 (2010)
19. Murray-Smith, R., Sbarbaro, D., Rasmussen, C., Girard, A.: Adaptive, cautious, predictive control with Gaussian process priors. In: Proceedings of 13th IFAC Symposium on System Identification. Rotterdam, Netherlands (2003)
20. Murray-Smith, R., Sbarbaro, D.: Nonlinear adaptive control using nonparametric Gaussian process prior models. In: Proceedings of IFAC 15th World Congress. Barcelona (2002)
21. Narendra, K., Parthasarathy, K.: Identification and control of dynamical systems using neural networks. IEEE Trans. Neural Networks **1**(1), 4–27 (1990)
22. Nguyen-Tuong, D., Peters, J.: Learning robot dynamics for computed torque control using local Gaussian processes regression. In: Symposium on Learning and Adaptive Behaviors for Robotic Systems, pp. 59–64 (2008)
23. Nguyen-Tuong, D., Seeger, M., Peters, J.: Real-time local GP model learning, chap. From Motor Learning to Interaction Learning in Robots, vol. 264, pp. 193–207. Springer (2010)
24. Petelin, D., Grancharova, A., Kocijan, J.: Evolving Gaussian process models for prediction of ozone concentration in the air. Simul. Model. Pract. Theory **33**, 68–80 (2013)
25. Petelin, D., Kocijan, J.: Control system with evolving Gaussian process model. In: Proceedings of IEEE Symposium Series on Computational Intelligence, SSCI 2011. IEEE, Paris (2011)
26. Quinonero-Candela, J., Rasmussen, C.E.: A unifying view of sparse approximate Gaussian process regression. J. Mach. Learn. Res. **6**, 1939–1959 (2005)
27. Rasmussen, C.E., Deisenroth, M.P.: Probabilistic inference for fast learning in control. In: Recent Advances in Reinforcement Learning, Lecture Notes on Computer Science, vol. 5323, pp. 229–242. Springer (2008)
28. Rasmussen, C.E., Kuss, M.: Gaussian processes in reinforcement learning. In: Thurn, S., Saul, L., Schoelkopf, B. (eds.) Advances in Neural Information Processing Systems conference. vol. 16, pp. 751–759. MIT Press (2004)
29. Rasmussen, C.E., Williams, C.K.I.: Gaussian Processes for Machine Learning. MIT Press, Cambridge (2006)
30. Sbarbaro, D., Murray-smith, R.: An adaptive nonparametric controller for a class of nonminimum phase non-linear system. In: Proceedings of IFAC 16th World Congress. Prague, Czech Republic (2005)
31. Sbarbaro, D., Murray-Smith, R., Valdes, A.: Multivariable generalized minimum variance control based on artificial neural networks and Gaussian process models. In: International Symposium on Neural Networks. Springer (2004)
32. Sbarbaro, D., Murray-Smith, R.: Self-tuning control of nonlinear systems using Gaussian process prior models. In: Murray-Smith, R., Shorten, R. (eds.) Switching and Learning in Feedback Systems. Lecture Notes in Computer Science, vol. 3355, pp. 140–157. Springer, Heidelberg (2005)
33. Seeger, M.: Low Rank Updates for the Cholesky Decomposition. University of California at Berkeley, Technical report (2008)
34. Seeger, M., Williams, C.K.I., Lawrence, N.D.: Fast forward selection to speed up sparse gaussian process regression. In: Ninth International Workshop on Artificial Intelligence and Statistics. Society for Artificial Intelligence and Statistics (2003)

35. Snelson, E., Ghahramani, Z.: Sparse Gaussian processes using pseudo-inputs. In: Neural Information Processing Systems (2005)
36. Wittenmark, B.: Adaptive dual control. In: Control Systems, Robotics and Automation, Encyclopedia of Life Support Systems (EOLSS), Developed under the auspices of the UNESCO. Eolss Publishers, Oxford, UK, Jan 2002

Part V
Novel Control Ideas and Variable-Structure Systems Control

So far as laws of mathematics refer to reality, they are not certain. And so far as they are certain, they do not refer to reality.

Albert Einstein

Chapter 25
Attenuation of Uncertain Disturbances Through Fast Control Inputs

Alexander B. Kurzhanski and Alexander N. Daryin

Abstract This paper introduces a new class of controls that ensure an effect similar to that produced by conventional matching conditions between control and disturbance inputs in a linear system, but now for a broader class of such inputs. Namely, this is due to an application of piecewise-constant control functions with varying amplitudes, generated by approximations of "ideal controls," which are linear combinations of delta functions and their higher order derivatives. Such a class allows to calculate feedback control solutions by solving problems of open-loop control, thus reducing the overall computation burden.

25.1 Introduction

The design of closed-loop control strategies is a cornerstone in mathematics of control under unknown disturbances. As is well known, such problem solutions require that controls depend not only on time (as under open-loop), but also on the online state of the system. But these are rather difficult to calculate. Here below, for a linear system , we indicate a new class of bounded controls (the so-called "fast controls") that effectively solve the problem of closed-loop target control under unknown disturbances with given hard geometric bounds. This is reached by reducing the problem of closed-loop control to simpler problems of open-loop control. The new class of controls is constructed through approximations of generalized impulse controls inputs that admit higher derivatives of delta functions [2, 12–14].

A.B. Kurzhanski · A.N. Daryin (✉)
Department of Computational Mathematics and Cybernetics,
Moscow State (Lomonosov) University, 119991 Moscow, Russia
e-mail: daryin@cs.msu.su; a.daryin@gmail.com
URL: http://sa.cs.msu.su/staff/daryin/en/

A.B. Kurzhanski
e-mail: kurzhans@mail.ru
URL: http://sa.cs.msu.su/staff/kurzhanski/en/

G.M. Dimirovski (ed.), *Complex Systems*, Studies in Systems, Decision and Control 55, DOI 10.1007/978-3-319-28860-4_25

25.2 The Problem

Consider the following linear system with control u and uncertain disturbance v:

$$\dot{x}(t) = A(t)x(t) + B(t)u(t) + C(t)v(t), \quad t \in [t_0, t_1]. \tag{25.1}$$

The vector dimensions are $x \in \mathbb{R}^n$, $u \in \mathbb{R}^m$, $v \in \mathbb{R}^k$, $m, k \leq n$ and the time interval $[t_0, t_1]$ is fixed in advance. The given matrix functions $A(t) \in \mathbb{R}^{n\times n}$, $B(t) \in \mathbb{R}^{n\times m}$, $C(t) \in \mathbb{R}^{n\times k}$ are assumed sufficiently smooth to allow our further constructions.

The disturbance $v(t)$ is a piecewise-continuous function subject to hard bound $v(t) \in \mathscr{Q}(t), t \in [t_0, t_1]$, where $\mathscr{Q}(t)$ is a set-valued function with values in conv $\mathbb{R}^k$—the class of nonempty convex compacts in $\mathbb{R}^k$. The function $\mathscr{Q}(t)$ is taken continuous in the Hausdorff metric. It could be, for example, defined by inequalities $\|v_i(t)\| \leq \nu_i$, $i = 1, \dots, k$.

The aim of the control is to steer the system to a given target set $\mathscr{M} \in \operatorname{conv} \mathbb{R}^n$ at a prespecified time t_1, despite the disturbances.

Considered here is the class of control functions described as follows. It is well known from theories of control under uncertainty and differential games [1, 5–7, 10] that if $B(t) \equiv C(t)$ and the control $u(t)$ belong to the same class as $v(t)$, then solutions to terminal control problems(like min-max over u, v of the distance to set $\mathscr{M} \in \operatorname{conv} \mathbb{R}^n$, at terminal time) are the same for both open-loop and closed-loop controls. Such requirements on control and disturbance inputs are known as "matching conditions," which may be such: if $u(t) \in \mathscr{P}(t)$, then $\mathscr{P}(t) = \alpha(t)\mathscr{Q}(t)$, $\|\alpha(t)\| \geq 1$, so that bounds on u and v are similar ("homothetic").

However, a completely different situation arises when no type of matching conditions is true. In this case the closed- loop control problem is much harder to solve than the open loop, and may require a significant increase in the computational burden.

Loosely speaking, we have the next.

Problem 25.1 Specify classes of controls that allow to reduce the closed-loop terminal min-max problem to problems of open-loop control.

It is known that in the case of conventional matching conditions between u and v the solutions may avoid the operation of convexification, which is crucial in the absence of such conditions and requires a larger computational burden. The *main point of the present paper* is that here we present a class of controls that produces a similar effect, and actually allows to *avoid convexification* for broader cases than conventional matching conditions. This is the class of piecewise-constant functions with varying amplitudes, generated by approximations of "ideal controls"—linear combinations of delta functions and their higher order derivatives. Such a class allows to calculate feedback controls by solving problems of open-loop control, avoiding operations of convexification.

25.3 Generalized Controls

Let the control input u be a generalized function (a distribution) of order s. The latter may be presented as the sum of generalized derivatives of functions $U_j(\cdot) \in BV([t_0, t_1]; \mathbb{R}^m)$ of bounded variation, namely [4, 13]:

$$u(t) = \sum_{j=0}^{s} \frac{d^{j+1} U_j(t)}{dt^{j+1}}, \quad U_j(\cdot) \in BV([t_0, t_1]; \mathbb{R}^m). \tag{25.2}$$

In particular, as indicated in [9], the optimal generalized control problem of steering the system to a prescribed state *in the absence of uncertainty* has the form

$$u(t) = \sum_{i=1}^{n} \sum_{j=0}^{s} h_{i,j} \delta^{(j)}(t - \tau_j), \tag{25.3}$$

where $\delta(t) = \chi'(t)$ is the delta function—the generalized derivative of the ("Heaviside") function $\chi(t) \in BV[t_0, t_1]$; vectors $h_{i,j} \in \mathbb{R}^m$ define the direction and the amplitude of the generalized impulses, τ_i are the times of these impulses.

Substituting control input (25.2) into the original differential equation (25.1), we come to the next impulse control system [9]:

$$dx(t) = A(t)x(t)dx(t) + \mathscr{B}(t)dU(t) + C(t)v(t)dt, \tag{25.4}$$

on $t \in [t_0, t_1]$, where

$$\mathscr{B}(t) = \left[L_0(t) \ \dots \ L_s(t)\right],$$

and matrix functions $L_j(t)$ are here defined by the recurrence relations

$$L_0(t) = B(t), \quad L_j(t) = A(t)L_{j-1}(t) - L'_{j-1}(t). \tag{25.5}$$

The new function $U(t) = \left[U_0(t) \ \dots \ U_s(t)\right] \in BV([t_0, t_1]; \mathbb{R}^{m(s+1)})$ is a *a generator of impulse control* through its derivatives, as shown in (25.3). Its aim is to ensure $x(t_1 + 0) \in \mathscr{M}$.

Higher order generalized impulses may increase the control possibilities in the sense that Range $\mathscr{B}(t) \supseteq$ Range $B(t)$ (here and further Range is the column space of a matrix).

Assumption 25.1 There exists an $s \leq n - 1$ such that Range $\mathscr{B}(t) \supseteq$ Range $C(t)$ for all $t \in [t_0, t_1]$.

This assumption holds if, for example, $A(t) \equiv A$, $B(t) \equiv B$, and $[A, B]$ is a controllable pair. In this case the minimum value of s coincides with the controllability index of the system.

We now replace the "ideal" impulse control in system (25.4) by physically realizable bounded functions. To do that we introduce a hard bound on the control input $\mathbf{u}(t) = dU/dt$: $\mathbf{u}(t) \in \mathscr{P}(t)$. Then system (25.4) acquires the form

$$\dot{x}(t) = A(t)x(t) + \mathscr{B}(t)\mathbf{u}(t) + C(t)v(t), \quad t \in [t_0, t_1]. \tag{25.6}$$

Here $\mathbf{u}(t) = \begin{bmatrix} u_0(t) & \dots & u_s(t) \end{bmatrix} \in \mathbb{R}^{m(s+1)}$, and the aim of the control is again $x(t_1) \in \mathscr{M}$.

It is known that if the *matching condition* holds[1]

$$(\mathscr{B}(t)\mathscr{P}(t) \dot{-} C(t)\mathscr{Q}(t)) + C(t)\mathscr{Q}(t) = \mathscr{B}(t)\mathscr{P}(t) \tag{25.7}$$

then the solution of feedback control problem simplifies significantly [6, 7]. This condition is equivalent to convexity of $f(\ell) = \rho(\ell \mid \mathscr{B}(t)\mathscr{P}(t)) - \rho(\ell \mid C(t)\mathscr{Q}(t))$—the difference in support functions for sets $\mathscr{B}(t)\mathscr{P}(t)$ and $C(t)\mathscr{Q}(t)$.

Our aim will be to match the bounds of control and disturbance in order to satisfy condition (25.7). With set $\mathscr{Q}(t)$ given, there exist at least the next two approaches:

1. Choose an appropriate $\mathscr{P}(t)$.
2. Choose $\mathscr{P}(t)$ such that $\mathscr{B}(t)\mathscr{P}(t) \dot{-} C(t)\mathscr{Q}(t) \neq \emptyset$. Then choose a set $\hat{\mathscr{Q}}(t) \supseteq \mathscr{Q}(t)$ such that the matching condition will hold.

25.4 Choosing Appropriate Constraints

25.4.1 General Case

Lemma 25.1 *For any set $\mathscr{N}(t) \in \operatorname{conv} \operatorname{Range} \mathscr{B}(t)$ there exists a set $\mathscr{P}(t) \in \operatorname{conv} \mathscr{R}^{m(s+1)}$, such that $\mathscr{B}(t)\mathscr{P}(t) = \mathscr{N}(t)$.*

Assumption 25.2 Matrices $A(t)$ and $B(t)$ are such that $\operatorname{rank} \mathscr{B}(t) \equiv \text{const}$. Furthermore, the indices of basis columns of $\mathscr{B}(t)$ may be chosen as piecewise-constant functions of t with a finite number of discontinuities.

Lemma 25.2 *Under Assumption 25.2 function $\mathscr{P}(t)$ may be chosen piecewise-continuous in t and upper semicontinuous in x with respect to inclusion.*

Then for the matching condition to hold, it is sufficient to set

$$\mathscr{N}(t) = \alpha C(t)\mathscr{Q}(t) + \mathscr{N}_0(t), \quad \alpha \geq 1,$$

where $\mathscr{N}_0(t) \in \operatorname{conv} \operatorname{Range} \mathscr{B}(t)$ is a parameter set-valued function that may be used for fine-tuning.

[1] Symbol $\dot{-}$ denotes the geometric (Minkowski) difference of the sets: $A \dot{-} B = \{x \mid x + B \subseteq A\}$.

25.4.2 Special Case

Consider a system with $A(t) \equiv A$, $B(t) \equiv B$ where the pair $[A, B]$ is completely controllable. We assume that the constraint on disturbance $\mathcal{Q}(t) \equiv \mathcal{Q}$ and the matrix $C(t) \equiv C$ are as described below.

In this case

$$\mathcal{B}(t) \equiv \mathcal{B} = \left[B\ AB\ \ldots\ A^s B\right].$$

Suppose the disturbance is $\mathbf{v}(t) = \left[v_1(t)\ \ldots\ v_r(t)\right]$ and matrix C is

$$C = \left[k_1 A^{j_1} B\ \ldots\ k_r A^{j_r} B\right], \quad 0 \le j_1 < j_2 < \ldots < j_r \le s; \quad k_i \in \mathbb{R}.$$

Suppose the constraint on disturbance $\mathcal{Q}$ is

$$\left\|v_1(t)\right\| \le 1, \quad \ldots, \quad \left\|v_s(t)\right\| \le 1.$$

Then the constraint on control may be chosen as

$$\left\|u_0(t)\right\| \le \mu_0, \quad \ldots, \quad \left\|u_s(t)\right\| \le \mu_s,$$

and the matching condition holds if

$$\mu_{j_1} \ge k_1, \quad \ldots, \quad \mu_{j_r} \ge k_r.$$

25.4.3 Example

Consider a three-body oscillating system [15]

$$\begin{cases} m\ddot{w}_1 = k(w_2 - 2w_1) + mv_1(t), \\ m\ddot{w}_2 = k(w_3 - 2w_2 + w_1) + mv_2(t), \\ m\ddot{w}_3 = k(w_2 - w_3) + mu(t) + mv_3(t), \end{cases} \tag{25.8}$$

that consists of a chain of linked weights of mass m connected by springs of stiffness k. Variables w_j are the displacements of the weights from equilibrium. Control u and disturbance v_j are the forces applied to the weights. We assume the hard bound on disturbance $\left\|v_j(t)\right\| \le \nu_j, j = \overline{1,3}$.

The matching condition for the system (25.8) does not hold since the control u only enters the last equation, whereas the disturbance is present in each of the equations.

Rewriting the system (25.8) in normal form (denoting $\omega = k/m$) we get

$$\begin{cases} \dot{x}_j = x_{3+j}, \quad j = \overline{1,3}; \\ \dot{x}_4 = \omega(x_2 - 2x_1) + v_1(t), \\ \dot{x}_5 = \omega(x_3 - 2x_2 + x_1) + v_2(t), \\ \dot{x}_6 = \omega(x_2 - x_3) + u(t) + v_3(t). \end{cases} \tag{25.9}$$

To fulfill condition Range $\mathscr{B}(t) \supseteq$ Range $C(t)$ it is necessary to apply distributions at least of order $s \geq 4$. In our example we choose $s = 5$. Then matrix $\mathscr{B}(t)$ will be

$$\mathscr{B} = \begin{bmatrix} 0 & 0 & 0 & 0 & 0 & \omega^2 \\ 0 & 0 & 0 & \omega & 0 & -3\omega^2 \\ 0 & 1 & 0 & -\omega & 0 & 2\omega^2 \\ 0 & 0 & 0 & 0 & \omega^2 & 0 \\ 0 & 0 & \omega & 0 & -3\omega^2 & 0 \\ 1 & 0 & -\omega & 0 & 2\omega^2 & 0 \end{bmatrix}.$$

To match the bounds on u and v we first perform a linear substitution of variables:

$$\hat{u}_1 = u_1 - \omega u_3 + 2\omega^2 u_5, \quad \hat{u}_3 = \omega u_3 - 3\omega^2 u_5, \quad \hat{u}_3 = \omega^2 u_5,$$
$$\hat{u}_2 = u_2 - \omega u_4 + 2\omega^2 u_6, \quad \hat{u}_4 = \omega u_4 - 3\omega^2 u_6, \quad \hat{u}_6 = \omega^2 u_6.$$

Then the system (25.9) takes the form

$$\begin{cases} \dot{x}_1 = x_4 + \hat{u}_6(t), & \dot{x}_4 = \omega(x_2 - 2x_1) + +\hat{u}_5(t) + v_1(t), \\ \dot{x}_2 = x_5 + \hat{u}_4(t), & \dot{x}_5 = \omega(x_3 - 2x_2 + x_1) + \hat{u}_3(t) + v_2(t), \\ \dot{x}_3 = x_6 + \hat{u}_2(t), & \dot{x}_6 = \omega(x_2 - x_3) + \hat{u}_1(t) + v_3(t). \end{cases}$$

We further choose the bounds on the controls as

$$\|\hat{u}_1(t)\| \leq \alpha_3 \nu_3, \quad \|\hat{u}_3(t)\| \leq \alpha_2 \nu_2, \quad \|\hat{u}_5(t)\| \leq \alpha_1 \nu_1, \quad \alpha_j \geq 1.$$

Controls $\hat{u}_2(t)$, $\hat{u}_4(t)$, $\hat{u}_6(t)$ may be bounded by an arbitrary convex set. In particular, we may set $\hat{u}_2(t) = \hat{u}_4(t) = \hat{u}_6(t) = 0$ in order to preserve the original physical sense (the control is a force which acts only on the velocities, but not on the displacements).

25.4.4 Constraints for Disturbance

Let function $\mathscr{P}(t)$ satisfy the following assumptions:

1. $\mathscr{B}(t)\mathscr{P}(t) \dot{-} C(t)\mathscr{Q}(t) \neq \emptyset$;
2. $\mathscr{B}(t)\mathscr{P}(t)$— is a generating set [11, p. 324] in the space Range $\mathscr{B}(t)$.

Recall that $\mathscr{X} \in \operatorname{conv} \mathbb{R}^n$ is a generating set if for any set $\mathscr{Y} \subseteq \mathbb{R}^n$ satisfying $\mathscr{X} \dot{-} \mathscr{Y} \neq \emptyset$, there exists a set $\mathscr{Z} \in \operatorname{conv} \mathbb{R}^n$ such that $\mathscr{X} \dot{-} \mathscr{Y} + \mathscr{Z} = \mathscr{X}$. Since $\mathscr{X} \dot{-} \mathscr{Z} + \mathscr{Z} = \mathscr{X}$ in this case, the matching condition holds for sets $\mathscr{X}$ and $\mathscr{Z}$.

Then by Lemma 25.1 we replace the set $\mathscr{Q}(t)$ with $\hat{\mathscr{Q}}(t) \supseteq \mathscr{Q}(t)$ such that (25.7) is true.

Next we indicate how one can choose a generating set for the constraint on control.

For $n = 2$ any nonempty convex compact set is generating [11, Theorem 4.2.6].

For $n \geq 3$, let the Assumption 25.1 hold. Choose basis columns of $\mathscr{B}(t)$—we assume that these are first q columns (otherwise we renumber them). Then $\mathscr{B}(t) = \left[\mathscr{B}_1(t)\ \mathscr{B}_2(t)\right]$, where $\mathscr{B}_1(t) \in \mathbb{R}^{n \times q}$ is a matrix of full rank. The control may be represented as $\mathbf{u}(t) = \left[\mathbf{u}_1(t)\ \mathbf{u}_2(t)\right]$, $\mathbf{u}_1(t) \in \mathbb{R}^q$. Choose constraint on $\mathbf{u}(t)$ in the following way:

1. constraint on $\mathbf{u_1}$ is one of:

   (a) $\left\|(\mathbf{u}_1)_j\right\| \leq \mu_j$;

   (b) $\left\|\mathbf{u}_1\right\| \leq \mu$.

2. $\mathbf{u_2} = 0$ (temporarily).

Then by [11, Theorems 4.2.3, 4.2.4, 4.2.7] the set $\mathscr{B}(t)\mathscr{P}(t)$ is generating. Numbers μ_j (or number μ) should be chosen such that

$$\mathscr{B}(t)\mathscr{P}(t) \dot{-} C(t)\mathscr{Q}(t) \neq \emptyset.$$

After the extended set $\hat{\mathscr{Q}}(t)$ is chosen, the zero constraint on $\mathbf{u}_2(t)$ may be replaced with an arbitrary constraint $\mathbf{u}_2(t) \in \mathscr{P}_2(t)$.

25.5 Control Inputs for the Original System

The suggested approach allows us to find a feedback control for system (25.6), so that then, for a certain realization of $v(t)$, one may calculate the control trajectory $\mathbf{u}(t)$. After that it is necessary to indicate the corresponding control input for the original system (25.1).

It is not possible to apply representation (25.2) directly, since the smoothness (and even the continuity) of function $\mathbf{u}(t)$ is originally not guaranteed. To overcome this difficulty, we suggest to approximate the generalized controls using one of the following schemes:

1. In (25.3), replace the derivatives of delta functions by their bounded approximations. In this case we come to a system different from (25.6), for which it is necessary to apply the theory of the above.
2. Solve the control problem for the system (25.6), then approximate the realization of the control $\mathbf{u}(t)$ by functions sufficiently smooth to apply the representation (25.2).

25.5.1 First Scheme

Following [3, 8], we replace in (25.3) the derivatives of the delta function by their piecewise-constant approximations:

$$u(t) = \sum_{i=1}^{n} \sum_{j=0}^{s} h_{i,j} \Delta_h^{(j)}(t - \tau_j), \tag{25.10}$$

where $\Delta_h^{(0)}(t) = h^{-1}\mathbf{1}_{[0,h]}(t)$,

$$\Delta_h^{(j)}(t) = h^{-1}\left(\Delta_h^{(j-1)}(t) - \Delta_h^{(j-1)}(t-h)\right). \tag{25.11}$$

Note the following properties of the approximations.

1. The weak* limit (as $h \to 0$) of $\Delta_h^{(j)}(t)$ in the space of generalized functions of order j is $\delta^{(j)}(t)$.
2. Recurrence relations (25.5) lead to the next explicit form of these functions:

$$\Delta_h^{(j)}(t) = h^{-(j+1)} \sum_{i=0}^{j} (-1)^i C_j^i \mathbf{1}_{[ih,(i+1)h]}(t).$$

The Cauchy formula for system (25.6) is

$$x(\vartheta) = X(\vartheta, t_0)x_0 + \sum_{j=0}^{s} \int_{t_0}^{\vartheta} X(\vartheta, t)L_j(t)u_j(t)dt \\ + \int_{t_0}^{\vartheta} X(\vartheta, t)C(t)v(t)dt. \tag{25.12}$$

Note that functions $L_j(t)$ from (25.5) are defined by relations

$$L_j(t) = (-1)^j X(t, t_0)[X(t_0, t)B(t)]^{(j)}.$$

We then represent these as convolutions with derivatives of the delta function:

$$L_j(t) = X(t, t_0) \int_{\mathbb{R}} X(t_0, \tau)B(\tau)\delta^{(j)}(\tau - t)d\tau,$$

After that we pass to approximations (25.11):

$$M_h^{(j)}(t) = \int_{t}^{t+(j+1)h} X(t, \tau)B(\tau)\Delta_h^{(j)}(\tau - t)d\tau. \tag{25.13}$$

Theorem 25.1 *Matrix functions* $\mathscr{M}_h^{(j)}(t)$ *satisfy recurrence relations*

$$M_h^{(j)}(t) = h^{-1}(M_h^{(j-1)}(t) - X(t, t+h)M_h^{(j-1)}(t+h)),$$
$$M_h^{(0)}(t) = h^{-1}\int_t^{t+h} X(t,\tau)B(\tau)d\tau.$$

In particular, for $A(t) \equiv A$, $B(t) \equiv B$

$$M_h^{(j)} = h^{-j}(I - e^{-Ah})^j M_h^{(0)}, \quad M_h^{(0)} = h^{-1}\left[\int_0^h e^{At}dt\right]B.$$

Theorem 25.2 *Let the matrix function* $A(t)$ *be continuous, and* $B(t)$ *be* $s+1$ *times continuously differentiable. Then functions* $M_h^{(j)}(t)$ *will converge to* $L_j(t)$ *uniformly on* $[t_0, t_1]$, $j = 0, \dots, s$ *as* $h \to 0$.

Corollary 25.1 *Under the stated conditions, the matrix function*

$$\mathscr{M}_h(t) = \left(M_h^{(0)}(t)\ \dots\ M_h^{(s)}(t)\right)$$

converges to $\mathscr{B}(t)$ *uniformly on* $[t_0, t_1]$ *as* $h \to 0$.

Corollary 25.2 *If* rank $\mathscr{B}(t) \equiv n$, *then for sufficiently small* $h > 0$ *one also has* rank $\mathscr{M}_h(t) \equiv n$.

Substituting into (25.12) the functions $L_j(t)$ by $M_h^{(j)}(t)$ we get

$$x_h(\vartheta) = X(\vartheta, t_0)x_0 + \sum_{j=0}^{s}\int_{t_0}^{\vartheta} X(\vartheta, t)M_h^{(j)}(t)u_j(t)dt$$
$$+ \int_{t_0}^{\vartheta} X(\vartheta, t)C(t)v(t)dt. \tag{25.14}$$

This is the Cauchy formula for system

$$\dot{x}_h(t) = A(t)x_h(t) + \mathscr{M}_h(t)\mathbf{u}(t) + C(t)v(t). \tag{25.15}$$

Theorem 25.3 *Trajectories* $x_h(t)$ *of system* (25.15) *converge uniformly to the trajectory* $x(t)$ *of the system* (25.6) *with* $h \to 0$ *over* $[t_0, t_1]$.

Theorem 25.4 *Let* $u(t) \equiv 0$, $v(t) \equiv 0$ *for* $t \in (\vartheta, \vartheta + (s+1)h]$. *Then*

$$x_h(\vartheta + (s+1)h) = x(\vartheta + (s+1)h),$$

where $x(t)$ is the trajectory of the original system (25.1) *with control*

$$u_h(t) = \sum_{j=0}^{s} \int_{t_0}^{t} \Delta_h^{(j)}(t-\tau)u_j(\tau)d\tau. \tag{25.16}$$

Note that $u_h(t)$ is non-anticipative: it depends only on values of $\mathbf{u}(\tau)$ for $\tau \le t$, i.e., it may be calculated using only the information available by time t.

Theorems stated above provide the following scheme for calculating control inputs in the original system:

1. Fix $h > 0$ and consider system (25.15).
2. Apply one of the approaches to choose the bounds on control and disturbance, with $\mathscr{B}(t)$ replaced by $\mathscr{M}_h(t)$. (Due to Corollary 25.2, if Range $\mathscr{B}(t) = \mathbb{R}^n$, then Range $\mathscr{M}_h(t) =$ Range $\mathscr{B}(t)$.)
3. For system (25.15) with chosen constraints design a feedback control $\mathscr{U}(t,x)$.
4. Find the realization of control trajectory $\mathbf{u}(t)$.
5. Using (25.16), find the control input for the original system (25.1). (Since $u_h(t)$ depends only on the past values of $\mathbf{u}(t)$, it may be calculated online.)

25.5.2 Second Scheme

Here we briefly describe the second scheme of calculating the control input for the original system. Let $\mathbf{u}(t) = \left[u_0(t) \ \dots \ u_s(t)\right]$ be the realization of the control of system (25.6). We approximate it by convolving with sufficiently smooth functions $\hat{\mathbf{u}}(t) = \left[\hat{u}_0(t) \ \dots \ \hat{u}_s(t)\right]$:

$$\hat{u}_j(t) = h^{-1} \int_{t_0}^{t_1} K_j((t-\tau)/h)u_j(\tau)d\tau.$$

The convolution kernels $K_j(t)$ should satisfy the following requirements: $K_j(t) = 0$ for $t < 0$; $K_j(t) \ge 0$ for $t \ge 0$; $K_j(t)$ is j times continuously differentiable; they satisfy the normalization condition: $\int_0^\infty K_j(t)dt = 1$.

One may select $K_j(t)$, for example, as the following piecewise-polynomial functions:

$$K_j(t) = \mathbf{1}_{[0,1]}(t)C_j(t(1-t))^{j+1}, \quad C_j = \frac{(2j+3)!}{((j+1)!)^2}.$$

The control $\hat{\mathbf{u}}(t)$ corresponds to the next control input for the original system (25.1):

$$\hat{u}(t) = \sum_{j=0}^{s} \hat{u}_j^{(j)}(t) = \sum_{j=0}^{s} h^{-(j+1)} \int_{t_0}^{t_1} K_j^{(j)}((t-\tau)/h)u_j(\tau)d\tau.$$

This approximation has the following properties.

1. $\hat{u}_j(t) \to u_j(t)$ almost everywhere when $h \to 0$.
2. Trajectories $\hat{x}(t)$ of the system (25.1) under control $\hat{u}(t)$ coincide with the trajectories of the system (25.6) under control $\hat{\mathbf{u}}(t)$. The latter converge pointwise to the trajectories $x(t)$ of the system (25.6) under control $\mathbf{u}(t)$.
3. $\hat{u}(t)$ depends only on values of $\mathbf{u}(\tau)$ for $\tau \le t$, i.e., it may be calculated using only the information available by time t.

25.5.3 Example

Consider system

$$\begin{cases} \dot{x}_1(t) = x_2(t) + v_1(t), \\ \dot{x}_2(t) = u(t) + v_2(t), \end{cases}$$

with hard bound on disturbance as $\|v_1\| \le \mu_1$, $\|v_2\| \le \mu_2$.

For this system we have

$$\mathscr{B}(t) = \begin{bmatrix} 0 & 1 \\ 1 & 0 \end{bmatrix}, \quad \mathscr{M}_h(t) = \begin{bmatrix} h/2 & 1 \\ 1 & 0 \end{bmatrix}.$$

To apply the first scheme, we make a linear change of variables: $\hat{u}_1(t) = hu_1(t)/2 + u_2(t)$, $\hat{u}_2(t) = u_1(t)$, which leads to system (25.15) of form

$$\begin{cases} \dot{x}_{h1}(t) = x_{h2}(t) + \hat{u}_1(t) + v_1(t), \\ \dot{x}_{h2}(t) = \hat{u}_2(t) + v_2(t). \end{cases}$$

Here one may choose the following constraint on control: $\|\hat{u}_1\| \le \nu_1$, $\|\hat{u}_2\| \le \nu_2$, where $\nu_j \ge \mu_j$.

Applying the second scheme we get a system (25.6) of form

$$\begin{cases} \dot{x}_1(t) = x_2(t) + u_1(t) + v_1(t), \\ \dot{x}_2(t) = u_2(t) + v_2(t), \end{cases}$$

where the hard bound on control may be also chosen as $\|u_1\| \le \nu_1$, $\|u_2\| \le \nu_2$, with $\nu_j \ge \mu_j$.

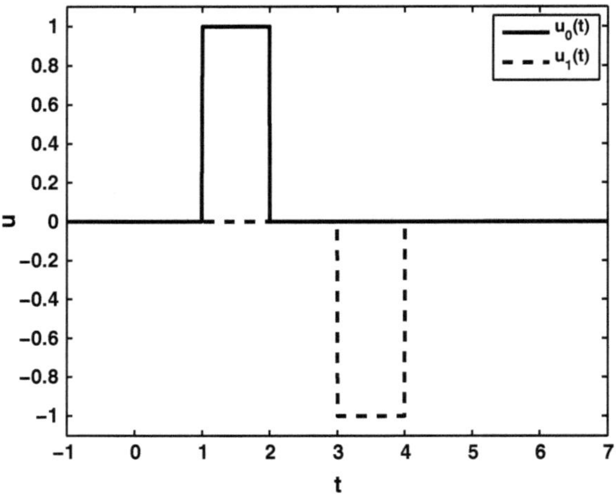

Fig. 25.1 Control input $u_j(t)$ for the system (25.6)

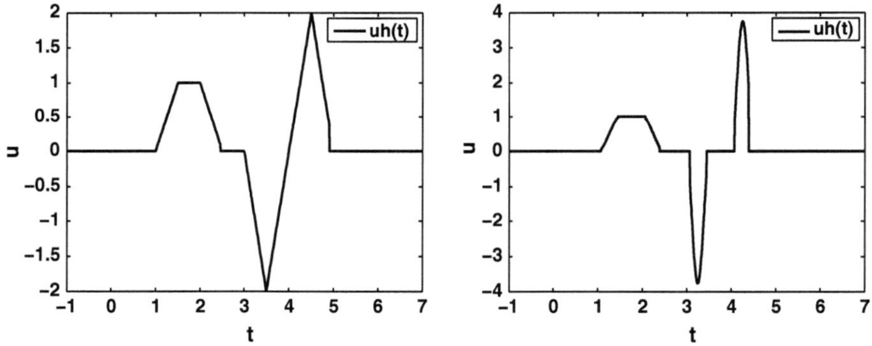

Fig. 25.2 Control input for the original system (25.1) calculated by the first (*left*) and the second scheme (*right*)

Suppose a control realization for system (25.6) is depicted in Fig. 25.1. Then Fig. 25.2 shows the control inputs for the original system (25.1) as calculated using both schemes. Here $t_0 = 0$, $t_1 = 5$, $h = 0{,}5$.

Acknowledgments This work is supported by the Russian Foundation for Basic Research (grants 12-01-00261-a, 12-01-31416-mol-a) and by the program State Support of the Leading Scientific Schools (grant NS-2239.2012.1).

References

1. Başar, T., Bernhard, P.: H^∞-Optimal Control and Related Minimax Design Problems. Birkhäuser, Boston (1995)
2. Bensoussan, A., Lions, J.L.: Contrôle impulsionnel et inéquations quasi variationnelles. Dunod, Paris (1982)

3. Dar'in, A.N., Kurzhanski, A.B.: Control synthesis in a class of higher-order distributions. Diff. Equ. **43**(11), 1479–1489 (2007)
4. Gel'fand, I.M., Shilov, G.E.: Generalized Functions. In: Properties and Operations, vol. 1. Academic Press, NY (1964)
5. Krasovski, N.N.: Rendezvous Game Problems. National Technical Information Service, Springfield (1971)
6. Krasovskii, N.N., Subbotin, A.I.: Game-theoretical control problems. (1988)
7. Kurzhanski, A.B.: Pontryagin's alternated integral in the theory of control synthesis. Proc. Steklov Inst. Math. **224**, 212–225 (1999)
8. Kurzhanski, A.B., Daryin, A.N.: Dynamic programming for impulse controls. Annu. Rev. Control **32**(2), 213–227 (2008)
9. Kurzhanski, A.B., Osipov, Y.S.: On controlling linear systems through generalized controls. Diff. Equ. **5**(8), 1360–1370 (1969) (In Russian)
10. Leitmann, G.: Optimality and reachability with feedback control. In: Avez, A., Blaquière, A., Marzollo, A. (eds.) Dynamical Systems and Microphysics: Geometry and Mechanics, pp. 119–141. Elsevier (1982)
11. Polovinkin, E.S., Balashov, M.V.: Elements of Convex and Strongly Convex Analysis. Fizmatlit, Moscow (2004) (In Russian)
12. Rockafellar, R.T., Wets, R.J.B.: Variational analysis. In: Grundlehren der mathematischen Wissenschaften, vol. 317. Springer (1998)
13. Schwartz, L.: Théorie des distributions. Hermann, Paris (1950)
14. Vladimirov, V.S.: Generalized Functions in Mathematical Physics. Mir Publishers, Moscow (1979)
15. Vostrikov, I.V., Dar'in, A.N., Kurzhanski, A.B.: On the damping of a ladder-type vibration system subjected to uncertain perturbations. Diff. Equ. **42**(11), 1524–1535 (2006)

Chapter 26
Sliding Manifold Design for Linear Systems with Scalar Unmatched Disturbances

Boban Veselić, Branislava Draženović and Čedomir Milosavljević

Abstract Although variable structure control systems (VSCS) are insensitive to so-called matched disturbances in ideal sliding mode (SM), they are vulnerable to the unmatched ones. This paper offers an efficient sliding manifold design method that minimizes the impact of unmatched disturbances onto SM dynamics. System sensitivity upon an unmatched constant external disturbance is evaluated through the steady-state vector norm. An infinite set of the sliding hyperplanes that minimize the chosen optimization criterion is determined. A way of selecting a manifold out of that set that provides adopted SM dynamics is also suggested. The proposed approach has been demonstrated on several numerical examples and investigated by computer simulations.

26.1 Introduction

Theoretical invariance to parameter perturbations and exogenous disturbances in ideal sliding mode (SM) is the main reason for the unabated interest in variable structure control (VSC) systems that lasts more than half a century. Conditions for such superior property were revealed and formulated in [1] by Draženović (1969), originally termed as invariance conditions, nowadays worldwide known as the

B. Veselić (✉)
Faculty of Electronic Engineering, University of Niš, Niš, Serbia
e-mail: boban.veselic@elfak.ni.ac.rs

B. Draženović
Faculty of Electrical Engineering, University of Sarajevo,
Sarajevo, Bosnia and Herzegovina
e-mail: brana_p@hotmail.com

Č. Milosavljević
Faculty of Electrical Engineering, University of Istočno Sarajevo,
Istočno Sarajevo, Bosnia and Herzegovina
e-mail: cedomir.milosavljevic@elfak.ni.ac.rs

G.M. Dimirovski (ed.), *Complex Systems*, Studies in Systems,
Decision and Control 55, DOI 10.1007/978-3-319-28860-4_26

matching conditions. This means that the SM dynamics is insensitive to the matched disturbances. However, unmatched disturbances have impact on the SM dynamics. In some cases SM along certain manifolds can result in severe dynamics deterioration under action of unmatched disturbances. This is the reason why SM control (SMC) systems with unmatched disturbances have attracted certain attention of the scientific community.

There are several approaches in dealing with unmatched disturbances within the context of SMC. Generally, they can be categorized into two groups according to the nature of the disturbances.

The first group is related to unmatched uncertainties of parametric nature, i.e., disturbances that occur due to parameter perturbations. A nonlinear control strategy providing motion both close to the sliding manifold and as close as possible to the ideal SM dynamics was proposed in [2] for a practical uncertainty class.

Dynamical approach to sliding manifold formulation is another aspect that can explicitly deal with unmatched uncertainties in SM. In [3] certain states were treated as inputs to the reduced order system that describes SM dynamics. Using an adaptive technique to design fictitious controllers for these inputs, the unmatched uncertainties can be tackled. This technique assumes certain constraints to the unmatched uncertainties as well as online estimation of the uncertainties present in the reduced order system. A similar concept of pseudo-control inputs into sliding function was utilized in [4, 5]. A new invariance condition in terms of LMI providing a linear sliding manifold that ensures SM dynamics to be stable and invariant to the matched and unmatched uncertainties was proposed in [6]. However, it is important to emphasize that the considered type of unmatched disturbances diminish as system states approach origin, so the asymptotic stability still can be achieved.

In case of other type of unmatched disturbances, which besides parametric contain external disturbances as well, asymptotic stability is never attained. A nonvanishing unmatched disturbance will force system trajectory to wander along the sliding manifold, not converging into the origin. The only way to handle this problem is to construct a sliding manifold that in some sense minimize SM dynamics sensitivity upon the unmatched disturbances. A way to choose sliding manifold in integral SMC systems that does not amplify the impact of an unmatched disturbance onto SM dynamics was proposed in [7]. Another appropriate sliding manifold selection [8] guarantees trajectory convergence into the minimal invariant ellipsoid for linear systems. State vector norm was used as a measure of SM sensitivity upon unmatched disturbances in [9]. Condition for a sliding hyperplane to minimize steady-state vector norm for linear scalar control systems subjected to a constant unmatched external disturbance was established. By imposing a single multiple eigenvalue that describes SM dynamics, the explicit solution was found.

This paper further studies minimization of a steady-state-based criterion function for linear systems under action of unmatched constant and slowly varying external disturbances. A new sliding manifold design method is developed that very efficiently finds an infinite set of sliding manifolds that minimize the optimization criterion. Furthermore, possibility of an arbitrary choice of SM dynamics under the minimization requirement is examined. Eigenvalue constraints are revealed in certain cases, for which a way to select and attain the feasible SM dynamics is suggested. The proposed sliding manifold design approach has been demonstrated on several numerical examples and verified by computer simulations.

The reminder of this paper is organized as follows. Section 26.2 analyzes impact of unmatched disturbances onto SM dynamics. A new sliding manifold design method is proposed in Sect. 26.3. Feasibility of arbitrary SM dynamics is investigated in Sect. 26.4. Demonstration of the developed method on three numeric examples is given in Sect. 26.5, whereas the simulation results are presented in Sect. 26.6. The paper ends with some concluding remarks and references.

26.2 SM in Systems with Unmatched Disturbances

Consider a linear dynamic system given by

$$\dot{x}(t) = Ax(t) + Bu(t) + jv(t), \tag{26.1}$$

where $x \in \Re^n$ is the state vector, $u \in \Re^m$ $(m < n)$ is the control vector, $A \in \Re^{n\times n}$ and $B \in \Re^{n\times m}$ are the system and input matrices, respectively, $j \in \Re^n$ is the disturbance input vector, and $v \in \Re$ represents a bounded external additive disturbance, $|v(t)| \le M < \infty$, $\forall t > 0$. The matching condition [1] implies that the disturbance acts through the control channel, that is, $\text{rank}[B \,\vdots\, j] = \text{rank}[B]$. It is assumed in the paper that this condition is not fulfilled in the considered system (26.1), i.e., $\text{rank}[B \,\vdots\, j] > \text{rank}[B]$. Also, it is assumed that the pair (A,B) is controllable and rank $(B) = m$.

Linear sliding manifolds, upon which SM is organized, are mainly used in SMC systems. A linear sliding manifold is defined by

$$\sigma(x) = Cx = 0, \qquad C \in \Re^{m\times n}, \quad \sigma = [\sigma_1, \sigma_2, \ldots, \sigma_m]^{\mathrm{T}}, \tag{26.2}$$

under restriction $\det(CB) \neq 0$, where σ is the switching function vector. The sliding manifold $\sigma = 0$ that is the intersection of m sliding hyperplanes $\sigma_i = 0$,$i = 1, \ldots, m$ includes the state space origin as the equilibrium point. To establish SM in a finite time it is sufficient to attain conditions $\sigma_i\dot{\sigma}_i \le -\eta_i|\sigma_i|$, $\eta_i > 0$, $i = 1, \ldots, m$ by an appropriate vector control, usually having discontinuous nature.

System dynamics in SM can be obtained using equivalent control method. Linear equivalent control [1] is a fictive linear control that provides sliding along the hyperplane (26.2). It can be determined from the well-known condition

$$\sigma = 0 \wedge \dot{\sigma}\big|_{u=u_{eq}} = 0 \quad \Rightarrow \quad u_{eq} = -(CB)^{-1}C(Ax + jv). \tag{26.3}$$

Substitution of (26.2) into (26.2) gives the full-order SM dynamics

$$\begin{aligned} \dot{x}(t) &= A_{eq}x(t) + j_{eq}v(t), \quad Cx(t) = 0, \\ A_{eq} &= PA, \quad j_{eq} = Pj, \quad P = I - B(CB)^{-1}C. \end{aligned} \tag{26.4}$$

It is obvious that the matrix P primarily determines SM behavior (26.4). To realize the impact of unmatched disturbances onto SM dynamics it is necessary to examine the properties of projector P.

Lemma 26.1 *Projector P has m zero eigenvalues and the remaining n-m eigenvalues are ones.*

Proof Suppose that λ is an eigenvalue of P, with the associated eigenvector v, so that $Pv = \lambda v$. Then $P^2 v = P(Pv) = P(\lambda v) = \lambda P v = \lambda^2 v$, indicating that P^2 has λ^2 eigenvalue associated with the unchanged eigenvector v. Since $P^2 = P$, identity $\lambda = \lambda^2$ must hold for each eigenvalue of P. The last equation has two solutions, $\lambda = 0$ and $\lambda = 1$, showing that the eigenvalues of P can take only two values, 0 or 1. It is shown in [10] that rank(P) = n-rank($B(CB)^{-1}C$) = n-rank(B). Since rank(B) = m, then rank(P) = n-m. Consequently, P has m zero eigenvalue, while the remaining n-m eigenvalues must be ones. □

Remark 26.1 Left and right eigenvectors that correspond to the zero eigenvalues of P are defined by C and B, respectively. It is obvious that $CP = 0$ and $PB = 0$.

Remark 26.2 Matrix A_{eq} has m zero eigenvalues ($CA_{eq} = CPA = 0$). Consequently, $\det(A_{eq}) = 0$ implying that A_{eq} is a singular matrix.

Remark 26.3 The remaining n-m nonzero eigenvalues of A_{eq} defines SM dynamics and should be stable.

Since rank$[B \mid j] \neq$ rank$[B]$, i.e., the disturbance is unmatched, $j_{eq} \neq 0$ and the equivalent disturbance $j_{eq}v(t)$ in (26.4) affects SM dynamics. Let $v(t)$ be an unknown constant scalar disturbance, i.e., $v(t) = Vh(t)$, $0 < V \leq M$ and $h(t)$ is the unity step function. Then the system trajectory will reach a nonzero steady state $x(\infty) = \lim_{t\to\infty} x(t) \neq 0$ along the manifold. By letting $t \to \infty$ in (26.4) under condition $\lim_{t\to\infty} \dot{x}(t) = 0$, it is obtained

$$\begin{aligned} A_{eq}x(\infty) &= -j_{eq}V, \\ Cx(\infty) &= 0, \end{aligned} \quad \Rightarrow \quad \begin{bmatrix} A_{eq} \\ \hline C \end{bmatrix} x(\infty) = \begin{bmatrix} -j_{eq}V \\ \hline 0 \end{bmatrix}. \tag{26.5}$$

System (26.5) consists of $n + m$ equations with n unknowns. Since $\text{rank}(A_{eq}) = n - m$, a unique solution for $x(\infty)$ exists. To solve this matrix equation with respect to $x(\infty)$, Eq. (26.5) is left multiplied by an $nx(n+m)$ matrix $\left[A_{eq}^{\mathrm{T}} \mid C^{\mathrm{T}}\right]$. Then it follows

$$\left[A_{eq}^{\mathrm{T}} \mid C^{\mathrm{T}}\right]\begin{bmatrix} A_{eq} \\ \hline C \end{bmatrix} x(\infty) = \left[A_{eq}^{\mathrm{T}} \mid C^{\mathrm{T}}\right]\begin{bmatrix} -j_{eq}V \\ \hline 0 \end{bmatrix} \Rightarrow$$
$$\Rightarrow (A_{eq}^{\mathrm{T}}A_{eq} + C^{\mathrm{T}}C)\,x(\infty) = -A_{eq}^{\mathrm{T}}j_{eq}V. \tag{26.6}$$

Now it is easy to find the steady state

$$x(\infty) = -(A_{eq}^{\mathrm{T}}A_{eq} + C^{\mathrm{T}}C)^{-1}A_{eq}^{\mathrm{T}}j_{eq}V, \tag{26.7}$$

which, apart from A, B, and j, depends also on matrix C explicitly and implicitly through A_{eq}.

However, a more elegant approach in determining the steady state using the properties of the projector P is proposed in the following lemma.

Lemma 26.2 *Steady state of system* (26.4) *on the sliding manifold* (26.2) *can be expressed as*

$$x(\infty) = A^{-1}(Bk - jV), \tag{26.8}$$

with $k \in \Re^m$ being some vector and under assumption that the matrix A is nonsingular.

Proof Equation (26.5) that defines the steady state may be rewritten as

$$P(Ax(\infty) + jV) = 0, \quad Cx(\infty) = 0. \tag{26.9}$$

According to equality $P(Ax(\infty) + jV) = 0$ it can be concluded using property $PB = 0$ that the vector $Ax(\infty) + jV$ may be expressed as $Ax(\infty) + jV = Bk$, where k is a constant vector. Under assumption that A is invertible, $x(\infty)$ may be found as (26.8). □

Remark 26.4 Expression (26.8) is free of matrix C, but $x(\infty)$ is on the sliding manifold so $Cx(\infty) = 0$ must also hold.

Let $x_0(\infty)$ denote normalized steady state that is obtained as $x_0(\infty) = x(\infty)/V$.

26.3 Linear Sliding Manifold Design

As established in the previous section, SM dynamics is affected by unmatched disturbances. Consequently, the steady state differs from the origin and depends on the matrix C. It is reasonable then to search for the sliding manifolds that provide minimal steady-state error for given unmatched disturbance. Sensitivity of the SM dynamics upon unmatched disturbances is evaluated here by means of Euclid vector norm $||x|| = (x^T x)^{1/2}$, which represents state distance from the origin. Therefore, the proposed optimization criterion may be expressed in a form

$$J = x^T(\infty)x(\infty) = ||x(\infty)||^2. \tag{26.10}$$

Substitution of (26.7) into criterion function (26.10) gives

$$J(C) = j_{eq}^T A_{eq}(A_{eq}^T A_{eq} + C^T C)^{-2} A_{eq}^T j_{eq} V^2. \tag{26.11}$$

Matrix C can be decomposed as $C = [c_1\ c_2 \cdots c_m]^T$ where vectors $c_i \in \Re^{1\times n}$, $i = 1, \ldots, m$ represent rows of C. To minimize a vector-valued scalar function $J(c_1, \cdots, c_m)$ it is necessary that m gradient vectors be zero, that is,

$$\nabla J(c_1, \ldots, c_m) = \frac{\partial J}{\partial c_i} = \begin{bmatrix} \frac{\partial J}{\partial c_{i1}} & \cdots & \frac{\partial J}{\partial c_{in}} \end{bmatrix}^T = 0,\ i = 1, \ldots, m. \tag{26.12}$$

This requirement imposes an nxm-dimensional homogeneous system of nonlinear equations with nxm unknowns. Besides trivial solution $C = 0$, such system will have an infinite number of solutions. Forming and solving the system (26.12) are not always so straightforward, since difficult differentiations and calculations might be involved.

However, the result given by Lemma 26.2 simplifies the problem to optimization with respect to vector k.

Theorem 26.1 *A dynamic system* (26.1) *in SM along the sliding manifold* (26.2), *subjected to an unmatched step disturbance* $v(t) = Vh(t)$, *will have a unique steady state*

$$\begin{aligned} x_{\min}(\infty) &= x_{\min 0}(\infty) \cdot V, \\ x_{\min 0}(\infty) &= A^{-1}[B(B^T A_s B)^{-1} B^T A_s - I]j, \end{aligned} \tag{26.13}$$

which ensures minimization of (26.10). $A_s = (A^{-1})^T A^{-1}$ *is a regular symmetric matrix. Also, for* $n > 2$ *there is an infinite set of sliding manifolds defined by*

$$\begin{aligned} C &= C_0[h \mid I_{n-1}], \quad C_0 \in \Re^{m\times(n-1)}, \\ h &= -\frac{1}{x_{1\min 0}(\infty)}[x_{2\min 0}(\infty) \quad \cdots \quad x_{n\min 0}(\infty)]^T, \end{aligned} \tag{26.14}$$

where matrix C_0 *may be freely chosen.* C_0 *determines SM dynamics that provides minimal steady state* (26.13).

Proof Expansion of (26.10) using (26.8) gives

$$J = k^T B^T A_s Bk - k^T B^T A_s jV - j^T A_s BkV + j^T A_s jV^2. \tag{26.15}$$

Minimization of (26.15) with respect to k can be done by solving $\partial J/\partial k = 0$ for $k = k_{min}$, giving the condition

$$B^T A_s B k_{min} - B^T A_s j V = 0. \qquad (26.16)$$

Since rank$[B \mid j] \neq$ rank$[B]$, solution $Bk_{min} = jV$ of (26.16) is rejected. The other solution is $k_{min} = (B^T A_s B)^{-1} B^T A_s j V$ that according to (26.8) gives (26.13). To prove that the steady state (26.13) ensures minimum of (26.10), it is necessary to show that Hessian $H = \partial^2 J/\partial k^2$ is a positive definite matrix. The second derivative of (26.15) with respect to k gives $H = 2B^T A_s B = 2B^T (A^{-1})^T A^{-1} B$. Since H can be expressed as $H = G^T G$, where $G = \sqrt{2} A^{-1} B$, it can be easily verified that H has all real nonnegative eigenvalues using singular value decomposition. Hence, H is a positive definite symmetric matrix indicating that the attained extremum is minimum.

To reach the minimal steady state (26.13) in SM, sliding manifold must contain the given point. Therefore,

$$Cx_{min}(\infty) = Cx_{min0}(\infty) \cdot V = 0 \quad \Rightarrow \quad Cx_{min0}(\infty) = 0 \qquad (26.17)$$

must hold. Since $x_{min0}(\infty)$ is known, to satisfy (26.17) the elements of one column of C must be constrained, i.e., expressed in terms of the remaining elements of C and the elements of $x_{min0}(\infty)$. This means that only the remaining $n - 1$ columns of C can be freely chosen. For example, if the first column of C is expressed in terms of others, then C can be partitioned as $C = [c_1 \mid C_0]$, $C_0 \in \Re^{m \times (n-1)}$, $c_1 \in \Re^{m \times 1}$. According to (26.17) it follows

$$c_1 = C_0 h, \quad h = \begin{bmatrix} \dfrac{-x_{2min0}(\infty)}{x_{1min0}(\infty)} & \cdots & \dfrac{-x_{nmin0}(\infty)}{x_{1min0}(\infty)} \end{bmatrix}^T. \qquad (26.18)$$

Then $C = [C_0 h \mid C_0]$, which confirms (26.14). □

Geometric interpretation of Theorem 26.1 is that the sliding manifold defined by C must contain the origin and the predetermined $x_{min0}(\infty)$, in order to provide minimal steady-state error under action of an unmatched step disturbance. Obviously, an infinite set of matrices C can be found that satisfies (26.17), defining an infinite number of sets of m hyperplanes whose intersections contain the obtained point, which is stable minimum.

Remark 26.5 The result from Theorem 26.1 can also be applied in case of slowly varying disturbances $v(t)$, when the SM dynamics is much faster than the disturbance change. In that case, after the transient phase, the assumption of the conducted analysis $\dot{x}(t) \to 0$ is valid.

26.4 Sliding Mode Dynamics

Optimization criterion (26.10) (i.e., steady-state error minimization) imposes on system the additional algebraic constraint in matrix C design. Besides providing desired SM dynamics, matrix C should satisfy relation (26.17) as well. Therefore, it is important to investigate the impact of such algebraic constraint onto feasibility of desired SM dynamics.

The reduced-order SM dynamics can be assessed by transforming the system into regular form [11]. Therefore, consider a coordinate change $x = T\bar{x}$. If the input matrix is partitioned as

$$B = \begin{bmatrix} B_1 \\ \hline B_2 \end{bmatrix}, \quad B_1 \in \Re^{(n-m)\times m}, \quad B_2 \in \Re^{m\times m}, \quad \det(B_2) \neq 0, \tag{26.19}$$

the transformation matrix T is given by

$$T = \left[\begin{array}{c|c} I_{n-m} & -B_1 B_2^{-1} \\ \hline 0 & B_2^{-1} \end{array}\right]^{-1}. \tag{26.20}$$

Linear transformation of system (26.1) results in the regular form

$$\begin{aligned} &\dot{\bar{x}}(t) = \bar{A}\bar{x}(t) + \bar{B}u(t) + \bar{j}v(t), \\ &\bar{x} = \begin{bmatrix} \bar{x}_1 \\ \hline \bar{x}_2 \end{bmatrix}, \bar{x}_1 \in \Re^{n-m}, \bar{x}_2 \in \Re^{m}, \bar{j} = T^{-1}j = \begin{bmatrix} \bar{j}_1 \\ \hline \bar{j}_2 \end{bmatrix}, \\ &\bar{A} = T^{-1}AT = \left[\begin{array}{c|c} A_{11} & A_{12} \\ \hline A_{21} & A_{22} \end{array}\right], \bar{B} = T^{-1}B = \begin{bmatrix} 0 \\ \hline I_m \end{bmatrix}. \end{aligned} \tag{26.21}$$

The switching function may be rewritten using the state transformation as $\sigma = Cx = CT\bar{x} = \bar{C}\bar{x}$, $\bar{C} = CT = [\bar{C}_1 \mid \bar{C}_2]$, $\bar{C}_1 \in \Re^{m\times(n-m)}$, $\bar{C}_2 \in \Re^{m\times m}$. In the SM, it holds $\bar{C}_1\bar{x}_1 + \bar{C}_2\bar{x}_2 = 0$, so the last m coordinates can be expressed as $\bar{x}_2 = -\bar{C}_2^{-1}\bar{C}_1\bar{x}_1$. Hence, system dynamics in SM can be described by reduced order system

$$\begin{aligned} &\dot{\bar{x}}_1(t) = (A_{11} - A_{12}\bar{C}_2^{-1}\bar{C}_1)\bar{x}_1(t) + \bar{j}_1 v(t) = A_r(\bar{C})\bar{x}_1(t) + \bar{j}_1 v(t), \\ &\bar{x}_2(t) = -\bar{C}_2^{-1}\bar{C}_1\bar{x}_1(t). \end{aligned} \tag{26.22}$$

If the pair (A, B) is controllable then the same holds for the pair (A_{11}, A_{12}) and the eigenvalues of the system matrix $A_r(\bar{C}) = A_{11} - A_{12}\bar{C}_2^{-1}\bar{C}_1$ can be adjusted by selection of matrix $\bar{C}$, i.e., matrix C. However, Matrix C cannot be entirely arbitrary chosen, since it must satisfy (26.17).

Let the transformation matrix T be partitioned as

$$T = \begin{bmatrix} T_1 \\ T_2 \end{bmatrix}, \quad T_1 \in \Re^{1\times n}, \quad T_2 \in \Re^{(n-1)\times n}. \tag{26.23}$$

Taking into account (26.14), it follows

$$\overline{C} = CT = C_0[h \mid I_{n-1}] \cdot \begin{bmatrix} T_1 \\ \hline T_2 \end{bmatrix} = C_0(hT_1 + T_2) = C_0 S\,. \tag{26.24}$$

Matrix $S = (hT_1 + T_2) \in \Re^{(n-1)\times n}$ can also be partitioned as $S = [S_1 \mid S_2]$, $S_1 \in \Re^{(n-1)\times(n-m)}$ and $S_2 \in \Re^{(n-1)\times m}$, which gives $\overline{C} = C_0 S = [C_0 S_1 \mid C_0 S_2] = [\overline{C}_1 \mid \overline{C}_2]$. Then, the system description in SM (26.22) can be rewritten as

$$\begin{aligned} \dot{\bar{x}}_1 &= [A_{11} - A_{12}(C_0 S_2)^{-1} C_0 S_1]\bar{x}_1 + \bar{J}_1 v = A_r(C_0)\bar{x}_1 + \bar{J}_1 v, \\ \bar{x}_2 &= -(C_0 S_2)^{-1} C_0 S_1 \bar{x}_1. \end{aligned} \tag{26.25}$$

Obviously from (26.25), SM dynamics, which is defined by *n-m* eigenvalues of matrix $A_r(C_0)$, can be set by selection of matrix C_0. Therefore, realization of the desired SM dynamics can be understood as a problem of providing *n-m* predefined eigenvalues of the reduced order system (26.25) by means of $m(n-1)$ adjustable parameters of C_0. Three different cases should be considered in the analysis of the given design problem.

26.4.1 Case 1 < m < n − 1

This case may be considered as the regular one. The task is to ensure *n-m* desired eigenvalues of (26.25) using pole placement technique. Obviously, the design problem has extra degrees of freedom, since there are *mn-m* adjustable parameters in C_0. To solve the obtained system of equations it is necessary to arbitrary chose $n(m-1)$ parameters of C_0. The remaining *n-m* parameters are determined by solving the system. Consequently, it may be concluded that in this case a desired eigenvalue spectrum can be freely adopted and there are infinite number of matrices C that provide the chosen spectrum.

26.4.2 *Case* m = n − *1*

This case may be considered as the singular one. Namely, for $m = n - 1$ matrices C_0 and S_2 become square, i.e., $C_0, S_2 \in \Re^{(n-1)\times(n-1)}$. Then the product $(C_0S_2)^{-1}C_0$ in (26.25) is equal to S_2^{-1}, which leads to simplification of (25) into

$$\begin{aligned}\dot{\bar{x}}_1 &= (A_{11} - A_{12}S_2^{-1}S_1)\bar{x}_1 + \bar{J}_1 v,\\ \bar{x}_2 &= -S_2^{-1}S_1\bar{x}_1.\end{aligned} \tag{26.26}$$

From (26.26) it is apparent that SM dynamics does not depend on C_0, indicating that SM dynamics cannot be chosen at all. This means that any matrix C from the infinite set satisfying (26.17) will give fixed SM dynamics described by (26.26).

This result also has a geometrical explanation. For $m = n - 1$ the sliding manifold is a line. Only a unique sliding line can be fitted through the origin and the predetermined $x_{\min 0}(\infty)$. Hence, SM dynamics is also predetermined and cannot be chosen.

26.4.3 *Case* m = *1*

This is a scalar control case. SM is organized along a single sliding hyperplane in an $n - 1$ dimensional subspace, offering $n - 1$ degrees of freedom in eigenvalue allocation. Due to the algebraic constraint (26.17) on C, sliding hyperplane must contain the line connecting the origin and $x_{\min 0}(\infty)$. This requirement reduces the number of degrees of freedom by one, disabling completely free choice of the eigenvalue spectrum. Hence, $n - 2$ eigenvalues can be arbitrary chosen, while the remaining one is constrained by others.

To determine the relationship between the eigenvalues let the eigenvalue spectrum be denoted as $\lambda \in \{p_1, p_2, \ldots, p_{n-1}\}$. Elements of $\bar{C}$ can be analytically determined using (26.22) by comparing two characteristic equations

$$\begin{aligned}(s-p_1)(s-p_2)\cdots(s-p_{n-1}) &= 0,\\ \det(sI - A_{11} + A_{12}\bar{C}_2^{-1}\bar{C}_1) &= 0.\end{aligned} \tag{26.27}$$

The elements of $\bar{C}$ are obtained as functions of the eigenvalues

$$\bar{C} = [\,\xi_1(p_1, \ldots, p_{n-1}) \quad \cdots \quad \xi_n(p_1, \ldots, p_{n-1})\,]. \tag{26.28}$$

The following theorem shows how to select a desired eigenvalue spectrum and how to obtain C_0.

Theorem 26.2 *For the dynamic system described in Theorem* 26.1, *a feasible SM eigenvalue spectrum can be formed from n* − *2 arbitrary chosen eigenvalues along with the last eigenvalue that satisfies condition*

$$\det[S^{\mathrm{T}} \mid \overline{C}^{\mathrm{T}}] = 0, \tag{26.29}$$

where $\bar{C}$ *is formed according to* (26.28). *The switching function vector C in* (26.14), *which minimizes* (26.10) *and provides the feasible SM dynamics, is defined by*

$$C_0 = \bar{C}_1 S_1^{-1}. \tag{26.30}$$

Proof Since in (26.28) $\bar{C}$ is expressed in terms of the desired eigenvalues, the design task is to solve (26.24) with respect to unknown C_0. It is more convenient to rewrite it in the transposed form $S^{\mathrm{T}} C_0^{\mathrm{T}} = \bar{C}^{\mathrm{T}}$. This is a system of n linear equations with n − 1 unknowns. This overdetermined system will have a unique solution only if $\operatorname{rank}(S^{\mathrm{T}}) = \operatorname{rank}[S^{\mathrm{T}} \mid \overline{C}^{\mathrm{T}}] = n - 1$. It is easy to verify that the obtained condition implies (26.29).

Equation (26.24) can be rewritten in the partitioned form as

$$\begin{aligned} C_0 S_1 &= \bar{C}_1, \\ C_0 S_2 &= \bar{c}_n. \end{aligned} \tag{26.31}$$

Since the rank of system (26.24) or (26.31) is n − 1, one equation is a linear combination of the remaining n − 1 equations. Having in mind that $\operatorname{rank}(S_1) = n - 1$, the last equation of (26.31) may be disregarded. Hence, C_0 can be calculated using upper equation of (26.31) as (26.30). □

Remark 26.6 If the constrained eigenvalue is obtained unstable, then another spectrum of n − 2 eigenvalues must be chosen.

Remark 26.7 For an even n − 1, if all complex eigenvalues are desired, only the real part of the last conjugated pair may be specified. The corresponding imaginary part should be determined to satisfy condition (26.29).

26.5 Illustrative Examples

The developed sliding hyperplane design method has been tested on three numerical examples of controllable systems (26.1), subjected to an unmatched disturbance $v(t) = h(t)$ ($V = 1$).

Example 26.1 Consider a system whose model (26.1) is given by

$$A=\begin{bmatrix}-1 & -3 & 1 & 5\\ 1 & -2 & 1 & -2\\ 1 & 2 & -3 & 2\\ -3 & -4 & -1 & -5\end{bmatrix},\quad B=\begin{bmatrix}2 & 1\\ 1 & 1\\ -1 & 3\\ 2 & 4\end{bmatrix},\quad j=\begin{bmatrix}1\\ 0\\ 1\\ -1\end{bmatrix}. \tag{26.32}$$

This is the regular case since $m < n - 1$. Using the result from Theorem 26.1, (26.13) gives the steady state $x_{\min}(\infty)=\frac{-1}{4148}[823 \quad 74 \quad 225 \quad 834]^{\mathrm{T}}$, whose norm is minimal $||x_{\min}(\infty)||=\sqrt{689/8188}$. Now, matrix C needs to be determined that enables reaching the minimal steady state and provides desired SM dynamics characterized, for example, by two eigenvalues $\lambda\in\{-1,-2\}$. Both requirements will be fulfilled if matrix C_0 is found that secures the desired spectrum of the reduced order system (26.25). The corresponding matrices are

$$A_{11}=\begin{bmatrix}\frac{17}{10} & 1\\ \frac{12}{5} & 0\end{bmatrix},\ A_{12}=\begin{bmatrix}\frac{239}{10} & \frac{206}{5}\\ \frac{24}{5} & \frac{32}{5}\end{bmatrix},\ C_0=\begin{bmatrix}c_{12} & c_{13} & c_{14}\\ c_{22} & c_{23} & c_{24}\end{bmatrix},$$
$$S=[S_1 \mid S_2]=\frac{1}{823}\left[\begin{array}{cc|cc}-74 & 823 & 675 & 749\\ -225 & 0 & -1273 & 2244\\ -824 & 0 & -22 & 2458\end{array}\right]. \tag{26.33}$$

Obviously, C_0 has six adjustable parameters, so four of them may be freely chosen. Let $c_{13}=c_{24}=1$ and $c_{14}=c_{23}=0$. By comparing the desired and actual characteristic equations, the remaining two parameters can be easily determined as $c_{12}=10.76$ and $c_{22}=4.24$. Finally, matrix C is calculated according to (26.14) as

$$C=\begin{bmatrix}-1.24 & 10.76 & 1 & 0\\ -1.39 & 4.24 & 0 & 1\end{bmatrix}. \tag{26.34}$$

As a verification of the implemented design, it can be easily proven that the determined C satisfies (26.17), and the related equivalent matrix

$$A_{eq}=\begin{bmatrix}-14.045 & -0.82 & -4.13 & -10.087\\ -3.58 & -0.47 & -0.88 & -5.0035\\ 21.11 & 4.01 & 4.36 & 41.32\\ -4.398 & 0.837 & -2.02 & 7.15\end{bmatrix} \tag{26.35}$$

of system (26.4) has the eigenvalue spectrum $\lambda\in\{0,0,-1,-2\}$.

Example 26.2 Consider a system (26.1) defined by

$$A=\begin{bmatrix}-1 & 4 & 1\\ 1 & -2 & 1\\ 1 & 2 & -3\end{bmatrix},\ B=\begin{bmatrix}2 & 1\\ 1 & 1\\ -1 & 3\end{bmatrix},\ j=\begin{bmatrix}1\\ 0\\ 1\end{bmatrix}. \tag{26.36}$$

Since $m = n - 1$, this example belongs to the singular case, Using (26.13), $x_{\min}(\infty) = \frac{1}{116}\begin{bmatrix} 5 & -16 & 3 \end{bmatrix}^{\mathrm{T}}$ is obtained, having minimal norm $||x_{\min}(\infty)|| = \sqrt{5/232}$. SM dynamics is entirely predefined by the minimal steady state and there is no degree of freedom left, as given by (26.26). The corresponding matrices are

$$A_{11} = \tfrac{-5}{2},\ A_{12} = \begin{bmatrix} \tfrac{9}{2} & 1 \end{bmatrix},\ S = \begin{bmatrix} S_1 \mid S_2 \end{bmatrix} = \frac{1}{5}\left[\begin{array}{c|cc} 16 & 37 & 21 \\ -3 & -11 & 12 \end{array}\right]. \tag{26.37}$$

The eigenvalue of the reduced order system (26.26) is calculated as $\lambda = -116/27$. Matrix C is obtained according to (26.14) as

$$C = \begin{bmatrix} \frac{1}{5}(16c_{12} - 3c_{13}) & c_{12} & c_{13} \\ \frac{1}{5}(16c_{22} - 3c_{23}) & c_{22} & c_{23} \end{bmatrix}. \tag{26.38}$$

Every matrix C from the infinite set of matrices defined by (26.38) will provide minimal steady-state vector norm and will result in the predefined SM dynamics. In deed, application of (26.38) into (26.4) gives the equivalent matrix

$$A_{eq} = \frac{1}{135}\begin{bmatrix} -50 & 160 & -30 \\ 160 & -512 & 96 \\ -30 & 96 & -18 \end{bmatrix}, \tag{26.39}$$

which has no adjustable parameters and whose eigenvalues are $\lambda \in \{0, 0, -116/27\}$.

Example 26.3 Consider a scalar control system (26.1) with

$$A = \begin{bmatrix} -1 & 0 & 1 \\ 1 & -2 & 1 \\ 0 & 0 & -3 \end{bmatrix},\ B = \begin{bmatrix} 0 \\ 1 \\ -1 \end{bmatrix},\ j = \begin{bmatrix} 1 \\ 0 \\ 1 \end{bmatrix}. \tag{26.40}$$

According to (26.13) the steady state $x_{\min}(\infty) = \frac{1}{9}\begin{bmatrix} 7 & 10 & -2 \end{bmatrix}^{\mathrm{T}}$ is obtained with minimal norm $||x_{\min}(\infty)|| = \sqrt{17}/3 = 1.374$. An infinite number of hyperplanes, defined by (26.14), can be constructed that minimize the steady-state vector norm. The next step is to select an appropriate sliding hyperplane, out from that infinite set, providing desired feasible SM dynamics. Sliding motion is described by (26.22), where the corresponding matrices are

$$\begin{gathered} A_{11} = \begin{bmatrix} -1 & 0 \\ 1 & -2 \end{bmatrix},\ A_{12} = \begin{bmatrix} -1 \\ 0 \end{bmatrix},\ \overline{C} = \begin{bmatrix} \overline{C}_1 \mid \overline{C}_2 \end{bmatrix} = \begin{bmatrix} \bar{c}_1 & \bar{c}_2 \mid \bar{c}_3 \end{bmatrix} \\ S = [S_1 \mid S_2] = \left[\begin{array}{cc|c} \frac{-10}{7} & 1 & 1 \\ \frac{2}{7} & 0 & -1 \end{array}\right]. \end{gathered} \tag{26.41}$$

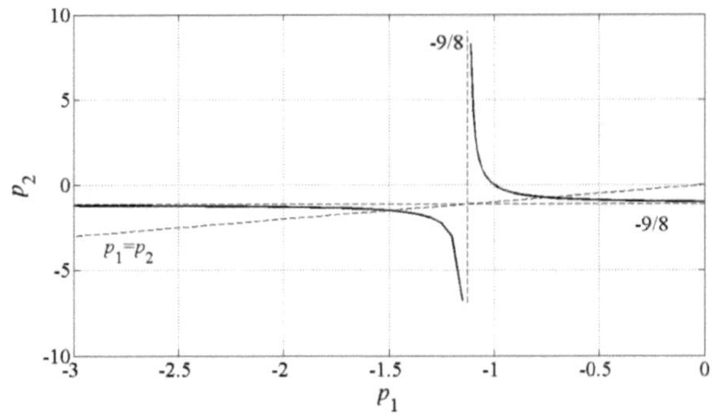

Fig. 26.1 Dependence between real eigenvalues defining SM dynamics for the considered third-order system

$\bar{C}$ that provides the eigenvalue spectrum $\lambda \in \{p_1, p_2\}$ is determined using (26.27) as

$$\overline{C} = [(p_1 + p_2 + 3)\bar{c}_3 \quad (2 + p_1)(2 + p_2)\bar{c}_3 \mid \bar{c}_3]\,. \tag{26.42}$$

Notice that $\bar{c}_3$ may be arbitrary chosen since it does not influence SM dynamics, so let $\bar{c}_3$ be chosen as $\bar{c}_3 = 1$. Feasible SM dynamics can be defined by detecting the constraint between the eigenvalues using condition (26.29) from Theorem 26.2, which gives

$$p_2 = -9(1 + p_1)/(9 + 8p_1), \quad p_1 \neq -9/8. \tag{26.43}$$

Graphical representation of the eigenvalues constraint (26.43) is depicted in Fig. 26.1. By including the constraint (26.43) into (26.42), matrix C_0 can be found as $C_0 = \bar{C}_1 S_1^{-1}$. Finally, C is obtained by virtue of (26.14) in the following form:

$$C = \begin{bmatrix} 2(3 + 2p_1)^2/(9 + 8p_1) \\ -(2 + p_1)(9 + 7p_1)/(9 + 8p_1) \\ -(27 + p_1(31 + 7p_1))/(9 + 8p_1) \end{bmatrix}^{\mathrm{T}}. \tag{26.44}$$

To check validity of the designed SM dynamics, eigenvalues of matrix A_{eq} should be inspected for C given by (26.44). The resulting eigenvalue spectrum is $\lambda \in \{0, p_1, -9(1 + p_1)/(9 + 8p_1)\}$, verifying the determined eigenvalue constraint.

Remark 26.8 The steady states with associated minimal norms for all three examples obtained using Theorem 26.1 have been confirmed by conducting optimization described by (26.11) and (26.12).

26.6 Simulation Results

Analytically obtained results from Example 26.3 have been verified by digital simulations. A very simple control law $u = -(CB)^{-1}(CAx + \alpha \operatorname{sgn} \sigma)$ is employed for SM realization in the system described by (26.1), (26.2) and (26.40), which is subjected to a step external unmatched disturbance $v(t) = h(t-6)$. Switching gain $\alpha = 3$ provides reaching and existing conditions in the presence of the disturbance. It is well known that such control law with dominant switching component may induce chattering in real systems with unmodeled dynamics. The reason for its application here is that the attained SM as much as possible should resemble an ideal one, which is the assumption of the conducted analysis.

Two sliding hyperplanes have been designed that minimize steady-state vector norm using the proposed method. The first one provides real eigenvalues that define SM dynamics, with the chosen $p_1 = -2$. Constraint (26.43) gives the other eigenvalue $p_2 = -9/7$ and the resulting sliding hyperplane vector is determined using (26.44) as $C = [-2/7 \quad 0 \quad -1]$. In the second hyperplane design, complex eigenvalues have been imposed. Only the real part of the conjugated eigenvalues has been chosen to be −1. The pair that satisfies (26.43) is $p_{1,2} = -1 \pm j\sqrt{2}/4$. Sliding hyperplane that guarantees the prescribed SM dynamics is defined by $C = [1 \quad -9/8 \quad -17/8]$, according to (26.44).

State space system trajectories for those two sliding hyperplanes are depicted in Fig. 26.2. From initial state $x(0) = [1 \quad -1 \quad 0]^{\mathrm{T}}$, system trajectories reach and slide along the corresponding planes into the origin. After the action of the unmatched disturbance, phase points are driven out of the origin along the respective planes into an identical steady state, which belongs to the intersection of the hyperplanes. Data analysis has confirmed that identical minimal steady state is reached as calculated in Example 26.3. The associated state coordinates time responses are given in Fig. 26.3.

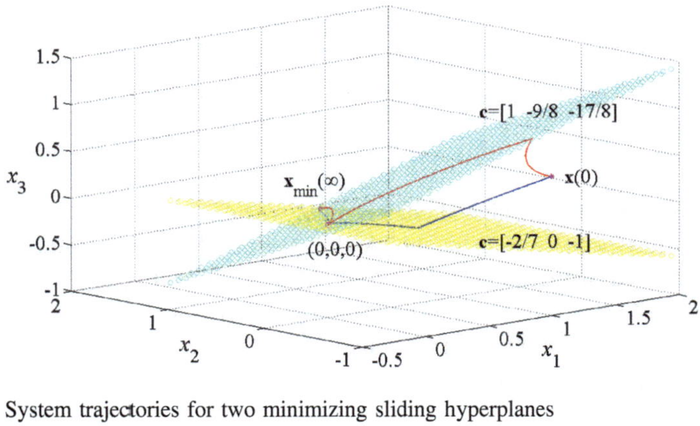

Fig. 26.2 System trajectories for two minimizing sliding hyperplanes

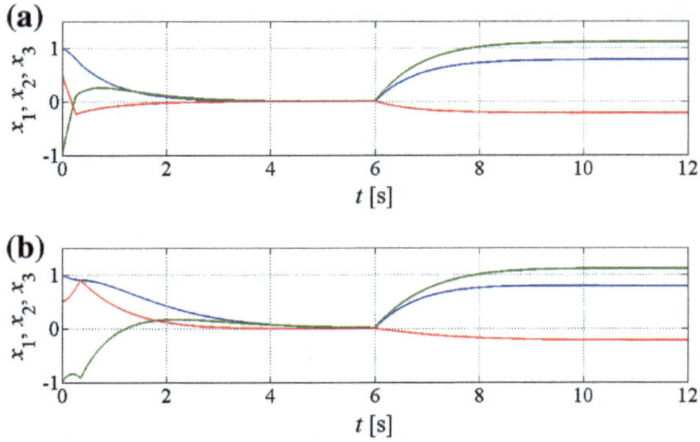

Fig. 26.3 State response for the minimizing sliding hyperplanes: **a** $C = [-2/7\ 0\ -1]$; **b** $C = [1\ -9/8\ -17/8]$

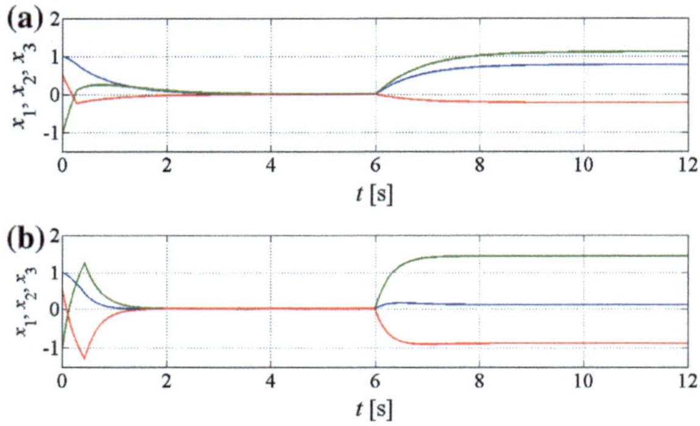

Fig. 26.4 State response for different sliding hyperplanes: **a** $C = [-2/7\ 0\ -1]$ (minimizing); **b** $C = [-3\ -1\ -2]$

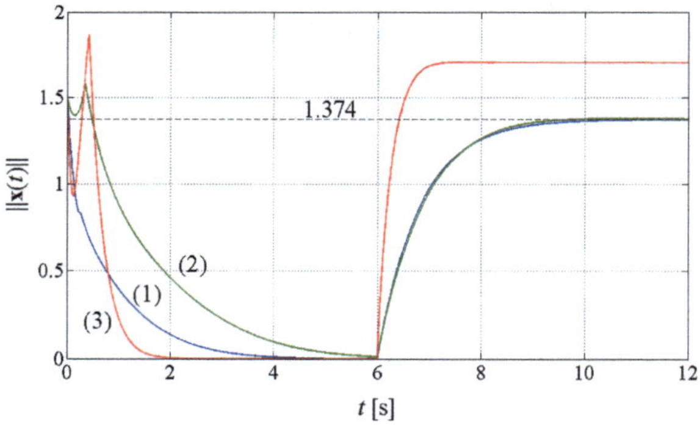

Fig. 26.5 State vector norms for different sliding hyperplanes: (*1*) $C = [-2/7\ 0\ -1]$ (minimizing); (*2*) $C = [1\ -9/8\ -17/8]$ (minimizing); (*3*) $C = [-3\ -1\ -2]$

To prove that the achieved steady-state vector norm is minimal, another hyperplane has been examined, which is constructed in the traditional way by imposing desired SM dynamics. $C = [-3 \quad -1 \quad -2]$ corresponds to the desired double eigenvalue $p_1 = p_2 = -3$. The resulting steady state and its norm have been calculated as $x(\infty) = [0.11 \quad 1.44 \quad -0.89]^{\mathrm{T}}$ and $||x(\infty)|| = 1.7$. This comparison confirms that the traditionally designed sliding hyperplane, having much faster SM dynamics, provides significantly larger steady-state vector norm than the minimal value accomplished by the hyperplanes constructed according to the proposed method. Figure 26.4 shows the state responses of the minimizing and not-minimizing sliding hyperplanes, and the state vector norms for all three hyperplanes are evaluated in Fig. 26.5.

26.7 Conclusion

The proposed linear sliding manifold design method very efficiently finds an infinite set of sliding manifolds minimizing the steady-state vector norm in SM, under action of unmatched constant external disturbances. The developed method avoids necessity for complex and difficult differentiations and calculations in solving the given optimization problem. The only prerequisite for the proposed method application is that the system matrix A is invertible.

Three cases have been recognized in prescribing SM dynamics. The regular case $(1 < m < n - 1)$, when SM dynamics can be freely selected; the singular case $(m = n - 1)$, when SM dynamics is predetermined by the minimal steady state and cannot be chosen at all; and the scalar control case $(m = 1)$, when a constraint arises between the eigenvalues, which can be easily discovered by the derived condition.

The presented numerical examples and simulation results have confirmed validity of the proposed design method by achieving both minimal steady-state vector norm and the prescribed SM dynamics.

Acknowledgments This work was supported by Project Grant III44004 (2011–2015) financed by the Ministry of Education, Science and Technological Development of the Republic of Serbia.

References

1. Draženović, B.: The invariance conditions in variable structure systems. Automatica **5**, 287–295 (1969)
2. Spurgeon, S.K., Davies, R.: A nonlinear control strategy for robust sliding mode performance in the presence of unmatched uncertainty. Intl. J. Control **57**(5), 1107–1123 (1993)
3. Kwan, C.M.: Sliding mode control of linear systems with mismatched uncertainties. Automatica **31**(2), 303–307 (1995)
4. Jang, S.H., Kim, S.W.: A new sliding surface design method of linear systems with mismatched uncertainties. IEICE Trans. Fundam. **E-88A**(1), 387–391 (2005)
5. Wen, C.C., Cheng, C.C.: Design of sliding surface for mismatched uncertain systems to achieve asymptotical stability. J. Franklin Inst. **345**, 926–941 (2008)
6. Choi, H.H.: An LMI-based switching surface design method for a class of mismatched uncertain systems. IEEE Trans. Autom. Control **48**(9), 1634–1638 (2003)
7. Castanos, F., Fridman, L.: Analysis and design of integral sliding manifolds for systems with unmatched perturbations. IEEE Trans. Autom. Control **51**(5), 853–858 (2006)
8. Polyakov, A., Poznyak, A.: Invariant ellipsoid method for minimization of unmatched disturbances effects in sliding mode control. Automatica **47**(7), 1450–1454 (2011)

9. Veselić, B., Peruničić-Draženović, B., Milosavljević, Č.: A sliding hyperplane design method for a class of linear systems with unmatched disturbances. In: Proceedings of the IEEE 11th International Workshop on Variable Structure Systems, Mexico City, Mexico, pp. 142–147 (2010)
10. El-Ghezawi, O.M.E., Zinober, A.S.I., Billings, S.A.: Analysis and design of variable structure systems using a geometric approach. Intl. J. Control **38**(3), 657–671 (1983)
11. Utkin, V.I.: Sliding Modes in Control and Optimization. Springer, Berlin (1992)

Chapter 27
Sliding Mode Based Anti-Lock Braking System Control

Dragan S. Antić, Darko B. Mitić, Zoran D. Jovanović, Staniša Lj. Perić, Marko T. Milojković and Saša S. Nikolić

Abstract In this paper we consider different continuous- and discrete-time sliding mode control (SMC) techniques in the control of antilock braking system (ABS). Having in mind that ABS is characterized by nonlinear and uncertain dynamics, SMC is a right choice for its control because of its robust characteristics. The survey of continuous-time SMC algorithms based on nonlinear models of ABS is given first. Then, the discrete-time nonlinear model of ABS is derived, and the overview of existing discrete-time SMC techniques is presented. The experimental results are given to show the effectiveness of analyzed SMC methods.

27.1 Introduction

In order to prevent abruptly stopping of vehicle, an antilock braking system (ABS) is used in most modern vehicles. ABS is an electronic system which monitors and controls wheel slip during vehicle braking, helping secure stopping of

D.S. Antić (✉) · D.B. Mitić · Z.D. Jovanović · S.Lj.Perić · M.T. Milojković · S.S. Nikolić
Faculty of Electronic Engineering, Department of Control Systems,
University of Niš, Aleksandra Medvedeva 14, 18000 Niš, Republic of Serbia
e-mail: dragan.antic@elfak.ni.ac.rs

D.B. Mitić
e-mail: darko.mitic@elfak.ni.ac.rs

Z.D. Jovanović
e-mail: zoran.jovanovic@elfak.ni.ac.rs

S.Lj.Perić
e-mail: stanisa.peric@elfak.ni.ac.rs

M.T. Milojković
e-mail: marko.milojkovic@elfak.ni.ac.rs

S.S. Nikolić
e-mail: sasa.s.nikolic@elfak.ni.ac.rs

G.M. Dimirovski (ed.), *Complex Systems*, Studies in Systems, Decision and Control 55, DOI 10.1007/978-3-319-28860-4_27

the vehicle in that way. It also reduces stopping distances on slippery road surfaces by limiting wheel slip and minimizing lockup, since rolling wheels have much more traction than locked ones. Vehicle stability and steerability are improved by reducing wheel slip, since stability increases as wheel slip decreases. If wheels stop completely during a sudden braking, the control over vehicle is lost and it can skid in an undesirable direction. ABS does not allow wheels to be stiffened and thus enable driver to normally operate with the vehicle, although the brake pedal is pressed to the end.

The first researches in ABS dated from 1930s and were conducted by Bosch, who has been developing it since then. In 1964, Teldix GmbH, with financial help of Bosch, constructed ABS 1 that successfully stopped wheels without locking up. The ABS 2 was model that was first commercially used in 1978 in a series car production of S-class Mercedes. Over the next 20 years, ABS has been significantly improved, and nowadays it represents the standard equipment of modern vehicles.

ABS does not require any additional effort to be applied to nearly all types of vehicles and can be successfully integrated into hydraulic and air brake systems (including air over hydraulic). In former systems, ABS consists of a hydraulic modulator (control valve), control electronics (central unit), and sensors mounted on the wheels. Sensors constantly measures braking power and when braking is so intensive that may block wheels, controller in central electronic unit sends the command to hydraulic system to open the electromagnetic valves. This reduces the pressure of oil in hydraulic system, and consequently the forces on the brake disks. As soon as a wheel rotating starts, sensor sends this information to controller and it closes solenoid valves providing maximum braking force until new wheel relock.

The main objective of controller is to provide the best adhesion of a wheel to a surface. This is performed by controlling a road adhesion coefficient. This coefficient is defined as a proportion between the friction force, generated during acceleration and braking phase, and the normal load of vehicle. It is shown that this coefficient is in nonlinear dependence on wheel slip, defined as a relative speed difference between the wheel and the vehicle. In the most papers, ABS controller is designed to regulate wheel slip so that the road adhesion coefficient has a maximum value. In [1], the regulator problem is analyzed by considering the constant value of desired wheel slip, whereas another approach based on real-time estimation of wheel slip value, with maximal road friction, is presented in [2]. In [3], it is shown that the optimal value of wheel slip should be in the range from 0.08 to 0.3.

ABS mathematical model is uncertain and has highly nonlinear structure, and, although this system has been used for several decades, it has not been completely developed yet. System nonlinearities are reflected in complex braking dynamic characteristics. Besides that, the system parameters change, and many external disturbances are not predictable. That is why sliding mode control (SMC) methods seem to be the right choice in the control of ABS.

SMC is a part of the well-known and studied class of discontinuous control systems [4–6]. Sliding mode arises when system state is forced to move along a predefined sliding surface, determined by the so-called switching function. In sliding mode, system becomes robust to parameter perturbations and external disturbances, and its dynamics is of the lower order and known in advance. The main SMC drawback is the existence of chattering. It occurs as a consequence of the high-frequency control signal, exciting system nonmodeled dynamics. It could potentially wear out ABS mechanical parts.

Many papers deal with the implementation of continuous-time SMC in ABS. Digital simulation results of ABS with conventional SMC, where hydraulic brake dynamics is neglected during the design, is shown in [7]. Traditional SMC enhanced by a gray system theory is proposed in [1, 8]. In these papers, a gray predictor is used to estimate the angular wheel and linear vehicle velocities. The wheel slip control through the engine torque control is given in [9]. The moving sliding surface is used to ensure that the system state is always on sliding surface. The chattering problem is discussed in [10], and the integral switching function is introduced to cope with this phenomenon. The integral sliding surface is also used in design of SMC for hybrid electric brake system [11]. The combination of SMC and the sliding mode observer is elaborated in [12]. In [13], SMC, allowing the maximum value of wheel–road friction force during the braking phase, without a priori knowledge of optimal slip, is discussed. This approach treats ABS SMC design in different manner than the concepts based on separation principle. The adaptive SMC of vehicle traction is considered in [14]. The proposed control method uses the difference between vehicle and wheel velocities (denoted as slip velocity) as controlled variable instead of the wheel slip representing the relative difference. The adaptive law, estimating road adhesion coefficient, is combined with the traditional SMC. The overall system stability is proved using Lyapunov theory. The integration of SMC and pulse-width modulation (PWM) method, realized using computer software as opposed to hardware, is given in [15]. This results in quasi-continuous control of ABS wheel slip. In [16], the previous approach is analyzed in detail and it is compared with traditional SMC of wheel slip. SMC concept is also implemented in hybrid electric vehicle brake system control [17], where a permanent magnet synchronous motor, mounted on wheel shafts via gears, serves as a generator producing a negative torque to the rear wheels and recovering kinetic energy into electric energy. In [18], the novel nonlinear control with integral feedback is compared with SMC. For overall vehicle stability enhancement, the traditional SMC is also used in wheel slip control [19], while the linear quadratic regulator is exploited for yaw moment controller. The traditional SMC of magneto rheological brake system is presented in [20]. The SMC with the neural networks and the moving sliding surface is considered in [21]. The paper [22] deals with a combination of fuzzy control (FC) and SMC. The fuzzy rules are used to obtain the parameters of SMC with constant plus proportional rate reaching law [5]. A survey of several continuous-time SMCs of ABS is given in [23]. In [24], ABS control using sliding mode optimization is discussed.

The realization of SMC in discrete time gives the quasi-sliding motion [25] as a result, and generates chattering in the $O(T)$ vicinity of a sliding surface (T is a sampling period). The survey of discrete-time SMC systems is presented in [26]. The discrete-time SMC algorithms for ABS may be classified in the group of input/output-based algorithms. Two approaches to design of discrete-time SMC of nonlinear plant exist so far: one based on discrete-time representation of linearized plant model [27], using the combination of (generalized) minimum variance control and discrete-time SMC [28–30], and the solution that uses the input/output discrete-time nonlinear plant models in discrete-time SMC design [31]. The latter approach is implemented for ABS control in [32]. The control algorithm for ABS, proposed in [33], combines control law presented in [30], for linear plants, with [31], for nonlinear plants, resulting in the robust control with chattering alleviation, thanks to the filtering of relay control component [32, 34, 35]. At the same time, high system steady-state accuracy is provided by introducing the modeling error estimation.

The paper is organized as follows. After short description of ABS given in Sect. 27.2, the continuous-time ABS model and the most used SMC techniques are presented in Sect. 27.3. Discrete-time ABS model and existing discrete-time SMC algorithms are discussed in Sect. 27.4. Section 27.5 contains experimental results of some implemented continuous-time and discrete-time SMC laws. Concluding remarks are given in the last section.

27.2 ABS Description and Graphical Model

ABS experimental setup, manufactured by Inteco, is depicted in Fig. 27.1 [36]. It consists of two rolling wheels. The lower vehicle-road wheel represents vehicle motion and it has smooth surface which can be covered by desired material in order to animate different road surface. The upper wheel models the vehicle wheel and it permanently remains in a rolling contact with the lower wheel, as it is the case with the real-world situation. This wheel is in connection with the balance lever and it is equipped in a tire.

The disk brake system, mounted on the upper wheel, is connected with the brake lever via hydraulic coupling. The brake lever is driven by the small DC motor via steel cord. We would like to point out that the steel cord causes strong nonlinearity

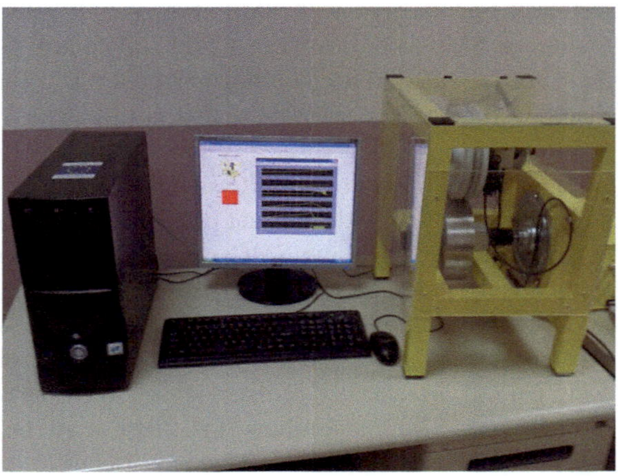

Fig. 27.1 ABS experimental framework (Inteco)

and limitation of control input signal on 40 % of its maximum nominal value. The lower wheel is connected to the big flat DC motor whose task is to accelerate the wheel, while in the braking phase its power supply is switched off. Both DC motors are controlled by pulse-width modulation (PWM) signals of the 3.5 kHz frequency. To measure the wheel angles, two rotary encoders are installed with the accuracy of $2\pi/2048 = 0.175°$. The wheel angular velocities are not measured and they have to be estimated. The simplest Euler formula is used with the sample time period of 0.5 ms.

The power interface amplifies the control signals which are transmitted from the PC to the DC motor. It also converts the encoders pulse signals to the digital 16-bit form to be read by the PC. The PC is equipped with the RT-DAC4/USB multi-purpose digital I/O board communicating with the power interface. The whole logic necessary to activate and read the encoder signals and to generate the appropriate sequence of PWM pulses to control the DC motors is configured in the Xilinx® chip of the RT-DAC4/USB board. All functions of the board are accessed from the ABS Toolbox which operates directly in the MATLAB®/Simulink® and the Real Time Windows Target (RTWT) toolbox environment.

We have to make several assumptions to describe a mathematical model of ABS. First, we observe the behavior of only one wheel (most of the vehicles have four wheels) and a mutual influence of vehicle wheels is neglected. Second, a rolling resistance force is ignored, as it is very small due to braking and, finally, a longitudinal dynamics of vehicle is only considered whereas the lateral and the vertical motions are not taken into consideration. An ABS graphical model is shown in

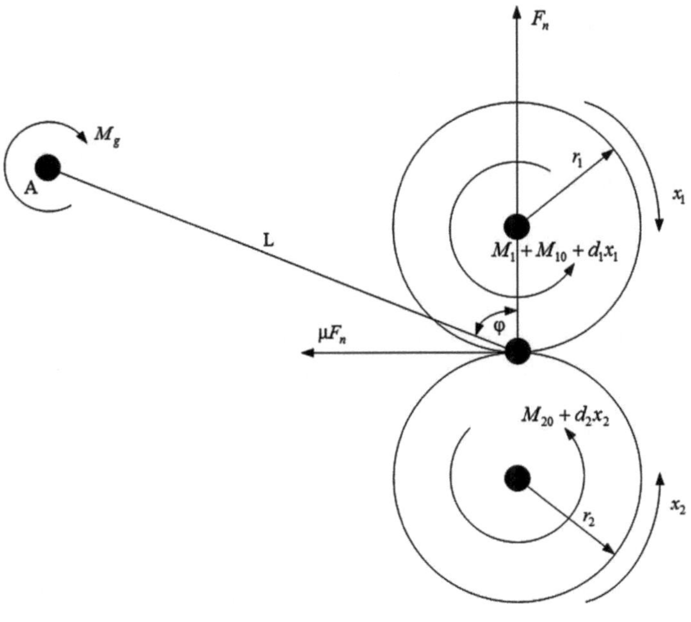

Fig. 27.2 ABS graphical model

Fig. 27.2. Despite this model simplification, it preserves the fundamental characteristics of real system.

Notice that the vehicle velocity is equal to the angular velocity of the lower wheel multiplied by the radius of this wheel, while the wheel velocity is equal to the angular velocity of the upper wheel multiplied by its radius. According to Fig. 27.2 there are three torques acting on the upper wheel: the braking torque M_1, the friction torque in the upper bearing, and the friction torque among the wheels, as well as two torques acting on the lower wheel—the friction torques in the lower bearing and among the wheels. Besides, two more forces are present on the lower wheel: the gravity force of the upper wheel and the pressing force of the shock absorber. In the next section, the mathematical model of ABS in continuous-time domain is derived in detail.

27.3 Continuous-Time SMC of ABS

27.3.1 Continuous-Time ABS Model

Using the Newton's second law, the upper wheel dynamics is described as

$$J_1\dot{x}_1 = F_n r_1 s\mu(\lambda) - d_1 x_1 - s_1 M_{10} - s_1 M_1, \tag{27.1}$$

where J_1 is the moment of inertia, d_1 is the viscous friction coefficient and M_{10} is the static friction of upper wheel. $\mu(\lambda)$ is the coefficient of proportion called the road adhesion coefficient. In a similar manner, the motion of lower wheel can be defined as

$$J_2\dot{x}_2 = -F_n r_2 s\mu(\lambda) - d_2 x_2 - s_2 M_{20}, \tag{27.2}$$

where J_2 is the moment of inertia, d_2 is the viscous friction coefficient, and M_{20} is the static friction of the lower wheel, and

$$s = \operatorname{sgn}(r_2 x_2 - r_1 x_1), \quad s_1 = \operatorname{sgn}(x_1), \quad s_2 = \operatorname{sgn}(x_2). \tag{27.3}$$

The friction force is assumed to be proportional to the normal pressing force F_n (the upper wheel acting on the lower wheel), and in many papers, F_n is considered to be a constant and its variations are treated as nonmodeled dynamics. In that case, (27.1) and (27.2) completely describe the dynamics of simplified quarter vehicle model. The normal force F_n can be derived from the sum of torques in the point A (Fig. 27.2) as

$$F_n = \frac{\left(M_g + s_1 M_1 + s_1 M_{10} + d_1 x_1\right)}{L(\sin\varphi - s\mu(\lambda)\cos\varphi)}, \tag{27.4}$$

where M_g represents the gravitational and shock absorber torques acting on the balance lever, L is the distance between the contact point of wheels and the rotational axis of balance lever and φ is the angle between the normal in the contact point and the line L.

The controlled variable is the wheel slip, so that it is of great importance to describe its dynamics in a proper manner. In normal operating conditions, the angular velocity of wheel would match the forward angular velocity of vehicle. During the braking and acceleration phases these velocities differ one from other, and their difference is called the wheel slip λ, and it is generally defined by

$$\lambda = \begin{cases} \frac{r_2x_2 - r_1x_1}{r_2x_2}, r_2x_2 \geq r_1x_1, x_1 \geq 0, x_2 \geq 0, \\ \frac{r_1x_1 - r_2x_2}{r_1x_1}, r_2x_2 < r_1x_1, x_1 \geq 0, x_2 \geq 0, \\ \frac{r_2x_2 - r_1x_1}{r_2x_2}, r_2x_2 < r_1x_1, x_1 < 0, x_2 < 0, \\ \frac{r_1x_1 - r_2x_2}{r_1x_1}, r_2x_2 \geq r_1x_1, x_1 < 0, x_2 < 0, \\ 1, x_1 < 0, x_2 \geq 0, \\ 1, x_1 \geq 0, x_2 < 0, \end{cases} \tag{27.5}$$

for all quarter vehicle model operating conditions. A zero value of wheel slip means that wheel and vehicle velocities are equal, and when the wheel slip is equal to one vehicle wheel is not rotating anymore and skidding on the road surface, i.e., vehicle is no longer steerable.

The road adhesion coefficient $\mu(\lambda)$ is nonlinear function of wheel slip and other physical variables, and one of its models can be defined by the following equation:

$$\mu(\lambda) = \frac{w_4\lambda^p}{a + \lambda^p} + w_3\lambda^3 + w_2\lambda^2 + w_1\lambda. \tag{27.6}$$

Taking into account (27.1)–(27.4), the ABS model can be rewritten in the following form:

$$\begin{aligned} \dot{x}_1 &= S(\lambda, x_1, x_2)(c_{11}x_1 + c_{12}) + c_{13}x_1 + c_{14} + (c_{15}S(\lambda, x_1, x_2) + c_{16})s_1(x_1)M_1\,, \\ \dot{x}_2 &= S(\lambda, x_1, x_2)(c_{21}x_1 + c_{22}) + c_{23}x_1 + c_{24} + c_{25}S(\lambda, x_1, x_2)s_1(x_1)M_1\,, \end{aligned} \tag{27.7}$$

where

$$S(\lambda) = \frac{s\mu(\lambda)}{L(\sin\varphi - s\mu(\lambda)\cos\varphi)}, \tag{27.8}$$

and

$$\begin{aligned} &c_{11} = (r_1d_1)/J_1,\ c_{12} = \left((s_1M_{10} + M_g)r_1\right)/J_1,\ c_{13} = -d_1/J_1, \\ &c_{14} = -(s_1M_{10})/J_1,\ c_{15} = r_1/J_1,\ c_{16} = -1/J_1,\ c_{21} = -(r_2d_1)/J_2, \\ &c_{22} = -\left((s_1M_{10} + M_g)r_2\right)/J_2,\ c_{23} = -d_2/J_2, \\ &c_{24} = -(s_2M_{20})/J_2,\ c_{25} = -r_2/J_2. \end{aligned} \tag{27.9}$$

After acceleration of vehicle to a certain velocity, the brakes are applied. The wheel speed decreases then and the force acting on the wheel increases, causing the slippage between the wheel and the road surface. Vehicle speed will be greater than wheel speed, i.e., $r_2x_2 \geq r_1x_1$ and $x_1 > 0$, $x_2 > 0$, and the wheel slip is

$$\lambda = \frac{r_2x_2 - r_1x_1}{r_2x_2}, \tag{27.10}$$

according to (27.5).

The main goal of ABS control design is to maintain wheel slip at a particular level, where the corresponding road adhesion coefficient, i.e., the friction force, has maximum value. That is why we need to determine quarter vehicle model with wheel slip as a controlled variable. Differentiating (27.10) and taking in consideration that

$$\lambda = 1 - \frac{r_1x_1}{r_2x_2} \Rightarrow x_1 = \frac{r_2}{r_1}(1-\lambda)x_2, \tag{27.11}$$

and $s = s_1 = s_2 = 1$, the ABS λ-model can be completely obtained from (27.7) in the following form:

$$\lambda = f(\lambda, x_2) + g(\lambda, x_2)M_1, \quad x_2 \neq 0, \tag{27.12}$$

where

$$\begin{aligned} f(\lambda, x_2) = &-\left((S(\lambda)c_{11} + c_{13})(1-\lambda) + \frac{r_1}{r_2x_2}(S(\lambda)c_{12} + c_{14})\right) \\ &+ \frac{(1-\lambda)}{x_2}\left(\left(S(\lambda)c_{21}\frac{r_2}{r_1}(1-\lambda) + c_{23}\right)x_2 + S(\lambda)c_{22} + c_{24}\right), \end{aligned} \tag{27.13}$$

$$g(\lambda, x_2) = -\frac{r_1}{r_2x_2}\left(c_{15}S(\lambda) + c_{16} - \frac{r_2}{r_1}c_{25}S(\lambda)(1-\lambda)\right). \tag{27.14}$$

27.3.2 Continuous-Time SMC Algorithms

In this section we elaborate the implementation of four sliding mode techniques in ABS control. The mathematical background of the SMC theory is given briefly. First, traditional and integral SMCs are described, thereafter the SMC based on constant plus proportional rate reaching law is discussed and, finally, SMC with exponential reaching law is considered.

The control design is based on the wheel slip dynamics described by (27.12) and the control algorithms consist of two terms: the so-called equivalent control M_1^{eq}

[37] and a reaching component $M_1^r(\sigma)$ ensuring a system state to hit a sliding surface from any initial point and to keep it moving on it

$$M_1 = M_1^{eq} + M_1^r(\sigma), \tag{27.15}$$

where σ is a switching function. Since the system is of the first order, the switching function is selected as

$$\sigma = \lambda - \lambda_r, \tag{27.16}$$

where λ_r is a reference wheel slip. The main control design objective is to find control providing $\sigma = 0$ and, consequently, $\lambda = \lambda_r$. For system motion in sliding mode, the equivalent control term is calculated from the condition $\dot{\sigma} = 0$, i.e.,

$$\dot{\sigma} = 0 \Rightarrow \lambda = 0 \Leftrightarrow f(\lambda, x_2) + g(\lambda, x_2) M_1^{eq} = 0, \tag{27.17}$$

assuming that $\lambda_r = \text{const}$. Equation (27.17) yields the equivalent control brake torque in the following form:

$$M_1^{eq} = -g(\lambda, x_2)^{-1} f(\lambda, x_2), \tag{27.18}$$

and it is calculated on the basis of nominal system parameters. Since the nominal values of $f(\lambda, x_2)$ and $g(\lambda, x_2)$ are $\hat{f}(\lambda, x_2)$ and $\hat{g}(\lambda, x_2)$, respectively, the equivalent control brake torque implemented in real control is

$$\hat{M}_1^{eq} = -\hat{g}(\lambda, x_2)^{-1} \hat{f}(\lambda, x_2), \tag{27.19}$$

and, consequently, (27.15) becomes

$$M_1 = \hat{M}_1^{eq} + M_1^r(\sigma). \tag{27.20}$$

For the sake of simplicity, we denote $f(\lambda, x_2) = f$, $\hat{f}(\lambda, x_2) = \hat{f}$, $g(\lambda, x_2) = g$, and $\hat{g}(\lambda, x_2) = \hat{g}$. We assume that $|\hat{f}| < \hat{F}$, $|f - \hat{f}| < \varepsilon_f$ and $|1 - g/\hat{g}| < \varepsilon_g < 1$, where $\hat{F}$, ε_f, and ε_g are the positive real constants, as well as $g/\hat{g} > 0$.

In traditional SMC control algorithm, the reaching component is selected as [8, 23, 38]

$$M_1^r(\sigma) = -\hat{g}^{-1} M_c \operatorname{sgn}(\sigma), \tag{27.21}$$

and the brake torque becomes

$$M_1 = \hat{M}_1^{eq} - \hat{g}^{-1} M_c \operatorname{sgn}(\sigma). \tag{27.22}$$

The parameter M_c is chosen to satisfy the reaching and existence condition of sliding mode

$$\sigma\dot{\sigma} < -\eta|\sigma|, \ \ \eta > 0, \tag{27.23}$$

providing that the final reaching time is

$$t_r < \frac{|\sigma(0)|}{\eta}. \tag{27.24}$$

Since $\dot{\sigma} = \lambda$, the substitution of (27.22) in (27.12), taking into account (27.19) and (27.23), gives

$$\frac{g}{\hat{g}}\left(\frac{\hat{g}}{g}(f - \hat{f})\sigma + \frac{\hat{g}}{g}\hat{f}\left(1 - \frac{g}{\hat{g}}\right)\sigma - M_c|\sigma|\right) < \eta|\sigma|. \tag{27.25}$$

The sliding motion in system defined by (27.12) with the control (27.22) will be established if M_c is chosen according to the following inequality:

$$M_c > \max\left(\left|\frac{\hat{g}}{g}\left((f - \hat{f}) + \hat{f}\left(1 - \frac{g}{\hat{g}}\right) + \eta\right)\right|\right). \tag{27.26}$$

Since

$$1 - \varepsilon_g < g/\hat{g} < 1 + \varepsilon_g \Rightarrow 1/(1 + \varepsilon_g) < \hat{g}/g < 1/(1 - \varepsilon_g), \tag{27.27}$$

Equation (27.26) can be rewritten as

$$M_c > \frac{\varepsilon_f + \hat{F}\varepsilon_g + \eta}{1 - \varepsilon_g}. \tag{27.28}$$

The reaching component of SMC is the discontinuous control signal with high-speed switching in the neighborhood of sliding surface. In that way, the system trajectory is forced to move always toward the sliding surface and to chatter along it. The chattering is an undesirable phenomenon in SMC systems, since it may excite nonmodeled system dynamics. The simplest way to alleviate the chattering problem is to replace the *sgn* function with continuous one

$$h(\sigma) = \frac{\sigma}{|\sigma| + \rho}, \tag{27.29}$$

where $\rho \geq 0$, and the control brake torque can be redefined then with

$$M_1 = \hat{M}_1^{eq} - \hat{g}^{-1}M_c h(\sigma). \tag{27.30}$$

In order to improve the system accuracy, the integral switching function

$$\sigma=(\lambda-\lambda_r)+c_1\int(\lambda-\lambda_r)dt, \tag{27.31}$$

is introduced in control law (27.22) forming the so-called integral SMC in that way. This control algorithm may influence on chattering reduction [10]. As in the previous case, the equivalent control brake torque is determined by

$$\hat{M}_1^{eq}=-\hat{g}^{-1}(\hat{f}+c_1(\lambda-\lambda_r)). \tag{27.32}$$

SMC with integral switching function has the same form as (27.22), and M_c is also selected according to (27.28) in order to establish sliding motion in system. When sliding mode exists in system, it becomes robust to parameter uncertainties and external disturbances. This robustness is lost when the discontinuous function (sgn(σ)) is replaced by continuous one (27.29) in the control signal.

In the case of SMC with constant plus proportional rate reaching law [5], the switching function is also defined by (27.16) and its dynamics is determined by the following equation:

$$\dot{\sigma}=\lambda=-M_p\sigma-M_c\text{sgn}(\sigma);\ M_p,M_c>0, \tag{27.33}$$

By substituting (27.33) in (27.12), the control brake torque is given by (27.20), where $\hat{M}_1^{eq}$ is defined by (27.19) and the reaching component of SMC is

$$M_1^r(\sigma)=-\hat{g}^{-1}(M_p\sigma+M_c\text{sgn}(\sigma)). \tag{27.34}$$

The parameters M_p and M_c are selected in accordance with the reaching and existence condition of sliding mode

$$\sigma\dot{\sigma}<-\delta\sigma^2-\eta|\sigma|;\ \ \delta,\ \eta>0. \tag{27.35}$$

The reaching time is then determined as

$$t_r<\frac{\ln((\delta|\sigma(0)|+\eta)/\eta)}{\delta}. \tag{27.36}$$

The implementation of (27.20), with (27.19) and (27.34), in (27.12), gives the reaching and existence condition of sliding mode in the following form:

$$\sigma\dot{c}=\frac{g}{\hat{g}}\left(\frac{\hat{g}}{g}(f-\hat{f})\sigma+\frac{\hat{g}}{g}\hat{f}\left(1-\frac{g}{\hat{g}}\right)\sigma-M_p\sigma^2-M_c|\sigma|\right)<-\delta\sigma^2-\eta|\sigma|, \tag{27.37}$$

which is fulfilled if M_c is chosen according to (27.28), while M_p satisfies

$$M_p > \frac{\delta}{1-\varepsilon_g}. \tag{27.38}$$

The proportional component of reaching law M_p makes the reaching phase faster in relation to traditional SMC, thus enabling selection of smaller values of the constant component M_c. In that manner, the chattering in ABS is rather reduced [39].

In order to alleviate the chattering phenomenon, the SMC with exponential reaching law is proposed in [40]. The constant term of reaching law is replaced by the exponential function adapting to the switching function variations, so that its dynamics is defined by

$$\dot{\sigma} = \frac{M_c}{N(\sigma)}\mathrm{sgn}(\sigma),\ M_c > 0, \tag{27.39}$$

and

$$N(\sigma) = \gamma + (1-\gamma)e^{-\beta|\sigma|^{\gamma}}, \tag{27.40}$$

where γ is a strictly positive constant less than one, r is a strictly positive integer, and β is a strictly positive number, causing $N(\sigma)$ to be also strictly positive function of σ and $\gamma < N(\sigma) < 1$. If we choose $\gamma = 1$ Eq. (27.39) is transformed to the constant rate reaching law

$$\dot{\sigma} = -M_c\mathrm{sgn}(\sigma),\ M_c > 0, \tag{27.41}$$

resulting in traditional SMC (27.22).

The control brake torque is also defined by (27.20), where $\hat{M}_1^{eq}$ is given by (27.19) and the reaching component of SMC is obtained from (27.12) and (27.39) as

$$M_1^r(\sigma) = -\hat{g}^{-1}\frac{M_c}{N(\sigma)}\mathrm{sgn}(\sigma). \tag{27.42}$$

The reaching and existence conditions of sliding mode can be written now in the following form:

$$\sigma\dot{\sigma} < -\frac{\eta}{N(\sigma)}|\sigma|,\ \eta > 0, \tag{27.43}$$

determining the reaching time as

$$t_r = \frac{1}{\eta}\left(\gamma|\sigma(0)| + (1-\gamma)\int\limits_0^{|\sigma(0)|} e^{-\beta|\sigma|^{\gamma}} d\sigma\right). \tag{27.44}$$

It can be easily shown, that the reaching time of SMC with exponential reaching law is less than the reaching time of the traditional SMC given by (27.24). The controller design for system with uncertain parameters is also considered in [40], but only for $g = \hat{g} = 1$. In this paper we treat the general case when both functions f and g are uncertain, as it is already done before for traditional SMC and SMC with constant plus proportional rate reaching law.

Based on (27.12), (27.19), (27.20), and (27.42), the condition (27.43) can be rewritten as

$$\sigma\dot{\sigma} = \frac{g}{\hat{g}}\left(\frac{\hat{g}}{g}(f-\hat{f})\sigma + \frac{\hat{g}}{g}\hat{f}\left(1-\frac{g}{\hat{g}}\right)\sigma - \frac{M_c}{N(\sigma)}|\sigma|\right) < -\frac{\eta}{N(\sigma)}|\sigma|. \tag{27.45}$$

The inequality (27.45) is satisfied if

$$\frac{M_c - \frac{\hat{g}}{g}\eta}{N(\sigma)} > \frac{\hat{g}}{g}\left((f-\hat{f}) + \hat{f}\left(1-\frac{g}{\hat{g}}\right)\right), \tag{27.46}$$

i.e.,

$$M_c > \max\left(\left|\frac{\hat{g}}{g}\left(N(\sigma)\left((f-\hat{f}) + \hat{f}\left(1-\frac{g}{\hat{g}}\right)\right) + \eta\right)\right|\right), \tag{27.47}$$

resulting in

$$M_c > \frac{\gamma(\varepsilon_f + \hat{F}\varepsilon_g) + (1-\gamma)e^{-\beta|\sigma|^{\gamma}}(\varepsilon_f + \hat{F}\varepsilon_g) + \eta}{1-\varepsilon_g}, \tag{27.48}$$

according to (27.27), (27.40), and the initial assumption on system parameter uncertainties. If (27.48) is fulfilled then the reaching condition of sliding mode is valid in the region determined by

$$|\sigma| > W, \tag{27.49}$$

where

$$W = \sqrt[r]{\frac{\ln\left(\frac{(1-\gamma)(\varepsilon_f + \hat{F}\varepsilon_g)}{M_c(1-\varepsilon_g) - \gamma(\varepsilon_f + \hat{F}\varepsilon_g) - \eta}\right)}{\beta}}, \tag{27.50}$$

is a width of boundary layer directly tuned by the parameter β and it exists if and only if

$$M_c > \frac{\gamma\left(\varepsilon_f + \hat{F}\varepsilon_g\right) + \eta}{1 - \varepsilon_g}. \tag{27.51}$$

SMC based on the exponential reaching law provides smaller finite reaching time with smaller values of control parameter M_c in comparison to the traditional SMC.

27.4 Discrete-Time SMC of ABS

In order to implement discrete-time SMC techniques in control of ABS, the discretization of continuous-time ABS model is performed first in this section. Then, the survey of existing control algorithms is briefly presented.

27.4.1 Discrete-Time ABS Model

We use the Euler's forward method to derive ABS discrete-time model of (27.12), which results in

$$\lambda_{k+1} = f_k^d + g_k^d M_{1k}, \tag{27.52}$$

where $f_k^d = \lambda_k + Tf_k(\lambda, x_2)$, $g_k^d = Tg_k(\lambda, x_2)$, T is a sampling period, and f_k, g_k, and M_{1k} are sampled values of f, g, and M_1 at time instant k. f_k^d and g_k^d represent smooth nonlinear functions, and it is assumed that the function g_k^d is strictly positive and bounded away from zero. The system must have stable and convergent zero dynamics, as well [41]. Notice that the plant model is linear in relation to the control input M_{1k}.

Due to parameter uncertainties, we use the nominal model of ABS in design of digital SMC

$$\lambda_{k+1} = \hat{f}_k^d + \hat{g}_k^d M_{1k}, \tag{27.53}$$

with $\hat{f}_k^d$ and $\hat{g}_k^d$ denoting nominal values of the functions f_k^d and g_k^d, respectively. The modeling error is defined by

$$\varepsilon_{k+1} = \lambda_{k+1} - \lambda_{k+1} = f_k^d - \hat{f}_k^d + \left(g_k^d - \hat{g}_k^d\right) M_{1k}, \tag{27.54}$$

and it satisfies

$$\max|\varepsilon_k| < E, \ \forall k. \tag{27.55}$$

Equation (27.52) can be rewritten now as

$$\lambda_{k+1} = \varepsilon_{k+1} + \hat{f}_k^d + \hat{g}_k^d M_{1k}. \tag{27.56}$$

27.4.2 Discrete-Time SMC Algorithms

The switching function σ_k of discrete-time SMC algorithms corresponds to the signal error e_k

$$\sigma_k = e_k = \lambda_k - \lambda_k^{ref}, \tag{27.57}$$

since ABS model is of the first order in discrete-time domain as well.

In [31], Munoz et al. propose the discrete-time SMC algorithm for nonlinear systems. It is slightly modified for implementation in ABS control, and given as

$$M_{1k} = -\frac{1}{\hat{g}_k^d}\left(\hat{f}_k^d - \lambda_{k+1}^{ref} - \sigma_k + \beta \mathrm{sgn}(\sigma_k)\right), \tag{27.58}$$

where $\lambda_{k+1}^{ref} = \lambda_k^{ref} = \text{const.}$ and $\beta = \text{const.}$, providing switching function dynamics

$$\sigma_{k+1} = \sigma_k - \beta \mathrm{sgn}(\sigma_k) + c_0 \varepsilon_{k+1}. \tag{27.59}$$

The quasi-sliding motion stability is ensured if

$$\beta > \max|c_0 \varepsilon_{k+1}|. \tag{27.60}$$

To obtain better system steady-state accuracy, the control law given in [35] is applied for ABS control as

$$M_{1k} = -\frac{1}{\hat{g}_k^d}\left(\hat{f}_k^d - \lambda_{k+1}^{ref} + \frac{\alpha T}{1 - z^{-1}} \mathrm{sgn}(\sigma_k)\right), \tag{27.61}$$

giving switching function dynamics

$$\sigma_{k+1} = \sigma_k - \alpha T \, \mathrm{sgn}(\sigma_k) + c_0(\varepsilon_{k+1} - \varepsilon_k), \tag{27.62}$$

with $\alpha = \text{const.}$ Notice that the chattering is alleviated since the high-frequency relay component of the control signal is passed through the digital integrator acting as a low-pass filter.

The existence of quasi-sliding mode is provided if [33]

$$\alpha T > \max|c_0(\varepsilon_{k+1} - \varepsilon_k)|, \tag{27.63}$$

where $\max|c_0(\varepsilon_{k+1} - \varepsilon_k)|$ determines system steady-state accuracy.

To improve the system steady-state accuracy one-step delayed modeling error estimator is introduced in (27.61)

$$\varepsilon_k = \lambda_k - \hat{f}^d_{k-1} - \hat{g}^d_{k-1} M_{1k-1}, \tag{27.64}$$

yielding SMC law

$$M_{1k} = -\frac{1}{\hat{g}^d_k}\left(\hat{f}^d_k - \lambda^{ref}_{k+1} + \frac{\alpha T}{1 - z^{-1}}\operatorname{sgn}(\sigma_k) + \varepsilon_k\right). \tag{27.65}$$

The switching function dynamics is now defined as

$$\sigma_{k+1} = \sigma_k - \alpha T \operatorname{sgn}(\sigma_k) + c_0(\varepsilon_{k+1} - 2\varepsilon_k + \varepsilon_{k-1}). \tag{27.66}$$

The control parameter α is selected in accordance with [33]

$$\alpha T > \max|c_0(\varepsilon_{k+1} - 2\varepsilon_k + \varepsilon_{k-1})|, \tag{27.67}$$

to provide stable switching function dynamics and reach quasi-sliding mode.

27.5 Experimental Results

The experimental setup consists of ABS braking mechanism and open architecture software environment for real-time control experiments [36]. The ABS framework supports real-time design and implementation of advanced control methods using MATLAB and Simulink environment in the solution of braking control problems.

This system has been used in all experiments with different control methods employed. Due to the lack of space, only wheel slip time responses in real time are given. The reference wheel slip is considered to be constant and it is chosen as $\lambda_r = 0.2$. The parameters of used ABS mathematical model are presented in Table 27.1, depicted below.

The wheel slip response of ABS with control (27.22) is presented in Fig. 27.3, for M_c = 8. The use of (27.29) to alleviate chattering has no effects anymore (Fig. 27.4), especially if we know that with *h*(σ) instead of *sgn* function system loses its robust characteristics.

Real-time experimental result of ABS with SMC, defined by (27.22), (27.31) and (27.32), without and with chattering reduction are depicted in Figs. 27.5 and 27.6, respectively, for M_c = 8. The use of (27.29) has no significant influence on

Table 27.1 Parameters of ABS model

Name	Value	Units
r_1	99.5×10^{-3}	m
r_2	99×10^{-3}	M
J_1	7.5281×10^{-3}	kgm^2
J_2	25.603×10^{-3}	kgm^2
d_1	1.2×10^{-4}	kgm^2/s
d_2	2.25×10^{-4}	kgm^2/s
M_{10}	3×10^{-3}	Nm
M_{20}	93×10^{-3}	Nm
M_g	19.6181	Nm
φ	65.61	o
w_1	−0.04240011450454	
w_2	0.00000000029375	
w_3	0.03508217905067	
w_4	0.40662691102315	
L	0.370	m
p	2.09945271667129	
a	0.00025724985785	

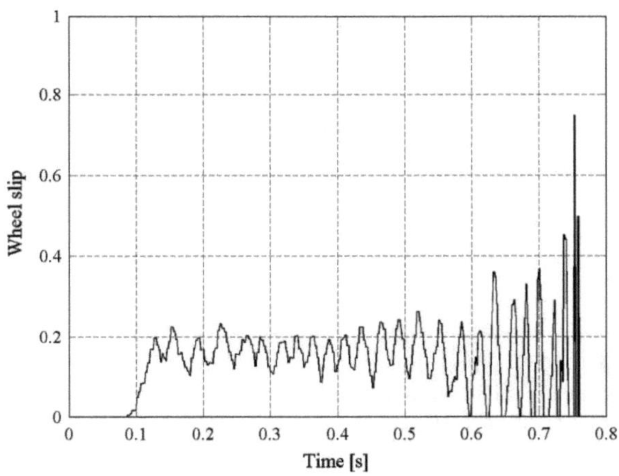

Fig. 27.3 Wheel slip response with control (27.22)

chattering reduction. Having this in mind, we do not utilize (27.29) in next two SMC control algorithms, as well as the integral switching function.

Figure 27.7 presents the ABS wheel slip response with SMC specified by (27.19), (27.20) and (27.34). To cope with nonmodeled dynamics, M_p and M_c are chosen to be 8 and 6, respectively. This SMC algorithm improves the system accuracy in comparison with the previously elaborated results.

Finally, SMC with exponential reaching law is implemented in control of ABS. The parameters of the controller are: $M_c = 8$, $\gamma = 1$, $r = 1$, and $\beta = 500$. The

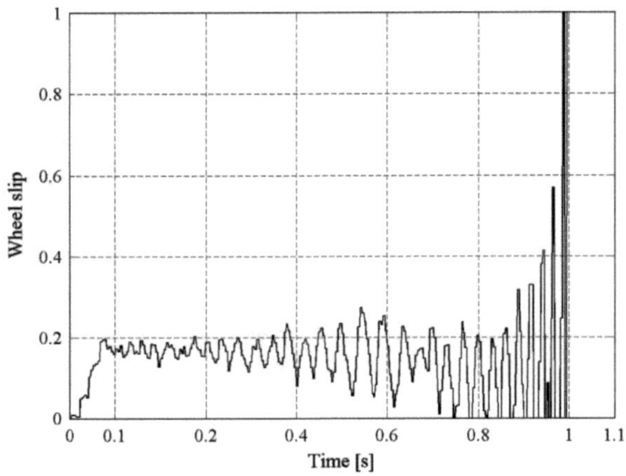

Fig. 27.4 Wheel slip response with control (27.30)

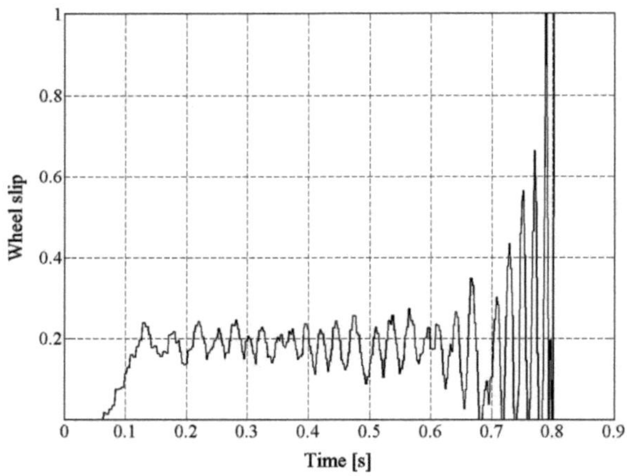

Fig. 27.5 Wheel slip response with control defined by (27.22), (27.31) and (27.32)

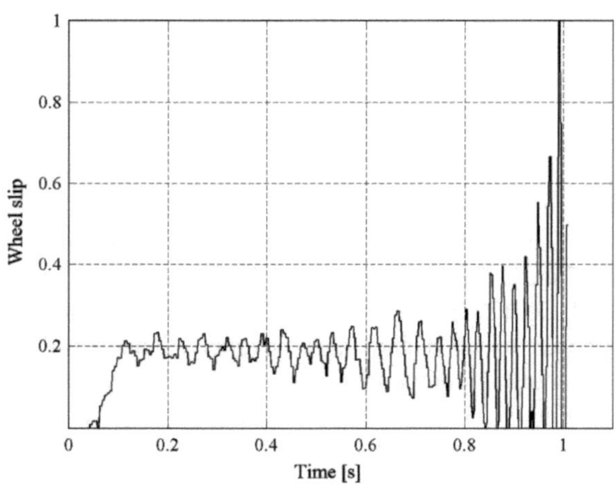

Fig. 27.6 Wheel slip response with control defined by (27.30), (27.31) and (27.32)

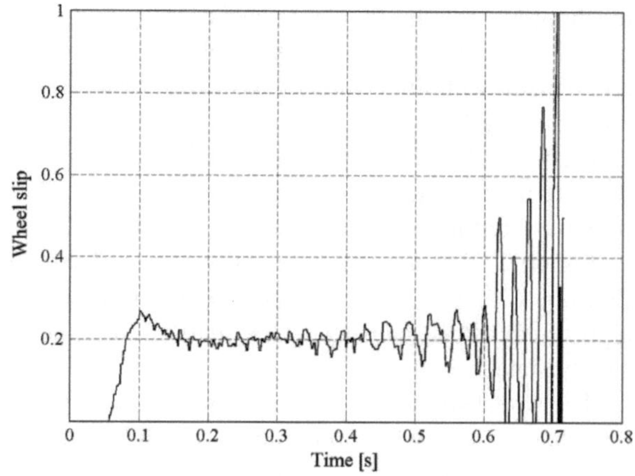

Fig. 27.7 Wheel slip response with control defined by (27.19) and (27.34)

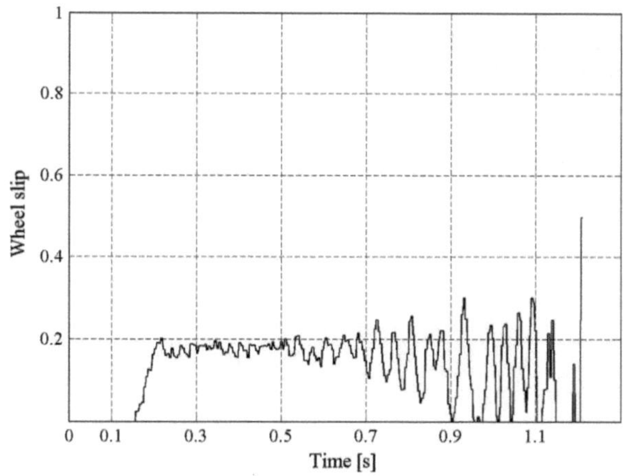

Fig. 27.8 Wheel slip response with control defined by (27.19) and (27.42)

experimental results (Fig. 27.8) are more similar to ABS responses with traditional SMC (with or without integral switching function) than with SMC based on constant plus proportional rate reaching law. This similarity can be explained by additional limitations of control signal introduced in real-time realization of control algorithms. Namely, in simulations the brake torque control signal is limited in the (0, 1) interval. This interval is too wide to be implemented in real-time control. Due to imperfect conveyance of DC motor torque to the hydraulic brake mechanism, realized by the steel cord, the implementation of maximum values of brake torque control signal (during saturation) produces very high oscillations in ABS responses, making wheel slip regulation impossible. This is caused by slow backward response of the hydraulic system when the control signal decreases. In order to make ABS responses more acceptable, the maximal brake torque control signal is limited on 40 % of its nominal value. Since traditional SMC and SMC with exponential reaching law require greater values of relay control term to guarantee the existence of sliding mode, it is comprehensible why their experimental results are worse in comparison with the results of ABS with SMC based on constant plus proportional rate reaching law.

Figures 27.9, 27.10, 27.11 contain the wheel slip responses, obtained using discrete-time SMC algorithms. The reference wheel slip is the same as in the previous continuous-time case. The sampling period T is 0.005 s.

Figure 27.9 represents the results of wheel slip response with control algorithm (27.32), (27.58). The parameter β, corresponding to αT in remaining control algorithms, is 0.1, causing wider quasi-sliding manifold and notable oscillations in steady state.

The wheel slip response of ABS with control (27.61) is shown in Fig. 27.10. The parameter α is 1, which provides quasi-sliding motion with smaller control efforts. The chattering phenomenon is alleviated thanks to the filtering of relay control term. Due to noise and nonmodeled dynamics, the system steady-state accuracy is

Fig. 27.9 Wheel slip response with control (27.58)

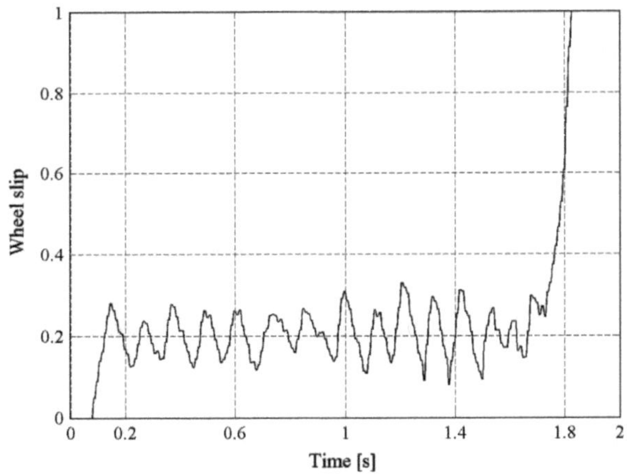

Fig. 27.10 Wheel slip response with control (27.61)

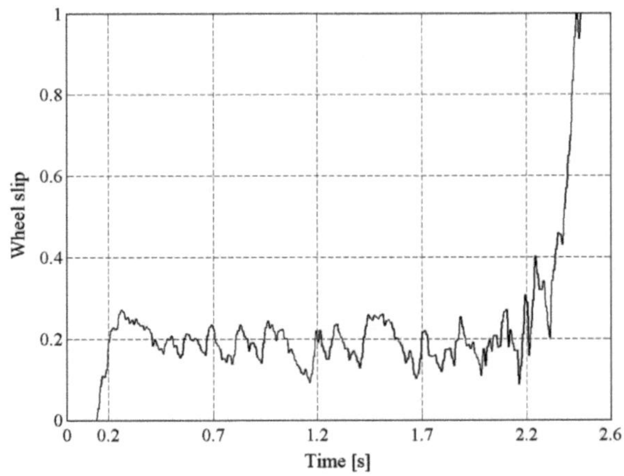

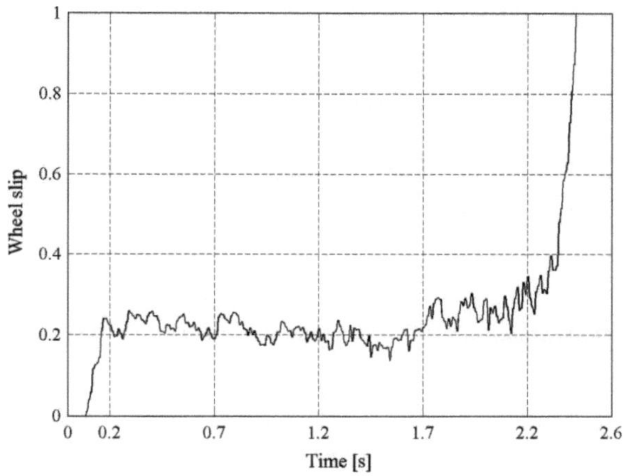

Fig. 27.11 Wheel slip response with control (27.65)

in wider boundaries than ones theoretically defined by $\sim \alpha T = 5 \times 10^{-3}$. Comparing to the previous case with control (27.58), the error signal is smaller.

The best wheel slip response of ABS with discrete-time SMC algorithms, presented in Fig. 27.11, is obtained using the digital SMC (27.65) first introduced in [33]. The parameter α is taken to be 10 times smaller than in the previous control law, i.e., $\alpha = 0.1$, and consequently $\alpha T = 5 \times 10^{-4}$.

The chattering is also alleviated thanks to the filtering of the relay control term by the digital integrator. The introduction of one-step delayed modeling error estimator signal in control law has the same effect as an implementation of one additional integrator in control [29]. This causes better system steady-state accuracy than with the implementation of two previously discussed algorithms.

27.6 Conclusions

The brief mathematical background of four continuous- and three discrete-time sliding mode control (SMC) techniques, used in antilock braking system control, is discussed here in this paper. The control algorithms are designed using continuous- and discrete-time nonlinear wheel slip models obtained from the quarter vehicle model. To illustrate the advantages of sliding mode control, a series of real-time experiments are performed for all control algorithms, showing that there are great agreements between the obtained theoretical and experimental results. Unfortunately, due to mechanical constraints, the braking torque control signal should be additionally limited to 40 % of its nominal value in order to get acceptable experimental results. This limitation introduces greater control signal saturation which favors continuous-time SMC with constant plus proportional rate reaching law and discrete-time SMC with one-step delayed modeling error estimator in comparison to other SMC laws.

Acknowledgments This paper was realized as a part of the projects "Studying climate change and its influence on the environment: impacts, adaptation and mitigation" (III 43007), "Development of new information and communication technologies, based on advanced mathematical methods, with applications in medicine, telecommunications, power systems, protection of national heritage and education" (III 44006), and "Research and Development of New Generation Wind Turbines of High-energy Efficiency" (TR 35005), financed by the Ministry of Education and Science of the Republic of Serbia within the framework of integrated and interdisciplinary research for the period 2011–2014.

References

1. Oniz, Y., Kayacan, E., Kaynak, O.: Simulated and experimental study of antilock braking system using grey sliding mode control. In: Proceedings of the IEEE International Conference on Systems, Man and Cybernetics SMC 2007, pp. 90–95 (2007)
2. Tanelli, M., Piroddi, L., Savaresi, S.M.: Real-time identification of tire-road friction conditions. IET Control Theory Appl. **3**, 891–906 (2009)
3. Zanten, A., Erhardt, R., Lutz, A.: Measurement and simulation of transients in longitudinal and lateral tire forces. SAE Pap. **99**, 300–318 (1990)
4. Hung, J.Y., Gao, W., Hung, J.C.: Variable structure control: A survey. IEEE Trans. Industr. Electron. **40**, 2–22 (1993)
5. Gao, W., Hung, J.C.: Variable structure control of nonlinear systems: a new approach. IEEE Trans. Industr. Electron. **40**, 45–55 (1993)
6. Utkin, V.I.: Sliding mode control design principles and application to electric drives. IEEE Trans. Industr. Electron. **40**, 23–36 (1993)
7. Hamzah, N., Sam, M. Y., Basari, A. A.: Enhancement of driving safety feature via sliding mode control approach. In: Proceedings of Fourth International Conference on Computational Intelligence, Robotics and Autonomous Systems, pp. 116–120 (2007)
8. Kayacan, E., Oniz, Y., Kaynak, O.: A grey system modeling approach for sliding-mode control of antilock braking systems. IEEE Trans. Industr. Electron. **56**, 3244–3252 (2009)
9. Chun, K., Sunwoo, M.: Wheel slip control with moving sliding surface for traction control system. Int. J. Automot. Technol. **5**, 123–133 (2004)
10. Harifi, A., Aghagolzadeh, A., Alizadeh, G., Sadeghi, M.: Designing a sliding mode controller for slip control of antilock brake systems. Transp. Res. Part C: Emerg. Technol. **16**, 731–741 (2008)
11. Song, J.: Performance evaluation of a hybrid electric brake system with a sliding mode controller. Mechatronics **15**, 339–358 (2005)
12. Unsal, C., Kachroo, P.: Sliding mode measurement feedback control for antilock braking systems. IEEE Trans. Control Syst. Technol. **7**, 271–281 (1999)
13. Drakunov, S., Ozguner, U., Dix, P., Ashrafi, B.: ABS control using optimum search via sliding modes. IEEE Trans. Control Syst. Technol. **3**, 79–85 (1995)
14. El Hadri, A., Cadiou, J. C., M'sirdi, N. K.: Adaptive sliding mode control of vehicle traction. In: Proceedings of the 15th Triennial IFAC World Congress, Barcelona, Spain, 21–26 July 2002
15. Wu, M., Shih, M.: Using the sliding-mode PWM method in an anti-lock braking system. Asian J. Control **3**, 255–261 (2001)
16. Wu, M., Shih, M.: Simulated and experimental study of hydraulic anti-lock braking system using sliding-mode PWM control. Mechatronics **13**, 331–351 (2003)
17. Yu, X., Li, G.: Hybrid electric vehicle brake system control metrics design and application. In: Proceedings of the International Symposium on Intelligent Information Systems and Applications, pp. 174–178 (2009)

18. Mirzaeinejad, H., Mirzaei, M.: A novel method for non-linear control of wheel slip in anti-lock braking systems. Control Eng. Pract. **18**, 918–926 (2010)
19. Zheng, S., Tang, H., Han, Z., Zhang, Y.: Controller design for vehicle stability enhancement. Control Eng. Pract. **14**, 1413–1421 (2006)
20. Park, E.J., Stoikov, D., Falcao, L., Suleman, A.: A performance evaluation of an automotive magnetorheological brake design with a sliding mode controller. Mechatronics **16**, 405–416 (2006)
21. Jing, Y., Mao, Y., Dimirovski, G. M., Zheng, Y., Zhang, S.: Adaptive global sliding mode control strategy for the vehicle antilock braking systems. In: Proceedings of the 28th American Control Conference, pp. 769–773 (2009)
22. Mitić, D., Antić, D., Perić, S., Milojković, M., Nikolić, S.: Fuzzy sliding mode control for anti-lock braking systems. In: Proceedings of 7th International Symposium on Applied Computational Intelligence and Informatics, pp. 217–222 (2012)
23. Antić, D., Nikolić, V., Mitić, D., Milojković, M., Perić, S.: Sliding mode control of anti-lock braking system: an overview. Facta Universitatis Series: Automatic Control and Robotics **9**, 41–58 (2010)
24. Utkin V. I., Chang H.-C.: Sliding mode control for automobile applications. In: Sabanovic, A., Fridman, L., Spurgeon, S. (eds.): Variable Structure Systems: From Principles to Implementation. IEE Control Engineering Series, vol. 66, pp. 319–332. The Institution of Electrical Engineers, London, United Kingdom (2004)
25. Milosavljević, Č.: General conditions for the existence of a quasi-sliding mode on the switching hyperplane in discrete variable structure systems. Autom. Remote Control **46**, 307–314 (1985)
26. Milosavljević, Č.: Discrete-time VSS. In: A Sabanovic, A., Fridman, L., Spurgeon, S. (eds.): Variable Structure Systems: From Principles to Implementation. IEE Control Engineering Series, Vol. 66, pp. 99–129 The Institution of Electrical Engineers, London, United Kingdom (2004)
27. Corradini, M. L., Orlando, G.: Discrete variable structure control for nonlinear systems.In: 3rd European Control Conference, pp. 1465–1470 (1995)
28. Furuta, K.: VSS type self-tuning control. IEEE Trans. Ind. Electron. **40**, 37–44 (1993)
29. Mitić, D.: Digital variable structure systems based on input-output model. Ph.D. Thesis (in Serbian) University of Niš, Faculty of Electronic Engineering, Niš, Republic of Serbia (2006)
30. Mitić, D., Milosavljević, Č.: Sliding mode based minimum variance and generalized minimum variance controls with $O(T^2)$ and $O(T^3)$ accuracy. Electr. Eng. **86**, 229–237 (2004)
31. Munoz, D., Sbarbaro, D.: An adaptive sliding-mode controller for discrete nonlinear systems. IEEE Trans. Industr. Electron. **47**, 574–581 (2000)
32. Tan, H.S., Tomizuka, M.: Discrete-time controller design for robust vehicle traction. IEEE Control Syst. Mag. **10**, 107–113 (1990)
33. Mitić, D., Perić, S., Antić, D., Jovanović, Z., Milojković, M., Nikolić, S.: Digital sliding mode control of anti-lock braking system. Adv. Electr. Comput. Eng. **13**, 33–40 (2013)
34. Mitić, D., Milosavljević, Č., Veselić, B.: One approach to I/O based design of digital sliding mode control for nonlinear plants. Electronics **8**, 64–67 (2004)
35. Mitić, D., Milosavljević, Č., Peruničić-Draženović, B., Veselić, B.: Digital sliding mode control design based on I/O model of nonlinear plants.In: 9th International Workshop on Variable Structure Systems, pp. 57–62 (2006)
36. Inteco: The laboratory Anti-lock Braking System controlled from PC-User's Manual. www.inteco.com.pl
37. Draženović, B.: The invariance conditions in variable structure systems. Automatica **5**, 287–295 (1969)
38. Antić, D., Nikolić, V., Mitić, D., Milojković, M., Perić, S.: Sliding mode control of anti-lock braking systems: An overview. In: Proceedings of X Triennial International SAUM Conference on Systems, Automatic control and measurements, pp. 41–48 (2010)

39. Mitić, D., Antić, D., Perić, S., Milojković, M., Nikolić, S.: Sliding mode control of anti-lock braking system based on reaching law method. In: Proceedings of XLVI International Scientific Conference on Information, Communication and energy systems and technologies, vol. 2, pp. 387–390 (2011)
40. Fallaha, C.J., Saad, M., Youssef, K.H., Al-Haddad, K.: Sliding-mode robot control with exponential reaching law. IEEE Trans. Industr. Electron. **58**, 600–609 (2011)
41. Chen, F.C., Khalil, H.: Adaptive control of a class of nonlinear discrete-time systems using neural networks. IEEE Trans. Autom. Control **40**, 791–797 (1995)

Chapter 28
Switching Frequency Optimization of DC/AC Inverters Using Sliding Mode

Khalifa Al Hosani and Vadim I. Utkin

Abstract It is common that a DC/AC converter for three-phase load is designed for controlling two variables (for example, speed and flux of AC motor). An additional degree of freedom can be utilized to minimize the switching frequency, which depends on the voltage of the load neutral point. In this chapter, a methodology of switching frequency minimization is proposed in the framework of the modified hysteresis control. The load neutral point voltage is selected as the third variable to be controlled. First, the tracking system algorithm is developed and then optimization with the switching frequency as a criterion is performed by a proper choice of the reference input for the neutral point voltage. The system accuracy is determined by the width of hysteresis loop and is the same for any switching frequency.

28.1 Problem Statement and Modeling of DC/AC Inverter

Consider the ideal-switch model of DC/AC inverter shown in Fig. 28.1 [1, 2]. The switches $S_i, i = 1, \ldots, 6$ are to be controlled such that phase currents i_a, i_b, and i_c track balanced sinusoidal reference currents $i_a^* = i_{\max}\sin(\omega t)$, $i_b^* = i_{\max}\sin\left(\omega t + \frac{2\pi}{3}\right)$, and $i_c^* = i_{\max}\sin\left(\omega t + \frac{4\pi}{3}\right)$ respectively with minimum possible switching frequency .

K. Al Hosani (✉)
Electrical Engineering Department, The Petroleum Institute, Abu Dhabi, United Arab Emirates
e-mail: khalhosani@pi.ac.ae

V.I. Utkin
Department of Electrical and Computer Engineering, The Ohio State University, Columbus, OH 43210, USA
e-mail: utkin@ece.osu.edu

G.M. Dimirovski (ed.), *Complex Systems*, Studies in Systems, Decision and Control 55, DOI 10.1007/978-3-319-28860-4_28

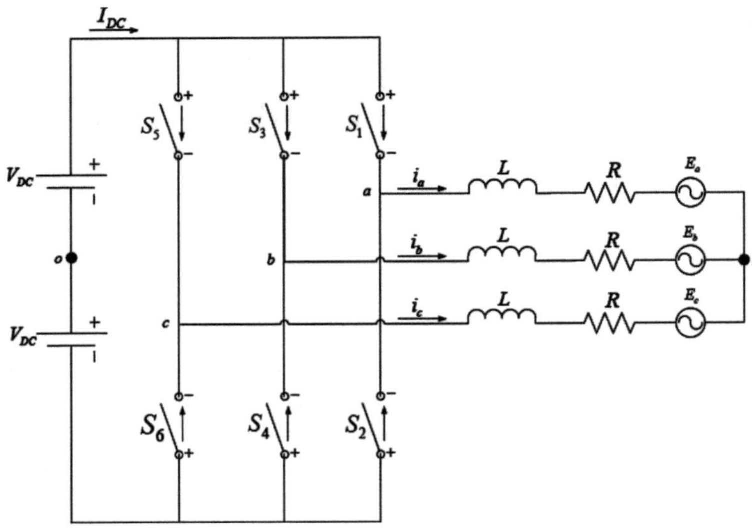

Fig. 28.1 Ideal-switch model of DC/AC inverter

It is assumed that all phase currents i_a, i_b, and i_c are measured. Binary switching functions $H_i \in \{0,1\}, i=1, \ldots, 6$ are defined for switches $S_i, i=1, \ldots, 6$ such that $H_i = 1$ if switch S_i is closed and $H_i = 0$ otherwise. Writing the voltage loop equations for the circuit in Fig. 4.1 results in the following equations

$$L\frac{di_a}{dt} + Ri_a + E_a = (H_1 - H_2)V_{DC} - v_n, \tag{28.1}$$

$$L\frac{di_b}{dt} + Ri_b + E_b = (H_3 - H_4)V_{DC} - v_n, \tag{28.2}$$

$$L\frac{di_c}{dt} + Ri_c + E_c = (H_5 - H_6)V_{DC} - v_n, \tag{28.3}$$

where E_a, E_b, E_c are the back electromotive force EMF voltage generated by the load or machine, and v_n is the voltage of the neutral point n. It is assumed that all voltages in this chapter are defined with respect to the middle point of DC bus (point o in Fig. 28.1).

The DC–AC inverter shown in Fig. 28.1 can be represented using switching matrix representation as in Fig. 28.2. The following conditions are imposed on switching functions $H_i \in \{0,1\}, i=1, \ldots, 6$

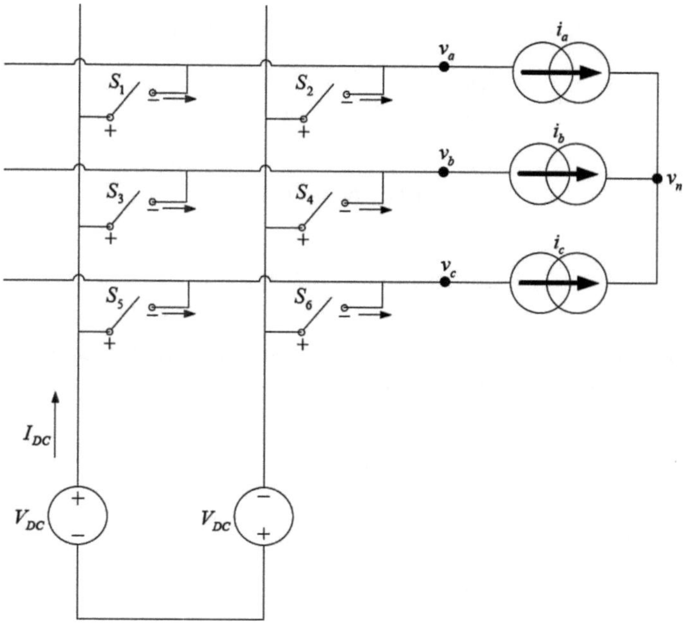

Fig. 28.2 Switching matrix representation of DC/AC inverter

$$H_1 + H_2 = 1, \tag{28.4}$$

$$H_3 + H_4 = 1, \tag{28.5}$$

$$H_5 + H_6 = 1. \tag{28.6}$$

Conditions (28.4)–(28.6) simply state that at least one switch at each phase must be closed at any instance for current in each phase to be continuous and no more than one switch at each phase should be closed simultaneously to prevent any voltage short circuiting. We now define new variables T_a, T_b, and T_c as

$$T_a = H_1 - H_2 = \begin{cases} +1 & \text{for} \quad (H_1, H_2) = (1, 0) \\ -1 & \text{for} \quad (H_1, H_2) = (0, 1), \end{cases} \tag{28.7}$$

$$T_b = H_3 - H_4 = \begin{cases} +1 & \text{for} \quad (H_3, H_4) = (1, 0) \\ -1 & \text{for} \quad (H_3, H_4) = (0, 1), \end{cases} \tag{28.8}$$

$$T_c = H_5 - H_6 = \begin{cases} +1 & \text{for} \quad (H_5, H_6) = (1, 0) \\ -1 & \text{for} \quad (H_5, H_6) = (0, 1). \end{cases} \tag{28.9}$$

Using (28.7)–(28.9), Eqs. (28.1)–(28.3) can be rewritten as

$$L\frac{di_a}{dt}+Ri_a+E_a=T_aV_{DC}-v_n, \tag{28.10}$$

$$L\frac{di_b}{dt}+Ri_b+E_b=T_bV_{DC}-v_n, \tag{28.11}$$

$$L\frac{di_c}{dt}+Ri_c+E_c=T_cV_{DC}-v_n. \tag{28.12}$$

Summing Eqs. (28.10) and (28.11) and using the fact that phase currents are balanced, i.e., $i_a+i_b+i_c=0$, the neutral point voltage is found to be

$$v_n=\frac{1}{3}(T_a+T_b+T_c). \tag{28.13}$$

The line voltages can be written as

$$v_a=T_aV_{DC}, \tag{28.14}$$

$$v_b=T_bV_{DC}, \tag{28.15}$$

$$v_c=T_cV_{DC}. \tag{28.16}$$

Thus, using (28.13)–(28.16), the line-to-neutral voltages are

$$v_{an}=\left(+\frac{2}{3}T_a-\frac{1}{3}T_b-\frac{1}{3}T_c\right)V_{DC}, \tag{28.17}$$

$$v_{bn}=\left(-\frac{1}{3}T_a+\frac{2}{3}T_b-\frac{1}{3}T_c\right)V_{DC}, \tag{28.18}$$

$$v_{cn}=\left(-\frac{1}{3}T_a-\frac{1}{3}T_b+\frac{2}{3}T_c\right)V_{DC}. \tag{28.19}$$

Equation (28.10) can be rewritten in matrix form as

$$\frac{di}{dt}=K+\frac{V_{DC}}{3L}B_1U, \tag{28.20}$$

where $i=[\,i_a \quad i_b \quad i_c\,]^T$, $K=\frac{1}{L}[\,-Ri_a-E_a \quad -Ri_b-E_b \quad -Ri_c-E_c\,]^T$,

$U=[\,T_a \quad T_b \quad T_c\,]^T$, and $B_1=\begin{bmatrix} 2 & -1 & -1 \\ -1 & 2 & -1 \\ -1 & -1 & 2 \end{bmatrix}$

The problem can now be reformulated as follows: control U is to be selected such that currents $i = [\, i_a \quad i_b \quad i_c \,]^T$ track reference currents $i^* = [\, i_a^* \quad i_b^* \quad i_c^* \,]^T$ with minimum possible switching. Since matrix B_1 is singular, two currents out of the there currents $i = [\, i_a \quad i_b \quad i_c \,]^T$ can be controlled independently. The remaining third current is automatically tracked because of the fact that phase currents are balanced, i.e., $i_a + i_b + i_c = 0$. Thus, we have one superfluous degree of freedom, although currents are equal to the desired values, suggesting that motion of the system is not unique. Thus, we can use the extra degree of freedom to control one more state variable [3]. For example, neutral voltage v_n can be equal to a desired time function. In the next section, control algorithm leading to current tracking and switching frequency minimization through proper selection of v_n is presented. In the sequel we apply the state-of-the-art developments in the theory and application of sliding-mode control and variable structure systems (e.g., see references [6–13]) in order to derive switching frequency optimization using sliding-mode technique [10].

28.2 Sliding-Mode Control Algorithm

Let surface $s = [\, s_a \quad s_b \quad s_3 \,]^T = 0$ be defined as

$$s_a = i_a - i_a^*, \tag{28.21}$$

$$s_b = i_b - i_b^*, \tag{28.22}$$

$$s_3 = \int \left(v_n^* - v_n \right) dt, \tag{28.23}$$

where $i_a^* = i_{\max} \sin(\omega t)$, $i_b^* = i_{\max} \sin\left(\omega t + \frac{2\pi}{3}\right)$, and $i_c^* = i_{\max} \sin\left(\omega t + \frac{4\pi}{3}\right)$ which results in $i_a^* + i_b^* + i_c^* = 0$. v_n^* is a reference voltage selectively chosen (as will be shown later) to minimize switching frequency.

If sliding mode exists and is enforced along the sliding surface $s = 0$, then all three components of s will be equal to zero and the problem will be solved. However, the switching along each surface component depends on switching function of all three switching surfaces. The system first needs to be decoupled such that the switching in each surface is controlled independently from the other two surfaces [4, 5]. The time-derivatives of the components of vector s are given by:

$$\frac{ds_a}{dt} = \frac{di_a^*}{dt} + \frac{R}{L} i_a - \frac{E_a}{L} - \frac{V_{DC}}{3L} (2T_a - T_b - T_c), \tag{28.24}$$

$$\frac{ds_b}{dt} = \frac{di_b^*}{dt} + \frac{R}{L} i_b - \frac{E_b}{L} - \frac{V_{DC}}{3L} (-T_a + 2T_b - T_c), \tag{28.25}$$

$$\frac{ds_3}{dt} = v_n^* - \frac{V_{DC}}{3}(T_a + T_b + T_c). \tag{28.26}$$

Equations (28.24)–(28.26) can be written in a matrix form as

$$\frac{ds}{dt} = F - \frac{V_{DC}}{3L} BU, \tag{28.27}$$

where $s = \begin{bmatrix} s_a & s_b & s_3 \end{bmatrix}^T$, $F = \begin{bmatrix} \frac{di_a^*}{dt} + \frac{R}{L}i_a - \frac{E_a}{L} & \frac{di_b^*}{dt} + \frac{R}{L}i_b - \frac{E_b}{L} & v_n^* \end{bmatrix}^T$,

$$B = \begin{bmatrix} 2 & -1 & -1 \\ -1 & 2 & -1 \\ L & L & L \end{bmatrix}, \text{ and } U = \begin{bmatrix} T_a & T_b & T_c \end{bmatrix}^T.$$

The following coordinate transformation is introduced:

$$\bar{s} = B^{-1}s. \tag{28.28}$$

Using (28.27) and (28.28), the time derivative of $\bar{s}$ is

$$\frac{d\bar{s}}{dt} = \bar{F} - \frac{V_{DC}}{3L}U, \tag{28.29}$$

where $\bar{F} = B^{-1}F = \frac{1}{3}\begin{bmatrix} \frac{di_a^*}{dt} - \frac{R}{L}i_a - \frac{E_a}{L} + \frac{v_n^*}{L} \\ \frac{di_b^*}{dt} - \frac{R}{L}i_b - \frac{E_b}{L} + \frac{v_n^*}{L} \\ -\left(\frac{di_a^*}{dt} + \frac{di_b^*}{dt}\right) + \frac{R}{L}(i_a + i_b) + \frac{(E_a + E_b)}{L} + \frac{v_n^*}{L} \end{bmatrix}$, and $\bar{s} = \begin{bmatrix} \bar{s}_1 \\ \bar{s}_2 \\ \bar{s}_3 \end{bmatrix}$.

Now, U is selected as

$$U = \begin{bmatrix} T_a \\ T_b \\ T_c \end{bmatrix} = \begin{bmatrix} \text{sign}(\bar{s}_1) \\ \text{sign}(\bar{s}_2) \\ \text{sign}(\bar{s}_3) \end{bmatrix}. \tag{28.30}$$

Elements of $\bar{F}$ are bounded. Thus, $\frac{V_{DC}}{3L}$ should be large enough to ensure that $\bar{s}\frac{d\bar{s}}{dt} < 0$ condition is satisfied all the time [6, 7] and thus sliding mode is enforced along the switching line $\bar{s} = 0$ (and thus along $s = 0$) after finite time.

Consider now the special case when $E_a = E_b = E_c = 0$. For this case,

$$\frac{d\bar{s}_1}{dt} = \frac{1}{3}\frac{di_a^*}{dt} - \frac{R}{3L}i_a + \frac{v_n^*}{3L} - \frac{V_{DC}}{3L}\text{sign}(\bar{s}_1), \tag{28.31}$$

$$\frac{d\bar{s}_2}{dt} = \frac{1}{3}\frac{di_b^*}{dt} - \frac{R}{3L}i_b + \frac{v_n^*}{3L} - \frac{V_{DC}}{3L}\text{sign}(\bar{s}_2), \tag{28.32}$$

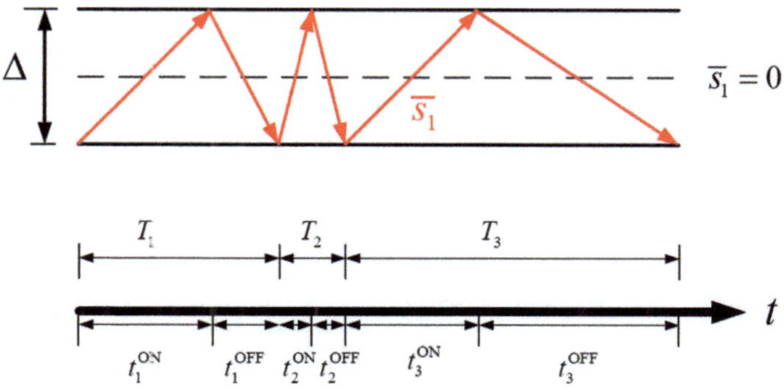

Fig. 28.3 Switching along the switching line $\bar{s} = 0$

$$\frac{d\bar{s}_3}{dt} = -\frac{1}{3}\left(\frac{di_a^*}{dt} + \frac{di_b^*}{dt}\right) + \frac{R}{3L}(i_a + i_b) + \frac{v_n^*}{3L} - \frac{V_{DC}}{3L}\operatorname{sign}(\bar{s}_3). \tag{28.33}$$

Assuming that the switching function sign is implemented with a hysteresis loop of width Δ as shown in Fig. 28.3, the switching frequency in the first phase ($\bar{s}_1$) is given by

$$T = \frac{1}{f} = t_{ON} + t_{OFF} = \frac{2\Delta\left(\frac{1}{3}\frac{di_a^*}{dt} - \frac{R}{3L}i_a + \frac{v_n^*}{3L}\right)}{\left(\frac{1}{3}\frac{di_a^*}{dt} - \frac{R}{3L}i_a + \frac{v_n^*}{3L}\right)^2 - \left(\frac{V_{DC}}{3L}\right)^2}. \tag{28.34}$$

Thus, switching frequency in the first phase is proportional to the term $\left(\frac{1}{3}\frac{di_a^*}{dt} - \frac{R}{3L}i_a + \frac{v_n^*}{3L}\right)^2 - \left(\frac{V_{DC}}{3L}\right)^2$. Similarly, switching frequencies in the second and third phases are proportional to $\left(\frac{1}{3}\frac{di_b^*}{dt} - \frac{R}{3L}i_b + \frac{v_n^*}{3L}\right)^2 - \left(\frac{V_{DC}}{3L}\right)^2$ and $\left(-\frac{1}{3}\left(\frac{di_a^*}{dt} + \frac{di_b^*}{dt}\right) + \frac{R}{3L}(i_a + i_b) + \frac{v_n^*}{3L}\right)^2 - \left(\frac{V_{DC}}{3L}\right)^2$ respectively. The ultimate goal here is to minimize the sum of switching frequencies in all three phases. Let f_1, f_2, and f_3 be defined as

$$f_1 = \frac{1}{3}\frac{di_a^*}{dt} - \frac{R}{3L}i_a, \tag{28.35}$$

$$f_2 = \frac{1}{3}\frac{di_b^*}{dt} - \frac{R}{3L}i_b, \tag{28.36}$$

$$f_3 = -\frac{1}{3}\left(\frac{di_a^*}{dt} + \frac{di_b^*}{dt}\right) + \frac{R}{3L}(i_a + i_b). \tag{28.37}$$

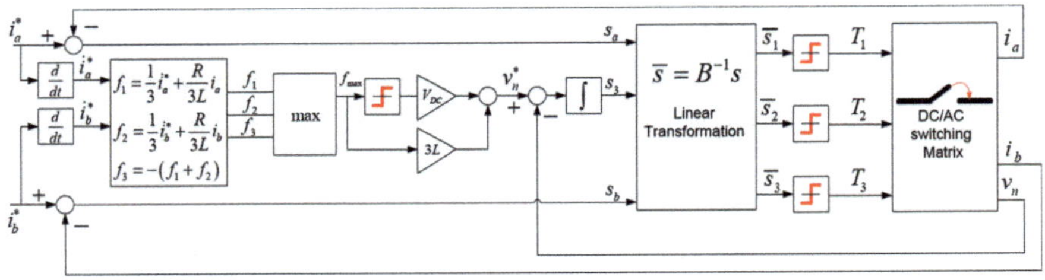

Fig. 28.4 Block diagram of control of DC/AC inverter with switching frequency minimization algorithm

The switching frequency in each phase takes a maximum value if $f_i + \frac{v_n^*}{3L} = 0, i = 1, 2, 3$ and decreases with increasing $\left|f_i + \frac{v_n^*}{3L}\right|$. Thus, v_n^* is to be selected such that $\sum_{i=1}^{3}\left|f_i + \frac{v_n^*}{3L}\right|$ is maximized, but without violating sliding mode enforcement condition of any of the three phases, i.e., $\left|f_i + \frac{v_n^*}{3L}\right| < \frac{V_{DC}}{3L}$. According to this discussion, the following algorithm to select v_n^* is proposed:

1. Select two of the three functions f_1, f_2, and f_3 such that they have the same sign (say for example f_j, f_k). The third function (say f_l) is necessarily equal in magnitude to the sum of the other two but with different sign.
2. Among the selected two functions $\mathrm{f_j}$, $\mathrm{f_k}$, select the one with the maximum magnitude $\mathrm{f_{max}} = \max\left(\left|\mathrm{f_j}\right|, \left|\mathrm{f_k}\right|\right)$.
3. Select $\mathrm{v_n^*}$ as $\mathrm{v_n^*} = \mathrm{V_{DC}}\mathrm{sign}(\mathrm{f_{max}}) - 3\mathrm{Lf_{max}}$

Figure 28.4 shows a block diagram of the overall system along with the switching frequency optimization algorithm's steps involved

Table 28.1 Parameter values for simulation in Figs. 28.5, 28.6, 28.7, 28.8, 28.9, 28.10, 28.11, 28.12 and 28.13

Parameter	Value
L	12 mH
R	0.06 Ω
V_{DC}	650 V
$i_{\max}$	18 $\sqrt{2}A$
ω	10π rad/sec

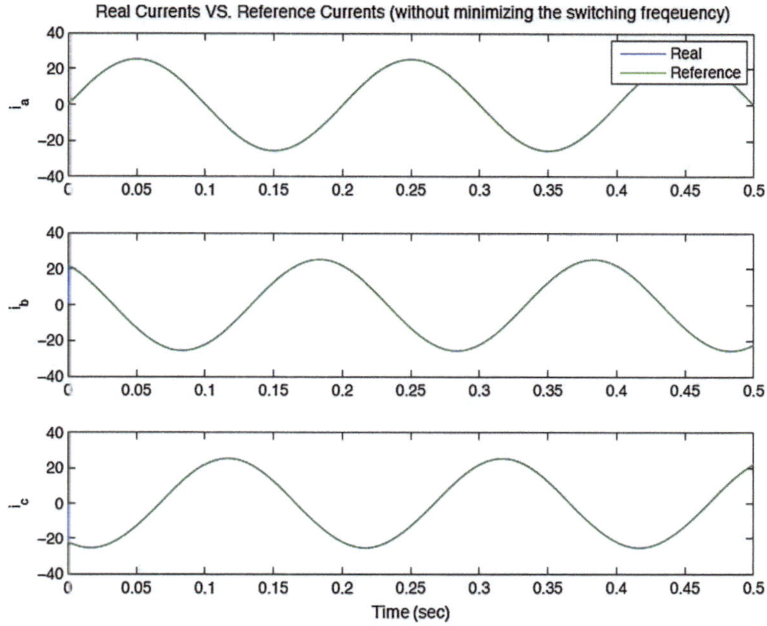

Fig. 28.5 Currents are tracked for the case without switching frequency minimization

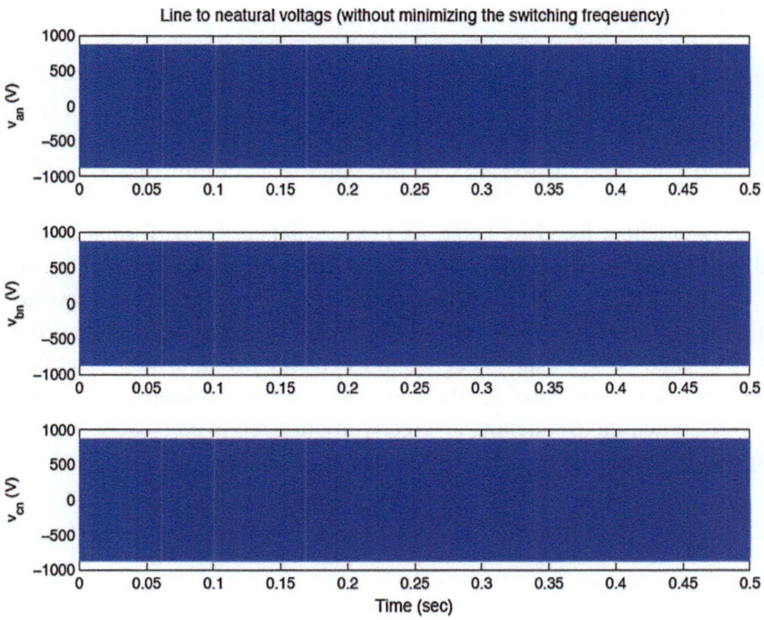

Fig. 28.6 Line-to-neutral voltages for the case without switching frequency minimization

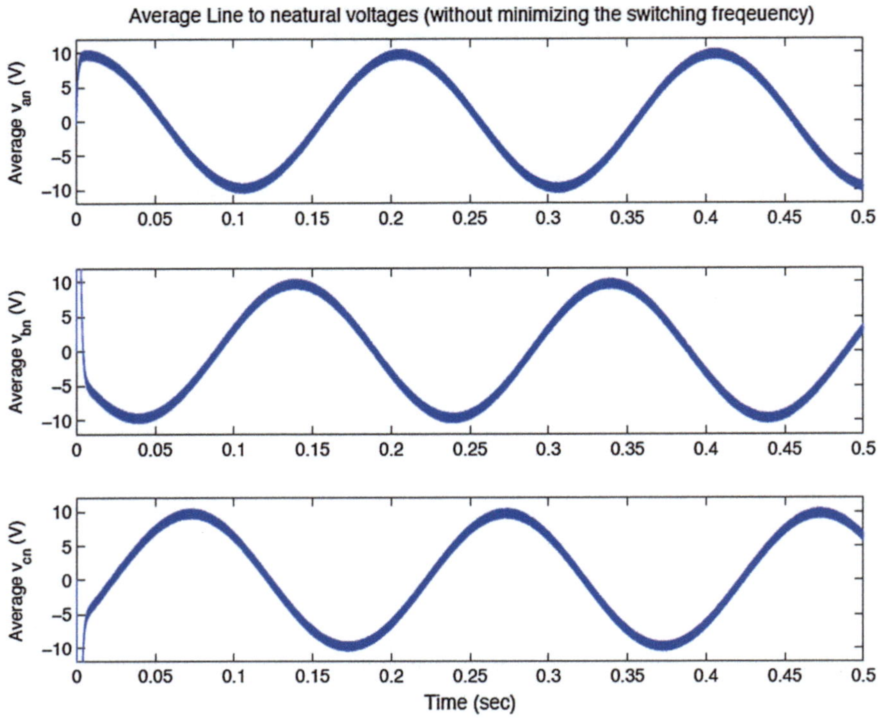

Fig. 28.7 Averaged line-to-neutral voltages for the case when no switching frequency minimization is performed

28.3 Simulation Result

To demonstrate the effectiveness of the control algorithm, the DC/AC inverter system with governing Eqs. (28.1)–(28.3) and controller as in Fig. 28.4 is simulated with the parameters listed in Table 28.1. To evaluate the performance of the proposed optimization algorithm, two cases are presented here. The first case (Figs. 28.5, 28.6, 28.7 and 28.8) appears when no switching frequency minimization is performed. By this we mean, $v_n^* = 0$. The second case (Figs. 28.9, 28.10, 28.11, 28.12and 28.13) takes place when frequency minimization is performed. As can be seen from simulation results, current tracking is established for both cases. However, switching frequency is greatly minimized when using the suggested sliding-mode algorithm.

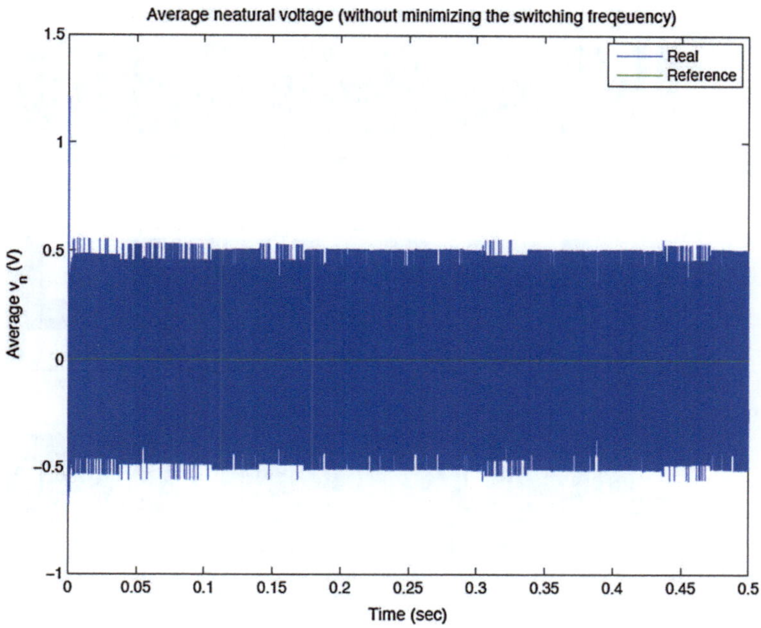

Fig. 28.8 Average neutral point voltage for the case when no switching frequency minimization is performed

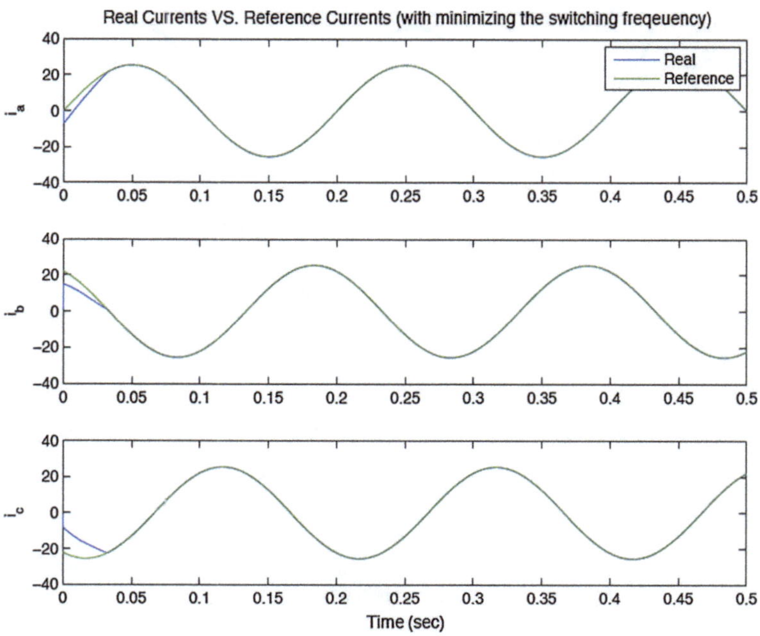

Fig. 28.9 Line currents when switching frequency is minimized

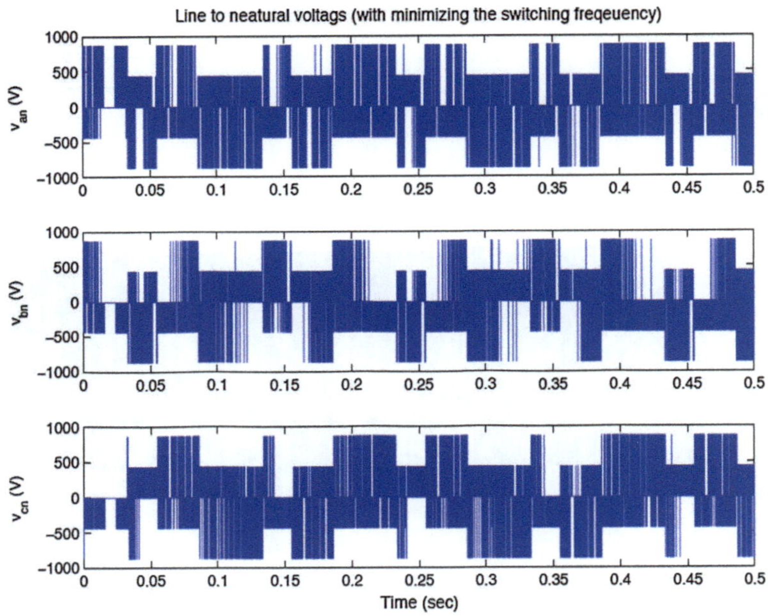

Fig. 28.10 Line-to-neutral voltages when switching frequency minimization is performed

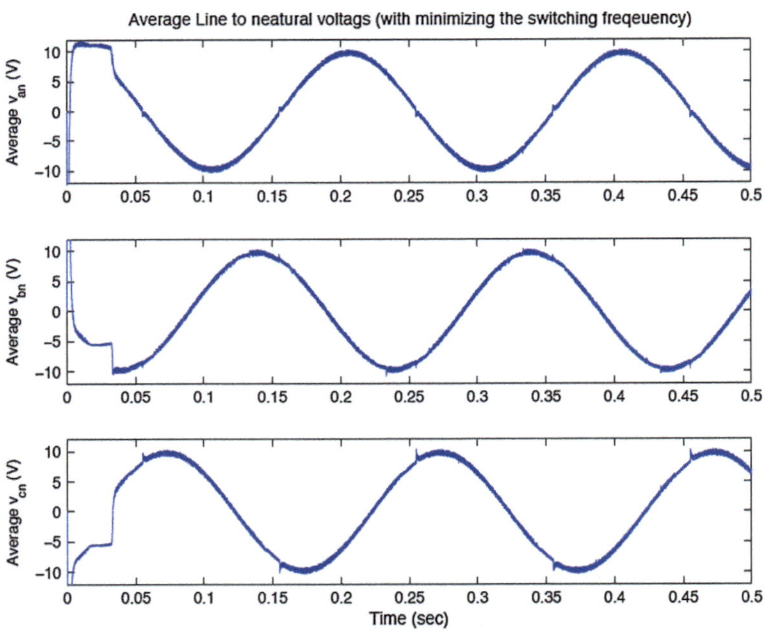

Fig. 28.11 Average line-to-neutral voltages when switching frequency minimization is performed

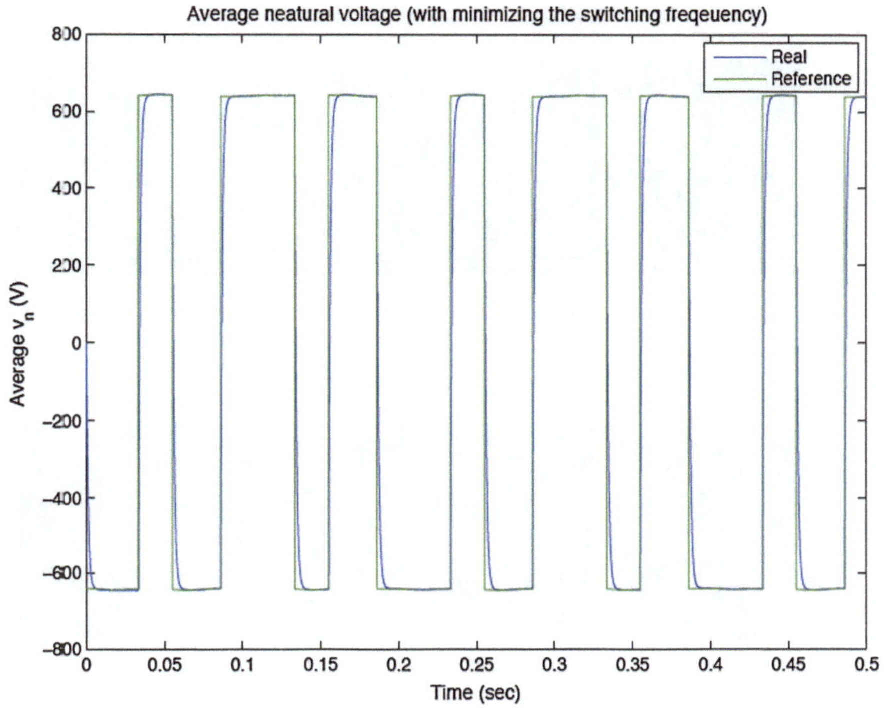

Fig. 28.12 Average neutral point voltage when switching frequency minimization is performed

28.4 Conclusion

The high level of heat losses for high power DC/AC converters is a crucial problem. In contrast to DC/DC converters with inverse proportional dependence between frequency and ripple amplitude, there exists one additional degree of freedom in DC/AC converters: indeed, DC/AC converter for three-phase load is designed for controlling two variables (for example, speed and flux of AC motor). An additional degree of freedom can be utilized to minimize the frequency preserving accuracy (or ripple amplitude) at the same level.

In this chapter, a methodology of switching frequency minimization is proposed in the framework of multidimensional sliding-mode control. The load neutral point voltage is selected as the third variable to be controlled. First, the tracking system algorithm is developed and then the optimization with the frequency as a criterion is performed by a proper choice of the reference input for the neutral point voltage. The system accuracy is determined by the width of hysteresis loop and is the same for any frequency.

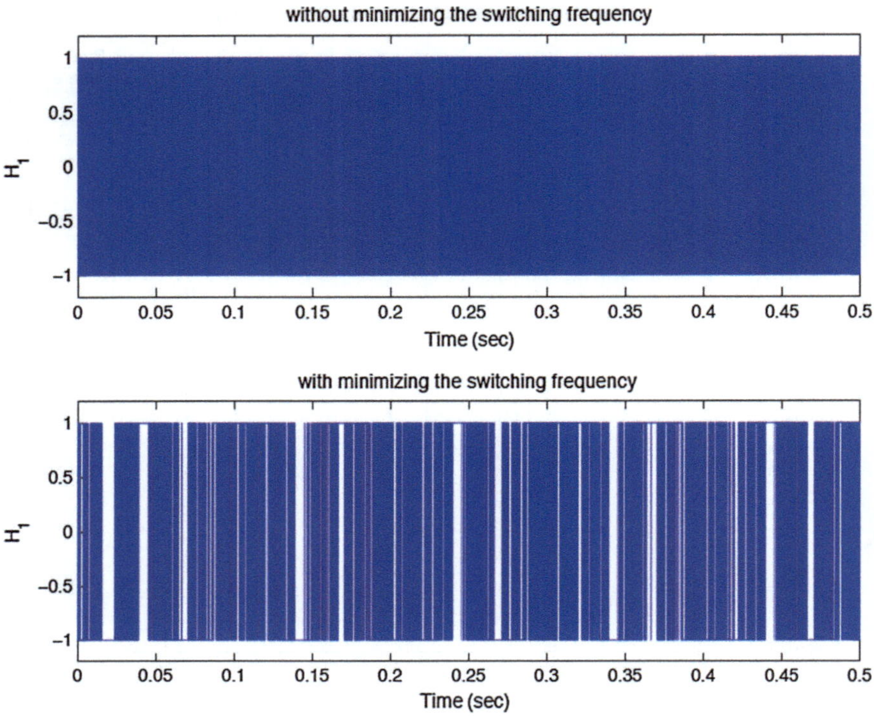

Fig. 28.13 Averaged line-to-neutral voltages for the case when no switching frequency minimization is performed. Comparison of the switching in phase a when no switching frequency optimization algorithm is used with that when the algorithm is utilized

References

1. Krein, P.T.: Elements of Power Electronics, 1st edn. Oxford University Press, Oxford (1997)
2. Mohan, N., Undeland, T.M., Robbins, W.P.: Power Electronics: Converters, Applications, and Designs, 3rd edn. Wiley, New York (2003)
3. Yan, W.: Multilevel sliding mode control in hybrid power systems. Ph.D. Dissertation, The Ohio State University, Columbus, OH, USA (2007)
4. Utkin, V.I.: Sliding Modes and their Application in Variable Structure Systems. MIR Publishers, Moscow (1978)
5. Utkin, V.I., Guldner, J., Shi, J.: Sliding Mode Control in Electro-Mechanical Systems (2nd edn.). Taylor & Francis, CRC Press, Boca Raton (2009)
6. Barbashin, E.A.: Introduction to the Theory of Stability (in Russian). Nauka, Moscow (1967)
7. Emelyanov, S.V. (ed.): Theory of Variable Structutre Systems (in Russian). Nauka, Moscow (1970)
8. Utkin, V.I.: VSS—research, status and primise. Autom. Remote Control **44**(9), 1105–1120 (1984)
9. Utkin, V.I.: Siding mode control design principle and application to electric drives. IEEE Trans. Ind. Electron. **40**(1), 23–36 (1993)
10. Utkin, V.I.: Sliding Modes in Control and Optimization. Springer, Berlin (1992)

11. Spurgeon, S.K., Davies, R.: A nonlinear control strategy for robust sliding mode performance in the presence of unmatched uncertainty. Int. J. Control **57**(5), 1107–1123 (1993)
12. Young, K.D., Utkin, V.I., Ozguner, U.: A control engineer's guide to sliding mode control. IEEE Trans. Control Syst. Technol. **7**(3), 328–342 (1999)
13. Sabanovic, A., Fridman, L.M., Spugeaon, S.K. (eds.): Variable Structure Systems from Principles to Implementation. The IET Press, London, UK (2004)

Chapter 29
Discrete-Time Sliding-Mode Servo Systems Design with Disturbance Compensation Approach

Čedomir Milosavljević, Branislava Draženović and Boban Veselić

Abstract This paper presents a chattering-free discrete-time sliding mode control algorithm with a new combined disturbance compensator, which is based on switching function measurement only. The complete system behaves as a high-accuracy tracking system with an excellent compensation of matched disturbances. Properties of the proposed design method are demonstrated on a velocity and a positional servo system.

29.1 Introduction

Sliding mode (SM) control (SMC) in variable structure systems is recognized as an efficient design method providing a robust performance with respect to the parameter uncertainties and external disturbances [1, 2]. Due to the implementation of microprocessors in control systems and the use of computers in controller design starting in seventies of the previous century, synthesis and application of discrete-time (DT) control systems, and computer-based control become widespread topics. The following question was crucial for its implementation: what happens with the continuous-time (CT) SMC systems when the DT signal processing is applied? Research [3] was one of the first to deal with the problem of

Č. Milosavljević (✉)
Faculty of Electrical Engineering, University of Istočno Sarajevo, Istočno Sarajevo, Bosnia and Herzegovina
e-mail: cedomir.milosavljevic@elfak.ni.ac.rs

B. Draženović
Faculty of Electrical Engineering, University of Sarajevo, Sarajevo, Bosnia and Herzegovina
e-mail: brana_p@hotmail.com

B. Veselić
Faculty of Electronic Engineering, University of Niš, Niš, Serbia
e-mail: boban.veselic@elfak.ni.ac.rs

G.M. Dimirovski (ed.), *Complex Systems*, Studies in Systems, Decision and Control 55, DOI 10.1007/978-3-319-28860-4_29

implementing discretization into SMC. The most important results of his research were presented in [4], where the term *quasi-sliding mode* has been introduced. Later, simular results were found by Potts and Yu [5], which introduced the term *pseudo-sliding mode*.

The contemporary approaches to digital algorithms realizations have introduced some new terms: *discrete-time sliding mode* (DSM) and *discrete-time equivalent control* [6–8]. Gao et al. [9] proposed *reaching law method* that controls dynamics in the reaching phase of the SM hyperplane. Bartolini et al. [10] proposed DSM control algorithm with a saturation function and Bartoszewicz [11] introduced a nonstationary sliding surface in way to constraint control amplitude. Combining the reaching law [9] and the results in [6], a new chattering-free control algorithm was proposed in [12], having bounded control in the reaching phase. In [13, 14] comparative studies of some Discrete-Time Sliding-Mode (DTSM) control algorithms are given.

One of the most alluring features of SMC is a possibility to achieve absolute insensitivity to matched disturbances. It is important to note that DSMs cannot even theoretically provide invariance neither to parameter changes nor to external disturbances. The main reason is that the control stays constant between two consecutive samples, and thus it cannot react to disturbances and parameter variations which act on the system all the time. It is sometimes forgotten that even in real Continuous-Time Sliding-Mode (CTSM) absolute invariance is not possible due to limited frequency of the actuators. However, a strong robustness is feasible, since in DSM it is easy to estimate and compensate the reasonably predictable disturbances. Results in Su et al. [15] rely on the assumption that disturbances are represented by functions that do not change much between sampling instants. The problem of disturbance observation or estimation has been the topic of many publications, some of which are: [13, 16–22]. The approach [20] is based on SM active dynamical disturbance observer, which is experimentally verified for positional systems with DC and AC motors in [19, 21, 22].

A novel approach to find a present disturbance estimate from their previous samples is proposed in [23]. The negative value of this estimate is then used as compensational control which partially cancels the real disturbance effects in the sampling moments for a linear controllable system with matched disturbances. This paper contains a theoretically explained extension of disturbance estimation and compensation principle proposed in [13, 19] where two control terms exist: a nonlinear term, with limited control amplitude, and a linear one equal to DT equivalent control. The linear term of control is chattering-free and acts alone in the vicinity of sliding hyperplane. The estimation and compensation of constant-type disturbance are applied in this linear mode. It has been established in [19] that the proposed disturbance compensator increases the control system type by one. For example, positional system can perfectly track quadratic parabola references and fully reject constant load disturbances. This high-accuracy tracking system structure was used in [20] as an auxiliary system for tracking error signal between controlled process and its nominal model. It is established that the control signal of the auxiliary system is identical to unknown output disturbance whose action is reduced to the plant input. Similar disturbance compensation structure was proposed by Abidi [24] and Abidi and Šabanović [25].

The previous value of the disturbance is calculated in [23] from the present switching function value using the recursive procedure. The estimate of the present time value of the disturbance is obtained from the saved previous values and used to obtain the compensational control. The equivalent control and the compensational control are both linear in state. For many typical slow disturbances, such control eliminates the need of switching type control, which is necessary to maintain SM in the presence of unknown disturbances and produces chattering. Besides estimation of constant-type disturbances, a method for estimation of ramp and quadratic parabola type disturbances was proposed as well. The obtained estimators of constant, ramp, and quadratic parabola disturbances become dynamical elements which contain discrete-time integrators of the first-, the second-, and the third-order, respectively. In the sequel, these compensators will be named compensators of the first-, the second-, and the third-order.

Further investigations in this area [26, 27] established that a combination of the first- and the second-order compensators gives much better dynamic behavior than the second-order compensator alone in control of the first-order plants. Furthermore, such compensator enlarges the control system type by two, providing design of high-accuracy tracking system. Unfortunately, this approach may lead to instability even in control of second-order plants.

This work is intended to study possibilities of application of combined disturbance compensator, which contain parallel connection of the first- and the second-order compensator, for design of DTSM controllers for the first- and second-order plants. Then, the obtained results will be applied in design of velocity and positional servo systems, whose controlled plants may be approximated by the first- and second-order models respectively. Influence of unmodeled dynamics will be of particular interest.

The next section presents the model of the controlled system, the way to obtain previous values of the disturbance, and linear disturbance estimators based on the previous values of the disturbance for some typical cases.

29.2 Preliminaries

Consider the following continuous-time controlled plant

$$\dot{\mathbf{x}}(t) = (\mathbf{A} + \Delta\mathbf{A}(t))\mathbf{x}(t) + (\mathbf{b} + \Delta\mathbf{b}(t))u(t) + \mathbf{d}v(t), \tag{29.1}$$

where $\mathbf{x}(t) \in \Re^n$, $u(t)$, $v(t) \in \Re$ are, respectively, vector of the state, control, and external disturbance. The pair $(\mathbf{A},\mathbf{b})$ is controllable; the matrices **A, b,** and **d** are of appropriate dimensions. The objective is to design a DTSM controller.

Assumption 29.1 Elements of **A, b, d,** and upper and lower bounds of $v(t)$ are known, the parameter variations $(\Delta\mathbf{A}, \Delta\mathbf{b})$ and external disturbance are bounded and matched [28], i.e., $\text{rank}[\mathbf{b}|\Delta\mathbf{A}|\Delta\mathbf{b}|\mathbf{d}] = \text{rank}(\mathbf{b})$ is satisfied, and the plant (29.1) is of minimum phase.

For DSM design purpose, model (29.1) has to be discretized. Usually, there are two options: (*i*) to apply traditional discrete-time state-space model with shift operator and (*ii*) to use δ-state-space model [29]. In this paper the δ-model is used for the controller design, since it is well suited for higher sampling frequencies [30], which provides better accuracy of DSM control system [15, 12]. Fast sampling is of great interest in robotic applications for providing high-speed performance of motion [31].

DT mathematical model of the plant (29.1) in delta form is

$$\delta\mathbf{x}(k) \; \overset{\Delta}{=} T^{-1}(\mathbf{x}((k+1)T) - x(kT)) = \mathbf{A}_\delta \mathbf{x}(kT) + \mathbf{b}_\delta(u(kT) + d(kT)), \qquad (29.2)$$

where

$$\mathbf{A}_\delta = T^{-1}(\mathbf{A}_d - \mathbf{I}); \mathbf{b}_\delta = T^{-1}\mathbf{b}_d; \delta \bullet (kT) = T^{-1}(\bullet((k+1)T) - \bullet(kT)),$$
$$d(kT) = [d_v(kT) + d_A(kT) + d_B(kT)]T^{-1}; d_v(kT) = \int_0^{\mathrm{T}} e^{\mathbf{A}T}\mathbf{d}v((k+1)T - \tau)\mathrm{d}\tau,$$
$$d_A(kT) = \int_0^{T} e^{\mathbf{A}T}\Delta\mathbf{A}((k+1)T - \tau)\mathbf{x}((k+1)T - \tau)d\tau; d_B(kT) = \int_0^{T} e^{\mathbf{A}T}\Delta\mathbf{b}(k+1)Td\tau,$$

$k = 0,1,2,\ldots$ $\mathbf{A}_\mathrm{d}$ and $\mathbf{b}_\mathrm{d}$ are matrices from the traditional discrete-time state-space model obtained using MATLAB® command [Ad,bd] = c2d(A,b,T).

Remark 29.1 If **(A,b)** is a controllable pair, the pair $(\mathbf{A}_\delta,\mathbf{b}_\delta)$ is also controllable for almost all choices of *T*.

Remark 29.2 For notational convenience, in the following $\bullet(k)$ stands for $\bullet(kT)$.

Remark 29.3 In general, if matching conditions hold for CT system (29.1) it does not necessarily follow that the same conditions hold in DT systems (29.2), because the zero-order-hold (ZOH) in the control channel does not change its output value between the sample instants. However, the corresponding error is $O(T^2)$ [15] if a first-order disturbance compensator is used. Therefore it is reasonable to choose the sampling time *T* as small as possible.

First, DT equivalent control for a nominal (undisturbed) system, $d(k) = 0$, will be found. Let DSM be made using the switching function as

$$g(k) = \mathbf{c}_\delta \mathbf{x}(k), \mathbf{c}_\delta \in \Re^{1\times n}. \qquad (29.3)$$

Vector $\mathbf{c}_\delta$ defines the control system behavior in DSM. Its design will be given later. The switching function dynamics becomes

$$\delta g(k) \;=^{\Delta} T^{-1}(g(k+1)-g(k)) = \mathbf{c}_\delta \delta \mathbf{x}(k). \tag{29.4}$$

Solving (29.4) for $g(k + 1)$ and replacing $\delta\mathbf{x}(k)$ from (29.2) it is obtained

$$g(k+1) = g(k) + T\mathbf{c}_\delta^{\mathrm{T}} \mathbf{A}_\delta \mathbf{x}(k) + T\mathbf{c}_\delta \mathbf{b}_\delta u(k). \tag{29.5}$$

The DSM is defined with relation [6–8]

$$g(k+1) = 0, \forall g(k). \tag{29.6}$$

Using (29.5) and (29.6), and an assumption that $\mathbf{c}_\delta \mathbf{b}_\delta = 1$ usually applied in SMC, the DT equivalent control is equal to

$$u(k) = u_{eq}(k) = -T^{-1} g(k) - \mathbf{c}_\delta \mathbf{A}_\delta \mathbf{x}(k). \tag{29.7}$$

From (29.7) it is evident that for small T and $g(k) \neq 0$ the control $u(k)$ may be very high. Since in real systems control magnitude is bounded, the control (29.7) was modified in [12] as follows

$$u(k) = -\mathbf{c}_\delta^{\mathrm{T}} \mathbf{A}_\delta \mathbf{x}(k) - \min\{\sigma, T^{-1}|g(k)|\}\mathrm{sgn}(g(k)), \tag{29.8}$$

where $\sigma = const > |d(k)|$ takes into account the limit of the control magnitude.

To complete DSM control design, the vector $\mathbf{c}_\delta$ of the switching function (29.3) needs to be determined. There are a few approaches. Here is given a comprehensive approach [32] oriented to use MATLAB software for conventional control system design via method of pole placement. Let a desired pole spectrum in continuous-time domain be

$$\Lambda = [\lambda_1, \lambda_2, \ldots, \lambda_{n-1}, 0], \;\mathrm{Re}\{\lambda_i\} < 0, i = 1, 2, \ldots, n-1. \tag{29.9}$$

One zero pole in the desired spectrum reflects algebraic constraint (29.6) related to the application of SMC technique.

The corresponding spectrum in δ-domain becomes

$$\Lambda_\delta = T^{-1}(e^{\Lambda T} - 1). \tag{29.10}$$

For a conventional pole placement design, state feedback vector is obtained by MATLAB as

$$\mathbf{K}_\delta = \mathrm{place}(A_\delta, b_\delta, \Lambda_\delta). \tag{29.11}$$

The vector $\mathbf{c}_\delta$ is then obtained as [32]

$$\mathbf{c}_{\delta} = [\,K_{\delta} \quad 1\,] * \mathrm{pinv}([\,A_{\delta} \quad b_{\delta}]) \,. \tag{29.12}$$

It was shown in [12] that the system (29.2) with the control (29.8) is robustly stable for bounded disturbances. Unfortunately, an error appears even if a constant type of disturbance is applied. To handle this error, an additional integral action was proposed in [19]. The introduced control component $u_I(k)$ is proportional to the integral of the switching function g(k), i.e.,

$$u_I(k) = \alpha T^{-1} g(k) + u_I(k-1),\ 0 < \alpha \leq 1. \tag{29.13}$$

Control (29.13) upgrades the system type by one with respect to the reference as well as to load disturbance. Stability of the resulting enhanced control system needs to be checked.

Unfortunately, the control component (29.13) deteriorates the excellent dynamical characteristic of the system with control (29.8) such as one step reaching of the hyperplane in the linear control mode. To prevent this drawback, the control component (29.13) should be activated in linear control mode only.

Mathematical analysis of the introduction of additional integral control component (29.13) was not given in [19], but it will be done in the next section. Moreover, the main idea will be extended by application of disturbance estimation, done by extrapolation using the measurements of the switching function $g(k)$. In such way, the system type and disturbance rejection capability are significantly improved.

29.3 Disturbance Compensators Synthesis

Now, consider the disturbed system (29.2) taking into account Remark 29.3,

$$\delta\mathbf{x}(k) \ \ {}^{\Delta}{=} T^{-1}(\mathbf{x}(k+1) - x(k)) = \mathbf{A}_{\delta}\mathbf{x}(k) + \mathbf{b}_{\delta}(u(k) + d(k)). \tag{29.14}$$

The control is equal to the equivalent control (7) of the nominal system, and assumption $\mathbf{c}_{\delta}\mathbf{b}_{\delta} = 1$ holds. Then, $g(k + 1)$ becomes

$$g(k+1) = Td(k) \text{ or } d(k-1) = T^{-1} g(k). \tag{29.15}$$

The similar results have been obtained by Chan [16] where a compensation algorithm was proposed.

To apply the new approach, the value of the disturbance $d(k)$ should be reasonably well predicted from a finite number of its previous values, which is the case with typical disturbances in a real industrial environment.

Theoretical value of the equivalent control for the system (29.14), $u_{eqd}(k)$ is

$$u_{eqd}(k) = u_{eq}(k) - d(k). \qquad (29.16)$$

The procedure used to obtain (29.16) is the same as one used to obtain (29.7). However, such control is not feasible, since the present value of the disturbance is not known. Because of that, the real disturbance term here is replaced by a compensational term denoted as $u_{com}(k)$. Its role is to suppress the disturbance influence on the system state. Now, the total control is

$$u(k) = u_{eq}(k) + u_{com}(k). \qquad (29.17)$$

The switching function dynamics is then

$$\delta g(k) = \mathbf{c}_\delta \mathbf{A}_\delta \mathbf{x}(k) + (u_{eq}(k) + u_{com}(k) + d(k)). \qquad (29.18)$$

Using (29.7), the future value of the switching function is

$$g(k+1) = Tu_{com}(k) + Td(k), \qquad (29.19)$$

whereas the present value of the switching function becomes

$$g(k) = Tu_{com}(k-1) + Td(k-1). \qquad (29.20)$$

Since the value of the switching function can be measured, and the value of the previous compensational control is known, it is possible to estimate the previous value of the disturbance as

$$d_{est}(k-1) = T^{-1}g(k) - u_{com}(k-1). \qquad (29.21)$$

The subscript "est" is added to emphasize that due to system parameters uncertainties this theoretical value may differ from the real one. If the disturbance is slowly changing, (29.21) may be used as a reasonable good estimate of the present disturbance's value.

The corresponding compensational control is

$$u_{com}(k) = -d_{est}(k-1) = -(T^{-1}g(k) - u_{com}(k-1)). \qquad (29.22)$$

To obtain a better dynamics, especially to prevent chattering in the system with unmodeled dynamics, it is not advisable to use full compensational control (29.22) but [19]

$$u_{com}(k) = -(\alpha T^{-1}g(k) - u_{com}(k-1));\ 0 < \alpha \leq 1. \qquad (29.23)$$

This is a recursive definition of the compensational control, and its initial value has to be set. With this compensational control the future value of the switching function will depend on the difference between the real and the estimated value of the disturbance

$$g(k+1) = T(d(k) - d_{est}(k-1)). \quad (29.24)$$

Obviously, the better disturbance estimation is used, and the system behavior is closer to the disturbance-less (nominal) system.

The reasonable question is: how this additional compensational control affects the stability of the nominal system? First, consider the system with the theoretical equivalent control (29.16). The system has the disturbance $d(k)$ as an input, and the switching function $g(k)$ as an output. The Z-transform is applied to Eqs. (29.19) and (29.23) assuming zero initial conditions, giving

$$zg(z) = Tu_{com}(z) + Td(z)). \quad (29.25)$$

$$u_{com}(z) = -\frac{\alpha}{T(1-z^{-1})}g(z). \quad (29.26)$$

Replacing (29.26) in (29.25) and solving for $g(z)$ yields

$$g(z) = \frac{T(z-1)}{z(z-(1-\alpha))}d(z). \quad (29.27)$$

It is obvious that $g(z)$ will be stable if $0 < \alpha \leq 1$. This result was obtained in [19]. For $\alpha = 0$ the switching function dynamics is equivalent to the dynamics of the system without disturbance compensation ($g(k) = Td(k-1)$). For $\alpha = 1$ switching function dynamics may be identified as one step delayed difference of disturbance scaled with factor T^2. In the later case the switching function response has an undesirable pulse deviation.

For a faster disturbance, the estimate may be obtained as an extrapolation based on more previous disturbance values, memorized in a buffer. The estimate obtained by extrapolation is denoted as $d_{app}(k)$. In general, the compensational control is given as

$$u_{com}(k) = -d_{app}(k). \quad (29.28)$$

In this paper only slowly varying disturbance such as step and ramp functions are considered. The estimates and the resulting compensational controls are given below.

The step disturbance estimate:

$$d_{app_1}(k) = d(k-1). \quad (29.29)$$

The ramp disturbance estimate is a linear extension of the two previous values:

$$d_{app_2}(k) = 2d(k-1) - d(k-2). \tag{29.30}$$

The compensational control for each case depends on the value of the switching function only. By replacing (29.29) and (29.30) in (29.21), and then by applying Z-transform and solving obtained equations for $u_{com}(z)$, the following is obtained:

$$(a) \quad u_{com_1}(z) = -\frac{\alpha_1}{T}\frac{z}{(z-1)}g(z), \tag{29.31}$$

$$(b) \quad u_{com_2}(z) = -\frac{\alpha_2}{T}\frac{(2z-1)z}{(z-1)^2}g(z). \tag{29.32}$$

It may be seen that compensational control (29.32) contains a series of two DT integrators. These integrators deal well with ramp-type disturbances.

Introducing compensation control (29.32) in (29.25), the value of the switching function in Z-domain is

$$g(z) = \frac{T}{z}\frac{(z-1)^2}{(z-1)^2 + \alpha_2(2z-1)}d(z). \tag{29.33}$$

For $\alpha_2 = 0$ $g(z) = Tz^{-1}d(z)$, i.e., $g(z) = Td(k-1)$ that agrees with the relation (29.15). For $\alpha_2 = 1$ the relation (29.33) becomes

$$g(z) = \frac{T^3}{z}\frac{(z-1)^2}{z^2T^2}d(z), \tag{29.34}$$

which indicates that the switching function dynamics is one step delayed second-order difference of disturbance $d(k)$ scaled by factor T^3. This introduces two undesirable pulses in the reaching process of the sliding hyperplane. It is not difficult to establish that two poles of (29.33) are stable and conjugate-complex which maximum of imaginary part is 0.5 for $\alpha_2 = 0.5$.

From the previous explanation, it can be concluded that the given type of extrapolations does not endanger the switching function stability for $0 < \alpha_{1,2} \leq 1$, but the reaching dynamics will be changed since unwanted pulses will arise. The number of these pulses in the nominal system is equal to the number of integrators introduced to obtain compensation control. To prevent this drawback, which may cause instability if there are unmodeled dynamics, an adaptation similar to one proposed in [19] may be applied.

Further investigations in this area [26, 27] established that the better solution to compensate complex disturbances is to use parallel connection of compensators

(29.31) and (29.32) instead of only compensator (29.32), which generally covers capabilities of compensator (29.31).

In the next section, the capabilities of compensator (29.32) and combined compensator as parallel connection of (29.31) and (29.32) will be discussed via pole placement. First-order dynamical process will be used as controlled plant.

29.4 DTSM Control of the First-Order Plants

In case of a first-order plant, relations (29.1) and (29.2) become

$$\dot{x}(t) = (a + \Delta a)x(t) + (b + \Delta b)u(t) + v(t), \tag{29.35}$$

$$\delta x(k+1) = a_\delta x(k) + b_\delta(u(k) + d(k)), \tag{29.36}$$

$$a_\delta = T^{-1}(e^{aT} - 1); b_\delta = T^{-1}\int_0^T e^{a\tau} b\mathrm{d}\tau = \frac{b}{a}a_\delta; d(k) = T^{-1}\int_0^T e^{a\tau}[v((k+1)T - \tau) + \Delta A((k+1) - \tau)x((k+1) - \tau) + \Delta b(k+1)]\mathrm{d}\tau. \tag{29.37}$$

Switching function dynamics is

$$g(k+1) = g(k) + Tc_\delta a_\delta x(k) + T(u(k) + d(k)), \tag{29.38}$$

and the equivalent control becomes

$$u_{eq}(k) = -T^{-1}g(k) - c_\delta a_\delta x(k) = -(c_\delta T^{-1} + k_{eq})x(k). \tag{29.39}$$

Here, since condition $\mathbf{c}_\delta \mathbf{b}_\delta = 1$ holds and c and b are scalars, then

$$c_\delta = b_\delta^{-1} = \frac{aT}{b(e^{aT} - 1)}. \tag{29.40}$$

In the z-domain, nominal plant dynamics with ZOH is described as

$$X(z) = \frac{b}{a}\frac{e^{aT} - 1}{z - e^{aT}}U(z). \tag{29.41}$$

SM control with disturbance compensator (29.31) in linear mode is described by

$$U(z) = U_{eq}(z) + U_{com1}(z) = -\frac{a}{b}\frac{(e^{aT}+\alpha_1)z - e^{aT}}{(e^{aT}-1)(z-1)}X(z). \tag{29.42}$$

The open-loop transfer function is

$$W(z) = \frac{(e^{aT}+\alpha_1)z - e^{aT}}{(z-e^{aT})(z-1)}. \tag{29.43}$$

Therefore, the characteristic equation of the control system, compensated by disturbance compensator (29.31), becomes

$$1 + W(z) = 0 \Rightarrow z(z-1+\alpha_1) = 0. \tag{29.44}$$

It is clear that the system has two poles. One is $z = 0$ which is consequence of SMC approach. The second pole, $z = 1-\alpha_1$, is introduced by disturbance compensator. It is obvious that by choosing $\alpha_1 = 1$, i.e., if full integral action of disturbance compensator (29.31) is applied, the system gets another pole at zero. In SMC system such double pole at zero may cause excitation of unmodeled dynamics. Therefore, the better solution is to choose $0 < \alpha_1 < 1$ to prevent chattering.

If the disturbance compensator described by (29.32) is used, then the characteristic polynomial becomes

$$z[z^2 - 2(1-\alpha_2)z + 1 - \alpha_2] = 0. \tag{29.45}$$

The system poles are

$$z_1 = 0; z_{2,3} = 1 - \alpha_2 \pm j\sqrt{\alpha_2(1-\alpha_2)}. \tag{29.46}$$

Pole z_1 is due to SMC. The other two poles are introduced by disturbance compensator. For $\alpha_2 = 1$, i.e., when full compensator is used, $z_{2,3} = 0$. Therefore the obtained control system has three zero (dead-beat) poles that present very fast modes. These modes probably may cause chattering, since in real dynamical system, described by the first-order model, unmodeled dynamics usually exists. For $0 < \alpha_2 < 1$ poles $z_{2,3}$ are conjugate-complex. Such poles introduce undesirable damped oscillation in the control system.

To avoid this oscillation and to assure compensation of complex disturbances, parallel connection of both compensators is here proposed. Such structure is named *combined disturbance compensator.*

By using the same procedure, characteristic equation of the system with the combined compensator is obtained as

$$z[z^2 - (2-\alpha_1-2\alpha_2)z + 1 - \alpha_1 - \alpha_2] = 0. \tag{29.47}$$

It is obvious that for $\alpha_1 = 0$ (29.47) becomes (29.45).

According to (29.47) one pole is again $z_1 = 0$. The other two poles are given by

$$z_{2,3} = 0.5\left[(2 - \alpha_1 - 2\alpha_2) \pm j\sqrt{4\alpha_2 - (\alpha_1 + 2\alpha_2)^2}\right]. \tag{29.48}$$

Here there exists an additional degree of freedom to adjust the system poles to prevent undesired damped oscillation and chattering excitation. Eigenvalues (29.48) should be real positive and not near zero. Then the condition

$$\alpha_2 = 0.5(1 - \alpha_1 - \sqrt{1 - 2\alpha_1}) \tag{29.49}$$

must be satisfied.

It may be concluded from (29.49) that the real value of α_2 exists only for $0 < \alpha_1 \leq 0.5$ that gives the range of α_2 as $0 < \alpha_2 \leq 0.25$. Then, from (29.48), real poles introduced by combined compensator are in the range $1 > z_{2,3} \geq 0.5$.

Replacing α_2 from (29.49) into (29.48) and solving it with respect to α_1 one obtains

$$\alpha_1 = 0.5\left[1 - (2z_{2,3} - 1)^2\right]. \tag{29.50}$$

Taking $z_{2,3}$ in the range $0.5 \leq z_{2,3} < 1$, gains α_1 and α_2 can be obtained using (29.50) and (29.49), respectively. Figure 29.1 gives graphical representation of (29.50) from which α_1 can be calculated for desired $z_{2,3}$.

In such a way, the compensation effectiveness of constant and ramp-type disturbances is achieved without appearance of any damped oscillation in the nominal system. This prevents chattering excitation if unmodeled dynamics exists in the considered system modeled by the first-order dynamics.

From the conventional control theory it can be easily concluded that the obtained control system is of type two, since its control loop contains two pure integrators. The both integrators are introduced by disturbance compensator (29.32). Therefore, this control system can track ramp-type references without steady state error and quadratic parabola type with a constant error. It is not difficult to obtain relations for

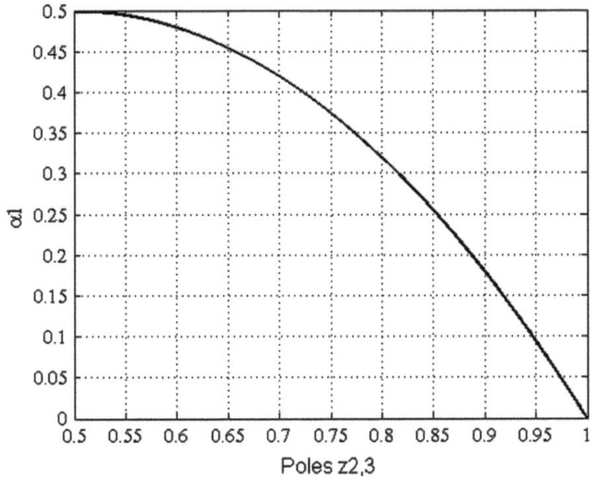

Fig. 29.1 Graphical presentation of the relation (29.50)

Table 29.1 Steady state errors of the analyzed control system

With respect to:	Disturbance compensator type		
	Combined	1st order	No dist. comp.
References	0 for $p = 0{,}1$ $-(a_\delta T^3/\alpha_2)r_0, p=2$	0 for $p = 0$ $-(a_\delta T^2/\alpha_1)r_0, p=1$	$-a_\delta T r_0, p=0$
Disturbances	0 for $q = 0{,}1$ $(b_\delta T^3/\alpha_2)d_0, q=2$	0 for $q = 0$ $(b_\delta T^2/\alpha_1)d_0, q=1$	$b_\delta T r_0, q=0$

steady state accuracy with respect to reference and load disturbance. These relations are given in Table 29.1, where reference input and load disturbance are respectively defined by relations

$$r(t) = r_0 t^p/p!, \quad d(t) = d_0 t^q/q!. \tag{29.51}$$

From these tables it is evident that the analyzed control system with combined disturbance compensator can track ramp-type reference and fully reject ramp-type disturbances. Tracking of quadratic parabola references and rejecting the same type of disturbances are more effective if sampling time T is smaller and α_2 is higher. Since there are limitations in choosing α_2 more realistic way to increase system accuracy is to choose smaller sampling time T. From the given tables it can be concluded that the steady state accuracy of the system with the combined compensator is of $O(T^3)$ order. For the system with first-order compensator the steady state accuracy is $O(T^2)$ order, which agrees with the results given in [33].

29.4.1 Velocity Servo System Design

A controlled plant, which consists of a DC motor and a power module, for the purpose of velocity control design is defined by a second-order dynamical model

$$\begin{aligned} \frac{\mathrm{d}\omega}{\mathrm{d}t}(t) &= -\frac{B}{J}\omega(t) + \frac{c}{J}i_r(t) - \frac{1}{J}M_L(t), \\ \frac{\mathrm{d}i_r}{\mathrm{d}t}(t) &= -\frac{R_r}{L_r}i_r(t) - \frac{c}{L_r}\omega(t) + \frac{k}{L_r}u(t), \end{aligned} \tag{29.52}$$

where ω is the angular velocity in rad/s, B is the coefficient of viscous friction in Nm/(rad/s), J is the moment of inertia of the motor with a load in kgm^2, c is the universal motor constant, M_L is the load torque in Nm, R_r, and L_r are respectively the rotor resistance in Ω and the inductance in H, i_r is the rotor current in A, k is the gain of a power module in V/V, and *u(t)* is the control voltage on the input of the power module in V.

Electrical time constant of a motor is usually smaller than the mechanical one. Moreover, it may by further decreased by using inner current control loop. Therefore, it is common to neglect electrical subsystem dynamics in servo system design, i.e,. the second relation in (29.52) may be treated as algebraic equation

$$0 = -R_r i_r(t) - c\omega(t) + ku(t) \Rightarrow i_r(t) = \frac{k}{R_r}u(t) - \frac{c}{R_r}\omega(t). \tag{29.53}$$

Substitution of i_r from (29.53) into the first relation of (29.52) gives

$$\dot{\omega}(t) = -a\omega(t) + b(u(t) - v(t)) \tag{29.54}$$

where

$$a = \frac{BR_r + c^2}{JR_r};b = \frac{kc}{JR_r}, v(t) = \frac{M_L(t)}{Jb}. \tag{29.55}$$

Tracking error dynamics in the signal error space

$$e(t) = r(t) - \omega(t) \tag{29.56}$$

is described as

$$\begin{aligned} \dot{e}(t) &= -ae(t) - b(u(t) - d(t)), \\ d(t) &= v(t) + b^{-1}ar(t) + b^{-1}\dot{r}(t), \end{aligned} \tag{29.57}$$

where $d(t)$, $|d(t)| \leq d_m < \infty$, is the sum of disturbance $v(t)$ and disturbances caused by the reference signal $r(t)$ and its first derivative.

Model (29.57) is in the form of (29.35). Therefore, the previously obtained results are valid for this velocity servo system. Only the influence of unmodeled inertial dynamics needs to be investigated. Investigation here will be conducted on a velocity servo system with DC motor with the following parameters:

$R_r = 1\Omega$, $L_r = 2.5$ mH, $J = 1.1810$ $a = 26$; $b = 654$.

The neglected inertial dynamics has the time constant of about 6 % comparing to the dominant one. Let the sampling time be $T = 1$ ms. Parameters of the discrete-time model (29.36) become

$$a_\delta = -25.6649; \mathrm{b}_\delta = 645.571.$$

Parameters of the controller are

$\mathrm{k_{eq}} = -0.039755; \mathrm{c}_\delta = 0.001549.$

Based on the recommendation, the combined compensator will be applied. Let the double pole introduced by the compensator be located at $z_{3,4} = 0.9472$. Using Fig. 29.1 it is obtained $\alpha_1 = 0.1$ which gives from (49) $\alpha_2 = 0.0027864$.

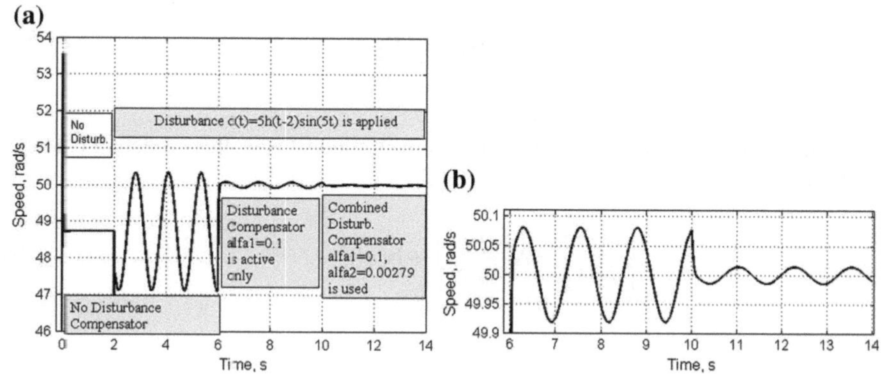

Fig. 29.2 Velocity regulation of 50 rad/s for different disturbance compensation. **a** detail around desired velocity and **b** zoomed detail when disturbance compensators are active

To verify the theoretically obtained results, simulations of the designed velocity servo system with the unmodeled dynamics of the electrical subsystem are given in Fig. 29.2. The system reference is 50 rad/s. In the time interval 0–2 s neither disturbance is present nor is the disturbance compensator applied. As seen from Fig. 29.2a, there is a fast overshoot due to unmodeled dynamics. Also, the control system without disturbance compensator has a steady state error caused by disturbance component $b^{-1}ar(t)$ in (29.57). At the time instant 2 s disturbance $v(t) = 5\,h(t\text{-}2)\sin(5t)$ is applied. During time interval 2–6 s the system is without disturbance compensator. The average value of the output stays as in the previous time interval, but amplitude of deviation from the desired value is significant. During time interval 6–10 s the first-order compensator is applied. The average value of the output velocity is equal to the desired. The amplitude deviation from the desired value is much smaller than in the previous time interval. At time 10 s the second-order disturbance compensator is switched on. Therefore, during time interval 10–14 s the combined disturbance compensator is active. As is obvious, in this time interval deviation from the desired velocity is very small. To underline that this complex disturbance cannot be fully rejected, the zoomed detail from Fig. 29.2a is given in Fig. 29.2b. It is clear that the combined compensator has exhibited the theoretically predicted behavior.

29.5 DTSM Control of a Positional Servo System

Design of a second-order DTSM control system with the proposed disturbance compensator will be presented here. This section is oriented only to positional servo system design whose plant is a DC servo motor with electronic power amplifier. The plant model is given by (29.52) supplemented by equation

$$\frac{d\theta}{dt}(t) = \omega(t), \tag{29.58}$$

where θ is angular shaft position in rad.

With assumption that electrical subsystem is much faster then mechanical one, model of the plant becomes

$$\begin{bmatrix} \theta(t) \\ \dot{\omega}(t) \end{bmatrix} = \mathbf{A} \begin{bmatrix} \theta(t) \\ \omega(t) \end{bmatrix} + \mathbf{b}(u(t) - f(t)), \tag{29.59}$$

where

$$\mathbf{A} = \begin{bmatrix} 0 & 1 \\ 0 & -a \end{bmatrix}; \mathbf{b} = \begin{bmatrix} 0 \\ b \end{bmatrix}; f(t) = \frac{M_L(t)}{Jb}; \tag{29.60}$$

Denoting a desired reference signal of a servo system with $\theta_r(t)$, the servo system dynamics in tracking error space,

$$e(t) = \theta_r(t) - \theta(t), \tag{29.61}$$

becomes

$$\dot{\mathbf{e}}(t) = \mathbf{A}\mathbf{e}(t) - \mathbf{b}(u(t) - d(t)), \ \mathbf{e}(t) = [e(t) \ \dot{e}(t)]^{\mathrm{T}}, \tag{29.62}$$

where

$$d(t) = f(t) + \frac{1}{b}(\dot{\theta}_r(t) + a\theta_r(t)) \tag{29.63}$$

is overall disturbance.

Block diagram of the considered DTSM servo system with the proposed combined disturbance compensator is given in Fig. 29.3. As was established in [19], the

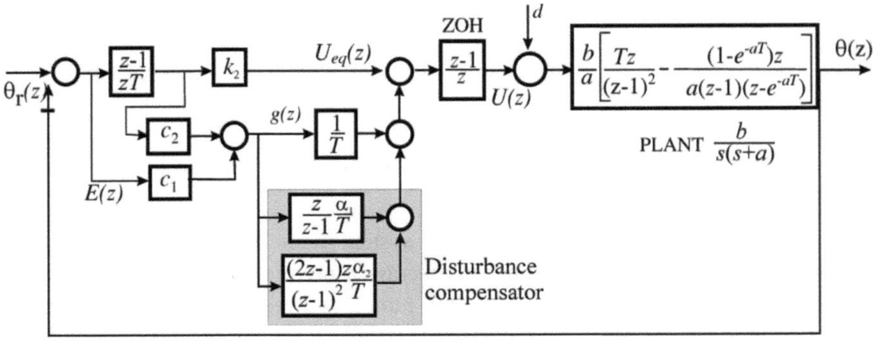

Fig. 29.3 Block diagram of the positional servo system with combined disturbance compensator

system without any disturbance compensator has a steady state error for ramp-type reference if no load is applied. This is due to the disturbance term $(a/b)\theta_r(t)$ in (29.63) caused by the reference signal. Since matching conditions in discrete-time are not satisfied, an error occurs.

Load disturbance and plant parameters uncertainties are matched for model (29.62). But, rigorously speaking, in practical servo systems with electrical motors, the mentioned assumptions about disturbances are not valid. First, the second-order mathematical model (29.62) of the plant is an approximation of the third-order model. Second, discretization corrupt matching conditions as well.

Positional servo system futures designed for model (29.62) will be examined in this section by simulations including unmodeled dynamics. For controller design it is necessary to define servo system dynamics by desired pole spectrum Λ and to choose an adequate sampling time T.

Let servo system be designed with the same DC servo motor from the previous section. Let desired dynamics be defined by the poles in continuous-time domain given by $\Lambda = [0, -50]$ and let the sampling time be $T = 1$ ms. In δ-discrete-time domain of model (29.62), corresponding to (29.2), the relevant matrices are obtained as

$$\mathbf{A}_\delta = \begin{bmatrix} 0 & 0.9871 \\ 0 & -25.6649 \end{bmatrix}; \mathbf{b}_d = \begin{bmatrix} 0.0032 \\ 650.57 \end{bmatrix}.$$

Using relations (29.10)–(29.12) following is obtained:

$$\begin{aligned} &\Lambda_\delta = [0, -48.77]; \\ &\mathrm{K}_\delta = [0, k_2] = [0, 0.03579]; \\ &\mathbf{c}_\delta = [c_1, c_2] = [0.0755, 0.0015]. \end{aligned} \tag{29.64}$$

Parameters of the disturbance compensator α_1 and α_2 depend on disturbance type. For constant-type disturbances only first-order disturbance compensator ($\alpha_1 > 0; \alpha_2 = 0$) need to be used. For systems with variable disturbances it is recommended to use combined disturbance compensator ($\alpha_1 > 0; \alpha_2 > 0$, $\alpha_1 > \alpha_2$). In any case, values of α_1 and α_2 have influence on the system stability. For the systems with unmodeled dynamics these parameters need to be tuned to obtain satisfactory stability margins and to prevent damped oscillations or chattering.

For the controller operating in linear mode, the transfer functions in z-domain becomes

$$G_c^{\alpha_1}(z) = \frac{A_1 z^2 + B_1 z + C_1}{z(z-1)T^2} \tag{29.65}$$

$$A_1 = (1+\alpha_1)(c_1 T + c_2) + k_2 T; B_1 = -(1+\alpha_1)c_2 - c_1 T - c_2 - 2k_2 T; C_1 = c_2 + k_2 T$$

for the system with the first-order disturbance compensator, and

$$G_c^{\alpha_{1,2}}(z) = \frac{A_2 z^3 + B_2 z^2 + C_2 z + D_2}{z(z-1)^2 T^2} \tag{29.66}$$

$$A_2 = (c_1 T + c_2)(1 + \alpha_1 + 2\alpha_2) + k_2 T; B_2 = -((c_1 T + c_2)(2 + \alpha_1 + \alpha_2) - c_2(1 + \alpha_1 + \alpha_2) - 3k_2 T; C_2 = c_1 T + c_2(3 + \alpha_1 + \alpha_2) + 3k_2 T; D_2 = -c_2 - k_2 T$$

for the system with the combined compensator. Equation (29.66) is valid as well for the system with the second-order disturbance compensator only after replacing $\alpha_1 = 0$ in it.

Characteristic equation of the nominal system with the first-order disturbance compensator is of fourth order, whereas for the system with the combined compensator or the system with the second-order disturbance compensator is of the fifth order.

Let the disturbance compensator parameters be chosen: $\alpha_1 = 0.1$, $\alpha_2 = 0.01$. Numerically obtained closed-loop poles for the given system parameters are shown in Table 29.2.

As can be seen from Table 29.2, all poles of the closed-loop system in any of the given cases are located inside the unit circle in the z-plane. Therefore, the given systems are stable in the nominal case. The pole z = 0.9524 is near desired pole ($z = \exp(-50T) = 0.9512$). This pole is dominant in the cases 1, 2, and 4. But in the case 3, conjugate-complex poles 0.9898 ± 0.0996i take over the role of dominant poles. In this case the nominal system dynamics becomes pure damped oscillatory.

Figure 29.4a gives simulation results of designed servo system in the nominal case. As can be seen, the combined disturbance compensator gives the best results comparing to the first-order or the second-order compensators if they are used separately. Moreover, if only the second-order disturbance compensator is used in the nominal system, undesirable damped oscillations occur (Fig. 29.4a, b). Furthermore, in the system with unmodeled dynamics chattering arises if the second-order disturbance compensator is applied alone, but the system with the combined disturbance compensator stays robust, Fig. 29.4c.

Steady state accuracy of the positional servo system with the considered disturbance compensators may be obtained in conventional way. Based on block diagram from Fig. 29.3, for different disturbance compensation, steady state errors are given in Table 29.3 for the reference tracking and disturbance rejection, defined by (29.51). Therefore, the proposed positional servo system with the combined

Table 29.2 The closed-loop poles for different disturbance compensators

Case	Disturbance compensator type	The closed-loop poles
1	$\alpha_1 = 0$, $\alpha_2 = 0$	0.2479 ± 0.6786i; 0.9525
2	$\alpha_1 = 0.1$, $\alpha_2 = 0$	0.2717 ± 0.7109i; 0.9013; 0.9523
3	$\alpha_1 = 0$, $\alpha_2 = 0.01$	0.253 ± 0.681i; 0.9898 ± 0.0996i; 0.9524
4	$\alpha_1 = 0.1$, $\alpha_2 = 0.01$	0.2771 ± 0.714i; 0.94 ± 0.0795i; 0.9524

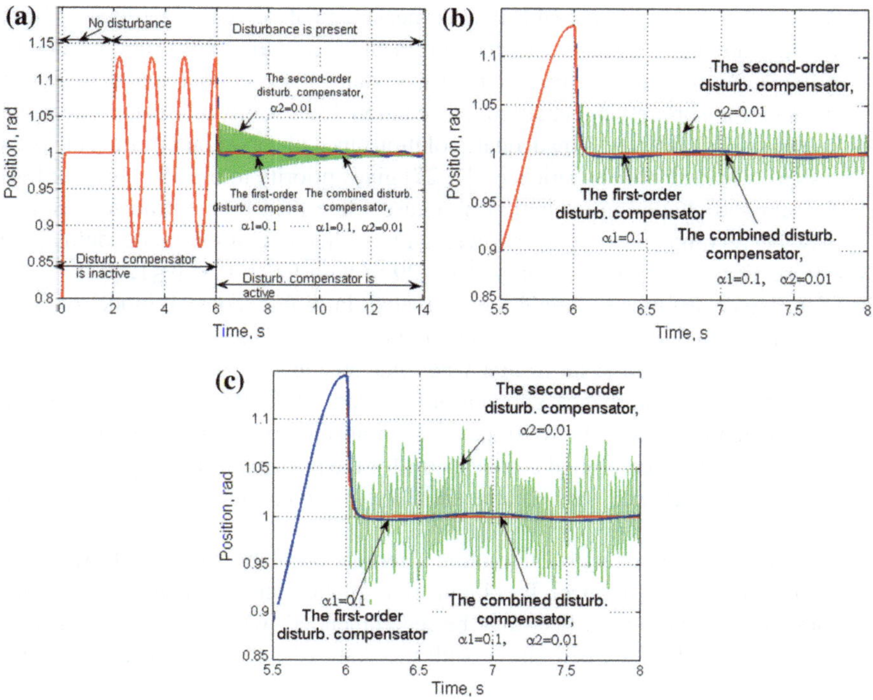

Fig. 29.4 Positioning of 1 rad without and with disturbance $d(t) = 5\,h(t-2)\sin(5t)$ with different disturbance compensation. **a** Nominal system; **b** Detail from Fig. 1*a;* **c** Detail if unmodeled dynamics is introduced

Table 29.3 Steady state error of positional servo system

With respect to:	Disturbance compensator type		
	Combined	1st order	No dist. comp.
References	0 for p = 0, 1, 2 $r_0 aT^3/bc_1\alpha_2, p=3$	0 for p = 0, 1 $r_0 aT^2/bc_1\alpha_2, p=2$	0 for p = 0 $r_0 aT/bc_1, p=1$
Disturbances	0 for q = 0, 1 $d_0 T^3/k\alpha_2 c_1, q=2$	0 for q = 0 $d_0 T^2/k\alpha_1 c_1, q=1$	$d_0 T/kc_1, q=0$

disturbance compensator may track references of cubic parabola type and suppress load disturbances up to quadratic parabola type, both with a finite error.

29.6 Some Practical Recommendations

For a correct implementation of the proposed DTSM control design it is necessary to follow some recommendations:

1. One of the problems in realization is measurement of state coordinates. SMC systems are more sensitive to measurement noises than conventional control systems. A possible solution is to use a state observer that makes a bypass channel for suppressing chattering excitation [33]. Conventional proportional Luenberger's observer is inadequate solution since observation error occurs if the system is loaded. Our experience [21, 27] gives priority to use PI^2 observer [34].
2. To suppress chattering excitation in a case when measurement noises occur, it may be recommended to use lower gain in linear control mode defined by (29.8). That is, to replace gain T^{-1} in (29.8) with βT^{-1}, $0 < \beta \leq 1$, [22].
3. Disturbance estimator should be activated in linear control mode only by resetting initial conditions on integrators to zero. According to (29.8), the resetting is triggered at an instant when $g(k)$ becomes $|g(k)| < \sigma T$.
4. The first-order disturbance compensator is very effective. In most of practical implementation it may be used alone as an absolutely adequate solution.
5. Parameters of the combined compensator for the positional servo system need to be carefully chosen to obtain satisfactory dynamics without chattering excitation.
6. In positional systems with DC motors a serious problem is Coulomb friction as an unmatched disturbance. To obtain high-quality servo system a disturbance compensation algorithm [21] may be recommended, which is based on DTSM control principle described above with the first-order disturbance compensator.

29.7 Conclusions

This chapter presents a method to achieve a high-quality control system design using DTSM approach with disturbance estimation and compensation. The employed control algorithm contains nonlinear and linear mode. Nonlinear mode bounds control signal in order to adapt it to permissive input of the controlled object and to obtain maximum permitted rise-time in the response. Linear control mode provides ideal discrete-time sliding mode in a nominal system or quasi-sliding mode in a real system. In the linear control mode it is easy to estimate matched disturbances by measurement of the switching function only. Using up to two previous samples of the switching function, it was established that constant- or ramp-type disturbances can be fully estimated by extrapolation. Other slow varying disturbances can be well enough estimated also. Three types of disturbance compensators are proposed on the basis of the used disturbance estimation method: the first-order, the second-order, and the combined compensator. Performance analysis of the control systems, in case of first-order and second-order plants, has established that combined disturbance compensator offers an additional possibility in tuning of system dynamics. Namely, the second-order disturbance compensator introduces undesirable slowly decaying oscillations in disturbance compensation, which in presence of a significant unmodeled dynamics may lead to chattering excitation.

Simulation results of the systems having first-order plant, such as velocity servo systems, and second-order plants, whose typical representative is positional servo system, showed significant improvements in control quality in comparison to conventional discrete-time and traditional DTSM control systems.

The proposed design method is not limited only to servo systems with DC motors. The method can be used for control of induction motors as well, and many other industrial processes which may be approximated by the first- or the second-order dynamical model without significant unmodeled inertial dynamics.

Acknowledgments The third author acknowledges support from the Ministry of Education and Science of the Republic of Serbia under Project Grant III44004.

References

1. Emel'yanov, S.V.: A method to obtain complex regulation laws using only the error signal or regulated coordinate and its first derivatives (in Russian). Avtomatika i Telemekhanika **18**(10), 873–885 (1957)
2. Utkin, V.I.: Sliding Modes in Control and Optimization. Springer, Berlin (1992)
3. Milosavljević, Č.: Some problems of the discrete variable structure systems control law realization. Ph.D. Dissertation (in Serbo-Croatian), University of Sarajevo, Faculty of Electrical Engineering, Sarajevo, (1982)
4. Milosavljević, Č.: General conditions for the existence of quasi-sliding mode on the switching hyper-plane in discrete variable structure systems. Autom. Remote Control **46**, 307–314 (1985)
5. Potts, R.B., Yu, X.: Discrete variable structure systems with pseudo-sliding mode. J. Aust. Math. Soc. Ser. B 365–376 (1991)
6. Bučevac, Z.: Design of digital discrete control systems with sliding mode. Ph.D. Dissertation (in Serbo-Croatian). University of Belgrade, Faculty of Mechanical Engineering, Belgrade (1985)
7. Salihbegović, A.: Contribution to analysis and synthesis of discrete realized systems with switched control. Ph.D. Dissertation (in Serbo-Croatian). University of Sarajevo, Faculty of Electrical Engineering, Sarajevo (1985)
8. Drakunov, S.V., Utkin, V.I.: On discrete-time sliding mode. In: Proceedings of the IFAC Symposium on Nonlinear Control Systems Design, Capry, Italy, pp. 484–489. CNR Commision for IFAC and the IFAC, Rome, IT, (1989)
9. Gao, W., Wang, Y., Homaifa, A.: Discrete-Time variable structure control systems. IEEE Trans. Ind. Electron. **42**, 117–122 (1995)
10. Bartolini, G., Ferrara, A., Utkin, V.I.: Adaptive sliding mode control in discrete-time systems. Automatica **31**(5), 769–773 (1995)
11. Bartoszewicz, A.: Discrete-Time quasi-sliding mode control strategies. IEEE Trans. Ind. Electron. **45**(4), 633–637 (1998)
12. Golo, G., Milosavljević, Č.: Robust discrete-time chattering-free sliding mode control systems. Syst. Control Lett. **41**, 19–28 (2000)
13. Milosavljević, Č.: Discrete-Time VSS. Chapter V in: Variable structure systems: from principles to implementation. In: Šabanovic, A., Fridman, L.M., Spurgeon, S.K. (eds.) pp. 99–127. The IEE Press, London, (2004)
14. Milosavljević, Č., Peruničić-Draženović, B., Mitić, D., Veselić, B.: Sampled data quasi-sliding mode control strategies. In: The IEEE International Conference on Industrial Technology ICIT '05, Mumbai, India, pp. 2640–2645. The IEEE Piscataway, NJ, 15–17 December 2006

15. Su, W.C., Drakunov, S.V., Özguner, Ü.: Implementation of variable structure control for sampled data systems. In: Garofalo, F., Glielmo, L.E. (eds.) Robust Control via Variable Structure and Lyapunov Techniques, pp. 87–106. Springer, London (1996)
16. Chan, C.Y.: Robust discrete-time sliding mode controller. Syst. Control Lett. **23**, 371–374 (1994)
17. Chen, X., Komada, S., Fukuda, T.: Design of nonlinear disturbance observer. IEEE Trans. Industr. Electron. **47**(2), 429–437 (2000)
18. Park, K.-B.: Discrete-Time sliding mode controller for linear time-varying systems with disturbances. In: ICASE: The Institute of Control Automation and Systems Engineers, Korea, **2**(4), 244–247 (2000)
19. Milosavljević, Č., Draženović, B., Veselić, B., Mitić, D.: A new design of servomechanisms with digital sliding mode. Electr. Eng. **89**(3), 233–244 (2007)
20. Veselić, B.: Application of digital sliding modes for coordinated tracking of complex trajectories. Ph.D. Dissertation (in Serbian). University of Niš, Faculty of Electronic Engineering, Niš (2006)
21. Veselić, B., Peruničić-Draženović, B., Milosavljević, Č.: High-Performance position control of induction motor using discrete-time sliding mode control. IEEE Trans. Ind. Electron. **55** (11), 3809–3817 (2008)
22. Veselić, B., Peruničić-Draženović, B., Milosavljević, Č.: Improved discrete-time sliding mode positional control using Euler velocity estimation. IEEE Trans. Ind. Electron. **57**(11), 3840–3847 (2010)
23. Lješnjanin, M., Draženović, B., Milosavljević, Č. Veselić, B.: Disturbance compensation in digital sliding mode. In: International Conference on Computer as a Tool. EUROCON '2011, CD, Paper 171 (2011)
24. Abidi, K.: Sliding mode control for high-precision motion control systems. MSc Thesis, Graduate School of Engineering and Natural Sciences, Sabanci University, Istanbul (2004)
25. Abidi, K., Šabanović, A.: Sliding mode control for high precision motion of Piezo-stage. IEEE Trans. Ind. Electron. **54**(1), 629–637 (2007)
26. Milosavljević, Č., Peruničić, B., Veselić, B.: Discrete-Time sliding mode based high performance velocity servo system design. In: Proceedings of the 16th International Symposium on Power Electronics Ee-2011, Novi Sad, Serbia, CD, Paper T4-3.6, 26–28 October (2011)
27. Milosavljević, Č., Peruničić-Draženović, B., Veselić, B.: Discrete-Time velocity servo system design using sliding mode control approach with disturbance compensation. IEEE Trans. Ind. Inf. **9**(2), 920–927 (2013)
28. Draženović, B.: The invariance conditions in variable structure systems. Automatica **5**, 287–295 (1969)
29. Yu, X., Wang, B., Li, X.: Computer-Controlled variable structure systems; the state of arts. IEEE Trans. Ind. Inf. **8**(2), 197–205 (2012)
30. Midleton, R.H., Goodwin, G.C.: Improved finite word length characteristics in digital control using delta operators. IEEE Trans. Autom. Control **31**, 1015–1021 (1986)
31. Shao, X., Sun, D.: Development of a new robot controller architecture with FPGA-based IC design for improved high-speed performance. IEEE Trans. Ind. Inf. **3**(4), 312–321 (2007)
32. Draženović, B., Milosavljević, Č., Veselić, B.: Comprehensive approach to sliding mode design and analysis linear systems. In: Bandyopdhyay, B., Janardhanan, S., Spurgeon, S.K. (eds.) Advances in Sliding Mode Control, Concept, Theory and Implementation. Lecture Notes in Control and Information Sciences, vol. 440, Chapter 1. Springer-Verlag, Berlin (2013)
33. Young, K.D., Utkin, V.I., Özgüner, U.: A control engineer's guide to sliding mode control. IEEE Trans. Control Syst. Theory **7**(3), 328–342 (1999)
34. Naumović, M.B.: Some Approaches to velocity estimation in digital controlled DC drives. In: Proceedings of the 6th International Conference on Systems, Automatic Control, and Measurements, Niš, SFR Yugoslavia, pp. 140–144 (1998)

Chapter 30
Pragmatic Design Methods Using Adaptive Controller Structures for Mechatronic Applications with Variable Parameters and Working Conditions

Stefan Preitl, Radu-Emil Precup, Zsuzsa Preitl, Alexandra-Iulia Stînean, Claudia-Adina Dragoş and Mircea-Bogdan Rădac

"The PID controller can be said to be 'the bread and the butter' of the control engineering".

(K.-J. Åström)

Abstract This chapter treats two pragmatic design methods for controllers dedicated to mechatronic applications working under variable conditions; for such applications adaptive structures of the control algorithms are of great interest. Basically, the design is based on two extensions of the modulus optimum method and of the symmetrical optimum method (SO-m): the Extended SO-m and the double parameterization of the SO-m (2p-SO-m). Both methods are introduced by the authors and they use PI(D) controllers that can ensure high control performance: increased value of the phase margins, improved tracking performance, and efficient

---

S. Preitl (✉) · R.-E. Precup · A.-I. Stînean · C.-A. Dragoş · M.-B. Rădac
Department of Automation and Applied Informatics, "Politehnica" University of Timisoara, Bd. V. Parvan 2, 300223 Timisoara, Romania
e-mail: stefan.preitl@aut.upt.ro

R.-E. Precup
e-mail: radu.precup@aut.upt.ro

A.-I. Stînean
e-mail: alexandra-iulia.stinean@aut.upt.ro

C.-A. Dragoş
e-mail: claudia.dragos@aut.upt.ro

M.-B. Rădac
e-mail: mircea.radac@aut.upt.ro

Z. Preitl
Siemens A.G., Erlangen, Germany
e-mail: zsuzsap@yahoo.com

G.M. Dimirovski (ed.), *Complex Systems*, Studies in Systems, Decision and Control 55, DOI 10.1007/978-3-319-28860-4_30

disturbance rejection. A short and systematic presentation of the methods and digital implementation aspects using an adaptive structure of the algorithms for industrial applications are given. The application deals with a cascade speed control structure for a driving system with continuously variable parameters, i.e., electrical drives with variable reference input, variable moment of inertia and variable disturbance input.

List of Abbreviations

SO-m	Symmetrical Optimum method
Mo-M	Modulus Optimum method
ESO-m	Extended Symmetrical Optimum method
2p-SO-m	Double parameterization of the SO-m
2-DOF	Two Degree of Freedom
VMI	Variable Moment of Inertia
t.f.	Transfer function
c.a.	Control algorithm
C-VR-MI-LD	Continuously Variable Reference, Moment of Inertia and Load Disturbance
DC-m, BLDC-m	DC-motors, Brush-Less DC-motors
MM	Mathematical Model
CS	Control Structure
CCS	Cascade Control Structure

30.1 Introduction: The Design Methods

The basic ideas of frequency domain optimization—based on the modulus optimum conditions—are synthesized in [1, 2] as:

$$M_r(\omega) = |H_r(j\omega)| \approx 1 \text{ , for values of } \omega \geq 0 \text{ as large as possible} \quad (30.1.1)$$

$$M_{v1,v2}(\omega) = |H_{v1,v2}(j\omega)| \approx 0 \text{ for values of } \omega \geq 0 \text{ as large as possible.} \quad (30.1.2)$$

Decomposing the expressions of $M_r(\omega)$, $M_{v1}(\omega)$, $M_{v2}(\omega)$ into Mc-Laurin series, the design conditions can be derived on the basis of the following requirements:

$$M_r(0) = 1, \left.\frac{\mathrm{d}^{\nu}|M_r(\omega)|}{\mathrm{d}\omega^{\nu}}\right|_{\omega=0} = 0 \qquad \text{(a)} \qquad (30.1.3)$$

$$M_{v1,v2}(0) = 0, \left.\frac{\mathrm{d}^{\nu}|M_{v1,v2}(\omega)|}{\mathrm{d}\omega^{\nu}}\right|_{\omega=0} = 0 \qquad \text{(b)} \qquad (30.1.4)$$

They are called modulus optimum (MO) conditions. Considering the classical control structure given in Fig. 30.1, the basic relations (neglecting the measurement noise) are expressed as

$$y(s) = H_0(s)S(s)r(s) + S(s)v_1(s) + P(s)S(s)v_2(s) \text{ and } H_0(s) = C(s) \cdot P(s) \tag{30.1.5}$$

$$u(s) = C(s)S(s)r(s) - C(s)S(s)v_1(s) - H_0(s)S(s)v_2(s), \tag{30.1.6}$$

$$\varepsilon(s) = S(s)r(s) - S(s)v_1(s) - P(s)S(s)v_2(s) \text{ and } r(s) = F(s)r_0(s) \tag{30.1.7}$$

$$S(s) = \frac{1}{1 + C(s) \cdot P(s)}, \ T(s) = \frac{C(s) \cdot P(s)}{1 + C(s) \cdot P(s)}, \ S(s) + T(s) = 1 \tag{30.1.8}$$

where $H_r(s)$, $H_{v1}(s)$ $H_{v2}(s)$, $H_0(s)$, $S(s)$, $T(s)$ are the main transfer functions (t.f. s) and characteristic functions of the system.

Many design methods based on the modulus optimum method (MO-m) and symmetrical optimum method (SO-m) conditions have been developed in various variants. Such examples include the basic approach due to Kessler [3, 4] and other ones developed afterwards and reported in [5–14].

The MO-m is applied in two classical variants:

- The first variant is based on determining domains of variation of controller parameters that satisfy the imposed requirements, finally determining "the best solution". This variant requires huge amount of calculation.
- The second variant is based on direct tuning relations. Applying this variant is closer to engineering practice.

This chapter treats two extensions related to the second variant, introduced by the authors in [6, 9] and focused on obtaining better dynamics of the control structure, enhancement of robustness, and enlarging of area of applications. The efficiency of controller tuning methods—regarding the basic variant but other tuning methods as well—can be proved in the time domain and in the frequency domain. The methods found various extensions in fuzzy control and so on, and they are currently used in several applications.

This chapter is structured as follows. In Sect. 30.2 a short overview of MO-m and SO-m in their practical version is synthesized. Section 30.3 is focused on two pragmatic extensions of SO-m (proposed by the authors); the local conclusions are

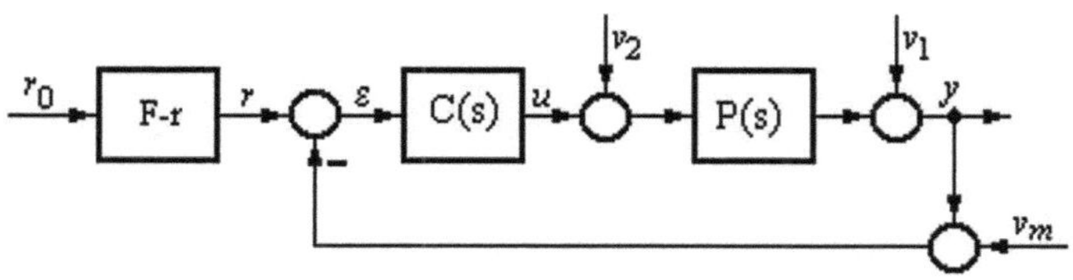

Fig. 30.1 Basic structure of the control loop

focused on versatility of the design, and its applicability for systems with plants characterized by variable parameters. Some possibilities to enhance the basic performances are given. Implementation in the form of a variable structure controller with bumpless transfer of the control signal is given. Section 30.4 offers an application example for systems with variable parameters: a control structure (CS) for a mechatronic system with variable parameters, the control solution for a winding system. Detailed analysis of the performances—including sensitivity analysis aspects—and the versatility of the solution prove its applicability. Section 30.5 is dedicated to some concluding remarks.

30.2 Pragmatic Forms of the Modulus Optimum Method and of the Symmetrical Optimum Method

The MO-m and the SO-m are dedicated mainly to servo systems [1–4] where they prove to be efficient [15]. Both methods are characterized by the fact that in the open-loop t.f. $H_0(s)$ (30.1.5), the result is a single (in case of MO-m) or a double pole (in case of SO-m) in origin and the parameters of the controllers (PI(D)-type) can be computed—and, if it is necessary, recalculated online—by means of compact formulas.

Basically, the MO-m and the SO-m design situations correspond to benchmark-type model for the plant and typical controllers—of PI(D) type—eventually, extended with lag components (L) and—in extension—with reference filters; they are synthesized in Table 30.1—for MO-m—and Table 30.2—for SO-m (T_Σ includes the effects of small time constants of the plant).

Mainly, the SO-m [1–4] is applicable—with some restrictions due to the resulting small phase reserve—to plants having a pole in origin (the main case for the positioning systems) characterized by a t.f. expressed in Table 30.2 for S0-1 and SO-2, when the use of a PI or a PID controller (having the t.f. expressed in Table 30.2 for S0-1 and SO-2; in this last case the pole-zero cancelation is applied $T_r' = T_1$).

k_p is the process gain, T_1 characterizes the mechanical time constant, T_2 and T_Σ characterize smaller time constants, having $T_1 \gg T_2 > T_\Sigma$

Table 30.1 The main design cases for MO-m (Kessler's variant)

Case	Plant, $P(s)$	Controller type $C(s)$	
0	1	2	3
MO-1.1 (a)	$\frac{k_p}{1+sT_\Sigma}$	I	$\frac{k_r}{s}$
MO-2.1 (b)	$\frac{k_p}{(1+sT_\Sigma)(1+sT_1)}$	PI	$\frac{k_r}{s}(1+sT_r)$, $T_r = T_1$
MO-3.1 (c)	$\frac{k_p}{(1+sT_\Sigma)(1+sT_1)(1+sT_2)}$ $T_1 > T_2 > T_\Sigma$	PID	$\frac{k_r}{s}(1+sT_r)(1+sT_r')$ $T_r = T_1$; $T_r' = T_2$

Table 30.2 The main design cases for SO-m (Kessler's variant)

Case	Plant, $P(s)$	Controller type $C(s)$	
0	1	2	3
SO-1 (a)	$\frac{k_p}{s(1+sT_\Sigma)}$	PI	$\frac{k_r}{s}(1+sT_r)$
SO-2 (b)	$\frac{k_p}{s(1+sT_\Sigma)(1+sT_1)}$ $T_\Sigma < 0.2T_1$	PID (-L1)	$\frac{k_r}{s}(1+sT_r)(1+sT_r')$, $T_r'=T_1$ $\frac{k_r}{s}(1+sT_r)\frac{(1+sT_r')}{(1+sT_f)}$ $T_r'=T_1$; $T_c'/T_f \approx (10\ldots20)$
SO-3 (c)	$\frac{k_p}{s(1+sT_\Sigma)(1+sT_1)(1+sT_2)}$ $T_1 > T_2 > T_\Sigma$, $T_\Sigma < 0.2T_1$	PID2-L2	$\frac{k_r}{s}(1+sT_r)\frac{(1+sT_r')(1+sT_d)}{(1+sT_f')(1+sT_f)}$ $T_r'=T_2$; $T_r'/T_f' \approx (10\ \ldots\ 20)$ $T_d=T_2$; $T_d/T_f \approx (10\ \ldots\ 20)$

Accordingly, the open-$H_0(s)$ and closed-loop t.f.s (with respect to the reference input r) $H_r(s)$ can be expressed as

$$H_0(s) = \frac{k_r k_p(1+sT_r)}{s^2(1+sT_\Sigma)},\ T_r > T_\Sigma.\ \text{and}\ H_r(s) = \frac{b_0 + b_1 s}{a_0 + a_1 s + a_2 s^2 + a_3 s^3}, \tag{30.2.1}$$

$(b_0 = a_0,\ b_1 = a_1,\ a_0 = k_r k_p,\ a_1 = k_r k_p T_r,\ a_2 = 1,\ a_3 = T_\Sigma)$ and the modulus

$$M_r(\omega) = |H_r(j\omega)| = \sqrt{\frac{a_0^2 + a_1^2\omega^2}{a_0^2 - (2a_0a_2 - a_1^2)\omega^2 - (2a_1a_3 - a_2^2)\omega^4 + a_3^2\omega^6}}. \tag{30.2.2}$$

The parameters of the "optimal controller" are obtained imposing in (30.2.2) the conditions

$$2a_0a_2 = a_1^2,\ 2a_1a_3 = a_2^2, \tag{30.2.3}$$

in the form of

$$k_c = \frac{1}{8k_pT_\Sigma^2},\ T_c = 4\,T_\Sigma\,(T_c' = T_1), \tag{30.2.4}$$

with $H_{r\,\mathrm{opt}}(s)$ having $a_0 = 1$, $a_1 = 4T_\Sigma$, $a_2 = 8T_\Sigma^2$, $a_3 = 8T_\Sigma^3$, and $b_0 = 1$, $b_1 = 4T_\Sigma$. The "optimal performance" guaranteed by the SO-m—viz. $\sigma_1 \approx 43\,\%$ (overshoot), $t_s \approx 16.5T_\Sigma$ (settling time), $t_1 \approx 3.1T_\Sigma$ (first settling time) and a relatively small phase margin, $\phi_r \approx 36°$ (which is the main drawback of SO-m)—are seldom acceptable, Fig. 30.2. Mainly in cases of plants with variable parameters, the retuning of the controller is strongly recommended.

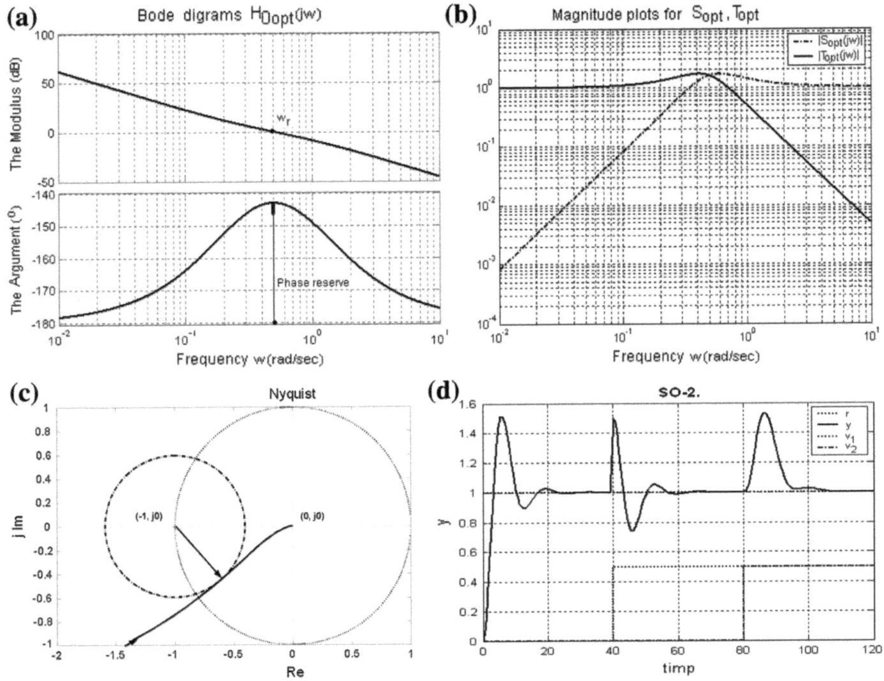

Fig. 30.2 The SO-m case: significant diagrams in frequency and time domain (for disturbance, only the case SO-1 is illustrated (*d*))

30.3 Extensions of the Symmetrical Optimum Method

Two efficient ways for performance enhancement—based on two extensions of the SO-m—have been introduced by the authors in [6, 9, 10], with focus on controller design based on benchmark-type models of the plant. They are synthesized as follows:

- the extended form of the symmetrical optimum method (ESO-m) [6], and
- the double parameterization form of the symmetrical optimum method (2p-SO-m) [9, 10].

Both methods employ the generalized form of Eq. (30.2.3)

$$\beta^{1/2}a_0a_2 = a_1^2,\ \beta^{1/2}a_1a_3 = a_2^2, \tag{30.3.1}$$

where β is the design parameter, whose value can be chosen by the designer, in correlation with desired performance indices. The methods are focused to fulfill an increased value for the phase margin, good (better) tracking performances, and efficient disturbance rejection.

30.3.1 The Extended Symmetrical Optimum Method (ESO-m)

The ESO-m [6, 16, 17] is dedicated mainly to controller tuning for positioning and tracking systems operating under continuously variable reference and plant parameters, characterized by the t.f.s in the forms given in Table 30.2(a)–(c). Applying the generalized form of the optimization relations (30.3.1), the compact parameter tuning equations are

$$k_c = \frac{1}{k_P \beta^{3/2} T_\Sigma^2},\ T_c = \beta T_\Sigma\ (T_c' = T_1), \tag{30.3.2}$$

and as a result the characteristic t.f.s $H_0(s)$ and $H_r(s)$ will obtain the forms (30.3.3), which lead to significantly improved performance (see, for example, the phase margin characteristics).

$$H_0(s) = \frac{1 + \beta T_\Sigma s}{\beta^{3/2} T_\Sigma^2 s^2 (1 + T_\Sigma s)} \text{ and } H_r(s) = \frac{1 + \beta T_\Sigma s}{\beta^{3/2} T_\Sigma^3 s^3 + \beta^{3/2} T_\Sigma^2 s^2 + \beta T_\Sigma s + 1}. \tag{30.3.3}$$

Figure 30.3 offers the main control system performance indices as function of the design parameter β [6]; the recommended domain for β is $4 < \beta \leq 9\,(16)$. The main advantage is that the increase of the phase margin (accompanied with the decreasing of ω_c) also leads to increase of the settling time. The reference behaviors can be corrected using adequate reference filter F-r.

Extensive analyses of the method and experimental results are also presented also in [16] and [19]. Important supplementary aspects deal with:

- *Sensitivity function analysis*. Based on the relation of $S_0(s)$ the maximum sensitivity value M_{s0} and its inverse M_{S0}^{-1} can be calculated. Figure 30.4a–c illustrate the Nyquist diagrams calculated for $\beta = 4$, 9, 16; the $M_{S0} = f(\beta)$ circles and the values of M_{S0}^{-1} are also marked. The curves point out the increase of robustness when the value of β is increased.

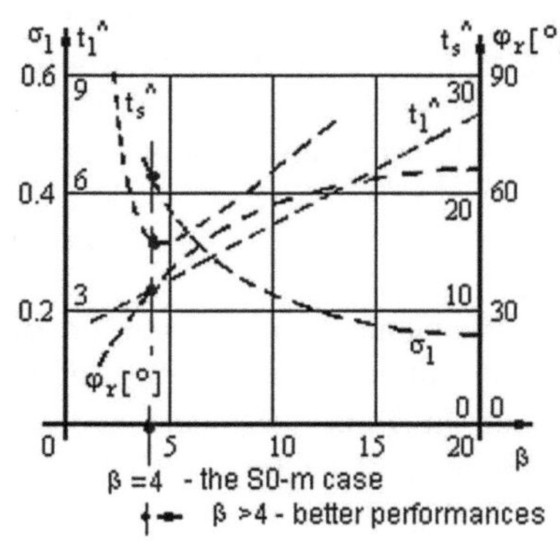

Fig. 30.3 Performance indices, σ_1, $\hat{t}_s = t_s/T_\Sigma$, $\hat{t}_1 = t_1/T_\Sigma$ and ϕ_r [°] versus β

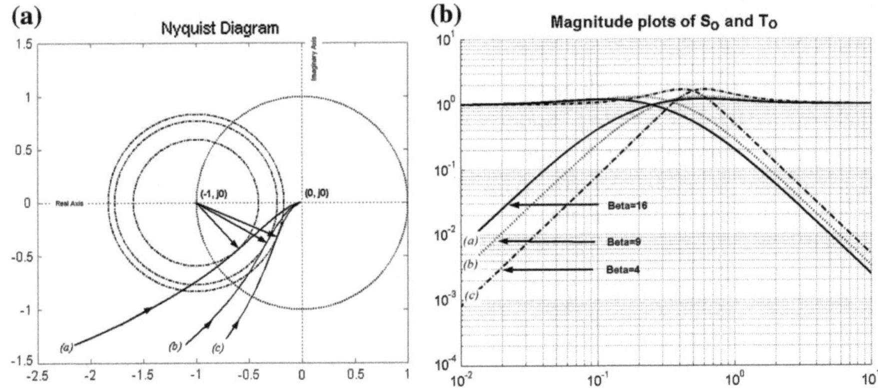

Fig. 30.4 **a** Nyquist curves and M_{S0}^{-1} *circles*, **b** Magnitude plot of the $M_P(\omega)$ for $\beta = 4, 9, 16$

- *Magnitude plot of the complementary sensitivity function*, $T_0(s)$; the dependences are depicted in Fig. 30.4b, having the characteristic values $M_{p\max(\beta=4)} \approx 1.6823$, $M_{p\max(\beta=9)} \approx 1.2990$ and $M_{p\max(\beta=16)} \approx 1.1978$.

The method offers good support to controller design for plants with variable input and continuously variable parameters, for example, variation of k_P in a relatively large domain $[k_{P\min}, k_{P\max}]$ or/and T_1 (the greatest time constant) the possibility for online (re)computing the controller's parameters (accepting an adequate value of β, which ensures a minimum guaranteed phase margin [6]. This situation is imposed by electrical drives with variable moment of inertia (VMI) treated in Sect. 30.4.

30.3.2 The Double Parameterization of the SO-m (2p-SO-m)

The double parameterization of SO-m, referred to as 2p-SO-m [9, 10], is dedicated mainly to driving systems (speed control) characterized by the t.f.s in the forms MO-2 and MO-3 given in Table 30.3, without an integral component, having $T_1 \gg (T_2 > T_\Sigma)$ which characterize plants with variable parameters, for example driving systems with VMI. The method is based on the optimization conditions (30.3.1) using an additional parameter m defined as

$$m = T_\Sigma / T_1 \; (T_\Sigma / T_1 \ll 1). \qquad (30.3.4)$$

The parameter m is a measure of the magnitude of the large time constant T_1 that points out mainly the moment of inertia. Accepting the controller-plant combination given in Table 30.3, and applying the indicated pole-zero cancelation, the t.f. computations lead to

Table 30.3 The basic situations (see also Table 30.1)

Case	$H_p(s)$	$H_c(s)$
0	1	2
2p-SO-m-1 and MO-2.1	$\frac{k_p}{(1+sT_\Sigma)(1+sT_1)}$	$\frac{k_c}{s}(1+sT_c)$, $T_c=T_1$
2p-SO-m-2 and MO-3.1	$\frac{k_p}{(1+sT_\Sigma)(1+sT_1)(1+sT_2)}$ $T_1>T_2>T_\Sigma$	$\frac{k_c}{s}(1+sT_c)\frac{(1+sT_c')}{(1+sT_f)}$ $T_c'=T_2$; $(T_c'/T_f'\approx 10)$ $\frac{k_c}{s}(1+sT_c)(1+sT_c')$, $T_c'=T_2$
2p-SO-m-3 (not detailed)	$\frac{k_p}{(1+sT_\Sigma)(1+sT_1)(1+sT_2)(1+sT_3)}$ $T_1>T_2>T_3>T_\Sigma$, $T_\Sigma/T_1<0.2$	$\frac{k_c}{s}(1+sT_c)\frac{(1+sT_c')(1+sT_d)}{(1+sT_f')(1+sT_f)}$ $T_c'=T_2$; $(T_c'/T_f'\approx 10)$ $T_d=T_3$; $(T_d/T_f\approx 10)$

$$L(s)=H_c(s)\,H_p(s)=\frac{k_ck_p(1+sT_c)}{s(1+sT_1)(1+sT_\Sigma)},\quad H_r(s)=\frac{L(s)}{1+L(s)},\tag{30.3.5}$$

$$H_r(s)=\frac{k_ck_p+sk_ck_pT_c}{k_ck_p+s(1+k_ck_pT_c)+s^2(T_1+T_\Sigma)+s^3T_1T_\Sigma},\tag{30.3.6}$$

(a proportional-derivative with lags (PDL-3)), having the coefficients a_ν, b_μ:

$$\begin{aligned}&a_0=k_ck_p,\quad a_1=1+k_ck_pT_c,\quad a_2=T_1+T_\Sigma,\quad a_3=T_1T_\Sigma\\&b_0=k_ck_p,\quad b_1=k_ck_pT_c.\end{aligned}\tag{30.3.7}$$

The condition (30.3.1) is imposed in relation with the notation (30.3.4). The main situations for interest are characterized by values of $m<(\ll)1$. The substitution of (30.3.7) in the second parameterization (30.3.4) yields

$$\beta^{1/2}k_ck_p(T_1+T_\Sigma)=(1+k_ck_pT_c)^2\ (a)\quad \beta^{1/2}(1+k_ck_pT_c)\,T_1T_\Sigma=(T_1+T_\Sigma)^2\ (b).\tag{30.3.8}$$

Finally, the characteristic t.f.s $H_0(s)$ and $H_r(s)$ will obtain the optimized forms

$$H_{0\,\mathrm{opt}}(s)=\frac{1+\beta T_{\Sigma m}s}{\beta^{3/2}T_\Sigma\frac{m}{(1+m)^2}s(1+sT_1)(1+sT_\Sigma)}\ \text{with}\ T_\Sigma'=\frac{T_\Sigma}{(1+m)},\tag{30.3.9}$$

$$H_{r\,\mathrm{opt}}(s)=\frac{(1+\beta T_{\Sigma m}s)}{\beta^{3/2}T_{\Sigma}^{\prime 3}s^3+\beta^{3/2}T_{\Sigma}^{\prime 2}s^2+\beta T_{\Sigma}^{\prime}s+1}=\frac{(1+\beta T_{\Sigma m}s)}{(1+\beta^{1/2}T_{\Sigma}^{\prime}s)[1+(\beta-\beta^{1/2})T_{\Sigma}^{\prime}s+\beta T_{\Sigma}^{\prime 2}s^2]}. \tag{30.3.10}$$

The compact tuning equations are

$$k_c=\frac{(1+m)^2}{\beta^{3/2}k_pT_1\frac{T_\Sigma}{T_1}m}\,(1+m),\ \ T_c=\beta T_\Sigma\frac{[1+(2-\beta^{1/2})m+m^2]}{(1+m)^3}\ \text{ or } T_c=\beta T_{\Sigma m}. \tag{30.3.11}$$

A. Performance analysis in time domain. The main system performances in time-domain regarding the reference input are synthesized in Fig. 30.5a, b. Elaborated useful conclusions are given in [10].

Values for the performance indices regarding a step disturbance are synthesized in Table 30.4. Mainly, the 2p-SO-m ensures efficient disturbance–rejection for a special case of servo system applications with "great and variable" moment of inertia.

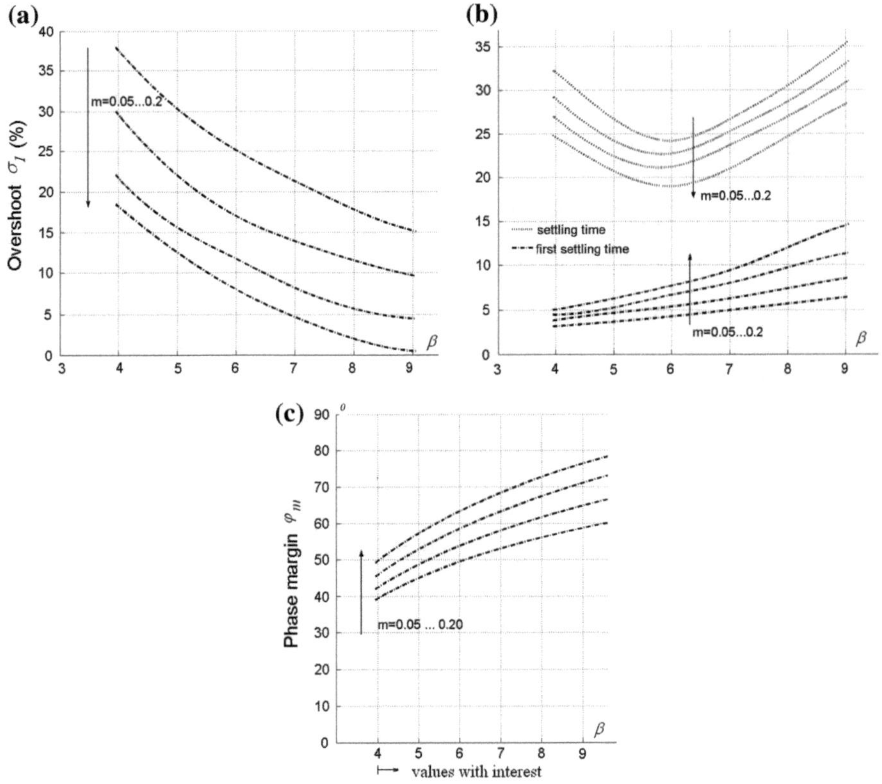

Fig. 30.5 System performance regarding the reference input; $\sigma_{1,r}, t_{s,r}, t_{1,r}=f(\beta)$, and the phase-margin curves, with m—parameter

Table 30.4 Comparison of performance indices for a step-form load disturbance

			2p-SO-m The value of β					
m		MO-m	4	5	6	7	8	9
0.05	$\hat{t}_{s,d2}$	*45,5*	**9.2**	**11.1**	**13.0**	**14.9**	**17.5**	**19.8**
	$\sigma_{1,d2}$	*9.3*	**7.7**	**8.7**	**9.7**	10.5	11.4	12.3
0.10	$\hat{t}_{s,d2}$	*28.7*	**10.6**	**12.6**	**14.5**	**17.1**	**19.6**	**23.4**
	$\sigma_{1,d2}$	*15.7*	**15.3**	17.4	19.1	20.8	22.1	2352
0.15	$\hat{t}_{s,d2}$	*19.7*	**15.2***	**17.9***	**13.9**	**16.7**	**19.7**	22.7
	$\sigma_{1,d2}$	*21.3*	22.9	25.4	28.3	30.1	32.1	34.0
0.20	$\hat{t}_{s,d2}$	*17.6*	**17.6***	**13.1**	**16.1**	19.5	22.4	26.8
	$\sigma_{1,d2}$	*25.9*	29.7	32.8	36.1	38.1	40.8	42.7

Using (30.3.9), the phase margin ϕ_r can be expressed as (Fig. 30.5c)

$$\phi_r = \mathrm{arctg}(\beta T_{\Sigma m}\omega_c) - \mathrm{arctg}(T_1\omega_c) - \mathrm{arctg}(T_\Sigma\omega_c) + \pi/2. \qquad (30.3.12)$$

The 2p-SO-m is recommended for servo systems (speed control) characterized by great differences between the large and the small time constants ($0.05 < m \le 0.2$) and high performance imposed regarding load disturbances.

Comparing the control system performance indices to those ensured by the MO-m, using for β values in the of domain of $4 < \beta \le 9$, the effects of load disturbances are lower maximum values faster rejection as shown in Table 30.4.

The 2p-SO-m is easily applicable to an analytic redesign of the controller, ensuring the "on-line" recalculation of the controller parameters, based on crisp relations.

B. Performance analysis in frequency domain. Using m and β as parameters, the Bode diagrams are illustrated in Fig. 30.6.

- ***Sensitivity function analysis***. The calculated maximum sensitivity value M_{s0} and its inverse M_{s0}^{-1} are calculated and presented only for $\beta \le 9$ (Table 30.5).

Remark: The dashed values are in the recommended domain or strictly close to it.

- ***Magnitude plot of the complementary sensitivity function***. For m and β as parameters, the graphics of $M_P(\omega) = |H_{ro}(j\omega)|$ are calculated and presented in Fig. 30.7 using the maximum value $M_{p\max}$ given in Table 30.6. The main conclusion, interpreted for example as in [1], is that an increased value of β leads to the decrease of the value of $M_{p\max}$ and the system becomes less and less oscillatory.

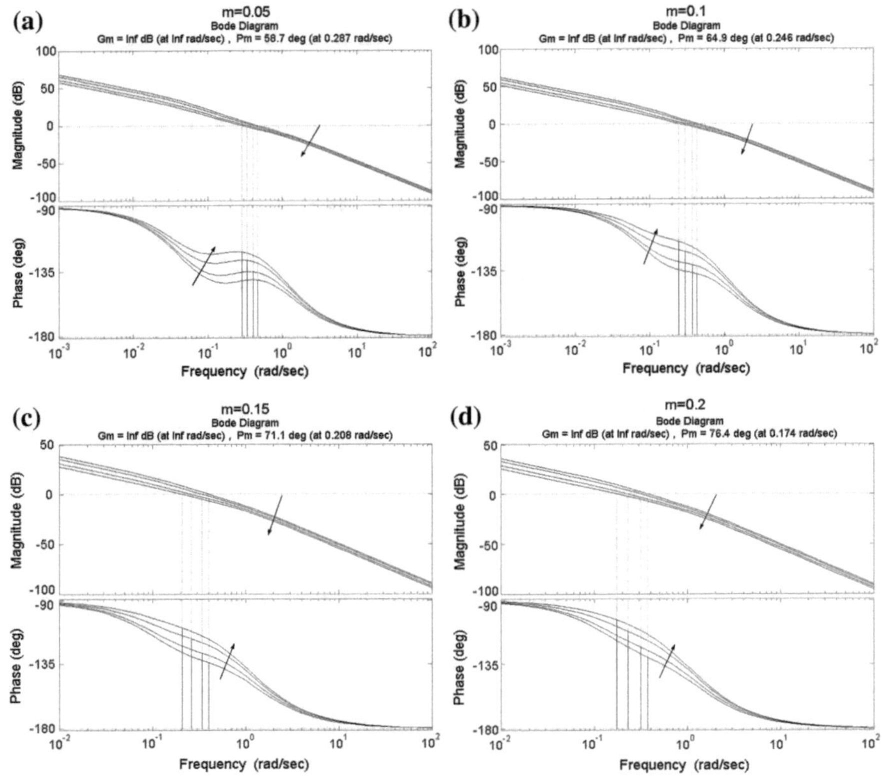

Fig. 30.6 Bode diagrams for m and β parameters

Table 30.5 The maximum value $M_{p\ max}$

	The value of β						
m	*4*	*5*	*6*	*7*	*8*	*9*	*12*
0.05	*1.573*	*1.415*	*1.321*	*1.257*	*1.211*	*1.176*	*1.104*
0.10	*1.456*	*1.303*	*1.210*	*1.147*	*1.102*	*1.067*	*1.008*
0.15	*1.343*	*1.199*	*1.114*	*1.058*	*1.023*	*1.004*	*0.998*
0.20	*1.241*	*1.113*	*1.042*	*1.006*	*0.999*	*0.998*	*0.997*

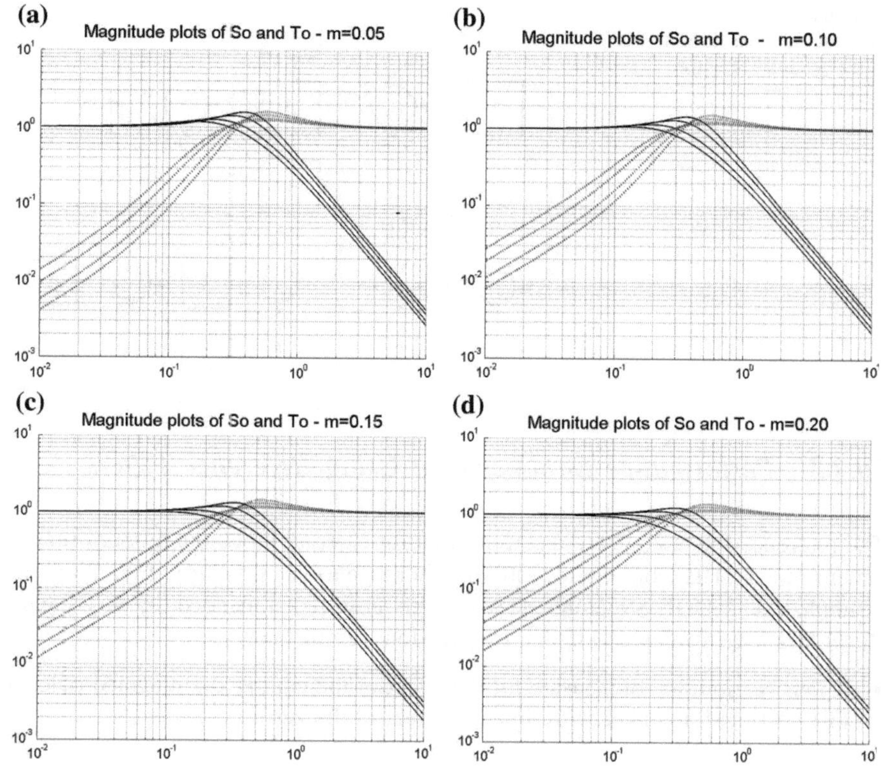

Fig. 30.7 The sensitivity and complementary sensitivity function, $\{m, \beta\}$ parameters

Table 30.6 The values for M_{s0} and M_{S0}^{-1}

		M_{s0}/M^{-1}_{s0}					
m	β	4	5	6	7	8	9
.05	M_{s0}	1.602	1.45	1.36	1.303	1.263	1.235
	M^{-1}_{s0}	**0.624**	**0.690**	**0.735**	**0.767**	**0.792**	**0.810**
.10	M_{s0}	1.529	1.385	1.302	1.248	1.212	1.185
	M^{-1}_{s0}	**0.654**	**0.722**	**0.768**	**0.801**	**0.825**	**0.844**
.15	M_{s0}	1.464	1.330	1.255	1.206	1.172	1.149
	M^{-1}_{s0}	**0.683**	**0.752**	**0.797**	**0.829**	**0.853**	**0.870**
.20	M_{s0}	1.406	1.285	1.217	1.172	1.143	1.122
	M^{-1}_{s0}	0.711	0.778	0.822	0.853	0.875	0.891

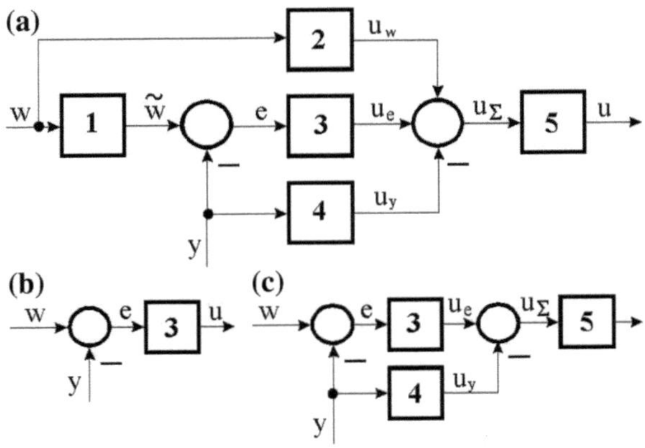

Fig. 30.8 Typical controller structures and particular forms of the modules, see for example, [19]

30.3.3 Performance Enhancement Using Reference Filters

The reference filters $F_r(s)$ are recommended for external input–output performance enhancement. The controllers with non-homogeneous structure, Fig. 30.8, for example the 2-DOF structures [18–20] are also recommended in this regard. Regarding use of reference filters, two versions are of interest:

- *A first version*, to compensate for the effect of the complex-conjugated poles in (30.3.10) and, together with this, the effect of the zero:

$$F_r(s) = \frac{1 + (\beta - \beta^{1/2})T_\Sigma s + \beta T_\Sigma^2 s^2}{(1 + \beta T_\Sigma s)(1 + sT_f s)}. \qquad (30.3.13)$$

Consequently, the control system behavior in the relation $r \to r_1 \to y$ becomes aperiodical with the main performance indices $\sigma_1 = 0$ and $t_s \approx (3\ldots5)(\beta - 1)T_\Sigma$ (and ϕ_r according to Fig. 30.1):

$$\tilde{H}_r(s) = \frac{1}{(1 + \beta T_\Sigma s)(1 + sT_f)}. \qquad (30.3.14)$$

- *A second version* of filter can be used to compensate for only the effect of the zero in (30.3.10); accordingly:

$$F_r(s) = \frac{1}{1 + \beta T_\Sigma s}. \qquad (30.3.15)$$

The control system behavior in the relation $r \rightarrow r_1 \rightarrow y$ is given by the following t.f. and the closed-loop system has an oscillatory behavior only for $\beta < 9$:

$$\tilde{H}_r(s) = \frac{1}{(1+\beta^{1/2}T_\Sigma s)[1+(\beta-\beta^{1/2})T_\Sigma s+\beta T_\Sigma^2 s^2]}. \tag{30.3.16}$$

Similar types of filters are used in case of the 2p-SO-m, having the t.f.s derived from relations (30.3.5)–(30.3.9) and β and m as parameters.

30.3.4 Variable Structure for the Controller with Bumpless Switch of the Control Algorithms

The design approaches presented in previous sections are expressed in continuous time. The discretized form of the PI(D) algorithms can be obtained using, for example, the well-known classical methods [1, 2]. The discrete t.f. of the controller results in the following form exemplified for a first order controller:

$$H_C(z^{-1}) = \frac{q_0 + q_1 z^{-1}}{p_0 + p_1 z^{-1}}, \tag{30.3.17}$$

In such cases, the bumpless switching between two or more control algorithms (c.a.s) needs the re-updating of the tuning parameters.

The notations q_ν^i, $\nu = 0, 1$ are used for the coefficients of the nominator of the discrete t.f. and $i = 1 \ldots m$, and m for the number of c.a.s (here $m = 3$). If the controller operates on the basis of c.a. (1) and it switches to c.a. (2), next c.a. (2) switches to c.a. (3), the algorithms are

$$\begin{aligned} &u_{mk} = q_1^{(m)} \cdot x_{1k}^{(m)} + q_0^{(m)} \cdot x_{2k}^{(m)}, \quad m = 1, 2, 3, \quad \text{c.a.}(m) \\ &\varepsilon_k^{(m)} = \varepsilon_k, \quad \varepsilon_k = r_k - y_k \end{aligned} \tag{30.3.18}$$

$x_{1k}^{(m)} = x_{2,k-1}^{(m)}$, with values which must be calculated. Since

$$x_{2k}^{(m)} = \varepsilon_k - p_1^{(m)} \cdot x_{1k}^{(m)}, \quad m = 1, 2, 3. \tag{30.3.19}$$

Then, the c.a. (1), c.a. (2) and c.a. (3) given in (30.3.18) can be transformed into:

$$u_{mk} = q_1^{(m)} \cdot x_{1k}^{(m)} + q_0^{(m)} \cdot \varepsilon_k - q_0^{(m)} \cdot p_1^{(m)} \cdot x_{1k}^{(m)}. \tag{30.3.20}$$

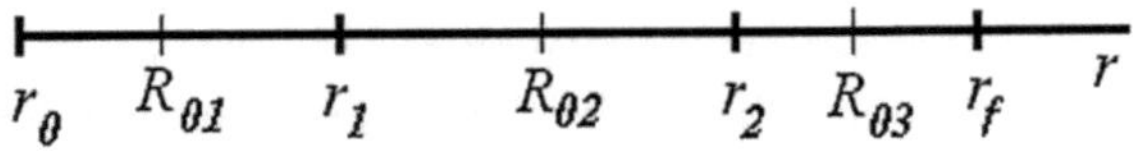

Fig. 30.9 Detailed block diagram regarded as the c.a.s switch

Imposing the bumpless switching condition $u_{2k} = u_{1k}$ and next $u_{3k} = u_{2k}$, leads to

$$\begin{aligned} x_{1k\,\mathrm{nec}}^{(n+1)} &= \left(q_1^{(n)} x_{1k}^{(n)} + q_0^{(n)} x_{2k}^{(n)}\right) / \left(q_1^{(n+1)} - q_0^{(n+1)} p_1^{(n+1)}\right) \\ &\quad - \left[q_0^{(n+1)} / \left(q_1^{(n+1)} - q_0^{(n+1)} p_1^{(n+1)}\right)\right] \varepsilon_k, \\ x_{2,k-1\,\mathrm{nec}}^{(n+1)} &= x_{1k\,\mathrm{nec}}^{(n+1)}, \quad n = 1, 2. \end{aligned} \tag{30.3.21}$$

The switching conditions must be connected and correlated to the changes in the plant, according to Fig. 30.9 and Eq. (30.3.21) considered in relation with the switching program

$$\begin{aligned} &\text{IF } (r < r_1) \text{ THEN c.a.}(1) \\ &\text{ELSE IF } (r < r_2) \text{ THEN c.a.}(2), \\ &\text{ELSE c.a.}(3), \end{aligned} \tag{30.3.22}$$

where r is the variable parameter which imposes the switch condition, r_1, r_2, r_3 are the switching values (included in the switching conditions) r_0—the initial value and r_f—the final value of r and R_{01}, R_{02}, R_{03} are the values for which the controllers are developed.

The block diagram of the controller is given in Fig. 30.10.

30.3.5 The Automatic Tuning/Retuning Steps

Imposing the requirements regarding variable reference tracking, load disturbance, rejection and robustness (also a minimum phase margin) and plant parameter changes (m is recalculated), the area of usable of the extended design methods and the value for the parameter β and can be adopted adequately.

In the field of electric drives, where demanding requirements are often met, the procedure for a systematic tuning/retuning of PI(D) controller parameters has to solve three issues:

(1) For choosing the tuning procedure it is necessary to end up in a control loop which achieves robust performance in terms of reference tracking and output disturbance rejection.
(2) Decide the optimal PI(D) type controller for the controlled plant; decide whether the controlled process needs I, PI, or PID control; if the D part has to

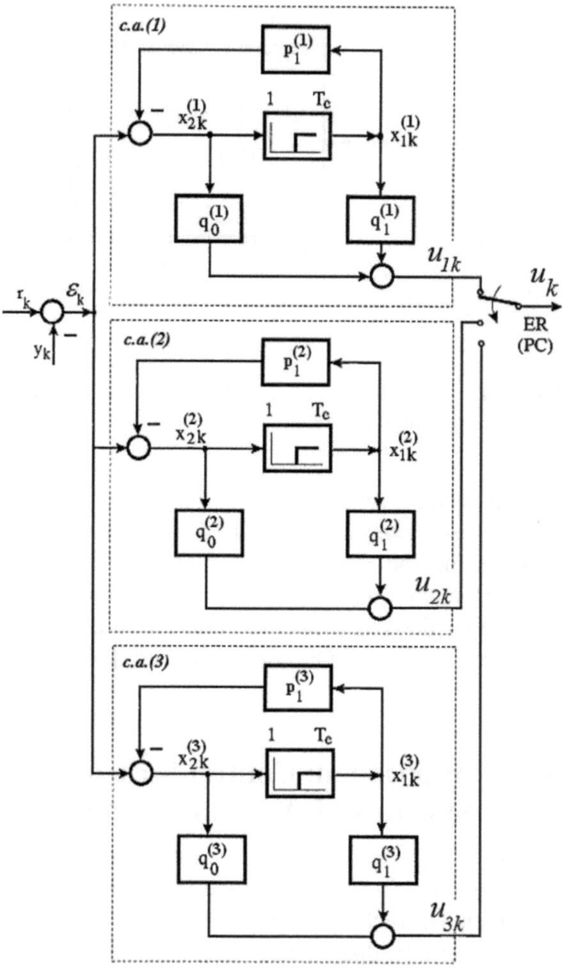

Fig. 30.10 Detailed block diagram of controller switching

be added only in the "feed-back" part of the algorithm, a non-homogenous structure results.

(3) Find the most important disturbances in the plant (external and/or parametric) and determine the effect in plant-parameter values (measured or estimated). Define an analytic form of parameter-changing.

(4) Impose the switch conditions, retune the controller's parameters based on the new model of the process and ensure a bumpless switching algorithm.

Due to the variations of the plant parameters, modifications in the controllers occur. Therefore, the method can be considered as adaptive, and the adaptive controller should have the benefit of taking into account such variations and retune its parameters.

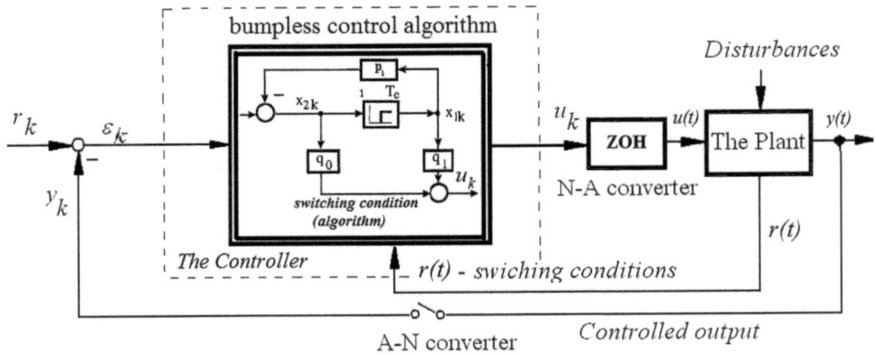

Fig. 30.11 Block diagram of control system with PI controller with bumpless switching between two or more control algorithms

If it is necessary, a stability analysis test can be applied for adjacent domains (worst cases analysis) using Kharitonov's method for a linear approach or Liapunov's method for a nonlinear approach. The switching structure presented in Fig. 30.9 and in Fig. 30.11 is applicable without difficulties to the classical (PI) case or to other derived structures. Illustrative examples are a Takagi–Sugeno PI type fuzzy controller, 2-DOF controllers, etc. [20–22].

30.4 Application: Control Structure for Mechatronic System with Variable Parameters

The essential objectives of some mechatronic applications is to ensure good reference signal tracking with small settling time with zero or small overshoot, good load (external) disturbance rejection, and reduced sensitivity with regard to parameter (internal) changes in a given domain and heavy operating regimes [15, 23–26].

Taking into account these objectives, the presented design methods ES0-m and 2p-SO-m—and based on it—different modern control solutions can be recommended and successfully applied; for example, "robust control algorithms" having the parameters permanently adapted to the variation of the plant parameters and load disturbances, adaptive fuzzy controllers (Takagi–Sugeno type), adaptive sliding-mode controllers, etc. Mainly such algorithms are based on the analytical or estimated model of the plant functioning in continuously variable conditions.

A classical application refers to electric drive systems with continuously variable reference, moment of inertia and load disturbance (abbrev. C-VR-MI-LD) for which, the variability of the parameters depends on the evolution of the plant. Representative cases are the driving systems (particularly electrical drive) with

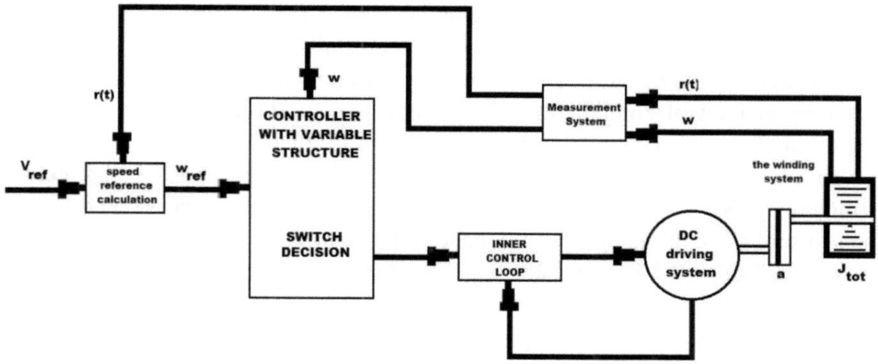

Fig. 30.12 The cascade control structure

DC-motors or BLDC-motors, [27–33] that wrap strip of various alloys on a drum (strip winding) as shown in Sect. 30.4.2.

The winding of the strip leads to the variation of the drum radius and thus the variation of the moment of inertia, which obviously changes the plant parameters and the overall system behaviors.

The speed control for the drive can be achieved using a cascade control structure (abbrev. CCS), used in the inner loop classical control solutions [27] and in the external loop PI(D) controllers (or controllers derived from it) with adaptable parameters. The functional structure of the CCS is illustrated in Fig. 30.12.

30.4.1 Steady-State Conditions and Anti-Windup Reset Measure

The mathematical model (MM) of a BLDC-m in the symmetrical operating mode [27–30] is very close to the MM of the DC-m; this leads to some similarities of the control solutions and of their design. The main (external) control loop design can be based on linearized equivalent second- or third-order benchmark-type t.f.s—see the plant t.f.s $P(s)$ in Tables 30.1, 30.2 and 30.3—connected to the operating points. So, the application became a classical control design case, with permanently adapted controller parameter based on crisp relations, see Sect. 30.3.

Table 30.7 Steady-state values of output and of control error for different values q_0

$r(s)$	y_∞			ε_∞		
	$q_0 = 0$	$q_0 = 1$	$q_0 = 2$	$q_0 = 0$	$q_0 = 1$	$q_0 = 2$
$\frac{1}{s} r_\infty$	$\frac{k_0}{1+k_0} r_\infty$	$1 \cdot r_\infty$	$1 \cdot r_\infty$	$\frac{1}{1+k_0} r_\infty$	$0 \cdot r_\infty$	$0 \cdot r_\infty$
$\frac{1}{s^2} r_\infty$	∞	∞	∞	∞	$\frac{1}{k0} r_\infty$	$0 \cdot r_\infty$

The first case, related to the ESO-m, is applied to processes with the t.f.s $P(s)$ characterized by an integral component (benchmark-type, Table 30.2).

The second case refers to speed control applications for processes with the t.f.s $P(s)$ without integral components (Table 30.3, the 2p-SO-m case), for which the condition $T_1 \gg 4T_\Sigma$ concerning the plant's t.f. is fulfilled [31].

The third case corresponds to time-variable reference input speed control structures (CS), where the applications require small control errors and so, the presence of a second integral component in the controller t.f. and, finally, in $H_0(s)$, $q_0 = 2$ in Table 30.7 is necessary (q_0 is the number of integral components). The subscript ∞ associated to a certain variable, points out the steady-state value of that variable, y and ε_∞ are the steady-state value of controlled output y and of control error ε.

In the first two cases an 1-DOF PI(D) controller can be used. The extension of the controller with an additional integral component, Fig. 30.13 is recommended only in the third case. Therefore, the controller can be characterized as an I+PI(D) with L structure and an anti-windup-reset (AWR) measure is recommended and pointed out in Fig. 30.13.

30.4.2 Application: The Strip Winding System with Variable Moment of Inertia. Bench-Mark Type Model for Controller Design

The C-VR-MI-LD application refers to a DC-m (it can also be a BLDC-m) [27–32], with a short (rigid) coupling with a rolling drum, which wraps a strip with thickness h and density ρ. The functional diagram of the application is presented in Fig. 30.14, where: a—the transmission parameter which characterizes the speed reduction unit, ω_f—the angular velocity (rolling drum) [rad/s], v_T—the linear velocity (rolling drum) [m/s], J_m, J_T—the moment of inertia of the DC-m and of the (rolling) drum [kgm^2], $J_{tot}(t)$—the moment of inertia of the whole system, [kgm^2], $r(t)$—the drum radius with strip wrapped on it [m], and f_h—the resistance force of the strip [N].

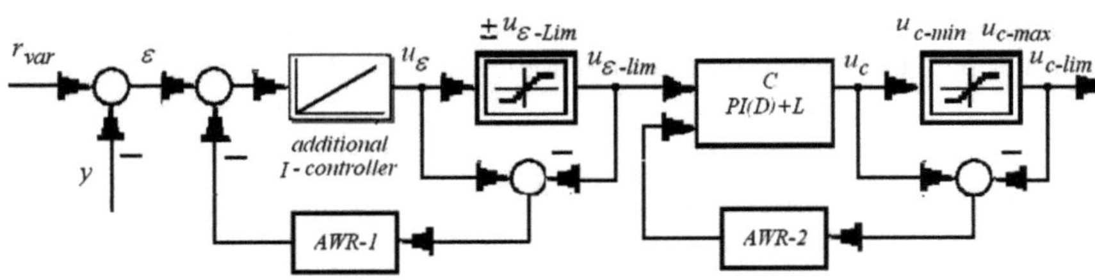

Fig. 30.13 Double integrating PI(D) controller structure with double AWR measure

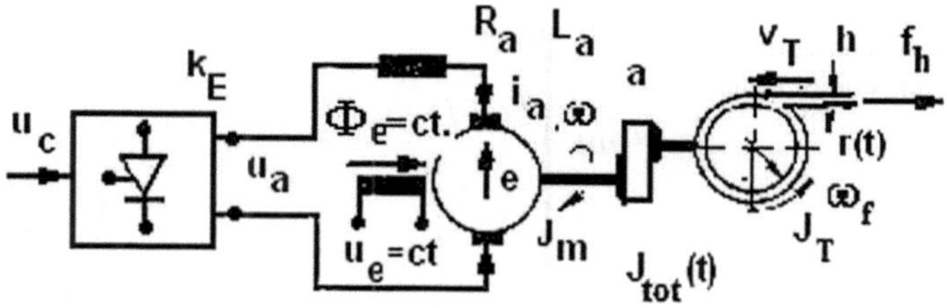

Fig. 30.14 The functional diagram of a DC electric drive system with VMI

The variation of the angular velocity of the drum and the total inertia of the system can be described by

$$a = \omega_f / \omega, \qquad J_{\text{tot}}(t) = J_m + a^2 \cdot J_T(t). \tag{30.4.3}$$

If h is sufficiently small, the drum radius variation and the variation of the moment of inertia of the drum can be approximated as (l—the drum width):

$$\frac{\mathrm{d}r(t)}{\mathrm{d}t} = \frac{h}{2 \cdot \pi} \cdot \omega_f(t) = \frac{h}{2 \cdot \pi} \cdot a \cdot \omega(t), \qquad J_T(t) = \rho \cdot \pi \cdot l \cdot r^4(t)/2. \tag{30.4.4}$$

This leads to the following extended MM of the electric drive system with VMI:

$$\begin{aligned} \frac{\mathrm{d}f_h(t)}{\mathrm{d}t} &= C \cdot a \cdot r(t) \cdot \omega(t) - C \cdot v_T(t), \\ \frac{\mathrm{d}i_a(t)}{\mathrm{d}t} &= -\frac{R_a}{T_a} \cdot i_a(t) - \frac{k_e}{L_a} \cdot \omega(t) + \frac{k_E}{L_a} \cdot u_c(t), \\ \frac{\mathrm{d}\omega(t)}{\mathrm{d}t} &= \frac{k_m}{J_{\text{tot}}(t)} \cdot i_a(t) - \frac{1}{J_{\text{tot}}(t)} \cdot \left[\frac{\mathrm{d}J_{\text{tot}}(t)}{\mathrm{d}t}\right] \cdot \omega(t) - \frac{a \cdot r(t)}{J_{\text{tot}}(t)} \cdot f_h(t) - \frac{k_f}{J_{\text{tot}}(t)} \cdot \omega(t), \end{aligned} \tag{30.4.5}$$

where C is the elasticity constant of the strip material. Using Eq. (30.4.3), Fig. 30.15 describes the variation of the moment of inertia (J) and also the variation of the mechanical time constant (T_m) versus drum radius (r).

Overall, the model is nonlinear considering the change of the system parameters, due mainly to changes in the moment of inertia of the drum (J_T). Accepting a constant value for the resistance force of the strip, f_h, the linear velocity (rolling drum) has an imposed constant value:

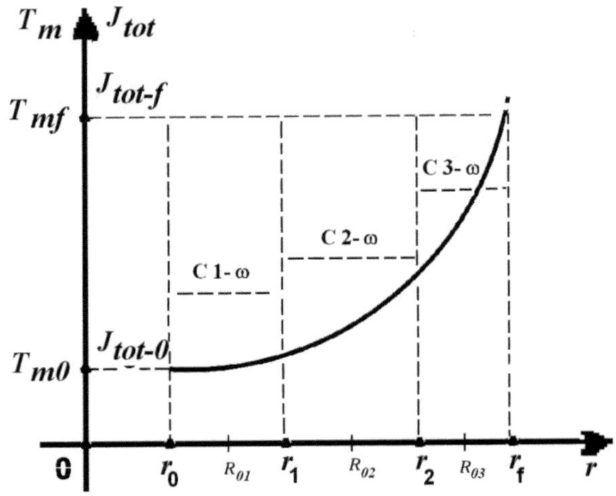

Fig. 30.15 The inertia and mechanical time constant variation as function of drum radius

$$v_T(t) \cong \text{const}, f_h(t) = \text{const} \qquad (30.4.6)$$

Accepting some simplifying assumptions and using the classical linearization technique connected to fixed value of radius, simplified benchmark-type MMs outlined in Tables 30.1, 30.2 and 30.3 can be obtained. They will be used in the CS design. The online retuning of the controller parameters is connected to three fixed value of radius for which the parameters of the three digital controllers are calculated using the ESO-method, with $\beta = 9$ (ensuring the phase margin φ_r around 55°).

The simulation block diagram given in Fig. 30.16 is developed on the basis of Eqs. (30.4.3)–(30.4.6). Other solutions can employ Takagi–Sugeno PI-Fuzzy Controllers (TS-PI-FCs) speed controllers C1-ω, C2-ω and C3-ω [32]; the solution can ensure good control system performance and compensation for plant nonlinearities. The implementation of the bumpless switching PI control algorithm (c.a.) follows the steps presented in Sect. 30.3.4 assisted by Eqs. (30.3.18)–(30.3.22).

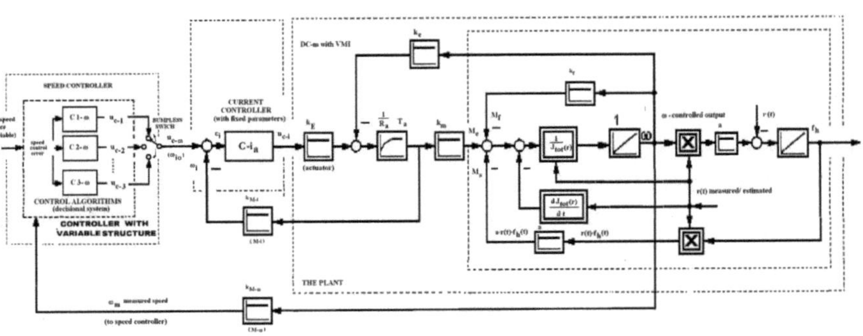

Fig. 30.16 Cascade control structure for C-VR-MI-LD driving system with a switching logic for a speed controller

30.4.3 Application: The Case Study and Simulation Results

This section offers simulation results for two possible CCS versions, which use a classical PI inner controller with fixed parameters calculated using the MO-m [1].

In the first version the speed controller comprises three digital PI controllers with fixed values for the parameters calculated relative to three significant values of the total moment of inertia reduced to the electric motor shaft, $J_{tot}(t)$: $J_{tot,R01}$, $J_{tot,R02}$, $J_{tot,R03}$, regarding three linearization points of the radius, $R_{01} = 0.0175$ m, $R_{02} = 0.0315$ m and $R_{03} = 0.05$ m and with bumpless switching of the control signal, Fig. 30.16.

In the second version, the CCS contains only one PI speed controller with fixed parameter values designed relative to three mentioned values of the total moment of inertia. The variants are summarized in Table 30.8.

- The first version of CCS. In design step the ESO-method was applied; using a sampling time $T_e = T_{a/4} = 0.00025$ s. The parameters of the digital c.a.s are:

$$\begin{aligned}
&C-i_a\colon q_{0i}=1.55,\ \ q_{1i}=-1.47, p_{0i}=1, p_{1i}=-1,\\
&C_{R_{01}}-\omega\colon q_0^{(R_{01})}=0.0492,\ \ q_1^{(R_{01})}=-0.0483, p_0^{(R_{01})}=1, p_1^{(R_{01})}=-1,\\
&C_{R_{02}}-\omega\colon q_0^{(R_{02})}=0.134,\ \ q_1^{(R_{02})}=-0.132, p_0^{(R_{02})}=1, p_1^{(R_{02})}=-1,\\
&C_{R_{03}}-\omega\colon q_0^{(R_{03})}=0.15,\ \ q_1^{(R_{03})}=-0.149, p_0^{(R_{03})}=1, p_1^{(R_{03})}=-1.
\end{aligned} \tag{30.4.7}$$

Imposing a constant value for the linear velocity v_f and recalculating permanently the reference speed $\omega_0(r(t))$, the simulation results are presented in Fig. 30.17. Based on these results it can be concluded that a PI speed controller with variable parameter values and bumpless transfer of the control ensure good control performances relating to the changes of the parameters over time. Since the application has continuously variable parameters, a sensitivity analysis in frequency domain conclusions can be useful.

Table 30.8 Combinations of plant parameters and controller parameters

		R_{01}/J_{01}	R_{02}/J_{02}	R_{03}/J_{03}
L1	*C-ω Optimal for R_{01}*	*Case study 1.1* $R_{01}, C_{R01}-\omega$	*Case study 1.2* $R_{02}, C_{R01}-\omega$	*Case study 1.3* $R_{03}, C_{R01}-\omega$
L2	*C-ω Optimal for R_{02}*	*Case study 2.1* $R_{01}, C_{R02}-\omega$	*Case study 2.2* $R_{02}, C_{R02}-\omega$	*Case study 2.3* $R_{03}, C_{R02}-\omega$
L3	*C-ω Optimal for R_{03}*	*Case study 3.1* $R_{01}, C_{R03}-\omega$	*Case study 3.2* $R_{02}, C_{R03}-\omega$	*Case study 3.3* $R_{03}, C_{R03}-\omega$

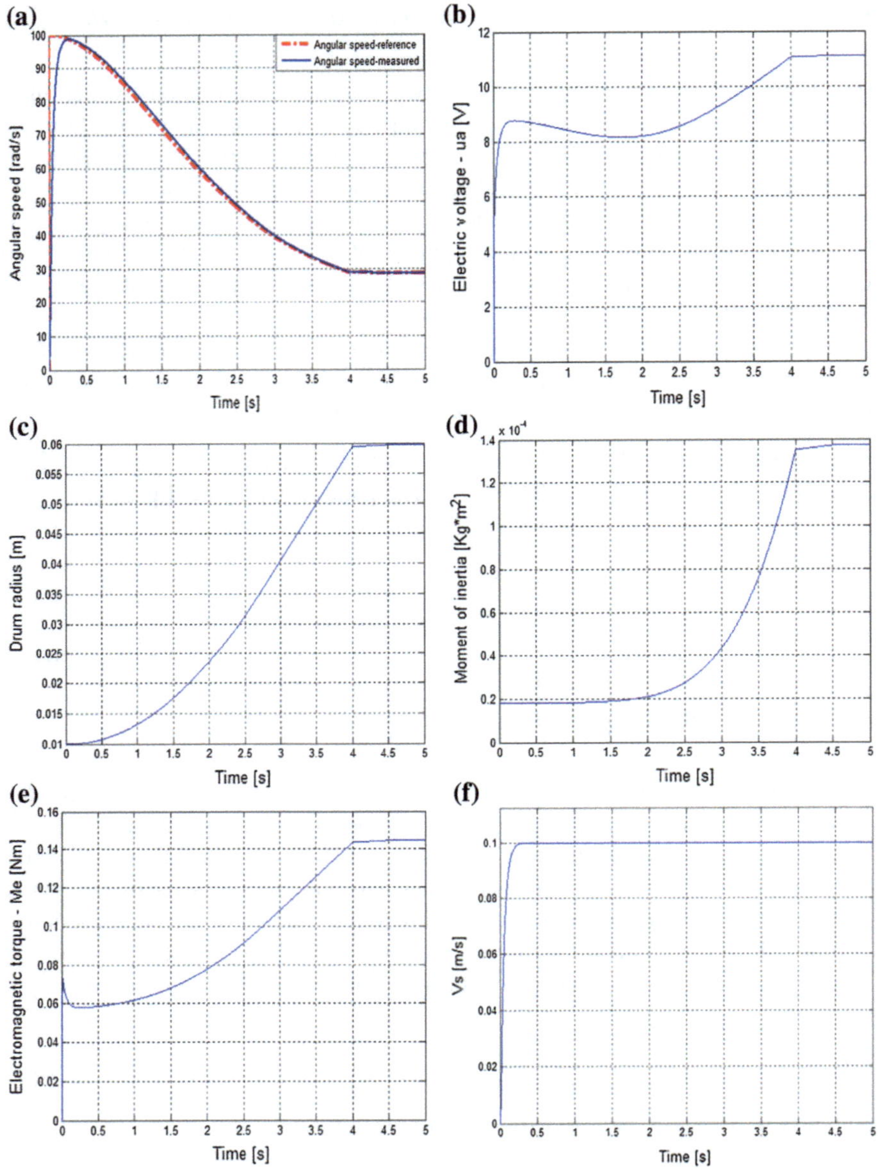

Fig. 30.17 Simulation results for C-VR-MI-LD driving system-first version of CCS: angular speed versus time (reference and measured) for the DC drive system: (**a**), electric voltage versus time for the DC drive system with VMI (**b**), drum radius versus time for the DC drive system with VMI (**c**), moment of inertia for the DC drive system with VMI (**d**), electromagnetic torque versus time for the DC drive system with VMI (**e**), linear speed of the drum versus time for the DC drive system with VMI (**f**)

- The second version of CCS. The parameters of the PI speed controllers were calculated using the ESO-m [6] ($\beta = 9$) for the same values of the radius resulting the same values of the digital c.a.s (30.4.7). In order to compare the performance of various controller-process combinations, summarized in Table 30.8 the commonly used descriptors were used: (a) step response of the angular speed versus time; (b) Bode characteristics; Fig. 30.18 presents only results for the CCS with PI current controller for the L2 and L3 combinations in Table 30.8 (*C-ω Optimal for* R_{02}, *C-ω Optimal for* R_{03}). According to [32] good performances are provided by case studies 2.1–2.3 with optimal case 2.2 and case studies 3.1–3.3 with optimal case 3.3 because in terms of settling times, $C_{R02} - \omega$ is favorable for R_{02} and least favorable for R_{01}, R_{03} and $C_{R03} - \omega$ is favorable for R_{03} and least favorable for R_{01}, R_{02}. The results are characterized by the following performance indices:

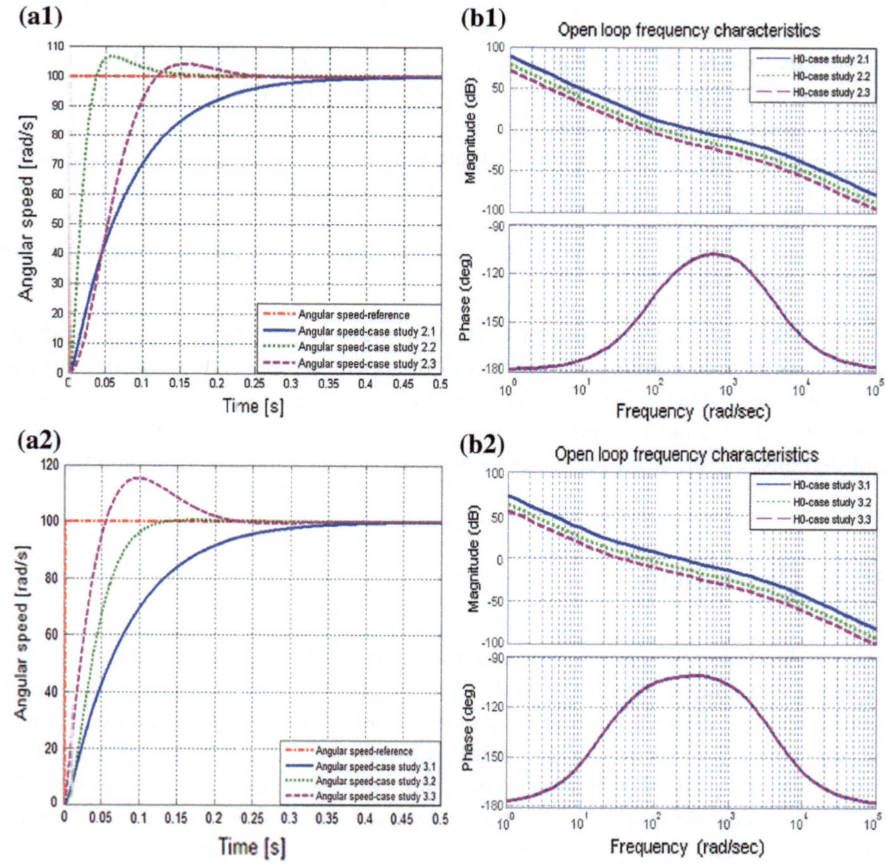

Fig. 30.18 Simulation results for C-VR-MI-LD driving system-second version of CCS: **a** step response of the angular speed versus time; **b** Bode characteristics

- *Case study 1.1*: the cutting frequency $\omega_t = 141$ rad/s, the phase margin $\varphi_r = 60°$, resonance peak $M_r = 1.28$ dB, bandwidth $\Lambda_b = 198$ rad/s and resonant frequency $\omega_r = 117$ rad/s.
- *Case study 1.2*: the cutting frequency $\omega_t = 60$ rad/s, the phase margin $\varphi_r = 39°$, resonance peak $M_r = 4.17$ dB, bandwidth $\Lambda_b = 90$ rad/s and resonant frequency $\omega_r = 55$ rad/s.
- *Case study 1.3*: the cutting frequency $\omega_t = 37$ rad/s, the phase margin $\varphi_r = 27°$, resonance peak $M_r = 7.1$ dB, bandwidth $\Lambda_b = 56$ rad/s and resonant frequency $\omega_r = 35$ rad/s.
- ***Case study 2.1***: the cutting frequency $\omega_t = 337$ rad/s, the phase margin $\varphi_r = 74°$, resonance peak $M_r = 0.0214$ dB, bandwidth $\Lambda_b = 431$ rad/s and resonant frequency $\omega_r = 225$ rad/s.
- ***Case study 2.2***: $\omega_t = 117$ rad/s, $\varphi_r = 61°$, $M_r = 1.4$ dB, $\Lambda_b = 161$ rad/s and $\omega_r = 90$ rad/s.
- ***Case study 2.3***: the cutting frequency $\omega_t = 62$ rad/s, the phase margin $\varphi_r = 47°$, resonance peak $M_r = 3.59$ dB, bandwidth $\Lambda_b = 92$ rad/s and resonant frequency $\omega_r = 49$ rad/s.
- ***Case study 3.1***: the cutting frequency $\omega_t = 201$ rad/s, the phase margin $\varphi_r = 78°$, resonance peak $M_r = 1.78$ dB, bandwidth $\Lambda_b = 233$ rad/s and resonant frequency $\omega_r = 185$ rad/s.
- ***Case study 3.2***: the cutting frequency $\omega_t = 65$ rad/s, the phase margin $\varphi_r = 70°$, resonance peak $M_r = 1.4$ dB, bandwidth $\Lambda_b = 83$ rad/s and resonant frequency $\omega_r = 34$ rad/s.
- ***Case study 3.3***: $\omega_t = 58.4$ rad/s, $\varphi_r = 58°$, $M_r = 1.25$ dB, $\Lambda_b = 81$ rad/s and $\omega_r = 50$ rad/s.

The design of the conventional controllers with fixed parameters is possible in such cases by determining the variation of the operating conditions and developing the controller for an adequately justified case. The results for the second study substantiate new control solutions that ensure the possibility to avoid the worst cases and demonstrate the need for controllers with variable parameters in variable CSs.

30.5 Conclusions

The mechatronic applications with continuously variable operating conditions (for example, variable reference, variable load disturbance, and variable moment of inertia) require adjustment of the conditions in which the controller parameters must be adaptable. To ensure good control performance, recalculation of controller parameters and providing bumpless switching between more control algorithms (in our case three) are required. To recalculate the controller parameters relations should be as simple as possible but well justified. The choice of the number of algorithms and the conditions for calculating the controller parameters are problems which need to be solved by the designer.

Based on the practical version of the SO-method, the chapter presents two extensions—the ESO-m and the 2p-SO-m—focused on benchmark-type plant models, which enable generalizations of the optimization conditions and based on it, compact design relations can be given. The presented extensions enlarge significantly the areas of application and usefulness of the SO method specific for mechatronic system applications, and they ensure better control system performance. The control design is discussed in continuous time, but the results can be easily implemented in quasi-continuous digital version using, for example, the approach given in [2].

Section 30.4 has presented a typical mechatronic application dedicated to servo systems with continuously variable conditions. In particular it is the case of a DC drive system that wraps on a drum strip of various alloys. The variation of the drum radius determines the modification of the moment of inertia, which obviously changes the plant parameters and the overall system behaviors. The basic idea of process control is to maintain a constant linear velocity of the wrapped strip by changing the angular velocity of the drum. The results presented support the benefits of design methods and their application to processes with variable parameters.

Extensions of the presented methods, i.e., Takagi–Sugeno fuzzy controller extension, the nonhomogeneous variant of the controller, are possible application themes. They can be accompanied by several modeling and application-oriented approaches [33–43].

Acknowledgments This work was supported by a grant in the framework of the Partnerships in priority areas—PN II program of the Romanian National Authority for Scientific Research ANCS, CNDI - UEFISCDI, project number PN-II-PT-PCCA-2011-3.2-0732, by a grant of the Romanian National Authority for Scientific Research, CNCS - UEFISCDI, project number PN-II-ID-PCE-2011-3-0109. Also, the work was partially supported by the strategic grant POSDRU ID 77265 (2010) of the Ministry of Labor, Family and Social Protection, Romania, co-financed by the European Social Fund—Investing in People.

References

1. Åström, K.J., Hägglund, T.: PID Controllers Theory: Design and Tuning. Instrument Society of America, Research Triangle Park, NC (1995)
2. Föllinger, O.: Regelungstechnik. Elitera Verlag, Berlin (1985)
3. Kessler, C.: Das symetrische Optimum. Regelungstechnik **6**(11), 395–400 (1958)
4. Kessler, C.: Das symetrische Optimum. Regelungstechnik **6**(12), 432–436 (1958)
5. Loron, L.: Tuning of PID controllers by the non-symmetrical optimum method. Automatica **33**(1), 103–107 (1997)
6. Preitl, S., Precup, R.-E.: An extension of tuning relations after symmetrical optimum method for PI and PID controllers. Automatica **35**(10), 1731–1736 (1999)
7. Vrancic, D., Peng, Y., Strmcnik, S.: A new PID controller tuning method based on multiple integrations. Control Eng. Pract. **7**(5), 623–633 (1999)
8. Vrancic, D., Strmcnik, S., Juricic, D.: A magnitude optimum multiple integration tuning method for filtered PID controller. Automatica **37**(9), 1473–1479 (2001)

9. Preitl, Z.: Improving disturbance rejection by means of a double parameterization of the symmetrical optimum method. Sci. Bull. “Politehnica” Univ. Timisoara, Trans. Autom. Comput. Sci. **50**(64), 25–34 (2005)
10. Preitl, Z.: Model Based Design Methods for Speed Control Applications. Editura Politehnica, Timisoara, Romania (2008)
11. Vrančić, D., Strmčnik, S., Kocijan, J., de Moura Oliveira, P.B.: Improving disturbance rejection of PID controllers by means of the magnitude optimum method. ISA Trans. **49**(1), 47–56 (2010)
12. Papadopoulos, K.G., Mermikli, K., Margaris, N.I.: Optimal tuning of PID controllers for integrating processes via the symmetrical optimum criterion. In: Proceedings of 19th Mediterranean Conference on Control and Automation (MED 2012), Corfu, Greece, pp. 1289–1294 (2011)
13. Papadopoulos, K.G., Mermikli, K., Margaris, N.I.: On the automatic tuning of PID type controllers via the magnitude optimum criterion. In: Proceedings of 2012 IEEE International Conference on Industrial Technology (ICIT 2012), Athens, Greece, pp. 869–874 (2012)
14. Papadopoulos, K.G., Margaris, N.I.: Extending the symmetrical optimum criterion to the design of PID type-p control loops. J. Process Control **22**(1), 11–25 (2012)
15. Isermann, R.: Mechatronic Systems: Fundamentals. Springer, Berlin, Heidelberg, New York (2005)
16. Preitl, S., Precup, R.-E.: Cross optimization aspects concerning the extended symmetrical optimum method. Preprints of PID'00 IFAC Workshop, Terrassa, Spain, pp. 223–228 (2000)
17. Preitl, S., Precup, R.-E.: Linear and fuzzy control extensions of the symmetrical optimum method. In: Kolemisevska-Gugulovska, T., Stankovski, M.J. (eds.) Proceedings COSY 2011 of the Special International Conference on Complex Systems: Synergy of Control, Computing & Communication, Ohrid, Macedonia, 16–20 September. The ETAI Society, Skopje, MK, pp. 59–68 (2011)
18. Precup, R.-E., Preitl, S.: Development of some fuzzy controllers with non-homogenous dynamics with respect to the input channels meant for a class of systems. In: Proceedings of European Control Conference (ECC'99), Karlsruhe, Germany, paper index F56, 6 pp (1999)
19. Preitl, S., Precup, R.-E., Preitl, Z.: Control Structures and Algorithms, vols. 1 and 2. Editura Orizonturi Universitare, Timisoara, Romania (2009) (in Romanian)
20. Preitl, S., Precup, R.-E.: Development of TS fuzzy controllers with dynamics for low order benchmarks with time variable parameters. In: Proceedings of 5th International Symposium of Hungarian Researchers on Computational Intelligence, Budapest, Hungary, pp. 239–248 (2004)
21. Preitl, S., Preitl, Z., Precup, R.-E.: Low cost fuzzy controllers for classes of second-order systems. Preprints of 15th World Congress of IFAC (b'02), Barcelona, Spain, paper index 416, 6 pp (2002)
22. Precup, R.-E., Hellendoorn, H.: A survey on industrial applications of fuzzy control. Comput. Ind. **62**(3), 213–226 (2011)
23. Koch, C., Radler, O., Spröwitz, A., Ströhla, T., Zöppig, V.: Project course ‘Design mechatronic systems’. In: Proceedings of IEEE International Conference on Mechatronics (ICM 2006), Budapest, Hungary, pp. 69–72 (2006)
24. Hehenberger, P., Naderer, R., Schuler, C., Zeman, K.: Conceptual design of mechatronic systems as a recurring element of innovation processes. In: Proceedings of 4th IFAC Symposium on Mechatronic Systems (MECHATRONICS 2006), Heidelberg, Germany, pp. 342–347 (2006)
25. Pabst, I.: An approach for reliability estimation and control of mechatronic systems. In: Proceedings of 4th IFAC Symposium on Mechatronic Systems (MECHATRONICS 2006), Heidelberg, Germany, pp. 831–836 (2006)
26. Su, Y., Mueller, C.: Smooth reference trajectory generation for industrial mechatronic systems under torque saturation. In: Proceedings of 4th IFAC Symposium on Mechatronic Systems (MECHATRONICS 2006), Heidelberg, Germany, pp. 896–901 (2006)

27. Boldea, I.: Advanced electric drives. Ph.D. courses. "Politehnica" Univ. Timisoara, Timisoara, Romania (2010–2011)
28. Nasar, S.A., Boldea, I.: Electric Drives. Electric Power Engineering Series, 2nd edn. CRC Press, Boca Raton (2005)
29. Yedamale, P.: Brushless DC (BLDC) Motor Fundamentals. Application Note 885, Microchip Technology Inc., Chandler, AZ (2003)
30. Baldursson, S.: BLDC motor modelling and control—A Matlab/Simulink implementation. M. Sc. Thesis, Institutionen för Energi och Miljö, Göteborg, Sweden (2005)
31. Grimble, M.J., Hearns, G.: Advanced control for hot rolling mills. In: Frank, P.-M. (ed.) Advances in Control: Highlights of ECC'99, pp. 135–170. Springer, London (1999)
32. Stînean, A.-I., Preitl, S., Precup, R.-E., Dragoş, C.-A., Petriu, E., Rădac, M.-B.: Choosing a proper control structure for a mechatronic system with variable parameters. Preprints of 2nd IFAC Workshop on Convergence of Information Technologies and Control Methods with Power Systems (ICPS'13), Cluj-Napoca, Romania, paper index 29, 6 pp (2013)
33. Škrjanc, I., Blažič, S., Matko, D.: Direct fuzzy model-reference adaptive control. Int. J. Intell. Syst. **17**(10), 943–963 (2002)
34. Baranyi, P., Tikk, D., Yam, Y., Patton, R.J.: From differential equations to PDC controller design via numerical transformation. Comput. Ind. **51**(3), 281–297 (2003)
35. Zhao, J., Dimirovski, G.M.: Quadratic stability of a class of switched nonlinear systems. IEEE Trans. Autom. Control **49**(4), 574–578 (2004)
36. Nakashima, T., Schaefer, G., Yokota, Y., Ishibuchi, H.: A weighted fuzzy classifier and its application to image processing tasks. Fuzzy Sets Syst. **158**(3), 284–294 (2007)
37. Vaščák, J.: Approaches in adaptation of fuzzy cognitive maps for navigation purposes. In: Proceedings of 8th International Symposium on Applied Machine Intelligence and Informatics (SAMI 2010), Herl'any, Slovakia, pp. 31–36 (2010)
38. Lian, J., Zhao, J., Dimirovski, G.M.: Integral sliding mode control for a class of uncertain switched nonlinear systems. Eur. J. Control **16**(1), 16–22 (2010)
39. Milojković, M., Nikolić, S., Danković, B., Antić, D., Jovanović, Z.: Modelling of dynamical systems based on almost orthogonal polynomials. Math. Comput. Modell. Dyn. Syst. **16**(2), 133–144 (2010)
40. Tikk, D., Johanyák, Z.C., Kovács, S., Wong, K.W.: Fuzzy rule interpolation and extrapolation techniques: criteria and evaluation guidelines. J. Adv. Comput. Intell. Intell. Inf. **15**(3), 254–263 (2011)
41. Angelov, P., Yager, R.: A new type of simplified fuzzy rule-based systems. Int. J. Gen Syst. **41**(2), 163–185 (2012)
42. Triharminto, H.H., Adji, T.B., Setiawan, N.A.: 3D dynamic UAV path planning for interception of moving target in cluttered environment. Int. J. Artif. Intell. **10**(S13), 154–163 (2013)
43. Wang, Y., Yang, Y., Zhao, Z.: Robust stability analysis for an enhanced ILC-based PI controller. J. Process Control **23**(2), 201–214 (2013)

Index

G.M. Dimirovski (ed.), *Complex Systems*, Studies in Systems, Decision and Control 55, DOI 10.1007/978-3-319-28860-4

MIX
Papier aus verantwortungsvollen Quellen
Paper from responsible sources
FSC® C105338

If you have any concerns about our products,
you can contact us on
ProductSafety@springernature.com

In case Publisher is established outside the EU,
the EU authorized representative is:
Springer Nature Customer Service Center GmbH
Europaplatz 3, 69115 Heidelberg, Germany

Printed by Libri Plureos GmbH
in Hamburg, Germany